DIESEL FUNDAMENTALS AND SERVICE
THIRD EDITION

FRANK J. THIESSEN
DAVIS N. DALES

Prentice Hall
Upper Saddle River, New Jersey   Columbus, Ohio

Library of Congress Cataloging-in-Publication Data
Thiessen, F. J.
  Diesel fuel injection, electrical and electronic systems,
and engine repair / Frank J. Thiessen, Davis N. Dales.
—3rd ed.
    p.   cm.
  Rev. ed. of: Diesel fundamentals / D.N. Dales. 2nd. ed.
1985, c1986.
  Includes index.
  ISBN 0-13-539736-7 (alk. paper)
  1. Diesel motor. I. Dales, D. N.   II. Diesel
fundamentals.   III. Title.
TJ795.T43 1997
621.43'6—dc20   96-11581
                                                CIP

Cover photo: Courtesy of MACK Trucks Inc.
Editor: Ed Francis
Production Editor: Alexandrina Benedicto Wolf
Cover Designer: Brian Deep
Production Buyer: Deidra M. Schwartz
Marketing Manager: Danny Hoyt

The book was set in Times Roman by the Clarinda Company
and was printed and bound by Quebecor Printing/Semline.
The cover was printed by Phoenix Color Corp.

 © 1997 by Prentice-Hall, Inc.
Simon & Schuster/A Viacom Company
Upper Saddle River, New Jersey 07458

All rights reserved. No part of this book may be reproduced,
in any form or by any means, without permission in writing
from the publisher.

Earlier editions ©1986, 1992 by Prentice-Hall, Inc.

Printed in the United States of America

10 9 8 7 6 5 4 3

ISBN 0-13-539736-7

Prentice-Hall International (UK) Limited, *London*
Prentice-Hall of Australia Pty. Limited, *Sydney*
Prentice-Hall Canada Inc., *Toronto*
Prentice-Hall Hispanoamericana, S.A., *Mexico*
Prentice-Hall of India Private Limited, *New Delhi*
Prentice-Hall of Japan, Inc., *Tokyo*
Simon & Schuster Asia Pte. Ltd., *Singapore*
Editora Prentice-Hall do Brasil, Ltda., *Rio de Janeiro*

PREFACE

Diesel engines are widely used in transportation in trucks, buses, railway locomotives, boats, and ships. In the construction and mining industries diesel engines power earth movers, ore carriers, bulldozers, backhoes, cranes, and trenchers. Diesel engines are also used to drive generators for electrical power, irrigation pumps, compressors, and other stationary equipment.

The widespread use of the diesel engine as a power source provides ample opportunity for a great number of rewarding careers in the use, maintenance, and service of diesel-powered equipment. Good pay is available in these positions, and the ongoing technological changes make a career in the diesel industry personally rewarding and challenging.

This third edition of *Diesel Fundamentals and Service* provides students with a sound basis upon which these careers can be built. The function, design, operation, diagnosis, service, and repair of the various systems and components are discussed in a straightforward and easily understood manner. The text is generously illustrated to help in explaining systems, components, their diagnosis, service, and repair. Illustrations are tied in bold type to the text. Each chapter has an introduction, a set of performance objectives, review questions, and test questions. The text is designed to be used to prepare the student for the writing of certification tests in the areas covered.

PLEASE NOTE: This text is not a service manual and should not be used as such. Always refer to the appropriate service manual for actual procedures and specifications for any specific engine.

CERTIFICATION TESTING

This book may be used to prepare for certification testing in the following areas: Induction, Exhaust, and Turbocharger Systems; Battery, Starting, and Charging Systems; Cooling and Lubrication Systems; Diesel Fuel Injection Systems including Multiplunger Injection Pumps, Distributor Injection Pumps, High-Pressure Fuel Injection Lines and Injection Nozzles; Unit Injector Fuel Systems; Mechanical Governor Systems; Electronic Fuel Injection Control Systems; Engine Diagnosis, Performance Testing, and Tune-up; and Cylinder Heads and Valves.

ACKNOWLEDGMENTS

The authors are deeply grateful to the many companies and individuals who have so generously provided the information and assistance required to produce this third edition of Diesel Fundamentals and Service. Without their help this book would not have been possible. Our thanks also to our colleagues and peers for their valued input. Thank you as well to the expert reviewers who took the time to examine and critique the work in manuscript form.

The ongoing advancement in diesel technology and electronic control systems has resulted in better engine performance, better fuel economy, lower exhaust emissions, greater durability, and an even longer engine life. Information on these improvements, the new control systems and their components, how they function, and how to diagnose and service them has been generously made available by the sales and service people in each of a large number of companies. Our sincere appreciation is extended to the following companies and the many individuals associated with them who were willing to spend their valuable time with us. AMBAC International Inc., American Trucking Association, Association of Diesel Specialists, Case International, Caterpillar, Inc., Cummins Engine Company, Inc., Cummins Mid-Canada Ltd., Deere and Company, Detroit Diesel Corporation, Dieseltech Division of Pritchard Engineering Co. Ltd., Donaldson Company Inc., Ford Motor Company, Garrett Corporation, General Motors Corporation, Imperial Oil Limited, Jacobs Manufacturing Company, Kent Moore Division, SPX Corporation, L. S. Starrett Company, Lucas CAV Ltd., Lucas Hartridge Inc., MAC Tools Inc., MACK Trucks Inc., Midwest Detroit Diesel-Allison, Navistar International Transportation Corporation, Peterbilt Motors Company, a Division of PACCAR, Robert Bosch Canada Ltd., Robert Bosch GmbH, Sioux Tools Inc., Snap-On Tools Corporation, Stanadyne Diesel Systems, Tobin-Arp Manufacturing Company, and Van Norman Machine Company.

A special thank you is extended to Prentice Hall editor Ed Francis, assistant editor Kim Gundling, production editor Alex Wolf, freelance copy editor Barbara Liguori, and the staff at Prentice Hall for their dedicated work.

BRIEF CONTENTS

1. DIESEL ENGINE OPERATION AND CLASSIFICATION 1
2. ENGINE PERFORMANCE, POWER, AND EFFICIENCY 19
3. INDUCTION, EXHAUST, AND TURBOCHARGER SYSTEM PRINCIPLES 29
4. INDUCTION, EXHAUST, AND TURBOCHARGER SYSTEM SERVICE 43
5. COOLING SYSTEM PRINCIPLES AND SERVICE 60
6. LUBRICATION SYSTEM PRINCIPLES AND SERVICE 83
7. ELECTRICAL PRINCIPLES AND SERVICE 103
8. BATTERY PRINCIPLES AND SERVICE 130
9. CHARGING SYSTEM PRINCIPLES AND SERVICE 147
10. STARTING SYSTEM PRINCIPLES AND SERVICE 174
11. DIESEL FUELS AND FUEL SUPPLY SYSTEMS 203
12. DIESEL FUEL INJECTION 217
13. GOVERNING FUEL DELIVERY 235
14. ELECTRONIC CONTROL OF DIESEL FUEL INJECTION—OPERATING PRINCIPLES 266
15. ROBERT BOSCH, AMBAC INTERNATIONAL, AND LUCAS CAV MULTIPLUNGER INJECTION PUMPS 279
16. V-MAC CONTROL SYSTEMS—ROBERT BOSCH AND AMBAC 311
17. DETROIT DIESEL MECHANICAL UNIT INJECTOR FUEL SYSTEM 324
18. DETROIT DIESEL ELECTRONIC CONTROL SYSTEMS 368
19. CUMMINS PT FUEL INJECTION SYSTEM 384
20. CUMMINS ELECTRONIC CONTROL SYSTEMS 424
21. CATERPILLAR FUEL INJECTION SYSTEMS 451
22. CATERPILLAR ELECTRONIC CONTROL SYSTEMS 512
23. HYDRAULIC ELECTRONIC UNIT INJECTOR (HEUI) FUEL SYSTEM 536
24. AMBAC INTERNATIONAL DISTRIBUTOR PUMPS 550
25. ROBERT BOSCH TYPE VE DISTRIBUTOR PUMP 568
26. STANADYNE (ROOSA MASTER) AND LUCAS CAV DISTRIBUTOR PUMPS 589
27. ENGINE DIAGNOSIS, PERFORMANCE TESTING, AND TUNE-UP 619
28. CYLINDER HEAD AND VALVE SERVICE 636

APPENDIX 671

INDEX 680

DIESEL FUNDAMENTALS AND SERVICE

Chapter 1 DIESEL ENGINE OPERATION AND CLASSIFICATION 1

Introduction 1
Performance Objectives 1
Terms You Should Know 1
Basic Engine Components 2
Energy Conversion 2
Four-Stroke-Cycle Diesel Engine Operation 5
Two-Stroke-Cycle Diesel Engine Operation 9
Valve Timing 9
Comparison of Two-Stroke-Cycle and Four-Stroke-Cycle Engines 10
Piston Positions 11
Bore, Stroke, and Displacement 12
Compression Ratios 13
Piston and Crank Pin Travel 14
Engine Speed 14
Engine Braking Systems 15
Engine Classification 15
Diesel Engine Systems 17
Review Questions 18
Test Questions 18

Chapter 2 ENGINE PERFORMANCE, POWER, AND EFFICIENCY 19

Introduction 19
Performance Objectives 19
Terms You Should Know 19
Work 19
Engine Power 20
Engine Torque and Brake Power 22
Engine Efficiency 22
Engine Balance 25
Cylinder Balance 26
Dynamometers 26
Review Questions 28
Test Questions 28

Chapter 3 INDUCTION, EXHAUST, AND TURBOCHARGER SYSTEM PRINCIPLES 29

Introduction 29
Performance Objectives 29
Terms You Should Know 29
Air Cleaner Function 31
Precleaners 31
Dry-Type Air Cleaners 31
Oil Bath Air Cleaner 32
Intake Manifold 33
Turbocharger and Blower Function 33
Turbocharger Operation 34
Air Shutdown Valve 35
Blower Design and Operation (Two-Stroke-Cycle Engine) 36
Intercoolers and Aftercoolers 37
Exhaust System Function and Components 37
Exhaust Emissions 39
Review Questions 41
Test Questions 41

Chapter 4 INDUCTION, EXHAUST, AND TURBOCHARGER SYSTEM SERVICE 43

Introduction 43
Performance Objectives 43
Terms You Should Know 43
Service Precautions 44
Intake, Exhaust, and Turbocharger Diagnosis 44
Dry-Type Air Cleaner Service 44
Cleaning the Dry Element 46
Oil Bath Air Cleaner Service 47
Turbocharger Service 47
Turbocharger Overhaul 50
Blower Service (Two-Stroke-Cycle Engine) 52
Exhaust System Service 55
Review Questions 58
Test Questions 58

Chapter 5 COOLING SYSTEM PRINCIPLES AND SERVICE 60

Introduction 60
Performance Objectives 60
Terms You Should Know 60
Liquid Cooling System Function 60
Antifreeze Coolant 71
Coolant Temperature Problems 71
Cooling System Diagnosis and Service 72
Water Pump Service 73
Review Questions 81
Test Questions 81

Chapter 6 LUBRICATION SYSTEM PRINCIPLES AND SERVICE 83

Introduction 83
Performance Objectives 83
Terms You Should Know 83
Lubrication System Components 83
Engine Oil Functions 85
Engine Lubricating Oils 86
Engine Oil Classification 87
Oil Pumps and Pressure Regulation 88
Oil Filters 91
Engine Oil Cooler 92
Oil Pressure Indicators 93
Crankcase Ventilation 94
Lubrication System Service 95
Review Questions 102
Test Questions 102

Chapter 7 ELECTRICAL PRINCIPLES AND SERVICE BASICS 103

Introduction 103
Performance Objectives 103
Terms You Should Know 103
What Is Electricity 104
Magnets and Magnetism 105
Producing Electricity 107
How Electricity Flows through a Conductor 107
How Electricity Is Measured 108
Electrical Circuits 109
Common Ground Principle 110
Electrical Wiring 111
Switches 114
Circuit Protection 117
Resistors 117
Capacitors 119
Transducers 120
Diodes 120
Transistors 121
Common Electrical Problems 122
Electrical Circuit Diagnosis and Service 123
Basic Test Equipment 124
Wiring Repairs 127
Review Questions 128
Test Questions 128

Chapter 8 BATTERY PRINCIPLES AND SERVICE 130

Introduction 130
Performance Objectives 130
Terms You Should Know 130
Battery Function and Construction 130
Multiple-Battery Systems 134
Battery Ratings 135
Battery Service Precautions 136
Jump-Starting (Negative-Ground System) 138
Battery Inspection, Testing, and Charging 138
Battery Selection and Installation 145
Review Questions 145
Test Questions 146

Chapter 9 CHARGING SYSTEM PRINCIPLES AND SERVICE 147

Introduction 147
Performance Objectives 147
Terms You Should Know 147
Charging System Function and Components 148
Types of Alternators 148
Alternator Components (Brush Type) 148
Alternator Operation 150
Voltage Regulation 154
Brushless Alternator 158
Charging System Service Precautions 159
Problem Diagnosis 159
Preliminary Inspection 161
Charging System Tests 161
Alternator Removal 166
Alternator Disassembly and Testing 166
Alternator Assembly 169
Review Questions 172
Test Questions 173

Chapter 10 STARTING SYSTEM PRINCIPLES AND SERVICE 174

Introduction 174
Performance Objectives 174
Terms You Should Know 174

Electric Starting System Components and Function 175
Electric Starting Motor Components 175
Starting Motor Operation 178
Starting Motor Circuits 180
Starting Motor Solenoid Circuits 181
Starting Motor Drives 182
Starter Protection Device 186
Electric Starting System Service 187
Starting Motor Bench Tests 189
Starter Disassembly 191
Parts Inspection and Replacement 191
Air-Starting System 195
Hydraulic Starting System 198
Review Questions 201
Test Questions 201

Chapter 11 DIESEL FUEL AND FUEL SUPPLY SYSTEMS 203

Introduction 203
Performance Objectives 203
Terms You Should Know 203
Diesel Fuel Quality 204
Alternative Fuels 207
Fuel Tank Design 209
Fuel Supply Pump Design and Operation 210
Fuel Filters 213
Bleeding Air from the Fuel System 215
Review Questions 215
Test Questions 215

Chapter 12 DIESEL FUEL INJECTION 217

Introduction 217
Performance Objectives 217
Terms You Should Know 217
Fuel Injection Principles 217

Combustion Chamber Design 218
Combustion Stages in a Diesel Engine 220
Injection Nozzle Function and Operation 221
Design and Application of Injection Nozzles 222
Diesel Fuel Injection Tubing 225
Injector Problems 226
Isolating a Faulty Injector Nozzle 226
Removing Injection Nozzles 226
Injection Nozzle Service 226
Installing Injection Nozzles 231
Review Questions 233
Test Questions 234

Chapter 13 GOVERNING FUEL DELIVERY 235

Introduction 235
Performance Objectives 235
Terms You Should Know 235
The Need for a Governor 235
Governor Terminology 238
Governor Types 238
Governor Functions 238
Mechanical Governors 240
Manifold-Pressure Compensator 256
Altitude-Pressure Compensator (ADA) 258
Electric Speed-Control Device 262
Pneumatic Governor 262
Review Questions 264
Test Questions 265

Chapter 14 ELECTRONIC CONTROL OF DIESEL FUEL INJECTION—OPERATING PRINCIPLES 266

Introduction 266
Performance Objectives 266
Terms You Should Know 266

Electronic Control Module (ECM) Operation 267

Electronic Control Module (ECM) Components 268

Input and Output Devices 270

Input Sensors and Their Function 271

Computer-Controlled Actuators (Output Devices) 272

Fault Codes 273

Features and Advantages of Electronic Controls 274

Review Questions 278

Test Questions 278

Chapter 15 ROBERT BOSCH, AMBAC INTERNATIONAL, AND LUCAS CAV MULTIPLUNGER INJECTION PUMPS 279

Introduction 279

Performance Objectives 279

Terms You Should Know 279

Injection Pump Design and Operation 280

Control-Rod Stops and Pressure Compensators 286

Pump Sizes 288

Design and Operation of the PF Pump 290

Injection Timing Control Device 290

Governor Types 291

Robert Bosch Electronic Diesel Control (EDC) System 292

Fuel Supply Pumps 292

Troubleshooting 296

Service Precautions 296

Preventive Maintenance 297

Injection Pump Removal 300

Injection Pump Flushing Procedure 300

Injection Pump Disassembly 301

Pump Phasing 302

Injection Pump Calibration 305

Injection Pump to Engine Timing 307

Bleeding the Injection System 309

Review Questions 309

Test Questions 310

Chapter 16 V-MAC ELECTRONIC CONTROL SYSTEMS— AMBAC AND ROBERT BOSCH 311

Introduction 311

Performance Objectives 311

Terms You Should Know 311

V-MAC Performance Features 311

V-MAC System Components and Function 313

Diagnostic and Troubleshooting Procedures 314

Procedure for No Active Codes, or Intermittent Problems 319

V-MAC Diagnostic Tools 322

Review Questions 323

Test Questions 323

Chapter 17 DETROIT DIESEL MECHANICAL UNIT INJECTOR FUEL SYSTEM 324

Introduction 324

Performance Objectives 324

Terms You Should Know 324

Unit Injector Function 325

Unit Injector Identification 326

Plunger, Barrel, and Helix Construction 326

Unit Injector Operation 327

Injector Timing 328

Needle Valve and Crown Valve Injectors 329

Governor Types 331

Fast-Idle Cylinder (Figure 17–16) 331

CONTENTS

Fuel Modulator (Figure 17–17)   333
Throttle Delay Mechanism   334
Tailored Torque (TT) Governor Operation   334
Unit Injector System Diagnosis and Service   335
Checking Fuel Flow   335
Injector Removal   338
Control Rack and Plunger Movement Test   339
Installing Fuel Injector in Tester
(Figures 17–24 and 17–25)   339
Valve Opening and Spray Pattern Test
(Figure 17–26)   340
Spray Tip Test (Figure 17–26)   341
Injector High-Pressure Test (Figure 17–26)   341
Pressure Holding Test (Figure 17–26)   341
Fuel Output Test   341
Injector Troubleshooting   342
Injector Disassembly   342
Cleaning Injector Parts   343
Inspecting Injector Parts   351
Injector Assembly   352
Checking Spray Tip Concentricity   354
Testing Reconditioned Injector   354
Injector Installation   354
Engine Tune-Up   355
Setting Injector Timing
(Two-Stroke-Cycle Engine)   355
Governor Gap Adjustment   357
Setting the Gap—Double-Weight Governor   359
Positioning Injector Rack Control Levers   360
Starting Aid Screw Adjustment   362
Maximum No-Load Engine Speed
Adjustment   363
Idle-Speed Adjustment   364
Buffer Screw Adjustment   364
Throttle Delay Adjustment   365
Fuel Modulator Adjustment (Figure 17–57)   366
Review Questions   366
Test Questions   367

Chapter 18 DETROIT DIESEL ELECTRONIC CONTROL SYSTEMS   368

Introduction   368
Performance Objectives   368
Terms You Should Know   368
DDEC Function and Development   369
DDEC III Operating Features   370
DDEC III Component Descriptions   370
DDEC Diagnosis and Testing   375
Electronic Unit Injector Replacement   378
Review Questions   382
Test Questions   382

Chapter 19 CUMMINS PT FUEL INJECTION SYSTEM   384

Introduction   384
Performance Objectives   384
Terms You Should Know   384
Fuel Pump Drive and Major Components   385
Fuel Pump Function and Operation   388
Fuel Supply Pump Operation   389
Governor Types and Operation   390
Air/Fuel Control (AFC)   393
Fuel Injector Operation   393
Cummins Injection Timing Control   398
Fuel System Diagnosis   403
Injector Service   403
Injector Adjustment
(Typical for L10 Series Engine)   410
Injection Timing
(Typical for L10 Series Engine)   414
Fuel System Checks and Adjustments   417
Fuel Pump Removal, Installation, and Testing
(Figures 19-72 to 19-74)   421
Review Questions   422
Test Questions   422

Chapter 20 CUMMINS ELECTRONIC CONTROL SYSTEMS 424

Introduction 424

Performance Objectives 424

Terms You Should Know 424

PACE and PT PACER Electronic Control Systems Components and Functions 425

PT PACER System Operation 427

Fuel Flow 427

Fault Codes 428

CELECT ECI (Electronically Controlled Injection) System Components and Function 429

CELECT System Basic Fuel Flow 437

CELECT Electronic Injector Operation 437

CELECT System Diagnosis and Service 439

Fault Code Diagnosis 439

Cummins Echeck Tool 440

Using the Compulink Tool 441

Pinpoint Testing 441

Cylinder Misfire Test 441

CELECT Injector Replacement 443

Injector and Valve Adjustment 446

Supply Pump and Shut-off Valve Tests 448

Review Questions 449

Test Questions 449

Chapter 21 CATERPILLAR FUEL INJECTION SYSTEMS 451

Introduction 451

Performance Objectives 451

Terms You Should Know 451

New Scroll Injection Pump Fuel System— Components and Fuel Flow (Figure 21–1) 452

Injection Pump Operation (Figure 21–3) 452

Governor Operation 452

Automatic Timing Advance Unit Operation (4MG3600-Up, 5KJ1-Up, 3ZJ1-Up) 458

New Scroll Fuel System Diagnosis and Service 462

Removal of Fuel Injection Pumps 462

Installation of Fuel Injection Pumps (Figure 21–25) 463

Finding the Top Center Compression Position for the No. 1 Piston 464

Checking Engine Timing with the 8T5300 Timing Indicator Group and the 8T5301 Diesel Timing Adapter Group 465

Engine Timing by the Timing Pin Method 4MG3600-Up, 5KJ1-Up, 3ZJ1-Up 467

Fuel Ratio Control and Governor Check (Later Engines) 468

Low-Idle Speed Adjustment 470

Checking the Set Point (Balance Point) 470

Fuel Ratio Control Adjustment (Later Engines) 472

Sleeve-Metering Fuel Injection System 473

Sleeve-Metering Fuel System Fuel Flow (Figure 21–40) 473

Sleeve-Metering Injection Pump Operation (Figures 21–42 and 21–43) 474

Governor Operation 475

Fuel Injection Nozzles 479

Sleeve-Metering System Diagnosis and Service 480

Finding the Top Center Compression Position for the No. 1 Piston 480

Checking Fuel Injection Pump Timing on the Engine 481

Fuel Setting 483

Load Stop Adjustment 484

Fuel Ratio Control Adjustment (Figure 21–69) 485

Adjustment of Crossover Levers 486

Low-Idle Adjustment (Figure 21–73) 487

Checking the Set Point (Balance Point) 487

xiv CONTENTS

Fuel Injection Pump Service 489

Checking Fuel Pump Calibration 489

Caterpillar Mechanical Unit Injector Fuel System (Figure 21–82) 493

Fuel Transfer Pump Operation 493

Unit Injector Operation 493

Control Rack Operation (Figures 21–87 and 21–88) 494

Governor Operation (Figures 21–89 to 21–92) 494

Mechanical Unit Injector System Diagnosis and Service 500

Fuel System Inspection 500

Checking Engine Cylinders Separately 500

Governor Low-Idle Adjustment (Figure 21–98) 500

Fuel Pressure Check (Figure 21–99) 501

Finding the Top Center Position for the No. 1 Piston (Figure 21–100) 501

Injector Synchronization (Rack Adjustment) 502

Fuel Setting 504

Fuel Timing 506

Review Questions 509

Test Questions 510

Chapter 22 CATERPILLAR ELECTRONIC CONTROL SYSTEMS 512

Introduction 512

Performance Objectives 512

Terms You Should Know 512

Programmable Electronic Engine Control (PEEC) System 513

Advantages and Features of the PEEC System 513

PEEC System Component Descriptions (Figures 22–1 to 22–3) 513

PEEC System Tests 517

3406E/3176B Electronic Control System (Figures 22–14 to 22–17) 521

Component Descriptions 522

Engine Protection System 526

System Fuel Flow and EUI Operation 526

Diagnostic Procedures 529

Troubleshooting with a Diagnostic Code 530

EUI Service 532

Throttle Position Sensor Test and Adjustment 533

Speed/Timing Sensor Removal, Installation, and Calibration 533

Electronic Injection Timing Check and Calibration 534

Review Questions 534

Test Questions 535

Chapter 23 HYDRAULIC ELECTRONIC UNIT INJECTOR (HEUI) FUEL SYSTEM 536

Introduction 536

Performance Objectives 536

Terms You Should Know 536

HEUI System Overview 537

HEUI Fuel System Components and Operation 537

Hydraulic Injection Control Pressure System 537

Rail Pressure Control Valve Operation 538

Hydraulic Electronic Unit Injector Components 542

Fuel Injector Operation 543

Electronic Control System 544

Glow Plug System 545

HEUI System Diagnosis and Service 545

Review Questions 548

Test Questions 548

Chapter 24 AMBAC INTERNATIONAL DISTRIBUTOR PUMPS 550

Introduction 550

Performance Objectives 550

Terms You Should Know 550

PS Model Pump Operation 552

PS Series Distributor Pump Timing 555

Governor Operation 558

Electronically Controlled Distributor Pump 559

Fuel System Troubleshooting 561

AMBAC Distributor Injection Pump Service 561

Pump Adjustments 563

Fuel Injection Pump Installation and Static Timing Procedure 564

Review Questions 566

Test Questions 567

Chapter 25 ROBERT BOSCH TYPE VE DISTRIBUTOR PUMP 568

Introduction 568

Performance Objectives 568

Terms You Should Know 568

Major Components of the VE Pump 569

VE Pump Operation 571

Governor Operation 573

Electronic Diesel Control (VE Pump) (Figure 25–19) 579

Injection Pump Removal (Typical) 580

Injection Pump Repairs 582

VE Injection Pump Timing 584

VE Injection Pump Installation 586

Review Questions 588

Test Questions 588

Chapter 26 STANADYNE (ROOSA MASTER) AND LUCAS CAV DISTRIBUTOR PUMPS 589

Introduction 589

Performance Objectives 589

Terms You Should Know 589

Stanadyne (Roosa Master) Distributor Injection Pumps 590

Stanadyne (Roosa Master) Injection Timing 595

Mechanical All-Speed Governor 596

Stanadyne Electronic Fuel Injection Pump 598

Stanadyne Distributor Pump Troubleshooting 599

Stanadyne Distributor Pump Testing and Calibration 599

General Test Procedure 602

Lucas CAV DPA Distributor Pump Design and Operation 603

Mechanical Governor Operation (Figure 26–27) 606

Timing Advance Unit Operation 606

Lucas DPA Pump Service 607

Pump Testing and Calibration 610

Review Questions 615

Test Questions 615

Chapter 27 ENGINE DIAGNOSIS, PERFORMANCE TESTING, AND TUNE-UP 617

Introduction 617

Performance Objectives 617

Terms You Should Know 617

Engine Diagnosis 617

Diesel Engine Tune-Up 627

Compression Testing 628
Cylinder Leakage Testing 630
Performance Testing with a Dynamometer 630
Review Questions 634
Test Questions 634

Chapter 28 CYLINDER HEAD AND VALVE SERVICE 636

Introduction 636
Performance Objectives 636
Terms You Should Know 636
Cylinder Head Removal Precautions 637
Cylinder Head Removal 638
Cylinder Head Disassembly 641
Cylinder Head Crack Detection 642
Cylinder Head Warpage 643
Valve Guide Service 644

Valve Bridge or Crosshead and Guide 645
Valve Seat Insert Replacement 646
Valve Seat Grinding 650
Valve Service 654
Valve Train Components Service 658
Injector Sleeve Testing and Replacement 660
Cylinder Head Assembly and Installation 660
Valve- and Injector-Operating Mechanism Service 664
Adjusting the Valve Train (Figures 28–79 to 28–82) 666
Review Questions 669
Test Questions 669

APPENDIX 671

INDEX 680

◆ CHAPTER 1 ◆

DIESEL ENGINE OPERATION AND CLASSIFICATION

INTRODUCTION

The diesel engine must produce the power needed to drive the vehicle or tractor under varying conditions of speed, load, and environment.

The size of the engine required is determined by the size and type of the vehicle or equipment and the load and performance requirements of the unit.

Both two-stroke-cycle and four-stroke-cycle engines are in common use. Multicylinder reciprocating piston diesel engines with two, three, four, five, six, eight, or more cylinders are produced.

Regardless of the type of engine being used, the demands placed on the engine to perform are rigorous. Speed and load requirements in many types of service are constantly changing during operation. Seasonal and regional temperature extremes are also encountered. All these factors are considered in engine design, operation, and service.

Diesel engines are used on such units as buses, trucks, tractors, mining equipment, construction equipment, agricultural equipment, railroad locomotives, ships and boats, power-generating units, and pumping and irrigation equipment.

To diagnose engine problems or service engines effectively, it is necessary to have a thorough understanding of the operating principles and construction features of the diesel engine.

PERFORMANCE OBJECTIVES

After thorough study of this chapter and the appropriate training models, you should be able to do the following:

1. Define energy conversion as it applies to the diesel engine.
2. List the basic components of a diesel engine.
3. Describe basic four-stroke-cycle and two-stroke-cycle diesel engine operation and compare the two.
4. Describe the relationship between piston and crank pin travel.
5. Describe and chart basic valve and port timing.
6. Define and calculate cylinder bore, piston stroke, displacement, and compression ratios.
7. Define diesel engine speed classifications.
8. List six diesel engine classifications.
9. List seven diesel engine systems.
10. Describe basic compression brake system operation.

TERMS YOU SHOULD KNOW

Look for these terms as you study this chapter, and learn what they mean.

piston block
cylinder connecting rod

crankshaft
reciprocate
camshaft
intake valve
exhaust valve
air/fuel ratio
four-stroke cycle
two-stroke cycle
TDC
BDC
intake stroke
compression stroke
power stroke
exhaust stroke
atmospheric pressure
combustion
expansion
blower
ports
scavenging
valve timing
port timing
bore
stroke
displacement
compression ratio
engine braking system
piston and crank pin travel
engine speed
engine classification
diesel engine systems

BASIC ENGINE COMPONENTS

The basic components of a diesel engine include a cylinder block, movable pistons inside the cylinders, connecting rods attached at the top end to the pistons and at the bottom to the offset portions of a crankshaft, a camshaft, pushrods and rocker arms to operate the two or four valves (intake and exhaust), and a cylinder head. A flywheel is attached to one end of the crankshaft. The other end of the crankshaft has a gear to drive the camshaft gear. The camshaft gear is twice as large as the crankshaft gear. This drives the camshaft at half the speed of the crankshaft on four-stroke-cycle engines. On two-stroke-cycle engines, the crankshaft and camshaft run at the same speed.

Study **Figures 1-1** to **1-7** to learn the names of engine parts and their functions.

ENERGY CONVERSION

The diesel engine is a device used to convert the chemical energy of diesel fuel into heat energy and usable mechanical energy. This conversion is achieved by combining the appropriate amounts of air and fuel and burning them in an enclosed cylinder at a controlled rate. A movable piston in the cylinder is forced down by the expanding gases of combustion.

The movable piston in the cylinder is connected to the top of a connecting rod. The bottom of the connecting rod is attached to the offset portion of a crankshaft. As the piston is forced down, this force is transferred to the crankshaft, causing the crankshaft to rotate. The reciprocating (back and forth or up and down) movement of the piston is converted to the rotary (turning) motion of the crankshaft, which supplies the power to drive the vehicle.

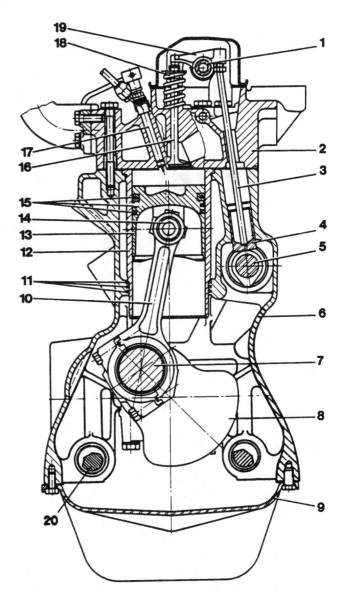

1—Rocker Arm Shaft
2—Cylinder Head
3—Push Rod
4—Cam Follower
5—Camshaft
6—Cylinder Block
7—Crankshaft
8—Crankshaft Counterweight
9—Oil Pan
10—Connecting Rod
11—Liner Packing Rings
12—Cylinder Liner
13—Piston
14—Piston Pin
15—Piston Rings
16—Valve
17—Fuel Injection Nozzle
18—Valve Spring
19—Rocker Arm
20—Balancer Shafts

FIGURE 1-1 Four-stroke-cycle engine cross section. (Courtesy of Deere and Company.)

High Efficiency Turbocharger — uses a pulsed-recovery exhaust manifold that provides increased heat flow contributing to fuel efficiency.

Parallel Ports — This unique configuration allows for very short intake and exhaust ports for efficient air flow, low pumping losses and reduced heat transfer.

Iron Crosshead Pistons — They allow the top ring to be placed much closer to the top of the piston. This reduces the dead volume above the top ring and improves fuel economy.

Gasket Eliminator — reduces engine service time since it is not necessary to get a separate gasket to complete a repair.

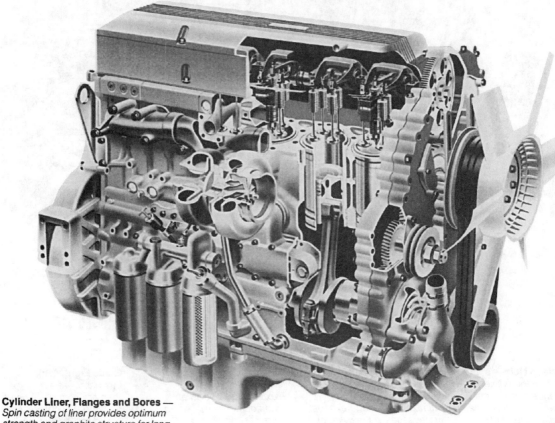

Cylinder Liner, Flanges and Bores — Spin casting of liner provides optimum strength and graphite structure for long life. Plateau honing minimizes piston ring break-in and allows quicker ring seal. Flanges at the liner upper ends seat in counterbores in the block deck and project slightly above it to compress the head gasket for a good seal. Cylinder bores feature replaceable, wet-type cylinder liners.

Isolators — reduce engine noise.

Crankshaft, Main and Rod Bearings — Crankshaft is forged induction hardened steel for high strength, and features computer positioned oil passages to promote a thick oil film in the highest loaded sections. Large main and rod bearings increase bearing life and tolerance to wear.

FIGURE 1–2 Detroit Diesel 60 series four-stroke-cycle diesel engine features. (Courtesy of Detroit Diesel Corporation.)

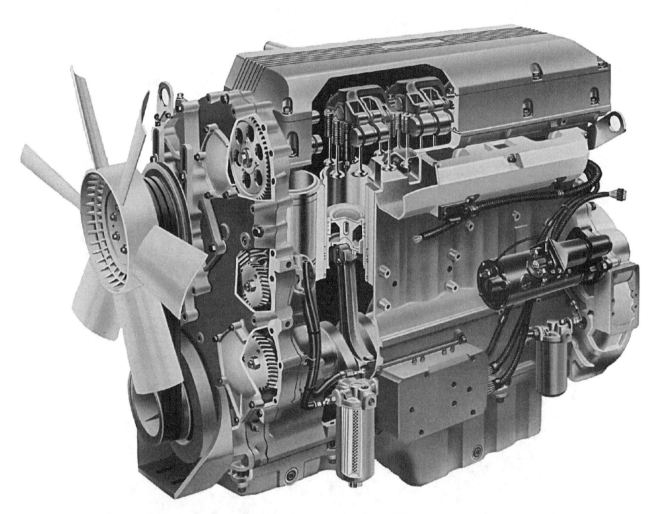

Fluid Weep Hole—is provided in the unlikely occurrence of an upper seal water leak. It will leak externally instead of internally to the crankcase. This also allows easy identification of a problem before damage can occur.

Overhead Camshaft—design eliminates parts, is easy to inspect and service and optimizes intake and exhaust air passages in the cylinder head for easier engine breathing, and minimizes valve train losses.

Eight Head Bolts per Cylinder—provide a uniform load on the gasket and liner to reduce stress on the liner flange and block counterbore.

Strong Cylinder Block—block is extensively ribbed and contoured for maximum rigidity and sound reduction, without excessive weight.

Redundant Internal Seals—provide an extra seal in the event of primary seal malfunction.

Grade Eight Metric Fasteners—are stronger than are commonly used on heavy-duty engines, thus improving gasket loads and decreasing likelihood of breaking. Flanged fasteners eliminate washers.

FIGURE 1–2 (continued)

- **Four-stroke-cycle design** — long, effective power strokes for more complete fuel combustion.
- **Aftercooler cools inlet charge temperature** — increasing power while reducing thermal stresses that can cause premature wear of pistons, rings and liners.
- **High torque rise** — offers superior lugging ability to keep moving through tough spots without downshifting.
- **High-pressure, direct-injection fuel system** — provides excellent fuel atomization for unmatched economy and is adjustment-free for superior reliability and durability.
- **Oil-cooled pistons and full-length, water-cooled cylinder liners** — provide maximum heat transfer for longer component life.
- **Engine oil cooler** — maintains optimum oil temperature for proper cooling and longer lubricant life.

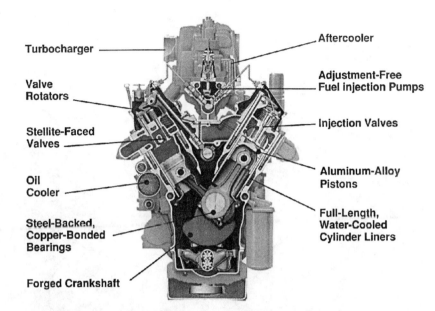

FIGURE 1–3 Caterpillar 3408 engine features. (Courtesy of Caterpillar Inc.)

In general an average air/fuel ratio for good combustion is about 15 parts of air to 1 part of fuel by weight. However, the diesel engine always takes in a full charge of air (since there is no throttle plate in most systems), but only a small part of this air is used at low or idle engine speeds. At idle the ratio of air to fuel may be as high as 100 to 1. At other speeds as much as 15% to 25% excess air is taken in. Air consists of about 20% oxygen, while the remaining 80% is mostly nitrogen. This means that for every gallon of fuel burned, the oxygen in 9000 to 10,000 gallons of air is required.

Much of the heat energy produced by the engine is lost through the exhaust, cooling, and lubrication systems, as well as through radiation, and is therefore not available to produce power. A further loss occurs through the friction of moving engine and drive train parts.

FOUR-STROKE-CYCLE DIESEL ENGINE OPERATION

The movement of the piston from its uppermost position (TDC, top dead center) to its lowest (BDC, bottom dead center) position is called a stroke. Many diesel engines operate on the four-stroke-cycle principle. A series of events involving four strokes of the piston completes one cycle. These events are: (1) the intake stroke, (2) the compression stroke, (3) the power stroke, and (4) the exhaust stroke. Two revolutions of the crankshaft and one revolution of the camshaft are required to complete one cycle **(Figure 1–8)**.

On the intake stroke the piston is pulled down in the cylinder by the crankshaft and connecting rod. During this time the intake valve is held open by the camshaft. The downward movement of the piston in the cylinder creates a low-pressure area (vacuum), so atmospheric pressure forces air past the intake valve into the cylinder. Atmospheric pressure is approximately 14.7 pounds per square inch (psi) [about 101.35 kilopascals (kPa)] at sea level. Pressure in the cylinder during the intake stroke is considerably less than this. The pressure difference is the force that causes the air to flow into the cylinder, since a liquid or a gas (vapor) will always flow from a high- to a low-pressure area. Atmospheric pressure and its effect on engine performance are described in more detail in Chapter 3.

As the piston is moved up by the crankshaft from BDC the intake valve closes. The air is trapped in the cylinder above the piston. Further piston travel compresses the air to approximately 1/20 of its original volume (approximately 20:1 compression ratio) by the

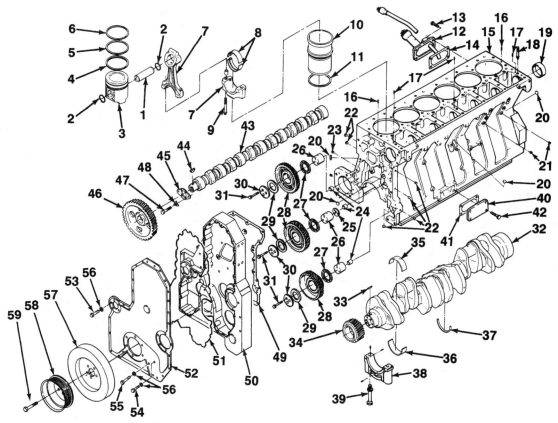

Ref. No.	Description	Qty.	Ref. No.	Description	Qty.	Ref. No.	Description	Qty.
1	Pin, Piston	6	22	Plug, Expansion (9.83 mm) [0.387 inch]	8	42	Capscrew, [10-1.50x20]	8
2	Ring, Retaining	12	23	Plug, Pipe [1/2 inch]	1	43	Camshaft	1
3	Piston	6	24	Dowel, Ring [19.23]	3	44	Key, Offset	1
4	Ring, Oil	6	25	Plate, Thrust Bearing Wear	1	45	Support, Camshaft Thrust	1
5	Ring, Compression	6	26	Shaft, Idler	3	46	Gear, Camshaft	1
6	Ring, Compression	6	27	Bearing, Thrust	3	47	Capscrew	2
7	Rod, Connecting	6	28	Assembly, Idler Gear	3	48	Lockplate	2
8	Bearing, Connecting Rod	12	29	Bearing, Thrust	3	49	Gasket, Gear Cover Housing	1
9	Capscrew, 12 point [14-1.5x91.25]	12	30	Gear Retainer	3	50	Gear Cover	1
10	Liner, Cylinder	6	31	Capscrew [10-1.50x60]	9	51	Gasket, Gear Cover	1
11	O-ring, Liner	6	32	Crankshaft	1	52	Cover, Gear	1
12	Housing, Dipstick	1	33	Key, Plain Woodruff	1	53	Capscrew [10-1.50x60]	12
13	Capscrew [10-1.50x20]	4	34	Gear, Crankshaft	1	54	Capscrew [10-1.50x30]	1
14	Gasket, Dipstick Housing	1	35	Bearing, Main (upper)	7	55	Capscrew [10-1.50x50]	1
15	Cylinder Block	1	36	Bearing, Main (lower)	7	56	Washer, Plain	15
16	Dowel, Pin [7/16x1 inch]	2	37	Bearing, Thrust (No. 4 Main)	4	57	Damper, Vibration	1
17	Dowel, Pin [6.000x12.00]	3	38	Cap, Main Bearing	7	58	Pulley, Crankshaft	1
18	Stud [10-1.50x50]	7	39	Capscrew and Washer, Main Bearing	14	59	Capscrew [14-1.50x50]	5
19	Bushing, Camshaft	7	40	Cover, Hand Hole	2			
20	Plug, Expansion [1.00 Dia]	3	41	Gasket, Hand Hole	2			
21	Plug, Pipe [1.4-18]	2						

FIGURE 1-4 Cylinder block components of Cummins L10 engine. (Courtesy of Cummins Engine Company, Inc.)

FOUR-STROKE-CYCLE DIESEL ENGINE OPERATION

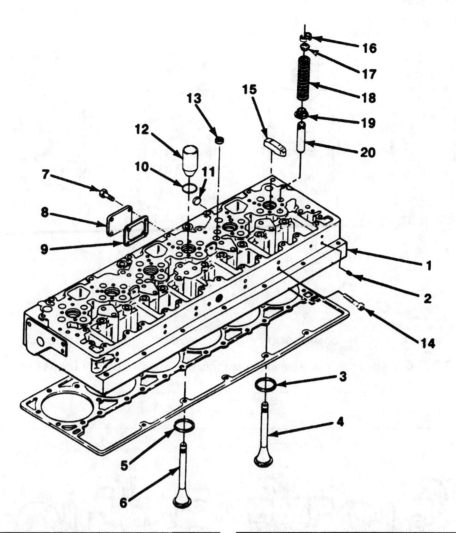

Ref. No.	Description	Qty.	Ref. No.	Description	Qty.
1	Cylinder Head	1	11	Expansion Plug	2
2	Pipe Plug	15	12	Injector Sleeve	6
3	Valve Insert	12	13	Expansion Plug	8
4	Valve, Exhaust	12	14	Fuel Passage Plug	5
5	Valve Insert	12	15	Crosshead	12
6	Valve, Intake	12	16	Valve Collet	24
7	Capscrew	4	17	Valve Spring Retainer	24
8	Cover Plate	1	18	Valve Spring	24
9	Cover Plate Gasket	1	19	Spring Guide	24
10	O-Ring	6	20	Valve Stem Guide	24

FIGURE 1–5 Cylinder head components of Cummins L10 engine. (Courtesy of Cummins Engine Company, Inc.)

time the piston reaches TDC. This completes the compression stroke. (Compression ratios vary from about 13:1 to about 22:1.)

Compressing the air to this extent creates a great deal of friction between the air molecules, which creates sufficient heat to ignite the fuel when it is injected. This temperature ranges from 800°F (426.6°C) to 1200°F (648.8°C). For this reason a spark ignition system is not required on a diesel engine.

The burning of the air/fuel mixture occurs at a controlled rate. Expansion of the burning mixture increases pressure. This increased pressure forces the

8 Chapter 1 DIESEL ENGINE OPERATION AND CLASSIFICATION

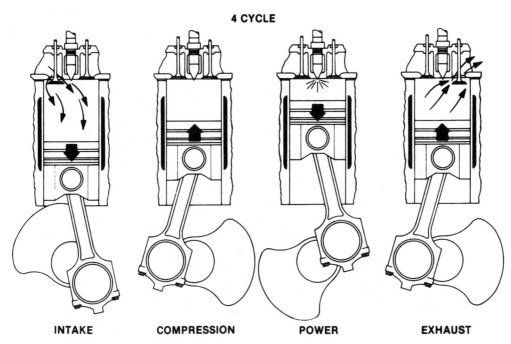

FIGURE 1–6 Four-stroke-cycle engine operation. (Courtesy of Detroit Diesel Corporation.)

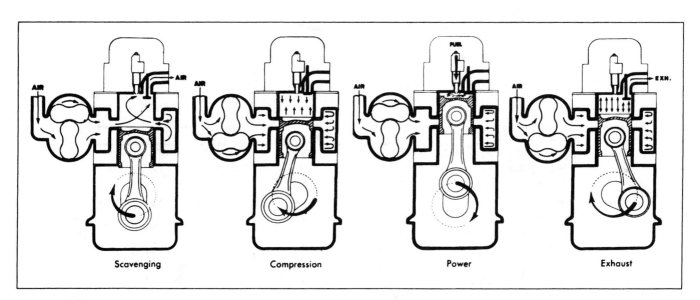

FIGURE 1–7 Two-stroke-cycle engine operation. (Courtesy of Detroit Diesel Corporation.)

piston down on the power stroke, causing the crankshaft to rotate. See Chapter 12 for a discussion of the four stages of diesel combustion.

At the end of the power stroke, the camshaft opens the exhaust valve, and the exhaust stroke begins. Remaining pressure in the cylinder and upward movement of the piston force the exhaust gases out of the cylinder. At the end of the exhaust stroke, the exhaust valve closes and the intake valve opens, continually repeating the entire cycle of events.

To start the engine, some method of cranking the engine is required to turn the crankshaft and cause piston movement. This is done by the starting motor when the key is turned to the start position. When sufficient air

VALVE TIMING 9

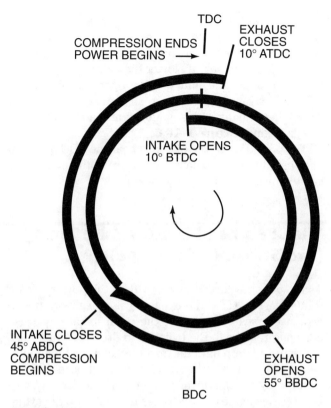

FIGURE 1–8 Valve timing diagram for a four-stroke-cycle engine. The diagram represents two crankshaft revolutions.

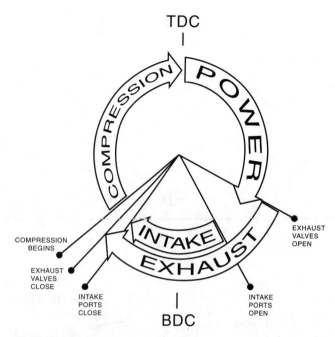

FIGURE 1–9 Two-stroke-cycle engine valve and port timing chart. The diagram represents 360° of crankshaft and camshaft rotation.

has entered the cylinders and fuel has been injected, the power strokes create enough energy to continue the crankshaft rotation. At this point the key is released and the starter is disengaged.

Sufficient energy is stored in the flywheel and other rotating parts on the power strokes to move the pistons and related parts through the other three strokes (exhaust, intake, and compression). The amount of fuel allowed to enter the cylinders determines the power and speed developed by the engine.

TWO-STROKE-CYCLE DIESEL ENGINE OPERATION

The two-stroke-cycle diesel engine completes all four events (intake, compression, power, and exhaust) in one revolution of the crankshaft or two strokes of the piston (**Figures 1–9** and **1–10**).

A series of ports or openings is arranged around each cylinder in such a position that the ports are open when the piston is at the bottom of its stroke. An air box running the length of the block surrounds the bottom of the cylinders. A blower forces air into the air box and cylinder through the open ports, expelling all remaining exhaust gases past the open exhaust valves and filling the cylinder with air. This is called *scavenging*.

As the piston moves up, the exhaust valves close, and the piston covers the ports. The air trapped above the piston is compressed by the upward-moving piston. Just before the piston reaches TDC, the required amount of fuel is injected into the cylinder. The heat generated by compression of the air ignites the fuel almost immediately. Combustion continues until the fuel injected has been burned. The pressure resulting from combustion forces the piston downward on the power stroke. When the piston is approximately halfway down, the exhaust valves are opened, allowing the exhaust gases to escape. Further downward movement uncovers the inlet ports, causing fresh air to enter the cylinder and expel the exhaust gases. The entire procedure is then repeated as the engine continues to run.

VALVE TIMING

Intake Valve Timing (Four-Stroke-Cycle Engine)

It is advantageous to have the intake valve begin to open before the piston reaches the TDC position on the exhaust stroke for two reasons. First, this ensures that the valve will be fully open to take advantage of the

Chapter 1 DIESEL ENGINE OPERATION AND CLASSIFICATION

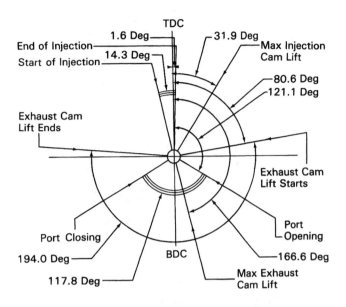

FIGURE 1–10 Series 92 Detroit Diesel two-stroke-cycle engine timing diagram. (Courtesy of Detroit Diesel Corporation.)

moves up on the exhaust stroke it forces the exhaust gases out of the cylinder. As the piston nears the top the velocity of the escaping exhaust gases reduces pressure in the cylinder to slightly lower than atmospheric pressure, and since the intake valve opens at about 15° BTDC, a scavenging effect is created. At about 10° to 15° ATDC, the exhaust valve is closed and a new cycle of events begins **(Figure 1–10)**.

The period when the piston is near the TDC position and the intake and exhaust valves are both open at the same time is known as *valve overlap*.

Intake Port and Exhaust Valve Timing (Two-Stroke-Cycle Engine)

As the piston is forced downward on the power stroke to about 110° ATDC the exhaust valve begins to open. Exhaust gases escape past the open exhaust valves, causing pressure in the cylinder to drop. The pressure continues to drop until the piston reaches a point about 130° to 135° ATDC. At this point the inlet ports are uncovered by the downward-moving piston. Air pressurized by the engine blower forces fresh air into the cylinder through the ports, which causes any remaining exhaust gases in the cylinder to be forced out past the open exhaust valves.

During this time the piston continues past the BDC position and starts to move upward. When it reaches a point about 48° ABDC, the inlet ports are closed by the upward-moving piston. A little later, about 63° ABDC, the exhaust valves are closed and the compression stroke begins. Just before the piston reaches TDC, about 20° BTDC, fuel is injected into the cylinder, causing ignition and expansion to take place. The cycle then repeats itself **(Figures 1–9 and 1–10)**.

rapid downward movement of the piston during the first 90° of crank rotation, when the greatest pressure drop in the cylinder occurs. Second, it helps overcome the *static inertia* (the tendency of the stationary air to remain stationary) of the air in the intake manifold and, as a result, starts air flow into the cylinder sooner.

It also helps engine breathing to keep the intake valve open until the piston has passed the BDC position. This allows the *kinetic inertia* (the tendency of air in motion to stay in motion) of the incoming air to fill the cylinder more completely.

Intake valve timing varies somewhat among the various makes and models of diesel engines. Not uncommon is an intake valve opening at 10° to 15° BTDC and an intake valve closing at 40° to 45° ABDC.

Exhaust Valve Timing

To ensure expulsion of exhaust gases, the exhaust valve opens at about 50° BBDC. Exhaust gas pressure in the cylinder begins to force the gases past the exhaust valve into the exhaust system. As the piston passes BDC and

COMPARISON OF TWO-STROKE-CYCLE AND FOUR-STROKE-CYCLE ENGINES

It could be assumed that a two-stroke-cycle engine with the same number of cylinders, the same displacement, compression ratio, and speed as a four-stroke-cycle engine would have twice the power, since it has twice as many power strokes. However, this is not the case, since both the power and compression strokes are shortened to allow scavenging to take place. The two-stroke-cycle engine also requires a blower, which takes engine power to drive.

About 160° of each 360° of crankshaft rotation is required for exhaust gas expulsion and fresh air intake (scavenging) in a two-cycle engine. About 415° of each

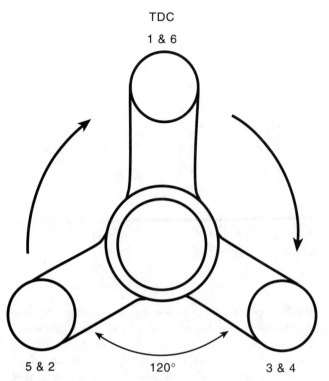

FIGURE 1–11 Operating pairs of a six-cylinder in-line four-stroke-cycle engine.

720° of crankshaft rotation in a four-stroke-cycle engine is required for intake and exhaust. These figures indicate that about 44.5% of crank rotation is used for the power-producing events in the two-stroke-cycle engine, while about 59% of crank rotation is used for these purposes in the four-stroke-cycle engine. Friction losses are consequently greater in the four-stroke-cycle engine. Heat losses, however, are greater in the two-stroke-cycle engine through both the exhaust and the cooling systems. In spite of these differences, both engine types enjoy prominent use worldwide.

PISTON POSITIONS

The engine operating principles just described apply to a single cylinder. On a multicylinder engine the other cylinders are at different stages of the same cycle of events. For example, in a six-cylinder four-stroke-cycle engine the pistons and connecting rods are connected to the crankshaft in pairs. Each pair is on the same plane, but the pairs are spaced 120° apart **(Figure 1–11)**. With a firing order of 1-5-3-6-2-4 each cylinder fires once in two turns (720°) of crankshaft rotation. When piston No. 1 is at TDC, so is piston No. 6; however, they are 360° apart in the four-stroke cycle of events. The other two pairs are each also 360° apart, as follows **(Figure 1–12)**:

No. 1—between compression and power strokes
No. 6—between exhaust and intake strokes
No. 5—coming up on the compression stroke
No. 2—coming up on the exhaust stroke
No. 3—going down on the intake stroke
No. 4—going down on the power stroke

On a six-cylinder engine with a firing order of 1-5-3-6-2-4 it is easy to identify the pairs by simply placing the first three numbers of the firing order over the last three. On an eight-cylinder engine with a firing order of 1-8-4-3-6-5-7-2, place the first four figures in the firing order over the last four. It is very helpful to know the firing order of an engine and what each cylinder is doing in relation to the others when adjusting the valves and unit injectors **(Figures 1–13 and 1–14)**.

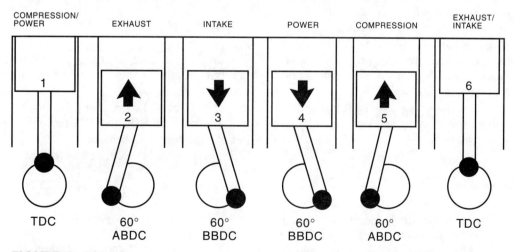

FIGURE 1–12 Operating relationship between cylinders in a six-cylinder engine with a firing order of 1-5-3-6-2-4.

$$\frac{1\ 5\ 3}{6\ 2\ 4}$$

$$\frac{1\ 8\ 4\ 3}{6\ 5\ 7\ 2}$$

FIGURE 1–13 Operating pairs of cylinders for a six-cylinder in-line engine (top) and a V-8 engine (bottom).

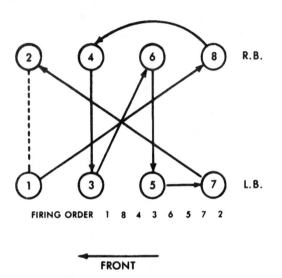

FIGURE 1–14 V-8 engine with a firing order of 1-8-4-3-6-5-7-2. (Courtesy of Detroit Diesel Corporation.)

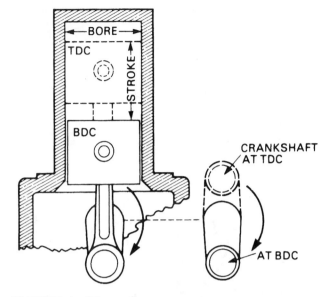

FIGURE 1–15 The distance the piston travels from its highest position in the cylinder (TDC, top dead center) to its lowest position in the cylinder (BDC, bottom dead center) is known as the *stroke*. The length of the stroke is determined by crankshaft design. The bore dimension is the cylinder diameter.

BORE, STROKE, AND DISPLACEMENT

A number of factors determine the ability of an engine to produce usable power. Some of these are determined by the manufacturer, such as cylinder bore, stroke, displacement, and compression ratio. Other factors such as force, pressure, vacuum, and atmospheric pressure also affect the power output of an engine. The amount of power an engine is able to produce is measured in several ways, as is the efficiency of an engine. All these factors, terms, and conditions must be properly understood to gain an understanding of their individual and combined effects on engine performance.

The following definitions are essential for understanding an engine's operating principles, its ability to produce power, and some of its limitations.

Cylinder Bore

Cylinder bore is the diameter of the engine's cylinder measured in inches or millimeters. Cylinder bore size is a major factor in determining engine displacement **(Figure 1–15)**.

Piston Stroke

The engine's stroke is the distance traveled by the piston from its BDC position to its TDC position measured in inches or millimeters. The distance traveled by the piston is determined by crankshaft design and is exactly twice the crank throw measurement.

The engine's stroke is also a major factor in engine displacement **(Figure 1–15)**.

Engine Displacement

The displacement of an engine is determined by cylinder bore diameter, length of stroke, and number of cylinders. Simply stated, it is the amount or volume of air pushed out of the cylinder (displaced) by one piston as it moves from BDC to TDC, multiplied by the number of cylinders in the engine.

Displacement is calculated as follows:

$$\pi r^2 \times \text{stroke} \times \text{no. of cylinders}$$

$$\pi = \frac{22}{7}$$

$$r^2 = \text{radius} \times \text{radius}$$

(radius = 1/2 of cylinder bore)

Therefore, a six-cylinder engine with a 4.50-in. bore and a 4.00-in. stroke has a displacement of:

$$\frac{22}{7} \times (2.00 \times 2.00) \times (4.00 \times 6) \quad \text{or} \quad \frac{22}{7} \times 4.00 \times 24$$
$$= 302 \text{ in.}^3$$

Another method of calculating displacement is to multiply 0.7854 × stroke × no. of cylinders. The proportionate area of a circle is 0.7854 that of a square with the same dimensions as the circle diameter.

In metric terms, the displacement of a six-cylinder engine with a 100.0 mm bore and a stroke of 80 mm is calculated as follows. First, since metric displacement is stated in cubic centimeters, it is necessary to convert the bore and stroke dimensions to centimeters.

$$100.0 \text{ mm} = 10.00 \text{ cm bore} \quad 80 \text{ mm} = 8.0 \text{ cm stroke}$$
$$\frac{22}{7} \times (5.0 \times 5.0) \times (8.0 \times 6) = \frac{22}{7} \times 25 \times 48$$
$$\text{or}$$
$$\frac{22}{7} \times 1200 = \frac{26,400}{7} = 3800 \text{ cm}^3 \text{ of displacement,}$$
$$\text{or 3.8 liters (L)}$$

The power an engine is able to produce depends very much on its displacement. Engines with more displacement are able to take in a greater amount of air on each intake stroke and can therefore produce more power. Engine displacement can be increased by engine design in three ways: (1) by increasing cylinder bore diameter, (2) by lengthening the stroke, and (3) by increasing the number of cylinders.

COMPRESSION RATIOS

The compression ratio of a diesel engine is much higher than that of a gasoline engine. This is true because only air is compressed. Compressing air at diesel compression ratios causes air molecules to collide rapidly with each other. The friction caused by these collisions creates heat. Temperatures of 1000°F (540°C) or higher can be reached depending on the compression ratio. This is hot enough to ignite the fuel when it is injected near the top of the compression stroke.

Compression ratios are determined by cylinder displacement and combustion chamber volume. To calculate compression ratio, divide the combustion chamber volume into the total cylinder volume. For example, if

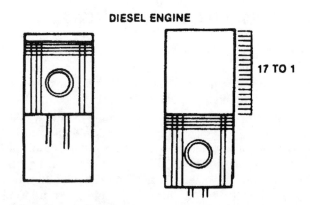

FIGURE 1-16 Typical diesel engine compression ratio.

the combustion chamber volume is 2 in.³ and the cylinder displacement is 32 in.³, the total cylinder volume is 34 in.³ **(Figure 1-16).**

$$34 \div 2 = 17:1$$

Metric Example. If the combustion chamber volume is 36 cm³ and the cylinder displacement is 650 cm³, the total cylinder volume is 686 cm³.

$$686 \div 36 = 19:1$$

These are typical compression ratios for diesel engines.

An engine with a 19:1 compression ratio will produce approximately 500 psi of compression pressure (3448 kPa). This rises to approximately 2000 psi of combustion pressure (13,790 kPa) when the piston reaches the TDC position.

The compression ratio of an engine can be changed in several ways. To raise the compression ratio, the combustion chamber volume can be reduced and the cylinder volume left unchanged, or the cylinder volume can be increased (increased bore or stroke) and the combustion chamber volume left unchanged.

To lower the compression ratio, the combustion chamber volume can be increased and the cylinder volume left unchanged, or the cylinder volume can be reduced (decreased bore or stroke) and the combustion chamber volume left unchanged.

Compression ratios in diesel engines must be high enough to create sufficient heat for good ignition and adequate power. Ratios range from about 11:1 to as high as 22:1 in some small high-speed diesel engines.

PISTON AND CRANK PIN TRAVEL

In general, the speed and distance traveled by the crank pin can be considered to be constant and in a uniform path at any given engine speed. This is not the case, however, with piston speed and travel.

When the movement of the crank pin is uniform and at a constant velocity, the speed and distance traveled by the piston connected to it varies due to crank pin and connecting rod angle. When the piston reaches the TDC position, its speed is zero. As it begins to move downward, its speed increases rapidly. At a point where the crank pin is about 63° ATDC, the piston has reached its maximum speed. This is the point at which the crank throw centerline and the connecting rod centerline form an angle of 90°. After this point, the piston speed decreases until it reaches zero at the BDC position. As the piston moves upward from this point its speed increases until it reaches its maximum at about 63° BTDC. From this point on the piston slows down until it reaches TDC again, where it once more comes to a stop.

Piston speed is normally stated as an average speed in feet per minute (ft/min) and can be calculated as follows:

$$\text{piston speed} = \frac{\text{stroke (in feet)} \times \text{rpm}}{2}$$

We divide by 2 because the piston travels the stroke distance twice (up and down) during each revolution.

It should also be noted that the distance traveled by the piston varies with the crank angle. Starting with the TDC position of the crank pin, the piston travels a greater distance during the first 90° of crank rotation than it does during the second 90°. As the crank pin continues past the BDC position through the third 90° of rotation, the piston travel is less than it is for the final 90° when it reaches the TDC position.

The force of combustion pressure has no rotating effect on the crankshaft when the piston is at the TDC position. As the crank pin passes the TDC position on the power stroke, the mechanical advantage through the angle of the connecting rod increases until it reaches its maximum at about 63° ATDC. This is the point where the crank pin and connecting rod form a 90° angle. Thereafter, the force advantage decreases rapidly while combustion pressure also decreases.

ENGINE SPEED

Diesel engines are often classified according to their speed capability. Low, medium, high, and super-high speed classifications are used. Both piston speed and rotating speed have been used as measures of engine speed capability.

Piston speed is expressed in feet per minute (or meters per minute), while rotating speed refers to crankshaft speed expressed in revolutions per minute (rpm). Both methods leave something to be desired. Rotating speed alone does not take into consideration the size and piston stroke of an engine. A long-stroke engine with a low rotating speed not exceeding 750 rpm may have quite a high piston speed. A high-speed short-stroke engine, on the other hand, has a relatively low piston speed. For these reasons an engine speed characteristic known as the *speed factor* is used.

The speed factor of an engine is determined as follows:

$$\text{speed factor} = \frac{\text{crankshaft rpm} \times \text{piston speed (in feet per minute)}}{100{,}000}$$

The division by 100,000 is an arbitrary step to produce a less cumbersome final figure. This allows all diesel engines to be assigned a speed factor between the figures 1 and 81. This results in the following classifications:

1. Engines with a speed factor of 1 to 3 are designated as low-speed engines.
2. Speed factors of 3 to 9 are medium-speed engines.
3. Speed factors of 9 to 27 are high-speed engines.
4. Speed factors of 27 to 81 are super-high-speed engines.

It should be noted that these classifications have been arbitrarily determined by multiplying each successively higher speed classification by a factor of 3 times the next lower speed category.

To find the speed factor of a six-cylinder engine with a 5-in. bore and a 6-in. stroke with a rated speed of 2500 rpm, proceed as follows. First, we find the mean piston speed of the engine from the stroke distance and the time it takes the piston to travel the stroke distance. The piston travels up once and down once during each revolution or two strokes of the piston. The distance traveled by the piston during one revolution is twice the stroke distance. Since piston speed is expressed in feet per minute and piston stroke is expressed in inches, we must convert inches to feet by dividing by 12. From this information we can calculate mean piston speed as follows:

$$\text{piston speed} = \text{rpm}\, \frac{2 \times \text{stroke}}{12}$$

Using the engine figures mentioned earlier, we calculate thus: $2500 \div 2 \times 6 \div 12 = 625$ ft/min. To obtain the speed factor, we calculate thus: rpm $\times$ piston speed $\div$ 100,000, or $2500 \times 625 \div 100,000 = 15.625$.

ENGINE BRAKING SYSTEMS

Vehicle retard systems that act on the engine include the Jacobs brake used by several engine manufacturers, the Mack Dynatard system used on some Mack trucks, and the Cummins C brake. All operate on the same general principle **(Figures 1–17 to 1–19)**.

The engine brake converts the engine cylinders into power-absorbing air compressors. The injector push rod is used to actuate a master piston that is connected to a slave piston positioned to open the exhaust valve at the end of the compression stroke on the Jacobs brake. This can happen only when the engine throttle is closed, the clutch is engaged, and a switch on the dash panel is in the ON position. The dash switch activates a solenoid that opens a valve to allow engine oil pressure to activate the brake. Jacobs' brake oil is instantly dumped through a separate passage when the dash switch is turned off.

Compressed air from the engine cylinders is vented through the exhaust valves near the end of the compression stroke, thereby preventing the stored energy of the compressed air from acting on the pistons and crankshaft on the downstroke of the piston.

Electronic control of engine compression braking is provided on engines with electronically controlled fuel systems **(Figure 1–20)**.

ENGINE CLASSIFICATION

Diesel engines can be classified in a number of different ways, depending on engine design.

1. *By cycles.* Two-stroke- and four-stroke-cycle engines are being used.

2. *By cooling systems.* Liquid-cooled engines and air-cooled engines are being used. Liquid-cooled engines are the most common in the diesel industry.

3. *By cylinder arrangement.* Engine block configuration or cylinder arrangement depends on cylinder block design. Cylinders may be arranged in a straight line one behind the other. The most common *in-line* designs are the four- and six-cylinder engines.

The V type of cylinder arrangement uses two *banks* of cylinders arranged in a 60° to 90° V design. The most common examples are those with two banks of three to eight cylinders each. The opposed engine uses two

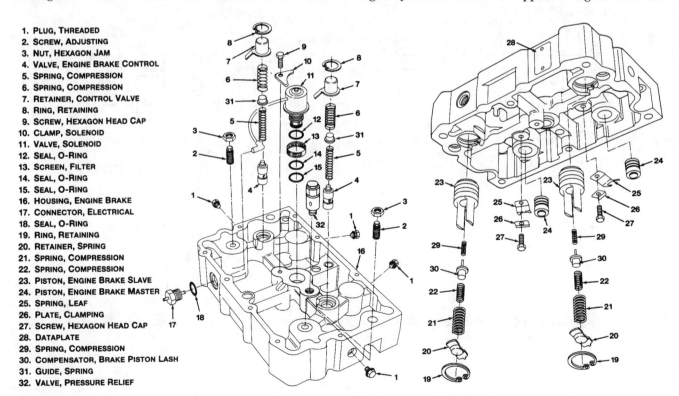

1. PLUG, THREADED
2. SCREW, ADJUSTING
3. NUT, HEXAGON JAM
4. VALVE, ENGINE BRAKE CONTROL
5. SPRING, COMPRESSION
6. SPRING, COMPRESSION
7. RETAINER, CONTROL VALVE
8. RING, RETAINING
9. SCREW, HEXAGON HEAD CAP
10. CLAMP, SOLENOID
11. VALVE, SOLENOID
12. SEAL, O-RING
13. SCREEN, FILTER
14. SEAL, O-RING
15. SEAL, O-RING
16. HOUSING, ENGINE BRAKE
17. CONNECTOR, ELECTRICAL
18. SEAL, O-RING
19. RING, RETAINING
20. RETAINER, SPRING
21. SPRING, COMPRESSION
22. SPRING, COMPRESSION
23. PISTON, ENGINE BRAKE SLAVE
24. PISTON, ENGINE BRAKE MASTER
25. SPRING, LEAF
26. PLATE, CLAMPING
27. SCREW, HEXAGON HEAD CAP
28. DATAPLATE
29. SPRING, COMPRESSION
30. COMPENSATOR, BRAKE PISTON LASH
31. GUIDE, SPRING
32. VALVE, PRESSURE RELIEF

FIGURE 1–17 Cummins C brake components. Solenoids act on two, four, or six cylinders depending on dash panel switch position. (Courtesy of Cummins Engine Company, Inc.)

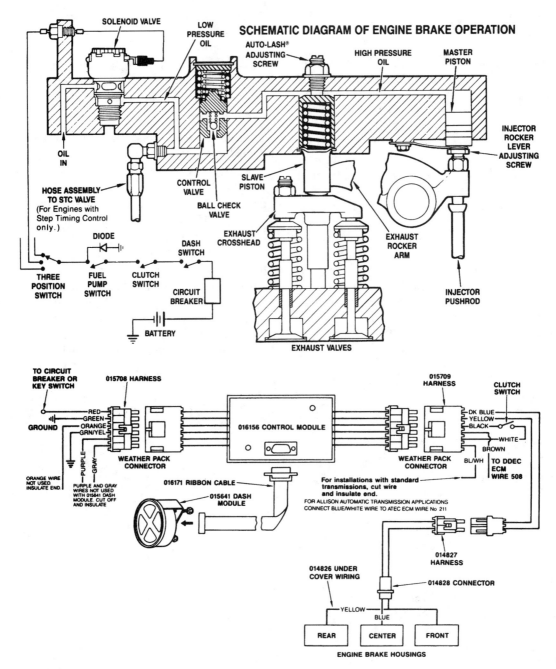

FIGURE 1-18 Typical Jacobs brake operation. (Courtesy of Cummins Engine Company, Inc.)

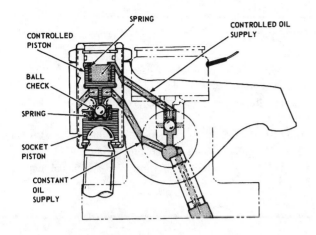

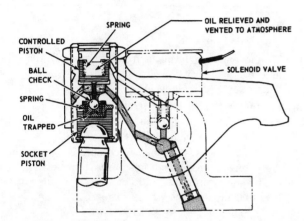

FIGURE 1–19 Mack Dynatard lash adjuster operation brake off (top) and brake on (bottom). (Courtesy of Mack Trucks Inc.)

banks of cylinders opposite each other with the crankshaft in between.

4. *By displacement.* Engine displacement is the amount of air displaced by the piston when it moves from BDC to TDC; it varies with cylinder bore size, length of piston stroke, and number of cylinders.

5. *By engine speed.* Engines are classified as low, medium, high, and super high speed.

6. *By application.* Industrial, highway, rail, agricultural, marine, or stationary use.

DIESEL ENGINE SYSTEMS

For a diesel engine to operate, the following systems must perform their functions in an efficient manner:

- The *fuel injection system* provides fuel to the cylinders.
- The *air induction system* provides the air to the cylinders.
- The *compression system* sufficiently compresses the air to provide a controlled rate of combustion and expansion. It is also used for engine braking.
- The *exhaust system* provides the means to efficiently dispose of the burned gases.
- The *lubrication system* reduces friction and wear and helps in cooling, sealing, and cleaning.
- The *cooling system* maintains the engine's most efficient operating temperature.
- The *ventilation system* removes harmful crankcase vapors.

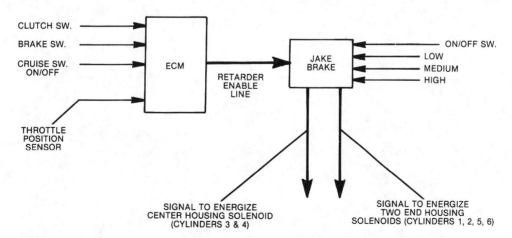

FIGURE 1–20 Electronic control system schematic for Jacobs brake. (Courtesy of Caterpillar Inc.)

REVIEW QUESTIONS

1. The diesel engine converts _____ energy of the fuel into _____ energy and to usable _____ energy.
2. What is the appropriate air/fuel ratio of a diesel engine?
3. List the major components of the diesel engine.
4. List and describe the four strokes of a four-stroke-cycle diesel engine.
5. Describe the basic operating principle of a two-stroke-cycle diesel engine.
6. The diesel engine compresses the air/fuel mixture on the compression stroke. (T) (F)
7. List the seven engine systems required for engine operation.
8. List six methods of classifying engine types and sizes.
9. Define engine bore, stroke, and displacement.
10. How is the compression ratio of an engine calculated?

TEST QUESTIONS

1. One of the functions of a diesel engine is to
   a. equalize weight distribution
   b. convert fuel energy into mechanical energy
   c. convert mechanical energy into horsepower
   d. convert kinetic energy into mechanical energy
2. In a four-stroke-cycle engine, the intake stroke follows the
   a. compression stroke
   b. power stroke
   c. exhaust stroke
   d. rate of flame propagation
3. Engine parts that convert reciprocating motion to rotary motion are the
   a. connecting rod and crankshaft
   b. piston and connecting rod
   c. crankshaft and flywheel
   d. crankshaft and timing chain
4. In a two-stroke-cycle diesel engine the power stroke occurs every
   a. four crankshaft revolutions
   b. two crankshaft revolutions
   c. crankshaft revolution
   d. every half crankshaft revolution
5. The movement of the piston from TDC to BDC is called a
   a. cycle   c. piston clearance
   b. stroke   d. half cycle
6. The purpose of the intake stroke is to create
   a. power   c. pressure
   b. a vacuum d. diesel flow
7. Technician A says the higher compression ratio in a diesel engine increases horsepower. Technician B says the higher compression ratio of a diesel engine creates heat to ignite the fuel. Who is correct?
   a. Technician A   c. both are correct
   b. Technician B   d. both are wrong
8. One of the main differences between a gasoline and a diesel engine is the method of
   a. starting   c. ignition
   b. valve timing   d. cycling
9. In the two-stroke-cycle diesel engine the function of the intake valve is taken over by the
   a. ports   c. blower
   b. sleeves   d. pistons
10. All diesel engines use which type of combustion?
    a. internal c. ignition
    b. instant   d. external

CHAPTER 2

ENGINE PERFORMANCE, POWER, AND EFFICIENCY

INTRODUCTION

The power an engine is able to produce is a measure of its performance. Engine power varies among the different makes and models of engines. Engine design is the determining factor in the maximum power an engine is able to produce. Engine torque is closely related to but is not the same as engine power. Engine wear eventually reduces engine power and efficiency. A dynamometer is used to test the power and efficiency of an engine.

PERFORMANCE OBJECTIVES

After thorough study of this chapter, you should be able to do the following:

1. Define the following terms:
   - work
   - power
   - horsepower
   - torque
   - SAE power
   - friction
   - inertia
2. Describe atmospheric pressure and its effects on engine performance.
3. Define the following types of engine efficiency:
   - mechanical
   - thermal
   - fuel
   - cycle
   - volumetric
   - scavenge
4. Define engine and cylinder balance.
5. Describe basic hydraulic dynamometer operation.

TERMS YOU SHOULD KNOW

Look for these terms as you study this chapter, and learn what they mean.

work	atmospheric pressure
engine power	mechanical efficiency
horsepower	thermal efficiency
kilowatt	scavenge efficiency
indicated power	fuel efficiency
friction power	cycle efficiency
SAE power	engine balance
engine torque	cylinder balance
engine brake power	dynamometer
volumetric efficiency	

WORK

Work is done when an applied force overcomes a resistance and moves through a distance. Work produces measurable results. When sufficient energy is expended through an application of force (push, pull, or twist) to overcome the resistance to motion of any particular object, movement is the result.

Pulling a 1 lb object a distance of 1 ft results in 1 ft-lb of work being done (if friction is ignored). In other words, force times distance equals work ($W = F \times D$).

ENGINE POWER

Power is the rate at which work is done. Power can also be defined as the ability to do a specific amount of work in a specific amount of time.

Engine power in the English system is stated in horsepower and in the metric system in kilowatts. Both systems are explained here. The formula for calculating power is $P = F \times D \div T$, where F is force, D is distance, and T is time.

Horsepower

James Watt, a Scottish inventor, observing the ability of a horse to do work in a mine, decided arbitrarily that this ability to do work was the equivalent of raising 33,000 lb of coal a distance of 1 ft in 1 minute. This became the standard measurement of a unit of power called horsepower (HP).

Expressed as a formula,

$$1 \text{HP} = 33{,}000 \text{ lb} \times 1 \text{ ft} \times 1 \text{ min}$$

This formula allows the horsepower of an engine to be calculated if certain factors are known, namely, the force produced by an engine and the distance through which that force moves in 1 minute.

A device known as the *prony brake* can be used to obtain these factors. Since the prony brake is a braking device, the output of an engine is stated in terms of brake power **(Figure 2–1)**.

A prony brake uses a drum attached to the engine flywheel. A contracting brake band surrounds the drum. The band can be tightened to increase the load on the engine. An arm is attached at one end to the band. The other end of the arm is connected to a scale through a knife edge device. This assures accuracy of arm length from the center of crankshaft rotation to the scale.

With the engine running, the band is slowly tightened. This causes the arm to exert pressure on the scale. The brake horsepower output of an engine can be calculated using Watt's formula for 1 HP and the prony brake test results by simply dividing the prony brake test results by Watt's formula.

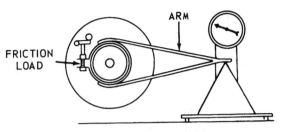

FIGURE 2–1 The prony brake uses a friction device to apply a load to an engine. An arm connected to the friction device deflects in proportion to the load applied and actuates the gauge, which indicates load in pounds or kilograms. Engine speed and load figures are then used to calculate brake horsepower.

Let's calculate the brake horsepower of a theoretical engine, assuming the following conditions:

- Engine speed: 2000 rpm
- Arm length: 3 ft (radius of circle arm would make if allowed to turn)
- Reading on scale: 100 lb

We use the formula

$$\frac{F \times D \times T}{33{,}000 \text{ lb} \times 1 \text{ ft} \times 1 \text{ min}} = \text{BHP}$$

To determine the circumference of a circle, we calculate $2\pi \times$ radius, or $2\pi r$, where π is 22/7.

Therefore,

$$F = 100 \text{ lb}$$
$$D = 3 \text{ ft} \times 2\pi \times 2000$$
$$T = 1 \text{ minute}$$

Substituting this information into the formula gives

$$\frac{2\pi \times 3 \times 2000 \times 100 \times 1}{33{,}000 \times 1 \times 1}$$
$$= \frac{3 \times 2000 \times 100 \times 1}{5250} = \frac{600{,}000}{5250} = 114 \text{ BHP}$$

Dividing 2π into 33,000 gives us the denominator 5250. Multiplying the remaining numerator figures results in 600,000. One horsepower is equal to 0.746 kilowatt (kW). Therefore, 114 BHP = 85 kW.

Indicated Power

Indicated power is the theoretical power an engine is able to produce. It is calculated from the following factors:

- P = mean effective pressure in the cylinder in pounds per square inch
- L = length of piston stroke in feet
- A = area of cylinder cross section in square inches
- N = number of power strokes per minute for one cylinder
- K = number of cylinders in the engine.

The formula for calculating indicated horsepower is, therefore,

$$IP = \frac{PLANK}{33{,}000}$$

Using this formula, it is possible to calculate the IP of an engine if the number of cylinders in the engine, the engine's bore and stroke, the engine's speed, and the mean effective pressure in the cylinder are known.

Friction Power

Friction power is the power required to overcome the friction of the various moving parts of the engine as it runs. Friction power increases as engine size and speed are increased.

The friction power of an engine can be calculated (if the indicated power and brake power are known) by subtracting the BP from the IP (IP − BP = FP).

SAE Power

SAE power is the power of an engine as determined by the Society of Automotive Engineers. Tests are performed under rigorously controlled conditions including the inlet air temperature, ambient temperature, humidity, and the like. A number of specific conditions, such as inlet air restriction and exhaust restriction, are also stated, since these are determining factors. Other factors and conditions must also be met.

SAE power is measured at the transmission output shaft, with all normal engine accessories mounted and operating. This includes the air cleaner and exhaust system.

Since a particular engine model may be used by a vehicle manufacturer for several different applications and may be equipped differently on different models, the SAE power for a particular engine varies depending on how it is equipped.

Engine Power—Kilowatts (kW)

In the metric system engine power is stated in kilowatts.

The power output of an engine is calculated, as stated earlier, by using the formula $P = W \div T$, where P is power, W is work, and T is time. We also know that Work = Force × Distance.

Force in the metric system is measured in newtons (N). Distance, for our purposes here, is measured in meters (m). Therefore, work can be expressed in terms of newton-meters (N · m).

Time is stated in minutes. To determine the power of an engine, we calculate force times distance divided by time ($F \times D \div T$).

The electrical unit for measuring work is the joule. One joule (J) is the equivalent of 1 ampere (A) of current under 1 volt (V) of pressure for 1 second. One joule is also the equivalent of 1 newton of force moving 1 meter of distance in 1 second (1 J = 1 N · m · s).

The watt is the unit of electrical power and is the equivalent of 1 joule per second. One kilowatt is 1000 watts.

To summarize:

$$1 \text{ newton-meter} = 1 \text{ joule (J)}$$
$$1 \text{ joule per second} = 1 \text{ watt (W)}$$
$$1000 \text{ watts} = 1 \text{ kilowatt (kW)}$$

It is possible to determine the power output in kilowatts of an engine by using a prony brake.

If, for example, the prony brake had a torque arm length of 1 meter, and the scale upon which it acted measured the applied force in newtons, the resultant power would be stated in newton-meters. The newton-meter output, at a given engine speed measured in revolutions per minute, could then be used to calculate engine power in kilowatts.

Assuming an engine speed of 2000 rpm, we obtain the engine speed per second by dividing 2000 by 60, since kilowatts are joules per second.

$$2000 \div 60 = 33.33$$

Assuming further that the applied force at the end of the torque arm is 1000 N and the length of the torque arm is 1 m, we calculate as follows:

$$2\pi \times 1 \times 33.33 = 209 \text{ kW}$$
or

$$\frac{2000}{60} \times 2\pi \times 1 = 209 \text{ kW}$$

or

$$2\pi \times 1 \times \frac{2000}{60} = 209 \text{ kW}$$

One kilowatt is 1.341 HP. Therefore, 209 kW = 280 HP.

Friction

A certain amount of force (push or pull) is required to slide one object over the surface of another. This resistance to motion between two objects in contact with each other is called *friction*. Friction increases with load. It requires more effort to slide a heavy object across a surface than it does to slide a lighter object over the same surface.

The condition of the two surfaces in contact also affects the degree of friction. Smooth surfaces produce less friction than rough surfaces. Dry surfaces cause more friction than surfaces that are lubricated or wet.

Residual oil clinging to the cylinder walls, rings, and pistons of an engine that has been stopped for some time will produce a greasy friction when the engine is started. Of course, as soon as the engine starts, the lubrication system supplies increased lubrication, which results in viscous friction.

Viscous comes from the word *viscosity*, which is a measure of an oil's ability to flow or its resistance to flow. Some energy is still required to slide a well-lubricated object over the surface of another. Although the layer of lubricant separates the two surfaces, the lubricant itself provides some resistance to motion. This is called *viscous friction*. Friction bearings provide a sliding friction action, while ball and roller bearings provide rolling friction. Rolling friction offers less resistance to motion than sliding friction.

Inertia

Inertia is the tendency of an object in motion to stay in motion or the tendency of an object at rest to stay at rest. The first can be called *kinetic inertia* and the latter *static inertia*. The moving parts of an engine are affected by kinetic inertia. A piston moving in one direction tries to keep moving in that direction because of kinetic inertia. The crankshaft and connecting rod must overcome this kinetic inertia by stopping the piston at its travel limit and reversing its direction. The static inertia of a vehicle that is stopped must be overcome by engine power to cause it to move.

ENGINE TORQUE AND BRAKE POWER

As an engine runs, the crankshaft is forced to turn by the series of pushes or power impulses imposed on the crank pins by the pistons and rods. This twisting force is called *torque*.

Engine torque and engine power are closely related. For instance, as we learned earlier, if we know the torque and speed of an engine, we can calculate its power.

Torque is equal to $F \times R$, where F is the force applied to the end of a lever and R is the length of the lever from the center of the turning shaft to the point on the lever at which force is being applied. R represents the radius of a circle through which the applied force would move if it moved through a complete revolution. Therefore, $T = F \times R$.

Engine torque is expressed in terms of pound-feet in the English system and in newton-meters in the metric system.

Maximum torque is produced in an engine when there is maximum pressure in the cylinders. Peak torque is therefore reached when there is maximum air and fuel delivery to the engine. This normally occurs at a somewhat lower engine speed than that at which maximum brake power is produced.

Engine torque drops off as engine speed increases to the point where cylinders take in less air. At this point, engine brake power is still increasing due to the increased number of power impulses per minute. Engine brake power drops off when the effect of an increased number of power impulses is offset by the reduced air intake of the cylinders.

Thus, engine torque and engine brake power are closely related to the volumetric efficiency of the engine (**Figures 2–2** and **2–3**).

ENGINE EFFICIENCY

The degree of engine efficiency is expressed in percentage figures resulting from a comparison of the theoretical power of an engine without any power losses, to the actual power available from the engine. Engine efficiency is measured in several ways.

Volumetric Efficiency

The amount of air an engine is able to take into the cylinder on the intake stroke compared to filling the cylinder completely with air at atmospheric pressure is known as the *volumetric efficiency* of an engine.

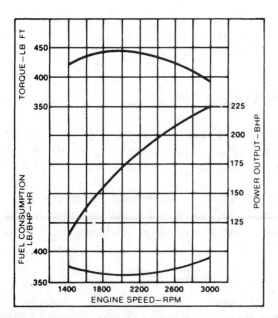

FIGURE 2-2 This chart shows the relationship between torque in pound-feet, brake horsepower, and fuel consumption over the operating speed range of a typical diesel engine. (Courtesy of Cummins Engine Company Inc.)

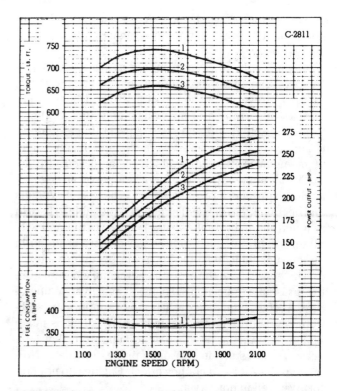

FIGURE 2-3 This chart shows an example of custom rating choices of a particular engine to suit torque and power requirements for particular applications. Rate of fuel delivery and engine governed speeds are tailored to provide desired performance. (Courtesy of Cummins Engine Company Inc.)

Another way to describe it would be to say that volumetric efficiency is the engine's ability to get rid of exhaust gases and take in the air as compared to the displacement of the engine.

The engine is not able to take in a 100% fill on each intake stroke because of design limitations. Such factors as valve and port diameters, manifold runner configuration, valve timing, engine speed, and atmospheric pressure all affect volumetric efficiency.

An engine running at 3000 rpm will have only half the time to fill the cylinder on each intake stroke as it would have at 1500 rpm. Since this is the case, volumetric efficiency drops as engine speed increases. As a result, engine torque also decreases (when engine speed exceeds a certain range).

An engine operating in an area that is 5000 ft above sea level will have less volumetric efficiency than the same engine at sea level because atmospheric pressure is lower at 5000 ft above sea level than at sea level. Since it is atmospheric pressure that forces the air into the cylinder, it is easy to see that there will be a corresponding decrease in volumetric efficiency as the altitude (at which an engine operates) increases.

To achieve 100% volumetric efficiency in an engine, it is necessary to use a blower to force the air into the cylinders. A turbocharger or a supercharger can be used for this. The turbocharger uses exhaust gas pressure to drive a turbine, and the turbine drives the blower. A supercharger uses a mechanically driven blower to do the job. Turbochargers and superchargers must have precise controls to prevent overcharging of the cylinders and consequent engine damage.

Effect of Atmospheric Pressure

Diesel engines require a great deal of air to produce the power required. It is air pressure (atmospheric pressure) that forces air into the engine (and the turbocharger or supercharger). Atmospheric pressure and the relative humidity of the air are important factors in how well an engine performs and the power it is able to produce. The atmosphere is a layer of air surrounding the earth's surface. This layer of air exerts a force against the earth's surface because of the earth's gravitational attraction. This force or pressure of the atmosphere against the earth's surface is called *atmospheric pressure*.

Atmospheric pressure is greatest at sea level, since there is more atmosphere above a given point at sea level than there is at a given point on a high mountain.

Chapter 2 ENGINE PERFORMANCE, POWER, AND EFFICIENCY

The air is, therefore, also less dense (air molecules are not packed together as tightly) at higher altitudes (**Figure 2–4**).

A 1 in.^{2} column of atmosphere at sea level weighs 14.7 lb. Atmospheric pressure at sea level is, therefore, 14.7 psi. However, at the top of a 10,000 ft high mountain, a 1 in.^{2} column of air weighs only 12.2 lb; therefore, atmospheric pressure is 12.2 psi at that altitude. It is important to recognize this fact, since the air intake of an engine is affected adversely by increased altitude.

The air temperature also has a bearing on an engine's ability to produce power. When air is heated, it expands and becomes less dense. The engine is not able to take in as much air on the intake stroke because of this and will, therefore, produce less power. Air density is stated in pounds per cubic foot or in kilograms per cubic meter.

The humidity of the air is the percentage of moisture the air is able to keep in suspension at a given temperature. At 100% humidity, the air cannot support any additional moisture. At 50% humidity, there is half as much moisture in the air as it is able to support at that temperature. Moisture in the air improves engine performance, since it has a cooling effect. Engines do not perform as well in hot, dry air.

Atmospheric pressure is measured with a barometer and is expressed in inches or millimeters of mercury. At sea level, 14.7 psi of atmospheric pressure results in 29.92 in. of mercury in the barometer. To convert inches of mercury to pounds per square inch of pressure, multiply the reading in inches of mercury (Hg) by 0.4912 (psi = in. Hg × 0.4912). In metric terms, barometric pressure at sea level is 101.35 kPa since 1 in. of mercury is equal to 3.38 kPa. 1 psi of pressure is equal to 6.895 kPa. Another unit of pressure measurement is the *bar*. One bar is equal to 0.986923 atmosphere.

A kilopascal is 1000 pascals, since the prefix *kilo* represents 1000. A pascal is equal to 1 newton of force applied over 1 square meter.

Mechanical Efficiency

Indicated power is the theoretical power an engine is able to produce. Brake power is the actual power delivered by an engine expressed in horsepower or kilowatts.

The formula for calculating the mechanical efficiency of an engine compares the brake power to the indicated power and is calculated by dividing BP by IP:

$$BP \div IP = \text{mechanical efficiency}$$

For example, an engine that produces 72 BHP and has an indicated power of 90 HP has a mechanical efficiency of 72/90, or 80%. This means that although 90 BHP is produced in the cylinders, only 72 BHP is being delivered at the flywheel (**Figure 2–5**).

Or, using the metric units, if an engine has an indicated power of 120 kW but delivers only 102 kW, the mechanical efficiency is 102/120, or 85%.

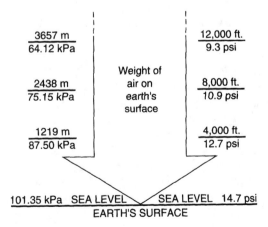

FIGURE 2–4 Atmospheric pressure is greatest at sea level and diminishes with altitude.

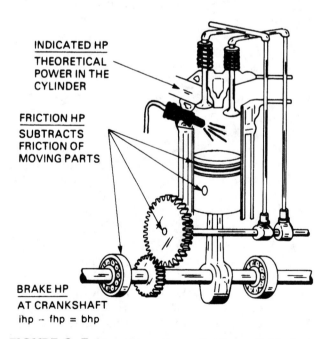

FIGURE 2–5 The power theoretically available in an engine is reduced by the friction of moving parts. Actual usable power at the crankshaft is calculated as follows: IHP − FHP = BHP.

Thermal Efficiency

The thermal efficiency of an engine is the degree to which the engine is successful in converting the energy of the fuel into usable heat energy or power. It is the heat energy in the cylinder that forces the pistons to move, which results in crankshaft rotation.

Scavenge Efficiency

The term *scavenge efficiency* is used to describe how thoroughly the burned gases are removed from the cylinder and how well it is filled with fresh air. The term is used for two-stroke-cycle engines, whereas the term *volumetric efficiency* is used for four-stroke-cycle engines.

It is desirable that the exhaust gases be removed as thoroughly as possible and that the cylinder be as completely filled with fresh air as possible.

Scavenge efficiency depends primarily on the precise location of the inlet ports, the design and arrangement of the exhaust valves and ports, and the efficiency of the intake and exhaust systems. Scavenge efficiency is reduced when intake or exhaust systems are restricted.

Fuel Efficiency

Fuel efficiency is actually the rate of fuel consumption over a distance traveled for highway vehicles or pounds of fuel consumed per brake horsepower hour. Fuel efficiency is dependent on all the foregoing factors as well as on vehicle weight, size, and load.

Diesel engine fuel efficiency is considerably better than that of the gasoline engine for several reasons. Diesel fuel has a higher heat value than gasoline. Also the compression ratio of the diesel engine is considerably higher than that of the gasoline engine.

Cycle Efficiency

Cycle efficiency is equal to engine output divided by fuel input, where output is expressed in units of power and input is expressed in units of heat value of the fuel consumed.

The diesel cycle is considerably more efficient than the gasoline engine cycle. Since diesel fuel has a higher heat value than gasoline and since diesel engine compression ratios are also higher, the combustion temperature of the diesel is considerably higher.

ENGINE BALANCE

Balance of rotating engine parts is achieved by equal distribution of mass radially around the axis of the rotating part. However, the reciprocating masses such as the pistons and connecting rods create an unbalanced condition that must be compensated for in engine design (**Figures 2–6** and **2–7**.)

These unbalanced forces are classified as primary and secondary forces. Primary unbalanced forces are equal in frequency to engine speed, whereas secondary force frequencies are at twice the engine speed. The primary forces are of greater consequence. The arrangement of the pistons and rods on the crankshaft determine the effect of these forces on the engine. These forces, known as *unbalanced couples*, tend to move the ends of the engine in an elliptical path.

Balance weights attached to the outer ends of camshafts, special weighted balance shafts, accessory gears, pulleys, or special balance gears are methods used to cancel the effects of unbalanced couples. The number, size, and position of the weights are designed to produce a couple that is equal in magnitude to and in a direction opposite the primary couple. This produces forces of the same magnitude and frequency, thereby canceling the effect of the unbalanced couple.

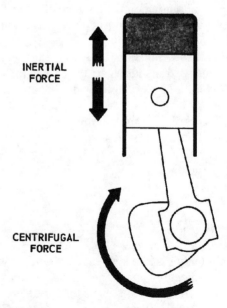

FIGURE 2–6 Unbalancing engine forces. (Courtesy of Deere and Company.)

Chapter 2 ENGINE PERFORMANCE, POWER, AND EFFICIENCY

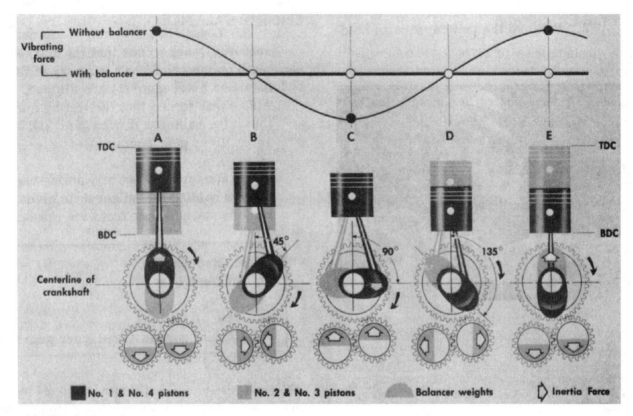

FIGURE 2–7 Engine balancer operation on a four-cylinder engine. (Courtesy of Deere and Company.)

CYLINDER BALANCE

The efficiency and smooth operation of an engine depend to a great degree on the balance of power output among all engine cylinders. Any conditions that result in reduced power output in one or more cylinders results in cylinder imbalance. The resulting uneven power impulses transmitted to the crankshaft create torsional vibrations as well as overall reduced power and torque.

Factors contributing to cylinder imbalance include incorrect valve adjustment, incorrect injector timing, unequal injector output, worn piston rings, leaking valves, and the like.

DYNAMOMETERS

The dynamometer is a reliable piece of equipment used to test all aspects of engine performance. By varying the engine speed and load, most operating conditions except for the weather can be simulated. The dynamometer is used to determine engine power output as well as to test engine systems under various speed and load conditions (**Figures 2–8** and **2–9**).

Engine Dynamometer. Used in a room especially equipped for the purpose in the diesel shop. It includes the dynamometer, provision for mounting the engine and connecting it to the dynamometer, an exhaust system, batteries for the starting and charging systems, a clean air supply, a fuel supply system, a heat exchanger for engine cooling, and the appropriate test instruments and gauges to monitor engine performance. In

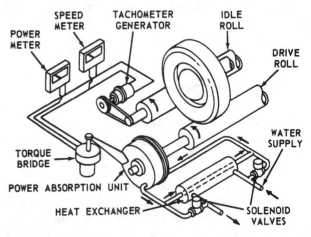

FIGURE 2–8 Major components of an in-floor dynamometer.

FIGURE 2-9 Vehicle must be securely anchored and blocked for dynamometer testing. (Courtesy of Cummins Engine Co. Inc.)

this system friction from the drive train components is not a factor, since they are not involved.

P.T.O. Dynamometer. Usually a two-wheeled portable unit used to test the performance of power take-off-equipped tractors in agriculture and industry. The unit is hitched to the tractor and the power shaft is connected to the tractor power take-off shaft. Test gauges are mounted in a panel on the unit. A water supply is needed for heat removal from the dynamometer.

Chassis Dynamometer. Equipped with floor-mounted drive rollers to accommodate highway tractor single- or tandem-drive axles. This unit is also usually installed in a special room for the purpose. No engine support systems are required with this unit, since only mobile units are tested. A set of test instruments and gauges are used to indicate vehicle performance. In this system drive train friction is a factor, and output is called *road horsepower* (**Figures 2-8** and **2-9**).

Steep grades, level roads, stop-and-go city driving, acceleration and deceleration, and a wide range of load conditions all can be simulated on a chassis dynamometer in the shop. With the proper diagnostic equipment connected to the vehicle during these tests, engine condition and performance can be accurately determined in a very few minutes. This kind of diagnosis and testing cannot be done on the road nor can it be done in the shop without a dynamometer. In addition to the test results obtained from the auxiliary test equipment, the dynamometer indicates vehicle or engine speed and power.

Most dynamometers, whether of the engine type or the chassis type, convert the torque and speed factors automatically to a brake power or road power reading on a dial.

The engine converts the heat energy of combustion to mechanical energy to drive a drive shaft or drive wheels. The drive shaft or drive wheels transfer this mechanical energy to the dynamometer by means of a shaft in the case of an engine dynamometer, and by means of rollers mounted in the shop floor in the case of a chassis dynamometer. This mechanical energy is transmitted to the dynamometer's power absorption unit, which converts the mechanical energy back to heat energy.

The power absorption unit, the torque bridge, and the connecting arm serve the same function as the prony brake. The torque bridge, however, converts the applied force to an electrical signal, which varies with the amount of force applied. This electrical signal provides a reading on the brake horsepower dial. The dynamometer also measures speed, which is indicated on a second dial.

Chassis, P.T.O., and engine dynamometers used in service shops are generally of the hydraulic type. The power absorption unit consists basically of two units: a drive unit and a driven unit. The drive unit is a drum with vanes attached to it internally. The driven unit is also a drum with vanes attached to its interior. The driven unit has an arm attached to it. The other end of the arm is connected to the torque bridge. The drive unit and driven unit are enclosed in a sealed housing, which can be filled with and emptied of fluid.

The fluid used in some dynamometers is water; in others it is oil. In either case, the amount of load applied is in direct proportion to the amount of fluid permitted to enter the power absorption unit. This is controlled by electrically operated solenoid valves. The solenoids are operated by a hand-held control device.

As fluid is allowed to enter the power absorption unit, the rotating drive member throws the fluid against the driven member, which is held by the connecting arm. As more fluid is allowed to enter the unit (more load applied), the force against the driven member is increased, which causes the arm to move slightly. This arm movement is converted to an electrical signal by the torque bridge, which then is indicated on the dial in brake horsepower. In this manner any combination of vehicle or engine speed and load can be observed.

As a load is applied to the dynamometer, the fluid in the power absorption unit heats up. This heat must be dissipated to prevent overheating. In the open water-type dynamometers, the absorption unit is connected to a cool-water pressure source that is constant (city main water) in order to keep the unit cool and dissipate the heat. The heated water is directed to the floor drain.

In the closed hydraulic-type of absorption unit, the oil is circulated through a heat exchanger, which is water cooled. This type requires less water.

Many dynamometers are equipped with an inertia flywheel, usually belt-driven from the rollers in the floor. The inertia flywheel can be used to simulate vehicle inertia during acceleration, deceleration, and coasting modes. It is useful in diagnosing engine and drive train problems. The flywheel can be engaged or disengaged by a manually operated lever.

When doing any type of dynamometer testing, make sure that the engine or vehicle is in a safe condition to be tested. Serious damage to the engine or vehicle can result from improper testing methods.

Be sure to follow all procedures recommended by the vehicle manufacturer and by the manufacturer of the equipment (dynamometer) being used.

It must be remembered that all test results observed during dynamometer testing are valid only for the conditions that existed at the time of the test, including engine and vehicle condition. See Chapter 10 for more on engine performance testing with a dynamometer.

REVIEW QUESTIONS

1. Describe atmospheric pressure.
2. What is a vacuum?
3. Define energy and work.
4. What is the formula for calculating power?
5. Engine power is measured in units of _____ or _____ .
6. How do you calculate engine brake power when indicated power and friction power are known?
7. What is kinetic inertia?
8. Define engine torque.
9. What is the mechanical efficiency of an engine with 72 BHP and 90 IHP?
10. Which engine has the higher thermal efficiency, the gasoline engine or the diesel engine?
11. Volumetric efficiency increases with engine speed. (T) (F)
12. What are the advantages of using a dynamometer for engine testing?

TEST QUESTIONS

1. If compression ratio equals cylinder volume plus combustion chamber volume divided by combustion chamber volume, what is the compression ratio of a cylinder with a 40 in^3 volume and a combustion chamber volume of 2.5 in^3?
   a. 16:1   c. 18:1
   b. 27:1   d. 17:1
2. Atmospheric pressure at sea level is
   a. 14.2 psi   c. 12.4 psi
   b. 14.7 psi   d. 17.4 psi
3. The potential or ability to do work is defined as
   a. work   c. horsepower
   b. power   d. energy
4. Technician A says a dynamometer is a device used to measure vehicle performance. Technician B says a dynamometer is a device used to simulate road and driving conditions. Who is correct?
   a. Technician A   c. both are correct
   b. Technician B   d. both are wrong
5. The amount of air an engine is able to take into a cylinder during the intake stroke, compared with filling the cylinder completely with air is known as
   a. fuel efficiency
   b. thermal efficiency
   c. volumetric efficiency
   d. mechanical efficiency
6. The resistance to motion between two objects in contact with each other is called
   a. viscosity   c. inertia
   b. friction   d. torque
7. Which of the following is *not* a horsepower rating?
   a. brake horsepower
   b. fractional horsepower
   c. indicated horsepower
   d. frictional horsepower
8. The prony brake is a _____ _____ to apply a load to an engine.
9. Atmospheric pressure is _____ at sea level.
10. A vehicle must be _____ _____ and _____ for dynamometer testing.

◆ CHAPTER 3 ◆

INDUCTION, EXHAUST, AND TURBOCHARGER SYSTEM PRINCIPLES

INTRODUCTION

The induction system must provide the engine with an adequate supply of clean air for good combustion (and for scavenging cylinders on two-stroke-cycle engines) for all operating speeds, loads, and operating conditions. Up to 1500 ft^3 of air per minute or more may be required, depending on engine size and load.

On a naturally aspirated four-stroke-cycle engine, the system includes the air cleaner, a precleaner (if used), the intake manifold, and the connecting tubing and pipes **(Figure 3–1)**. On the two-stroke-cycle, the system also includes a blower for scavenging air and for combustion.

On a turbocharged engine, additional air is supplied by means of a turbocharger, which is exhaust gas–driven **(Figure 3–2)**. On a supercharged engine a mechanically driven blower is used to supply additional air.

An air shutdown valve may be included to allow engine intake air to be shut off completely for emergency engine shutdown.

An intercooler or aftercooler may also be included in the induction system. Since cooler air is more dense, a greater amount of air is in fact supplied if the air is cooled. The intercooler is mounted to cool the intake air after it leaves the discharge side of the turbocharger and before it enters the engine (before it enters the blower on two-stroke-cycle diesels). The aftercooler is mounted in the two-stroke-cycle diesel engine block so that it will cool intake air after it leaves the blower and before it enters the cylinder ports.

PERFORMANCE OBJECTIVES

After adequate study of this chapter and sufficient practical experience on appropriate training models and with proper tools, equipment, and shop manuals, you should be able to do the following:

1. Describe the function, design, and operation of dry- and oil bath type air cleaners.

2. Describe the function, design, and operation of the exhaust driven turbocharger.

3. Describe the function, design, and operation of the two-stroke-cycle engine blower.

4. Describe the function and operation of the exhaust system and its components.

5. List and describe the five types of harmful diesel engine exhaust emissions.

TERMS YOU SHOULD KNOW

Look for these terms as you study this chapter, and learn what they mean.

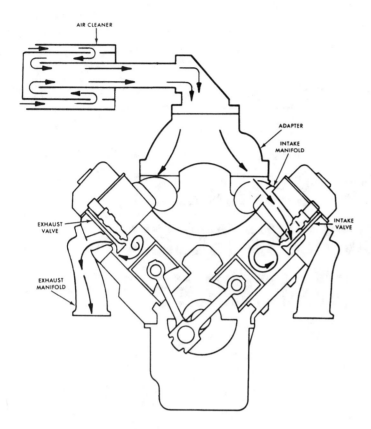

FIGURE 3–1 Naturally aspirated air intake system schematic. (Courtesy of Detroit Diesel Corporation.)

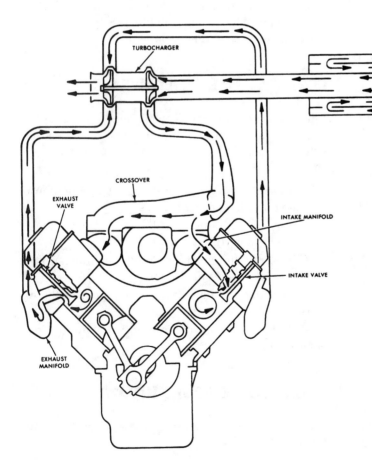

FIGURE 3–2 Air intake system schematic with turbocharger. (Courtesy of Detroit Diesel Corporation.)

induction system
naturally aspirated engine
turbocharged engine
air cleaner capacity
precleaner
dry-type air cleaner
air filter element
restriction indicator
oil bath air cleaner
intake manifold
turbocharger
turbine wheel
compressor wheel
air shutdown valve
blower
blower rotor
rotor timing
intercooler
aftercooler
air-to-air aftercooler
liquid-to-air aftercooler
exhaust system
exhaust backpressure
exhaust manifold
muffler
exhaust emissions
NO_x
HC
CO
PM
SO_2
particulate trap

AIR CLEANER FUNCTION

The air cleaner is designed to remove moisture, dirt, dust, chaff, and the like from the air before it reaches the engine. It must do this over a reasonable time period before servicing is required. The air cleaner also silences intake air noise.

If dirt is allowed to enter the engine cylinders, the abrasive effects will result in rapid cylinder and piston ring wear. If the air cleaner is not serviced at appropriate intervals for the conditions in which it must operate, it will become restricted and prevent an adequate air supply for complete combustion from reaching the cylinders. Incomplete combustion results in carbon deposits on valves, rings, and pistons, which in turn results in engine wear and oil consumption problems.

Air cleaner types include (1) precleaners, (2) dry types of various designs, and (3) oil bath types. Air cleaner capacity (cubic feet per minute or liters per minute) may have to be up to twice as great on a turbocharged engine as compared with the same engine naturally aspirated.

PRECLEANERS

Precleaners are mounted on the intake tube of the air cleaner on some diesel engine–powered equipment. The simplest precleaner consists of a screened hood at the top of the air cleaner inlet. Other precleaners include a spirally vaned drum, which causes incoming air to spin and to force dirt, which is heavier, to the outside. There the dirt falls into a dust cup or passes

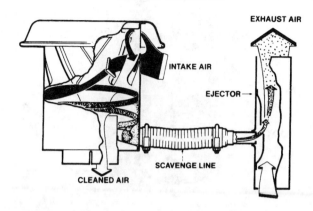

FIGURE 3–3 Precleaner with dirt ejector and scavenge line connected to exhaust system. (Courtesy of Donaldson Company Inc.)

through a scavenging line connected to the exhaust pipe, where it is ejected into the atmosphere by exhaust gases (**Figure 3-3**).

When dust in the transparent dust cup reaches the level indicated by a line on the cup, it is removed, emptied, and reinstalled.

DRY-TYPE AIR CLEANERS

Dry-type air cleaners may have one or more replaceable filter elements and may include primary and secondary or safety filtering elements. Most dry-type air cleaners also include a vaned type of precleaning device. The vanes may be part of the filter element or they may be part of the air cleaner housing (**Figures 3–4 and 3–5**).

As air enters the air cleaner, it passes over the vanes, which impart a swirling action to the air. The swirling action causes the heavier dust and dirt to be thrown outward centrifugally against the air cleaner housing, where it goes to the dust collector cup or bin. A one-way rubber discharge valve ejects dust and water from the dust cup into the atmosphere directly or through a scavenge line connected to the engine's exhaust. A one-way check valve in the exhaust scavenge line prevents engine exhaust gases from entering the air cleaner.

Air is further cleaned as it passes through the filter element or elements. Air cleaners that use primary and secondary filter elements have the advantage of protecting the engine in cases of primary filter element damage, since air passes through the primary element first and then through the secondary or safety element. Filtered air then passes through the air cleaner outlet and to the engine.

Chapter 3 INDUCTION, EXHAUST, AND TURBOCHARGER SYSTEM PRINCIPLES

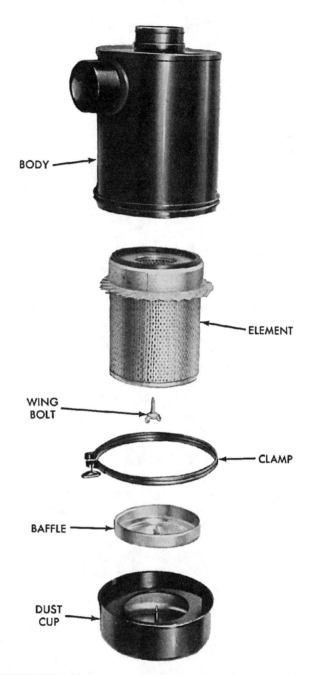

FIGURE 3-4 Dry-type air cleaner components. (Courtesy of Detroit Diesel Corporation.)

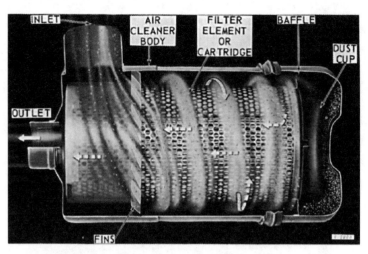

FIGURE 3-5 Air flow diagram of dry element–type of air cleaner. Fins cause circular flow and centrifugal separation of dirt from air. (Courtesy of Deere and Company.)

FIGURE 3-6 Air cleaner restriction indicator shows red when filter must be replaced. (Courtesy of Caterpillar Inc.)

An air cleaner restriction indicator that reflects intake air vacuum may be mounted near the outlet side of the air cleaner **(Figure 3-6)**. The restriction indicator may be of the self-contained type, which shows a red flag or card when restriction reaches filter replacement levels, or it may be of the vacuum-actuated light type. Another type is a gauge that continuously reads restriction in inches of water (H_2O) vacuum when the engine is in operation. The restriction indicator may be remote-mounted where it may be easily observed by the vehicle or engine operator.

OIL BATH AIR CLEANER

In the oil bath air cleaner, shown in **Figure 3-7,** air is drawn through the inlet and down through the center tube. At the bottom of the tube, the direction of air flow is reversed and oil is picked up from the oil reservoir cup. The oil-laden air is carried up into the separator screen, where the oil, which contains the dirt particles, is separated from the air by collecting on the separator screen.

TURBOCHARGER AND BLOWER FUNCTION

A low-pressure area is created toward the center of the air cleaner as the air passes a cylindrical opening formed by the outer perimeter of the central tube and the inner diameter of the separator screen. This low pressure is caused by the difference in air current velocity across the opening.

The low-pressure area, plus the effect of gravity and the inverted cone shape of the separator screen, causes the oil and dirt mixture to drain to the center of the cleaner cup. This oil is again picked up by the incoming air, causing a looping cycle of the oil; however, as the oil is carried toward another cycle, some of the oil overflows the edge of the cup, carrying the dirt with it. The dirt is deposited in the outer area surrounding the cup. Oil then flows back into the cup through a small hole located in the side of the cup. Above the separator screen, the cleaner is filled with a wire screen element, which removes any oil that passes through the separator screen. This oil also drains to the center and back into the pan. The clean air then leaves the cleaner through a tube at the side and enters the intake system.

INTAKE MANIFOLD

The intake manifold is of cast-iron or cast-aluminum alloy construction. It is designed to direct air from the air cleaner or turbocharger to each cylinder intake port on four-stroke-cycle engines. On the two-stroke-cycle engine, it directs air from the air cleaner or turbocharger to the blower inlet opening.

The intake manifold is equipped with an emergency air shutdown valve on some engines. Closing the air shutdown valve prevents air from reaching the cylinders, thereby stopping combustion in the cylinders, which stops the engine.

Inside diameters of manifold air passages must be of sufficient diameter to provide adequate air for combustion for all engine operating speeds and loads (**Figures 3–8 and 3–9**).

TURBOCHARGER AND BLOWER FUNCTION

Turbochargers and superchargers are designed to increase the amount of air delivered to each of the engine's cylinders. In general, superchargers are mechanically driven from the engine crankshaft, while turbochargers are driven by waste exhaust gases from

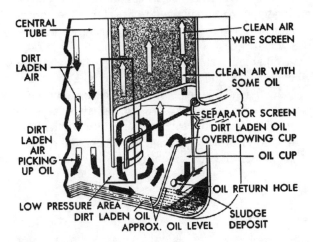

FIGURE 3–7 Oil bath air cleaner operation. (Courtesy of Detroit Diesel Corporation.)

the engine's exhaust system. Turbochargers are the most common. Two-stroke-cycle engines use a mechanically driven blower as well as a turbocharger. The blower is required to provide startup air and scavenging air on the two-stroke diesel.

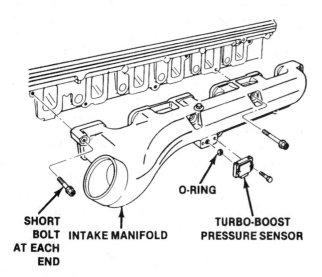

FIGURE 3–8 One-piece intake manifold. (Courtesy of Detroit Diesel Corporation.)

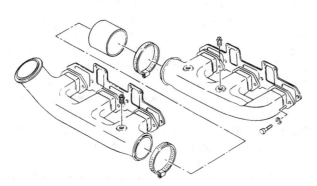

FIGURE 3–9 Two-piece air intake manifold for in-line engine. (Courtesy of Mack Trucks Inc.)

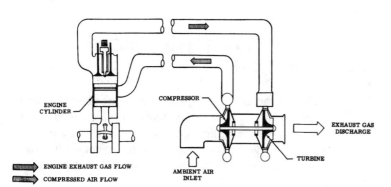

TURBOCHARGER OPERATION

The turbocharger compresses intake air to a density up to four times that of atmospheric pressure. This greater amount of dense air allows more fuel to be burned, thereby doubling the engine's power output. The turbocharger also reduces exhaust emissions and exhaust noise. It compensates for the less dense air encountered in higher altitude operation **(Figure 3–10)**.

A naturally aspirated engine has a limited supply of air for combustion. The air has only atmospheric pressure pushing it into the cylinders. A turbocharger provides pressurized air, which allows for more air to be packed into a cylinder for each firing. This provides more power and much better combustion efficiency. Thus, more power is realized from a given engine size, fuel economies improve, and emissions are reduced.

A centrifugal compressor pulls air through a rotating wheel at its center, accelerating the air to a high velocity, which flows radially outward through a shell-shaped housing. The air velocity is slowed after leaving the wheel, which converts velocity energy into pressure. This type of compressor is a high-speed device. Current turbochargers run at 80,000 to 130,000 rpm. A normal engine may run at only 2500 to 4000 rpm. Gear- or belt-driven compressors require engine power to drive them. Of the fuel energy available for an engine, about 40% is wasted in the exhaust. A turbocharger uses some of this waste energy to drive its compressor.

The centrifugal compressor wheel is attached to a shaft, which has a turbine wheel on the other end. This arrangement can be likened to a water wheel mounted in a flowing river, with the force of the water turning the wheel, which in turn powers machinery. Like the river, flowing exhaust gases drive the compressor wheel. The turbine wheel is enclosed by a shell-shaped housing much like the compressor section, but the flow is the reverse, or rapidly inward. Exhaust gases enter tangentially and flow toward the rotating wheel at the center. After flowing through the wheel, the gases exit at the center and continue through the exhaust pipe to

FIGURE 3–10 Schematic diagram of turbocharger operation. (Courtesy of Detroit Diesel Corporation.)

the atmosphere. The turbine works best at high speeds, which makes it a good match to the compressor section. The common shaft runs in sleeve bearings between the wheels.

These bearings are free-floating, having an oil film on both the inside diameter and outside diameter. The action of the oil flow and the shaft rotation causes the bearings to rotate at approximately one-third of the shaft speed. Each turbocharger has a *journal* bearing system and a *thrust* bearing system. There are system variations for different sizes of turbochargers. The compressor cover is attached to one end of the bearing housing, and the turbine housing is attached to the other end. The bearings are lubricated by engine oil. Only the wheels, shaft, and bearings rotate. The turbocharger unit can nearly double an engine's horsepower, can fit in a 1 ft^3 space, and weighs approximately 45 lb for a medium-size unit. Turbochargers will flow up to 1750 ft^3 of air per minute at pressure ratios up to four times atmospheric **(Figures 3–11 to 3–14)**.

An engine may be equipped with one or two turbochargers. In the dual turbocharger arrangement each unit provides boosted air to half of the engine's cylinders. When connected in a series, boosted air from the primary turbocharger is fed to the secondary turbocharger, where it is boosted to even higher pressure before going to the aftercooler and engine.

AIR SHUTDOWN VALVE

An air shutdown valve on two-stroke-cycle engines allows the air supply to be closed off to stop the engine when abnormal operating conditions require

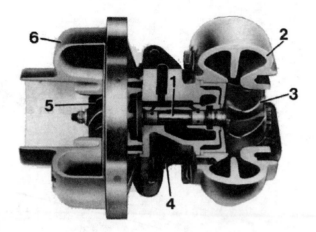

1—Shaft
2—Turbine Housing
3—Turbine Wheel
4—Center Housing
5—Compressor Wheel
6—Compressor Housing

FIGURE 3–12 Turbocharger cross section. (Courtesy of Deere and Company.)

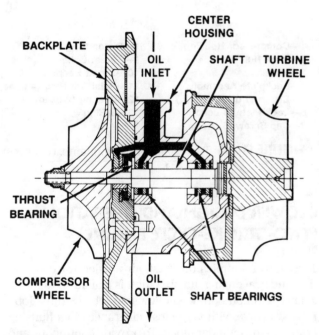

FIGURE 3–13 Turbocharger oil flow. (Courtesy of Detroit Diesel Corporation.)

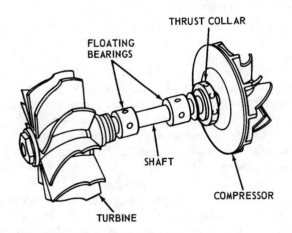

FIGURE 3–11 Rotating parts of a turbocharger. (Courtesy of Deere and Company.)

an emergency shutdown. It is located between the air cleaner and the blower. The major components are the housing, valve, and control linkage. When the valve is open, the air supply is not restricted. When the valve is closed, the air supply to the engine is completely shut off, thereby stopping the engine **(Figure 3–15)**.

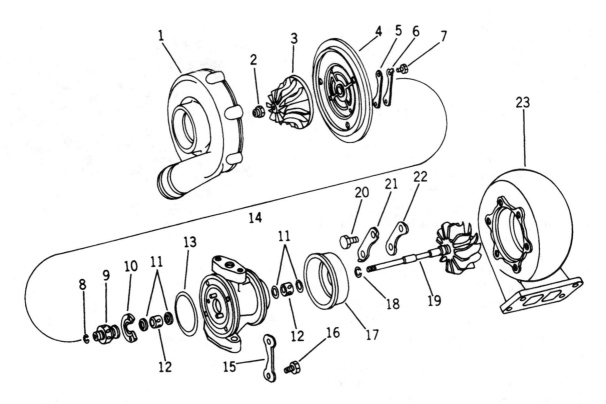

FIGURE 3-14 Turbocharger components. (Courtesy of Deere and Company.)

1—Compressor Housing
2—Lock Nut
3—Impeller
4—Backplate Assembly
5—Clamp (3 used)
6—Lock Plate (3 used)
7—Cap Screw (6 used)
8—Piston Ring
9—Thrust Collar
10—Thrust Bearing
11—Retaining Ring (4 used)
12—Bearing (2 used)
13—O-Ring
14—Center Assembly Housing
15—Lock Plate (2 used)
16—Cap Screw (4 used)
17—Wheel Shroud
18—Piston Ring
19—Turbine Wheel With Shaft
20—Special Bolt (6 used)
21—Lock Plate (3 used)
22—Clamp (3 used)
23—Turbine Housing

BLOWER DESIGN AND OPERATION (TWO-STROKE-CYCLE ENGINE)

The blower supplies the fresh air needed for combustion and scavenging. Its operation is similar to that of a gear-type oil pump. Two hollow two-lobe or three-lobe rotors revolve with very close clearances in a housing bolted to the cylinder block. To provide continuous and uniform displacement of air, the rotor lobes are helical- (spiral) shaped.

Two timing gears, located on the drive end of the rotor shafts, space the rotor lobes with a close tolerance; therefore, because the lobes of the upper and lower rotors do not touch at any time, no lubrication is required.

Oil seals located in the blower end plates prevent air leakage and also keep the oil used for lubricating the timing gears and rotor shaft bearings from entering the rotor compartment.

Each rotor is supported in the end plates of the blower housing by a roller bearing at the front end and a two-row preloaded radial and thrust ball bearing at the gear end (**Figures 3–16** and **3–17**).

The blower rotor is driven by the blower drive shaft, which is coupled to the rotor timing gear by means of a flexible drive hub.

The ratio between the blower speed and the engine speed, and the number of teeth in the blower drive gears and reduction gears, is dependent on engine type and size and whether it is naturally aspirated or turbocharged.

The blower rotors are timed by the two rotor gears at the rear end of the rotor shafts. This timing must be correct; otherwise, the required clearance between the rotor lobes will not be maintained.

Normal gear wear causes a decrease in the rotor-to-rotor clearance between the edges of the rotor lobes. Clearance between the opposite sides of the rotor lobes is increased correspondingly.

EXHAUST SYSTEM FUNCTION AND COMPONENTS

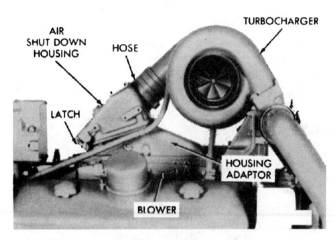

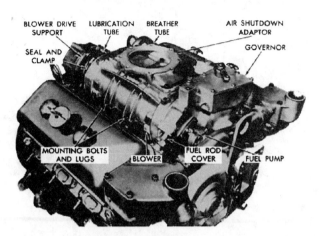

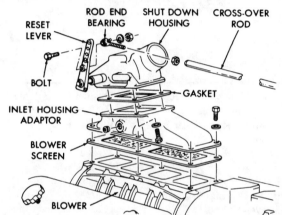

FIGURE 3–15 Air shutdown location (top) and parts detail (bottom). (Courtesy of Detroit Diesel Corporation.)

FIGURE 3–16 Blower mounting on V engine. (Courtesy of Detroit Diesel Corporation.)

The intercooler (or aftercooler) is designed to cool turbocharger boost air before it enters the engine. Both terms refer to the same unit, which is located between the turbocharger and engine at the output side of the turbocharger. Cooling is accomplished by using either ambient air or engine coolant as the cooling medium.

In the air-to-air aftercooler arrangement, a radiator-like device is mounted in front of the radiator. Turbocharged air is routed through the tubes of the cooler and back to the engine. Ambient air flowing over the aftercooler fins and tubes cools the turbocharged air as it flows through the cooler **(Figures 3–18 and 3–19)**.

Alternatively, engine coolant passes through a series of tubes (aftercooler unit) and back to the engine cooling system. Turbocharged air flows across the tubes (usually in the direction opposite liquid flow) and then back to the engine air intake **(Figure 3–20)**.

Worn blower components can result in blower noise and increased friction. Rotor-to-rotor contact abrasion can cause abrasives to be ingested into the engine, which in turn will cause piston, piston ring, and cylinder liner damage. Normal preventive maintenance and inspection procedures should prevent this kind of damage from occurring, since deterioration would be detected in time.

INTERCOOLERS AND AFTERCOOLERS

The intake air temperature increases considerably as a result of turbocharging. Compressing the air causes friction between the air molecules. Because friction creates heat, the air temperature rises. Boosted air temperature may reach 300°F (148.8°C) or higher. Higher air temperature means that the air is less dense, so less air is delivered to the engine.

EXHAUST SYSTEM FUNCTION AND COMPONENTS

The exhaust system is designed to collect the exhaust gases from the engine cylinders, direct them to the muffler, where exhaust noise is reduced, and discharge them into the atmosphere. In addition, exhaust gases may be used to drive a turbocharger for improved air induction for combustion. The exhaust may also be used to eject dirt and dust from the air cleaner or precleaner into the atmosphere. Exhaust gas–driven turbine cargo unloaders for certain materials are used on some trucks. On some older model highway vehicles an operator-controlled valve in the exhaust system is used for braking purposes **(Figures 3–21 to 3–24)**.

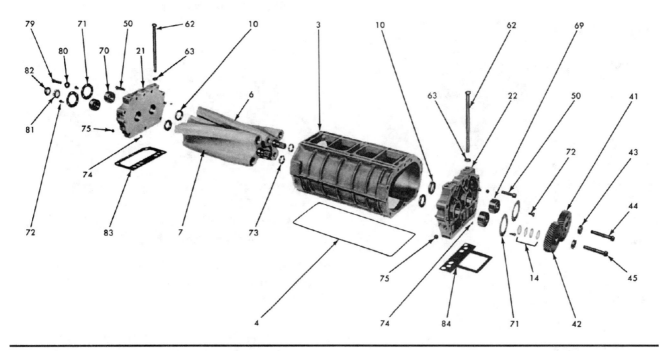

3.	Housing—Blower	22.	Plate—Rear End	62.	Bolt—Blower-to-Block	74.	Strainer—End Plate Oil Passage
4.	Gasket	41.	Gear—Rotor Drive L.H. Helix	63.	Washer—Special	75.	Plug—Pipe
6.	Rotor Assy.—R.H. Helix	42.	Bear—Rotor Driven R.H. Helix	69.	Bearing—Rotor Rear (Roller)	79.	Bolt
7.	Rotor Assy.—L.H. Helix	43.	Pilot—Rotor Gear	70.	Bearing—Rotor Front (Ball)	80.	Washer
10.	Oil Seal—Rotor Shaft	44.	Bolt—Drive Gear	71.	Retainer—Bearing	81.	Lock Washer
14.	Shims	45.	Bolt—Driven Gear	72.	Bolt—Bearing Retainer	82.	Nut
21.	Plate—Front End	50.	Bolt—Socket Head	73.	Sleeve—Oil Seal	83.	Gasket
						84.	Gasket

FIGURE 3–17 Typical blower components. (Courtesy of Detroit Diesel Corporation.)

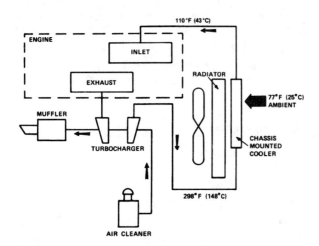

FIGURE 3–18 Air-to-air aftercooler schematic diagram. (Courtesy of Caterpillar Inc.)

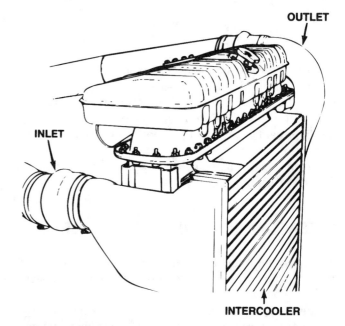

FIGURE 3–19 Air-to-air intercooler mounted in front of truck radiator. (Courtesy of Detroit Diesel Corporation.)

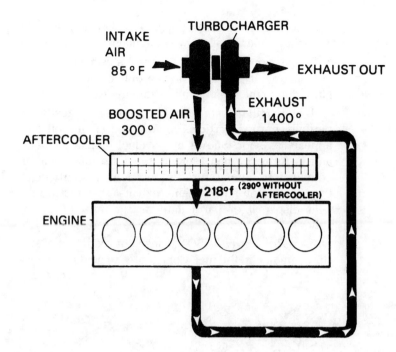

FIGURE 3-20 Typical intake air, boosted air, aftercooled air, and exhaust temperatures (85°F = 29.44°C, 300°F = 148.88°C, 218°F = 103.33°C, 1400°F = 760°C). (Courtesy of Cummins Engine Company Inc.)

Exhaust system components include the exhaust manifolds, exhaust pipe, muffler, muffler extension, and the connecting gasket and clamps. Exhaust pipes may include rigid steel tubing and flexible tubing. A hinged exhaust pipe cap is used on many vertical installations to prevent rain from entering the system. The cap closes automatically when the engine is shut off. A heat shield is used on many applications to prevent injury and component damage.

Exhaust system components must be of sufficient capacity to remove effectively exhaust gases produced by the engine at all operating speeds and loads.

Any restriction in the exhaust system through external damage or internal deterioration will affect the engine's performance.

EXHAUST EMISSIONS

Increasing concern over the effects of harmful exhaust emissions has resulted in increasingly stringent

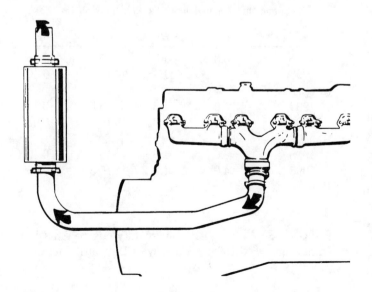

FIGURE 3-21 Typical exhaust system showing exhaust manifold, exhaust pipe, muffler, and extension. (Courtesy of Cummins Engine Company Inc.)

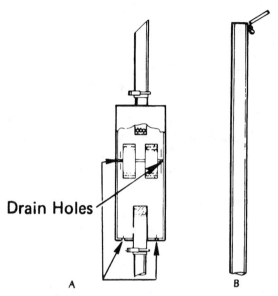

FIGURE 3–22 (A) Moisture drain holes in muffler exhaust system prevent moisture from entering engine. Hinged rain cap (B) prevents moisture from entering exhaust system. Cap opens only when the engine is running and closes automatically when the engine is shut off. (Courtesy of Cummins Engine Company Inc.)

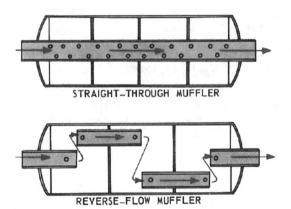

FIGURE 3–23 Muffler types. (Courtesy of Deere and Company.)

combustion. They are especially evident during cold ambient temperatures as white exhaust smoke.

3. *Carbon monoxide* (CO): results from incomplete combustion. It is not a serious problem with diesel engines due to their lean combustion process.

4. *Particulate matter* (PM): consists mostly of partially burned particles of fuel and engine oil and appears as soot. Shows up as black exhaust smoke and is formed when there is too little intake air or combustion temperature is too low. A particulate trap is used on some engine exhaust systems to prevent dispersal of particulate matter into the atmosphere.

5. *Sulfur dioxide* (SO_2): caused by oxidation of sulfur contained in the fuel. Contributes to acid rain.

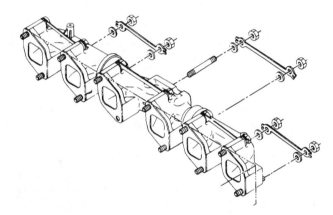

FIGURE 3–24 Exhaust manifold and related parts. (Courtesy of Mack Trucks Inc.)

U.S. EPA HEAVY-DUTY ENGINE EMISSION STANDARDS

(g/bhp-hr measured during the EPA Transient Test)

Model Year	NO_x	HC	CO	PM
1991	5.0	1.3	15.5	0.25
1993	5.0	1.3	15.5	0.10
(Urban Bus)				
1994	5.0	1.3	15.5	0.10
1994	5.0	1.3	15.5	0.05[a]
(Urban Bus)				
1998	4.0	1.3	15.5	0.10
1998	4.0	1.3	15.5	0.05[a]
(Urban Bus)				

[a]EPA can set at 0.07

FIGURE 3–25 U.S. EPA heavy-duty engine emission standards, 1991–1998. (Courtesy of Detroit Diesel Corporation.)

legislation designed to control and reduce their volume. The following emissions are of concern **(Figure 3–25)**.

1. *Oxides of nitrogen* (NO_x): form by a reaction between nitrogen and oxygen at higher temperatures in the combustion chamber. NO_x reacts with hydrocarbons in the atmosphere in the presence of sunlight to form ozone and photochemical smog.

2. *Hydrocarbons* (HC): include over 100 different hydrocarbon compounds resulting from incomplete

Factors Affecting Exhaust Emissions

The rated high-idle speed of an engine is an important factor affecting exhaust emissions. Engines designed to run at higher speeds have less time for combustion to take place than do lower speed engines. Consequently, injection must begin earlier for the combustion process, cylinder pressure, piston position, and crank angle to be at the most advantageous state and position to provide as much push on the piston as possible and still not provide excessive emissions. The duration and rate of fuel injection are precisely controlled to aid in achieving the best results in engine performance, economy, and low emissions. This process involves injector design and injection pressures. Injector design determines the spray pattern and the degree of fuel atomization achieved. Injection pressures generally have become higher in order to improve further the atomization of fuel as well as to provide a thorough mixing of fuel particles and air.

Electronically Controlled Fuel Injection

The electronic control of fuel injection systems has resulted in greatly reduced exhaust emissions. Very precise control of injection timing and fuel metering is achieved in response to a variety of sensors that closely monitor engine operation. Changes in the temperature of engine coolant, lubricating oil, fuel, and intake air are factors in determining the best injection timing and fuel metering. The altitude at which the engine operates and the turbocharger boost pressure also affect fuel metering and injection timing.

REVIEW QUESTIONS

1. The induction system must provide the engine with an adequate supply of _____.
2. Up to _____ ft^3 of air per minute or more may be required from the induction system.
3. What does the term *naturally aspirated* mean?
4. Turbochargers are _____ _____ driven.
5. An intercooler or aftercooler may be included in the _____.
6. What is the function of the air cleaner?
7. Oil bath air cleaners are not often used. (T) (F)
8. The intake manifold is of a _____ _____ or _____ _____ construction.
9. Superchargers are usually _____ driven.
10. Two-stroke-cycle engines use a mechanically driven blower as well as a _____.
11. Current turbochargers run at 80,000 to 130,000 rpm. (T) (F)
12. An engine may be equipped with _____ or _____ turbochargers.
13. What is the function of the blower on a two-stroke-cycle engine?
14. Explain the function of the (a) intercooler; (b) aftercooler.
15. Carbon monoxide emissions are a result of _____ _____.

TEST QUESTIONS

1. A supercharged or turbocharged engine will have
   a. volumetric efficiency problems
   b. no change in volumetric efficiency
   c. increased volumetric efficiency
   d. decreased volumetric efficiency
2. Technician A says a turbocharger is driven by exhaust gas. Technician B says a supercharger is driven by exhaust gas. Who is correct?
   a. Technician A   c. both are right
   b. Technician B   d. both are wrong
3. The speed and power of a diesel engine are controlled by varying the amount of
   a. air only   c. fuel only
   b. fuel and air   d. atmospheric pressure
4. Boost pressure in a turbocharger is controlled by
   a. air density   c. turbine speed
   b. altitude compensation d. engine vacuum

5. Diesel engines require an air filter with a capacity
   a. the same as that of a gasoline engine
   b. smaller than that of a gasoline engine
   c. larger than that of a gasoline engine
   d. three times as large as that of a gasoline engine
6. Technician A says some diesel engines have a muffler in the air cleaner inlet to reduce noise. Technician B says diesel engines do not use enough air to produce noise in the air inlet. Who is correct?
   a. Technician A   c. both are right
   b. Technician B   d. both are wrong
7. Many turbocharged engines cool the intake air to
   a. reduce manifold expansion
   b. prevent engine overheating
   c. increase volumetric efficiency
   d. reduce horsepower at high rpm
8. Exhaust manifolds are made of
   a. aluminum or steel
   b. steel or cast iron
   c. cast iron or aluminum
   d. aluminum or platinum
9. A substance or condition that does *not* produce smog is
   a. oxides of nitrogen
   b. unburned hydrocarbons
   c. sunshine and still air
   d. complete fuel combustion

CHAPTER 4

INDUCTION, EXHAUST, AND TURBOCHARGER SYSTEM SERVICE

INTRODUCTION

It is critical to efficient engine operation that the air intake system and the exhaust system function in a manner that will not limit the engine's ability to take air in or to expel exhaust gases. Any restriction or leakage in either system will affect this ability and result in serious performance problems.

PERFORMANCE OBJECTIVES

After thorough study of this chapter and sufficient practical work on the appropriate components, and with the necessary shop manuals, tools, and equipment, you should be able to do the following:

1. Follow the accepted general precautions as outlined in this chapter.
2. Diagnose the basic induction system and exhaust system problems according to the diagnostic charts provided.
3. Safely remove, disassemble, clean, inspect, and accurately measure all induction system and exhaust system components.
4. Recondition all components as required.
5. Correctly assemble, install, and adjust all system components according to specifications provided.
6. Perform necessary tests and inspections to determine the success of the service performed.

TERMS YOU SHOULD KNOW

Look for these terms as you study this chapter, and learn what they mean.

- air intake restriction indicator
- measuring air intake restriction
- noisy turbocharger
- turbine wheel damage
- compressor wheel damage
- turbine shaft end play
- turbine shaft radial movement
- priming the turbocharger
- blower rotor clearance
- rotor backlash
- rotor-to-end plate clearance
- rotor timing
- measuring exhaust backpressure
- exhaust smoke density
- opacity

Chapter 4 INDUCTION, EXHAUST, AND TURBOCHARGER SYSTEM SERVICE

SERVICE PRECAUTIONS

When servicing the induction and exhaust systems, observe the precautions given in the service manual. In addition, follow these precautions:

- Be careful not to damage parts by mishandling, improperly storing, or improperly using tools and equipment. Parts such as air cleaner filter elements and turbocharger parts are easily damaged. Use the proper tools in the recommended way and in the sequence given in the appropriate service manual.
- Be aware of the danger of hot engine components, particularly cooling systems and exhaust systems.
- Be careful of rotating parts when checking engine components with the engine running. The engine fan, belts, and the turbocharger rotating assembly are examples. Do not get clothing, hands, electrical cords, hoses, or wiping rags near rotating parts.
- Be aware of the danger of an opened induction system's ability to inhale dirt, rags, and paper while the engine is running. Make sure that there is no possibility that this can occur before the engine is started **(Figure 4–1)**. Any ingested material will seriously damage the blower, turbocharger, or engine.
- Always be sure that you know exactly how to stop an engine quickly in case of emergency.

DRY-TYPE AIR CLEANER SERVICE

No matter what system or method is used for determining air cleaner element service intervals, it should be geared to the amount of restriction present.

Restrictions are best recorded by a water manometer, a dial-type gauge calibrated in inches of water, or an air cleaner service indicator.

Normally, restrictions are measured at high idle, no load on naturally aspirated or supercharged diesel engines and at full load, wide-open throttle on turbocharged diesel engines.

FIGURE 4–1 Inlet shield helps prevent the turbocharger from ingesting foreign objects when the engine is operated with the air inlet piping disconnected. (Courtesy of Detroit Diesel Corporation.)

INTAKE, EXHAUST, AND TURBOCHARGER DIAGNOSIS

Trouble and Symptoms	Probable Causes
Engine lacks power	1,4,5,6,7,8,9,10,11,18,20,21,22,25,26,27,28,29,30,A
Black smoke	1,4,5,6,7,8,9,10,11,18,20,21,22,25,26,27,28,29,30,A
Blue smoke	1,4,8,9,19,21,22,32,33,34,36,A
Excessive oil consumption	2,8,17,19,20,33,34,36,A
Excessive oil, turbine end	2,7,8,16,17,19,20,22,32,33,34,36
Excessive oil, compressor end	1,2,4,5,6,8,9,16,19,20,21,33,36,A
Insufficient lubrication	15,16,22,23,24,31,36
Oil in exhaust manifold	2,7,19,20,22,28,29,30,33,34
Damaged compressor wheel	3,6,8,20,21,23,24,36
Damaged turbine wheel	7,8,18,20,21,22,34,36
Drag or bind in rotating assembly	3,6,7,8,13,14,15,16,20,21,22,31,34,36
Worn bearings, journals, bearing bores	6,7,8,13,14,15,16,20,23,24,31,35,36
Noisy	1,3,4,5,6,7,8,9,10,11,18,20,21,22,A
Sludged or coked center housing	2,15,17

DRY-TYPE AIR CLEANER SERVICE

KEY TO PROBABLE CAUSE CODE NOS.

Code No.	Probable Cause	Code No.	Probable Cause
1	Dirty air cleaner elements	20	Worn journal bearings
2	Plugged crankcase breathers	21	Excessive dirt buildup in compressor housing
3	Air cleaner element missing, leaking, not sealing correctly Loose connections to turbocharger	22	Excessive carbon buildup behind turbine wheel
		23	Too fast acceleration at initial start (oil lag)
		24	Too little warm-up time
4	Collapsed or restricted air tube before turbocharger	25	Fuel pump malfunction
5	Restricted/damaged crossover pipe turbocharger to inlet manifold	26	Worn or damaged injectors
		27	Valve timing
6	Foreign object between air cleaner and turbocharger	28	Burned valves
		29	Worn piston rings
7	Foreign object in exhaust system (if from engine, check engine)	30	Burned pistons
		31	Leaking oil feed line
8	Turbocharger flanges, clamps, or bolts loose	32	Excessive engine pre-oil
9	Inlet manifold cracked; gaskets loose or missing; connections loose	33	Excessive engine idle
		34	Coked or sludged center housing
10	Exhaust manifold cracked, burned; gaskets loose, blown, or missing	35	Oil pump malfunction
		36	Oil filter plugged
11	Restricted exhaust system	A.	Oil bath air cleaner
12	Oil lag (oil delay to turbocharger at start-up)	1.	Air inlet screen restricted
13	Insufficient lubrication	2.	Oil pull-over
14	Lubricating oil contaminated with dirt or other material	3.	Dirty air cleaner
15	Improper type lubricating oil used	4.	Oil viscosity low
16	Restricted oil feed line	5.	Oil viscosity high
17	Restricted oil drain line		
18	Turbine housing damaged or restricted		
19	Turbocharger seal leakage		

Courtesy of Airesearch Industrial Division of The Garrett Corporation.

Restrictions are measured in the air cleaner outlet tap, at a tap in the air transfer tube, or within the engine intake manifold.

As an air cleaner element becomes loaded with dust, the vacuum on the engine side of the air cleaner (at the cleaner outlet) increases. This vacuum is generally measured as "restriction in inches of water."

The engine manufacturer places a recommended limit on the amount of restriction that the engine will stand without loss in performance before the element must be cleaned or replaced.

The element in the air cleaner should be serviced only when the maximum allowable restriction has been reached. The element should not be serviced on the basis of visual observation because this will lead to overservice.

The excess handling that is a result of overservice can cause

1. element damage;
2. improper installation of element;
3. contamination from ambient dust; and
4. increased service cost, time, and material.

Mechanical gauges, warning devices, indicators, and water manometers are used to tell the operator when the air cleaner restriction reaches this recommended limit. These gauges and devices are reliable, but the water manometer is the most accurate and dependable.

To use the manometer, hold it vertically and fill both legs approximately half-full with water. Connect one of the upper ends to the restriction tap on the outlet side of the air cleaner by means of a flexible hose. Leave the other end open to the atmosphere (**Figure 4–2**).

Maximum restriction in the air cleaner occurs at maximum air flow. On a naturally aspirated or supercharged (not turbocharged) diesel, the maximum air flow occurs at maximum (high-idle) speed without regard to engine power. On a turbocharged diesel engine, the maximum air flow occurs only at maximum engine power.

With the manometer held vertically and the engine drawing maximum air, the difference in the height of

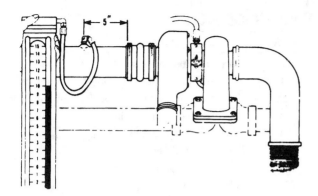

FIGURE 4–2 Measuring air inlet restriction with a water manometer. (Courtesy of Cummins Engine Company Inc.)

the water columns in the two legs, measured in inches, is the air cleaner restriction. Restriction indicators are generally marked with the restriction at which the red signal flag "locks up."

Most engine manufacturers suggest a maximum restriction of 20 to 30 in. of water for diesels. Exceeding these maximums affects engine performance.

CLEANING THE DRY ELEMENT

After removing the element or elements, inspect them for damage. Damaged elements should be replaced. If the element is not damaged, it must be cleaned or replaced as recommended by the manufacturer **(Figures 4–3 and 4–4)**. Some filter elements must not be washed, because this would cause damage.

Compressed air at 40 psi maximum may be used to remove dust and dirt from the filter element. Air should be directed at the "clean" side at least 1 in. away from the surface. The entire surface area must be cleaned.

FIGURE 4–3 Changing dry-type air filter element. (Courtesy of Cummins Engine Company Inc.)

FIGURE 4–4 Cleaning a dry air filter element by washing and rinsing. (Courtesy of Deere and Company.)

The following procedure is typical for washing a filter element:

1. Remove loose dirt from the element with compressed air or with a water hose.

CAUTION:

Compressed air, 40 psi maximum with nozzles at least 1 in. away from element; water hose, 40 psi maximum without nozzle.

2. Mix the proper amount of the recommended solution with the correct amount of warm or cold water.

3. Soak the element in the solution for 15 minutes. Do not soak for more than 24 hours.

4. Swish the element around in the solution to help remove dirt, or use a cleaning tank with the proper solution and recirculating pump.

5. Rinse the element from the "clean" side to the "dirty" side with a gentle stream of water (less than 40 psi) to remove all suds and dirt.

6. Dry the element before reuse. Warm air (less than 160°F) must be circulated. Do not use a light bulb to dry the element.

7. Inspect for holes and tears by looking through the element toward a bright light. Check for damaged gaskets or dented metal parts. Do not reuse damaged elements.

8. Protect the element from dust and damage during drying and storage.

9. Install the elements.

10. Check to make sure that all gaskets are sealing. Examine for dust trails, which indicate leaks. Check to be certain that the wing nut is tight.

11. Examine the clean air transfer tubing for cracks, loose clamps, or loose flange joints.

12. Check air compressor connections (if used) to be certain that these are leak-free.

13. Check ether fittings (if used) to be certain that no contaminants are entering through these connections.

OIL BATH AIR CLEANER SERVICE

1. Remove the oil cup from the cleaner by loosening the retaining band (or wing nuts). Empty the cup and wash it with solvent to remove all the sediment. Wipe the cup dry.

2. Remove the detachable screen by loosening the wing nuts and rotating the screen.

3. Remove the hood and clean it by brushing or by blowing out with compressed air. Push a lint-free cloth through the center tube to remove dirt or oil from the walls.

4. The fixed element should be serviced as operating conditions warrant. Remove the entire cleaner from the engine, soak the unit in the recommended cleaning solution to loosen the dirt, then flush and allow to dry thoroughly.

5. Clean and check all gaskets and sealing surfaces to ensure airtight seals.

6. Refill the oil cup *only* to the oil level marked. Use oil of the same grade as used in the engine crankcase. Do not overfill.

7. Install the removable screen in the housing and reinstall the housing.

8. Install the oil cup and the hood.

9. Check all the joints and tubes and make sure that they are airtight.

All oil bath air cleaners should be serviced as operating conditions warrant. At no time should more than ½ in. of "sludge" be allowed to form in the oil cup or the area used for sludge deposit, nor should the oil cup be filled above the oil level mark.

TURBOCHARGER SERVICE

Most turbocharger failures are caused by one of the three basic reasons: lack of lubricant, ingestion of foreign objects, and contamination of lubricant. Many turbochargers are removed needlessly because proper diagnostic procedures are not followed. The purpose of system troubleshooting is to identify the reason for failure so that repair can be made before installing a new unit **(Figure 4–5)**.

Common symptoms that may indicate possible turbocharger trouble are (1) lack of engine power, (2) black smoke, (3) blue smoke and excessive engine oil consumption, and (4) noisy operation of the turbocharger.

Engine Lacks Power; Black Exhaust Smoke

Both lack of engine power and black exhaust smoke can result from insufficient air reaching the engine and can be caused by restrictions to the air intake or air leaks in the exhaust or induction systems. The first step in troubleshooting any turbocharger trouble is to start the engine and listen to the sound the turbo system makes. As you become more familiar with this characteristic sound, you will be able to identify an air leak between the compressor outlet and engine, or an exhaust leak between engine and turbo, by a higher pitched sound. If the turbo sound cycles or changes in intensity, a plugged air cleaner, loose material in the compressor inlet ducts, or dirt buildup on the compressor wheel and housing is the likely cause.

After listening, check the air cleaner for a dirty element. If in doubt, measure for restrictions per the engine manufacturer's shop manual. Next, with the engine stopped, remove the ducting from air cleaner to turbo and look for dirt buildup or foreign object damage. Then check for loose clamps on compressor outlet connections and check the engine intake system for loose bolts, leaking gaskets, etc. Then disconnect the exhaust pipe and look for restrictions or loose material. Examine the engine exhaust system for cracks, loose nuts, or blown gaskets. Then rotate the turbo shaft assembly. Does it rotate freely? Are there signs of rubbing or wheel-impact damage? **(Figures 4–6 and 4–7)** Axial shaft play is end-to-end movement and radial shaft play is side-to-side. There is normally end-to-end play; however, if this play is sufficient to permit either of the wheels to touch the housing when the shaft is rotated by hand, then there is excessive wear. If none of these symptoms is present, the low-power complaint is not being caused by the turbocharger. Consult the engine troubleshooting procedures for the cause. (See Chapter 27 for engine diagnosis procedure.)

Oil Consumption; Blue Exhaust Smoke

Blue smoke is an indication of oil consumption and can be caused by either turbo seal leakage or other

TROUBLE SHOOTING CHARTS

CHART 1

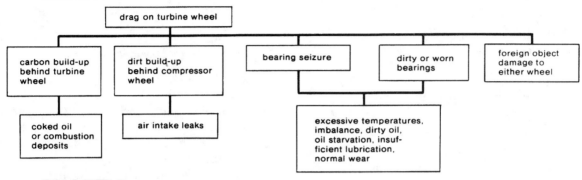

CHART 2

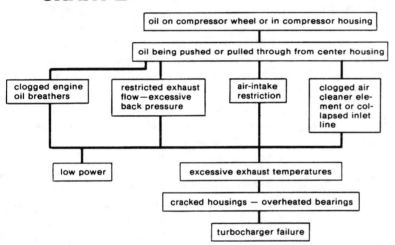

CHART 3

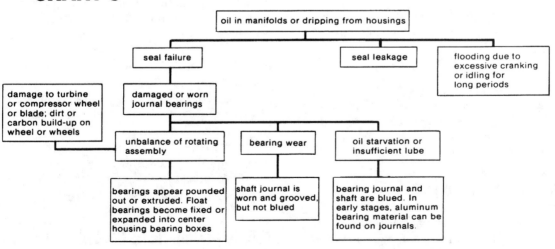

FIGURE 4–5 (Courtesy of Detroit Diesel Corporation.)

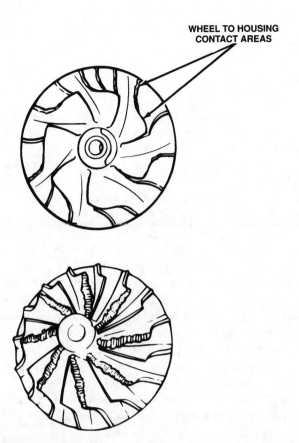

FIGURE 4–6 Typical compressor wheel damage. (Top) housing contact damage; (bottom) foreign object damage. (Courtesy of Detroit Diesel Corporation.)

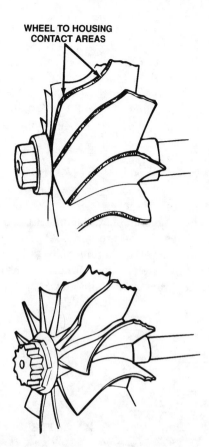

FIGURE 4–7 Typical turbine wheel damage. (Top) housing contact damage; (bottom) foreign object damage. (Courtesy of Detroit Diesel Corporation.)

internal engine problems. First, check the air cleaner for restrictions per the engine manufacturer's shop manual. Higher than normal air cleaner restriction can cause compressor oil seal leakage. Next, with the engine stopped, remove the turbo ducts and check the shaft assembly for free rotation, damage to wheels, or rubbing against housing walls. Next, check the oil drain line for restriction or damage, which can cause seal flooding and leakage. Also check for high crankcase pressure. If in doubt, measure crankcase pressure, which must be within the engine manufacturer's specifications. Finally, loosen the exhaust manifold duct and check for oil in the engine exhaust. If oil is present the engine must be repaired.

Oil Consumption without Exhaust Smoke

First check for air cleaner restrictions. Next, check the compressor discharge duct for loose connections, and check crankcase pressure, which must be within the manufacturer's specifications. Check the turbo shaft assembly for free rotation. Also check for evidence of wheel rubbing on housing walls. If there is rubbing, you should be able to feel it as you rotate the shaft while pulling or pushing on it. If no fault is found then the engine may be at fault.

Noisy Turbocharger

First, check all pressure connections for tightness—compressor discharge ducting, exhaust manifold, etc. Check the turbo shaft for looseness, and look for wheel rubbing or impact damage to blades from foreign material. If rubbing or impact damage is found, remove and replace the turbocharger. If a replacement turbocharger is necessary, it is important that the unit be prelubed and the oil system primed prior to installation. Check the oil supply line for damage or restriction.

Instruct the driver as to the proper shutdown procedure. If the driver shuts down from high speed, the turbo will continue to rotate after the engine oil pressure has dropped to zero, which may cause bearing damage. Also, advise about oil lag. Allow 30 seconds for oil flow to become established before running up high rpm. Ask the driver if the engine oil and oil filter are changed at recommended intervals; review the proper lube oil and filter change interval with the operator. Contaminated oil may cause sludge buildup.

TURBOCHARGER OVERHAUL

If the turbocharger requires overhaul, remove the turbocharger from the engine by disconnecting the intake and exhaust lines, the oil pressure and oil return lines, and the turbocharger mounting bolts **(Figure 4–8)**. The following disassembly, cleaning, and reassembly is provided as general information only. For actual procedures and specifications refer to the manufacturer's service manual.

Disassembly

1. Clean the exterior of the turbocharger with cleaning solvent to remove accumulated surface matter before disassembly.

2. Mark related positions of the compressor housing, center housing, and turbine housing with a punch or scribe prior to disassembly to assure reassembly in the same relative position **(Figure 4–9)**.

3. Remove V-band clamps, which hold the compressor and turbine housings to the center housing. Tap with a soft hammer if force is needed for removal. Exercise care in removal so that no damage occurs to the wheel blades.

4. Clamp a suitable socket or box-end wrench in a vise and place an extended hub on a shaft in the socket or wrench. Hold the center housing upright and remove the shaft locknut **(Figure 4–10)**.

5. Press the turbine wheel shaft from the compressor wheel using an arbor press. Remove and discard the turbine piston ring seal. The turbine wheel shroud that is not retained will fall free when the shaft wheel is removed.

6. Remove the thrust plate retaining bolts.

7. Insert a soft rod through the bearings and push the thrust plate from the center housing.

8. Remove the seal spacers and thrust collar from the thrust plate. Discard the piston ring and seal ring.

9. Remove the bearings, bearing washers, and retainers from the center housing.

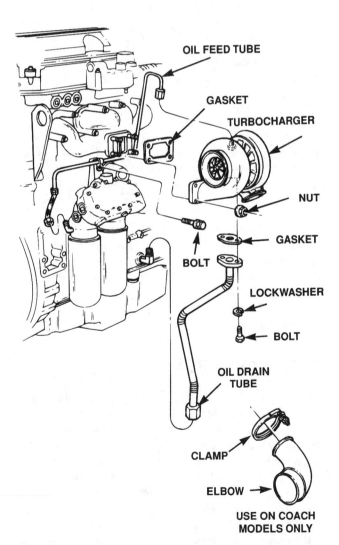

FIGURE 4–8 Turbocharger removal. (Courtesy of Detroit Diesel Corporation.)

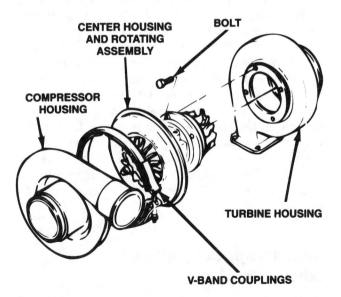

FIGURE 4–9 Major sections of a turbocharger. (Courtesy of Detroit Diesel Corporation.)

TURBOCHARGER OVERHAUL 51

FIGURE 4–10 Removing the turbine shaft locknut. (Courtesy of Detroit Diesel Corporation.)

Cleaning

Before cleaning, inspect all parts for signs of burning and rubbing or other damage that may not be evident after cleaning.

Clean all parts in a noncaustic cleaning solution. Use a bristle brush, a plastic blade scraper, and dry compressed air to remove surface accumulation. Completely remove all surface matter. Do not use an abrasive cleaning method, which might destroy or damage machined surfaces. Especially check the center housing cavity and remove all carbon.

Reassembly

1. Install the turbine outboard retainer with round shoulder toward the bearing.
2. Insert the turbine bearing and bearing washer and install with inboard bearing retainers, also with the round shoulder toward the bearings.
3. Install the piston ring on the shaft and place the shaft wheel upright.
4. Place the wheel shroud over the shaft and then guide the shaft through the center housing bore. Do not force the piston ring into place. A gentle rocking action will assist in seating the ring in the recess.
5. Install the compressor bearing and inboard thrust washer. Ensure that retaining pins engage the washer properly and that the washer is seated flat against the housing.
6. Install the thrust collar over the shaft and flat against the thrust washer.
7. Install the seal ring in the groove on the center housing.
8. Place the piston ring on the seal spacer and insert the spacer into the thrust plate bore with the piston ring forward.
9. Align the oil feed holes of the center housing and thrust plate and guide the assembly over the shaft until it bottoms in the housing recess. Keep finger pressure on the spacer to keep in place.
10. Install the thrust plate bolts and lock tabs. Torque to specifications and bend the lock tabs.
11. Place the center housing assembly in a suitable holding fixture and install the compressor wheel over the shaft. Pull the wheel down with the shaft locknut until it bottoms, and tighten to specifications (**Figure 4–11**).
12. Check the clearance between the backplate and turbine wheel (**Figure 4–12**). Check turbo shaft end play (**Figure 4–13**) and radial movement (**Figure 4–14**).

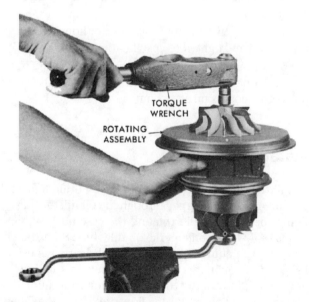

FIGURE 4–11 Tightening the turbine shaft locknut. (Courtesy of Detroit Diesel Corporation.)

52   Chapter 4 INDUCTION, EXHAUST, AND TURBOCHARGER SYSTEM SERVICE

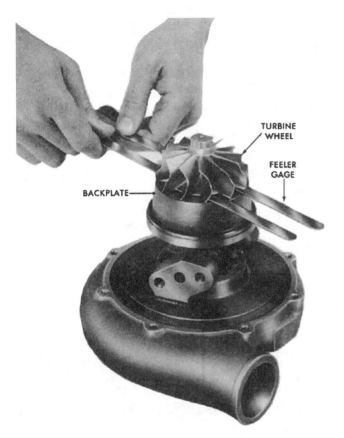

FIGURE 4–12 Checking turbine wheel to backplate clearance. (Courtesy of Detroit Diesel Corporation.)

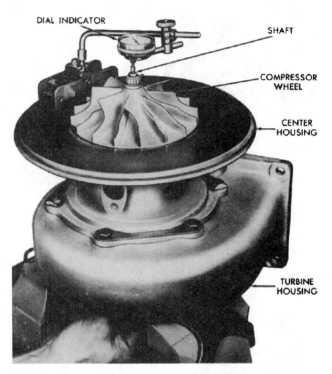

FIGURE 4–13 Checking shaft end play. (Courtesy of Detroit Diesel Corporation.)

13. Orient the compressor housing to the center housing. Install the V-band clamp and tighten to specifications.

14. Orient turbine housing to center housing. Install the V-band and tighten to specifications.

15. After assembly, check the unit for binding. If the unit is to be stored, lubricate it internally and install protective covers on all openings.

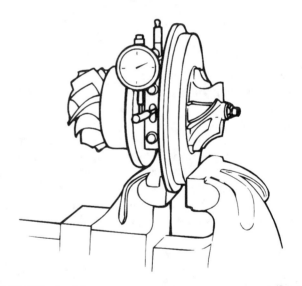

FIGURE 4–14 Checking radial movement of shaft. (Courtesy of Detroit Diesel Corporation.)

Installation

Install the turbocharger, connect the oil drain line, and prelube the turbocharger through the oil inlet fitting by filling the cavity while rotating the assembly by hand **(Figure 4–15)**. Connect the oil inlet line. Start the engine and run at idle until pressure lubrication is established. If in doubt, this can be confirmed by running the engine at idle with the turbocharger oil return line disconnected and observing the flow. Make sure that all intake and exhaust connections to the turbocharger are clean and tight.

BLOWER SERVICE (TWO-STROKE-CYCLE ENGINE)

The following service procedures are general and do not apply to any particular blower or engine. For specific procedures and specifications refer to the appropriate service manual for the unit being serviced.

FIGURE 4–15 Priming the turbocharger with oil. (Courtesy of Mack Trucks Inc.)

Inspection (on Engine)

For on-engine inspection, the air inlet housing and air shutdown housing must be removed. These components and controls should be inspected for damage as they are removed; damaged parts should be replaced.

Visually inspect the blower rotors for burrs or scratches. If these cause any interference between rotors, the blower should be removed and repaired. This requires disassembly and removing the burrs from the rotors to eliminate any interference.

Inspect the area around the seals on the inside of the housing and the blower rotors for signs of oil leakage. If oil is present, the blower will have to be removed for repairs and seal replacement.

A noisy blower when running at idle indicates a worn blower drive. With the engine stopped, grasp the right-hand rotor and attempt to turn it. It may move from 3/8 in. to 5/8 in., measured at the crown of the lobe. A springing action should be felt while doing this. When released from this spring pressure, the rotor should move back at least 1/4 in. If rotors cannot be moved in this manner or there is no spring action, the drive coupling must be replaced.

Inspect the rotor lobes for contact with each other, looking for wear patterns on the crown or root of the lobes. Check for rotor-to-housing contact at the ends of the housing or the housing itself. If any of these conditions is present, it may be caused by a loose shaft or worn bearings.

Excessive gear backlash will cause rotors to rub each other full length. Check backlash with a dial indicator; if excessive, drive gears will have to be replaced.

Blower Removal

Follow the manufacturer's service manual for blower removal, since procedures vary considerably among different models and accessory equipment. Be sure to have the blower adequately supported with the proper lift attachment before removing the mounting bolts from where they are supporting blower weight (**Figure 4–16**).

Disassembly and Cleaning

Cover the inlet and outlet openings of the blower and clean the outside of the blower with fuel oil. Dry the blower with compressed air. Make sure that no dirt or foreign matter enters the blower, since the blower lobes are easily damaged.

Use only recommended methods, tools, and pullers for disassembly. Do not clamp parts that are easily damaged in a vise. Follow the sequence and procedures given in the manufacturer's service manual for disassembly. Changes in design occur regularly, which means changes in procedures. Keep parts in order: do not mix bearings, etc.

After disassembly, wash all parts in clean fuel oil and dry them with compressed air.

Inspection

Examine the bearings for any indications of corrosion or pitting. Lubricate each bearing with light engine oil.

FIGURE 4–16 Removing blower from 6V engine. (Courtesy of Detroit Diesel Corporation.)

Then, while holding the bearing inner race to keep it from turning, rotate the outer race slowly by hand and check for rough spots.

The double-row ball bearings are preloaded and have no end play. A new double-row bearing will seem to have considerable resistance to motion when rotated by hand.

Examine the rotor shafts and the oil seal sleeves, if used, for wear.

Inspect the blower rotor lobes, especially the sealing ribs, for burrs and scoring. If the rotors are slightly scored or burred, they may be cleaned up with emery cloth.

Examine the rotor shaft serrations for wear, burrs, or peening. Also inspect the bearing contact surfaces of the shafts for wear and scoring.

Inspect the inside surface of the blower housing for burrs and scoring. If the inside surface of the housing is slightly scored or burred, it may be cleaned up with emery cloth.

Check the finished ends of the blower housing for flatness and burrs. The end plates must set flat against the blower housing.

The finished inside face of each end plate must be smooth and flat. If the finished face is slightly scored or burred, it may be cleaned up with emery cloth.

Examine the serrations in the blower rotor gears for wear and peening; also check the teeth for wear, chipping, or damage. If the gears are worn to the point where the backlash between the gear teeth exceeds 0.004 in. or are damaged sufficiently to require replacement, both gears must be replaced as a set.

Check the blower drive shaft serrations for wear or peening. Replace the shaft if it is bent.

Inspect the blower-drive coupling springs (pack) and the cam for wear.

Replace all worn or excessively damaged blower parts.

Clean the oil strainer in the vertical oil passage at the bottom of each blower end plate and blow out all oil passages with compressed air.

Assembly and Installation

Install new oil seals and sleeves. Install bearings, maintaining the same component relationship as originally, if parts are used again. Lubricate the seals and bearings with engine oil.

Install the blower rotors into the blower housing and align as specified in the service manual by positioning the housing over the rotors with the rear of the housing facing up. One end of the blower housing is marked "rear" on the outside face of the housing. Install the rear end plate. Make sure that bearings and seals are lubricated with engine oil. If necessary, tap the end plate lightly with a soft-face hammer. Install bearing retainers and tighten bolts to specifications. Install ball bearings on rotor shafts and front end plate. Make sure that bearings are lubricated with engine oil. Install bearing retainers and tighten bolts to specified torque.

Check rotor-to–end plate clearance with a feeler gauge. Check rotor-to-housing clearance. Make sure that clearances are within specifications (**Figures 4–17** and **4–18**).

FIGURE 4–17 Checking rotor-to-housing clearance (top) and rotor-to–end plate clearance (bottom). (Courtesy of Detroit Diesel Corporation.)

EXHAUST SYSTEM

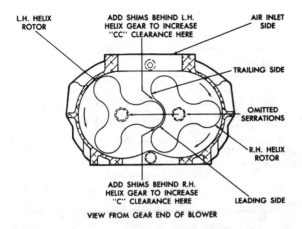

FIGURE 4–18 Location of shims to correct rotor lobe clearances. (Courtesy of Detroit Diesel Corporation.)

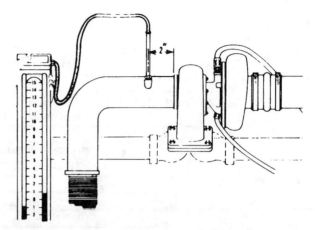

FIGURE 4–19 Measuring exhaust back pressure with a manometer. (Courtesy of Cummins Engine Company Inc.)

Install the blower rotor gears as specified in the shop manual. Install and align gears and shims as outlined in the manual. Tighten all bolts to specified torque. Check the backlash between rotor gears. Replace gears if the backlash exceeds specifications.

Time the blower rotors to provide the specified clearance between rotor lobes by adding or removing shims between the gears and the bearings until specified clearances are obtained. Clearance should be measured from both the inlet and outlet sides of the blower.

Attach the remaining accessories to the blower and install the blower, using new gaskets and appropriate gasket sealer (sparingly) to keep the gasket in place during blower positioning. Tighten all bolts as specified. Attach all remaining accessories, blower drive, governor, alternator, air compressor, and the like as well as fuel lines and control linkages as required.

Connect all coolant lines and fill the engine with coolant as outlined in the engine section of this text. Attach intake and exhaust connections.

Adjust all linkages as outlined in the appropriate service manual. Recheck the entire installation before starting the engine.

EXHAUST SYSTEM SERVICE

The entire exhaust system, manifolds, pipes, turbocharger connections, and mufflers should be inspected for leakage or external damage that could cause exhaust restriction. Any damaged component should be repaired or replaced. The system should also be checked for back pressure with a mercury or water manometer or an appropriately calibrated pressure gauge. This test reveals internal restriction that may otherwise not be evident **(Figure 4–19)**.

If exhaust back pressure exceeds service manual specifications, early engine failure and poor performance may be expected.

Start the engine and let it run until oil temperature reaches 140°F (60°C).

Take back pressure readings when the engine is developing its maximum horsepower at maximum engine speed.

Add the mercury reading in both columns for the final figure.

Example. If mercury is 1 in. (25.4 mm) high in the left column and 1 in. (25.4 mm) low in the right column, there is 2 in. (50.8 mm) of pressure. If the mercury is 1

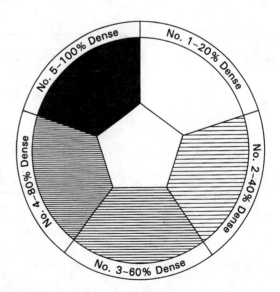

FIGURE 4–20 Exhaust smoke density chart. (Courtesy of Detroit Diesel Corporation.)

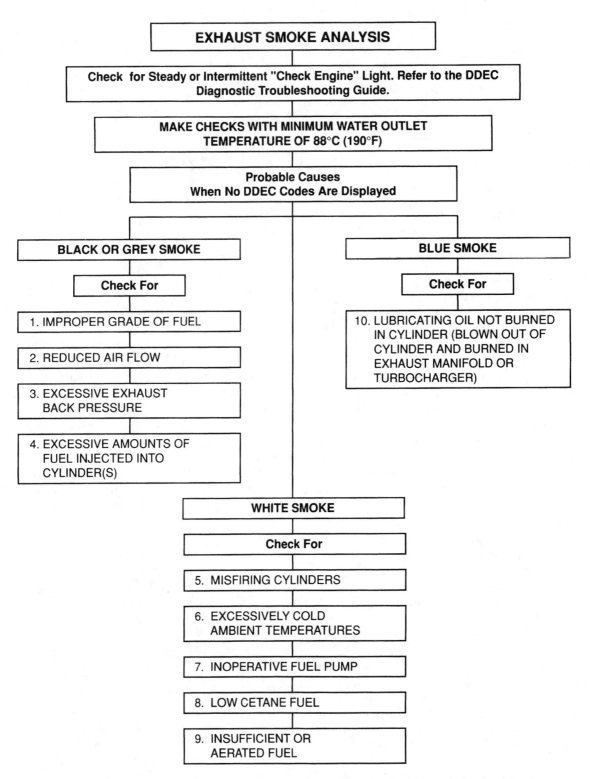

FIGURE 4–21 Smoke analysis chart for series 60 engine. (Courtesy of Detroit Diesel Corporation.)

SUGGESTED REMEDY

Black or Gray Smoke

1. Check for use of an improper grade of fuel. The use of low cetane fuel will cause this condition. Refer to "Fuel Specifications" in the appropriate service manual.
2. Reduced air flow to the engine cylinders is caused by a restricted intercooler, air cleaner, an air leak in the piping between the air cleaner and the intake manifold, or a faulty turbocharger. Check, clean and/or repair these items as necessary.
3. Excessive exhaust back pressure may be caused by faulty exhaust piping or muffler obstruction and is measured at the exhaust manifold outlet from the turbocharger with a manometer or suitable gauge. Replace faulty parts.
4. The fuel injector(s) may be injecting excessive amounts of fuel. Perform a cylinder cutout test under load, using a dynamometer, in an effort to determine if one or more injectors is causing problem.

White Smoke

5. Misfiring cylinders may be caused by improperly timed or malfunctioning injectors. To check injectors for proper operation, use the DDEC Reader Injector Cutout Procedure as outlined in the DDEC Troubleshooting Guide, 6SE477 or 6SE489. If no trouble codes are present, remove the valve rocker cover and set the intake and exhaust valve clearances and injector heights.
6. White smoke may also be the result of excessively cold ambient temperatures. Auxiliary engine heating devices may be employed. Also, when DDEC system/engine is in the cold start mode (increased idle speed) do not put DDEC system in the fast idle mode!
7. An inoperative or defective fuel pump may result in white smoke.
8. Low cetane fuel.
9. Perform a fuel flow test. If less than specified amount is returning, or if the fuel is aerated, check for no fuel or insufficient fuel.

Blue Smoke

10. Check for internal lubricating oil leaks and refer to "High Lubricating Oil Consumption" in Chapter 6.

FIGURE 4–22 (Courtesy of Detroit Diesel Corporation.)

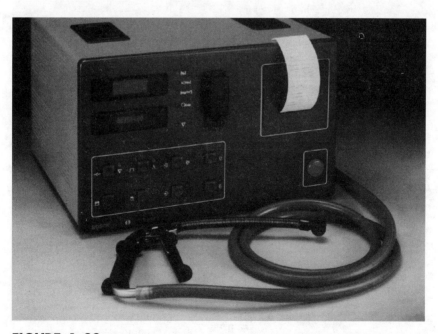

FIGURE 4–23 Exhaust smoke opacimeter. (Courtesy of Robert Bosch Canada Ltd.)

in. (25.4 mm) high in the right column, and 1 in. (25.4 mm) low in the left column, there is 2 in. (50.8 mm) of vacuum. Maximum permissible back pressure is 3 in. (76.2 mm) Hg or 40.7 in. (1 m) of water.

If exhaust restriction is excessive, the faulty component should be identified and replaced.

Analyzing Exhaust Smoke

Diesel engine exhaust smoke is analyzed in terms of opacity (denseness). Engine exhaust smoke is analyzed when the engine coolant has reached the specified operating temperature, usually about 160°F (70°C). The exhaust smoke is then analyzed under specified operating modes such as acceleration and lugging modes. The exhaust smoke emitted is compared to a smoke density chart held at arm's length. During the acceleration mode, for example, a 40% opacity may be the maximum acceptable, whereas during the lugging mode a 20% maximum opacity may be allowed. See the density chart illustrated in **Figure 4–20** and the smoke analysis charts in **Figures 4–21** and **4–22**.

Exhaust Smoke Opacity Meter

An exhaust smoke opacity meter provides fast, accurate measurement of exhaust smoke opacity. A clip-on sensor inserted into the exhaust pipe samples the exhaust for analysis by the tester. Tests are performed at the specified full load and acceleration modes. An integral printer provides a printed record of test data. Custom program modules are available for state-specific and lab applications. A memory module provides data on vehicle specific test sequences. Test results indicate percent opacity, light absorption, and mass concentration. Test results are compared with current specifications to determine whether standards are being met **(Figure 4–23)**.

REVIEW QUESTIONS

1. When servicing exhaust systems be aware of the danger of contacting _____ _____ .
2. Be careful of _____ parts when checking components while the engine is running.
3. While the engine is running, any open induction system has the ability to ingest dirt, rags, paper, and the like. (T) (F)
4. How is intake restriction measured when a dry-type air cleaner is used?
5. The _____ manometer is the most accurate instrument for measuring air intake restriction.
6. Some dry air filter elements can be washed. (T) (F)
7. After cleaning an oil bath air cleaner, refill the oil cup only to the _____ marked.
8. List three common reasons why turbochargers fail.
9. Axial shaft play is _____ _____ movement, and radial shaft play is _____ _____ .
10. What can cause blue exhaust smoke?
11. When disassembly of a turbocharger is required, be sure to mark all parts so as to _____ in _____ position.
12. Describe an on-engine blower inspection on a two-stroke-cycle engine.
13. Follow _____ _____ _____ instructions for blower removal.
14. List the steps in blower inspection after disassembly.
15. When reassembling the blower, be sure to install new _____ _____ and sleeves.

TEST QUESTIONS

1. An inlet shield should be used to help prevent the turbocharger from
   a. excessive speed
   b. ingesting foreign objects
   c. excessive oil consumption
   d. overheating
2. Air inlet restriction is measured with a
   a. vacuum gauge
   b. mercury manometer
   c. water manometer
   d. any of the above

3. When a dry-type air cleaner is serviced, it must be
   a. replaced
   b. washed
   c. blown out with air
   d. replaced *or* cleaned and reinstalled
4. Turbocharger compressor wheel damage may be caused by
   a. housing contact
   b. foreign objects
   c. lack of oil
   d. all of the above
5. When tightening the turbine shaft locknut, be sure to use a
   a. vise
   b. torque wrench
   c. torque wrench to proper specifications
   d. shop manual
6. Turbine wheel to backplate clearance is checked with
   a. a dial gauge
   b. feeler gauges
   c. a plastic gauge
   d. a torque wrench
7. Turbocharger shaft end play is checked with a dial gauge. (T) (F)
8. Radial movement of the shaft is checked with a feeler gauge. (T) (F)
9. After being rebuilt, the turbocharger must be primed with oil. (T) (F)
10. A blower rotor to housing and rotor to end plate is checked for wear with a feeler gauge. (T) (F)

◆ CHAPTER 5 ◆

COOLING SYSTEM PRINCIPLES AND SERVICE

INTRODUCTION

Proper cooling system operation is critical to optimum engine performance, good fuel economy, and long engine life. A faulty cooling system can result in serious engine damage if not corrected in time. Understanding the operation of the cooling system and its components is essential for accurately diagnosing and servicing system components. This chapter discusses the function, operation, diagnosis, and service of diesel engine cooling systems and their components.

PERFORMANCE OBJECTIVES

After thorough study of this chapter and sufficient practice on the appropriate engine components, you should be able to:

1. Describe the purpose, construction, and operation of the cooling system and its components.
2. Diagnose cooling system problems.
3. Recondition or replace the faulty components.
4. Test the reconditioned system to determine the success of the service performed.

TERMS YOU SHOULD KNOW

Look for these terms as you study this chapter, and learn what they mean.

conduction
convection
radiation
liquid cooling system
air cooling system
marine engine cooling system
keel cooler
fins
shroud
blower
damper valve
bypass
water pump
raw water pump
thermostat
radiator hose
water manifold
radiator
pressure cap
pressure relief valve
vacuum valve
reserve tank
radiator fan
variable pitch fan
viscous fan drive
fan drive clutch
fan braking
radiator shutters
coolant filter
coolant conditioner
marine heat exchanger
antifreeze coolant
ethylene glycol
freeze protection
V belt
poly V belt
belt tension
cooling system flushing
pressure cap testing
cooling system pressure testing

LIQUID COOLING SYSTEM FUNCTION

The diesel engine cooling system is designed to bring the engine to its most efficient operating temperature (as soon as possible after starting) and to maintain that temperature through all operating conditions. Some of the heat absorbed by the cooling system is used to heat the cab interior in cold weather and to keep windows clear of moisture and frost.

The cooling system of an engine relies on the principles of conduction, convection, and radiation. Heat is conducted from the metal surrounding the cylinders, from valves, and from cylinder heads to the coolant in the water jackets of the block and head. The hot coolant is forced out of the block and cylinder heads by the water pump to the radiator, where the heat is removed by convection. Some cooling of the engine takes place through radiation. Air flow around the engine carries this heat away.

Since about 30% of the heat energy of the burning fuel in an engine, as well as heat from friction, is absorbed by the cooling system, all components involved must be of sufficient capacity, and they must be in good operating condition.

There are two general classifications of diesel engine cooling systems: the *liquid system* and the *air cooling system*. Of these the liquid cooling system is by far the most common **(Figures 5–1 and 5–2)**.

The liquid system uses a liquid (water or antifreeze) as the medium to absorb engine heat and transfer it to the radiator or heat exchanger. A fan blows air across the radiator tubes and fins to dissipate the heat. A centrifugal water pump circulates the coolant through the engine and radiator. A flow control valve called a *thermostat* determines whether circulation is allowed through the radiator. The opening temperature range of the thermostat, approximately 160°–185° F (71°–85° C), maintains operating temperature in that range. Engine coolant is also used to control lubricating oil temperature by means of a liquid-to-liquid heat exchanger or oil cooler.

Marine Engine Cooling System

Cooling systems for marine engines are often equipped with two water pumps. One pump circulates coolant as described above but uses a liquid-to-liquid heat exchanger instead of a radiator. The other pump, known as the *raw water pump*, circulates raw sea water through separate passages in the heat exchanger to cool the engine coolant with sea water.

In the *keel cooling system*, the coolant is drawn by the water pump from the keel cooler and is forced through the engine oil cooler, cylinder block, cylinder heads, and exhaust manifold to the thermostat housings. A bypass from the thermostat housings to the inlet

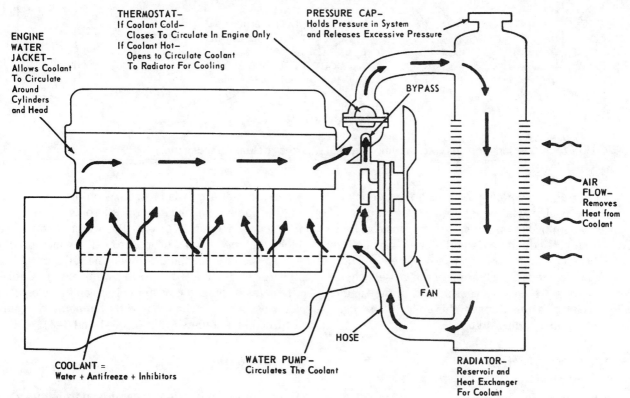

FIGURE 5–1 Major cooling system components and their functions. (Courtesy of Deere and Company.)

62 Chapter 5 COOLING SYSTEM PRINCIPLES AND SERVICE

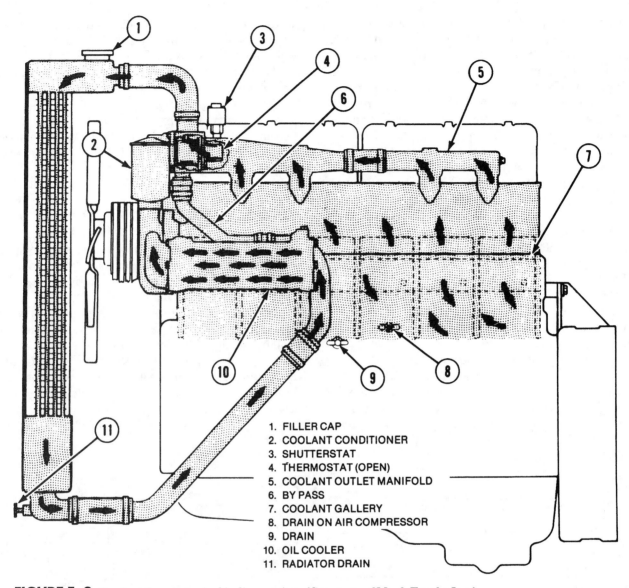

1. FILLER CAP
2. COOLANT CONDITIONER
3. SHUTTERSTAT
4. THERMOSTAT (OPEN)
5. COOLANT OUTLET MANIFOLD
6. BY PASS
7. COOLANT GALLERY
8. DRAIN ON AIR COMPRESSOR
9. DRAIN
10. OIL COOLER
11. RADIATOR DRAIN

FIGURE 5-2 Coolant flow in typical in-line engine. (Courtesy of Mack Trucks Inc.)

side of the water pump permits circulation of coolant through the engine while the thermostats are closed. When the thermostats are open, the coolant can flow through the keel cooling coils and then to the suction side of the water pump for recirculation. The keel cooling coils are located at the bottom of the vessel, where they are exposed to sea water. The heat of the engine coolant is transferred through the coils of the keel cooler to the surrounding water.

Coolant Circulation

In the liquid cooling system, the coolant circulates through the cylinder block, up through the cylinder head, and back through the bypass to the water pump when the thermostat is closed. When the temperature of the coolant reaches thermostat-opening temperature, coolant circulates from the water pump to the block and the cylinder head, through the open thermostat to the radiator inlet, through the radiator and the radiator outlet, and back to the water pump. Coolant circulates through the heater core at all times on some vehicles, but only when the heater temperature control is turned on in others **(Figures 5-1 and 5-2)**.

Water Pumps

The water pump is belt- or gear-driven from the crankshaft pulley. Water pump capacity must be sufficient to provide adequate coolant circulation. Centrifugal,

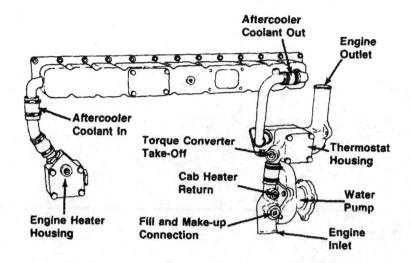

FIGURE 5-3 Water pump and related parts. (Courtesy of Cummins Engine Company, Inc.)

vane-type, nonpositive displacement pumps are commonly used. The impeller, shaft, fan hub, and pulley are supported in the water pump housing by one or more bearings. A water pump seal prevents coolant from leaking (**Figures 5-3 to 5-5**).

The water pump forces coolant into the engine block as the impeller rotates. Coolant enters the center area of the impeller from the radiator outlet and is thrown outward centrifugally to create a flow into the block. Coolant flow returns to the water pump through the bypass when the thermostat is closed, and through the radiator when the thermostat is open.

Raw (Sea) Water Pump

The raw water pump used on some marine diesel engines is similar in design to the conventional freshwater engine coolant pump in many ways (**Figure 5-6**). The shaft, bearings, seal, and housing are very similar to those found in other water pumps. The major difference is in the type of impeller used. The raw water pump usually uses a flexible vane-type impeller made from a special rubber compound and bonded to a steel hub that is mounted to the pump shaft. Since this type of pump is subjected to impurities such as salt and sand normally found in raw sea water, a conventional pump would not last in this type of application.

Water Manifold

Many multicylinder diesel engines are equipped with water manifolds. The water manifold is designed to ensure more even temperature control throughout the engine. Coolant heated by the cylinders at the rear of the engine must flow forward through the other cylinders on its way to the radiator. Consequently, the forward

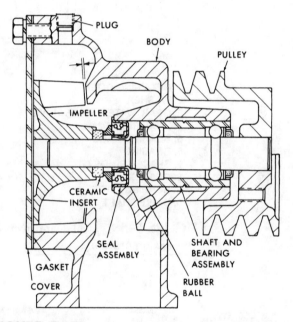

FIGURE 5-4 Cross section of a belt-driven water pump. (Courtesy of Detroit Diesel Corporation.)

cylinders run somewhat hotter than the others when not equipped with a water manifold. The water manifold collects water from each cylinder or pair of cylinders individually and routes it to the thermostats and radiator or heat exchanger. This results in equalization of cylinder temperatures. Coolant outlets in the cylinder head connecting to the water manifold are usually located near the exhaust valve port, where the hottest coolant is continually removed (see **Figures 5-2** and **5-3**).

Radiator Hoses and Clamps

Radiator hoses include the straight, molded, and flexible types. The straight and molded types must not be

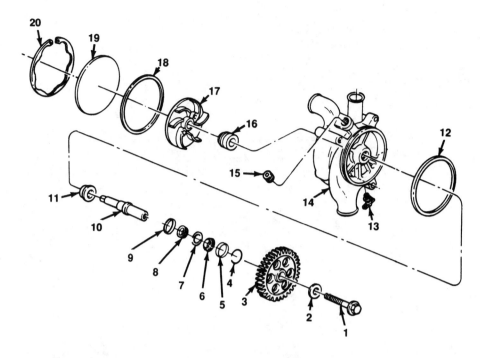

1. RETAINING BOLT
2. WASHER
3. GEAR, WATER PUMP DRIVE
4. SNAP RING
5. BEARING RACE
6. BEARING
7. SPACER RINGS (2)
8. BEARING
9. BEARING RACE
10. DRIVE SHAFT
11. OIL SEAL
12. O-RING, WATER PUMP HOUSING
13. DRAIN COCK
14. HOUSING, WATER PUMP
15. PIPE PLUG, WATER PUMP HOUSING
16. WATER SEAL
17. IMPELLER
18. O-RING, WATER PUMP COVER
19. WATER PUMP COVER
20. SNAP RING, WATER PUMP COVER

FIGURE 5–5 Typical gear-driven water pump components. (Courtesy of Detroit Diesel Corporation.)

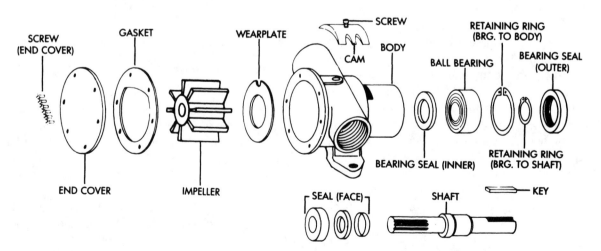

FIGURE 5–6 Raw water pump components for a marine engine. (Courtesy of Detroit Diesel Corporation.)

distorted when installed. The flexible hose can be bent as needed. The hose connecting the radiator to the inlet side of the water pump usually has a spiral wire support built in to prevent hose collapse due to low internal pressure. Hoses are required to make flexible connections between cooling system components.

Hoses are secured to their connections with hose clamps. A variety of designs are in use, with the most common being the geared type. Some engines use special silicone hoses that require hose clamps specially designed for them.

Radiator hoses can deteriorate both internally and externally. Hoses soften and disintegrate from contamination by lubricating or fuel oil. Silicone hoses are not affected in this way. Rubber hoses can swell and restrict coolant circulation. Material from damaged or deteriorated interior walls of rubber hoses can contaminate the coolant and impede thermostat operation.

Thermostats

The thermostat is a temperature-sensitive flow control valve located in the thermostat housing at the front of the engine. The thermostat remains closed until the engine reaches operating temperature. As the temperature increases, the thermostat opens. This allows coolant to be circulated through the radiator for cooling. When the engine coolant falls below operating temperature, the thermostat closes once again. Coolant circulation is restricted to the engine block and cylinder heads and the cab's interior heater when the thermostat is closed. A bypass provides the passage for coolant return to the pump (**Figures 5–7 to 5–9**).

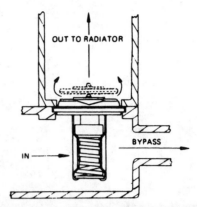

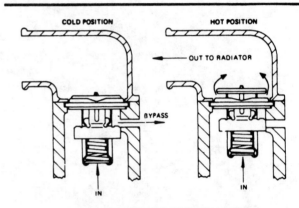

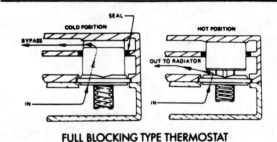

FIGURE 5–8 Various thermostat designs. (Courtesy of Detroit Diesel Corporation.)

Many V-type engines have two or more thermostats, one or two for each bank of cylinders.

Semiblocking thermostats are used in the rapid-warm-up cooling system.

In this warm-up system, enough coolant to vent the system is bypassed to the radiator top tank by means of a separate external deaeration line and then back to the water pump without going through the radiator cores. As the coolant temperature rises above 170°F, the thermostat valves start to open, restricting the bypass system, and permit a portion of the coolant to circulate through the radiator. When the coolant temperature reaches approximately 185°F, the thermostat valves are

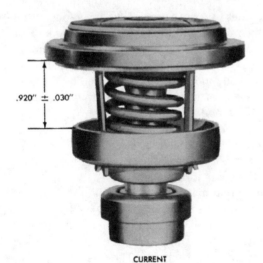

FIGURE 5–7 Example of wax pellet–type thermostat. (Courtesy of Detroit Diesel Corporation.)

66 Chapter 5 COOLING SYSTEM PRINCIPLES AND SERVICE

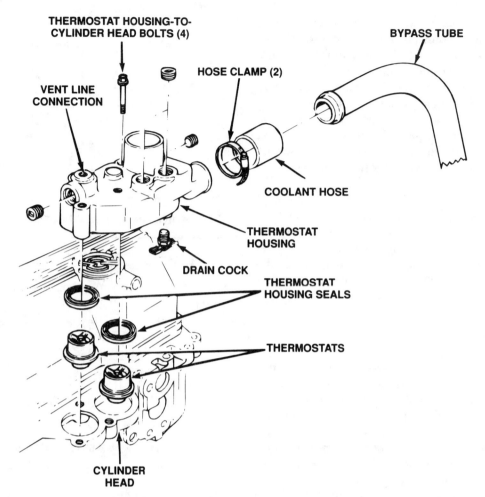

FIGURE 5-9 Thermostat mounting detail for series 60 engine. (Courtesy of Detroit Diesel Corporation.)

fully open, the bypass system is completely blocked off, and all of the coolant is directed through the radiator.

A defective thermostat that remains closed, or only partially open, will restrict the flow of coolant and cause the engine to overheat. A thermostat that is stuck in a full-open position may not permit the engine to reach its normal operating temperature. The incomplete combustion of fuel due to cold engine operation will result in excessive carbon deposits on the pistons, rings, and valves.

Properly operating thermostats are essential for efficient operation of the engine. If the engine operating temperature deviates from the normal range, the thermostats should be removed and checked.

Radiator and Pressure Cap

Radiators are designed to allow rapid heat dissipation and good air flow through the radiator core.

A radiator consists of two metal tanks connected to each other by a core consisting of a series of thin tubes and fins. Coolant flows from the inlet tank through the tubes to the outlet tank whenever the thermostat in the engine is open. The tubes and fins radiate heat from the hot coolant, and the air flow created by the fan, or ram air, dissipates the heat to the atmosphere (**Figures 5-1, 5-10, and 5-11**).

The inlet tank is equipped with a filler neck and radiator cap as well as an overflow tube. The overflow tube allows excess pressure to escape either to the ground or to the coolant reserve tank.

The radiator is made of aluminum or copper-brass. Both of these metals have good heat conductivity. A vertical-flow radiator has the inlet tank at the top and the outlet tank at the bottom. A horizontal-flow radiator has one tank at each side. The inlet tank is connected to the thermostat housing, while the outlet tank is connected to the water pump inlet. Hoses are used to make these connections.

LIQUID COOLING SYSTEM FUNCTION 67

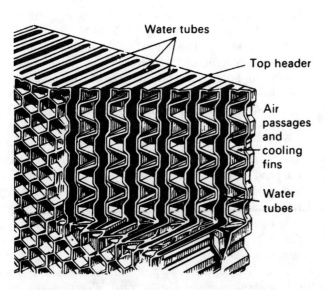
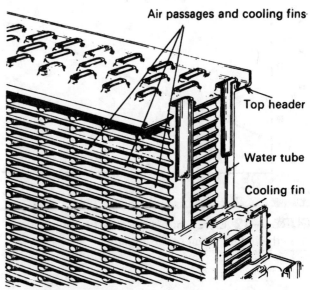

FIGURE 5-10 Two types of radiator core construction: honeycomb or cellular type (left); tube and fin (right). (Courtesy of Ford Motor Company.)

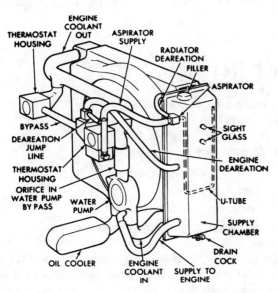

FIGURE 5-11 Cross-flow radiator and related components. (Courtesy of Detroit Diesel Corporation.)

The *radiator cap* incorporates two valves. The pressure relief valve limits pressure in the cooling system to a predetermined level. The pressure cap causes the system to become pressurized as a result of coolant expansion. Engine heat causes the coolant to expand. Pressurizing the cooling system raises the boiling point of the coolant by approximately 3.25°F (1.8°C) for every 1 psi (6.895 kPa) of pressure increase. This reduces the tendency for coolant to boil **(Figure 5-12)**.

A 10 psi (68.95 kPa) radiator pressure cap will increase the boiling point of a 50:50 solution of antifreeze from 230°F (110°C) to 262.5°F (128°C).

If the coolant expands sufficiently to cause system pressure above radiator cap design relief pressure, the pressure valve opens and allows coolant to escape, via the overflow tube, to the reserve tank, until pressure in the system is stabilized.

When the engine is shut off, the coolant cools down and contracts. This creates a low pressure in the cooling system and allows coolant to reenter from the reserve tank through the vacuum valve in the radiator pressure cap **(Figure 5-13)**. This prevents radiator hose collapse and allows the cooling system to remain full of coolant at all times. Reduced oxidation and rust formation are benefits of the constantly full cooling system.

Radiator capacity is determined by core size, thickness, and surface area. Engine size, number of accessories such as air conditioning, and type of service determine the radiator capacity of different equipment.

Radiator Fan

The fan is designed to provide sufficient air flow through the radiator core to ensure adequate cooling at all engine speeds and loads. The fan is bolted to a drive hub, which may be mounted on the water pump shaft or mounted and driven separately. Fans mounted on the water pump rely on water pump bearings for support, while fans mounted separately are supported by a hub and bearings of their own. A fan clutch or viscous

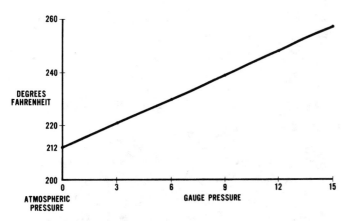

FIGURE 5-12 Effects of pressure in the cooling system on the boiling point of the coolant.

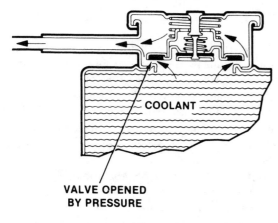

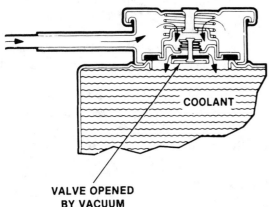

FIGURE 5-13 Radiator cap operation during coolant expansion (top) and contraction (bottom). (Courtesy of Detroit Diesel Corporation.)

fluid drive may be used to control fan operation according to engine need **(Figure 5-14)**.

Fan capacity is determined by the number of blades, total fan diameter, blade pitch, and fan speed. Fan efficiency can be increased by the use of a shroud around the circumference of the fan. This reduces air turbulence and air recirculation around the tips of the fan blades.

The pitch of the fan blades may be fixed or variable. Variable-pitch fans have flexible curved blades that tend to flatten out as fan speed increases. On highway vehicles, ram air adds to the air flow through the radiator at highway speeds. Variable-pitch fans and clutch-controlled fans reduce the power required to drive the fan, thereby reducing fuel consumption.

Fans may push or pull air through the radiator and may move air forward, sideways, or rearward in relation to the engine depending on equipment design and the manufacturer's preference. The fan may be direct-drive from an engine gear or shaft, or it may be driven by a belt from the crankshaft pulley.

Fan Drives

Fan drives are designed to achieve the following beneficial results.

1. Faster engine warm-up.
2. Reduced fuel consumption.
3. Reduced fan noise.
4. Increased fan drive belt life.
5. Improved engine temperature control.

Two types of fan drives are commonly used: the viscous drive and the clutch drive. Variations in design and operation exist in each type, but the operating principles are similar.

The thermo-modulated viscous fan drive is an integral unit with no external controls or control lines. It operates on the principle of transmitting drive torque from the input shaft to the fan through the shearing of a viscous silicone fluid film between the input and output plates of the unit. The input plate is connected to the fan drive shaft, and the output plate is connected to the fan. The plates and fluid are contained in a housing. An integral thermostatic control element reacts to changes in engine temperature and varies the thickness of the fluid film between the plates, thereby varying the fan speed. This unit requires no periodic service **(Figure 5-14)**.

about 185°F (85°C), the coolant sensor closes its contacts, completing the electrical circuit to the air control solenoid. This opens the air valve, applying air pressure (from the vehicle reservoir) to engage the fan clutch against spring pressure to drive the fan. When coolant temperature drops below sensor switch closing temperature, the switch opens and the solenoid cuts off air pressure to the fan clutch at the same time, venting air pressure from the clutch. The clutch spring disengages the clutch and the fan freewheels.

On vehicles or equipment with an electronic control system, the fan clutch may be controlled by the electronic control module (ECM) to control cooling as well as to provide as much as 30 HP of braking on deceleration.

Shutters and Shutter Controls

The radiator shutters provide the means for controlling air flow through the radiator to maintain proper engine operating temperature. The shutter consists of a series of vanes mounted in a frame and attached in front of the radiator. The vanes open and close by pivoting **(Figure 5–15)**.

The shutterstat in the air-operated system is mounted in the upper radiator hose and senses engine coolant temperature. The temperature-sensing element operates the needle-type air valve to control compressed air to the shutter cylinder.

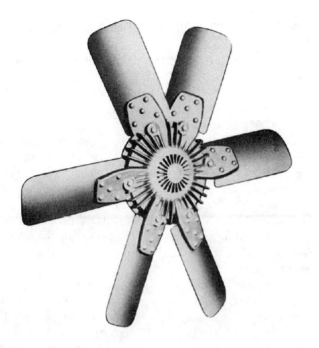

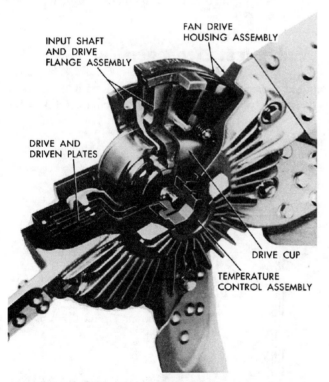

FIGURE 5–14 Thermo-modulated fan and fan drive assembly. (Courtesy of Detroit Diesel Corporation.)

Another design uses a single plate clutch released by spring pressure and applied by air pressure. Air pressure is controlled by a coolant-temperature-sensing switch, which in turn controls a solenoid-operated air control valve. As engine coolant temperature rises to

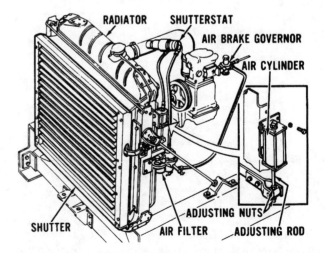

FIGURE 5–15 Air-operated type of radiator shutters. Automatically controlled shutters maintain relatively constant engine operating temperature. (Courtesy of Ford Motor Company.)

When the engine is below operating temperature, the control valve directs air to the shutter cylinder to close the shutter. Engine temperature then rises quickly because air cannot be drawn through the radiator. When the operating temperature of the control valve is reached, air is allowed to exhaust from the shutter cylinder, and the springs open the shutter.

The set temperature is maintained by the length of time the shutter remains open and closed, which varies, depending on engine load and air temperature. This oscillation of the shutter between fully open and fully closed is accompanied by a lag of approximately 7°F in the control valve to prevent excessive cycling.

In the vernatherm shutter control system, the vernatherm temperature-sensing element is located in the bottom radiator tank. When the coolant temperature rises to the correct operating temperature, the material in the element expands against a diaphragm, which in turn is connected to the operating piston. Continued expansion causes the diaphragm and piston to move against the push rod, which is connected to the shutter control linkage. This movement opens the shutter, allowing air to pass through the radiator, which causes the coolant temperature to drop. This cools the sensing element, causing the material in the element to contract. As a result, the piston returns to the original position once again to close the shutter.

The engine thermostat, fan, and shutter operate as a team to control the engine operating temperature (**Figure 5–16**). Each member of the team must operate at precise on and off coolant temperatures to maintain proper temperature control. Service manuals provide accurate and detailed specifications on temperature-sensing devices and controls.

Coolant Filters and Conditioners

Some engines are equipped with coolant filters and conditioners. They are designed to provide a cleaner cooling system, better heat dissipation, and improved heat transfer, thereby increasing engine efficiency and service life (**Figure 5–17**).

Both spin-on and canister (replaceable element) types are used. The filter element removes any impurities such as sand or rust suspended in the coolant. The filter element includes corrosion inhibitors and water softeners. The corrosion inhibitors dissolve in the coolant and form a protective film on the metal parts of the cooling system to prevent rust. Softening the water reduces scale deposits and maintains the coolant in an acid-free condition.

Periodic replacement of the chemically treated filter element is required to ensure effective cooling system protection. If engine coolant is drained and discarded, the filter must be replaced, since many of its protective elements will have been consumed in the discarded coolant.

Marine Heat Exchanger

Some marine engines are equipped with a liquid-to-liquid heat exchanger. The heat exchanger takes the place of the radiator on land-based engines (**Figure 5–18**).

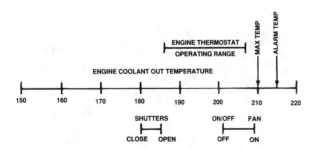

FIGURE 5–16 Temperature chart showing thermostat operating range, engine maximum temperature, alarm temperature, shutter operating range, and electronically controlled fan clutch operating range. (Courtesy of Detroit Diesel Corporation.)

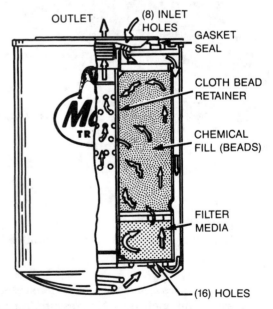

FIGURE 5–17 Coolant filter and conditioner. (Courtesy of Mack Trucks Inc.)

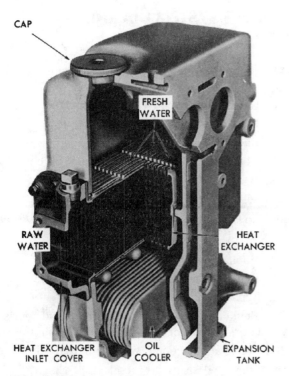

FIGURE 5–18 Cross section of heat exchanger for a marine engine. (Courtesy of Detroit Diesel Corporation.)

The unit consists of a housing with an expansion tank and a core of cells or tubes. Hot engine coolant circulates through the expansion tank and around the core cells back to the engine water pump and the engine. The core mounted inside the housing has separate inlet and outlet passages. Raw sea water is circulated through the core by the raw water pump. As engine coolant flows over the outside of the core, it is cooled by the cooler sea water inside the core.

Zinc electrodes are used to protect the heat exchanger from the corrosive effects of the electrolytic action of the raw sea water. One or more electrodes extending into the sea water in the heat exchanger may be used.

The expansion tank provides the means for filling the system and is provided with a filler cap. An overflow pipe connected to the expansion tank provides a vent to atmosphere or to a coolant reserve tank.

ANTIFREEZE COOLANT

Liquid coolant is the medium used to absorb heat while it is in the engine and transfer it to the radiator, where it is dissipated to the atmosphere. Although water is a satisfactory liquid to use for absorbing and transferring heat, it has several deficiencies. It has a relatively low boiling point and freezes readily. Also, inhibitors must be added to water to prevent rust and scale formation and for water pump seal lubrication. For these reasons, an ethylene glycol–based liquid is used for year-round service.

Ethylene glycol–based antifreeze coolant has a higher boiling point than water, it has the necessary inhibitors and additives required to retard the buildup of rust and scale, and it also has a water pump seal lubricant. Silicate inhibitors are added to prevent the corrosion of aluminum parts that may be used on some cooling systems.

A mix of 50% undiluted ethylene glycol antifreeze and 50% water will provide freeze protection to approximately −34°F (−36°C) and will have a boiling point of approximately 230°F (110°C) at 14.7 psi atmospheric pressure (101.4 kPa). A greater than 60% ethylene glycol content is not practical, since further increasing the antifreeze content can cause the antifreeze to thicken at low temperatures, restrict coolant circulation, and cause antifreeze in the engine to boil. **Figure 5–19** shows the degree of freeze protection provided by different mixes of antifreeze and water.

In operation, antifreeze additives and inhibitors tend to lose their effectiveness. For this reason, engine manufacturers and antifreeze manufacturers recommend changing the coolant at regular intervals as specified in the appropriate service manual.

COOLANT TEMPERATURE PROBLEMS

The average temperatures of engine components are relatively high, when compared with the boiling point of water; pistons will run as high as 500°F (260°C); exhaust valves, 1200°F (649°C); and the water side of the cylinder liners, 250°F (121°C). These temperatures are high enough to cause the water to boil. The flow of coolant in these areas is critical, because the water velocity must be high enough to prevent localized boiling, which would result in loss of heat transfer and hot spots. The engine design includes porting cylinder blocks in the water header to assure a high-velocity swirl around the cylinder liner and in the head areas.

It is necessary to maintain a continuous coolant circulation throughout the engine under all operating conditions. When the coolant flow stops, even for a short interval, the engine is immediately placed in danger. The piston rings will scuff and pistons will score or seize as quickly as 30 seconds after flow ceases, when operating at full throttle. Even though the coolant is present, local boiling at the point of maximum heat

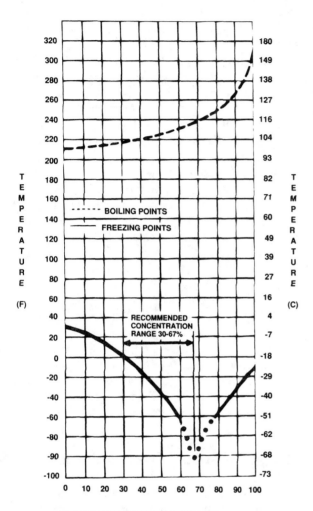

FIGURE 5–19 (Courtesy of Detroit Diesel Corporation.)

COOLING SYSTEM DIAGNOSIS AND SERVICE

Visual Inspection

1. Coolant level and freeze protection
2. Cleanliness of coolant
3. Radiator cap gasket
4. Hose and connections for leaks or hose collapse
5. Engine lubricating oil level

Note: Too low or too high an oil level will cause overheating and possible loss of the coolant.

6. Coolant in the oil and vice versa
7. Bent or damaged blades in the fan
8. Tension and condition of the belt
9. Bent core fins or accumulation of road debris in the radiator; damage or linkage wear in the shutters
10. Leaks at the cooler, water pump, truck heater, or other accessories

If any problems are indicated by these checks, determine the cause and make the appropriate correction.

Most cooling system malfunctions result in one of two conditions: overheating or overcooling.

Overheating

The obvious effects of severe overheating are seized pistons and resulting progressive damage. Intermittent loss of coolant flow can result in ring or piston scuffing and scoring, damage to the liner seals and the head gaskets, loosening of the injector sleeves, scuffed valve guides, and even cracked cylinder heads. This damage is often not obvious at the time of flow loss but later develops into an engine failure. In the case of piston and/or ring scuffing and scoring, this may show up as high oil consumption or excessive blowby several miles or hours later.

Continued operation with coolant temperatures above the recommended outlet temperatures can result in thermal fatigue of such parts as pistons, cylinder heads, and valves. High temperature causes the rubber parts to age prematurely, harden, and fail.

The resulting high lubricating oil temperatures will result in much thinner or lower viscosity oil and therefore higher oil consumption; higher than normal wear on the rings, bearings, and other parts; increased lacquer or varnish deposits; and reduced life of various assemblies.

transfer will occur; the temperature is high enough to destroy the lubricating oil film. The piston also expands as its temperature is raised, and scoring inevitably occurs.

The engine may continue to operate after brief periods of overheating, but the residual damage due to overheating will begin to affect various components. Liner O-ring life may be shortened because of accelerated heat-age hardening. Scuff marks on the liner or piston may lead to high oil consumption and can lead to excessive blowby and piston seizure at later dates. In many cases, failures such as high oil consumption are caused by cooling system malfunction.

Overcooling

Too much cooling may be the result of faulty thermostat or shutterstat operation, a leaking thermostat seal, too low a temperature setting, poor installation, or no thermostat. As a result, many parts will be operating at lower than normal temperature at greater than normal clearances. The lubricating oil temperature (even engines without an oil cooler) is directly related to coolant temperature; thus low coolant temperatures will result in low oil temperature.

Low oil temperatures reduce the effectiveness of oil detergency and the ability of filters to clean, and increase the possibility of water content and acidity. Several gases (e.g., sulfur dioxide, nitrous oxide) pass through the sump cavity, which contains water vapor. These gases combine with water to form acid. Low oil temperatures from overcooling allow the water to condense in the crankcase, causing increased acidity and internal formation of sludge, lacquer, and rust. Operating with extreme overcooling may result in excessive wear and/or poor engine-operating performance.

CAUTION:

Never allow an engine to idle for more than 15 minutes at a time. Long periods of idling can be harmful to an engine because the operating temperatures drop so low that the fuel may not burn completely, resulting in deposit on the valves, clogged injector spray holes, and sticking piston rings.

1. Periodically check the coolant temperature and level indicator.
2. Listen for changes in the fan noise at all speeds.
3. Listen for belt squealing, particularly during acceleration.
4. Always allow the engine to cool before shutdown. Operate the engine for at least 3 minutes, preferably 5 minutes, at idle speed. Much can be done on the road by easing off at the end of the run. This will prevent localized boiling and loss of coolant at shutdown.

The cooling system diagnostic chart lists common problems and their causes in the cooling system and related systems. You should have a thorough knowledge of the purpose, construction, and operation of the entire system and its components as well as the operating conditions to which the engine is subjected in order to make an intelligent diagnosis of the problem and its cause. If this is the case, the appropriate correction is obvious from the stated cause, and the necessary repair or service procedures should be made in accordance with service manual procedures and specifications.

WATER PUMP SERVICE

1. Inspect the water pump, idler pulley, and fan hub for wobble and evidence of grease or coolant leakage. Replace the damaged units.

2. Check the weep holes for evidence of continuous water leakage past the seal and make sure that the holes are clean and clear. Small streaks at the hole are normal if there is no coolant leakage. If leakage is present, the water pump must be removed and rebuilt or replaced with a new or rebuilt unit. Water pump and engine designs vary considerably. As a result, pump removal, service, and replacement procedures are also quite different. Refer to the appropriate service manual for specific procedures. Pump overhaul generally requires replacing the seals and bearings and in some cases the impeller and shaft as well. Carefully inspect the disassembled parts for wear or other damage and replace the faulty components. Assemble and install the pump as outlined in the service manual.

3. Check the pulley grooves for wear by the belt method. Push a new belt down in the groove. There should be 1/16 to 1/8 in. (0.06 to 0.13 mm) protrusion above the pulley groove (**Figure 5–20**). The belts should not bottom in the grooves.

Radiator Shroud Inspection

Shrouds should be checked periodically for proper clearance around the fan and for air leaks, cracks, and deterioration.

V-Belt Service

When using matched sets of V belts, be sure that the belts in the set are from a single manufacturer.

Note: Do not install or try to make a matched set of belts by mixing old belts with new belts or belts from different manufacturers.

If one belt of a matched set fails, remove all the belts in the set and install a new set of matched belts.

COOLING SYSTEM DIAGNOSIS

Problem	Cause
Loss of coolant due to external leakage	1. Leaking pipe plugs, such as core plugs, which seal off coolant passages. Core hole plugs are sometimes loosened by corrosion or vibration. 2. Loose clamps, faulty hose and piping 3. Leaking radiator or cab heater core 4. Leaking radiator deaeration top tank or surge tank 5. Leaking gaskets due to improper tightening of capscrews, uneven gasket surfaces, or faulty gasket installation 6. Leaking drain cocks 7. Leaking water pump. Badly worn or deteriorated seals are the cause of leaks at the pump. Premature failure of the pump seals often results from suspended abrasive materials in the cooling system, excessive heat from lack of coolant, or cavitation 8. Leaks at engine or air compressor cylinder heat gasket 9. Leaks at upper cylinder liner counterbore 10. Leaking engine or auxiliary oil cooler 11. Leaking air intake aftercooler (intake air heater in some cases) 12. Leaking coolant-cooled exhaust manifold 13. Leaking water manifold and/or connections
Loss of coolant due to internal leakage	1. Leaking engine or air compressor cylinder head gasket. Coolant passes into the cylinders or the crankcase. 2. Cracked engine or air compressor cylinder head. Coolant passes into the cylinders and is blown out the exhaust or out the air compressor discharge. 3. Deteriorated, severed, or chafed liner packing; defective liner packing bore. Coolant passes into the engine crankcase. 4. Improperly seated or defective injector sleeves. Coolant can pass into the cylinder or crankcase and, in the case of heads having cylindrical injectors, coolant can enter the fuel system if it gets past the injector body O-ring. 5. Porous cylinder block or head casting. Coolant can pass into the crankcase via holes in the cooling system jacket. In the case of cylinder heads with internal fuel passages, coolant can enter the fuel system via holes in the wall of the fuel rifle, which adjoins coolant passages. 6. Cracked or porous water coolant exhaust and faulty manifold-to-head gaskets. Coolant passes into the cylinder or out through the exhaust system. 7. Leaking engine air intake aftercooler or intake air heater. Coolant passes into the cylinders on naturally aspirated engines and on turbocharged and supercharged engines when the intake manifold pressure is less than the coolant pressure.
Loss of coolant due to overflow	1. Slush freeze of frozen coolant resulting from insufficient antifreeze in the system or poor mixing of antifreeze and water 2. Dirt, scale, or sludge in the cooling system 3. Plugged radiator core 4. Combustion gas enters the cooling system and displaces the coolant, causing it to overflow
Engine overheating	1. Poor circulation of coolant caused by collapse of soft hose and restriction 2. Overfueling the engine 3. Radiator shutter malfunction or improper adjustment of the thermal controls 4. Incorrect adjustment or malfunction of variable pitch, electronically controlled, or modulating fans 5. Crankcase oil level too high. Crankshaft dips in oil and causes a corresponding increase in temperatures due to friction and parasitic load on the engine. Crank-

WATER PUMP SERVICE

shaft oil dipping can also be encountered when operating an engine beyond the angles for which the oil pan was originally designed.

6. Dirty engine exterior; heavy accumulations of dirt and grease can severely hinder normal heat dissipation through the exterior walls of all the engine components.
7. Air pressure in the cooling system. The following are the most common causes of air entrainment:
   a. Low coolant level due to a leaking water pump or leaks at hose and coolant accessories
   b. Insufficient venting of the coolant system or plugging of vent lines
   c. Leaking air intake aftercooler
   d. Inadequate cooling system deaerating top tank. Allows air entrainment of the coolant by failing to control the top tank coolant turbulence or by not maintaining the proper level of coolant
   e. Leaking engine or air compressor cylinder head gasket
   f. Improperly seated injector sleeves (where applicable)
8. Inadequate cooling capacity. This condition can be the result of the misapplication of any one or any combination of the following cooling system components:
   a. radiator
   b. top tank
   c. surge tank
   d. fan
   e. fan shroud
   f. water pump
   g. auxiliary coolers
   h. recirculation baffles
   i. fan speed
9. Afterboil. Coolant boils and overflows after the engine is abruptly shut down following heavy loading.
10. Improper fan belt tension

Coolant contaminated with combustion gases	1. Cracked cylinder head 2. Blown head gasket 3. Injector sleeves leaking at the bore (where applicable)
Coolant contaminated with fuel oil	1. Injector sleeves leaking at the bore (where applicable) 2. Porous cylinder head casting; the fuel enters the cooler system through holes in the wall of the fuel rifle, which adjoins the coolant passages.
Coolant circulating at high pump speed only	1. Badly deteriorated water pump impeller; impeller damage is primarily caused by corrosion and cavitation erosion. 2. Excessive impeller-to-body clearance 3. Loose water pump drive belts 4. Impeller slipping on the shaft 5. Cracked impellers
Fan or water pump belts breaking prematurely	1. Foreign material in drive 2. Shock or extreme overloads 3. Belt damaged on installation by localized stretching 4. Belts not properly matched 5. Pulleys misaligned 6. Pulleys nicked or rough 7. Guard or shield interference during operation

Continued.

COOLING SYSTEM DIAGNOSIS—CONT'D

Problem	Cause
Engine operating too hot or overheating when loaded and coolant is known to be at the proper level	1. An altered horsepower rating of an engine or new engine installation that exceeds the original design of the cooling system or the vehicle 2. Clogged radiator air passages 3. Damaged radiator core fins 4. Heat exchanger element contains heavy lime and scale deposits 5. Thermostat not opening fully 6. Fan shrouding missing, damaged, or improperly positioned with respect to the fan 7. Recirculation baffles on the sides of the radiator missing or damaged 8. Fan drive belts slip. Impeller slips on the water pump shaft. 9. Excessive heat load because the torque converter is operating below 0.3 speed ratio 10. Clogged coolant passages, radiator, and engine 11. Faulty automatic radiator shutters. Shutters open only partially or open too late. 12. Faulty thermatic or modulated fan drive. Fan engages too late or operates too slowly. 13. Faulty electronic fan clutch control. Fan clutch does not engage. 14. Faulty variable pitch fan. Fan operates with the blades at an insufficient pitch.
Engine coolant temperature too low	1. Thermostat stuck in open position or malfunctioning to allow premature opening 2. Thermostat seal lip deteriorating and hardening, allowing the coolant to bypass the closed thermostat and enter the radiator, heat exchanger, or keel cooler 3. Excessive bypassing of coolant to the radiator or heat exchanger, with the thermostat closed and properly sealed 4. Engine exposed to very low temperatures and high wind with a low load factor
Lack of temperature control	1. Defective thermostat. Thermostat fails to open at the proper coolant temperature range and/or does not open completely. 2. Defective thermostat seal. Coolant leaks by the seal and on to the water pump, thereby bypassing the radiator or heat exchanger. 3. Operating engine having a pressurized cooling system without a pressure cap, with a defective pressure cap or a defective self-contained relief valve
Clogged water filter element	1. Contamination of the coolant caused by soluble oil, engine lube oil, gear oil or hydraulic transmission fluid, engine fuel, or rust, scale, and lime 2. Contamination of the cooling system with dirty fill or makeup coolant 3. Use of antileak additive
Corrosion, rust and scale buildup occurring in the cooling system even though the water filter is serviced regularly	1. Insufficient water filter capacity 2. Improper water filter installation 3. Careless servicing of the water filter 4. Aeration of the coolant. Free oxygen and carbon dioxide contained in the air are highly corrosive.
Oil overheating even though the cooling system has adequate capacity and is free of defects	1. Clogged or restricted oil cooler oil passages 2. Restricted oil cooler oil supply and return lines 3. Clogged or restricted oil cooler passages due to dirt, rust, and scale 4. Restricted oil cooler supply and discharge lines
Cab heater not producing heat at normal engine operating temperatures	1. Valves in the supply and/or return coolant lines shut off. Improperly sized heater for the compartment size. 2. Restricted heater cooler lines 3. Plugged heater core 4. Air entrainment in the heater core caused by operating the engine with a low coolant level 5. Heater lines not installed across a pressure drop or in the hot coolant portion of the circuit

WATER PUMP SERVICE

BELT TENSION ADJUSTMENT

In order for the belt to drive its intended accessory properly, it must be properly tensioned (**Figures 5–21 and 5–22**). The belt must have sufficient tension to transmit the load of the accessory, or excessive slip will

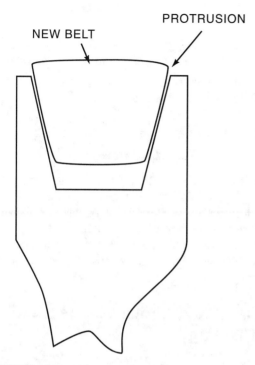

FIGURE 5–20 New belt must protrude above the pulley. If it does not, the pulley is worn and should be replaced.

FIGURE 5–21 Checking fan drive belt tension with a special gauge. (Courtesy of Cummins Engine Company Inc.)

1. All the belts of a matched set should be riding at approximately the same height in the groove. Differences between the belts should not exceed 1/16 in. (0.06 mm).

2. No object should rub against the belts. Check the heat shields, guards, etc. for proper clearance and stability at all engine speeds.

3. If the side walls of the belt become frayed, remove and replace with a matched set of belts. Check to determine the reason for fraying.

4. If the belts become saturated with oil, they should be replaced. Loss of friction surface will cause slippage.

5. Small shallow cracks in the cushion or at the top of cog may be ignored. If one or more cracks are deep, the belts must be replaced.

6. Check the tensioning of the belt. If the belt squeals, it is a sign of insufficient tension. Squealing most often occurs during engine acceleration.

7. Check the pulleys for damage or nicked grooves. Replace the pulley if the nick cannot be removed with a file.

8. Check for loose pulley shafts at the bearing. Replace worn or damaged bearings or shafts, which cause pulley movement or wobble.

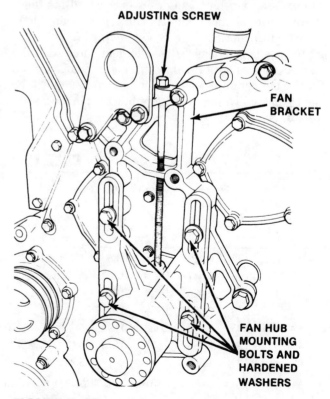

FIGURE 5–22 Example of fan hub mounting and belt tension adjustment screw. (Courtesy of Detroit Diesel Corporation.)

result. This slip glazes the belt, reducing its friction; more slippage results, causing premature belt failure.

Excessive tension will damage the accessory bearings, resulting in premature bearing failure and reduced belt life.

A new belt must be tensioned at a higher value than the belt that has been in operation for at least 15 minutes. This higher value allows for the "seating-in" effect and tension loss **(Figure 5–23)**.

Note: After the belt has seated in, check the tension to see that it is at the proper operating tension. The belt tension must be checked by using an approved belt tension gauge. This gauge must be properly calibrated for acceptable accuracy. Used belts (belts that have been run for at least 15 minutes) are tensioned at a lower value. Using the appropriate gauge, check and/or adjust belt to the tension as specified in the appropriate service manual.

Radiator Shutter Service

All moving parts of the shutters should be cleaned periodically. The shutterstat, filter-lubricator, and operating cylinder must be cleaned and inspected following the manufacturer's instructions.

The shutter-blade pivot points require cleaning with compressed air. Little lubricant is required for these pivots; use a small amount of nongumming low-temperature lubricant. *Never use engine oil.*

If installed, replenish the filter-lubricator with the correct fluid as prescribed by the shutter manufacturer.

The temperature-sensing unit or shutterstat must be set to operate within the ranges indicated in the appropriate service manual.

ADJUSTMENTS

1. Check to make sure that the shutters are completely sealed in the closed position.
2. Make sure that the shutters are completely open at the desired temperature setting.

CAUTION:

Shutters must sense a temperature in the coolant flow stream. This requirement prohibits putting the sensing unit in the top tank above the baffle. If a point other than the engine water outlet is used, the temperature setting must be adjusted to allow for the temperature drop that occurs between the water outlet and the sensing point.

Radiator and Coolant Inspection and Testing

1. Inspect the radiator for leaks. If there are no visible leaks, carefully remove the radiator cap and install a cooling system pressure tester.
2. Pressurize the system to the specified limit (usually stamped on the radiator cap).

CAUTION:

Do not exceed the limit, since this could cause leaks in the radiator or heater core to develop.

BELT TENSION CHART

Belt Size	Width Belt Top mm	Width Belt Top in.	Width Top of Pulley Groove mm	Width Top of Pulley Groove in.	Belt Tension "Initial"[a] Gauge Reading N	Belt Tension "Initial"[a] Gauge Reading lb	Belt Tension "Used"[b] Gauge Reading N	Belt Tension "Used"[b] Gauge Reading lb	Borroughs Gauge Numbers Old Gauge No.	Borroughs Gauge Numbers New Gauge No.
3/8	10.72	0.422	9.65	0.380	445 ± 22	100 ± 5	400 ± 22	90 ± 5	BT-33-95	BT-33-97
1/2	13.89	0.547	12.70	0.500	534 ± 22	120 ± 5	400 ± 44	90 ± 10	BT-33-95	BT-33-97
5V	15.88	0.625	15.24	0.600	534 ± 22	120 ± 5	400 ± 44	90 ± 10	BT-33-72-4-15	BT-33-72C
11/16	17.48	0.688	15.88	0.625	534 ± 22	120 ± 5	400 ± 44	90 ± 10	BT-33-72-4-15	BT-33-72C
3/4	19.05	0.750	17.53	0.690	534 ± 22	120 ± 5	400 ± 44	90 ± 10	BT-33-72-4-15	BT-33-72C
15/16	23.83	0.983	22.30	0.878	534 ± 22	120 ± 5	400 ± 44	90 ± 10	BT-33-72-4-15	BT-33-72C
8K	27.92	1.099			800 ± 22	180 ± 5	489 ± 44	110 ± 10		BT-33-109

MEASURE TENSION OF BELT FARTHEST FROM THE ENGINE

[a] "INITIAL" BELT TENSION is for a new belt.
[b] "USED" BELT TENSION is for a belt which has more than 30 minutes of operation at rated speed of engine.

FIGURE 5–23 Typical belt tension chart. (Courtesy of Caterpillar Inc.)

3. With the pressure applied, inspect the radiator again for leaks. If leakage is present the radiator must be removed, cleaned, and repaired.

4. Test the freeze protection of the coolant with an antifreeze tester. Inspect the coolant withdrawn with the tester for any signs of contamination. Hold the tester in a level position to prevent the float from sticking. Observe the freeze protection of the coolant. Inspect the coolant withdrawn with the tester for any sign of contamination by oil or other foreign matter. The coolant should normally be a translucent red or green. Drain and flush the system if the coolant is contaminated. Strengthen the freeze protection if it is below requirements. Follow service manual instructions for these procedures.

Pressure Cap Testing

There are several different cap testers that can be used to check the cap. The cap is turned on to the end of the tester, and a pump built into the tester is used to apply pressure. A pressure gauge, which is a part of the tester, indicates the pressure as it is applied. A pressure cap that tests more than 2 psi above or below its required value should be replaced **(Figure 5-24)**.

Adapters for long or short caps usually come with the cap pressure tester, which permits the use of the tester during cylinder block pressure checks.

Check the filler neck to make sure that the cap is sealing when installed. The vacuum relief valve in the cap should be checked for seal when seated and for adequate spring tension to hold a seal in position. A vacuum check should be made to verify reseating of the vacuum relief valve.

Thermostat Service

The thermostat is usually replaced if cooling system diagnosis indicates that it may be faulty. The following procedure may be used to test it. Immerse the thermostat and a test thermometer in a container of water **(Figure 5-25)**. Heat the water to the thermostat's specified opening temperature. Note the temperature at which the valve begins to open. Continue to heat the water to the specified temperature at which the valve should be fully open. If the thermostat does not operate within the specified temperature range (usually stamped on the thermostat body), it must be replaced. To replace the thermostat, clean the gasket surfaces and use new seals and gaskets **(Figures 5-26 and 5-27)**. You may have to index the thermostat to a certain position in the housing. The sensor element must be placed to face in the proper direction (toward the hotter engine coolant).

Cleaning and Flushing the System

The cooling system should be cleaned and flushed if the coolant is contaminated or is due to be changed as a result of scheduled maintenance. Cleaning and flushing must be done only as recommended in the appropriate service manual. For thorough cleaning and testing, the radiator and heater core may have to be removed and serviced by professionals. The following cleaning and flushing procedure is typical.

1. Drain all the coolant from the cooling system.

2. Add the recommended commercial radiator cleaning compound and fill the cooling system with water.

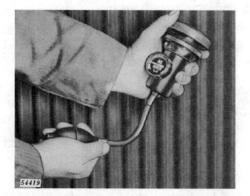

FIGURE 5-24 Pressure testing a radiator cap. The same tester is used to pressure test the cooling system. (Courtesy of Deere and Company.)

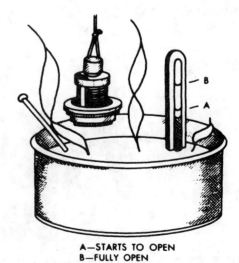

A—STARTS TO OPEN
B—FULLY OPEN

FIGURE 5-25 Testing a thermostat. (Courtesy of Detroit Diesel Corporation.)

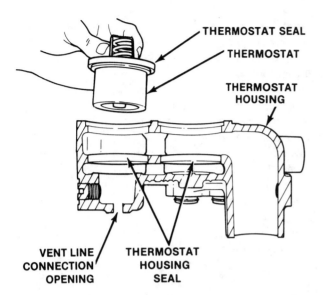

FIGURE 5–26 Thermostat and seal replacement. (Courtesy of Detroit Diesel Corporation.)

3. Run the engine at a high idle for 30 minutes. Keep constant watch on the temperature. The thermostat must be open for at least 15 minutes.

Note: Temporary covering of the radiator may be required to obtain the proper coolant temperature.

4. Drain the system completely and allow the engine to cool.

5. Add neutralizer and fill with water.

6. Run the engine with the thermostat open for 5 minutes after the coolant temperature reaches 180°F (85°C).

7. Drain the complete cooling system and allow the engine to cool.

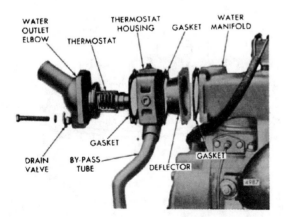

FIGURE 5–27 Thermostat and related parts assembly. (Courtesy of Detroit Diesel Corporation.)

8. Flush with clean water and refill the system with antifreeze solution.

9. Run the engine at high idle until the thermostat opens, and check for leaks. If none are found, shut down the engine.

BACKFLUSHING CAUTION

Backflushing of the radiator and block may be attempted if chemical cleaning does not satisfactorily clean the radiator.

CAUTION:

Backflushing can loosen scale formation, causing the cooling system to clog at a later date during operation.

Since the cooling system may clog from the material loosened during flushing, it is usually better, if the radiator condition warrants, to take the radiator to a qualified radiator repair shop.

Changing the Coolant Filter/Conditioner

CANISTER-TYPE ELEMENT

1. Close the shut-off valves on the inlet and drain lines. Unscrew the drain plug at the bottom of the housing.

2. Remove the cover capscrews and lift off the cover; discard the gasket and canister.

3. Remove the new canister from the package; install the canister in the housing.

4. Install the cover.

5. Reinstall the drain plug and open the shut-off valves in the inlet and drain lines.

SPIN-ON ELEMENT

1. Close the shut-off valves on the inlet and drain lines.

2. Unscrew the element and discard.

3. Install a new element and tighten until the seal touches the filter head. Tighten an additional one-half to three-fourths turn. Open the shut-off valves.

Heat Exchanger

Heat exchangers require inspections (at engine rebuild) of the interior tube bundles to ensure against buildup on the tubes or clogging from oil sludge, which reduces heat transfer. Marine units and some industrial heat exchangers have electrolytic protection provided by a zinc plug. These plugs have a sacrificial anode that is meant to be consumed while protecting other metallic portions of the system.

These plugs should be removed and inspected at recommended intervals. If the zinc anode has deteriorated to less than approximately half the original diameter, it should be replaced. If these are neglected, the raw or sea water could react with and consume the heat exchanger or oil cooler elements.

Filling the Cooling System

After all the necessary flushing and cooling system repairs are made, the system should be filled with the recommended coolant as follows (general procedure):

1. Close drain cocks and install drain plugs.
2. Turn the heater temperature control to high.
3. Add coolant to the system until the radiator remains full.
4. Add additional coolant to the required level in the coolant reserve tank.
5. Observe the sight glass to make sure that the system is completely deaerated.
6. Install the radiator cap.
7. Start the engine and run until operating temperature is reached.
8. Switch the heater blower fan on. If the heater produces heat, the heater core is full; if not, accelerate the engine several times to remove air lock from the heater until heat is produced.
9. Switch the engine off.
10. Correct coolant level in the reserve tank.
11. Make sure that there are no coolant leaks.

REVIEW QUESTIONS

1. The cooling system removes _____ _____ from the engine.
2. The cooling system maintains the engine at its most _____ _____ _____ _____ .
3. Many multicylinder diesel engines are equipped with _____ manifolds.
4. A raw sea water pump is used on _____ engines.
5. The cooling system of an engine relies on the principles of _____ , _____ , and _____ .
6. About _____ _____ of the heat developed by the burning fuel is absorbed by the cooling system.
7. The water pump is usually _____ or _____ driven.
8. The water pump impeller throws coolant _____ to create flow in the engine block.
9. When the thermostat is closed, coolant flows back to the water pump through the _____ .
10. The thermostat remains _____ until the engine reaches _____ operating temperature.
11. Thermostats have either a _____ valve or a _____ valve.
12. The radiator _____ is connected to the water pump _____ .
13. The radiator cap has a _____ _____ valve and a _____ valve.
14. Coolant returns to the radiator from the _____ when the engine cools down.
15. Pressurizing the coolant system _____ the boiling point of the coolant.
16. A good antifreeze solution usually consists of _____ % water and _____ % ethylene glycol.

TEST QUESTIONS

1. Ethylene glycol is
   a. the same as alcohol antifreeze
   b. permanent antifreeze
   c. not recommended
   d. a solution of antifreeze and water

Chapter 5 COOLING SYSTEM PRINCIPLES AND SERVICE

2. How much engine heat is actually used to produce power?
   a. 60%
   b. 35%
   c. 75%
   d. 50%

3. A hydrometer is used for checking
   a. speed
   b. distance
   c. hydro
   d. freezing protection

4. Pressurizing a cooling system
   a. raises the boiling point
   b. lowers the boiling point
   c. does not affect the boiling point
   d. keeps the coolant from evaporating

5. A 50% solution of water and antifreeze will give protection to
   a. −90°F
   b. −40°F
   c. −34°F
   d. 0°F

6. The term *permanent* when used with antifreeze refers to
   a. continuous use during summer and winter
   b. its low freezing point
   c. its boiling point with reference to that of water
   d. its mixing readily with water

7. The cooling system removes heat from the engine by the following methods:
   a. conduction, convection, radiation
   b. radiation, evacuation, precipitation
   c. evacuation, conduction, convection
   d. convection, precipitation, evacuation

8. Technician A says antifreeze is used to provide freeze protection. Technician B says antifreeze is used to provide a higher boiling point. Who is correct?
   a. Technician A
   b. Technician B
   c. both are correct
   d. both are incorrect

9. The primary purpose of the pressure cap is to
   a. prevent coolant leaks
   b. prevent air leaks
   c. reduce the cooling system pressure
   d. increase the coolant boiling point

10. Technician A says the cooling system should be pressure tested to determine engine internal leakage. Technician B says the cooling system should be pressure tested to determine external leakage. Who is correct?
    a. Technician A
    b. Technician B
    c. both are correct
    d. both are incorrect

11. If 1 psi pressure is applied to a cooling system, the boiling point
    a. decreases by 3.25°F
    b. decreases by 7°F
    c. increases by 7.5°F
    d. increases by 3.25°F

12. When the cooling system thermostat is closed, water circulates mainly through
    a. the radiator
    b. the engine
    c. both engine and radiator
    d. the bypass and the engine

13. Technician A says the thermostat should be installed with the thermal element toward the radiator. Technician B says the thermostat should be pressure tested to determine if it is operating properly. Who is correct?
    a. Technician A
    b. Technician B
    c. both are correct
    d. both are incorrect

14. Removing the radiator cap on a hot engine will usually result in a loss of coolant. This occurs because
    a. coolant pressure pushes the coolant out
    b. reduced pressure allows the coolant to boil
    c. the hot coolant has expanded and will push out of the radiator neck
    d. the pump has caused rapid coolant circulation

15. The vacuum-operated radiator shutter system consists of a
    a. shutter assembly, vacuum power cylinder, and shutterstat
    b. vacuum valve and shutter assembly
    c. mechanical linkage, vacuum valve, and shutterstat
    d. shutter assembly only

16. On the air-operated radiator shutter system, a thermostatically controlled air valve is called a
    a. control valve
    b. shutter control
    c. shutter assembly
    d. shutterstat

17. The automatic radiator shutter controls engine temperature by
    a. opening shutters when the engine is cold
    b. regulating air flow through the radiator
    c. opening shutters when the engine is warm
    d. partially opening shutters when the engine is cold

18. On a thermostat-operated shutter system, as the coolant reaches operating temperatures
    a. the power element expands and opens the shutters by working against the spring tension
    b. the power element expands and opens the shutters by working with the spring tension
    c. the power element contracts and opens the shutters by working with the spring tension
    d. the spring alone opens the shutters

19. One of the main advantages of radiator shutters is
    a. the fan uses less horsepower because of less air flow
    b. slower engine warm-up and less variation in operating temperature
    c. faster engine warm-up and less variation in operating temperature
    d. less maintenance

CHAPTER 6

LUBRICATION SYSTEM PRINCIPLES AND SERVICE

INTRODUCTION

Engine oils are the lifeblood of the diesel engine. Their quality and performance are often taken for granted, without full appreciation of the severe service conditions under which engine oils must function. Lubrication must be accomplished despite highly oxidizing conditions, very high temperatures, and the presence of large amounts of contaminants. The high output of today's engines, coupled with extended drain intervals, has increased the severity of the conditions under which the oil must perform.

Extensive research and in-service field testing are required to develop oils meeting the increasing demands of today's diesel engines.

Engine oils are formulated from selected lube basestocks and fortified with the right additives to provide the performance level required.

PERFORMANCE OBJECTIVES

After thorough study of this chapter and sufficient practice on the appropriate engine components, you should be able to:

1. Describe the purpose, construction, and operation of the lubrication system and its components.
2. Diagnose lubrication system problems.
3. Recondition or replace faulty components.
4. Test the reconditioned system to determine the success of the service performed.

TERMS YOU SHOULD KNOW

Look for these terms as you study this chapter, and learn what they mean.

- engine oil
- lubrication
- additives
- detergents
- dispersants
- oxidation inhibitors
- corrosion inhibitors
- rust inhibitors
- foam depressant
- pour point depressant
- viscosity index improver
- SAE viscosity classification
- multigrade oil
- API service classification
- oil pan
- oil pump
- pressure regulator
- scavenge pump
- oil filter
- oil cooler
- oil pressure indicator
- crankcase ventilation

LUBRICATION SYSTEM COMPONENTS

Although the lubrication systems of different engines vary considerably in design, the components used are quite similar. System components include the following (**Figure 6–1**):

1. Engine oil lubricant
2. Oil sump or oil pan (contains engine oil)
3. Oil pump (produces oil flow and oil pressure)

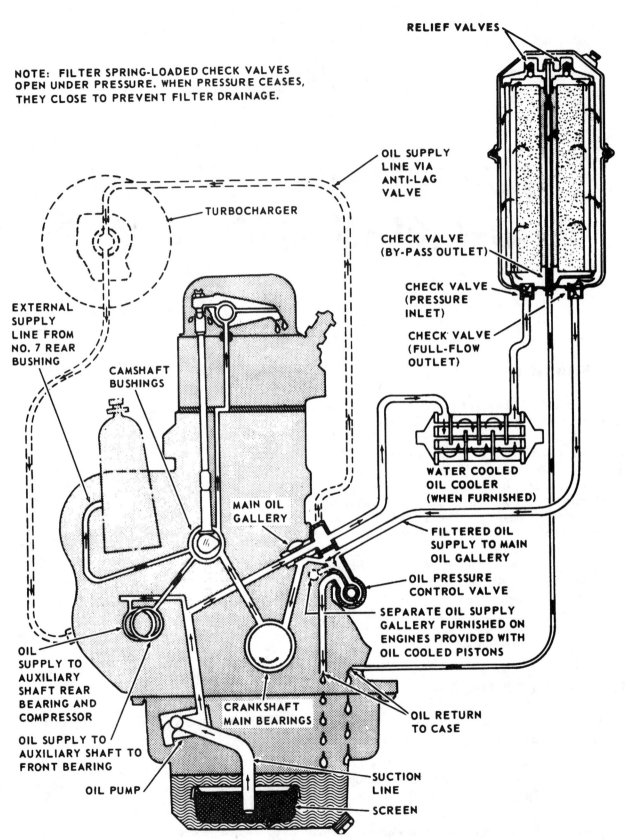

FIGURE 6-1 Lubrication system and oil circulation in a Mack truck engine. (Courtesy of Mack Trucks Inc.)

4. Pressure regulator (regulates engine oil pressure)
5. Inlet screen and pickup tube (screens oil flowing into pump)
6. Oil cooler (cools engine oil to avoid overheating)
7. Oil pressure indicator (indicates engine oil pressure)
8. Oil galleries (internal passages in engine block and head that distribute oil to engine parts)
9. Oil filter (removes foreign particles from engine oil)

ENGINE OIL FUNCTIONS

The engine oil must perform many direct functions without causing any negative impact in other areas of engine performance.

LUBRICATION

The oil must provide a fluid film between all moving engine parts to reduce friction, heat, and wear. Friction and wear are caused by metal-to-metal contact of the moving parts. Wear is also caused by acidic corrosion, rusting, and abrasion from the contaminants carried in the oil.

SEALING

High combustion pressures are encountered. Piston rings require an oil film between ring and liner and between ring and piston groove to seal against these high pressures and to prevent blowby, the escape of combustion gases past the piston rings.

COOLING

The engine oil is largely responsible for piston cooling. This is accomplished by direct heat transfer through the oil film to the cylinder walls and by oil sprayed at the underside of the pistons. Heat is carried by the oil from the undercrown and skirt to the crankcase. Thermally stable oils are required to withstand the high temperatures encountered.

DEPOSIT CONTROL

Rings must maintain freedom of movement to function properly and to maintain a good seal. Deposit buildup in the ring grooves and on piston lands must therefore be controlled.

VARNISH CONTROL

Engine parts, particularly the pistons, must be kept free of varnish to ensure performance and proper cooling.

SLUDGE CONTROL

High- and low-temperature sludge-forming contaminants must be held in suspension and not allowed to drop out and accumulate. The larger particles are removed by the filter. Sludge and abrasives are removed with the oil when drained.

BEARING PROTECTION

Oil breakdown and corrosive combustion products can cause bearing corrosion. The additives in the oil counteract this action by minimizing breakdown, neutralizing blowby products, and helping to form a protective film.

RUST CONTROL

Engine components including valve lifters, valve stems, rings, and cylinder walls, are subjected to severe rust-promoting conditions, particularly in winter stop-and-go driving. Rusting is controlled by oil formulation.

WEAR CONTROL

Wear occurs through metal-to-metal contact, acidic corrosion, rusting, and abrasive action of the oil's contaminant load. Metal-to-metal contact is controlled by proper viscosity selection and use of film-forming compounds. Acidic corrosion and rusting are controlled by formulation of the oil, while the abrasive wear is controlled by air and oil filtration and oil drain intervals.

SCUFF PROTECTION

High peak pressures occur in such areas as valve train mechanisms, particularly the camshaft lobes. Antiweld- or antiseize-type additives are required to minimize this type of wear.

CONTROL OF COMBUSTION CHAMBER DEPOSITS

Deposits, including oil-derived ones, accumulate in the combustion chamber, increasing the compression ratio and creating hot spots.

Combustion chamber deposits increase exhaust emissions. Oils must be formulated to reduce such deposits.

CONTROL OF VALVE DEPOSITS

Some higher ash oils tend to create deposits on exhaust valves in some severe types of service. This tendency must be minimized.

The success of the lubrication system in performing all these functions depends on a number of factors and conditions. There must be an adequate supply of good-quality lubricant delivered to all moving engine parts under sufficient pressure to provide hydrodynamic lubrication for rotating parts and oil adhesion to surfaces subject to sliding friction. Therefore:

- The oil and filter must be changed at regular intervals.
- The engine must operate at its most efficient temperature.
- Engine oil temperatures must not be excessively hot or cold.

ENGINE LUBRICATING OILS

Diesel engine lubricating oils fall into two basic categories: petroleum-based oils and synthetic oils. Petroleum-based oils, however, contain a variety of additives; so in fact they, too, are partly synthetic. Some of the major additives include those described here.

METALLIC DETERGENTS

These are metallic ash-containing compounds having a detergent/dispersant action in controlling deposits and keeping engine parts clean. They will clean up existing deposits as well as disperse particulate contaminants in the oil. As the performance of ashless dispersants continues to improve, they are replacing the metallic detergents, thus reducing the ash level of the finished oil. Metallic detergents are the major contributor of the base number required for acidic corrosion control. They are also good antiwear, antiscuff, and antirust additives.

ASHLESS DISPERSANTS

These are ashless organic compounds having a detergent/dispersant action in controlling deposits and keeping the engine parts clean. Their cleanup action on existing deposits is much slower and less effective than that of the metallic detergents. However, their dispersancy is much more effective in suspending in the oil potential carbon-forming deposits and, in particular, low-temperature sludge. Improvements in their diesel performance over the past few years have increased their use in combination with the metallic detergents in lower ash formulations.

OXIDATION INHIBITORS

These prevent oxygen from attacking the lubricant base oil. Without inhibitors the oil would react with oxygen, eventually solidifying in service or turning acidic and causing bearing corrosion.

BEARING CORROSION INHIBITORS

Bearing corrosion is the result of acid attack on the oxides of the bearing metals. The acids involved originate either from the blowby combustion gases or from oxidation of the crankcase oil. Acidic corrosion is controlled by the addition of inhibitors, which form protective barrier films on the bearing surfaces.

RUST INHIBITORS

Rusting results from oxygen attack on the metal surface and usually occurs in thin film areas such as hydraulic lifters or push rods. It is controlled by the addition of an inhibitor to the oil formulation.

ANTIWEAR

Wear results from metal-to-metal contact, acidic corrosion, and contaminant or dirt load. Metal-to-metal contact is overcome by the use of film-forming compounds. The acidic corrosion, originating mainly from acidic blowby gases, is neutralized by the use of alkaline additives.

Reserve alkalinity is required to control corrosive wear. The type and sulfur level of the fuel will determine the alkalinity reserve required. The use of highly alkaline or "overbased" additives provides high alkaline reserve where needed.

FOAM DEPRESSANTS

Detergent/dispersant-type oils tend to entrain air, which, when rapidly released, causes foaming. Foam depressants are added to control release of entrained air, thus eliminating this problem.

POUR POINT DEPRESSANTS

Base oils contain hydrocarbons that tend to solidify or crystallize into waxy materials at lower temperatures. Use of pour point depressants in the oil formulation modifies the wax crystal structure, resulting in a lower pour point and, in some instances, improved low-temperature fluidity. (*Pour point* is the lowest temperature at which the oil will flow when tested under prescribed conditions.)

VISCOSITY INDEX IMPROVERS

Petroleum oils thin out with increasing temperature. Viscosity index is a measure of this rate of viscosity change. The addition of a viscosity index improver slows down the rate of "thinning"; thus, the oil remains thicker at the engine operating temperature. Viscosity index improvers are used extensively to formulate multigrade oils.

Engine oils play a significant role in providing improved fuel economy. Additives such as colloidal graphite, colloidal molybdenum disulfide, and soluble friction modifiers are being used. Lighter viscosity grades and multigrades provide greater fuel efficiency. Lighter grades require improved wear performance of the oil.

ENGINE OIL CLASSIFICATION

Engine oils are classified by viscosity and performance as follows.

SAE Viscosity Classification

The Society of Automotive Engineers (SAE), the American Society for Testing and Materials (ASTM), and the American Petroleum Institute (API) classify engine oils by SAE viscosity numbers, commonly called *viscosity grades* or *SAE grades*.

The W grades are defined by cold cranking simulator viscosity (in centipoises) at 0°F or −18°C, a value related to ease of cranking an engine in cold weather. The other grades are defined by kinematic viscosity (in centistokes) at 212°F or 100°C, to define high-temperature wear protection.

The SAE system is based solely on viscosity; other factors of oil quality or performance are not considered.

In the viscosity test a measured amount of oil is brought to the specified temperature (usually 100°C or 212°F). The length of time in seconds required for a specified volume of oil to flow through a small orifice in an instrument such as a saybolt or kinematic viscometer is recorded.

MULTIGRADE OILS

The viscosity of an oil changes with temperature. At low temperatures the oil is thick, and its viscosity is high. As temperature increases, the oil becomes thinner, and its viscosity decreases. A sluggish oil makes engine starting difficult and delays warm-up lubrication, while excessively thin oils give poor lubrication and high oil consumption.

The temperature-viscosity relationship of typical SAE 10W, 40, and 10W-40 grades may be described as follows. The 10W grade is control tested at 0°F or −18°C. The 40 grade is controlled at 212°F or 100°C. The multigrade oil passes the specification for both the 10W and 40 grades; they therefore can be called 10W-40. The reduced rate of thinning out with increasing temperature is due to the viscosity index improver additive used.

The 15W-40 and 10W-30 multigrade oils provide the cranking ease and reliability required. The SAE grade required is determined by referring to the SAE chart provided in the manufacturer's service manual (**Figures 6–2** and **6–3**).

API Service Classification

The API, the ASTM, and the SAE have classified and described engine oils in terms intended to aid equipment makers in recommending proper oils and consumers in selecting them.

The engine oil service classification is divided into an S series covering engine oils generally sold for use in passenger cars and light trucks (mainly gasoline engines), and a C series for oils for use in commercial, farm, construction, and off-highway vehicles (mainly diesel engines). An oil can meet more than one classification, e.g., SE/CC. The system is open-ended so that

SAE Viscosity Grade: 15W-40
API Classification: CF-4
Military Specification: Mil-L-2104E
HT/HS Viscosity: 3.7 cP min.

FIGURE 6–2 SAE grade markings on oil containers and what they mean. HT/HS means high temperature/high speed. (Courtesy of Detroit Diesel Corporation.)

Chapter 6 LUBRICATION SYSTEM PRINCIPLES AND SERVICE

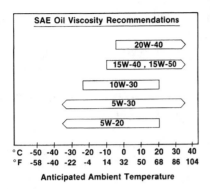

FIGURE 6–3 Typical engine oil viscosity recommendations. (Courtesy of Cummins Engine Company Inc.)

CA—Light-duty operation
CB—Moderate-duty operation
CC—Light-duty turbocharged engine operation
CD—Severe-duty operation
CE—Low-speed/high-load and high-speed/high-load operation
CF—On-highway, heavy-duty truck engine operation

FIGURE 6–4 Diesel engine oil service classifications.

new categories can be added in either series as needed **(Figure 6–4)**.

Synthetic Oils

Synthetic oils are made from synthetic basestocks and fortified with additives similar to those used with conventional petroleum basestocks. The advantage of the synthetic basestock is that it can be made to meet the required performance characteristics.

Properly formulated synthetic oils can outperform conventional oils in oil life and the control of oil thickening in service. They also have extremely good low-temperature performance and may be used when extremely low temperature starting is anticipated. Advantages of synthetic oils include the following:

- Higher viscosity stability over a wider operating temperature range
- Reduced effects of oxidation (reduced oil thickening)
- Reduced wear and increased load-carrying ability
- Reduced loss through evaporation
- Reduced crankcase oil temperatures
- Reduced oil consumption
- Fewer engine deposits
- Increased fuel economy

Engine Oils Data Book

The publication *EMA Lubricating Oils Data Book for Heavy Duty Automotive and Industrial Engines* is available from the Engine Manufacturers Association, 401 N. Michigan Ave., Chicago, Illinois, 60611-4267. The publication lists brand-name lubricants distributed by worldwide oil distributors and shows their performance levels, viscosity grades, and sulfated ash contents.

OIL PUMPS AND PRESSURE REGULATION

The engine oil pump must provide a continuous supply of oil at sufficient pressure and in sufficient quantity to provide adequate lubrication at all times to the entire engine. The oil pump is mounted on the engine block on either the outside or inside. It picks up oil from the reserve in the oil pan through the inlet screen and pickup tube. The oil is forced out of the pump outlet to the pressure regulator valve, which limits maximum oil pressure, and to the main oil gallery **(Figures 6–5 to 6–8)**.

Two types of oil pumps are used: the rotor type and the gear type. One of the rotors or gears is gear-driven by the crankshaft.

Since there is no direct path or opening for oil to flow through between the pump inlet and outlet, the pump is a positive displacement pump. Bearing clearances and metered oil holes restrict the flow of oil from the pump. This results in back pressure and pressure buildup. To limit maximum pressure, a pressure regulator valve is incorporated in the oil pump. Excess oil pressure is vented to the pump inlet or to the oil pan. Thus, engine oil pressure is dependent on pressure regulator valve–spring tension and restriction to flow on the outlet side of the pump. It follows, then, that excessive bearing clearances will cause a drop in oil pressure. A sticking regulator valve or damaged regulator valve spring will also affect oil pressure.

Internal wear in the oil pump, allowing oil to bypass back to the inlet side, will cause oil pressure to drop correspondingly. Oil pumps are designed to have sufficient reserve capacity to allow for normal wear.

Scavenge Pump

Many diesel engines used in equipment required to operate on fairly steep grades use an additional oil pump known as a scavenge pump. It is designed and posi-

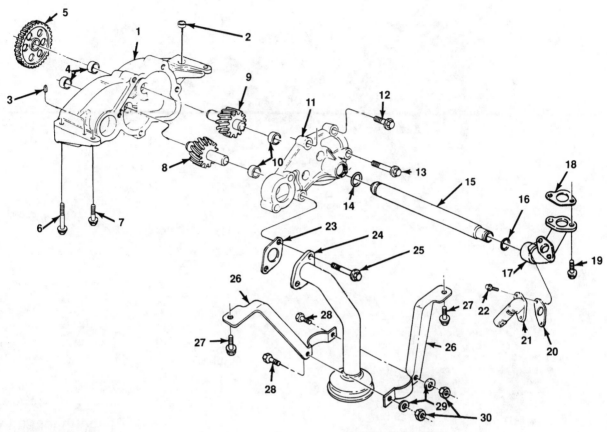

1. Body, Oil Pump
2. Dowel, Oil Pump Locating
3. Dowel, Oil Pump Locating
4. Bushing, Oil Pump Housing (2)
5. Gear, Oil Pump Drive
6. Bolt, Pump Body to Block (2)
7. Bolt, Pump Body to Block (2)
8. Driven Shaft and Gear Assembly
9. Drive Shaft and Gear Assembly
10. Bushing, Oil Pump Cover
11. Cover, Oil Pump Housing
12. Bolt, Oil Pump Cover (2)
13. Bolt, Oil Pump Cover (2)
14. Ring, Seal
15. Tube, Oil Pump Outlet
16. Ring, Seal
17. Elbow, Oil Pump Outlet
18. Gasket, Oil Pump Outlet Elbow
19. Bolt, Elbow-to-Cylinder Block (2)
20. Gasket, Relief Valve-to-Elbow
21. Relief Valve
22. Bolt, Relief Valve-to-Elbow (2)
23. Gasket, Inlet Pipe-to-Pump
24. Pipe and Screen Assembly, Inlet
25. Bolt, Inlet Pipe-to-Oil Pump (2)
26. Bracket, Inlet Pipe (2)
27. Bolt, Inlet Pipe Bracket-to-Cylinder Block (2)
28. Bolt, Bracket-to-Inlet Pipe (2)
29. Washer, Bracket-to-Inlet Pipe (2)
30. Nut, Bracket-to-Inlet Pipe (2)

FIGURE 6–5 External gear-type oil pump and related parts for a series 60 engine. (Courtesy of Detroit Diesel Corporation.)

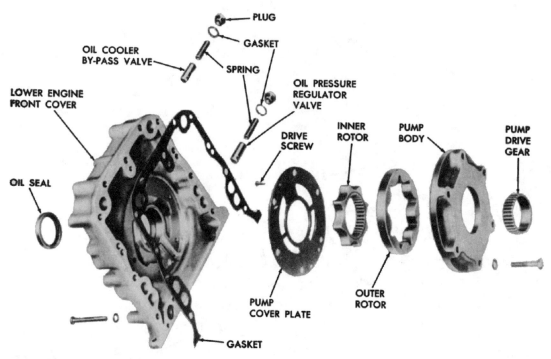

FIGURE 6–6 Rotor type of oil pump with internal splined pump drive. (Courtesy of Detroit Diesel Corporation.)

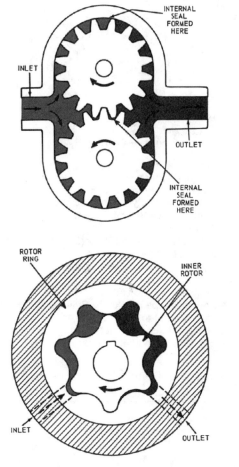

FIGURE 6–7 Engine oil pump operation. (Top) gear pump; (bottom) rotor pump. (Courtesy of Deere and Company.)

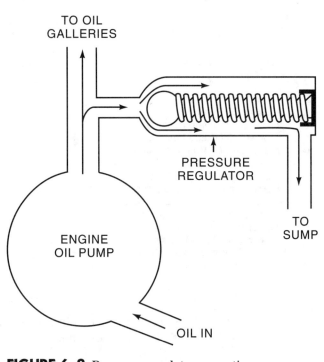

FIGURE 6–8 Pressure regulator operation.

OIL FILTERS 91

tioned to ensure that the main oil pump is always supplied with adequate lubricating oil regardless of equipment attitude. Depending on oil pan design, oil may accumulate at the front or rear in the pan. Without a scavenge pump, the main oil pump would be starved of oil when the equipment is operated at certain steep angles. The scavenge pump ensures that oil is returned to the main sump and oil pump at all times. Scavenge oil pump design is similar to main oil pump design.

OIL FILTERS

Oil filters are designed to trap foreign particles suspended in the oil and prevent them from getting to engine bearings and other parts (**Figures 6–9** and **6–10**). Most engines use the full-flow filtering system in which all the oil delivered by the pump must pass through the filter before reaching the engine bearings (**Figure 6–11**).

Should the filter be neglected and become completely clogged, a bypass valve allows oil to flow directly from the oil pump to the bearings (**Figure 6–12**). Another valve prevents the oil from draining out of the filter when the engine is stopped. This ensures oil delivery to engine parts immediately after the engine is started.

Other engines may use the shunt type or bypass type of filtering system (**Figure 6–13**). Filters may be mounted directly on the engine block or on a special adapter with no external oil lines, or they may be remote mounted, requiring external oil lines.

Some engines have only one lubricating oil filter; others may have two or more.

Filter elements come in three types: the surface type, the depth type, and the combination type. The *surface type* is usually of the pleated paper bellows design; foreign matter collects on the surface of the paper element as the lubricating oil flows through the filter element. The *depth type* consists of a perforated canister filled with a fibrous material. Foreign matter is trapped

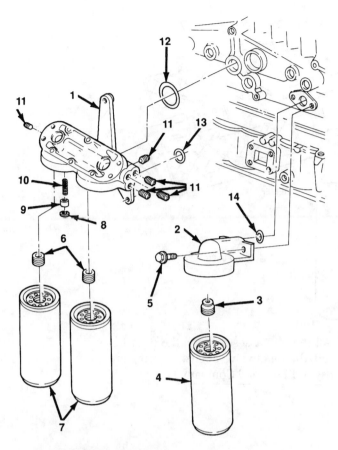

1. OIL FILTER ADAPTOR
2. BYPASS OIL FILTER ADAPTOR
3. INSERT, BYPASS FILTER-TO-ADAPTOR
4. OIL FILTER, BYPASS
5. BOLT, ADAPTORS-TO-BLOCK (6)
6. INSERT, FULL-FLOW FILTER-TO-ADAPTOR
7. OIL FILTER, FULL-FLOW
8. SNAP RING
9. VALVE, BYPASS
10. SPRING, BYPASS
11. PIPE PLUGS (5)
12. O-RING, ADAPTOR-TO-BLOCK
13. O-RING, ADAPTOR-TO-BLOCK
14. O-RING, ADAPTOR TO BLOCK

FIGURE 6–9 Oil filters and related parts. (Courtesy of Detroit Diesel Corporation.)

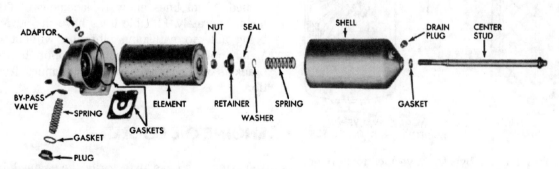

FIGURE 6–10 Exploded view of cartridge type of full-flow oil filter. (Courtesy of Detroit Diesel Corporation.)

92 Chapter 6 LUBRICATION SYSTEM PRINCIPLES AND SERVICE

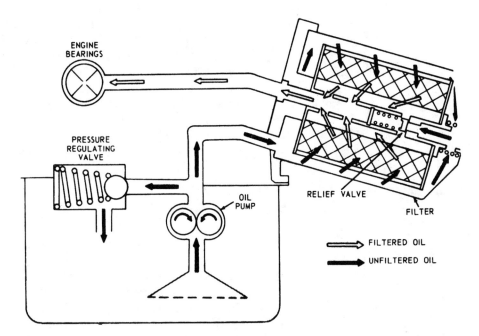

FIGURE 6–11 Full-flow filtering system. All the oil passes through the filter before it reaches the bearings. If filter is plugged, the relief valve opens, allowing unfiltered oil to lubricate the bearings. (Courtesy of Deere and Company.)

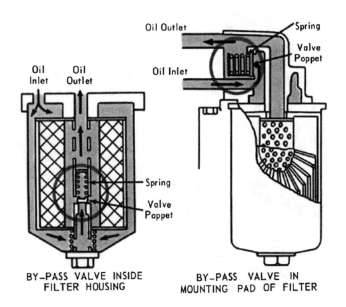

FIGURE 6–12 Filter bypass valve locations. (Courtesy of Deere and Company.)

in this element at various depths in the fibrous material as the oil flows through **(Figure 6–14)**. The combination type of filter element combines the surface and depth designs in one element. Another type of oil filter is the *centrifugal type* **(Figure 6–15)**. Lubricating oil forced against a rotor causes the rotor to spin. Heavier foreign particles are separated from the oil in this manner and deposited on the filter wall while the oil drains back to the sump.

Lubricating oil lines include flexible synthetic rubber and braided lines as well as tubular metal lines. Flexible lines usually have the fittings permanently attached. Metal lines allow replacement of fittings or lines separately. Flexible lines have the advantage of being able to withstand vibration without damage. Metal tubing requires adequate support by means of brackets on longer lines to prevent damage from vibration.

ENGINE OIL COOLER

To perform its job satisfactorily, the engine lubricating oil must be kept within the proper temperature range. If

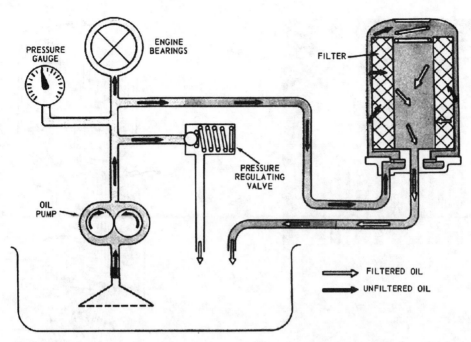

FIGURE 6–13 Bypass type of oil filtering system. Engine bearings are fed from the oil pump. Some oil is directed from the main gallery to the filter. Filtered oil is returned to the sump. (Courtesy of Deere and Company.)

the oil is too cold, it does not flow freely. If it is too hot, it is not able to support the bearing loads, it cannot carry away enough heat, and too great an oil flow can result. If this happens, oil pressure can drop below specified limits and oil consumption can become excessive.

During engine operation, the lubricating oil absorbs a considerable amount of heat that must be dissipated by an oil cooler.

The oil cooler consists of a housing containing a series of plates or tubes (see oil coolers in **Figures 6–1** and **6–2**). The cooler is connected to the engine cooling system and to the lubricating system. This requires two inlets and two outlets. One inlet and one outlet are connected to the cooler housing, while the other inlet and outlet are connected to the oil cooler core plates or tubes. In the typical tube-type cooler, coolant from the engine water pump flows through a passage in the oil cooler; it then passes through the tubes of each section of the oil cooler, back to the outlet passage, and into the water jackets in the engine block. Engine oil from the engine oil pump enters a passage in the oil cooler, passes through the oil filter, then around the tubes of the oil cooler, back through the outlet passage, and on to the oil galleries in the cylinder block (**Figure 6–16**).

A bypass valve is provided to permit engine oil to flow to the engine lubricating system should the oil cooler become plugged.

OIL PRESSURE INDICATORS

Oil pressure indicators inform the operator whether the lubrication system is functioning normally. Direct-reading gauges that indicate actual oil pressure are used in some instances. These are either pressure-sensitive direct-acting gauges or electric gauges that respond to a pressure-sensitive variable-resistance sending unit switch (**Figure 6–17**).

Oil warning lights that indicate inadequate oil pressure when they light are also used (**Figure 6–18**).

Some engines are equipped with automatic alarms or automatic shutdown devices, which are triggered when oil pressure drops below specified levels.

Chapter 6 LUBRICATION SYSTEM PRINCIPLES AND SERVICE

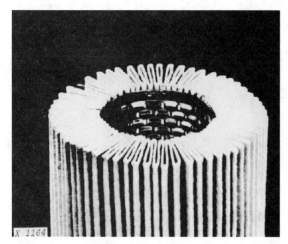

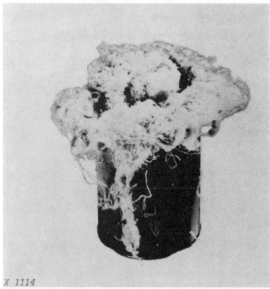

FIGURE 6-14 Pleated paper filter element (top) and cotton waste filter (bottom). (Courtesy of Deere and Company.)

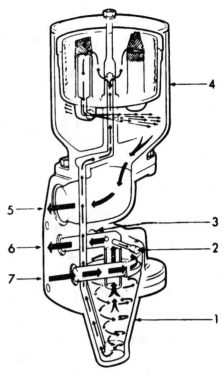

FIGURE 6-15 Phantom view of centrifugal type of oil cleaner: 1. cyclone, 2. oil outlet, 3. oil outlet, 4. cleaner housing and rotor, 5. clean oil outlet to sump, 6. clean oil outlet to engine bearings, 7. oil inlet. (Courtesy of Mack Trucks Inc.)

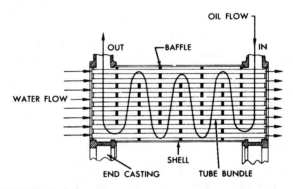

FIGURE 6-16 Oil cooler schematic (tube type) showing oil flow and coolant flow. (Courtesy of Detroit Diesel Corporation.)

CRANKCASE VENTILATION

Crankcase ventilation is required to remove harmful crankcase vapors and to prevent pressure buildup in the crankcase. Combustion byproducts, condensation, and contaminants are removed by the ventilation system. Clean air enters the crankcase from a breather cap usually equipped with a filter or in some cases from the air cleaner **(Figure 6-19)**.

A completely open system includes a vent pipe and an open filtered air inlet. Harmful vapors are simply vented to the atmosphere. Other systems route the crankcase vapors to the air induction system. One method used is an internal pipe or tube leading from the rocker cover to the air inlet.

Another system uses a valve mounted on top of the rocker cover. The valve regulates flow from the rocker chamber to the air manifold.

Crankcase pressure is caused by combustion gases that get past the piston rings into the crankcase; this is

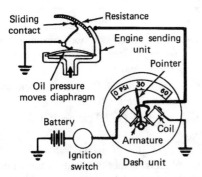

FIGURE 6-17 Variable-resistance oil pressure sending unit and balanced-coil dash gauge unit. (Courtesy of Deere and Company.)

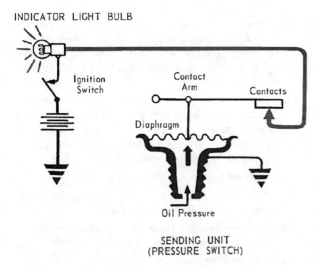

FIGURE 6-18 Oil pressure indicator light circuit. When oil pressure drops too low, the contacts close to complete the circuit to the lamp. (Courtesy of Deere and Company.)

called *blowby*. If the engine is in good condition, blowby will be minimal. Worn rings and cylinders will increase blowby.

LUBRICATION SYSTEM SERVICE

For lubrication system problem diagnosis and corrective steps, refer to the diagnostic chart **(Figure 6-20)**. The following service operations are common to lubrication system service.

Checking Oil Level and Condition

It is important that the engine oil level be checked in a consistent manner in order to be able to determine ac-

FIGURE 6-19 Crankcase ventilation valve controls flow of crankcase vapors from valve cover to intake manifold. (Courtesy of Caterpillar Tractor Company.)

curately the level of oil in an engine. It is equally important that the engine oil level not be allowed to fall below the recommended level and that the crankcase not be overfilled.

Several methods of checking crankcase oil levels are recommended by vehicle manufacturers. One manufacturer includes an oil level check with the engine running at idle. One side of the dipstick is marked for the proper oil level with the engine running, and the other side indicates the oil level with the engine stopped. When the engine oil level is checked, the vehicle should be level, the engine should be at operating temperature, and the engine should be stopped for at least 20 minutes to allow time for oil drainback. Oil levels should be kept between the add-oil and full marks on the dipstick. Overfilling may cause oil to be thrown around by the crankshaft, which can cause foaming and increased oil consumption as too much oil is thrown up onto the cylinder walls. Too low a lubricating oil level will cause higher oil temperatures and consequently reduced lubricating effectiveness.

Oil Consumption, Oil Leaks

A certain amount of oil consumption is normal. It is therefore normal that oil may have to be added between oil changes.

Problem	Cause	Correction
Low lubricating oil pressure	1. Oil leak–line gasket, etc. Insufficient lubricating oil level.	1. Check oil level and add make-up oil as required. Oil must be to recommended specifications. Check for oil leaks.
	2. Wrong oil viscosity	2. Drain lubricating oil. Change oil filters and fill with oil meeting specifications.
	3. Defective oil pressure gauge	3. Check operation of oil gauge. If defective, replace gauge.
	4. Dirty oil filter(s)	4. Check operation of bypass valve for the filter. Install new oil filter elements. Clean or install new oil cooler core. Drain oil from engine and install oil meeting specifications.
	5. Lubricating oil diluted with fuel oil	5. Check fuel system for leaks. Make necessary repairs. Drain diluted lubricating oil. Install new filter elements and fill crankcase with recommended oil meeting specifications.
	6. Defective oil pump relief valve	6. Remove valve, check for seat condition and sticking relief valve–spring tension and cap. Check assembly parts. The use of incorrect parts will result in improper oil pressure. Make necessary repairs or install new relief valve.
	7. Incorrect meshing of oil pump gears	7. Check mounting arrangement. If engine has been rebuilt, check for proper gear ratio combination of the oil pump–driven gear and drive gear. Incorrect gear combinations will result in immediate gear failure and possible engine damage.
	8. Excessive clearance between crankshaft and bearings	8. Overhaul engine and replace worn defective parts.
Oil in cooling system	1. Defective oil cooler O-ring(s)	1. Disassemble and replace O-ring(s).
	2. Defective oil cooler core	2. Remove oil cooler. Disassemble and repair/replace oil cooler core.
	3. Blown head gasket	3. Replace head gasket.
Coolant in lubricating oil	1. Defective oil cooler core	1. Disassemble and repair/replace oil cooler core.
	2. Blown head gasket	2. Replace head gasket.
	3. Defective water pump oil seal(s)	3. Remove water pump; disassemble and replace defective parts.
	4. Cylinder sleeve seals failure	4. Replace cylinder sleeve seals.
Excessive oil consumption	1. External oil leaks	1. Check engine for visible signs of oil leakage. Look for loose, stripped oil drain plug, broken gaskets (cylinder head cover, etc.), front and rear oil seal leakage. Replace all defective parts.
	2. Clogged crankcase breather	2. Remove obstruction.
	3. Excessive exhaust back pressure	3. Check exhaust pressure and make necessary corrections.
	4. Worn valve guides	4. Replace valve guides.
	5. Air compressor passing oil	5. Repair or replace air compressor.
	6. Failure of seal rings in turbocharger	6. Check inlet manifold for oil and make necessary repairs.
	7. Internal engine wear	7. Overhaul engine.

FIGURE 6–20 Lubrication system diagnosis and correction chart.

To determine whether a vehicle is using an excessive amount of oil, the method of checking the oil level must be applied consistently over a period of time and mileage.

Oil consumption may be the result of worn piston rings, worn valve guides, excessive bearing clearance, or oil leakage. If an oil consumption problem exists, any oil leakage must be corrected first before the condition of the engine is blamed. If no oil leakage exists, a thorough diagnosis of the engine's mechanical condition should be performed to determine the cause. A wet and dry compression test can help determine whether the piston rings or valve guides may be at fault.

Oil and Filter Change Intervals

The oil and filter change intervals should be followed as outlined in the owner's manual or appropriate service manual. The SAE rating (viscosity) and the service rating of the oil used should comply with the manufacturer's recommendations.

The frequency of oil and filter changes may vary, depending on the type of service that the vehicle is required to provide. Under severe operating conditions such as dusty conditions, extensive idling, frequent short trips (especially in cold weather), and sustained high-speed driving in hot weather, more frequent oil changes may be required. Oil and filter change intervals may be based on number of miles or kilometers driven, number of hours of service, or analysis of engine oil samples established, over a period of time. Engine manufacturer's or lubricating oil supplier's recommendations should be followed to ensure long engine service life **(Figure 6–21)**.

Oil Contamination

Excessive oil dilution requires changing the engine oil and filters and correcting the cause of oil dilution. This may require adjusting or correcting any fuel delivery system problems resulting in excessive fuel delivery to the cylinders, including injector operation.

Coolant leakage past cylinder liner seals or head gaskets may cause sludge buildup and bearing damage and must be corrected.

Engine Oil Analysis

Periodic analysis of the engine oil can provide valuable information about engine operation and condition. Chemical analysis includes such methods as UV/visible

Maximum Allowable Oil Drain Intervals (Normal Operation)

Service Application	Oil Drain Interval
Highway truck & motor coach	15,000 miles (24,000 km)
City transit coach	6000 miles (10,000 km) or 1 year[a]
Pick-up and delivery, stop and go, short trip	6000 miles (10,000 km)

[a]Whichever comes first

FIGURE 6–21 Example of oil drain interval recommendations. Always refer to the manufacturer's service manual for specific engines and types of engine service. (Courtesy of Detroit Diesel Corporation.)

WARNING LIMITS

	ASTM Designation	Series 60
Pentane insolubles mass %	D-893	1.00
Carbon (soot) content mass %, max.	E-1131	1.50
Viscosity at 100°F, SUS		
% Max. increase	D-445	40
% Max decrease	& D-2161	15
Total base number (TBN) min.	D-664	1.00
Total base number (TBN) min.	D-2896	2.00
Water content (dilution), vol. %, max.	D-85	.30
Flash point, °F, max. reduction	D-92	40.0
Fuel dilution, vol. %, max.	a	2.50
Glycol dilution, ppm, max.	D-2982	1000
Iron content, ppm, max.	b	150
Sodium content, ppm, max. allowed over lube oil baseline	b	50
Boron content, ppm, max. allowed over lube oil baseline	b	20
Copper content, ppm, max.	b	90
Lead, ppm, max.	b	100
Tin, ppm, max.	b	40
Silicon, ppm, max.	b	20
Chrome, ppm	b	15

[a]No ASTM procedure designation.
[b]Elemental analyses are conducted using either emission or atomic absorption spectroscopy. Neither method has an ASTM designation.

FIGURE 6–22 Typical used engine oil analysis specifications (series 60 DDC engine). (Courtesy of Detroit Diesel Corporation.)

spectrophotometry, atomic absorption spectroscopy, and infrared analysis. These tests can be performed only in a properly equipped chemical lab.

The results of these tests monitored over a period of time can provide valuable engine wear rate data from which the need for engine repair can be predicted with considerable accuracy. This, in turn, can prevent costly on-the-job engine breakdown. Tests indicate wear of engine parts in the form of fine metallic particles found in the oil. Based on the amount and types of metal identified, degree of wear of bearings, rings, and cylinder walls can be determined (**Figure 6–22**).

Oil Pressure Testing

If for any reason an oil pressure problem is suspected, a master test gauge with a range of 0 to 100 psi (0 to 689.5 kPa) should be used to verify actual oil pressure produced by the engine (**Figure 6–23**).

Generally, procedures include bringing the engine oil to operating temperature. The engine is then shut off and the test gauge installed at the point indicated in the service manual (usually a plug in the main oil gallery). The engine is restarted and readings are taken at specified engine speed. These readings are then compared with those provided in the shop manual.

If the oil pressure is too high, the problem is usually a stuck oil pump pressure relief valve. The relief valve should be removed and polished with crocus cloth to correct this condition. Relief valve–spring pressure should also be checked at this time.

If the oil pressure is too low, the cause may be any of the following:

- worn oil pump
- excessive bearing clearances (camshaft or crankshaft)
- weak or broken pressure relief valve spring
- relief valve stuck in the open position
- excessive oil dilution
- plugged oil pickup screen
- air leak into oil pump inlet

If the oil pump is worn, it is usually replaced. The oil pump must be removed and checked to determine its condition.

If the engine bearing clearances are excessive, they must be corrected to bring clearances to specified limits. This requires replacement of the bearings and may require replacing the camshaft and crankshaft as well. If the relief valve or spring is at fault, it should be cleaned or replaced.

If the oil pickup tube or screen is faulty, it should be cleaned or replaced.

Oil Pump Service

If the oil pump is suspected of being faulty, it must be removed and checked for wear. Both gear- and rotor-type pumps are measured similarly to determine if wear is excessive (**Figure 6–24**).

The relationship of the pump parts to one another must be maintained during this procedure if the pump is to be used again. Do not mark pump parts with a center punch. Use a felt pen or chalk to mark parts for correct reassembly. Some pumps have the gears or rotors marked during manufacture. Procedures for measuring pump clearances are illustrated, and results should be compared with limits specified in the service shop manual.

The oil pump should be assembled properly, lubricated, and primed to ensure immediate lubrication on engine start-up. The correct thickness of the cover plate gasket must be used (where applicable) to ensure correct gear or rotor tolerances and adequate oil pump pressure. Install the pump and oil pan as outlined in the appropriate service manual, and tighten all bolts to specified torque and proper sequence (**Figure 6–25**).

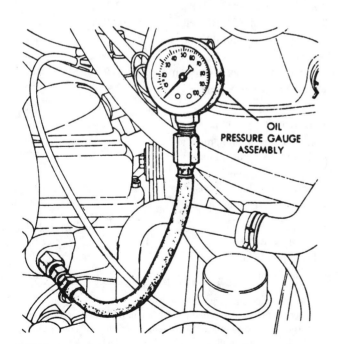

FIGURE 6–23 Using a master oil pressure gauge to check lubrication system operating pressure.

LUBRICATION SYSTEM SERVICE

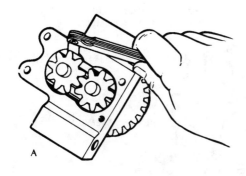

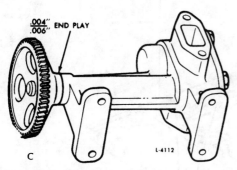

FIGURE 6–24 Measuring oil pump clearances. (A) Measuring gear-to-housing clearance with a feeler gauge; (B) measuring gear-to–cover plate clearance with a Plastigage®, (C) typical oil pump drive shaft end play requirement. [(A) and (B) Courtesy of Deere and Company; (C) Courtesy of Detroit Diesel Corporation.]

Crankcase Ventilation Service

A plugged or restricted crankcase ventilation system will contribute to oil leaks and oil consumption by causing the crankcase to be pressurized. The ventilation system should be inspected and cleaned to make sure that there is no restriction in either the air inlet or the vent outlet. Any filters in the system should be cleaned or replaced, as well as the valve, if so equipped.

Oil Cooler Service

Remove and clean the oil cooler core according to the manufacturer's recommendations. This may include removing the cooler and immersing the core in the recommended chemical solution for cleaning. Thorough flushing with water is necessary after chemical cleaning, or a chemical neutralizing flushing solution may be used. Pressure test the cooler as recommended in the service manual (**Figures 6–26** to **6–28**).

If engine failure has resulted in the circulation of metallic particles in the lubricating oil and oil cooler, the best procedure is to replace the cooler core. This will ensure that no metal particles will be recirculated from the oil cooler into the lubrication system after the engine overhaul.

All gaskets, lines, and fittings must be properly sealed and installed. Tighten all connections and bolts to the recommended torque values given in the service manual.

Priming the Lubrication System

After completing all required lubrication system service, fill the crankcase with the recommended amount and type of lubricating oil. The entire lubrication system, all lines, filters, coolers, etc., can be filled and primed by using a pressure tank connected to the main oil gallery to charge the system. The pressure tank has a reservoir for lubricating oil, a connecting line and valve to the engine, and a valve for pressurizing with compressed air. After connecting the line to the engine, open the valve, allowing oil to enter the lubrication system. Continue this process until system pressure is reached (engine lubrication pressure). This procedure ensures immediate lubrication when the engine is started.

After the pressure tank is removed and the engine oil gallery plug is replaced, the engine can be started. Lubricating oil pressure should be observed to make sure that it is within specifications. The entire system should be checked for leaks during engine warm-up and again after engine operating temperature has been reached.

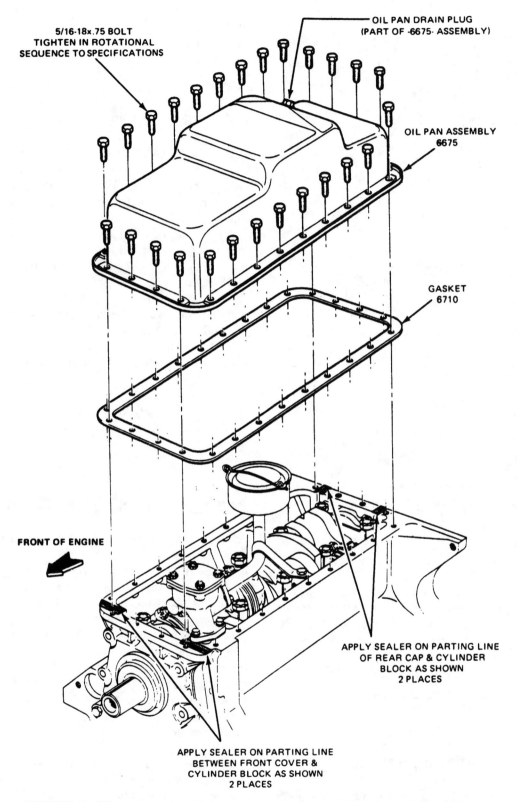

FIGURE 6–25 Typical oil pump and pan installation. (Courtesy of Ford Motor Company.)

FIGURE 6–26 Pressure testing oil cooler core for leaks. (Courtesy of Mack Trucks Inc.)

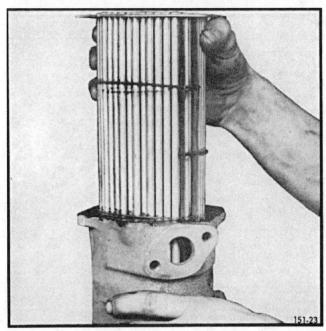

FIGURE 6–27 Installing oil cooler tube bundle. (Courtesy of Mack Trucks Inc.)

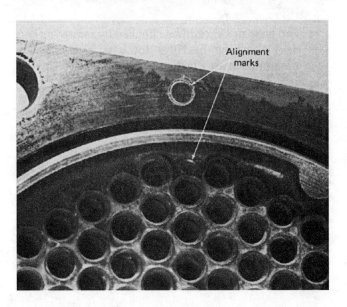

FIGURE 6–28 Oil cooler core and housing alignment marks. (Courtesy of Cummins Engine Company Inc.)

REVIEW QUESTIONS

1. Engine oil helps provide a seal. (T) (F)
2. _____ is largely responsible for piston cooling.
3. Hydrodynamic lubrication relies on the _____ of the engine oil.
4. Viscosity ratings of engine oils designate the _____ and _____ of the oil.
5. List five common additives used in the production of engine oils.
6. A 5W-30 oil has the _____ _____ viscosity of a _____ oil.
7. Synthetic oils are made from _____ compounds.
8. The oil pump may be mounted on the inside or outside of the _____ or on the front end of the _____.
9. Oil pump designs include the _____ and _____ types.
10. To limit maximum oil pressure, a _____ is used.
11. A worn oil pump should be _____.
12. Oil pump wear can be measured with a _____ _____.
13. Modern diesel engines use the _____ _____ type of oil filter.

TEST QUESTIONS

1. The lubrication system performs the following task(s) in the engine:
   a. cleans the engine
   b. cools engine parts
   c. seals engine parts
   d. all of the above
2. Hydrodynamic lubrication is
   a. hydraulic lifter lubrication
   b. hydraulic brake lubrication
   c. hydraulic pump lubrication
   d. wedging of oil between a bearing and a journal
3. Sludge is a formation of
   a. water, dirt, and oil
   b. water, fuel, and dirt
   c. fuel, oil, and water
   d. dirt, fuel, and water
4. Technician A says that varnish is caused by excessively high lubricating oil temperatures. Technician B says varnish causes hydraulic lifters to stick. Who is correct?
   a. Technician A
   b. Technician B
   c. both are correct
   d. both are incorrect
5. Technician A says sludge is caused by short-trip, cold-weather driving. Technician B says that varnish is caused by long-trip, high-speed, high-temperature driving. Who is correct?
   a. Technician A
   b. Technician B
   c. both are correct
   d. both are incorrect
6. Technician A says that the viscosity ratings and SAE ratings are the same. Technician B says that the service ratings and API ratings refer to the same rating. Who is correct?
   a. Technician A
   b. Technician B
   c. both are correct
   d. both are incorrect
7. Oil pump types include
   a. gear type, rotor type, and chain type
   b. chain type, belt type, and gear type
   c. rotor type, gear type, and belt type
   d. rotor type and gear type
8. Excessive internal oil pump wear causes
   a. high oil pressure
   b. high oil temperature
   c. low oil pressure
   d. low oil temperature
9. Engine oil should be changed
   a. every 5000 miles (8000 km)
   b. at least once a month
   c. at intervals specified in the service manual
   d. whenever the filter is changed
10. Filling the crankcase above the specified level with oil
    a. lets you drive longer before changing oil
    b. can cause aeration and lubricant leakage
    c. has no effect on vehicle operation
    d. sells more oil

CHAPTER 7

ELECTRICAL PRINCIPLES AND SERVICE BASICS

INTRODUCTION

The electrical system's alternator produces electrical energy, stores it in the battery in chemical form, and delivers it on demand to the various vehicle or equipment electrical systems and components. Voltage in the different systems may range from as low as 0.5 to more than 100 V. A great many of the components on vehicles and equipment are electrically or electronically operated. A thorough understanding of how electricity acts, how it is put to use, how it is controlled, and what can cause electrically operated systems to malfunction is essential for the intelligent diagnosis and service of electrical systems and components.

PERFORMANCE OBJECTIVES

After thorough study of this chapter and sufficient practice with the appropriate electrical components and test equipment, you should be able to:

1. Describe electricity.
2. Describe the characteristics of permanent magnets and electromagnets.
3. Describe the characteristics of a magnetic field.
4. Describe electromagnetic induction.
5. Define volts, amperes, ohms, and Ohm's law.
6. Explain the operating characteristics of series, parallel, and series–parallel electrical circuits.
7. Calculate resistance in a circuit.
8. Describe electrical wiring, wiring harness, terminals, and connectors.
9. Describe different types of manual and automatic electrical switches, solenoids, and relays.
10. Describe different types of fuses, fusible links, and circuit breakers.
11. Describe different types of resistors and capacitors.
12. Describe different types of transducers.
13. Describe diodes and transistors.
14. List common electrical problems.
15. Use basic electrical test equipment to test electrical components.

TERMS YOU SHOULD KNOW

Look for these terms as you study this chapter, and learn what they mean.

atom	potential difference
electricity	current
magnetism	amperes
polarity	resistance
electromagnet	ohms
magnetic field	Ohm's law
electromagnetic induction	electrical power
volts	watts
electromotive force	series circuit

parallel circuit
series–parallel circuit
conductor
insulator
fuse
fusible link
circuit breaker
resistor
potentiometer
rheostat
thermistor
capacitor
transducer
sensor
diode
transistor
conductors
solid wire
stranded wire
insulation
wire size
wire harness
clips
retainers
boots
straps
tubing
terminals
connectors
crimping

soldering
locking device
wiring diagram
manual switch
automatic switch
push-pull
rotary
toggle
slider
thermal switch
vacuum switch
pressure switch
relay
solenoid
open
short
ground
feedback
jumper wire
test lamp
self-powered test light
circuit breaker
voltmeter
ohmmeter
ammeter
multimeter
VOM
digital multimeter
voltage drop

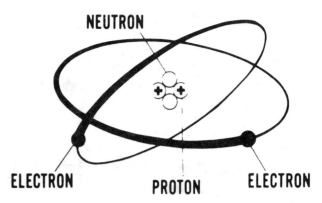

FIGURE 7–1 The principle of electricity can best be explained when the structure of an atom is understood.

WHAT IS ELECTRICITY?

Electricity behaves according to certain reliable rules that produce predictable results and effects. The best way to explain electricity is to look at the atom. Simply stated, the atom is composed of electrons orbiting around a nucleus. The electrons have a negative electrical charge, and the protons (which are part of the nucleus) have a positive charge. The positive charge of the nucleus attracts the negatively charged electrons and keeps them in orbit **(Figure 7–1)**.

The electrons in an electrical wire are loosely held by the nucleus and can be dislodged by an external force such as the voltage from a battery or an alternator. When electrons are knocked out of their orbit in an electrical conductor, electrical current (a flow of electrons) is produced. This electron flow or current is the result of electrons being "pushed" into the conductor at one end and out of the other as long as a potential difference exists between the two ends. Electrons move from one atom to the next as in a chain reaction. Current will continue to flow as long as the voltage from the battery or alternator is applied and the opposing voltage or resistance is less. Electricity is the flow of electrons **(Figure 7–2)**.

The electrons in an insulator are very tightly held together and are very hard to dislodge. That is why the insulation around a current-carrying conductor (electri-

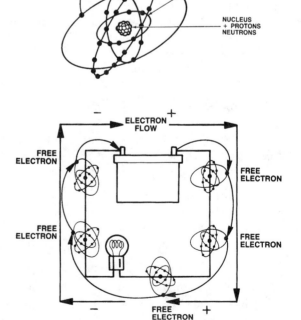

FIGURE 7–2 Electricity is the flow of free electrons.

cal wire) prevents any electrons from escaping through the insulation.

MAGNETS AND MAGNETISM

Magnetism produces almost all the electricity used in our homes, offices, and industries. The alternator in your car or truck uses magnetism to produce the electricity to run the electrical systems and keep the battery fully charged.

There are two kinds of magnets: natural and artificial. Natural magnets are found in the earth in the form of magnetite, which will attract pieces of iron and steel. If you suspend these materials on a string, they will align themselves with the earth's magnetic poles.

Artificial magnets can be produced by putting pieces of iron, steel, or certain alloys of aluminum and nickel (called alnico) in an intense magnetic field. These substances will acquire all the magnetic properties of magnetite, but they will be stronger magnets. Softer metal magnets eventually lose their magnetic properties. The harder magnets made of steel or alnico tend to retain their magnetic properties indefinitely.

All magnets have polarity. Like the earth, they have a north and a south pole. Like poles (N–N or S–S) repel each other. Unlike or opposite poles (N–S) attract each other (**Figures 7–3** and **7–4**). This is an important characteristic of magnets that is applied to many of the equipment's electrical components.

Magnetic Field

Magnetic force is invisible. The only way we know that it exists is by the effects it produces. The action of a magnet can best be explained by recognizing that invisible lines of force (magnetic field) leave the magnet at one end and reenter the magnet at the other end. These

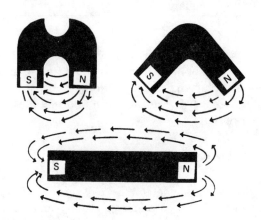

FIGURE 7–3

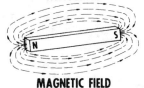
MAGNETIC FIELD

UNLIKE POLES ATTRACT

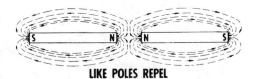
LIKE POLES REPEL

FIGURE 7–4 (Courtesy of General Motors Corporation.)

invisible lines of force are also called *magnetic flux lines*. The shape of the space they occupy is called a *flux pattern*. The number of lines of flux per inch or centimeter is called *flux density*. The stronger the magnetic field, the greater the flux density. Flux lines are continuous and unbroken; they do not cross each other. From experiments we know that a magnetic field has force and direction.

In the "conventional" theory of current direction, current is said to flow from positive to negative. This theory can be used to determine the direction of the magnetic field. Using the right hand with the thumb pointing toward negative and the fingers wrapped around the conductor, the fingers indicate the direction of the magnetic field. With the arrival of the electronic age, the direction of current was determined to be from negative to positive. In this case the left hand can be used with the thumb pointing toward positive. The direction of the magnetic field is indicated by the fingers wrapped around the conductor (**Figure 7–5** and **7–6**).

Electromagnets

An electromagnet is made by wrapping a soft-iron core with a coil of insulated wire. When one end of the wire is connected to an electrical current and the other end is grounded, the iron core acquires all the magnetic properties of a natural magnet but is much stronger. The iron core concentrates the magnetism in an area surrounding the electromagnet (**Figure 7–7**). Ignition

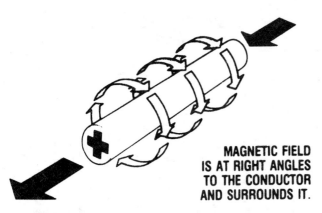

FIGURE 7–5 The magnetic field around a conductor.

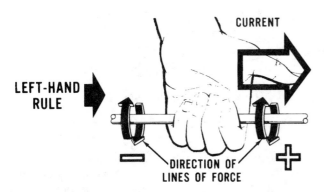

FIGURE 7–6 (Courtesy of General Motors Corporation.)

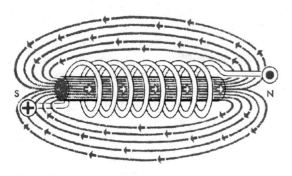

FIGURE 7–7 Use of an iron core in a coil increases the field strength in the core and forms an electromagnet. (Courtesy of Deere and Company.)

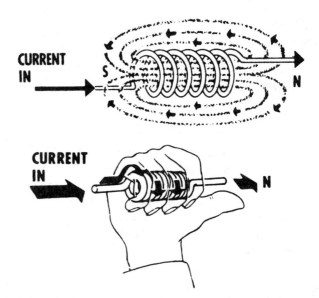

FIGURE 7–8 When a current-carrying conductor is wound in the form of a coil, the magnetic lines of force are concentrated inside the coil, making a stronger magnetic field. The polarity (direction) of these magnetic lines of force can be established by using the right hand with the fingers pointing in the direction of current in the coil winding. The thumb then points to the north pole. (Courtesy of General Motors Corporation.)

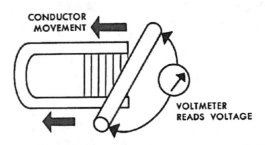

FIGURE 7–9 Moving a conductor across a magnetic field induces a current in the conductor. (Courtesy of Deere and Company.)

coils, solenoids, and relays (explained later) are examples.

To establish the direction of the magnetic field inside an electromagnet a similar method can be used. With the fingers wrapped around the coil and pointing in the direction of current flow, the thumb will point in the direction of the magnetic field inside the electromagnet **(Figure 7–8)**. This principle is applied to solenoids to determine which direction the solenoid plunger will move when energized. The solenoid on a starter is an example. As soon as the current to the solenoid is stopped, the magnetic lines of force collapse, and the solenoid plunger is returned to its normal position by spring pressure.

Electromagnetic Induction

Electricity and magnetism are closely related. Magnetism can be used to produce electricity, and electricity can be used to produce magnetism **(Figure 7–9)**. This is why alternators produce electricity, and electric motors use magnetism to produce power. When a magnetic field is passed across a conductor, voltage is induced in the conductor, and current begins to flow. When a lamp is connected to the ends of that conductor, current flows, lighting the lamp. This is called *electromagnetic induction*. This principle is used to make alternators work.

PRODUCING ELECTRICITY

An electrical current can be formed under certain conditions by the following forces: friction, chemical, heat (thermal), pressure (piezoelectric), and magnetism (induction).

Friction

When two materials are rubbed together, frictional contact between them will actually rub some of the electrons from their orbits. Thus, there is a transfer of electrons to one of the materials, giving it a *static negative charge*, and a lack of electrons in the other material, giving it a *static positive charge*. When these materials are brought close to a grounded object (one at zero voltage), a spark will jump between the materials and the grounded object.

Pressure

Perhaps the best known use of electricity produced by pressure is the crystal in the arm of a record player or in the diaphragm of a crystal microphone. Crystals of certain materials, such as quartz or Rochelle salt, develop a small electric charge when pressure is applied to them. When sound waves strike the diaphragm of a microphone, mechanical pressure is transferred to the crystal. This causes the crystal to flex and bend, producing a small voltage at its surface. This voltage is at the same frequency and amplitude as the incoming sound. This current is then amplified. This phenomenon is called a *piezoelectric effect*.

Heat

A direct conversion of heat to electricity can be accomplished by heating a bimetallic junction of twisted wires made from two dissimilar metals like copper and iron. This type of junction is called a *thermocouple*.

Producing electricity in this manner is called *thermal conduction*. Since thermocouples cannot produce large amounts of electrical current, they are used as heat-sensing units. The amount of electrical current generated is dependent on the difference in temperature between the bimetallic junction of the thermocouple and the opposite ends of the wire. The greater the temperature difference, the greater the flow of current and voltage produced.

HOW ELECTRICITY FLOWS THROUGH A CONDUCTOR

Keep in mind these six rules of electrical behavior:

1. All electrons repel each other.
2. All like charges repel each other. (Negatively charged objects repel other negatively charged objects; positively charged objects repel other positively charged objects.)
3. Unlike charges attract each other. (Positively charged objects attract negatively charged objects and vice versa.)
4. Electrons flow in a conductor only when affected by an electromotive force (EMF).
5. A voltage difference is created in the conductor when an EMF is acting on the conductor. Electrons flow only when a voltage difference exists between the two points in a conductor.
6. Current tends to flow to ground in an electrical circuit. *Ground* is defined as the area of lowest voltage. Electrical current moves through a conductor to ground in an attempt to reach a balance or equilibrium with the ground voltage (which is zero).

When electrons are set into motion, they display a variety of behaviors. The behavior of electrons moving through a conductor accounts for the many things electricity can do. By understanding how electricity behaves, you will be able to understand the function and operation of the various diesel engine electrical systems. This will aid you in diagnosing electrical problems.

Let us turn our attention to a single copper atom in a conductor. The copper atom has a single electron in its outermost orbit. This electron is not tightly held by the nucleus, and it can easily be freed by an electromotive force (EMF).

Once an electron escapes from its orbit, it is free to move—possibly colliding with other atoms in the conductor. As the free electron approaches the outer orbit of another copper atom, its electrostatic force starts to interact with the electron in orbit, repelling it. At the same time this electron is being repelling, the nucleus of that atom is attracting the free electron into its orbit.

The free electron now enters the copper atom's orbit, replacing the ejected electron. As more and more electrons collide with other atoms in the conductor, an electrical current begins.

Once EMF is applied, it causes electrons to be freed from their orbits. This starts a chain reaction between the electrons and atoms in the conductor, causing the freed electrons to move away from the electromotive force. This effect is called *electron drift* and accounts for the flow of electrons through a conductor. Whenever electrons flow or drift in mass, an electrical current is formed **(Figure 7–10)**.

HOW ELECTRICITY IS MEASURED

Electricity is measured in volts and amperes. Resistance to electrical flow is measured in ohms. The electrical power required to operate a light or electric motor is stated in watts. These terms and their relationship to one another must be understood to be able to diagnose and service the equipment's electrical systems and their components.

Voltage (Volts)

Voltage is an electrical pressure or electromotive force. In diesel engines this voltage is applied by the battery and alternator. Voltage can be described as a potential difference (in electrical pressure). The potential difference between the two posts on a 12-volt (V) battery is normally about 12.6 V. Voltage is measured with a voltmeter. The symbol for voltage is *V*. For electromotive force the symbol is EMF or *E*. They represent the same force. One volt is equal to 1 ampere (explained later) of current across 1 ohm (explained later) of resistance when 1 watt of power is being consumed **(Figure 7–11)**.

Current (Amperes)

Current is the rate of electron flow. Electron flow (or current) increases as voltage increases provided that

FIGURE 7–10 Electron flow in a conductor. (Courtesy of Deere and Company.)

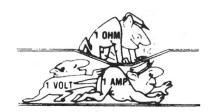

FIGURE 7–11 One volt is equal to 1 ampere of current across 1 ohm of resistance.

resistance remains constant. Electrical current is measured in amperes with an ammeter. One ampere (A) is equivalent to the current produced by 1 volt when applied across a resistance of 1 ohm. Another term for amperes is *intensity* of current. The symbol for current intensity is *I*.

Resistance (Ohms)

Electrical resistance is opposition to electron flow. It is measured in ohms with an ohmmeter. One ohm is the resistance that allows 1 ampere of current to flow when 1 volt is applied. The letter *R* is the symbol for resistance. The Greek capital letter omega (Ω) is the symbol for ohms. The resistance of an electrical wire increases as its length is increased and as its temperature is increased. The diameter or cross-sectional area of the wire is also a factor. Wires with a greater cross-sectional area have less resistance. This can be compared to the flow of water in a pipe or hose: the larger the inside diameter of the hose or pipe, the greater the flow of water. A smaller diameter hose or pipe has greater resistance to the flow of water; therefore, the flow is reduced. In the same way if the hose or pipe is flattened, the resistance to flow is increased, resulting in less flow.

Ohm's Law

When any two values in an electrical circuit are known, the third can be calculated by using Ohm's law **(Figure 7–12)**. Ohm's law can be expressed in several ways.

You already know that many electrical circuits operate at battery voltage (12 V). If you know that a circuit operates at 3 A, you can calculate the normal resistance in the circuit as follows:

$$12 \text{ V} \div 3 \text{ A} = 4 \text{ }\Omega$$

If resistance in the circuit is too high, 6 Ω for example, then there will be less current available to operate the

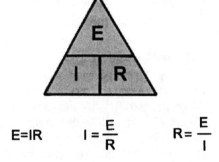

E= VOLTAGE IN VOLTS
I= CURRENT IN AMPERES
R= RESISTANCE IN OHMS

FIGURE 7-12 Ohm's law. (Courtesy of Deere and Company.)

electrical device in the circuit. The current available can be calculated as follows:

$$12\,V \div 6\,\Omega = 2\,A$$

This is a 33% reduction in current. Reduced current causes lights to dim and electric motors to run more slowly or not at all. The cause for the increased resistance, for example, loose or corroded connections or a faulty electrical device, must be found and corrected.

Electrical Power

The rate of work done by electricity is called electrical power and is measured in watts (W). If the voltage and current values are known, the power in watts can be calculated simply by multiplying the number of volts times the number of amperes: $V \times A = W$. For example, a 12 V starting system using 150 A would use 1800 W. The proper amount of power can be delivered only if voltage, current, and resistance values are as they should be.

ELECTRICAL CIRCUITS

Every electrical system requires a complete circuit to function. A complete circuit provides an uninterrupted path for electricity to flow from its source through all circuit components and back to the electrical source. Whenever the circuit is interrupted (broken), electricity will not flow. This interruption can be in the form of a switch or an open (broken) wire.

There are three basic types of electrical circuits: series, parallel and series–parallel. An electrical system may have one or more of these circuit types.

Series Circuit

A series circuit provides only one path for current to flow from the electrical source through each component and back to source. If any one component fails, the entire circuit will not function. Total resistance in a series circuit is simply the sum of all the resistances in the circuit. For example, a series circuit with a light and two switches will have a total resistance of 4 Ω if the light has a resistance of 2 Ω and each switch has a resistance of 1 Ω. Total resistance = 2 + 1 + 1 = 4 Ω. Another example is shown in **Figure 7-13.**

Parallel Circuit

A parallel circuit provides two or more paths for electricity to flow. Each path has several resistances (loads) and operates independently or in conjunction with the other paths in the circuit. If one path in the parallel circuit does not function, the other paths in the circuit are not affected. One example of this is the headlight circuit: If one headlight burns out, the other headlight will still function. To calculate the total resistance in a parallel circuit, the following equation may be used:

$$R = \frac{1}{\frac{1}{R_1} + \frac{1}{R_2} + \frac{1}{R_3}} \quad \text{or} \quad \frac{1}{R} = \frac{1}{R_1} + \frac{1}{R_2} + \frac{1}{R_3}$$

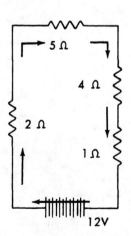

FIGURE 7-13 This series circuit has a total resistance of 12 Ω. (Courtesy of General Motors Corporation.)

depending on the number of resistances involved. If R_1, R_2, and R_3 are 4, 6, and 8 Ω, respectively, the total resistance can be calculated as follows:

$$R = \frac{1}{1/4 + 1/6 + 1/8}$$
$$= \frac{1}{6/24 + 4/24 + 3/24}$$
$$= \frac{1}{13/24} \quad \text{or} \quad 1 \div \frac{13}{24}$$
$$= 1 \times \frac{24}{13} \quad \text{or} \quad 1.85 \text{ Ω}$$

In a parallel circuit the total resistance is always less than the resistance in any single device in the circuit.

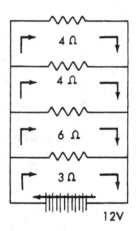

FIGURE 7–14 This parallel circuit has a total resistance of 1 Ω. (Courtesy of General Motors Corporation.)

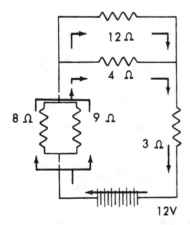

FIGURE 7–15 Example of a series–parallel circuit. (Courtesy of General Motors Corporation.)

This is because there is more than one path for electricity to follow. **Figure 7–14** gives another example.

Series–Parallel Circuits

A series–parallel circuit combines the series and parallel circuits. A series–parallel circuit is shown in **Figure 7–15**. In a headlight circuit the headlight and dimmer switches are in series, while the headlights are in parallel with each other. To calculate the total resistance in a series–parallel circuit, first calculate the equivalent series resistance of each parallel circuit, then calculate the resistance of the "new" series circuit as described earlier. Thus, for the two parallel resistances in **Figure 7–15**,

$$R_1 = \frac{1}{1/8 + 1/9} \quad \text{and} \quad R_2 = \frac{1}{1/4 + 1/12}$$
$$= \frac{1}{9/72 + 8/72} \qquad\qquad = \frac{1}{3/12 + 1/12}$$
$$= \frac{1}{17/72} \qquad\qquad = \frac{1}{4/12}$$
$$= 1 \times \frac{72}{17} \qquad\qquad = 1 \times \frac{12}{4}$$
$$= 4.2 \text{ Ω} \qquad\qquad = 3 \text{ Ω}$$

The resistance of the circuit can then be calculated using the formula for series resistances:

$$R_{\text{circuit}} = 4.2 \text{ Ω} + 3 \text{ Ω} + 3 \text{ Ω}$$
$$= 10.2 \text{ Ω}$$

COMMON GROUND PRINCIPLE

In some illustrations and wiring diagrams a return or ground wire is shown as the method to connect the circuit or component to the negative or ground terminal of the battery. However, this is not always the case with electrical circuits and components. It is not always necessary and would require more wires than are actually used. Instead manufacturers use the common ground method, which uses the vehicle or equipment frame or body as part of the return circuit. A starting motor, for example, is bolted directly to the engine while the engine is grounded by a ground strap. The internal circuits of the starter are grounded to the starter frame and through it to the engine and ground. A separate ground wire for the starter therefore is not needed. Other components may be mounted in plastic and must have a ground wire to connect them to ground. Some manufacturers have adopted a system of common ground terminals at various points in the vehicle or equipment to provide good grounds and keep ground wires to a minimum.

ELECTRICAL WIRING

Electrical wires may be one solid single strand or a number of smaller wires twisted together to form a stranded wire. Wires are usually stranded for more flexibility in diesel equipment wiring **(Figure 7–16)**.

Wire diameter is specified in gauge sizes or in square millimeters (mm^2). Gauge sizes use numbers to indicate size. The smaller the gauge number, the larger the wire diameter. The larger the metric number designation, the larger the wire cross section **(Figure 7–17)**.

Large-diameter wires are required for circuits subject to high current (amperes). Smaller-diameter wires are used for low current.

Heavy insulation is required for high-voltage wires (e.g., spark plug wires), and lighter insulation is used for low-voltage circuits. The uninsulated side of electrical system circuits uses the equipment frame and body for a return path. The insulated side of the circuit is sometimes called the *live side*.

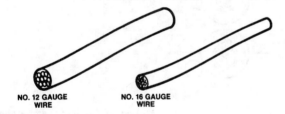

FIGURE 7–16 Stranded wire.

CABLE CONVERSION CHART

Metric Size (mm^2)	Gauge Size (AWG)
0.5	20
0.8	18
1.0	16
2.0	14
3.0	12
5.0	10
8.0	8
13.0	6
19.0	4

FIGURE 7–17 Wire size comparison chart. (Courtesy of General Motors Corporation.)

Microprocessor control system wiring may contain special design features not found in standard wiring. Environmental protection is used extensively to protect electrical circuits. Special twisted and twisted/shielded cable is essential to system performance. Additionally, special high-density connections are used in many of the wiring harnesses.

The current and voltage levels in this system are very low. Three types of cable construction are used: straight wire, twisted wires, and twisted/shielded wires. Unwanted induced voltages are prevented from interfering with computer operation. Environmental deterioration by moisture, rust, and corrosion is prevented for the same reason.

Wiring diagrams using a variety of symbols are used in service manuals. Examples are shown in **Figures 7–18** to **7–20**.

Wire Harness

Wire harnesses are an assembled group of wires that branch out to the various electrical components of a vehicle. Groups of insulated wires are wrapped together with tape or inserted in insulating tubing to form a harness. There are several more complex wiring harnesses as well as a number of simple harnesses. The engine compartment harness and the underdash harness are examples of a complex harness. Lighting and accessory circuits use a more simple harness.

The wiring harness makes for easier assembly and replacement and requires fewer supporting clips or clamps. A harness, however, makes it harder to locate a problem in the wires enclosed in it. In this case wires can be traced by using the wire's color code.

Supporting Clips and Devices

Proper routing and support of electrical wire is achieved with various types of insulated clips, retainers, boots, straps, and tubing. All of these are mounted to avoid any tension or stretching of the wiring. Without these devices wiring could chafe or come in contact with exhaust systems or moving parts.

Wire Terminals and Connectors

A variety of terminals and connectors are used to connect the wiring to the different electrical components. This includes round connectors, spade or blade-type connectors, junction blocks, and bulkhead connectors. Connectors and terminals may be attached to the wiring by soldering or by crimping or both. Soldering

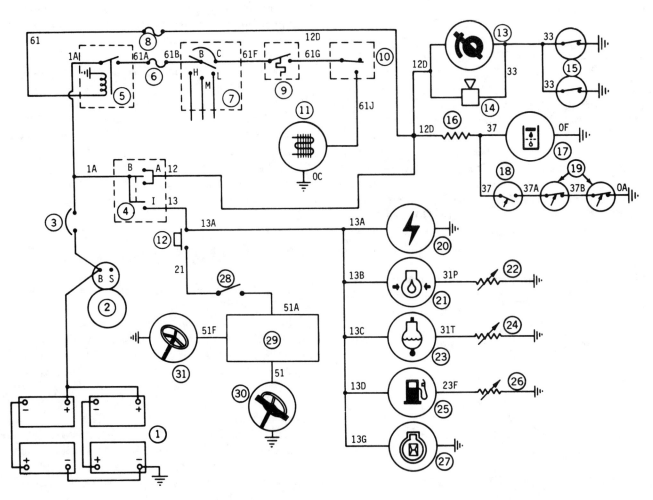

1—Batteries
2—Starting Motor
3—Main Circuit Breaker
4—Key Switch
5—Main A/C Relay
6—Pressurizer Fuse
7—Pressurizer Switch
8—Air-Heater Fuse
9—Thermostat
10—Thermal Switch
11—Plugged Condenser Indicator
12—Start Switch
13—Low Brake Pressure Indicator
14—Low Brake Pressure Buzzer
15—Low Brake Pressure Switches
16—Resistor
17—Filter Restriction Indicator
18—Test Switch
19—Filter Restrictor Indicator Sending Units
20—Voltmeter
21—Engine Oil Pressure Gauge
22—Engine Oil Pressure Sending Unit
23—Engine Coolant Temperature Gauge
24—Engine Coolant Temperature Sending Unit
25—Fuel Gauge
26—Fuel Sending Unit
27—Hourmeter
28—Neutral Start Switch
29—Emergency Steering Logic Board
30—Primary Steering Failed Indicator
31—Emergency Steering Failed Indicator

FIGURE 7-18 Typical loader instrument panel wiring diagram. (Courtesy of Deere and Company.)

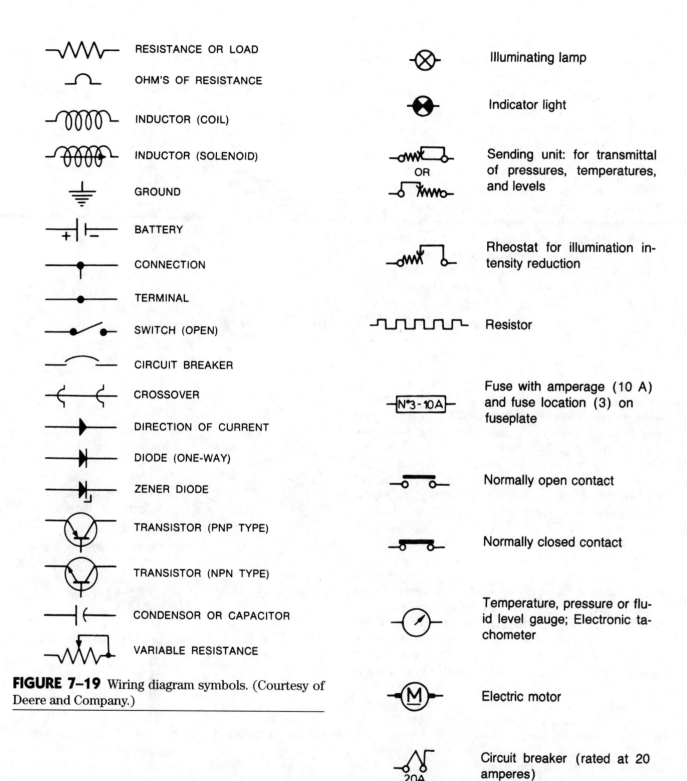

FIGURE 7-19 Wiring diagram symbols. (Courtesy of Deere and Company.)

FIGURE 7-20 Electrical symbols used in Mack Truck electrical service manuals. (Courtesy of Mack Trucks Inc.)

fuses the solder to the wire and the terminal, making a good electrical connection as well as a good mechanical connection. Crimping squeezes the terminal tightly around the wire, actually partially biting into the wire. Crimping requires that both the terminal and the wire end be "electrically" clean to provide a good electrical connection. A good mechanical connection is not necessarily a good electrical connection.

Spade or blade terminals, junction blocks, and bulkhead connectors usually have locking devices or tabs to ensure that the connections do not come apart from vibration during vehicle operation. To disconnect these devices requires disengaging the locking device before pulling them apart. Pulling on the wires without unlocking the connection can cause the wires to be pulled out of their terminals. This would require repair or replacement of the harness (**Figures 7–21** to **7–23**).

SWITCHES

Electrical switches are used to open and close electrical circuits. Some of these are operated manually, whereas others operate automatically. Manually operated switches include the following types: push-pull, toggle, turn, and slider. Manually operated switches are used to operate headlamps, radios, tape players, speakers, heaters, air conditioners, windshield wipers, speed control, power seats, power door locks, glow plugs, the starting system, trip and fuel calculators, and the like. Some of these are switched on manually but switch off automatically; others must be switched on and off manually.

Automatic switches include those controlled by heat, pressure, vacuum, solenoids, and relays. Heat-sensitive switches are used for coolant temperature indicators controlled by a thermal sending unit in contact with engine coolant. They may control cold engine temperature indicator lights as well as hot engine indicators. An example of a pressure-sensitive switch is the oil pressure indicator sending unit or switch screwed into a main engine oil gallery. With the engine off, there is no oil pressure: the switch is closed. When the start switch is turned on, the oil pressure indicator light goes on. When the engine is started and engine oil pressure rises above approximately 8 to 12 psi (55 to 82 kPa), the switch contacts separate, opening the circuit; then the dash indicator light goes out.

A *relay* is a switch that opens and closes an electrical circuit conducting relatively high current controlled by a circuit of relatively low current value. This reduces the need for long, heavy electrical wiring in many

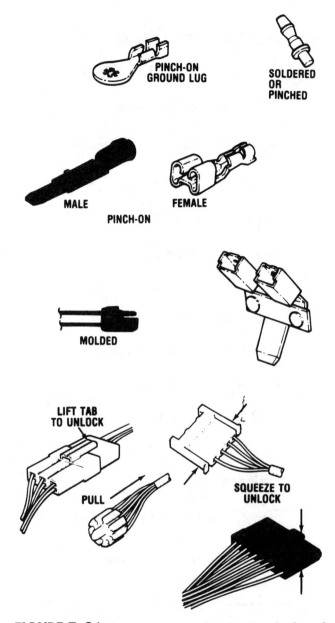

FIGURE 7–21 Common types of wiring terminals and connectors.

cases, since only light wire is used to connect the relay to the actuating device or switch. Starter relays are good examples of the relay type of switch. When the control circuit switch is closed, the relay coil winding becomes an electromagnet that causes the armature of the relay to be drawn toward the coil, closing the relay points or contacts. One of the relay contacts is connected to the power source and the other to the device to be operated. Closing the relay points actuates the starting motor.

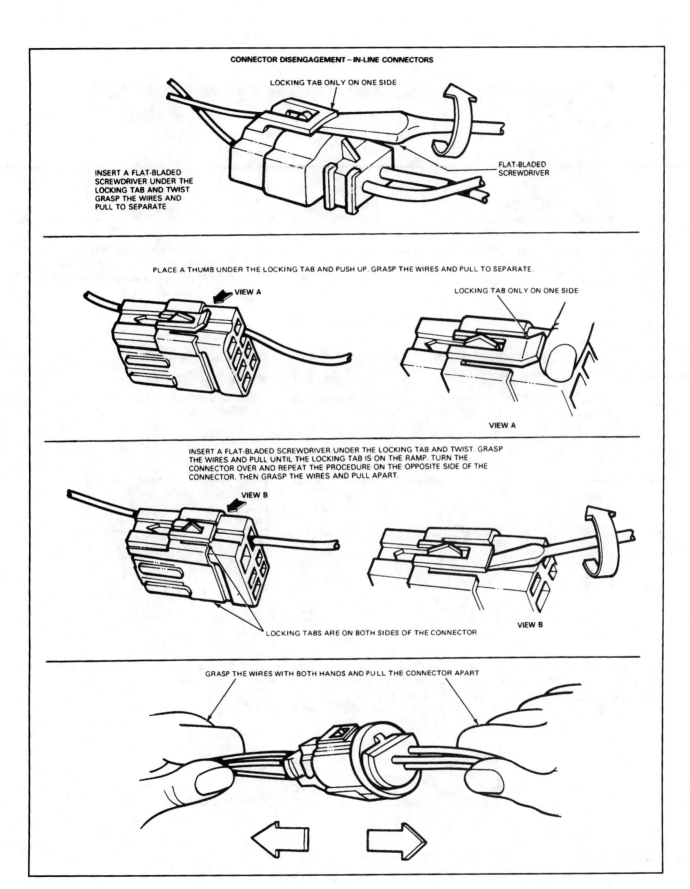

FIGURE 7-22 Typical disconnect procedures for multiple connectors. (Courtesy of Ford Motor Company.)

The 40 pin V-MAC module connector and the serial communication port connector are Deutsch connectors. For terminal removal and replacement, follow the instructions listed below.

Contact Insertion

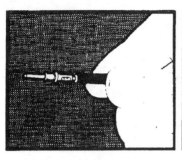 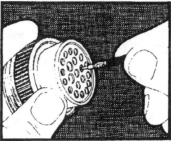 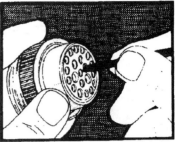

Grasp contact approximately 1 inch (25.4mm) behind the contact crimp barrel.

Hold connector with rear grommet facing you.

Push contact straight into connector grommet until a positive stop is felt. A slight tug will confirm that it is properly locked in place.

Figure 12

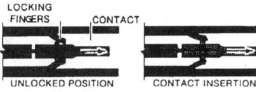 

Contact Removal

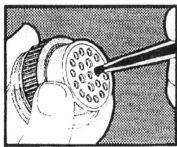 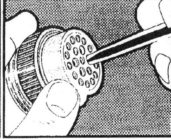 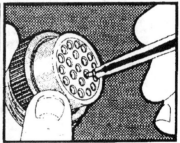

Figure 13

With rear insert toward you, snap appropriate size extractor tool over the wire or contact to be removed.

Slide tool along wire into the insert cavity until it engages contact and resistance is felt.
NOTE: Do not twist or insert tool at an angle.

Pull contact-wire assembly out of connector.

FIGURE 7-23 Deutsch connector contact insertion and removal instructions. Deutsch connectors are used on many heavy-duty vehicles and equipment. (Courtesy of Mack Trucks Inc.)

CIRCUIT PROTECTION

Fuses and Fusible Links

A chain is only as strong as its weakest link; when a load is applied, the weakest link will break.

In the same way, a fuse or a fusible link is the weakest point electrically in an electric circuit. It is needed in order to protect wiring and other components in the circuit from damage due to overloading of the circuit. Circuit overload can occur due to mechanical overload of the electrical device (i.e., windshield wiper motor, heater motor) or to shorts or grounds in the circuit (**Figures 7-24 and 7-25**).

Because of their lower current capacity, fuses and fusible links are designed to "blow" or "burn out" at a predetermined value, depending on the circuit capacity they are designed to protect. One type of fuse is the cylindrical glass type with the fusible link visible in the glass and connected at each end to a metal cap. The metal capped ends snap into place between two spring clip connectors in the fuse holder. Another type is enclosed in transparent plastic and has two blade terminals that plug into corresponding connectors in the fuse holder. Fuse capacity ranges anywhere from about 3 to 30 A. A failed fuse is easily identified by the gap in the wire visible in the fuse. The cause for fuse failure should be determined and corrected before fuse replacement. Replacement fuses should never exceed original fuse capacity.

A *fusible link* is a short piece of wire of smaller diameter than the wire the circuit is designed to protect. When the circuit is overloaded, the link burns in two before damage can occur to the rest of the circuit. Fusible links are identifiable in the wiring harness by color code or by an attached tag. Fusible links are insulated in the same way as the rest of the circuit. A failed fusible link can often be identified by heat-damaged insulation or exposed wire (**Figure 7-26**). They are used in such circuits as charging and lighting systems (see **Figure 7-25**).

Circuit Breakers

Circuit breakers, like fuses and fusible links, are designed for circuit protection. The circuit breaker is more costly but has the advantage of opening and closing the circuit intermittently. In a headlight circuit, for example, the circuit breaker allows headlights to go on and off, which allows the driver to safely pull over to the side and stop. A fuse or fusible link failure, in this circuit, would cause the lights to go out completely, leaving the driver in the dark.

A circuit breaker has a pair of contact points, one of which is attached to a bimetal arm. The arm and contacts are connected in series in the circuit.

When circuit overload current heats the bimetal arm, the arm bends to open the contacts, stopping electrical flow in the circuit. When the arm cools, the contacts close again, energizing the circuit once more. This action continues until the circuit is switched off or repaired.

Voltage Limiter

The instrument voltage regulator is designed to limit voltage to the instrument panel gauges. Power to the voltage limiter is supplied when the start switch is in the ON position. Voltage is limited to approximately 5 V at the instrument gauges on some vehicles.

The voltage limiter consists of a bimetal arm, a heating coil, and a set of contact points enclosed in a housing. Two terminals provide connections in series into the circuit. When the switch is turned on, the heating coil heats the bimetal arm, causing it to bend and open the contacts. This disconnects the voltage supply from the heating coil as well as from the circuit. When the bimetal arm cools sufficiently, the contacts close and the cycle repeats itself. The rapid opening and closing of the contacts results in a pulsating voltage at the output terminal averaging approximately 5 V.

The voltage limiter protects the instrument gauges against high voltage surges and prevents erroneous gauge readings caused by voltage fluctuations.

RESISTORS

Resistors are devices used in electrical circuits to reduce current and voltage levels from those supplied by

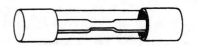
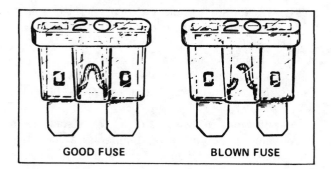

FIGURE 7-24 Typical fuse types.

118 Chapter 7 ELECTRICAL PRINCIPLES AND SERVICE BASICS

1. Fusible Link (3.0 mm²)
2. Headlamp Dimmer Switch
3. White Wire (8 mm²)
4. Engine Glowplug Circuit
5. Headlamp—Upper Beam—Left
6. Headlamp—Upper Beam—Right
7. Headlamp—Lower Beam—Left
8. Headlamp—Lower Beam—Right
9. Headlamp—Main
10. Taillamps, I.D. Lamps, Clearance Lamps
11. Stop Lamps, Wiper And Washer, Engine Stop Circuit, And Horn
12. Turn Signal And Hazard Warning Circuits, Domelamp
13. AC Compressor Circuit
14. AC Condensor Fan Circuit
15. Engine Glow Plugs Relay Control Circuit
16. Starter Relay Control Circuit
17. Cigar Lighter, Radio
18. Back Up Lamp Circuit, Exhaust Brake, Drain Heater
19. Voltmeter, Fuel Gage, Coolant Temperature Gage, Oil Pressure Warning Lamp, Low Air Pressure Warning Lamp, Parking Indicator Lamp, And Charging System Lamp
20. Starter Relay
21. 2 Speed Axle (Circuit Breaker In The Motor)
22. Heater Blower Motor Circuit
23. White Wire (5 mm²)
24. White Wire (5 mm²)
25. Green Wire (3 mm²)
26. Fusible Link (1.0 mm²)
27. Fusible Link (1.25 mm²)
28. Engine Control Switch

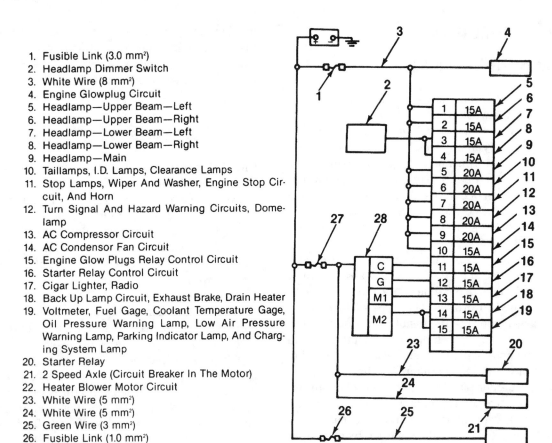

FIGURE 7-25 Fuse block diagram. (Courtesy of General Motors Corporation.)

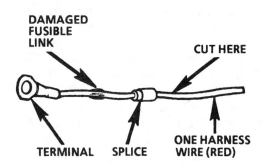

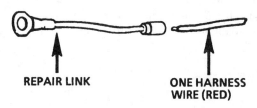

FIGURE 7-26 Typical fusible link repair. (Courtesy of General Motors Corporation.)

the power source. They are used to protect devices or circuits designed to operate at a lower current level than that supplied by the battery or charging system. They are also used to control current and voltage levels produced by charging systems and to control light intensity.

Resistors provide opposition to electron flow. This opposition causes the electrons to work harder to try to get through. The increased electron activity generates heat. Since some of the electrical energy is used up to produce heat, the voltage through the resistor is at a reduced level.

Several types of resistors are used in diesel equipment. These include fixed-value resistors, variable resistors, ballast resistors, and thermistors (**Figures 7-27 to 7-29**).

Fixed-value resistors maintain a constant resistance value once operating temperature has been reached. A manually operated rheostat or variable resistor (as in the dash light control switch incorporated in the headlamp switch) inserts more or less resistance into the circuit to dim or brighten the dash lights as the

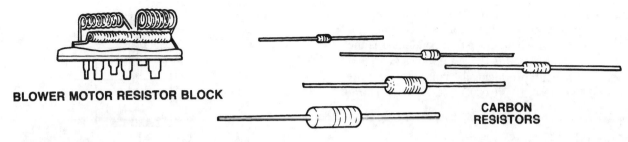

FIGURE 7-27 Typical resistors.

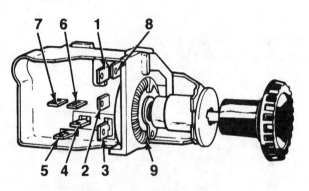

1. Terminal 1 — Power from Battery
2. Terminal 2 — To Instrument Panel Lamps
3. Terminal 3 — To Front Marker Lamps (C5,C6,C7)
4. Terminal 4 — To Taillamps
5. Terminal 5 — Power from Fuse Block
6. Terminal 6 — To Headlamp Dimmer Switch
7. Terminal 7 — Not Used
8. Dome Lamp Terminal (C5, C6, C7 and P6 Motorhome)
9. Rheostat — Instrument Panel Lamps

FIGURE 7-28 Headlamp switch with rheostat. (Courtesy of General Motors Corporation.)

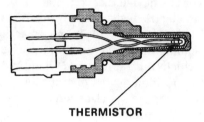

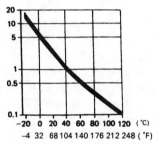

FIGURE 7-29 A thermistor is used in a coolant temperature sensor. As coolant temperature increases, thermistor resistance decreases. Resistance on left of chart is in kilohms (kΩ). (Courtesy of Robert Bosch Canada Ltd.)

switch knob is turned. A *ballast resistor* is a wire coil housed in a ceramic block to regulate temperature changes. The resistance of a ballast resistor increases with increased current and decreases with decreased current. Increased current causes the resistor to heat up, which, in turn, increases resistance.

The resistance value of a *thermistor* (a type of resistor) varies with temperature. As the temperature of the thermistor increases, its resistance decreases. It is used in charging systems to vary charge voltage with ambient temperature change and in cooling system temperature sensors.

CAPACITORS

A *capacitor* (sometimes called a *condenser*) is a device that is used in an electrical circuit to temporarily store an electrical charge until it is needed to perform its job or until it can safely be dissipated if it is not to be used.

The typical capacitor consists of several thin layers of electrically conductive material, such as metal foil, separated by thin insulating material known as *dielectric* material. Alternate layers of foil are connected to one terminal of the capacitor. The other layers of foil are connected to ground. The entire assembly is rolled up tightly and enclosed in a metal cylinder. The unit is completely sealed and moisture proof. The metal container is the ground connection, and a pigtail lead provides the other connection. The capacitor is connected in parallel with the circuit. Any surge of current (excess electrons) enters the capacitor and is stored on the capacitor plates.

Capacitors of various types and sizes are used in electrical circuits to collect and dissipate stray or unwanted current. This prevents the unwanted current from interfering with other electrical functions. A radio-suppressor type of capacitor in the alternator is a typical example. The capacity or capacitance of a condenser is measured in units called *farads* (F). A farad is a charge of 1 ampere for 1 second producing a 1 volt potential difference. A microfarad (μF) is 0.000001 (1/1,000,000) farad.

TRANSDUCERS

A transducer is a device that converts another form of energy to an electrical signal. MAP (manifold absolute pressure) sensors are used to provide a varying voltage signal depending on intake manifold pressure to the computer, which controls injection timing and fuel rate.

A throttle position switch or transducer is used to send a varying voltage signal, dependent on throttle position, to the electronic control unit or computer. The computer uses this information to increase or decrease fuel delivery accordingly.

DIODES

A diode is a solid-state (completely static) device that allows current to pass through itself in one direction only (within its rated capacity). Acting as a one-way electrical check valve, it allows current to pass in one direction and blocks it in the other direction **(Figure 7–30)**.

The silicon wafer is chemically treated to produce either a positive or negative diode. Diodes may be encased in noncorrosive heat-conductive metal with the case acting as one lead and a metal wire connected to the opposite side of the wafer as the other lead. The unit is hermetically sealed to prevent the entry of moisture. This type of diode is used in AC charging system alternators. A minimum of six diodes are used—three positive diodes and three negative diodes to provide full-wave rectification (changing alternating current to direct current). Many charging systems use more than six diodes.

Other diodes used in electronic systems are much smaller and may be sealed in epoxy resins with two leads for connection into the circuit. Diodes in computers may be very tiny in comparison with the more visible charging system diode.

Negative diodes are identified by a black paint mark, a part number in black, or a black negative sign. Positive diodes are similarly identified in red or with a red positive sign.

The manner in which the metallic disk is installed in the diode assembly determines whether the diode is negative or positive. (Inverting the disk in a positive diode would make it a negative assembly.) This disk is

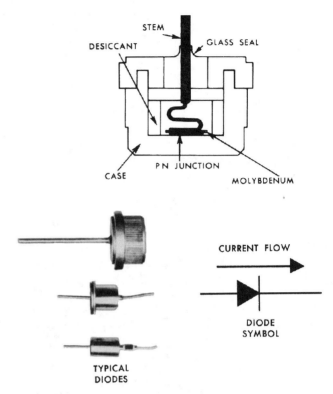

FIGURE 7–30 Typical diodes and diode symbol showing direction of current. (Courtesy of General Motors Corporation.)

only .008 in. to .010 in. thick and approximately ⅛ in. square, depending upon current rating.

Some rectifier assemblies contain diodes that are exposed, while others have them built in. Those with built-in diodes contain only the wafer portion of the diode.

The silicon crystal material for diodes and transistors is processed or "doped" by adding other material to it. Phosphorus or antimony may be used to produce a negative or N type material. These materials have five electrons in the outermost orbit of their atoms, compared with four for silicon. Thus, the crystal of the N material has one extra or free electron. The free electron can easily be made to move through the material when voltage is applied. Electrons are considered to be negative current carriers.

Boron or indium may be used to treat silicon crystal to produce a positive or P type material. These elements have only three electrons in the outermost orbit of their atoms. This leaves a shortage of one electron in the crystal of P type material. This shortage or vacancy is called a *hole*. Holes are considered to be positive current carriers.

A diode consists of a very thin slice of each material, P type and N type, placed together. The area where the two materials meet is called the *junction*. When the N material side of the diode is connected to a negative current supply, such as the battery negative terminal, and the P material side is connected to the positive battery terminal, the diode will conduct current. This happens because the negative battery terminal has an excess of electrons that repel the electrons in the diode toward the positive side. At the same time, the positive holes in the P material move toward the N side. This interchange of electrons and holes occurs at the junction of the N and P material in the diode. Connecting a diode in this manner is called *forward bias*.

When a diode is connected in the opposite manner (reverse bias) it will not conduct current. It cannot do so since the N material side of the diode is connected to the positive battery terminal and the P material side to the negative battery terminal. The electrons in the N material are attracted to the positive battery terminal side away from the diode junction. At the same time, the holes in the positive diode material are attracted to the negative battery terminal side of the diode away from the junction area. This in effect creates an open circuit, which cannot conduct current.

Of course, these conditions apply only if normal diode design voltage is not exceeded. When applied in reverse bias, excessive current will cause the bond structure to break down and allow reverse current, which causes the diode to be damaged. Diodes are designed with the necessary current and voltage capacity for the circuit in which they are to be used.

Excessive reverse current will destroy a diode due to excessive heat. A "blown" diode will not conduct current, resulting in an open circuit. Blown diodes must be replaced. A shorted diode will conduct current in both directions and must be replaced.

Light-emitting diodes (LEDs) are used for digital display of instrument panel gauges on some vehicles.

Zener Diode

The zener diode is a specially designed diode that conducts current like a normal diode but will also safely conduct current in a reverse direction when reverse current reaches the specified design voltage. A zener diode can prevent reverse current if it is below design voltage, but when reverse current reaches and exceeds design voltage, the zener diode will conduct reverse current. This type of diode is used in control circuits such as in the field circuit in an alternator **(Figure 7–31)**.

TRANSISTORS

A transistor is a solid-state switching device used to control current in a circuit. It operates like a relay ex-

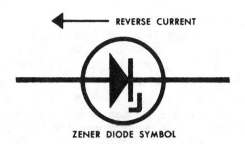

FIGURE 7–31 The zener diode will allow current in the reverse direction when specified voltage is imposed. (Courtesy of General Motors Corporation.)

cept that it has no moving parts. A relatively small current is used to control a larger current. The transistor either allows current to pass or stops it.

Transistors used in alternator applications are usually of the PNP type. This means that they are designed with a thin slice of N material sandwiched between two pieces of P material. The P material on one side is called the *emitter*, the N material in the middle is called the *base*, and the other P material is called the *collector*. NPN transistors are also produced but are not commonly used in diesel equipment applications.

The very thin slice of N type base material is attached to a surrounding metallic ring that provides the means for circuit connection. The emitter and collector material are also provided with circuit connections. The physical arrangement of the three pieces of material is such that the distance between the emitter and the collector is shorter than the distance between the emitter and the base. This feature results in the unique manner in which the transistor controls current.

A transistor is connected into a circuit in a manner that allows a low base–emitter current to control a larger collector–emitter current. A typical example of this is in the control module of an electronic voltage regulator.

When the base circuit is energized (by closing the ignition switch for example), a small base current is applied to the transistor emitter–base. This causes the electrons and holes in the emitter–base to act similar to the way they do in a diode. However, since the emitter is closer to the collector than it is to the base, most of the current is conducted by the emitter–collector section of the transistor. This is because electricity normally follows the path of least resistance.

The control current is called *base current*. The base circuit or current controls the emitter–collector current.

The same type of semiconductor material used in diodes is also used in transistors. The transistor, however, uses a second section of this material, which results in three terminals instead of two (as in the diode).

If, for example, the base circuit of a transistor is energized with 5 A of current, the transistor divides this current into base current and emitter–collector current. This division is known as the *current gain factor*. This factor varies with transistor design. The emitter–collector current may be 24 times that of the base current. In this example, therefore, base current would be 0.2 A, and emitter–collector current would be 4.8 A.

Transistors are used in electronic voltage regulators to control charging system voltage and in computers **(Figures 7–32 and 7–33)**.

COMMON ELECTRICAL PROBLEMS

Common electrical problems include incorrect resistance values, loose or corroded connections, electrical feedback, opens, shorts, and grounds. Many electrical problems are easily corrected if these basic problems are recognized and understood.

Incorrect Resistance

All electrical wiring and components have some resistance. The effective operation of these components depends on their resistance values being correct or as specified in the service manual. If a wire or electrical component has excessive resistance, electric current and power are reduced. If an electrical component has no resistance, it is inoperative and acts as if it were switched off.

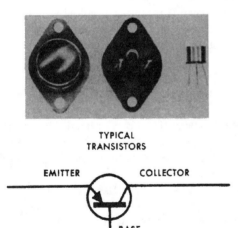

FIGURE 7–32 Typical transistor and transistor symbol. A small base current turns the transistor on, allowing a larger current from emitter to collector. (Courtesy of General Motors Corporation.)

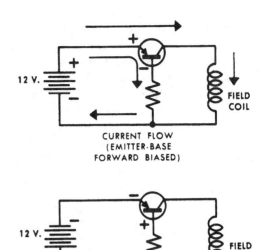

FIGURE 7–33 Diagrams showing transistor operation. (Courtesy of General Motors Corporation.)

Corroded and Loose Connections

Connections that are loose or corroded cause high resistance. High resistance creates an extra load and causes components to operate at reduced capacity or not at all. Heating due to arcing at a loose connection can cause corrosion of the terminal and high resistance. Water and road salt add to the corrosion. Loose or corroded connectors may be hard to spot; however, connections in a faulty circuit are easily cleaned and tightened to ensure a good electrical connection. If the locking tabs on an electrical connector are broken, vibration can cause electrical contact to be broken. Unplug, inspect, clean, and repair or replace any faulty connectors as required.

Electrical Feedback

Electrical feedback can cause lamps to light or accessories to operate when they are not supposed to. The most common cause of feedback is a blown or faulty fuse or fusible link. This can allow current to find a different electrical path, since some circuits are fed from several sources. Inspect and test suspected fuses or fusible links with an ohmmeter. Make sure that there is no current in the component being tested, because this would damage the ohmmeter. Make sure that all ground connections are clean and tight at all electrical components. A poor ground or an open bulb filament can cause other bulbs to light. Replace any faulty bulbs.

Voltage Drop

As current passes through a resistance, circuit voltage across it will drop. Total voltage drop in an electrical circuit will always equal available voltage at the source of electrical pressure.

Circuit resistance, if excessive at any point, will result in excessive voltage drop across that portion of the circuit. The voltage drop method is commonly used to determine circuit resistance. The voltage drop across a battery cable, for example, should not exceed 0.2 V per 100 A at 68°F (20°C).

Opens, Shorts, and Grounds

Electrical systems may develop an open circuit, a shorted circuit, or a grounded circuit. Each of these conditions will render the circuit more or less ineffective **(Figure 7–34)**.

OPENS

An *open circuit* is a circuit in which there is a break in continuity. For electricity to be able to flow there must be a complete and continuous path from the electrical source through the circuit back to the electrical source. If this path is broken, the condition is referred to as an open circuit. An open circuit, therefore, is no longer operational and acts the same as if it were switched off.

SHORTS

A *shorted circuit* is a circuit that allows current to bypass part of the normal path. An example is a shorted solenoid coil. Coil windings are normally insulated from each other; however, if this insulation breaks down and allows copper-to-copper contact between turns, part of the coil windings will be bypassed. In a starter pull-in winding, this condition will reduce the number of windings through which electricity will flow. If the short causes 50 windings of approximately 100 windings to be bypassed, this will reduce coil capacity by 50% and prevent starter engagement.

GROUNDS

A *grounded circuit* is a condition that allows current to return to ground before it has reached its intended destination. An example of this is a grounded light circuit. If the wire leading to the light has an insulation breakdown, allowing the wire to touch the frame or body, electricity will flow to ground at this point

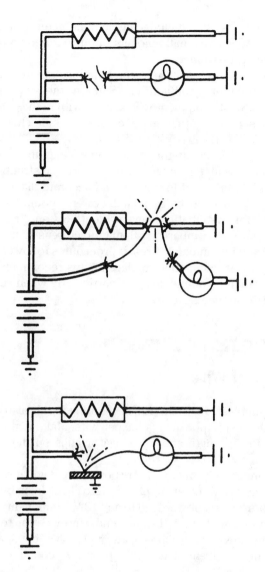

FIGURE 7–34 (Top) open electrical circuit, (middle) shorted circuit, and (bottom) grounded circuit. (Courtesy of General Motors Corporation.)

and return directly to the battery without reaching the light.

ELECTRICAL CIRCUIT DIAGNOSIS AND SERVICE

A number of common tools and instruments are used to diagnose electrical circuits. The most common are the test light, the jumper wire, the voltmeter, the ammeter, and the ohmmeter. The ammeter and voltmeter are often combined in a single piece of test equipment, which also may include a carbon pile resistor capable of applying varying loads to electrical circuits. The ohmmeter is often included in the multimeter, which is able to measure small amounts of current and voltage as well as resistance.

Variations of equipment combinations are available from different manufacturers. Some of these types of multiuse test equipment will be dealt with in the appropriate sections of this text. Examples include charging system and starting system testers.

In every case both the test equipment manufacturer's instructions and the engine manufacturer's instructions should be followed for proper and accurate diagnosis. Test sequences given by manufacturers should be followed for systematic problem identification. Test results must be compared with manufacturer's specifications and recommended repair procedures followed. In many cases, component replacement is recommended rather than component repair, particularly when solid-state components are involved.

BASIC TEST EQUIPMENT

Jumper Wire

The simplest electrical troubleshooting tool is also one of the most important—a jumper wire. Make it at least a meter in length and use alligator clips on the ends **(Figure 7–35)**.

Connect one end to battery positive and you have an excellent 12 V power supply. Use it to check lamp bulbs, motors, or as a power feed to any 12 V component. But be careful not to drop the other end; any place you touch on the engine is ground—battery negative; big sparks and high current will result, possibly "cooking" the jumper wire!

Self-Powered Test Light

The self-powered test light is used to check for continuity in a circuit or load device. The self-powered test light is connected and performs the same basic checks for continuity as an ohmmeter. It is also effective in checking for shorts to ground **(Figure 7–36)**.

CAUTION:

Never use a self-powered test light on a circuit that could be damaged by applying test light power. Check the service manual to avoid problems.

A short finder may also be used **(Figure 7–37)**.

Test Lamp

Sometimes you want to *look* for power, rather than *supply* it. That's when a *test lamp* is perfect. Just ground one side, and you can go to most "hot" 12 V points in the system and the lamp will light. But sometimes it won't light up fully with a hot circuit—for example, if you test the circuit *after* the voltage has "dropped" over a load. (See **Figure 7-35**).

Circuit Breaker

A circuit breaker reacts to excess current by heating, opening up, and cutting off the excess current. With no

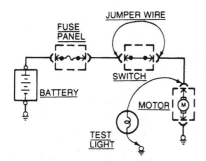

FIGURE 7–35 Using a jumper wire to check a switch. (Courtesy of General Motors Corporation.)

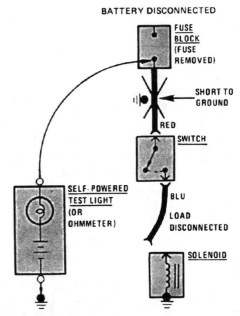

FIGURE 7–36 Checking for a short with a self-powered test light or ohmmeter. (Courtesy of General Motors Corporation.)

BASIC TEST EQUIPMENT 125

Voltmeter

A voltmeter is connected in parallel with a circuit—it reads directly in volts (**Figure 7–38**). In parallel the meter draws only a small current—just enough to sample the voltage. That's why you can "short" right across the battery terminals with a voltmeter without damaging it.

However, never try to check voltage by putting the meter in *series*. The voltmeter hookup should always parallel the circuit being measured because the high-resistance meter would disrupt the circuit in a series connection.

Closed Circuit Voltage. In the top part of **Figure 7–39**, the voltage at A is 12 V positive. There is a drop of 6 V over the 1.0 Ω resistor, and the reading is 6 V positive at B. The remaining voltage drops in the fan load, and the voltmeter reads zero (12 V negative) at C, indicating normal motor circuit operation.

Open Circuit Voltage. In the bottom part of **Figure 7–39**, we read the voltage in the same circuit as above but with *no* electricity flowing because there is an open circuit (broken wire or poor ground) at point X. The voltage at A, B, and C will be 12 V positive, indicating circuit continuity up to, but not through, point X. Remember, there is no voltage drop across a resistor or load if there is no electrical flow.

Ammeter

You measure current draw with an ammeter. But unlike the parallel voltmeter, you must put the ammeter in *series* with the load to read the current draw (**Figure 7–38**). That means disconnecting the load and reconnecting

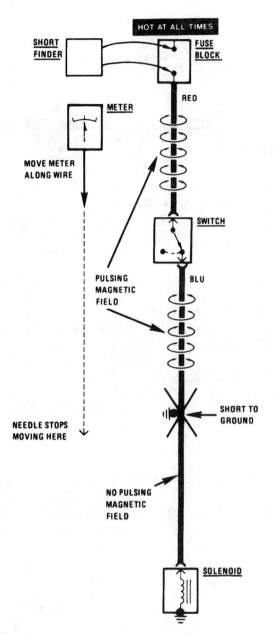

FIGURE 7–37 Using a short finder. (Courtesy of General Motors Corporation.)

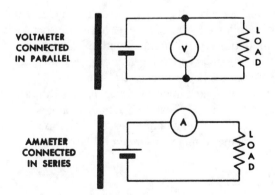

FIGURE 7–38 Always connect a voltmeter in parallel (top) and an ammeter in series (bottom). (Courtesy of General Motors Corporation.)

current flowing, the heating stops, and the breaker closes again and restores the circuit. If the high-current cause is still in the circuit, the breaker will open again. This cycling on and off will continue as long as the circuit is overloaded.

You can use a cycling breaker fitted with alligator clips in place of a fuse that keeps blowing. The circuit breaker will keep the circuit "alive" while you check the circuit for the cause of high current draw—usually a short.

126 Chapter 7 ELECTRICAL PRINCIPLES AND SERVICE BASICS

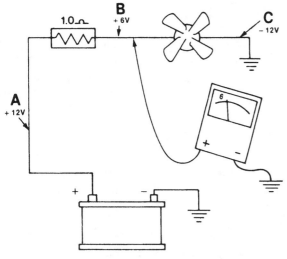

Closed circuit voltages.

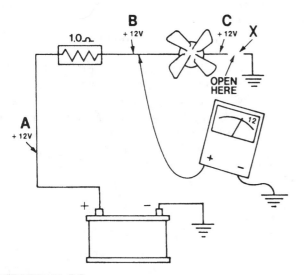

FIGURE 7-39 Using a voltmeter to isolate a problem in a motor circuit.

with all the current going through the meter. Polarity must be followed—the red lead going to the positive side. The induction type of ammeter with the clamp type of induction pickup is very convenient to use, since the wire being tested is not disconnected **(Figure 7-40)**.

Always use an ammeter that can handle the expected current, because excessive current can damage a meter. Also, never connect an ammeter across a circuit (parallel) or you may damage the meter or the circuit.

Ohmmeter

Another useful meter measures resistance. The ohmmeter is an excellent *continuity* checker, too. If you

FIGURE 7-40 Ammeter induction pickup clamp. (Courtesy of Kent Moore Tools Division, SPX Corporation.)

want to see if a wire is *continuous* or open inside a harness, clipping the ohmmeter leads to each end will show 0 Ω (if the wire is good).

Just remember *never* to use an ohmmeter in a hot circuit. Always be sure there is no power (voltage or current) in the circuit when using an ohmmeter to avoid damaging it **(Figure 7-41)**.

Ohmmeters have batteries for their power supply. For that reason, don't leave the ohmmeter connected for longer than necessary to read the scale, or the batteries will run down.

Digital Volt-Ohmmeter (DVOM)

A volt-ohmmeter with digital display and at least 10 MΩ (10,000 ohms) impedance (resistance) is required for

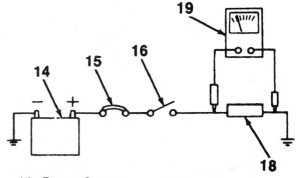

14. Power Source 18. Load
15. Circuit Breaker 19. Ohmmeter
16. Switch (Open)

FIGURE 7-41 An ohmmeter is used to measure resistance of electrical components. Component being tested must be isolated from any electrical current. (Courtesy of General Motors Corporation.)

testing many electronic components where very low voltages are present. This ensures that very little power will be drawn from the component being tested. Using a low-impedance meter will cause damage to sensitive electronic components.

Multimeter

A multimeter consists of a voltmeter, an ammeter, and an ohmmeter in a single unit. A control switch allows the user to select the test meter function desired. A variety of multimeter types are available with different volt, ohm, and resistance capacities for any application.

WIRING REPAIRS

The fast and easy way to repair electrical wiring is to use crimp-style terminals and connectors. Remove about ⅜ in. (10 mm) of insulation from the wire ends. Slip the terminal onto the wire end and crimp it tightly onto the wire with a crimping tool. This will form a good connection. Two pieces of wire can be joined together using a crimp-type connector in the same manner **(Figure 7–42)**. The connections can be soldered using a soldering gun and rosin-core solder. Heat the connection with the soldering gun until the solder flows freely into the connection. Allow the connection to cool, then tape it up with at least three layers of electrical tape. Never use acid-core solder for electrical repairs because of its corrosive effects.

To join braided wires together without using a crimp connector, spread the exposed braided ends apart, then push them together and twist them tightly together. Heat the connection with a soldering gun until the solder flows freely into the connection. Allow the connection to cool, then tape it up **(Figure 7–43)**.

A burned-out fusible link is replaced with a new one using similar procedures. Shielded twisted wiring used in computer control system wiring requires special procedures and special kinds of tape. Refer to the service manual for specific instructions.

WIRING TEST CHART

Type of Failure	Test Unit and Expected Results if Wiring Has Failed
Open (broken wire)	Ohmmeter—
	Infinite resistance at other end of wire
	Infinite to adjacent wire
	Infinite to ground.
	Voltmeter—
	Zero volts at other end of wire
Ground (bare wire touching frame)	Ohmmeter—
	Zero resistance to ground
	Infinite to adjacent wire
	May or may not be infinite to other end of wire
	Voltmeter—
	Instead of testing, normally look for blown fuse or tripped circuit breaker
Short (rubbing of two bare wires)	Ohmmeter—
	Zero resistance to adjacent wire
	Infinite to ground
	Zero to other end of wire
	Voltmeter—
	Voltage will be read on both wires

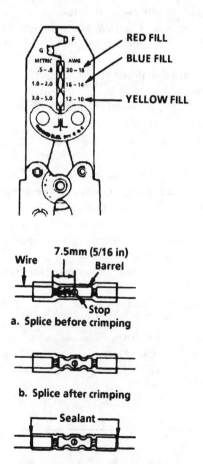

FIGURE 7–42 Crimping pliers (top) and insulated crimp connector wiring repair. Heating the crimp insulator shrinks and seals the insulation. (Courtesy of General Motors Corporation.)

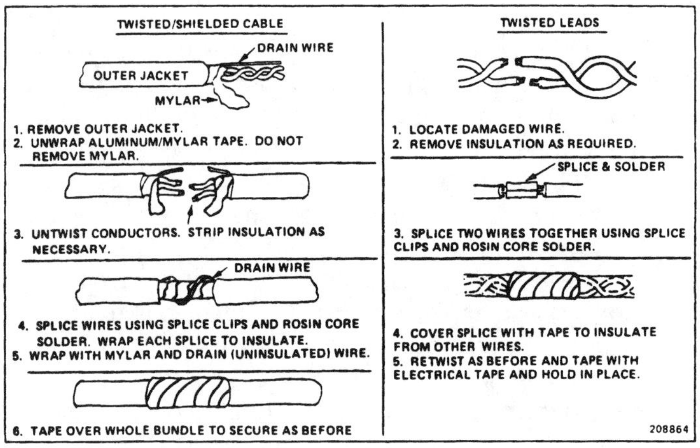

FIGURE 7–43 Typical wiring repair procedure. (Courtesy of General Motors Corporation.)

REVIEW QUESTIONS

1. Electricity behaves according to certain reliable _____ _____ that produce predictable _____ .
2. Electrons in an electrical wire can be dislodged by an external force such as an _____ or a _____ .
3. Electrons in an insulator are _____ bound together.
4. _____ produces almost all electricity.
5. Name two kinds of magnets.
6. Alnico is an alloy of _____ and _____ .
7. All magnets have _____ .
8. What are the two main parts of an electromagnet?
9. The only way we know that a magnetic field exists is by the _____ it produces.
10. From experiments we know that a magnetic field has _____ and _____ .
11. When a magnetic field passes over a conductor, a _____ is induced.
12. Electricity is measured in _____ and _____ .
13. Voltage is an electrical _____ .
14. One volt is equal to 1 _____ of current across 1 _____ of resistance when 1 _____ of power is consumed.
15. Current is the rate of _____ .
16. One ampere is equivalent to the current produced by 1 _____ when applied across a resistance of _____ _____ .
17. Resistance is measured in _____ .
18. State the three versions of Ohm's law using the letter symbols.
19. What is the formula for calculating electrical power when the values for amperes and volts are known?
20. What is the total resistance in a series circuit with three lamps of 2 Ω each?
21. In a parallel circuit if one bulb is burned out, the entire circuit is dead. (T) (F)
22. List six common electrical problems.
23. An ohmmeter is used to test for the presence of voltage in a circuit. (T) (F)
24. Why must a high-impedance digital voltmeter be used to test certain electronic components?
25. Voltage drop testing is another method of testing for excessive resistance. (T) (F)

TEST QUESTIONS

1. An atom consists of
   a. electrons, neutrons, and protons
   b. positive, negative, and neutral electrons
   c. positive electrons, negative protons, and neutrons
   d. positive protons, negative neutrons, and electrons
2. Technician A says that all electrons repel each other. Technician B says that unlike charges attract each other. Who is right?
   a. Technician A   c. both are right
   b. Technician B   d. both are wrong

3. Technician A says that $R = V \times I$. Technician B says that $I = E \times R$. Who is right?
   a. Technician A
   b. Technician B
   c. both are right
   d. both are wrong
4. Electrical power can be calculated as follows:
   a. $P = IR$
   b. $P = E - R$
   c. $P = E - I$
   d. $P = EI$
5. Total resistance in a circuit with three parallel resistances of 3, 4, and 6 Ω each is
   a. 7.5
   b. 0.75
   c. 1.33
   d. 1.67
6. An automotive lighting circuit consists of components connected in
   a. series
   b. parallel
   c. series–parallel
   d. all of the above
7. A switch suspected of being faulty may be checked with
   a. a jumper wire
   b. a test light
   c. an ohmmeter
   d. any of the above
8. Circuit resistance can be tested with
   a. a jumper wire
   b. a test light
   c. an ohmmeter
   d. any of the above
9. The invisible lines of force surrounding a current-carrying conductor are known as
   a. electromotive force
   b. counterelectromotive force
   c. repelling field
   d. magnetic field
10. Electrical feedback can cause
    a. lamps to light when they are not supposed to
    b. lamps to go off when they are not supposed to
    c. battery current to be reversed
    d. the alternator to produce alternating current
11. An electrical short
    a. allows current to bypass part of the normal circuit
    b. weakens the magnetic field
    c. reduces battery charging time
    d. reduces the length of wiring required
12. A multimeter is able to test
    a. watts, power, and volts
    b. volts, amperes, and ohms
    c. ohms, watts, and volts
    d. amperes, ohms, and power

CHAPTER 8

BATTERY PRINCIPLES AND SERVICE

INTRODUCTION

Chapter 7 discussed electrical principles and service. Voltage, current, resistance, and electrical power were also discussed. The battery is a source of voltage, current, and electrical power. This chapter discusses the function, construction, operation, and service of batteries. Studying Chapter 7 makes it easier to understand the material in this chapter.

PERFORMANCE OBJECTIVES

After thorough study of this chapter and sufficient practice with the appropriate components and equipment, you should be able to:

1. Describe the function of the battery.
2. Describe the construction features of the battery.
3. Describe battery operation during discharging and charging.
4. Diagnose battery problems.
5. Inspect, clean, and test a battery.
6. Charge a battery using a shop charger.
7. Replace a battery.

TERMS YOU SHOULD KNOW

Look for these terms as you study this chapter, and learn what they mean.

cell
element
positive plate
negative plate
separator
envelope
case
cell connector
caps
electrolyte
specific gravity
battery terminal
battery voltage
maintenance-free battery
cold-cranking rating
reserve capacity
battery stand
battery hold-down
battery cable
dead battery
hydrometer
state of charge
voltage test
cell test
drain test
battery charger
fast charger
slow charger
load test
quick-charge test
sulfated battery

BATTERY FUNCTION AND CONSTRUCTION

The battery performs the following tasks.

1. Provides all the electrical energy to the vehicle whenever the engine is not running

2. Operates the cranking motor, fuel injection control system, instrumentation, and other electrical devices during starting

BATTERY FUNCTION AND CONSTRUCTION

3. Provides extra electrical power whenever power requirements exceed the output of the charging system
4. Stores energy over long periods of time
5. Acts as an electric shock absorber or capacitor to absorb stray voltages from the vehicle's electrical systems

Battery Construction

The 12 V battery has six 2 V cells. Each cell is made up of a number of positive and negative plates separated by insulating separator plates. Negative and positive plates are arranged alternately in each cell. All the negative plates are connected to each other and so are the positive plates. This arrangement provides a positive cell connection and a negative cell connection. This assembly is submerged in a cell case full of battery electrolyte, which is 64% water and 36% sulfuric acid.

Each battery cell produces approximately 2 V, regardless of the number or size of plates per cell. Six of these 2 V cells arranged in a single battery case form a 12 V battery. The battery case is usually made of polypropylene. The case has built-in cell dividers and sediment traps. The six battery cells are connected in series. This means that the positive side of a cell is connected to the negative side of the next cell throughout all six cells. If the cells were connected in parallel (positive to positive and negative to negative), the battery would have only a 2 V potential **(Figures 8–1 to 8–5)**.

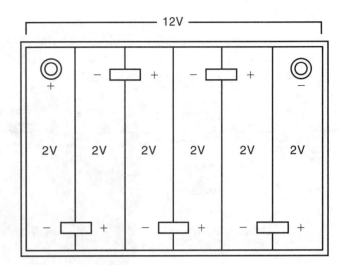

FIGURE 8–2 A 12 V battery has six 2 V cells connected in series.

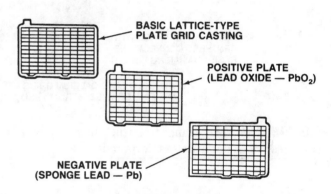

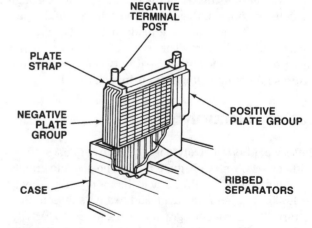

FIGURE 8–3 Battery cell construction. (Courtesy of Ford Motor Company.)

Cell Connectors

Cell connectors **(Figure 8–6)** are used to connect the cell elements in series, that is, the positive strap of one cell is connected to the negative strap of the adjacent

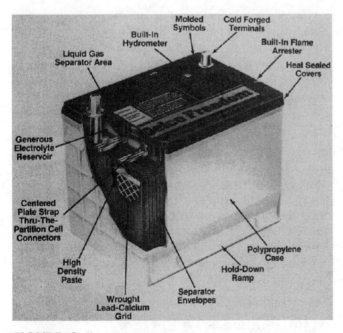

FIGURE 8–1 Maintenance-free battery with tapered post terminals. (Courtesy of General Motors Corporation.)

132 Chapter 8 BATTERY PRINCIPLES AND SERVICE

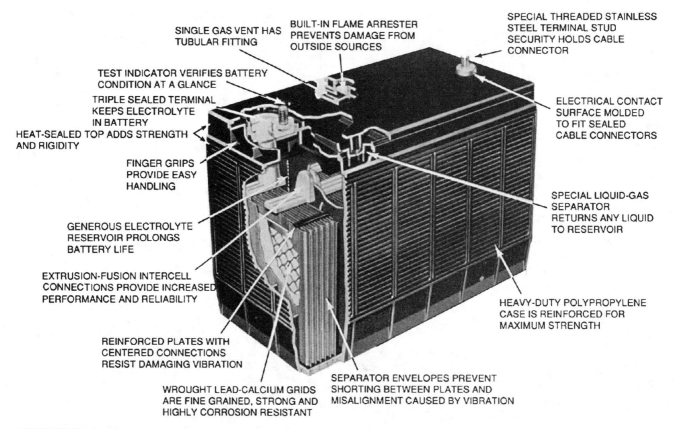

FIGURE 8–4 Maintenance-free truck battery features. Note threaded terminals. (Courtesy of General Motors Corporation.)

cell. Connections between the cells are either through the cell partitions in the case or over the top of the partition. The connections are made before the cover is placed on the battery. This type of construction not only provides an acid-tight seal between the cells but assures minimum voltage loss from cell to cell.

Battery Electrolyte (Acid)

Battery acid or electrolyte is a mixture of 36% sulfuric acid and 64% water **(Figure 8–7)**. The specific gravity of water is 1.000. Sulfuric acid has a specific gravity of 1.835. The sulfuric acid and water solution in a battery has a specific gravity of 1.275 ± 0.010. This makes it 1.275 times heavier than plain water. When the specific gravity of battery acid is too low, it may freeze in colder climates **(Figure 8–8)**. When the specific gravity is too high, the plate grids in the battery will be damaged. The specific gravity of a maintenance-free battery is checked through the built-in hydrometer visible at the top of the battery **(Figure 8–9)**. A squeeze bulb and float type of hydrometer is used on other batteries.

Battery acid is extremely corrosive and will cause skin burns if contacted. It will also burn holes in many kinds of clothing and damage metal and painted surfaces if allowed to come in contact with them.

Charging and Discharging

In operation, the battery is normally being partially discharged and recharged. There is a constant reversing of the chemical action taking place in the battery. The cycling of the charge and discharge modes slowly wears away the active materials on the battery cell plates. This eventually causes the battery positive plates to oxidize. When the oxidation reaches the point that there is insufficient active plate area to charge the battery, the battery is worn out and must be replaced **(Figures 8–10 and 8–11)**.

Maintenance-Free Battery

Maintenance-free batteries have several design features not always found on other batteries **(Figures**

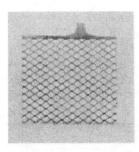

Wrought Lead-Calcium Grids...offer considerable strength while maintaining a very fine and consistent grain structure. Conventional lead-antimony battery grids and competitive lead-calcium cast grids are susceptible to attack by corrosion. Delco wrought grids are substantially resistant to grid corrosion, over-charge, gassing, water usage, self-discharge and thermal runaway.

Polypropylene Case...is a reinforced design, precisely tailored to support the battery elements. The case has beveled corners to reduce handling damage and breakage, while outside case bottom is waffled to prevent puncture. Polypropylene case material is exceptionally strong and durable and easily withstands road shock and vibration.

Envelope Separators Encapsulate Plates...replacing the traditional separator during element assembly. Envelope construction or encapsulation, improves vibration durability. It prevents "treeing" and internal shorting between the plates. The Delco element rests on a flat case bottom, as there is no need for a sediment chamber beneath the plates.

Exclusive Liquid-Gas Separator...has been built into the battery cover to prevent minute electrolyte losses by collecting the liquid and returning it to the main electrolyte reservoir of the battery. Although gassing is virtually eliminated, this vent also allows the battery to "breathe", especially during temperature changes and charging.

Centered Cast-on Plate Straps...used to connect plates...are stronger than the thinner gas-burned conventional connections. These straps are located near the center of the plates, which reduces the lever action movement resulting from road shock. In addition, the element is reinforced with thermoplastic anchors to further improve vibration durability.

Heat-Sealed Cover...is installed to the case at the factory after the battery is charged. This prevents future contamination and also adds to the strength and rigidity of the case-cover assembly. Permanent flame arrestor has been built-in to prevent an accidental explosion which could be caused by either sparks or flame from outside the battery.

FIGURE 8-5 Heavy-duty 12 V battery construction. (Courtesy of General Motors Corporation.)

8–4 and 8–5), including a larger electrolyte reserve capacity. Since all lead–acid batteries are subject to some vapor loss, water must be added periodically unless the reserve capacity for electrolyte is adequate for years of operation. Another feature is the use of calcium instead of antimony to strengthen the grid plates. The use of calcium reduces normal gassing. Instead of using separator plates between the positive and negative plates, each plate is encased in a porous fiberglass envelope. The envelopes prevent any material that may shed from the plates from causing a short between the plates and sediment collected in the bottom of the battery. Also provided is an expansion chamber to allow for internal expansion and contraction to occur. The top of the battery is sealed except for tiny indirect vapor vents. This reduces surface discharge and corrosion caused by electrolyte on the battery surface. A built-in hydrometer indicates the battery state of charge.

Battery Terminals and Polarity

The positive plate group in one end cell of a battery is connected to the positive battery external terminal. This terminal is identified in one of the following ways: POS, + sign, or red-colored terminal. The tapered positive post is larger in diameter than the negative post.

The negative plate group at the other end of the battery is connected to the external negative battery terminal. It can be identified as follows: NEG or − sign on terminal. The tapered negative post is smaller in diameter than the positive post. Proper battery polarity must always be adhered to when working with vehicle and equipment electrical systems.

The following are the most common types of battery terminals.

1. *Post or top terminals.* (See **Figure 8–1** for an example of post terminals on top of the battery.)

2. *Side terminals:* positioned in the sidewall of the container near the top edge. These terminals are

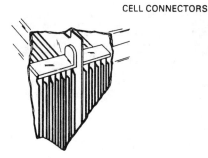

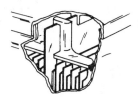

FIGURE 8–6 Types of battery cell connectors. (Courtesy of Ford Motor Company.)

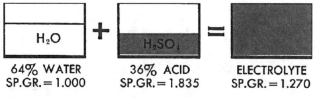

FIGURE 8–7 Makeup of battery electrolyte and specific gravities. (Courtesy of General Motors Corporation.)

Specific Gravity	Freezing Temperature
1280	−90°F (−67°C)
1250	−62°F (−53°C)
1200	−16°F (−20°C)
1150	+ 5°F (−15°C)
1100	+19°F (− 7°C)
1050	+27°F (− 2°C)

FIGURE 8–8 Specific gravity at which electrolyte will freeze.

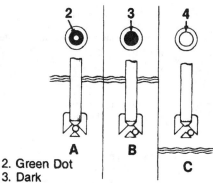

2. Green Dot
3. Dark
4. Clear
A. 65% or Above State of Charge
B. Below 65% State of Charge
C. Low Level Electrolyte

FIGURE 8–9 Built-in hydrometer operation. Note position of float ball in each case. (Courtesy of General Motors Corporation.)

threaded and require a special bolt to attach the cables. **(Figures 8–12 to 8–14)**.

Battery Cables

Battery cables must be of sufficient current-carrying capacity to meet all electrical loads. Normal 12 V cable size usually is 4 gauge (19 mm^2) or 6 gauge (13 mm^2) **(Figure 8–15)**. Various cable clamps and terminals are used to provide a good electrical connection at each end. Connections must be clean and tight to prevent arcing, corrosion, and high resistance.

MULTIPLE-BATTERY SYSTEMS

Most North American vehicles and equipment use a 12 V system. Some European ones and others use a 12 V charging system with a 24 V starting system **(Figure 8–16)**. Some earlier domestic systems were also of the 12/24 V design. The 12 V system is by far the most common. This system may have only one battery or it may have several connected in parallel **(Figures 8–17 and 8–18)**. Multiple battery systems provide the increased cranking power required for larger engines and cold temperatures. For example, when four 12 V, 600 cold-cranking ampere (CCA) batteries are connected in parallel, the total CCA is 2400 (4 × 600) while the voltage remains at 12 V. The total available power in this system is 28,000 W (12 × 2400), whereas a single 600 CCA battery has only 7200 W (12 × 600). This system uses a high-output starting motor, as described in Chapter 10.

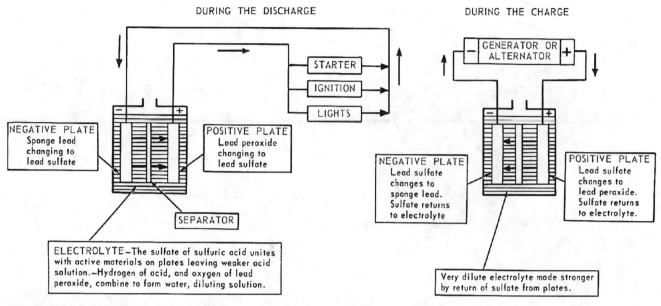

FIGURE 8-10 Chemical action in battery during discharge and charge cycles. (Courtesy of Deere and Company.)

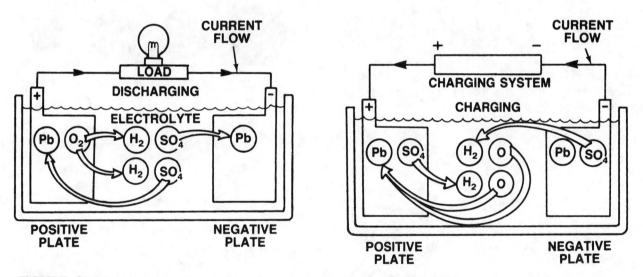

FIGURE 8-11 Chemical action in battery during discharge (left) and charge (right) cycles.

BATTERY RATINGS

Battery capacity ratings are established by the Battery Council International (BCI) and the Society of Automotive Engineers. Commonly used ratings are as follows:

1. *Cold-cranking amperes:* the load in amperes a battery is able to deliver for 30 seconds at 0°F (−17.7°C) without falling below 7.2 V for a 12 V battery.

2. *Reserve capacity:* the length of time in minutes a battery can be discharged under a specified load at 80°F (26.6°C) before battery cell voltage drops below 1.75 V per cell.

Factors that determine the battery rating required for a vehicle or equipment include engine size and type and climatic conditions under which it must operate. Battery power drops drastically as temperatures drop below freezing **(Figure 8-19)**. As the temperature drops much below freezing, the engine is harder to crank, owing to increased friction resulting from oil thickening.

136 Chapter 8 BATTERY PRINCIPLES AND SERVICE

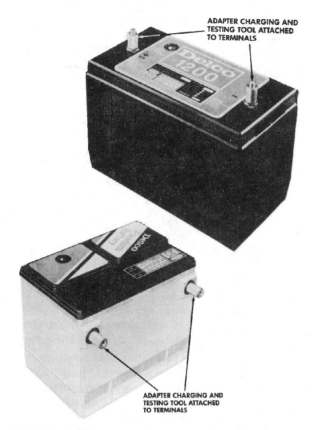

FIGURE 8–12 Charging and testing tool adapters on battery terminals protect posts. (Courtesy of General Motors Corporation.)

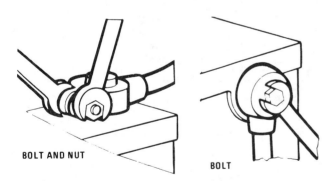

FIGURE 8–13 Common cable-to-battery connections.

BATTERY SERVICE PRECAUTIONS

1. Battery acid is extremely corrosive. Avoid contact with skin, eyes, and clothing. If battery acid should accidentally get into your eyes, rinse thoroughly with clean water and see your doctor. Contacted skin should be washed thoroughly with clean water; some baking soda with the wash will neutralize the

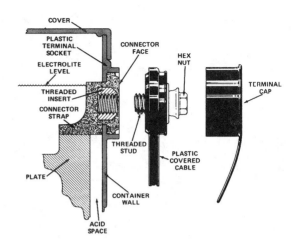

FIGURE 8–14 Battery side-terminal detail. (Courtesy of Battery Council International.)

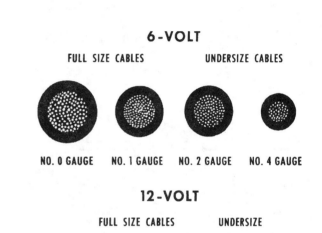

FIGURE 8–15 Battery cable sizes and cross sections.

action of the acid. Painted surfaces and metal parts are also easily attacked by acid, and contact should be avoided.

2. When making connections to a battery in or out of the vehicle, always observe proper battery polarity: positive to positive and negative to negative.

3. Avoid any arcing (sparks) or open flame near a battery. The battery produces highly explosive vapors that can cause serious damage if ignited.

4. When disconnecting battery cables, always remove the ground cable first; and when connecting cables, always connect the ground cable last. This helps prevent accidental arcing.

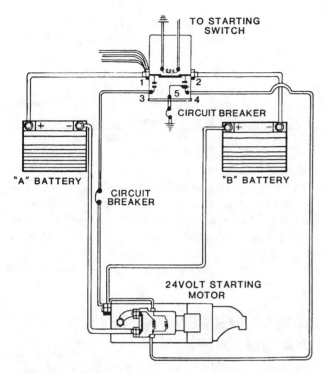

FIGURE 8–16 Series–parallel system with 24 V starting and 12 V charging systems. (Courtesy of Snap-on Tools Corporation.)

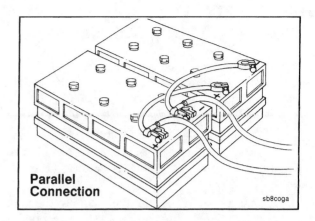

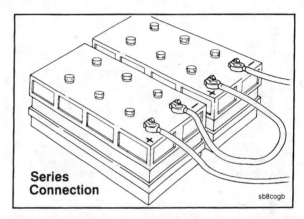

FIGURE 8–18 Parallel and series battery connections. (Courtesy of Cummins Engine Company, Inc.)

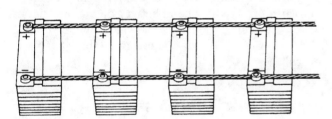

FIGURE 8–17 Parallel battery connections. (Courtesy of Snap-on Tools Corporation.)

5. Observe equipment manufacturer's instructions when charging batteries. Never allow battery temperature to exceed 125°F (51.6°C).

6. Some battery manufacturers place restrictions on the use of a booster battery to jump start. Follow battery manufacturer's recommendations. One manufacturer says: "Do not charge, test, or jump start this battery when the built-in hydrometer is clear or yellow: battery must be replaced."

7. Use the proper battery carrier to handle the battery; this avoids injury and possible battery damage.

8. Use the proper protective clothing (apron, gloves, and face shield) when handling batteries to ensure safety.

9. Do not weld or smoke near a battery charging or storage area.

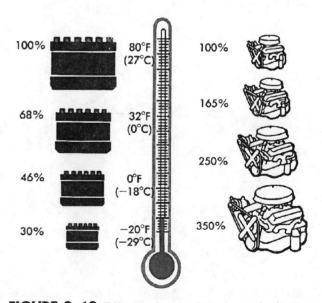

FIGURE 8–19 Effects of temperature on the battery and the engine when starting. (Courtesy of Deere and Company.)

JUMP-STARTING (NEGATIVE-GROUND SYSTEM)

1. Set the parking brake and place the transmission in PARK (NEUTRAL for manual transmission). Turn off the lights and all other electrical loads.
2. Check the built-in hydrometer. If it is clear or light yellow, replace the battery **(Figure 8–20)**.
3. Observe the correct battery voltage and polarity (same for both vehicle and booster battery) **(Figures 8–21 and 8–22)**.
4. Attach the end of one jumper cable to the positive terminal of the booster battery and the other end of the cable to the positive terminal of the discharged battery. Do not permit vehicles to touch each other because this could cause a ground connection and counteract the benefits of the procedure.
5. Attach one end of the remaining negative cable to the negative terminal of the booster battery, and the other end to a solid engine ground (such as a compressor bracket or generator mounting bracket) at least 18 in. from the battery of the vehicle being started. **(DO NOT CONNECT DIRECTLY TO THE NEGATIVE TERMINAL OF THE DEAD BATTERY.)**
6. Start the engine of the vehicle that is providing the jump-start and turn off all electrical accessories. Then start the engine with the discharged battery.
7. Reverse these directions exactly when removing the jumper cables. The negative cable must be disconnected first from the engine that was jump-started.

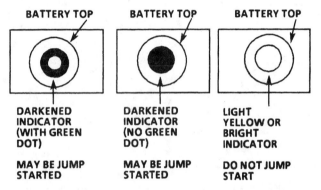

FIGURE 8–20 Built-in hydrometer readings indicate whether battery may be jump-started. (Courtesy of General Motors Corporation.)

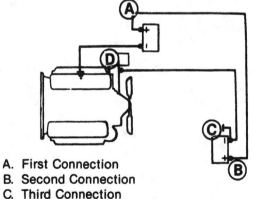

A. First Connection
B. Second Connection
C. Third Connection
D. Last Connection (Made on the Engine, Away from the Battery)

FIGURE 8–21 Jumper cable connections for a single-battery system. (Courtesy of General Motors Corporation.)

BATTERY INSPECTION, TESTING, AND CHARGING

Inspect for dirt on top of the battery. Look for corroded battery cable connections, evidence of gassing (battery top wet with acid), case damage, and a damaged or dirty battery stand and hold-down. These problems must be corrected for good battery operation. On batteries with cell caps, check the electrolyte level. Add distilled water if the level is below the fill ring. Never overfill.

Surface Discharge Test

A dirty or wet battery top can result in current loss and self-discharge across the top. This condition may result in low battery power and poor starting. To check for surface discharge, use a digital voltmeter on the low-voltage scale **(Figure 8–23)**. Connect the negative voltmeter lead to the battery negative post. Touch the surface area near the positive post with the positive voltmeter lead. If a voltage reading appears, battery current is being lost. The battery must be cleaned.

Cable Connection Test

To check for a poor connection between the battery cable and battery post, use a digital voltmeter to measure the voltage drop between the post and cable. Connect the negative lead to the cable end. With the ignition disabled (so the engine will not start), crank the engine while touching the positive lead to the battery terminal

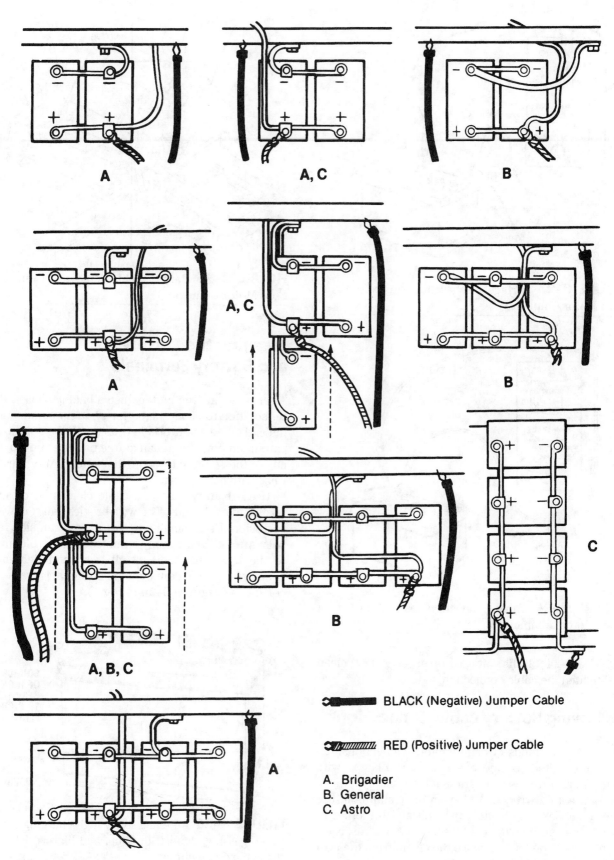

FIGURE 8–22 Typical jump-start cable connections for GMC trucks. (Courtesy of General Motors Corporation.)

140 Chapter 8 BATTERY PRINCIPLES AND SERVICE

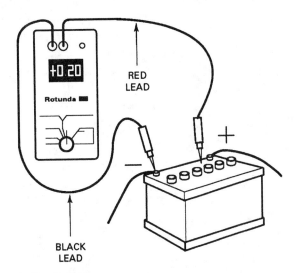

FIGURE 8-23 Surface discharge test. (Courtesy of Ford Motor Company.)

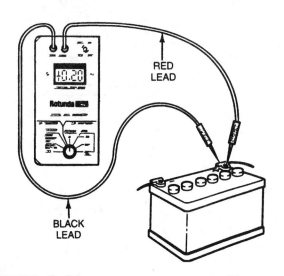

FIGURE 8-24 Cable connection test. (Courtesy of Ford Motor Company.)

(**Figure 8-24**). If the voltage drop exceeds 0.5 V, clean and tighten the cable connection.

Removing Battery Cable Connections

There are two kinds of battery cable connections: the flat ends bolted to side terminals, and clamps with bolts. To remove the side-terminal type, simply remove the bolt with a wrench. To loosen the clamp bolt, use two wrenches, one to hold the bolt and the other to loosen the nut. With the nut loosened a few turns, use a battery cable puller to remove the clamp from the post (**Figures 8-25, 8-13,** and **8-14**).

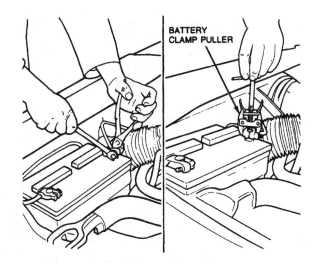

FIGURE 8-25 Battery cable removal. (Courtesy of Ford Motor Company.)

Cleaning the Battery and Battery Terminals

After removing the battery, use a baking soda and water solution to clean the battery, battery stand, and cable ends. Use a soft bristle brush with the solution. Be careful on non-maintenance-free batteries not to allow any of the solution to enter the battery. After cleaning, rinse with clear water.

Use a battery post and cable cleaner to clean the posts and cable ends. The wire bristles will remove any corrosion. Use a small wire brush to clean side terminals. After cleaning apply a thin coat of petroleum jelly to the terminals, then install them and tighten just enough for a good connection. Petroleum jelly prevents corrosion (**Figures 8-26** to **8-29**).

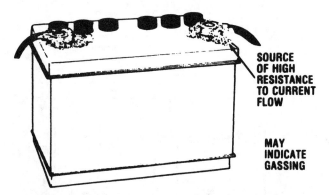

FIGURE 8-26 An excessively high charging rate can cause gassing, deposit buildup, and high resistance at the cable connections.

BATTERY INSPECTION, TESTING, AND CHARGING 141

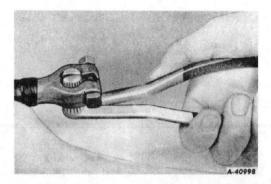

FIGURE 8–27 Using a cable clamp spreader. (Courtesy of Kent Moore, SPX Corporation.)

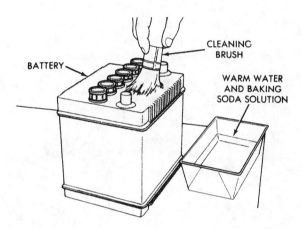

FIGURE 8–28 Cleaning the battery with a warm water and baking soda solution. (Courtesy of Ford Motor Company.)

Checking the State of Charge

The specific gravity of the battery electrolyte indicates its state of charge. This is measured with a battery hydrometer. Maintenance-free batteries have a built-in hydrometer with a sight glass. When the green dot is visible, the battery is fully charged. When the sight glass shows dark with no green dot, the battery is discharged. When the sight glass shows a light yellow color, the battery should be replaced **(Figure 8–9)**. On batteries with cell caps a glass-float hydrometer is used to check the specific gravity of the electrolyte. A fully charged battery in excellent condition has a specific gravity reading of 1.275 (± 0.010). To use the hydrometer, squeeze the air out of the rubber bulb and insert the tube into the battery electrolyte **(Figure 8–30)**. Slowly release the bulb to allow electrolyte to be drawn into the hydrometer. When the float has

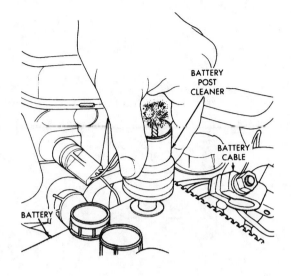

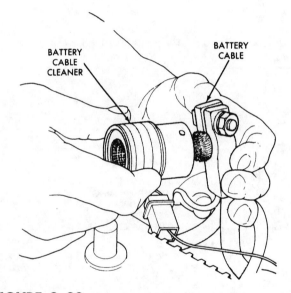

FIGURE 8–29 Cleaning the battery post and cable clamp.

risen but is not touching the top, stop releasing the rubber bulb and note the reading on the float at the electrolyte level. Note the temperature of the electrolyte as well. Calculate the temperature-corrected reading as shown in **Figure 8–31**. Test each cell in the same manner.

SAFETY CAUTION:

Never allow battery acid to drip on your skin or on any painted surface. It will burn your skin and damage paint.

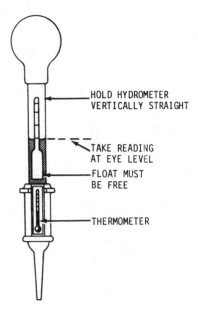

FIGURE 8–30 Battery hydrometer. (Courtesy of Deere and Company.)

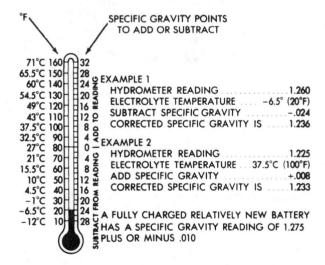

FIGURE 8–31 To accurately determine the state of charge of a battery, the reading must be temperature corrected as shown here.

Battery Voltage and State of Charge Test

1. With the start key off and no electrical loads on, connect the negative (−) lead of a voltmeter to the negative battery cable clamp.

NOTE: The range setting on the voltmeter should be at least 0 to 15.

2. Connect the positive (+) lead of the voltmeter to the positive battery cable clamp.

3. If the voltmeter reading is more than 12.4 V at 70°F (21°C), the battery voltage is acceptable. If the reading is 12.4 V or less, the battery needs charging **(Figures 8–32 and 8–33).**

Cell Voltage Test

A cell voltage test determines whether the battery is defective or just discharged. It can be performed only on batteries with individual cell caps. To perform the test, insert the cadmium tips into adjoining cell pairs, one tip in each cell. Start at one end of the battery and test each cell this way. Note the voltage of each cell as it is tested. If cell voltage varies more than 0.10 V between cells, the battery should be replaced. If cell voltage is even, recharge the battery and test it again.

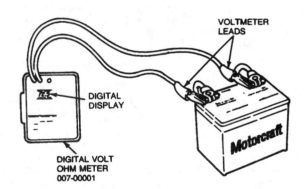

FIGURE 8–32 Checking battery open-circuit voltage. (Courtesy of Ford Motor Company.)

Open-Circuit Volts	Percent Charge
11.7 or less	0
12.0	25
12.2	50
12.4	75
12.6 or more	100

FIGURE 8–33 Open-circuit voltage and state-of-charge figures.

Quick-Charge Test

A quick-charge test can be performed to determine if the battery is sulfated. Charge the battery for 3 minutes at 30 to 40 A. Connect a voltmeter across the battery terminals and observe battery voltage during charging. If battery voltage exceeds 15.5 V on a 12 V battery, the battery is sulfated and must be replaced.

Charging the Battery

CAUTION: Disconnect the negative battery cable before making charger connections. Failing to do so may result in a high-voltage surge and damage to electronic parts.

Batteries below 40°F (5°C) do not readily accept a charge. It may take 4 to 8 hours for a battery to warm up enough to take a charge **(Figure 8–34)**. A completely discharged battery may also be slow to accept a charge. In this case the dead battery switch on the charger is used. However, the initial charge rate accepted is so low that it may not register on the charger ammeter. Follow the charger manufacturer's instructions on how to operate the charger.

There are two methods by which a battery may be charged: the automatic setting on chargers so equipped and the manual or constant-current setting. The automatic setting maintains a safe charging rate by automatically controlling the current and voltage to prevent spewing of electrolyte. It may take from 2 to 4 hours to charge a battery to a serviceable state. Several additional hours at 3 to 5 A are required to bring it to a fully charged state **(Figures 8–35 to 8–37)**.

The manual or constant-current setting is initially set at 30 to 40 A for approximately 30 minutes or as long as there is no major gassing or spewing of electrolyte. If gassing is excessive, the rate must be reduced.

Percent Charge	Specific Gravity	Charge Rate	Charging Time
100	1.280		
75	1.225	20	50 min
50	1.190	20	70 min
25	1.155	20	90 min
0	1.120	5	12 hrs

FIGURE 8–34 Typical charge times for batteries at different state of charge.

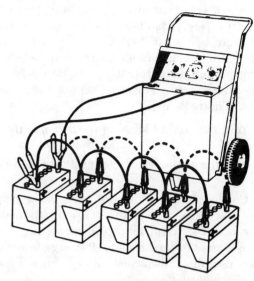

FIGURE 8–36 Charging a number of batteries at the same time. Batteries are connected in parallel, so the combined battery voltage remains at 12 V. (Courtesy of Ford Motor Company.)

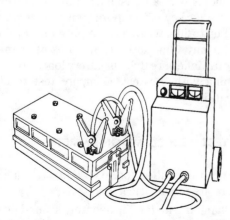

FIGURE 8–35 Fast charger connected to battery out of vehicle. (Courtesy of Cummins Engine Company, Inc.)

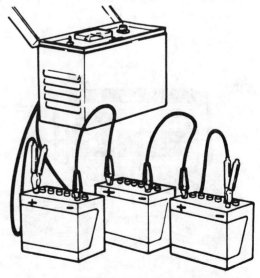

FIGURE 8–37 (Courtesy of Ford Motor Company.)

Battery Capacity Test (Load Test)

For this test the battery must be at or very near the fully charged state. Check the state of charge as outlined earlier. If necessary, charge the battery to bring it up to a full charge. Remove the surface charge by applying the specified load for the battery being tested for 10 to 15 seconds only, then allow the battery to recover for 2 minutes. Battery electrolyte temperature should be at 70°F (21°C). Never load test a battery with electrolyte temperature below 60°F (17°C). If the tester has an adjustment for temperature correction be sure to set it at the proper setting. Use a high-rate-discharge battery–starter tester combined with a voltmeter for this test **(Figure 8–38)**.

1. Turn the control knob on the tester to the OFF position.
2. Turn the voltmeter selector switch to the 10 or 20 V position.
3. Connect both positive test leads to the positive battery post and both negative test leads to the negative battery post. The voltmeter clips must contact the battery posts and not the high-rate-discharge tester clips. Unless this is done, the actual battery terminal voltage will not be indicated.
4. Turn the load control knob in a clockwise direction until the ammeter reads approximately half the cold-cranking ampere rating of the battery. A battery with a 400 CCA rating should be tested at 200 A (half the cold-cranking rating; see **Figure 8–39**).
5. With the ammeter reading the required load for 15 seconds, note the voltmeter reading. Avoid leaving the high-discharge load on the battery for periods longer than 15 seconds.

Cold-Cranking Amperes	Ampere-Hour Rating (Approx.)	Watt Rating	Load Test Amperes
200	35–40	1800	100
250	41–48	2100	125
300	49–62	2500	150
350	63–70	2900	175
400	71–76	3250	200
450	77–86	3600	225
500	87–92	3900	250
550	93–110	4200	275

FIGURE 8–39 Load test figures for various battery ratings.

6. If the voltmeter reading is 9.6 V at 70°F (21°C) or more, the battery has a good output capacity and will readily accept a charge, if required.
7. If the voltage reading obtained during the capacity test is below 9.6 V at 70°F (21°C) and the battery is fully charged, the battery is defective and must be replaced. If unsure about the battery's state of charge, charge the battery.
8. After the battery has been charged, repeat the capacity test. If the capacity test battery voltage is still less than 9.6 V, replace the battery. If the voltage is 9.6 V or more, the battery is satisfactory for service.

Battery Drain Test

The battery drain test will show whether battery current is being used with the ignition key off. Most computer-controlled cars have a small current drain to maintain computer memory, which must be taken into consideration. Vehicles with air suspension or load-leveling systems may have key-off temporary current drains ranging from 0.1 up to 20 A when the compressor is cycling. This can occur any time up to 70 minutes after the key is turned off. Make sure that all battery loads are off before proceeding with the test. Compare test results with specifications.

DRAIN TEST WITH CLAMP-ON AMMETER

1. Turn the ignition to the OFF position and ensure that there are no electrical loads.
2. Clamp the meter clip securely around positive or ground battery cable (all cables if two or more lead to post).

NOTE: Do not start vehicle with clip on cable.

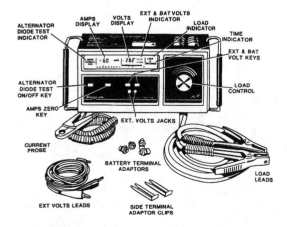

FIGURE 8–38 AVR tester. (Courtesy of Snap-on Tools Corporation.)

The current reading (current drain) should be less than 0.05 A. If it exceeds 0.05 A, it indicates a constant current drain that could cause a discharged battery. Possible sources of current drain problems are vehicle lamps (under hood, glove compartment, luggage compartment, etc.) that do not shut off properly.

If the drain is not caused by a vehicle lamp, remove fuses one at a time until the cause of the drain is located. If the drain is still undetermined, disconnect leads at the starter relay one at a time to find the problem circuit.

BATTERY SELECTION AND INSTALLATION

The correct shape, physical size, post location, and battery rating requirements must all be considered when replacing a battery. Use only the type, size, and rating that are equal to original specifications. Install the new battery in a clean tray. Make sure that the tray has no protrusions that could damage the battery case. Place the hold-down in position and bolt it in place. Lightly coat the terminals and cable ends with petroleum jelly, place them in position, and tighten the bolts.

REVIEW QUESTIONS

1. The battery provides all the electrical energy for the vehicle at all times. (T) (F)
2. The battery provides all the electrical energy to operate the starting motor to crank and to start the engine. (T) (F)
3. A battery cell is made up of _____ and _____ plates separated by _____ plates.
4. Each battery cell produces approximately _____ V.
5. A 12 V battery has _____ cells.
6. The 12 V battery has one _____ and one _____ terminal.
7. Battery terminals may be located at the _____ or the _____ of the battery.
8. In operation the battery is normally being partially _____ and _____ .
9. Currently, batteries are rated as to their _____ _____ ability and their _____ _____ _____ .
10. Battery electrolyte consists of approximately _____ % water and _____ % _____ .
11. Maintenance-free batteries have an _____ _____ to allow for terminal expansion and contraction.
12. A built-in _____ indicates the state of _____ in a maintenance-free battery.
13. Skin contact with battery acid should be avoided because it is extremely _____ .
14. Battery acid can be neutralized with a _____ _____ and water solution.
15. When disconnecting a battery, always disconnect the _____ cable first.
16. Avoid any sparks or open flame near a battery because the battery produces _____ .
17. Surface discharge can result from a _____ or _____ battery top.
18. To check for surface discharge, use a _____ set on the _____ scale.
19. To check for a poor connection between the battery _____ and battery _____ , use a _____ .
20. A fully charged battery has a specific gravity of _____ .
21. A quick-charge test can be used to determine if a battery is _____ .
22. Batteries below _____ °F (_____ °C) do not readily accept a charge.
23. A high-rate-discharge battery–starter tester is used to perform a battery _____ test.
24. A battery drain test can be made with an _____ or a _____ .
25. When performing a battery drain test, consideration must be given to possible _____ current drain.
26. When replacing a battery it must be of the proper _____ , _____ location and _____ .

TEST QUESTIONS

1. Battery electrolyte consists of approximately
   a. 36% sulfuric acid and 64% water
   b. 64% sulfuric acid and 36% water
   c. 34% muriatic acid and 64% water
   d. 64% muriatic acid and 36% water

2. A battery cell consists of
   a. positive plates, negative plates, sponge plates, and lead plates
   b. insulator plates, positive plates, negative plates, and electrolyte
   c. sulfuric acid, electrolyte positive plates, and negative plates
   d. sponge plates, lead plates, antimony plates, and cadmium plates

3. Battery cells are connected in
   a. series
   b. parallel
   c. series–parallel
   d. none of the above

4. A vehicle is brought into the shop with a dead battery. What should you do?
   a. replace the battery
   b. test the battery and charge it
   c. charge the battery and test it
   d. jump-start the vehicle

5. The following should be considered when replacing a battery:
   a. physical size and post location
   b. cold-cranking capacity and reserve capacity
   c. engine size and climatic conditions
   d. all of the above

6. When load testing a battery with a 400 CCA rating, apply a load of
   a. 400 A
   b. 40 A
   c. 20 A
   d. 200 A

7. To perform a battery surface discharge test, use a(n)
   a. ohmmeter
   b. ammeter
   c. wattmeter
   d. voltmeter

8. When performing a battery cable connection test, the voltage drop between the post and cable should not exceed
   a. 1 V
   b. 0.1 V
   c. 0.5 V
   d. 0.05 V

9. The open-circuit voltage of a fully charged battery at room temperature should be
   a. 12 V
   b. 12 to 12.2 V
   c. 12.2 to 12.4 V
   d. 12.6 V

10. After charging a battery just prior to a load test, first remove the
    a. overcharge
    b. surface charge
    c. undercharge
    d. subsurface charge

◆ CHAPTER 9 ◆

CHARGING SYSTEM PRINCIPLES AND SERVICE

INTRODUCTION

The charging system is an important component of a vehicle or equipment electrical system. If it fails to function properly, the battery will soon be discharged, and none of the electrical systems will be able to function. The design and operation of the charging system must be thoroughly understood in order to maintain and service it properly. This chapter deals with the function, design, operation, and service of the charging system and its components.

PERFORMANCE OBJECTIVES

After thorough study of this chapter and sufficient practice with the appropriate components and equipment, you should be able to:

1. Describe the function of the charging system.
2. List the major components of the charging system and state their function.
3. Describe alternator construction and operation.
4. List the different types of charge indicators.
5. Describe alternator output control.
6. List common charging system problems.
7. Perform a preliminary inspection of a charging system.
8. Perform charging system tests as required.
9. Test and replace alternator components.
10. Remove and replace an alternator.

TERMS YOU SHOULD KNOW

Look for these terms as you study this chapter, and learn what they mean.

charging system	Y stator
AC	delta stator
DC	field current
alternator	charge indicator
drive belt	brushless alternator
regulator	alternator noise
wiring harness	overcharging
rotor	undercharging
stator	voltmeter test
rectifier	surface charge
diode	base voltage
brush holder	no-load test
brushes	load test
slip rings	current output test
drive end frame	voltage output test
rectifier end frame	regulator bypass test
fan	full field current
pulley	circuit resistance test
integral regulator	rotor winding resistance
diode trio	continuity
single phase	stator winding resistance
three phase	diode test
capacitor	

CHARGING SYSTEM FUNCTION AND COMPONENTS

While the engine is running, the charging system provides the power needed to operate all of the vehicle's electrical systems and to charge the battery whenever needed. It does this by converting mechanical energy into electrical energy.

The major components of the charging system are (**Figure 9-1**):

1. *Alternator:* converts engine mechanical power to electrical power
2. *Alternator drive belt:* connects the crankshaft pulley to the alternator pulley to drive the alternator
3. *Voltage regulator:* controls the output voltage of the alternator
4. *Charge indicator:* dash-mounted ammeter, voltmeter, or indicator light informs the operator of charging system operation
5. *Battery:* provides current to energize the alternator initially; also acts as a voltage stabilizer
6. *Wiring harness:* connects charging system components to each other

TYPES OF ALTERNATORS

Manufacturers produce a variety of alternator designs for different applications with different voltage and current output. Output voltage may be 12, 24, 30 or 32 V. Current output ranges from around 40 A on the smaller units to 160 A on the larger heavy-duty units. Brush-type and brushless models are used. Alternators may be belt-driven or gear-driven. Belt-driven alternators are driven from a pulley on the engine crankshaft. The belt drive may be a single V, multiple V, or multirib belt. The gear-driven alternator is usually driven by a gear in the gear train at the front of the engine. Voltage regulators may be bolted on the outside of the alternator, or they may be mounted on the inside. Modern regulators are solid-state electronic devices, while older types were electromechanical.

Alternator manufacturers include General Motors Corporation, Ford Motor Company, Prestolite (Leece Neville), Robert Bosch, Motorola, Nippondenso, and others.

ALTERNATOR COMPONENTS (BRUSH TYPE)

The alternator consists of the following components, as illustrated in **Figures 9-2** and **9-3**.

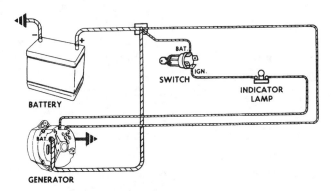

FIGURE 9-1 Charging system components and wiring. (Courtesy of General Motors Corporation.)

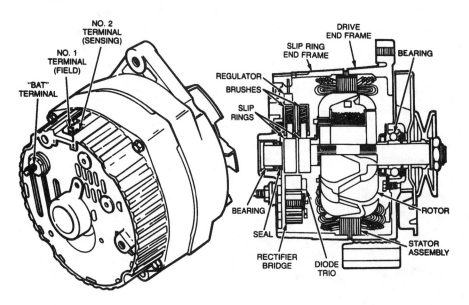

FIGURE 9-2 Major components of an alternator. (Courtesy of General Motors Corporation.)

ALTERNATOR COMPONENTS (BRUSH TYPE)

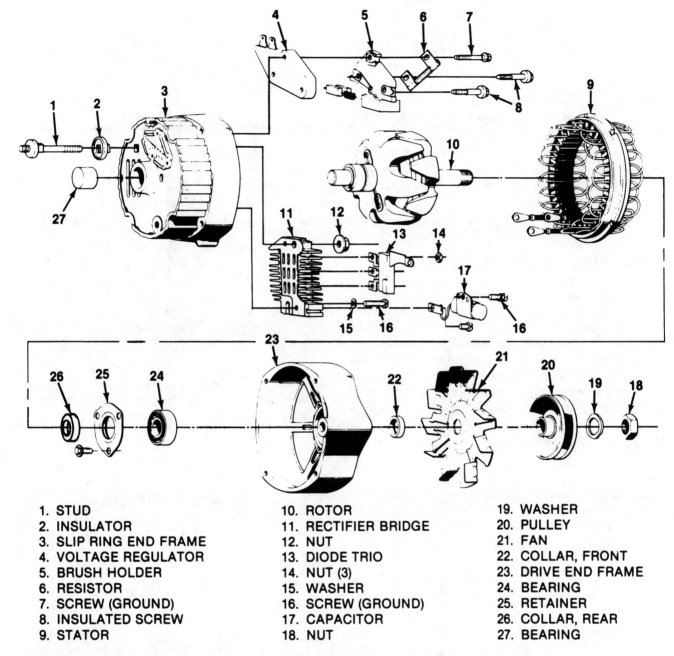

1. STUD
2. INSULATOR
3. SLIP RING END FRAME
4. VOLTAGE REGULATOR
5. BRUSH HOLDER
6. RESISTOR
7. SCREW (GROUND)
8. INSULATED SCREW
9. STATOR
10. ROTOR
11. RECTIFIER BRIDGE
12. NUT
13. DIODE TRIO
14. NUT (3)
15. WASHER
16. SCREW (GROUND)
17. CAPACITOR
18. NUT
19. WASHER
20. PULLEY
21. FAN
22. COLLAR, FRONT
23. DRIVE END FRAME
24. BEARING
25. RETAINER
26. COLLAR, REAR
27. BEARING

FIGURE 9–3 Components of a brush-type alternator with an integral regulator. (Courtesy of General Motors Corporation.)

1. *Rotor:* includes the shaft, a field winding, enclosed by two sets of magnetic claw poles (one north and the other south), and two slip rings. One slip ring is connected to one end of the field winding and the other to the opposite end (10 in **Figure 9–3**).

2. *Stator assembly:* three sets of stationary copper wire windings positioned on a circular stator core. The stator windings have output lead connections (9).

3. *Rectifier or bridge assembly:* diodes mounted in a heat sink or diode plate with electrical connections (11).

4. *Brush holder assembly:* housing with two brushes, brush springs, and wire leads. Current is fed from the battery to the slip rings to energize the field winding (5).

5. *Drive end frame:* front housing and bearing assembly with mounting provision (23).

6. *Rectifier or slip ring end frame:* rear housing and bearing assembly with mounting provision. The rectifier and brush holder assemblies are usually mounted in this housing (3).

7. *Fan and pulley:* includes fan for cooling, drive pulley, spacer, washer, and nut (21 and 20).

8. *Integral regulator:* voltage regulator mounted in rectifier end housing or in computer (4).

9. *Diode trio:* used on some alternators to feed the rotor field from stator winding output. Usually mounted in the heat sink or end frame (13).

ALTERNATOR OPERATION

Single-Phase Stator Operation

A simple alternator having a single loop of wire to represent the stator winding serves to illustrate how an alternating current is produced (**Figures 9–4** and **9–5**). An alternator of this type is called a single-phase alternator, regardless of the number of turns of wire in the stator winding. When all the windings in the stator are connected in series to form one continuous circuit, the term *single-phase* applies.

Three-Phase Stator Operation

The stator assembly in the alternator has three sets of windings. Each winding has one terminal or end that is independent of the others. The other end of each winding is connected to form an insulated junction called a Y connection or neutral junction. The single-phase AC

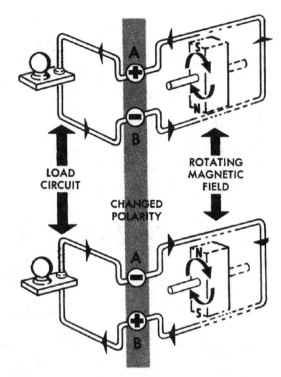

FIGURE 9–5 Alternators produce alternating (AC) current that must be rectified to direct current (DC). (Courtesy of General Motors Corporation.)

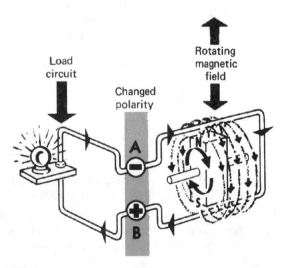

FIGURE 9–4 Single-phase alternator operation. (Courtesy of General Motors Corporation.)

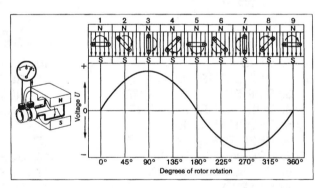

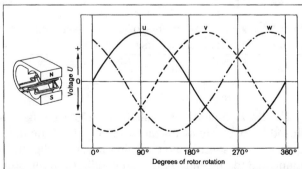

FIGURE 9–6 Single-phase AC voltage output (top) and three-phase output (bottom). (Courtesy of Robert Bosch Canada Ltd.)

ALTERNATOR OPERATION

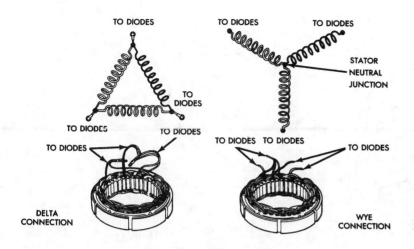

FIGURE 9-7 (Courtesy of General Motors Corporation.)

voltage is produced between any two of the open terminals. Combining these three single phases forms the three-phase-connected stator. This means that it produces three overlapping sets of current. Some alternators use a delta connection of the three phases **(Figures 9-6 and 9-7)**.

Rectifying the Current

Six silicone-diode rectifiers are used to rectify the AC output of the three-phase-connected alternator to direct current. Three of these positive diodes are mounted on a heat sink that is called a *positive rectifier assembly*. This rectifier assembly is isolated from the alternator end shield and connected to the alternator output BAT terminal of the alternator. The other three diodes have negative polarity and are mounted on a heat sink and are called a *negative rectifier assembly*. This rectifier assembly is mounted directly to the alternator end shield, which is a ground. Both rectifier assemblies are mounted in the airstream to provide adequate cooling of the diodes.

The diodes are connected into the alternator circuit between the stator winding and the battery and the vehicle electrical load. This arrangement provides a smooth direct current. The diodes also provide a blocking action to prevent the battery from discharging through the alternator. There are two diodes in each phase: one diode allows current in one direction, and the other diode in the same phase allows current in the opposite direction **(Figures 9-8 and 9-9)**.

A capacitor is connected from the output BAT terminal to ground. It is used to absorb any peak voltages and thus protects the rectifiers and helps reduce radio interference.

Energizing the Field

Battery current is fed to the field coil in the rotor through a brush and slip ring. The slip ring is connected to one end of the field coil. The other end of the field coil is connected to the other slip ring and a second brush. This brush is connected to ground to complete the circuit **(Figure 9-10)**. Current to the field coil is controlled by the voltage regulator or through the computer on computerized systems.

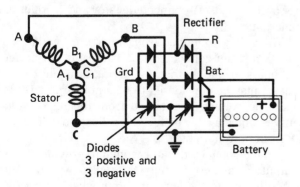

FIGURE 9-8 Diodes rectify alternating current. (Courtesy of General Motors Corporation.)

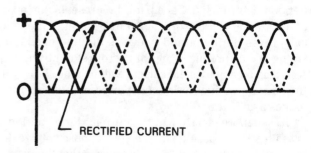

FIGURE 9-9 Rectified current output. (Courtesy of Ford Motor Company.)

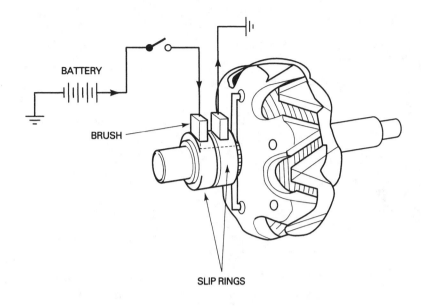

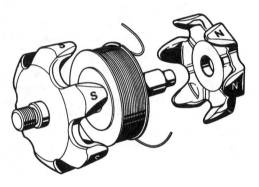

FIGURE 9-10 Current is fed to the rotor field winding through the brushes and slip rings. Note polarity of claw poles.

Voltage Production

This discussion follows the principle of positive voltage output. If desired, negative voltage output may be applied to the diagrams (**Figures 9-11** and **9-12**) and the discussion by reversing the direction of the arrows in the diagrams and by reversing the direction of current in the discussion. For convenience, the three AC voltage curves provided by the Y-connected stator for each revolution of the rotor have been divided into six periods, 1 through 6. Each period represents one-sixth of a rotor revolution, or 60°. An inspection of the voltage curves during period 1 reveals that the maximum voltage being produced appears across stator terminals BA. This means that the current will be from B to A in the stator winding during this period and through the diodes as illustrated.

To see more clearly why the current during period 1 is as illustrated, assume that the peak phase voltage developed from B to A is 16 V. This means that the potential at B is 0 V, and the potential at A is 16 V. Between periods 1 and 2, the maximum voltage being impressed across the diodes changes or switches from phase BA to phase CA.

Taking the instant of time at which this voltage is 16 V, the potential at A is 16 and at C is zero. Following the same procedure for periods 3 to 6, the current conditions can be determined, and they are shown in the illustrations. These are the six major current conditions for a three-phase Y-connected stator and rectifier combination.

The voltage obtained from the stator–rectifier combination when connected to a battery is not perfectly flat but is so smooth that, for all practical purposes, the output may be considered to be a nonvarying DC voltage. The voltage is obtained from the phase voltage curves and can be pictured as illustrated in **Figure 9-11** at bottom right.

A delta-connected stator wound to provide the same output as a Y-connected stator also will provide a smooth voltage and current output when connected to a six-diode rectifier. For convenience, the three-phase AC voltage curves obtained from the basic delta con-

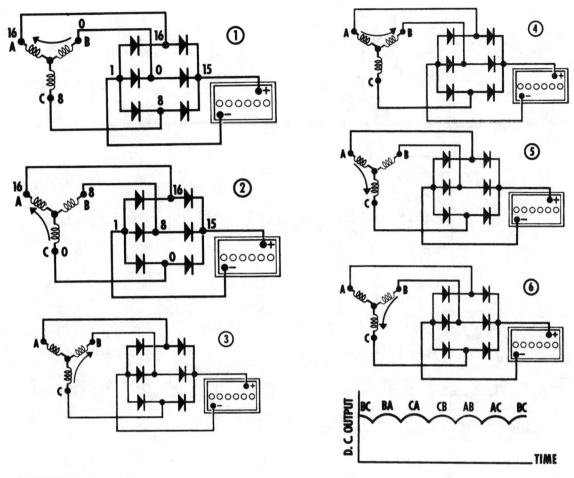

FIGURE 9-11 (Courtesy of General Motors Corporation.)

nection for one rotor revolution are reproduced here and have been divided into six periods.

During period 1, the maximum voltage being developed in the stator is in phase BA. To determine the direction of current, consider the instant at which the voltage during period 1 is at maximum, and assume this voltage to be 16 V. The potential at B is zero and at A is 16 V.

An inspection of the delta stator, however, reveals a major difference from the Y stator. Whereas the Y stator conducts current through only two windings throughout period 1, the delta stator conducts current through all three. The reason for this is apparent, since phase BA is in parallel with phase BC plus CA. Note that since the voltage from B to A is 16 V, the voltage from B to C to A must also be 16 V. This is true because 8 V is developed in each of these two phases.

During period 2, the maximum voltage developed is in phase CA, and the voltage potentials are shown in **Figure 9-12** at the instant the voltage is maximum. Also shown are the other phase voltages, and again, the current through the rectifier is identical to that for a Y stator, since the voltages across the diodes are the same. However, as during period 1, all three delta phases conduct current as illustrated in **Figure 9-12**.

Following the same procedure for periods 3 to 6, the current directions are shown. These are the six major current conditions for a delta stator.

Alternator Current Control

The alternator is self-limiting in its maximum current output. This occurs as the magnetic field produced by the current in the stator windings opposes in polarity and approaches in value the magnetic field provided by the rotor as the generator output increases. This causes the generator to limit its own output to a maximum value.

154 Chapter 9 CHARGING SYSTEM PRINCIPLES AND SERVICE

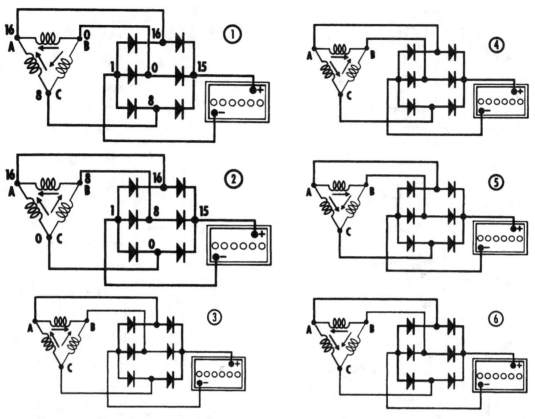

FIGURE 9-12 (Courtesy of General Motors Corporation.)

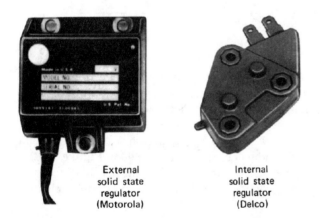

FIGURE 9-13 Examples of electronic voltage regulators. (Courtesy of Deere and Company.)

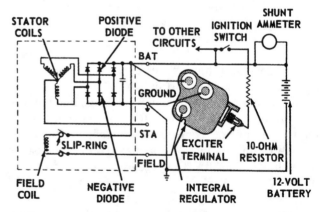

FIGURE 9-14 Charging system wiring diagram with integral regulator. (Courtesy of Ford Motor Company.)

VOLTAGE REGULATION

The voltage delivered by an alternator must be regulated to protect the charging circuit. This is accomplished with a vibrating-type electromechanical regulator, with a transistorized regulator, or currently on most systems with an integral electronic regulator that incorporates transistors, diodes, and resistors in an arrangement that accomplishes the same end (**Figures 9-13 to 9-20**).

VOLTAGE REGULATION 155

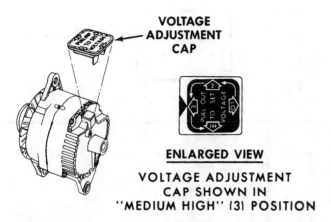

FIGURE 9-15 The voltage adjustment cap can be installed in any of four positions to achieve specified voltage output on some alternators. (Courtesy of General Motors Corporation.)

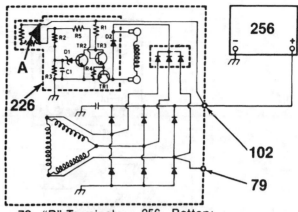

79. "R" Terminal 256. Battery
102. Output Terminal A. Voltage Adjustment
226. Voltage Regulator

FIGURE 9-16 Wiring diagram for 400 series alternator with adjustable voltage regulator. (Courtesy of General Motors Corporation.)

Voltage Output Factors

The following factors affect the magnitude of voltage generated:

1. Voltage will increase as the speed of the rotor is increased.
2. Voltage will increase as the strength of the rotor magnetic field is increased.
3. The strength of the rotor magnetic field may be increased by the following:
    a. The number of turns and type of wire used in the rotor.
    b. The air gap between the rotor poles and stator. Reducing the air gap increases the strength of the field.
    c. The voltage applied to the rotor through the slip rings and brushes.
4. Voltage will increase as the number of turns of wire in the stator winding is increased. This is because more conductors will be cut by the lines of force from the rotor.

Charge Indicators

A charge indicator is mounted in the dash to inform the driver about charging system operation. There are three kinds of charge indicators.

1. *Warning light:* light is off during normal charging system operation and on when a problem exists.
2. *Voltage indicator:* indicates alternator output voltage.
3. *Ammeter:* indicates alternator output in amperes.

CHARGING SYSTEM WARNING LIGHT

Most vehicles and equipment have a charging system warning light. When the starting switch is first turned on, current is provided to the alternator field through the charge indicator lamp. This turns the indicator lamp on. When the engine is started, the alternator begins charging. A circuit between the alternator output and the indicator lamp is connected to the indicator lamp on the side opposite the ignition switch connection. This allows alternator output current to flow to the lamp in a direction opposite that provided by the ignition switch. When opposing alternator output current reaches the specified level, the indicator lamp goes out **(Figure 9-19)**. Any time vehicle or equipment electrical components use more current than the alternator is able to produce, the lamp lights up.

When the start switch is turned on, battery current flows through the ignition switch, through the alternator indicator lamp and resistor to the voltage regulator I terminal. From there the current can be traced to the voltage limiter upper contacts, via internal connections, and out through the voltage regulator F terminal to the alternator FLD terminal (rotor coil). There is enough current through the circuit to prime the alternator so that when the engine is started, the magnetism in the field will cause the alternator to start generating voltage.

While engine rpm is increasing from cranking speed to idle speed, the alternator develops sufficient voltage

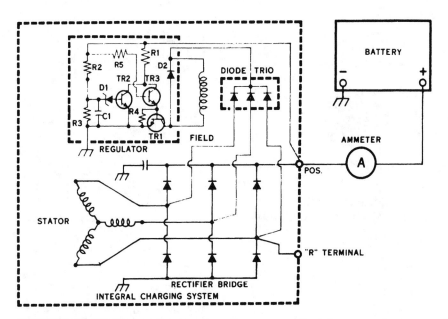

FIGURE 9-17 Wiring diagram for alternator with delta-wound stator and diode trio. (Courtesy of General Motors Corporation.)

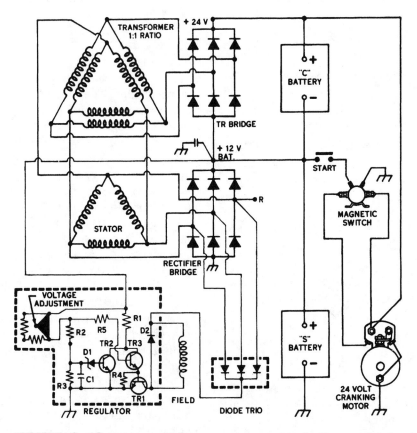

FIGURE 9-18 Wiring diagram for system with 24 V starting circuit and 12 V charging system. C, cranking battery; S, starting battery. The batteries are connected in series for 24 V starting. The transformer circuit eliminates the need for a series–parallel switch. (Courtesy of General Motors Corporation.)

VOLTAGE REGULATION 157

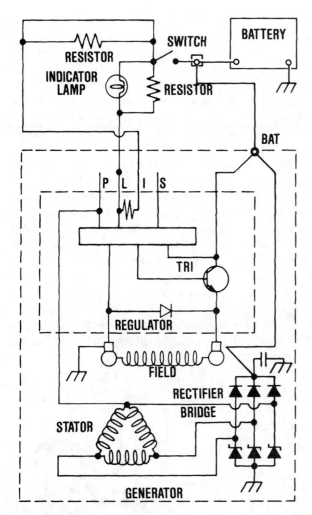

FIGURE 9–19 Delco model CS alternator wiring diagram. Compare with Figure 9–20. (Courtesy of General Motors Corporation.)

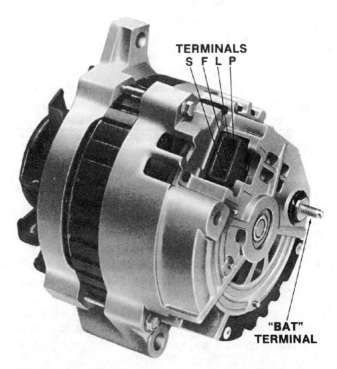

FIGURE 9–20 Wiring connections for a Delco model CS alternator. Compare with Figure 9–19. (Courtesy of General Motors Corporation.)

at its STA terminal to close the field relay contacts. This action turns out the alternator indicator lamp and applies battery voltage to the alternator field via the voltage regulator A terminal, the field relay contacts, and the voltage limiter upper contacts. Some units provide added field current through a diode trio (see **Figures 9–17** and **9–18**).

ELECTRONIC VOLTAGE REGULATOR

The electronic voltage regulator (**Figures 9–13** to **9–19**) is a solid-state unit with no moving parts or adjustment. It is serviced only by replacement. This regulator governs the electrical system voltage by limiting the output voltage that is generated by the alternator. The regulator does this by controlling the value of field current that it allows to pass through the field windings. Basically, the electronic regulator operates as a voltage-sensitive switch.

The electronic regulator contains several transistors, diodes, resistors, and a capacitor. A large transistor is placed in series with the alternator field winding and a control circuit that senses the system voltage. This control unit turns the transistor on and off many times a second to keep the field current and alternator output voltage at a proper level.

VOLTAGE INDICATOR

The voltage indicator is connected to the alternator output circuit either directly or through an induction pickup. Voltage output is an indicator of current output. When output voltage drops below a specified value, a charging system problem is indicated. Since battery voltage is approximately 12.6 V when fully charged, alternator voltage output must be greater than 12.6 V to maintain a fully charged battery.

AMMETER

The ammeter type of indicator shows alternator current output in amperes. With a normally operating charging system the ammeter needle will be to the right or positive side of the midpoint on the gauge while the battery is being charged. With a fully charged battery the needle will be very near the zero point on the gauge. When the battery is being discharged, the needle will be to the left or negative side of the gauge. A charging system problem is indicated if the ammeter needle is always on the left side. The ammeter is connected to the alternator output circuit either directly or through induction pickup (See **Figure 9-17**).

BRUSHLESS ALTERNATOR

The following description applies to the Delcotron type 400 alternator, which is typical of this design (**Figures 9–21** and **9–22**).

The Integral Charging System is a self-rectifying, brushless unit featuring a built-in voltage regulator. The only movable part in the assembly is the rotor, which is mounted on a ball bearing at the drive end, and a roller bearing at the rectifier end. All current-carrying conductors are stationary. These conductors are the field winding, the stator windings, the six rectifying diodes, and the regulator circuit components. The regulator and diodes are enclosed in a sealed compartment.

A fan located on the drive end provides airflow for cooling. Extra-large grease reservoirs contain an adequate supply of lubricant so that no periodic maintenance of any kind is required.

Only one wire is needed to connect the Integral Charging System to the battery, along with an adequate ground return. The specially designed output terminal is connected directly to the battery. A red output terminal is used on negative ground models and is to be connected only to battery positive. A black output terminal is used on positive ground models and is to be connected only to battery negative. An R terminal is provided for use in some circuits to operate auxiliary equipment.

The hex head bolt on the output terminal is electrically insulated; no voltage reading can be obtained by connecting to the hex head.

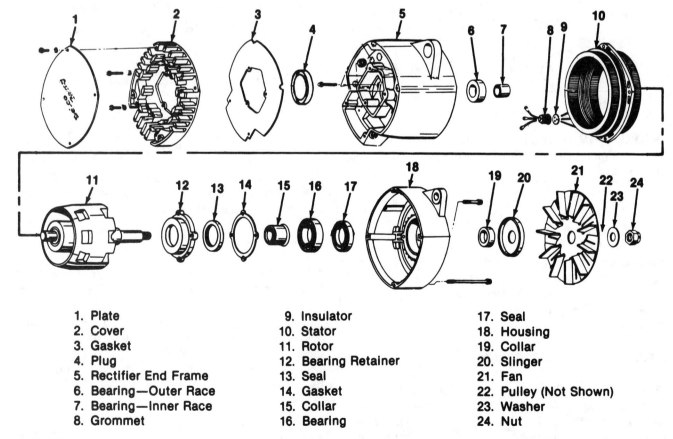

1. Plate
2. Cover
3. Gasket
4. Plug
5. Rectifier End Frame
6. Bearing—Outer Race
7. Bearing—Inner Race
8. Grommet
9. Insulator
10. Stator
11. Rotor
12. Bearing Retainer
13. Seal
14. Gasket
15. Collar
16. Bearing
17. Seal
18. Housing
19. Collar
20. Slinger
21. Fan
22. Pulley (Not Shown)
23. Washer
24. Nut

FIGURE 9–21 Components of a brushless alternator with an integral regulator. (Courtesy of General Motors Corporation.)

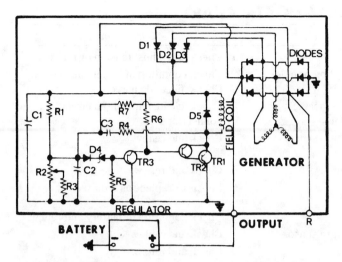

FIGURE 9-22 Wiring schematic of a brushless integral charging system. (Courtesy of General Motors Corporation.)

Operating Principles

The wiring diagram in **Figure 9-22** applies to generator models with an internal voltage adjustment potentiometer mounted on the voltage regulator. The generator rear cover has to be removed for access to this adjustment.

The basic operating principle for the generator is as follows:

As the rotor begins to turn, the permanent magnetism therein induces voltages in the stator windings. The voltages across the six diodes cause current to flow to charge the battery.

Current from the stator flows through the three diodes to resistor R6 and the base–emitter of TR2 and TR1 to turn these transistors on. Current also flows from the stator through the diode trio D1, D2, and D3, the field coil and transistor TR1, returning to the stator through the other three diodes. All stator current, except through the diode trio D1, D2, and D3, flows through the six diodes connected to the stator.

Current flow through R1, R2, and R3 causes a voltage to appear at zener diode D4. When the voltage becomes high enough due to increasing generator speed, D4 and the base–emitter of TR3 conduct current, and TR3 turns on. TR2 and TR1 then turn off, decreasing the field current, and the system voltage decreases. The voltage at D4 decreases, D4 and TR3 turn off, TR2 and TR1 turn back on, and the system voltage increases. This cycle then repeats many times per second to limit the system voltage as determined by the setting of the potentiometer R2, R3.

Capacitor C1 protects the generator diodes from high transient voltages and suppresses radio interference.

Resistor R5 prevents current leakage through TR3 at high temperatures.

Diode D5 prevents high transient voltages in the field coil when the field current is decreasing.

Resistor R7, capacitor C3, and resistor R4 all act to cause transistors TR2 and TR1 to turn on and off more quickly.

CHARGING SYSTEM SERVICE PRECAUTIONS

1. Always disconnect the negative battery cable from the battery before connecting a battery charger and before removing any charging system components.

2. Always observe battery polarity when making connections. Reversing polarity can damage electronic components.

3. Do not accidentally ground any electrical system components. Grounding an electrical terminal can cause damage. Never ground or short an electrical terminal unless instructed to do so by the service manual.

4. Do not operate the alternator on open circuit (disconnected output). Uncontrolled output may cause alternator damage.

5. Do not attempt to polarize the alternator unless directed to do so in the service manual. Damage will result if attempted.

6. Follow the service manual and equipment manufacturer's instructions to perform charging system tests.

7. Never remove either battery cable with the engine running. Serious electrical system damage could result.

PROBLEM DIAGNOSIS

Charging system problems result in one or more of the following symptoms.

1. *Dead battery:* will not crank the engine.

2. *Abnormal noise:* belt squeal, noisy alternator bearing, loose mounting bolts.

3. *Overcharged battery:* top of battery moist from gassing, corroded battery terminals. Premature light burnout and possible electronic system damage may result.

CHARGING SYSTEM DIAGNOSTIC CHART

Problem	Possible Cause	Correction
Battery low in charge or discharged	1. Loose or worn alternator drive belt 2. Defective battery not accepting or holding charge; electrolyte level low 3. Excessive resistance due to loose charging system connections 4. Defective battery temperature sensor (where fitted) 5. Defective voltage regulator 6. Defective alternator	1. Check and adjust tension or renew. 2. Check condition of battery and renew. Check, fill, and charge. 3. Check, clean, and tighten circuit connections. 4. Check and renew. 5. Check and renew. 6. Repair or replace as required.
Alternator charging at high rate (battery overheating)	1. Defective battery 2. Defective battery temperature sensor 3. Defective voltage regulator 4. Defective alternator	1. Check condition of battery and renew. 2. Check and renew. 3. Check and renew. 4. Repair or replace as required.
No output from alternator	1. Alternator drive belt broken 2. Loose connection or broken cable in charging system 3. Defective temperature sensor (where fitted) 4. Defective voltage regulator 5. Defective alternator	1. Renew and tension correctly. 2. Inspect system, tighten connections, and repair or renew faulty wiring. 3. Check and renew. 4. Check and renew. 5. Repair or replace as required.
Intermittent or low alternator output	1. Alternator drive belt slipping 2. Loose connection or broken cable in charging system 3. Defective temperature sensor (where fitted) 4. Defective voltage regulator 5. Defective alternator	1. Check and adjust tension or renew. 2. Inspect system, tighten connections, and repair or renew faulty wiring. 3. Check and renew. 4. Check and renew. 5. Repair or replace as required.
Warning light dimming and/or battery low	1. Faulty external charging circuit connections 2. Faulty rotor slip rings or brushes	1. Inspect system, clean and tighten connections. 2. Inspect and repair or renew.
Warning light going out or becoming brighter with increased speed	1. Faulty external charging circuit connections 2. Faulty rectifier or rectifying diodes	1. Inspect system, clean and tighten connections. 2. Check and renew.
Warning light normal but battery boiling	1. Defective voltage regulator 2. Faulty battery temperature sensor (where fitted)	1. Check and renew. 2. Check and renew.
Warning light normal but battery discharged	1. Defective voltage regulator 2. Faulty stator 3. Faulty rectifier or rectifying diodes	1. Check and renew. 2. Check and renew. 3. Check and renew.

FIGURE 9–23

Warning light illuminated continuously and/or flat battery	1. Loose or worn alternator drive belt 2. Defective surge protection diode (where fitted) 3. Defective isolation diodes (where fitted) 4. Faulty battery temperature sensor (where fitted) 5. Faulty rotor, slip rings, or brushes 6. Faulty voltage regulator 7. Defective stator 8. Defective rectifier or rectifying diodes	1. Check and adjust tension or renew. 2. Check and renew. 3. Check and renew. 4. Check and renew. 5. Inspect, repair, or renew. 6. Check and renew. 7. Inspect and renew. 8. Check and renew.
Warning light extinguished continuously and/or flat battery	1. Burned-out bulb 2. Alternator internal connections 3. Defective voltage regulator 4. Faulty rotor, slip rings, or brushes 5. Defective stator	1. Check and renew. 2. Inspect and test circuitry; repair or renew. 3. Check and renew. 4. Check, repair, or renew. 5. Check and renew.
Warning light flashing intermittently	1. Faulty external charging circuit 2. Alternator internal connections	1. Inspect circuit, clean and tighten connections, repair or renew faulty wiring. 2. Inspect and test circuitry repair or renew.
Warning light dimming continuously and/or flat battery	1. Defective rotor, slip rings, or brushes 2. Defective voltage regulator	1. Check, repair, or renew. 2. Check and renew.

FIGURE 9–23 *(cont'd)*

4. *Charge indicator reading is abnormal:* indicator light stays on, or gauge reading is always low.

See **Figure 9–23** for typical charging system problems, causes, and corrections. See **Figures 9–24** to **9–26** for typical charging system testers.

PRELIMINARY INSPECTION

Look for obvious signs of trouble—abnormal conditions that you can see or detect with your hands.

1. Check the alternator drive belt for proper tension, glazing, cracks, or other damage. Adjust the belt tension to specifications or replace the belt if necessary.
2. Inspect the battery and battery cables for loose connections, corroded terminals, or other damage. Check the battery state of charge at the built-in hydrometer.
3. Inspect the charging system wiring for loose or corroded connections and insulation damage. Check whether connections are tight by wiggling the wires.
4. With the engine running, check for any abnormal noises. A glazed or loose drive belt may squeal when accelerating. Check for alternator bearing noise with a stethoscope. Running the engine briefly with the belt removed will determine whether the belt or alternator is causing the noise.

CHARGING SYSTEM TESTS

Voltmeter Tests

Voltmeter testing of the charging system can be used to indicate voltage and current output. These tests are easily done and require only a voltmeter.

1. Turn the headlights on for 15 seconds to remove any surface charge from the battery.
2. Turn the voltmeter knob to the 20 V scale, or nearest thereto on 12 V systems.

3. With the engine and all accessories off, connect the positive voltmeter lead to the positive battery post and the negative lead to the negative post and note the voltmeter reading. It should be about 12.6 V with a fully charged battery in good condition. This is the base or open circuit voltage reading.

4. Start the engine and turn off all accessories. Run the engine at 1500 to 2000 rpm and note the voltmeter reading. This is the no-load test. The voltmeter reading should be a minimum of 0.5 V higher than base voltage. If the voltage increase is more than 3 V, the charging rate is probably too high. The regulator or wiring is faulty. If the voltage does not increase, the alternator, regulator, or wiring is faulty.

5. If the no-load test results are normal, a full-load test should be performed. Start the engine, turn all electrical accessories, lights, heater, and air conditioning blowers on high, and turn headlights on high beam, to apply a full load to the charging system. Run the engine at 2000 rpm and note the voltmeter reading. It should be 0.5 V higher than base voltage. This indicates that the charging system produces enough current to operate all the electrical devices and is able to charge the battery as well. If the load test voltage is not 0.5 V above base voltage, perform a regulator bypass test.

Current Output Test

1. Make sure that the battery is in good condition and is fully charged (see Chapter 8 for battery testing)

Charging System Analyzer

Now you can diagnose suspected charging system problems quickly and easily with this pocket-sized analyzer. It's designed for use on any vehicle with a 12 volt, positive or negative grounded alternator charging system. In seconds you can discover if the battery and regulator are working properly and single out an open or shorted rectifier, rotor field, stator or diode trio. Solid state design provides instant read-out; no meters or scopes to decipher. To use, simply attach the clip-on connectors to the alternator and start the engine—it's like a charging system test stand in your pocket!

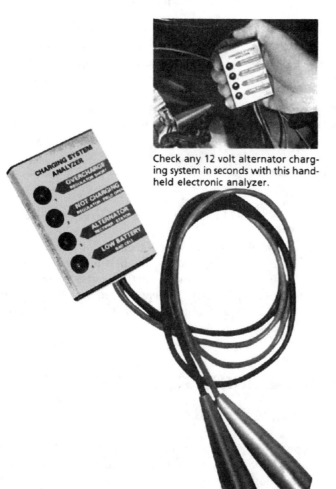

Check any 12 volt alternator charging system in seconds with this hand-held electronic analyzer.

REGULATOR — Checks out internal or conventional regulators.

RECTIFIER BRIDGE — Readout displays open or shorted heat sink.

STATOR/FIELD — Open/shorted field or stator winding shown.

DIODE TRIO — Check possible diode trio malfunction.

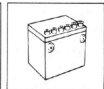

BATTERY — Determine low voltage or possible bad cell quickly.

FIGURE 9–24 Hand-held charging system analyzer allows quick diagnosis of system. (Courtesy of Kent Moore, SPX Corporation.)

CHARGING SYSTEM TESTS   163

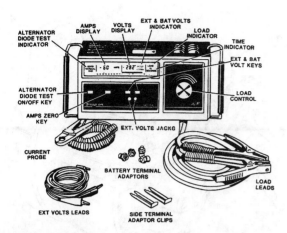

FIGURE 9–25 AVR tester. (Courtesy of Snap-on Tools Corporation.)

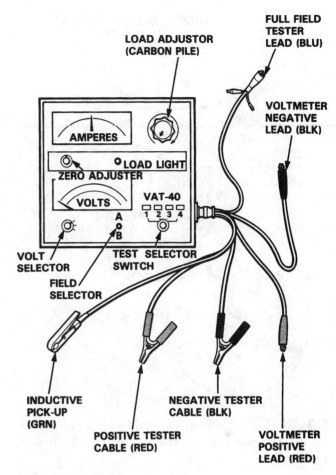

FIGURE 9–26 Charging system tester. (Courtesy of Sun Test Equipment.)

and that the alternator drive belt is in good condition and adjusted to proper tension.

2. Obtain the proper alternator current and voltage output specifications from the service manual. (Output may range from about 14.0 to 15.0 V).

3. Connect the tester according to the test equipment instructions.

4. Start the engine and turn off all accessories.

5. Increase engine speed and hold it at 2000 rpm.

6. Turn the load control knob to increase the load until the highest possible reading is obtained on the ammeter while maintaining a voltmeter reading of 12 V or slightly above 12 V.

7. Amperage reading should be within range of rated output. Add 10 to 15 A to the reading to compensate for engine operation. Turn the load control knob off as soon as you have taken the ammeter reading.

Regulator Bypass Test

When the charging system fails the output tests, a regulator bypass test should be done where possible. The general procedure is the same as for the current output test except that full field current is applied to the alternator. This bypasses the regulator and determines whether the alternator or regulator is at fault. The procedure is also known as *full fielding* an alternator. Full fielding makes the alternator produce maximum output.

Several methods are used to apply full field current to an alternator, depending on charging system design. One method is to ground the field with a screwdriver through an access hole at the back of the alternator. This method is used on many GM alternators with an internal regulator (**Figure 9–27**). Another method uses a jumper wire to connect the battery terminal to the field terminal at the back of the alternator (**Figure 9–28**). Always follow service manual procedures for full fielding an alternator. Never ground or connect any two terminals to each other unless instructed to do so by the service manual. Making the wrong connections can cause serious damage to the system.

Circuit Resistance Tests

Circuit resistance tests are used to pinpoint excessive resistance in the charging circuit. Too much resistance on the insulated side of the circuit or the grounded side results in reduced output. Corroded or loose terminals or plug-in connectors and poor wire insulation are common causes of too much resistance.

To perform a ground circuit resistance test, full field or load the alternator and connect a voltmeter to the

164 Chapter 9 CHARGING SYSTEM PRINCIPLES AND SERVICE

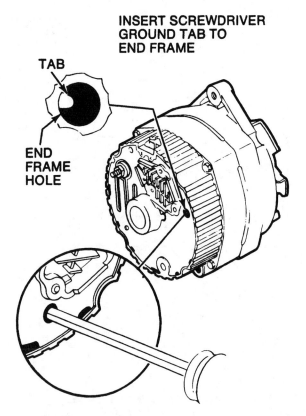

FIGURE 9-27 Grounding the field with a screwdriver on a Delco alternator. (Courtesy of General Motors Corporation.)

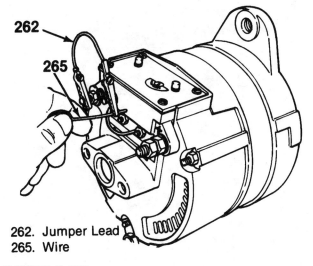

FIGURE 9-28 Full fielding a Leece Neville alternator. (Courtesy of General Motors Corporation.)

negative battery terminal and the alternator housing (**Figure 9-29**). The voltmeter should not read above 0.1 V. If it does, check for loose or corroded connections.

To perform an insulated circuit test, connect the tester as outlined in the service manual. Insulated circuit tests are not recommended on some charging systems. The general procedure is to connect a voltmeter across each insulated circuit connection with the charging system under load. Voltage drop across all connections should not exceed 0.5 V.

Voltage Drop (Circuit Resistance) Test (Typical) (Courtesy of Detroit Diesel Corporation)

1. With the engine *not* running, connect a carbon pile to the alternator output terminal and the alternator housing as shown in **Figure 9-30**. (The carbon pile must be in the OFF position.)

NOTE: Use care when connecting the carbon pile to the alternator output terminal not to allow the pile clamp to touch ground. The output terminal is at battery voltage.

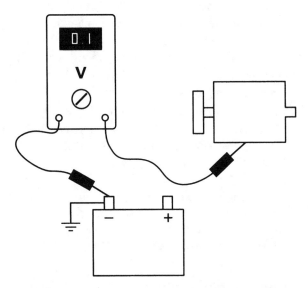

FIGURE 9-29 Ground circuit resistance test.

2. Connect a voltmeter across one of the batteries, red lead to positive and black lead to negative as shown at A in **Figure 9-30**.

3. With the engine *not* running, turn the knob of the carbon pile until the gauge on the carbon pile reads the alternator rated output in amps.

NOTE: The alternator's rated output is typically stamped on the alternator housing or on a name tag located on the alternator housing.

4. Read and record the battery voltage. Turn the carbon pile off.

5. Move the voltmeter to the alternator. Connect the red voltmeter lead to the alternator output terminal and the black voltmeter lead to the alternator housing as shown at B in **Figure 9–30**.

NOTE: Do not connect the voltmeter leads to the carbon pile leads.

6. Turn the carbon pile knob until the meter again reads alternator rated output in amps.

7. Read and record the voltage at the alternator. Turn the carbon pile off.

8. Subtract the voltage reading recorded at the alternator from the voltage reading obtained at the battery.

Voltage reading at battery	Voltage reading at alternator	Total voltage drop
_____ V −	_____ V =	_____ V

9. The result is a measure of the charging circuit voltage drop. The maximum allowable charging circuit voltage drop is 0.5 V for a 12 V system (1.0 V for a 24 V system).

10. If the voltage drop reading was below specification, the system is satisfactory. Go to the Alternator Output Test. If the reading was more than allowed, it will be necessary to determine if the voltage drop is located in the positive or negative side of the charging circuit. Go to the next step.

11. With the carbon pile still connected, connect a low-scale voltmeter (digital preferred) as shown in **Figure 9–31**.

12. Connect the red lead of the voltmeter to a battery positive terminal. Connect the black voltmeter lead to the alternator output terminal as shown in STEP 1, **Figure 9–31** (you may need a jumper wire to extend your voltmeter leads).

13. Turn the carbon pile knob until the gauge on the carbon pile reads the alternator rated output in amps.

14. Read and record the voltmeter reading. Turn the carbon pile off.

15. Connect the voltmeter leads to the negative side of the charging circuit as shown in STEP 2, **Figure 9–31**.

16. Connect the red voltmeter lead to the alternator housing and the black voltmeter lead to the battery negative terminal.

17. Turn the carbon pile knob until the gauge on the carbon pile reads the alternator rated output in amps.

18. Read and record the voltmeter reading. Turn the carbon pile off.

Positive circuit loss	Negative circuit loss	Total system loss
_____ V −	_____ V =	_____ V

19. The maximum allowable system loss is 0.5 V for a 12 V circuit (1.0 V for a 24 V system).

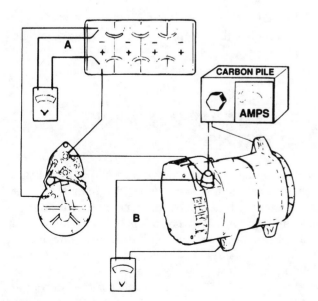

FIGURE 9–30 Hookup for voltage drop testing. (Courtesy of Detroit Diesel Corporation.)

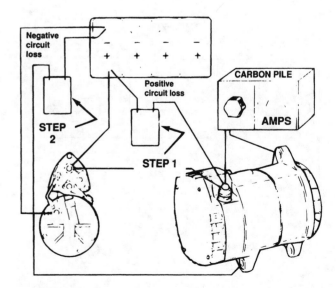

FIGURE 9–31 Voltage drop test. (Courtesy of Detroit Diesel Corporation.)

20. The preceding procedure should show which circuit has the excessive voltage loss. Remove the carbon pile and voltmeter. Repair or replace the problem portion of the circuit.

Alternator Output Test

1. The engine must be at shop temperature. Connect a charging–starting system analyzer, with a voltmeter and ammeter, to the vehicle. Connect the voltmeter leads to one of the batteries, observing proper polarity. If the analyzer has an inductive pickup, place it around the alternator output wire **(Figure 9–32)**.

2. If the charging–starting system analyzer does not have a carbon pile, connect a carbon pile across one of the batteries.

NOTE: On 24 V vehicles, connect the carbon pile across one 12 V battery. Connect the voltmeter across the normal 24 V battery connection.

If the wiring in the circuit is adequate, adjust the alternator drive belts to the specifications found in the service manual. Tighten both the alternator mounting bolts (adjusting rod and pivot point), and be sure all the charging circuit cables and connections are clean and tight. Then test the alternator output as follows:

3. With *all* electrical loads turned off, start the engine and accelerate to a fast idle. Observe the voltmeter. When the voltage stabilizes (does not increase) for 2 minutes, read and record the voltage. Voltage should not exceed 15 V (30 V for a 24 V system). If the voltage exceeds the maximum allowable by 1 V, the voltage regulator is defective. Remove the alternator for repair.

Regulator Service

When charging system tests determine that the regulator is at fault, it must be replaced. Some externally mounted electronic regulators are adjustable. An adjusting screw is turned or the cap is repositioned to increase or decrease the voltage setting. To replace an externally mounted regulator, disconnect the negative battery cable, disconnect the wiring from the regulator, and remove the mounting screws. To install the new regulator, reverse this procedure.

ALTERNATOR REMOVAL

1. Disconnect the negative battery cable at the battery.

SAFETY CAUTION

Failure to observe this step may result in injury from the hot battery lead at the alternator.

---

2. Disconnect the wiring leads from the alternator.
3. Loosen the adjusting bolts and move the alternator to provide slack in the belt.
4. Remove the alternator drive belt.
5. Remove the bolts that retain the alternator.
6. Remove the alternator from the vehicle or equipment.
7. Replace or test and repair the alternator.

ALTERNATOR DISASSEMBLY AND TESTING

Some alternators can be repaired, while others must be replaced as a unit. Check the service manual to determine whether repair is recommended. Usually, the pulley and fan are removed from the old alternator and installed on the new one. This involves removing the pulley nut (if so equipped) and then removing the pulley. On others, the pulley is a press fit on the rotor shaft and is removed with a puller. Refer to the appropriate service manual for specific disassembly procedures, which vary depending on alternator design. Mark each section of the alternator before disassembly to ensure

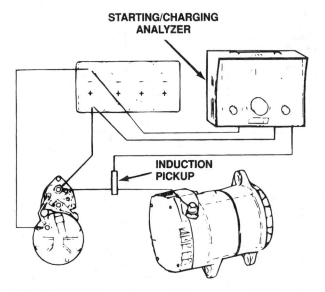

FIGURE 9–32 Alternator output test. (Courtesy of Detroit Diesel Corporation.)

ALTERNATOR DISASSEMBLY AND TESTING

proper alignment during assembly. Use a chisel or punch.

Stator Testing

The stator windings must be tested for open circuits or shorts to ground. To perform this test you must remove the stator leads from the rectifier for accurate results. If the leads are soldered, hold the lead with long-nose pliers and unsolder the connection. Do not blow away molten solder with compressed air. This could cause a short in the rectifier.

Touch each of the three leads with a continuity tester, ohmmeter, or tester light. If there is no continuity, replace the stator coil assembly. To check for a grounded stator, touch the test probes to one stator lead and the stator frame. If continuity is shown, the stator is grounded and must be replaced **(Figure 9–33)**.

Rotor Testing

The rotor field coil should be tested for continuity. Touch the tester probes to the slip rings. If there is no continuity, the field windings are open and the rotor must be replaced. Examine the slip rings for dirt or roughness. If the slip rings cannot be smoothed with light sanding, replace the rotor assembly. The rotor must also be tested for ground. Touch one test probe to a slip ring and the other to the rotor core. If there is no continuity, the coil is shorted to ground and the rotor must be replaced **(Figure 9–34)**.

Field Brush Testing

The field brushes should be inspected for excessive wear or damage, and replaced if needed. If visual inspection reveals no problem, test the brushes with an ohmmeter for continuity. If no continuity is observed, replace the brushes **(Figure 9–35)**.

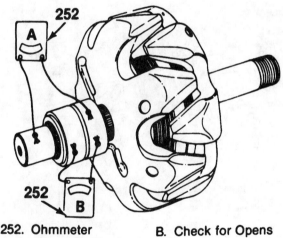

252. Ohmmeter B. Check for Opens
A. Check for Grounds

FIGURE 9–34 Testing the rotor with an ohmmeter. (Courtesy of General Motors Corporation.)

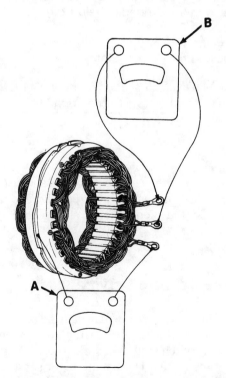

A. Ohmmeter Connections Checking for Grounds
B. Ohmmeter Connections Checking for Opens

FIGURE 9–33 Stator winding tests. (Courtesy of General Motors Corporation.)

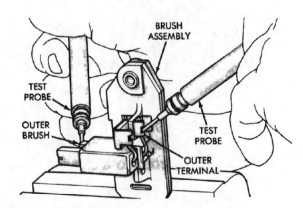

FIGURE 9–35 Field brush continuity test with ohmmeter.

Diode Testing

If a diode malfunctions, it can allow AC current to leave the alternator and possibly damage the vehicle's electrical system. There are two methods of testing diodes. If a special diode tester is being used, it is not necessary to disconnect the stator leads from the rectifier. If an ohmmeter or other continuity tester is being used, unsolder or unbolt the stator leads prior to testing.

Place one test probe on the diode heat sink and the other on the diode pin **(Figures 9–36** and **9–37)**. Reverse the probes and note the reaction. If the tester shows continuity one way but not the other, the diode is good. If there is no continuity, the diode is open. If

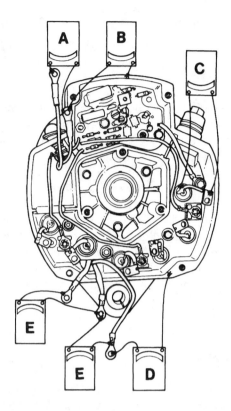

A. Checking for Shorts or Opens (Field Coil)
B. Checking for Grounds (Field Coil)
C. Checking Diodes
D. Checking for Grounds (Stator)
E. Checking for Opens (Stator)

FIGURE 9–37 Delco 25-SI alternator ohmmeter tests. (Courtesy of General Motors Corporation.)

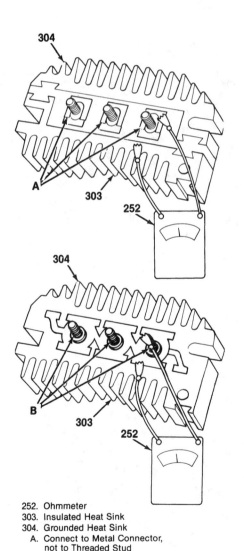

252. Ohmmeter
303. Insulated Heat Sink
304. Grounded Heat Sink
  A. Connect to Metal Connector, not to Threaded Stud
  B. Connect to Threaded Stud

FIGURE 9–36 Rectifier bridge tests. (Courtesy of General Motors Corporation.)

there is continuity both ways, the diode is shorted. Repeat this procedure on all pins in the positive and negative diodes. Test the diode trio as shown in **Figure 9–38**.

Alternator Bearing Service

The bearings in an alternator will usually be sealed ball bearings or needle bearings. Use the recommended bearing removal and installation tools to avoid damage to parts **(Figure 9–39)**. During alternator service, examine the bearings carefully for signs of wear or roughness when turning. If a needle bearing assembly is replaced, do not add any lubricant to the bearing. The bearing assembly will come prelubricated. Handle the assembly carefully so that dirt won't get into the needle bearing unit before installation.

ALTERNATOR ASSEMBLY

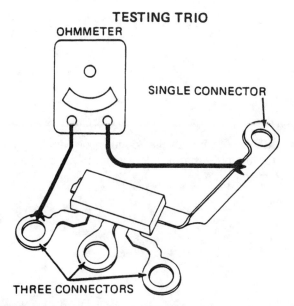

FIGURE 9-38 (Courtesy of General Motors Corporation.)

aligned. On some alternators the brushes must be held back against spring pressure with a small pin or Allen wrench. Push the brushes back far enough to allow the pin to be inserted through the holes (**Figure 9-40**). This keeps the brushes out of the way while the rotor is inserted into the end frame. On others the brushes may be installed from the outside after alternator assembly.

4. Install the through bolts, making sure that the three major components remain aligned and are fully seated against each other before tightening the through bolts.

5. Remove the wire or Allen wrench to allow the brushes to contact the slip rings.

6. Check to make sure that the rotor turns freely without any unusual noise.

7. Install any spacer, the fan, pulley, lock washer, and nut on the front of the shaft. Tighten the nut to specifications (**Figure 9-41**).

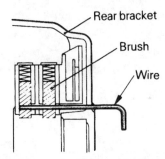

FIGURE 9-40 Wire holds brushes out of the way for alternator assembly. (Courtesy of General Motors Corporation.)

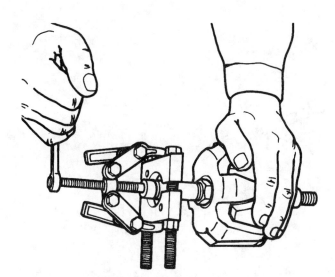

FIGURE 9-39 Using a puller and adapter to remove the alternator bearings. (Courtesy of General Motors Corporation.)

ALTERNATOR ASSEMBLY

Alternator assembly procedures vary with alternator design. Typically, the procedure is as follows.

1. Install front and rear bearings.
2. Install all the components in the slip ring end frame.
3. Assemble the slip ring end frame, stator, and drive end frame. Make sure that the aligning marks or pins are

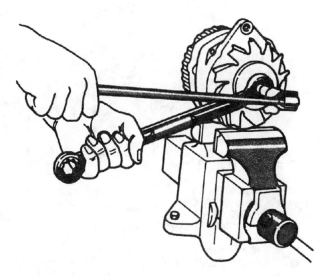

FIGURE 9-41 (Courtesy of General Motors Corporation.)

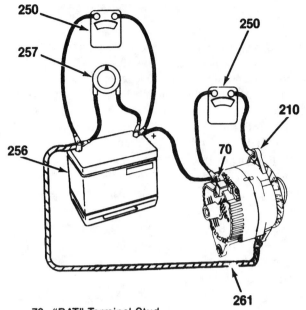

FIGURE 9-42 Bench test connections for a single-wire-terminal alternator. (Courtesy of General Motors Corporation.)

70. "BAT" Terminal Stud
210. Generator
250. Voltmeter
256. Battery
257. Carbon Pile
261. Ground Cable

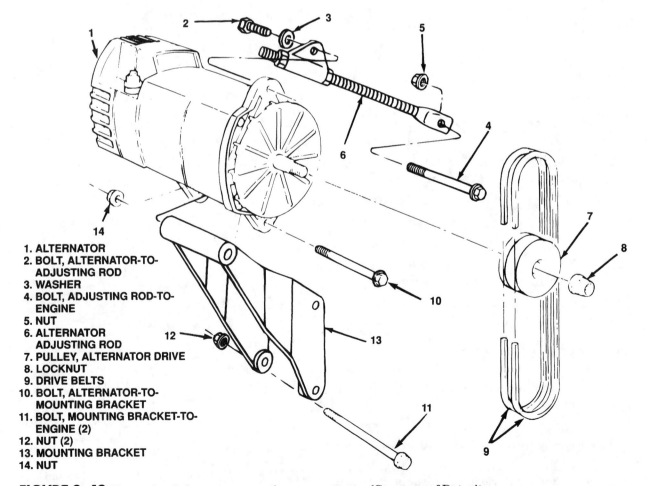

1. ALTERNATOR
2. BOLT, ALTERNATOR-TO-ADJUSTING ROD
3. WASHER
4. BOLT, ADJUSTING ROD-TO-ENGINE
5. NUT
6. ALTERNATOR ADJUSTING ROD
7. PULLEY, ALTERNATOR DRIVE
8. LOCKNUT
9. DRIVE BELTS
10. BOLT, ALTERNATOR-TO-MOUNTING BRACKET
11. BOLT, MOUNTING BRACKET-TO-ENGINE (2)
12. NUT (2)
13. MOUNTING BRACKET
14. NUT

FIGURE 9-43 Example of alternator mounting components. (Courtesy of Detroit Diesel Corporation.)

8. Test alternator output on a bench tester if available.

Alternator Bench Test

Bench testing the alternator may be done before disassembly to determine possible internal problems and to aid in diagnosis. It can also be done to verify alternator performance after repair and reassembly. The following procedure is typical for 12 V Delcotron alternators.

1. Mount the alternator securely in a suitable test bench and make the drive connection.

2. Using a fully charged battery to ensure accuracy, make the test connections as shown in **Figure 9–42** but leave the carbon pile disconnected until later.

3. Connect a resistor rated at 10 Ω and at least 6 W between terminal 1 on the alternator and the battery positive post (except on alternators with a single wire terminal). If the alternator was disassembled, the proper magnetism may have to be reestablished. To do this, make the normal alternator connections and momentarily make a "touch" connection with a jumper wire from the battery positive post to the alternator R terminal.

4. Start the test stand motor and increase the speed while observing the rise in voltage. If the voltage rises quickly above 15.5 V and does not change when the drive speed is changed, replace the voltage regulator.

5. If the voltage remains below 15.5 V, stop the motor and connect the carbon pile across the battery as shown.

6. Test the alternator at the specified speed while applying enough carbon pile load to produce maximum current output. Current output should be within

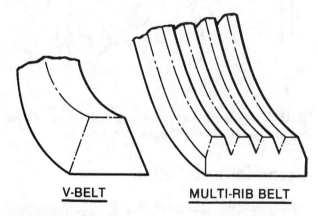

FIGURE 9–44 Alternator drive belt designs. (Courtesy of General Motors Corporation.)

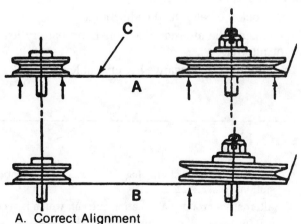

A. Correct Alignment
B. Incorrect Alignment
C. Cord Or Straight Edge

FIGURE 9–45 Pulley alignment is critical to belt operating life. (Courtesy of General Motors Corporation.)

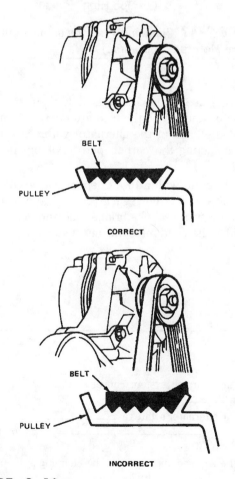

FIGURE 9–46 Correct and incorrect ribbed belt alignment. (Courtesy of Ford Motor Company.)

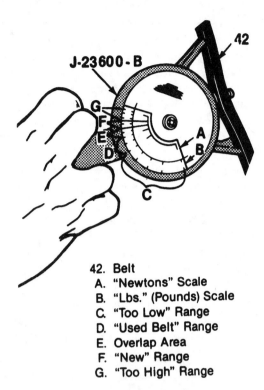

42. Belt
A. "Newtons" Scale
B. "Lbs." (Pounds) Scale
C. "Too Low" Range
D. "Used Belt" Range
E. Overlap Area
F. "New" Range
G. "Too High" Range

FIGURE 9–47 Checking V-belt tension. (Courtesy of General Motors Corporation.)

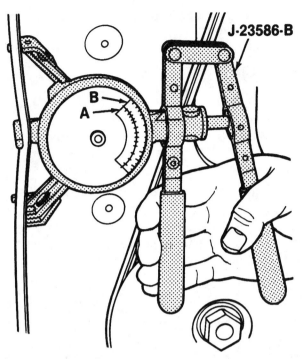

A. "Newtons"
B. "Lbs." (Pounds)

FIGURE 9–48 Checking multirib belt tension. (Courtesy of General Motors Corporation.)

10 A of rated output. If it is not, then ground the field by pushing in on the field ground tab through the hole at the back of the alternator with a screwdriver while adjusting the carbon pile to obtain maximum output.

7. If output within 10 A of specifications cannot be obtained, replace the regulator, check the field winding, diode trio, rectifier bridge, and the stator as described earlier, and retest the alternator.

Alternator Installation

1. Install the alternator to the mounting bracket with bolts, washers, and nuts **(Figure 9–43)**.
2. Install the drive belt **(Figures 9–44 to 9–46)**.
3. Tighten the belt to the specified belt tension. **(Figures 9–47 and 9–48)**.
4. Lock the belt tension adjustment.
5. Install the alternator terminal and battery leads to the alternator.
6. Connect the negative battery cable.

REVIEW QUESTIONS

1. The charging system converts _____ energy to _____ energy.
2. The alternator rotates a _____ _____ across _____ _____ to produce electricity.
3. The major components of the charging system are the _____, drive _____, _____ regulator, _____ indicator, storage _____, and wiring _____.
4. The three-phase stator has _____ separate stationary _____.
5. Alternators produce alternating current, which is rectified by _____ diodes to produce _____.
6. Each stator winding or phase is connected to _____ diodes.
7. Stator windings are connected to each other by the _____ or the _____ method.

8. The alternator is _____ _____ in its maximum current output.
9. The voltage regulator controls maximum _____ by controlling the _____ _____ value.
10. There are three kinds of charge indicators: the _____ _____, _____, and _____.
11. Computer-controlled charging systems have the _____ _____ located inside the computer.
12. Always disconnect the _____ battery cable first before performing charging system repairs.
13. List four common indicators of charging system problems.
14. If a booster battery or battery charger has been connected to the battery in reverse polarity, the _____ _____ protecting the _____ may be burned out.
15. Always test the _____ before proceeding with other charging system tests.
16. Battery base voltage should be about _____ V.
17. A no-load charging system test should result in an increase of at least _____ V over base voltage.
18. A charging system load test should show an increase of at least _____ V over base voltage.
19. A current output test maintains a voltmeter reading of _____ V while increasing the load until _____ current output is achieved.
20. During a regulator bypass test, full _____ _____ is applied to the alternator.
21. During a circuit resistance test the maximum voltage drop should not exceed _____ V.
22. The stator windings should be tested for _____ and _____.
23. The rotor field coil should be tested for _____ and _____.
24. A diode that has continuity both ways is _____.

TEST QUESTIONS

1. Alternator initial field current is fed from the battery to the
   a. diodes
   b. stator windings
   c. rotor windings
   d. heat sink
2. The alternator produces electrical current using the principle of
   a. mutual induction
   b. magnetic induction
   c. static induction
   d. solid-state induction
3. One of the most common alternator types uses a
   a. three-phase field winding
   b. phased-out field winding
   c. single-phase stator winding
   d. three-phase stator winding
4. The minimum number of diodes required in an alternator heat sink is
   a. 1
   b. 3
   c. 6
   d. 12
5. The alternator's current-producing winding is a
   a. rotating single-phase winding
   b. rotating three-phase winding
   c. stationary three-phase winding
   d. stationary single-phase winding
6. Alternating current is changed to DC current by the
   a. slip rings
   b. commutator
   c. transistor
   d. diodes
7. The alternator regulator controls
   a. current only
   b. voltage only
   c. current and voltage
   d. current, voltage, and reverse current
8. During a charging system load test the voltmeter reading should exceed base voltage by
   a. 5 V
   b. 0.5 V
   c. 0.05 V
   d. 1.5 V
9. During a charging system voltage output test the system should produce
   a. 13.5 to 14.7 V
   b. 12 to 13.5 V
   c. 10 to 12.5 V
   d. 12.5 to 12.8 V
10. During a regulator bypass test full field current is applied to the
    a. stator
    b. rectifier
    c. heat sink
    d. rotor
11. The alternator should be checked for
    a. opens
    b. shorts
    c. grounds
    d. all of the above
12. Diodes should be checked for
    a. shorts
    b. opens
    c. grounds
    d. all of the above
13. A charging system check should include checking the
    a. battery and cables
    b. alternator belt
    c. wiring connections
    d. all of the above

CHAPTER 10

Starting System Principles and Service

INTRODUCTION

Diesel engines commonly use electric or air-operated starting systems. A small gasoline engine was used on some earlier diesel engines. Both 12 and 24 V electric starting motors have been used. The 24 V system required a series–parallel switch to provide 24 V starting and 12 V for accessory operation. The advent of the high-torque 12 V starting motors saw the decline of 24 V systems. The air-powered starting system uses air pressure to power the starting motor. Air is supplied from a reservoir that is replenished by an air compressor when the engine is running.

PERFORMANCE OBJECTIVES

After studying this chapter thoroughly and with sufficient practical experience on adequate training models, and with the appropriate shop manuals, tools, and equipment, you should be able to do the following:

1. Describe the purpose, construction, and operation of the starting system and each of its components.
2. Diagnose starting system problems according to the diagnostic charts provided.
3. Safely remove, recondition, adjust, and replace any faulty starting system components in accordance with manufacturer's specifications.
4. Performance test the reconditioned starting system to determine the success of the repairs performed.

TERMS YOU SHOULD KNOW

Look for these terms as you study this chapter, and learn what they mean.

- battery
- battery cables
- ignition switch
- starter relay
- starter solenoid
- starter drive
- starting motor
- armature
- field
- brushes
- commutator
- magnetic field
- drive pinion
- permanent magnet starter
- series-wound
- shunt-wound
- compound-wound
- brushes
- brush holders
- drive housing
- field frame
- shift lever
- solenoid pull-in winding
- solenoid hold-in winding
- solenoid switch
- relay switch
- series–parallel switch
- overrunning clutch drive
- noise
- clicking
- buzzing
- humming
- whirring
- grinding
- current draw
- volt/ampere tester
- starter draw tester
- load control
- voltage drop tests
- armature tester
- growler
- armature open test
- armature ground test
- armature short test

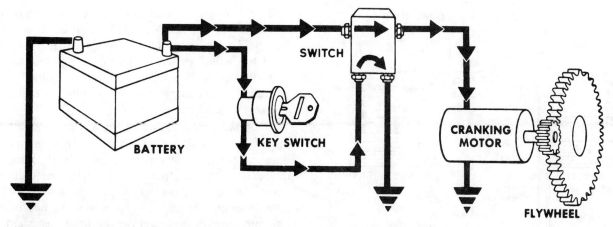

FIGURE 10-1 Starting system diagram. Note gear reduction and relay switch. (Courtesy of General Motors Corporation.)

field coil test
solenoid test
relay test
brush holder test

starter drive test
free-running test
air starter
hydrostarter

ELECTRIC STARTING SYSTEM COMPONENTS AND FUNCTION

The electric starting system includes the following components (**Figures 10-1 to 10-3**):

1. *Battery:* provides the electrical power to operate the starting motor. More than one battery may be used depending on the size of the engine. Larger engines require more cranking power and high-torque starting motors to crank over the engine.

2. *Starting motor:* uses battery electrical power to crank the engine over at sufficient speed to cause the engine to start.

3. *Key switch (Start switch):* ON/OFF switch that controls the starting system electrical circuit and the starting motor.

4. *Solenoid:* an electromagnetic switch that turns battery current to the starter on and off in response to the operator's use of the key start switch. The starter-mounted solenoid also shifts the starter drive pinion in and out of mesh with the starter ring gear.

5. *Starter protection device:* prevents starting motor engagement beyond a 30-second time period to prevent overheating of the starting motor. This prevents the starter from being engaged again for several minutes to allow for cool-down time. Another device cuts off power to the starting motor the instant the engine starts and prevents engaging

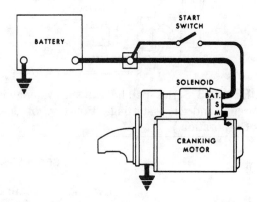

FIGURE 10-2 Basic electric control circuit for starter without relay-type solenoid. (Courtesy of Mack Trucks Inc.)

the starter again until the engine has come to a full stop.

6. *Cables and wiring:* Heavy electrical cables transmit the high current from the battery to the starting motor. Other wiring connects the key start switch and solenoid to each other for starting motor control.

ELECTRIC STARTING MOTOR COMPONENTS

The electric starting motor comprises the following major components (**Figures 10-4 and 10-5**):

1. *Field frame:* contains the field coils, armature, and brush assemblies. The drive housing is mounted on one end and the commutator end cap or plate on the other. The solenoid is mounted on the outside of the field frame.

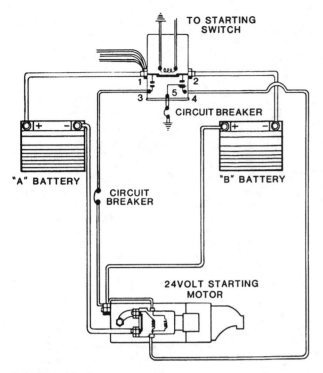

FIGURE 10–3 Series–parallel starting system with 12 V charging and 24 V starting. (Courtesy of Snap-on Tools Corporation.)

2. *Field coils:* The field coils are mounted on pole shoes that are bolted to the inside of the field frame. The field coils consist of a series of heavy copper-insulated windings wound around the pole shoes. As battery current is applied to these windings during starting they become strong electromagnets that create a strong magnetic field.

3. *Armature:* consists of many slotted steel disks mounted on a shaft with heavy copper-ribbon-type conductors fitted into the grooves formed by the laminated disks. Each ribbon forms a loop with the ends connected to separate copper bar segments that are insulated from one another. The bar segments that form the commutator are mounted on the armature shaft and are insulated from it.

4. *Brush assembly:* Each brush is a solid block (rectangle or square) containing a mixture of carbon, graphite, and copper. The brushes are mounted in spring-loaded brush holders that keep them in contact with the commutator. As the armature turns, the brushes slide over the commutator surface. As they slide over the commutator segments they switch power on and off to each copper-ribbon winding in the armature. Short, flexible electrical leads provide the electrical connections between the brushes and the starter battery terminal.

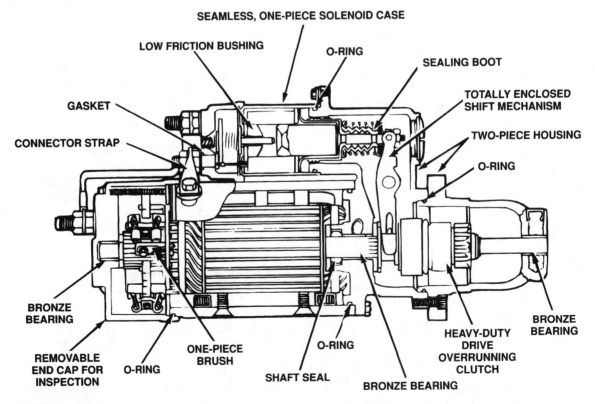

FIGURE 10–4 Model 42MT starting motor cross section. (Courtesy of Detroit Diesel Corporation.)

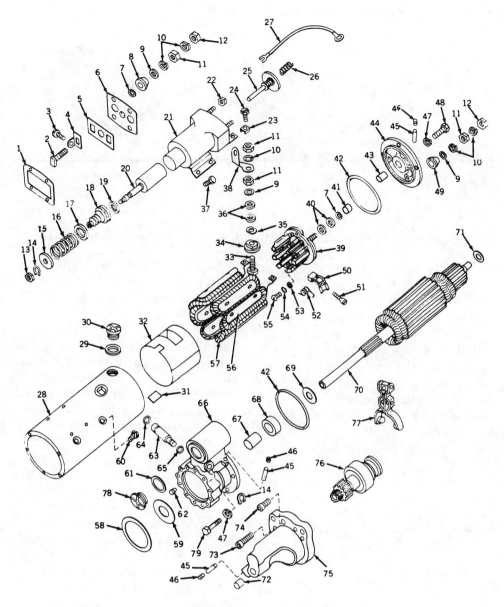

1—Terminal Plate Gasket	27—Solenoid Return Wire	54—Lock Washer (3 used)
2—Terminal Stud (2 used)	28—Field Frame	55—Machine Screw (3 used)
3—Machine Screw	29—Gasket (4 used)	56—Field Pole Shoe (6 used)
4—Solenoid Winding Terminal	30—Brush Plug (3 used)	57—Field Coil Assembly
5—Terminal Insulation	31—Field Coil Lead Insulator (2 used)	58—Gasket
6—Terminal Plate	32—Field Coil and Brush Lead Insulator	59—Brake Washer
7—Packing (3 used)	33—Field Terminal Stud	60—Pole Shoe Screw (12 used)
8—Insulator (2 used)	34—Insulator	61—Adjusting Hole Plug Gasket
9—Plain Washer (4 used)	35—Special Packing	62—Seal Plug
10—Lock Washer (7 used)	36—Insulating Washers (2 used)	63—Shift Lever Shaft
11—Jam Nut (5 used)	37—Solenoid Mounting Screw (4 used)	64—O-Ring
12—Hex. Nut (3 used)	38—Field Coil Connector	65—Small O-Ring
13—Self-Locking Nut	39—Brush Holder Assembly	66—Lever Housing
14—Snap Ring (2 used)	40—Insulating Washers (2 used)	67—Lever Housing Bushing
15—Spring Retainer	41—Insulator Bushing	68—Oil Seal
16—Solenoid Return Spring	42—O-Ring (2 used)	69—Spacer Washer
17—Spring Retainer	43—Commutator End Frame Bushing	70—Armature
18—Boot	44—Commutator End Frame	71—Thrust Washer
19—Washer	45—Lubricating Wick (3 used)	72—Drive End Housing Bushing
20—Solenoid Plunger	46—Plug (3 used)	73—Screw (5 used)
21—Solenoid Winding and Case Assembly	47—Lock Washer (9 used)	74—Special Screw
	48—Cap Screw (6 used)	75—Drive End Housing
22—Sealing Nut (4 used)	49—Insulator	76—Sprag Clutch Assembly
23—Lock Clip (2 used)	50—Brush (12 used)	77—Shift Lever
24—Machine Screw (2 used)	51—Machine Screw (12 used)	78—Adjusting Hole Plug
25—Contact Assembly	52—Brush Spring (12 used)	79—Cap Screw (7 used)
26—Contact Return Spring	53—Plain Washer (3 used)	

FIGURE 10–5 Exploded view of a typical starter. (Courtesy of Deere and Company.)

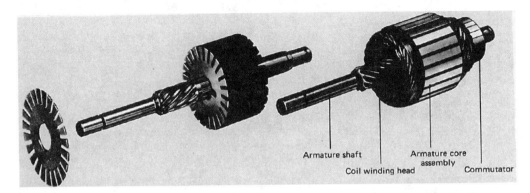

FIGURE 10-6 Starter armature detail. (Courtesy of Robert Bosch Canada Ltd.)

5. *Starter drive:* The starter drive is splined to the armature shaft and is able to slide back and forth as the starter is engaged and disengaged. The drive contains a one-way clutch that locks up during cranking and overruns as soon as the engine starts.

6. *Drive end housing:* The drive end housing supports one end of the armature in a bushing or bearing. A flange on the housing provides the means for mounting the starter on the engine. The starter drive operates inside the drive end housing.

7. *Commutator end cap or plate:* supports the commutator end of the armature in a bushing or bearing and closes the field frame at the commutator end.

8. *Intermediate bearing:* supports the armature between the starter drive and the armature windings.

STARTING MOTOR OPERATION

Magnetically, the cranking motor is made up of two parts, the armature and the field winding assembly **(Figures 10-6 and 10-7)**. The armature consists of a number of low-resistance conductors placed in insulated slots around a soft-iron core assembled onto the armature shaft. The commutator consists of a number of copper segments insulated from one another and from the armature shaft. The conductors are so connected to each other and to the commutator that electricity will flow through all the armature conductors when brushes are placed on the commutator and a source of current is connected to the brushes **(Figure 10-8)**. This creates magnetic fields around each conductor. Current through the field windings creates a powerful magnetic field. **Figure 10-9** shows the relationship between a magnetic field from a permanent magnet, a single current-carrying conductor, and the direction of the force that is exerted on the conductor. The magnetic lines of force pass from the north to the south pole as the arrows indicate. Current through the conductor is in the direction shown.

Using the right-hand rule, it can be seen that when the direction of the current through the conductor is as shown, the magnetic lines of force around the conductor will be in a counterclockwise direction, as indicated by the circular arrow around the conductor. Looking end on at the conductor in **Figure 10-10,** it can be seen that to the right of the conductor the magnetic field from the permanent magnet and the circular magnetic field around the conductor oppose each other. To the left of the conductor they are in the same direction. When a current-carrying conductor is located in a magnetic field, the field of force is distorted, creating a strong field on one side of the conductor and a weak field on the opposite side. The conductor will be forced to move in the direction of the weak field; therefore, in this instance, the conductor will be pushed to the right. The more current through the

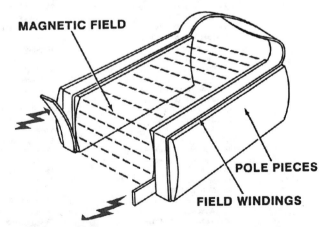

FIGURE 10-7 Pole pieces (shoes) support the field windings, which create a magnetic field. (Courtesy of Deere and Company.)

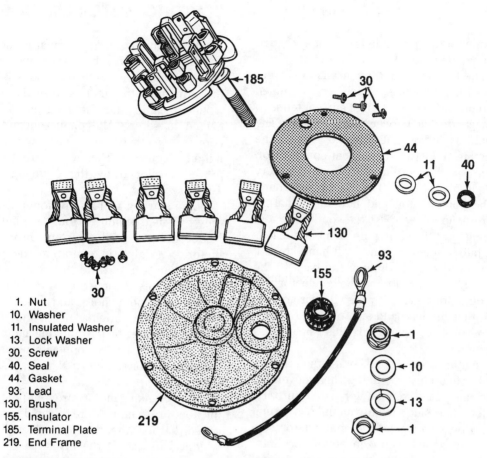

1. Nut
10. Washer
11. Insulated Washer
13. Lock Washer
30. Screw
40. Seal
44. Gasket
93. Lead
130. Brush
155. Insulator
185. Terminal Plate
219. End Frame

FIGURE 10–8 End plate and brush assembly components for a heavy-duty starter. (Courtesy of General Motors Corporation.)

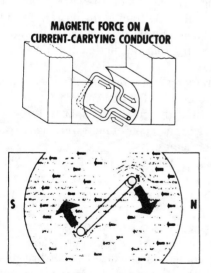

FIGURE 10–9 The action of magnetic fields causes the looped conductor to rotate.

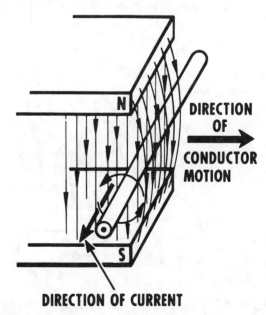

FIGURE 10–10 The magnetic field of a magnet and the magnetic field of current-carrying conductors are used in cranking motor operation.

conductor, the stronger is the force exerted on the conductors.

The application of this principle is shown in **Figure 10–11,** with a simple, one-turn armature electric motor. The magnetic field is created by means of current through the field coil windings, which are assembled around the two poles. The direction of current tends to increase the magnetic field between the two poles. The U-shaped armature winding that is placed between the two poles is connected to a two-segment commutator. In the position shown, current is from the battery through the right-hand brush, through the right-hand segment of the commutator, and into the armature winding, where it flows first past the south pole and then, in returning, past the north pole then into the left-hand commutator segment and left-hand brush, through the north pole field winding, and back to the battery. The magnetic fields around the conductor will be in the two directions shown by the circular arrows. It can be seen that the left-hand side of the armature winding will be pushed downward, thus imparting a clockwise rotation.

Since the armature winding and commutator are assembled together and must rotate together, movement of the winding causes the commutator also to turn. By the time the left-hand side of the winding has swung around toward the south pole, the commutator segments will have reversed their connections with respect to the brushes. Current is in the opposite direction with respect to the winding; since the winding has turned 180°, the force exerted on it still will tend to rotate it in a clockwise direction.

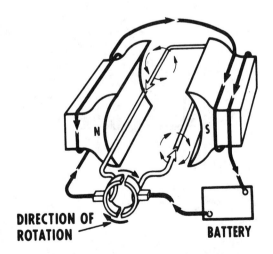

FIGURE 10–11 Armature loop and field windings are connected in series through the brushes and commutator to create magnetic fields that cause the armature to rotate.

STARTING MOTOR CIRCUITS

The field windings and the armature normally are connected in such a way that all the current that enters the cranking motor passes through both field windings and the armature. In other words, the motor is series wound, meaning that the fields and armature are connected in series. All conductors are heavy ribbon-copper types that have a very low resistance and permit a high current. The more current, the higher is the power developed by the cranking motor (**Figure 10–12**).

Some starting motors are four-pole units but have only two field windings, thus keeping resistance low. Notice the part of the current through this cranking motor. Note that in operation the poles with field coil windings have a north polarity at the pole shoe face. The lines of force pass through the armature, enter the plain pole shoes, and pass through the frame and back to the original north shoe to complete the magnetic circle. In all cranking motors the adjacent pole shoes must be of opposite polarity, so that in a four-pole motor there is a N, S, N, S sequence around the frame.

Other cranking motors are four-pole, four-field winding, four-brush units. Here the field windings are paired off so that half the current is through one set of field windings to one of the insulated brushes, while the other half of the current is through the other set of field windings to the other insulated brush. With four-field coil windings, it is possible to create more ampere turns and consequently stronger magnetic fields, thus producing cranking motors with greater torque or cranking ability. By tracing the current from the terminal, it can again be shown that the poles alternate S, N, S, N, providing four magnetic paths through the armature core.

The cranking motor designed for heavy-duty service uses 6 poles and 6 brushes; 24 V starters have 12 brushes. Here the current is split in three ways, one-third through each pair of field windings to one of the 3 insulated brushes.

Increasing the number of circuits through the cranking motor helps keep the resistance low so that a high current and a high horsepower can be developed.

As a rule, all the insulated brushes are connected together by means of jumper leads or bars, so that the voltage is equalized at all brushes. Without these equalizing bars there may be conditions that cause arcing and burning of commutator bars, eventually insulating the brush contact and preventing cranking.

STARTING MOTOR SOLENOID CIRCUITS

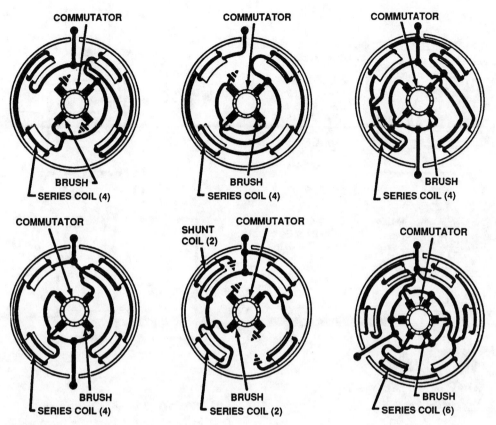

FIGURE 10-12 Starting motor circuit designs. (Courtesy of General Motors Corporation.)

STARTING MOTOR SOLENOID CIRCUITS

The solenoid switch on a cranking motor not only closes the circuit between the battery and the cranking motor but also shifts the cranking motor pinion into mesh with the engine flywheel ring gear. This is accomplished by means of a linkage between the solenoid plunger and the shift lever on the cranking motor. Solenoids are energized directly from the battery through the switch or in conjunction with a magnetic switch (**Figures 10-13** to **10-19**).

When the circuit is completed to the solenoid, current from the battery is through two separate windings, designated as the *pull-in* and the *hold-in* windings. These windings produce a magnetic field that pulls in the plunger, so that the drive pinion is shifted into mesh and the main contacts in the solenoid switch are closed, completing the cranking motor circuit.

Closing the main contacts in the solenoid switch at the same time shorts out the pull-in winding, since it is

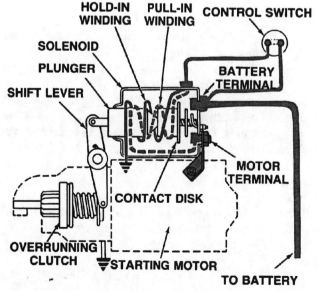

FIGURE 10-13 Starter solenoid wiring diagram. (Courtesy of Deere and Company.)

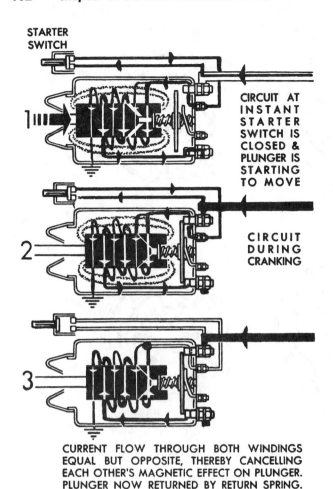

FIGURE 10-14 Starter solenoid operation. (Courtesy of General Motors Corporation.)

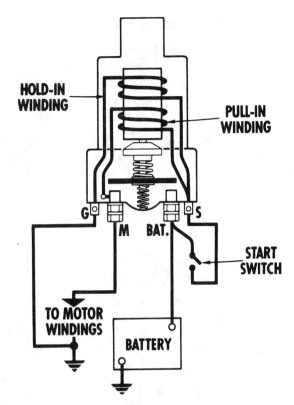

FIGURE 10-15 Heavy-duty starter solenoid circuit. (Courtesy of General Motors Corporation.)

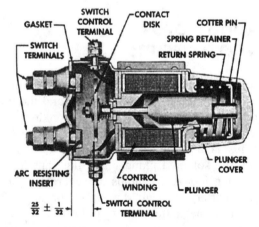

FIGURE 10-16 Magnetic starting motor switch cross section. (Courtesy of General Motors Corporation.)

connected across the main contacts. The heavy current draw through the pull-in winding occurs only during the movement of the plunger.

When the control circuit is broken after the engine is started, current no longer reaches the hold-in winding. The tension of the return spring then causes the plunger to return to the at-rest position. Low system voltage or an open circuit in the hold-in winding will cause an oscillating action of the plunger. The pull-in winding has sufficient magnetic strength to close the main contacts, but when they are closed, the pull-in winding is shorted out, and there is no magnetic force to keep the contacts closed. Check for a complete circuit of the hold-in winding as well as the condition of the battery whenever chattering of the switch occurs.

Whenever a solenoid is replaced on a cranking motor, it is necessary to adjust the pinion travel, with the exact clearance and method of adjustment varying among different motor designs.

STARTING MOTOR DRIVES

There are two common types of cranking motor drives in general use: the Bendix drive and the overrunning clutch.

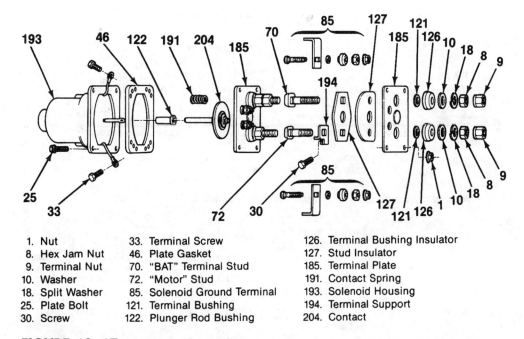

1. Nut	33. Terminal Screw
8. Hex Jam Nut	46. Plate Gasket
9. Terminal Nut	70. "BAT" Terminal Stud
10. Washer	72. "Motor" Stud
18. Split Washer	85. Solenoid Ground Terminal
25. Plate Bolt	121. Terminal Bushing
30. Screw	122. Plunger Rod Bushing
126. Terminal Bushing Insulator	
127. Stud Insulator	
185. Terminal Plate	
191. Contact Spring	
193. Solenoid Housing	
194. Terminal Support	
204. Contact	

FIGURE 10–17 Magnetic switch components. (Courtesy of General Motors Corporation.)

Bendix Drive

The Bendix drive **(Figure 10–20)** depends on inertia to provide meshing of the drive pinion with the engine flywheel ring gear.

The Bendix drive consists of a drive pinion, sleeve, spring, and spring-fastening screws. The drive pinion is normally unbalanced by a counterbalance on one side. It has screw threads on its inner bore. The Bendix sleeve, which is hollow, has screw threads cut on its outer diameter that match the screw threads of the pinion. The sleeve fits loosely on the armature shaft and is connected through the Bendix drive spring to the Bendix drive head, which is keyed to the armature shaft. Thus, the Bendix sleeve is free to turn on the armature shaft within the limits permitted by the flexing of the spring.

When the cranking motor switch is closed, the armature begins to revolve. The rotation is transmitted through the drive head and the spring to the sleeve, so that all these parts pick up speed with the armature. The pinion, however, being a loose fit on the sleeve screw thread, does not pick up speed along with the sleeve. In other words, the increased inertia of the drive pinion due to the effect of the counterbalance prevents it from rotating. The result is that the sleeve rotates within the pinion. This forces the drive pinion endwise along the armature shaft so that it goes into mesh with the flywheel teeth. As soon as the pinion reaches the pinion stop, it begins to rotate along with the sleeve and armature. This rotation is transmitted to the flywheel. The Bendix spring takes up the shock of meshing.

When the engine begins to operate, it spins the pinion at a higher speed than that of the cranking motor armature. This causes the pinion to rotate relative to the sleeve, so that the pinion is driven back out of mesh from the flywheel teeth. Thus, the Bendix drive auto-

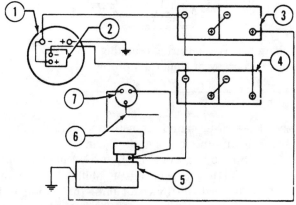

1. ALTERNATOR
2. TRANSFORMER-RECTIFIER (T-R) UNIT
3. CHASSIS BATTERIES
4. CRANKING BATTERIES
5. CRANKING MOTOR
6. WIRE TO KEY SWITCH
7. CRANKING MOTOR RELAY

FIGURE 10–18 A 12/24 V electrical system with cranking motor relay. (Courtesy of Mack Trucks Inc.)

184 Chapter 10 STARTING SYSTEM PRINCIPLES AND SERVICE

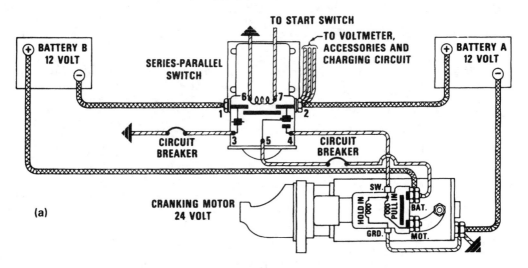

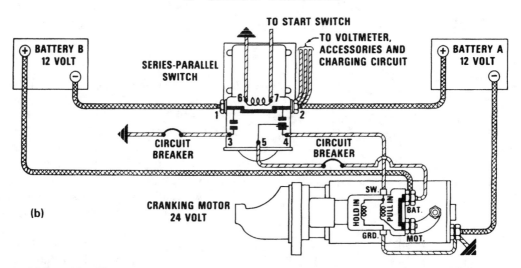

FIGURE 10–19 Typical 12/24 V system during charging (a) and during cranking (b). Note the difference in switch contact positions. (Courtesy of General Motors Corporation.)

matically meshes the pinion with the flywheel ring gear as soon as the engine starts. The spring-loaded antidrift pins prevent disengagement until the engine speed reaches approximately 350 rpm.

Some heavy-duty Bendix drives incorporate a friction clutch that is spring loaded and provides momentary slippage to cushion engagement.

Positork Drive

This drive unit provides positive engagement and remains engaged until the electrical circuit is deenergized. The starter cannot be activated until the pinion is in mesh with the flywheel ring gear.

When the starter control circuit is energized, the solenoid is activated, causing the shift lever to move the drive toward the flywheel until the pinion is in mesh. When the pinion is fully meshed, the starting motor switch contacts in the solenoid assembly are closed, completing the electrical circuit to the starting motor. When the engine starts, the flywheel turns the pinion faster than the armature and drive parts, causing the drive collar teeth to demesh from the pinion teeth. A lockout device prevents these teeth from reengaging as long as the pinion is meshed with the ring gear.

STARTING MOTOR DRIVES

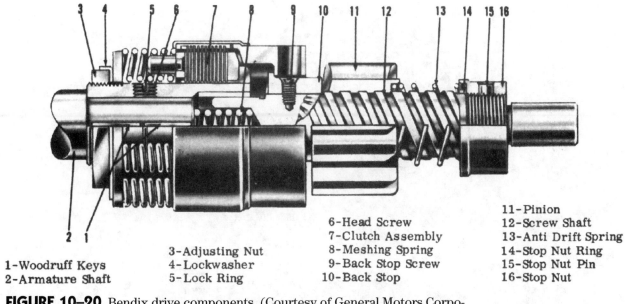

1—Woodruff Keys
2—Armature Shaft
3—Adjusting Nut
4—Lockwasher
5—Lock Ring
6—Head Screw
7—Clutch Assembly
8—Meshing Spring
9—Back Stop Screw
10—Back Stop
11—Pinion
12—Screw Shaft
13—Anti Drift Spring
14—Stop Nut Ring
15—Stop Nut Pin
16—Stop Nut

FIGURE 10–20 Bendix drive components. (Courtesy of General Motors Corporation.)

Sprag Clutch

The sprag clutch is the device that has made the solenoid-actuated type of starter feasible. It is a roller-type or sprag-type clutch that transmits torque in only one direction, turning freely in the other. In this way, torque can be transmitted from the starting motor to the flywheel but not from the flywheel to the starting motor **(Figure 10–21).**

A typical sprag clutch is shown in **Figure 10–22.** The clutch housing is internally splined to the starting motor armature shaft. The drive pinion turns freely on the armature shaft within the clutch housing. When the clutch housing is driven by the armature, the spring-loaded rollers are forced into the small ends of their tapered slots and wedge tightly against the pinion barrel. This locks the pinion and clutch housing solidly together, permit-

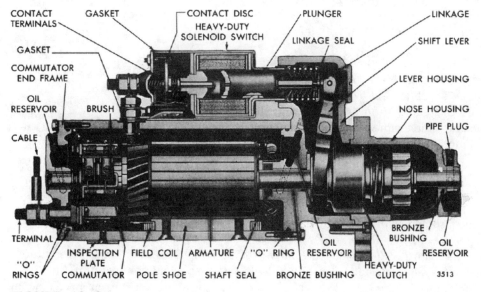

FIGURE 10–21 Starting motor with sprag clutch drive. (Courtesy of Detroit Diesel Corporation.)

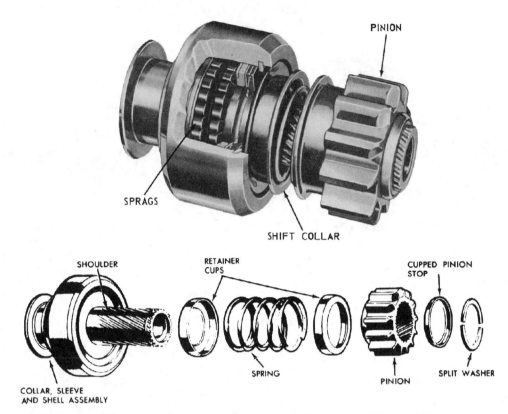

FIGURE 10-22 Starter drive with sprag clutch. (Courtesy of General Motors Corporation.)

ting the pinion to turn the flywheel and thus crank the engine.

When the engine starts, the ring gear begins to drive the pinion faster than the starter motor because of the pinion-to-ring gear reduction ratio. This action unloads and releases the clutch rollers, permitting the pinion to rotate freely around the armature shaft without stressing the starter motor.

The operator always should be careful not to reengage the cranking motor drive too soon after a false start. It is advisable to wait at least 5 seconds between attempts to crank. Burred teeth on the fly-wheel ring gear are an indication of attempted engagement while the engine is running.

STARTER PROTECTION DEVICE

In addition to the protection provided by the overrunning clutch in the starter drive, other devices are used to prevent damage to the starting motor. One such device uses a thermostatic sensor to detect field coil temperature **(Figure 10-23)**. The sensing switch is connected in series with the starter magnetic switch. The switch is normally closed to allow starting motor operation. When the starter is

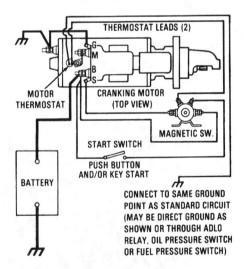

FIGURE 10-23 A thermostatic sensor circuit. (Courtesy of General Motors Corporation.)

operated to the point where the field coils reach the specified temperature, the thermostatic switch opens to prevent current from reaching the starter. The starter cannot be reengaged until it has cooled off enough to allow the thermostatic switch to close and restore the circuit.

ELECTRIC STARTING SYSTEM SERVICE

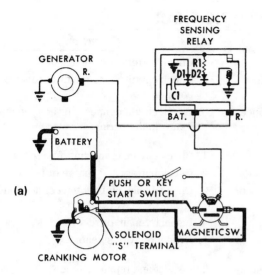

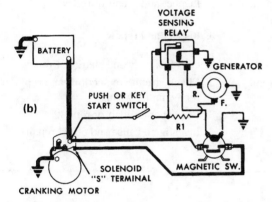

FIGURE 10–24 (a) Frequency-sensing and (b) voltage-sensing ADLO circuits. (Courtesy of General Motors Corporation.)

Another device used is the Automatic Disengagement and Lock-Out (ADLO) circuit, which can be used on solenoid-type starters. The system uses a relay that is either frequency-sensing or voltage-sensing. The relay contacts are normally closed to allow starting motor operation. As soon as the engine starts, voltage from the alternator energizes the relay and causes the relay contacts to open. With the relay contacts open, the starter cannot be engaged. This system not only disengages the starter immediately when the engine starts but also prevents the starter from being engaged while the engine is running **(Figure 10–24)**.

ELECTRIC STARTING SYSTEM SERVICE

General Precautions

When servicing the starter system, some general precautions should be observed as follows:

- Always disconnect the battery ground cable before disconnecting wiring from the starting motor or removing the starting motor.
- Always have the unit properly positioned on a hoist or safely supported for any work underneath.
- For any cranking tests, make sure that transmission is in NEUTRAL or PARK with the parking brakes applied. Follow directions to prevent engine starting during cranking motor tests.
- Do not wash or immerse electrical components in solvent; clean with compressed air only.

In general, cranking system diagnosis and service includes checking the following:

1. engine mechanical condition
2. battery and battery cables
3. starting control circuit
4. cranking motor current draw
5. cranking motor removal, cleaning, inspection, testing, overhaul, and installation, or replacing motor with a new or rebuilt unit

Preliminary Inspection

Always begin with a quick visual check of the supply circuits to note any obvious trouble sources such as corroded or loose connections. The supply circuit consists of the battery, battery cables, clamps, and connectors.

Many slow-turning starters have been corrected by simply cleaning the battery terminal posts and cable clamps. Inspect starter and ground cables for corrosion or damage. In checking the supply circuit, always begin with a visual inspection of the battery post and cable clamps.

Test the battery to make sure that it is in good condition and has a minimum specific gravity reading of 1.220, temperature corrected, and see that the battery passes the high-rate-discharge test described in the battery section.

See **Figure 10–25** for starting system problem diagnosis.

Engine Won't Crank Properly

There are several possible causes for this condition. Assuming that the battery checks out and has been eliminated as a possible cause, either battery power is being prevented from reaching the starter motor or the motor is defective and must be repaired or replaced. First perform the resistance (voltage drop) tests.

STARTING SYSTEM DIAGNOSTIC CHART

Problem	Possible Cause	Correction
Engine will not crank and starting motor relay or solenoid does not engage	1. Battery discharged 2. Key start switch, safety start switch, relay or solenoid inoperative 3. Starting circuit open or high resistance	1. Check battery and change or renew. 2. Check circuitry and repair or renew faulty components. 3. Check circuit connections and repair or renew faulty wiring.
Engine will not crank but starting motor relay or solenoid engages	1. Engine seized 2. Battery discharged 3. Defective starting motor connections or loose battery connections 4. Starting motor faulty 5. Relay or solenoid contacts burned	1. Check engine crankshaft free to turn. 2. Check battery and charge or renew. 3. Check, clean, and tighten connections. 4. Inspect, repair, or renew. 5. Renew relay or solenoid.
Starting motor turns but does not crank engine	1. Defective starting motor drive assembly 2. Defective solenoid or pinion engagement levers 3. Defective flywheel ring gear	1. Inspect and repair or renew. 2. Inspect and repair or renew. 3. Inspect and renew.
Engine cranks slowly	1. Discharged battery 2. Excessive resistance in starting circuit 3. Defective starting motor 4. Tight engine	1. Check battery and charge or renew. 2. Check circuit connections and repair or renew faulty wiring. 3. Inspect and repair or renew. 4. Investigate cause and effect repair.

FIGURE 10–25

Resistance (Voltage Drop) Tests

Excessive resistance in the circuit between the battery and starter motor will reduce cranking performance. The resistance can be checked by using a voltmeter to measure the voltage drop in the circuit while the starter motor is operated.

There are three checks to be made (**Figure 10–26**):

1. The voltage drop between the positive battery terminal and the fastening device at the other end of the positive battery cable.

2. The drop between the unit frame and the starter motor field terminal (or the field frame if there is no terminal). Disregard this test on starter motors used in 24 V insulated systems.

3. The drop between the negative battery terminal and the fastening device at the other end of the negative battery cable.

Each of the checks should show no more than 0.1 V drop with the starter motor cranking the engine.

If excessive voltage drop is found in any of these circuits, disconnect the cables, clean the connections carefully, and then reconnect the cables firmly in place. If this does not help, replace the cable. Broken strands in the cable that are not visible to the eye are probably creating excessive resistance.

NOTE: On some applications, extra-long battery cables are required due to the location of the batteries and starter motor. This may result in somewhat higher voltage drops than the above recommended 0.1 V. On such applications, the normal voltage should be established by checking several units. Then, when a voltage drop well above this normal figure is found, abnormal resistance will be indicated and correction can be made as already explained.

Starter Removal

1. Disconnect the battery ground cable.
2. Disconnect all the wiring connections from the starter.
3. Remove the starter mounting bolts and the starter from the engine.

VIEW A

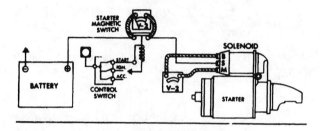

VIEW B

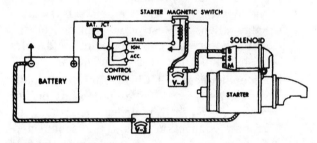

VIEW C

FIGURE 10–26 Checking starting circuit resistance using the voltage drop method. (Courtesy of General Motors Corporation.)

CAUTION:

The starter is quite heavy. Make sure you support it properly during removal and placement on the repair bench.

STARTING MOTOR BENCH TESTS

With the starting motor removed from the engine, the armature should be checked for freedom of rotation by prying the pinion with a screwdriver. Tight bearings, a bent armature shaft, or a loose pole shoe screw will cause the armature not to turn freely. If the armature does not turn freely, the motor should be disassembled immediately. However, if the armature does rotate freely, the motor should be given a no-load test before disassembly.

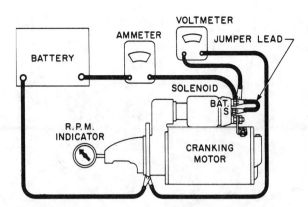

FIGURE 10–27 Test instruments connected to perform starter no-load test. (Courtesy of Mack Trucks Inc.)

Free-Speed Test

In the free-speed test (**Figures 10–27** and **10–28**), the starter motor is connected in series with a battery of the specified voltage and an ammeter capable of reading at least 700 A. A 1000 A capacity is preferable. An rpm indicator also should be used to measure the armature revolutions per minute.

Torque Test

The torque test requires the equipment illustrated in **Figure 10–29**. The starter motor is securely mounted and the brake arm hooked to the drive pinion. Then, when the specified voltage is applied, the torque can be computed from the reading on the scale. If the brake arm is 1 ft long as shown, the torque will be indicated directly on the scale in foot-pounds. A high-current-carrying variable resistance should be used so that the specified voltage can be applied. Many torque testers indicate the developed pounds-feet of torque on a dial.

WARNING:

Stand clear of the apparatus when current is applied. A broken arm can result if the brake arm slips.

The specifications are normally given at low voltages so that the torque and ammeter reading obtained

190 Chapter 10 STARTING SYSTEM PRINCIPLES AND SERVICE

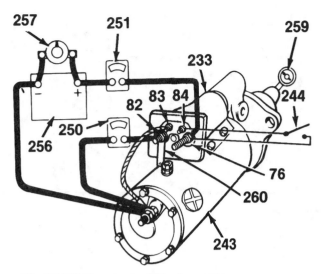

76. "BAT" Terminal
82. "MTR"
83. "GRD"
84. "SW"
233. Solenoid
243. Starting Motor
244. Switch
250. Voltmeter
251. Ammeter
256. Battery
257. Carbon Pile
259. RPM Indicator
260. Connector Strap

FIGURE 10–28 No-load test connections on starters with ground terminals. (Courtesy of General Motors Corporation.)

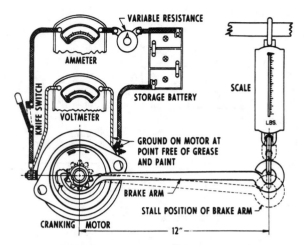

FIGURE 10–29 Starter torque test setup. (Courtesy of General Motors Corporation.)

will be within the range of the testing equipment available in the field.

Interpreting Test Results

1. Low free speed and high current draw with low developed torque may result from:
   a. Tight, dirty or worn bearings, bent armature shaft or loose field pole screw, which will allow the armature to drag.
   b. Shorted armature. Check further after the armature is disassembled by revolving the armature on a growler with a steel strip such as a hacksaw blade held above it. The blade will vibrate above the area of the armature core in which the short circuit is located.
   c. A grounded armature or field. Check by raising grounded brushes and insulating them from the commutator with cardboard, and then checking with a test lamp between the armature terminal and starter motor frame. If the test lamp lights, there is a ground, so raise the other brushes from the commutator and check fields and commutator separately to determine whether it is the fields or armature that is grounded (see step 1.d. below). Grounds in the armature can be detected after disassembly by use of a test lamp and test points. If the lamp lights when one test point is placed on the commutator with the other point on the core or shaft, the armature is grounded.
   d. Grounded field coil. Check after disassembly with a test lamp. If the starter has one or more coils normally connected to ground, the ground connection must be disconnected. Connect one lead of the test lamp to the field frame and the other lead to the field connector. If the lamp lights, at least one field coil is grounded, which must be repaired or replaced.
2. Failure to operate with high current draw:
   a. A direct ground in the switch, terminal, or fields.
   b. Frozen shaft bearings that prevent the armature from turning.
3. Failure to operate with no current draw:
   a. Open field circuit. Connect test lamp leads to the ends of field coils. If the lamp does not light, the field coils are open and must be replaced.
   b. Open armature coils. Inspect the commutator for badly burned bars. Running free speed, an open armature will show excessive arcing at the commutator bar that is open.
   c. Broken or weakened brush springs, worn brushes, improperly seated brushes, high mica on the commutator, or other causes that would prevent good contact between the

brushes and commutator. Any of these conditions will cause burned commutator bars.

4. Low free speed with low torque and low current draw:

a. An open field winding. Raise and insulate ungrounded brushes from commutator and check fields with test lamp.

b. High internal resistance due to poor connections, defective leads, dirty commutator and causes listed under Step 3 above.

5. High free speed with low developed torque and high current draw indicates shorted fields. There is no easy way to detect shorted fields, since the field resistance is already low. If shorted fields are suspected, replace the fields and check for improvement in performance.

Solenoid Test

Proper operation of the solenoid switch depends on maintaining a definite balance between the magnetic strengths of the pull-in and hold-in windings. The two windings may be tested separately to check their efficiency. Test the pull-in winding by connecting between the solenoid switch terminal and the solenoid motor terminal with a source of variable voltage (battery and variable resistance in series), and an ammeter. Connect a voltmeter between the same terminals. Adjust the variable resistance to secure the specified voltage and note the current draw. Test the hold-in coil in a similar manner except make connections between the solenoid switch terminal and the other small solenoid terminal. If the solenoid does not come up to specifications and corrections cannot be made by wire brushing the control disk, the solenoid should be replaced (**Figures 10–30 and 10–31**).

Magnetic Switch Test

Connect a test voltmeter between the switch control terminals and connect a source of variable voltage to these terminals. Where the magnetic switch has but one control terminal, make these connections between that terminal and the switch base. A battery and a variable resistance connected in series will be found to be a suitable source of variable voltage. Increase the voltage on the switch slowly and note the voltage required to close the contacts. Closing of the contacts is indicated by a click that can normally be heard. A more accurate way of checking the instant that the contacts close is to connect test lamp points between the two main switch terminals. The lamp will light as the contacts close.

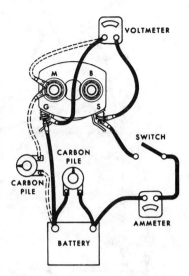

FIGURE 10–30 Checking solenoid hold-in and pull-in windings. (Courtesy of Mack Trucks Inc.)

With an ammeter connected in series with the magnetic switch windings, increase the voltage to the specified value and note the current draw.

STARTER DISASSEMBLY

1. Match mark the drive end housing, the field frame, and the commutator end cover for proper alignment during assembly later. Use a center punch or chisel to make the marks.

2. Remove the bolts that hold the commutator end cover and field frame together. On some starters the brushes must be disconnected before the end cover can be removed.

3. Remove the bolts that hold the drive end housing to the field frame, and remove the housing and armature from the field frame. The solenoid plunger will slide out of the solenoid during this procedure.

4. Remove the armature from the housing. On some units the lever housing is a separate part that must be removed before the armature can be withdrawn.

5. Remove the solenoid from the field frame.

PARTS INSPECTION AND REPLACEMENT

1. Disassemble the starter as outlined above.

2. Wipe the field coils, armature, commutator and armature shaft with a clean cloth. Wash the springs,

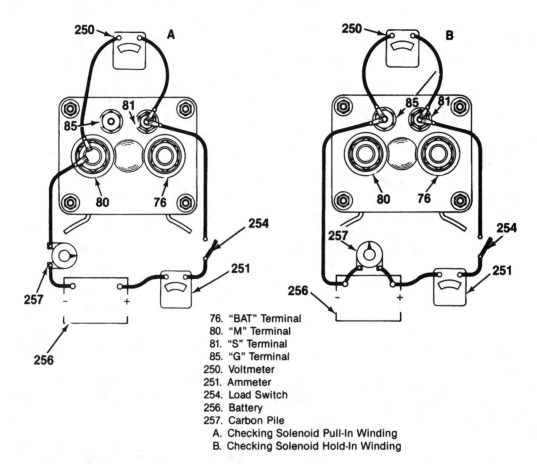

76. "BAT" Terminal
80. "M" Terminal
81. "S" Terminal
85. "G" Terminal
250. Voltmeter
251. Ammeter
254. Load Switch
256. Battery
257. Carbon Pile
A. Checking Solenoid Pull-In Winding
B. Checking Solenoid Hold-In Winding

FIGURE 10–31 Solenoid test connections on a heavy-duty starter. (Courtesy of General Motors Corporation.)

FIGURE 10–32 Checking armature runout. (Courtesy of Deere and Company.)

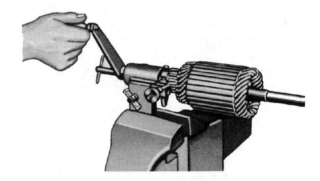

FIGURE 10–33 Cutting down the commutator. (Courtesy of Deere and Company.)

shims, thrust washer, trip collar, locking collar, and brush end plate assembly (not the brushes) in solvent and dry the parts.

3. Inspect the armature for broken or burned insulation and unsoldered connections. Inspect the commutator surface for grooves, and check for runout **(Figure 10–32)**.

4. If the surface of the commutator is rough or more than 0.002 in. (0.051 mm) out of round, turn it down. Do not undercut the mica **(Figure 10–33)**.

5. Check the armature windings with a growler tester **(Figures 10–34 to 10–36)**.

6. With the starter disassembled, check the field windings with a growler light for shorts to the housing or pole shoes. Connect one growler light lead to the starter housing; connect the other growler lead to each of the field coil brush leads in turn. A short circuit will

PARTS INSPECTION AND REPLACEMENT   193

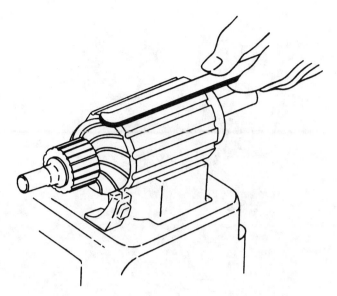

FIGURE 10–34 Checking the armature for short circuits. (Courtesy of Deere and Company.)

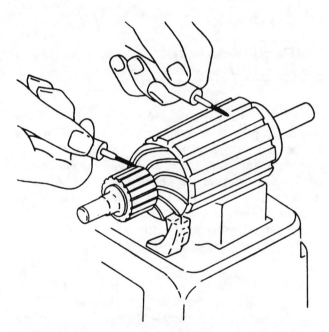

FIGURE 10–35 Checking the armature for grounds. (Courtesy of Deere and Company.)

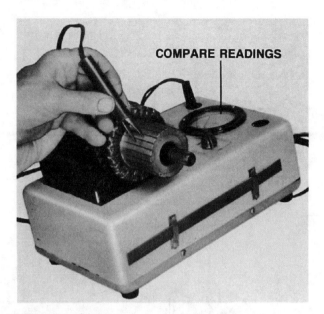

FIGURE 10–36 Testing the armature for opens across each pair of commutator bars. (Courtesy of Deere and Company.)

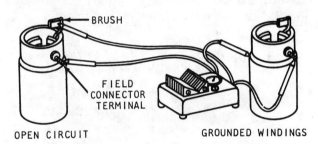

FIGURE 10–37 Checking the field coils for opens and grounds. (Courtesy of Deere and Company.)

light the growler light. Replace damaged coils **(Figure 10–37)**.

7. Examine the starter bushings for excessive wear. Replace worn or badly scored bushings.

8. Examine the contacts on the solenoid assembly and, if necessary, clean them with very fine sandpaper or crocus cloth.

9. The solenoid windings can be tested for internal shorts or open circuits with an ohmmeter. Connect the instrument across the windings and note the reading. The resistance of the pull-in winding and the hold-in winding should be as specified. A low reading generally indicates an internal short circuit, while no reading at all indicates an open circuit.

If the solenoid contacts are badly burned, or if the solenoid winding is shorted, replace the solenoid assembly, then adjust drive travel.

10. Inspect the pinion teeth for excessive wear or damage from improper engagement. If a new drive is to be installed, be sure it has the same number of teeth as the replaced component. Install the new drive on the armature shaft and check for free movement **(Figures 10–38** and **10–39)**. If necessary, lightly lap the pinion and armature shaft together, then adjust drive travel.

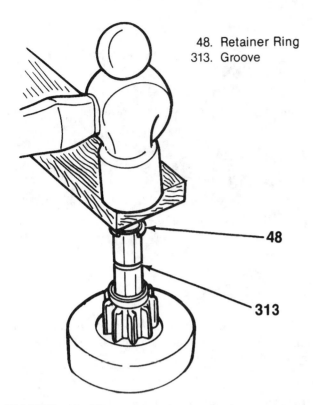

FIGURE 10–38 Installing the pinion gear retainer ring. (Courtesy of General Motors Corporation.)

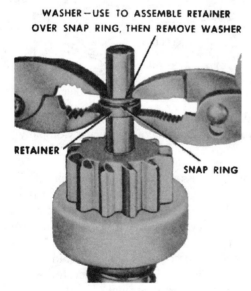

FIGURE 10–39 Installing retainer over snap ring. (Courtesy of Mack Trucks Inc.)

11. Check the tension of the brush springs for conformance to specifications.

12. Check the brushes for free movement in the brush holders. Inspect the brushes for excessive wear. Where brush leads are insulated, be sure that the insulation is not burned or worn **(Figure 10–40)**.

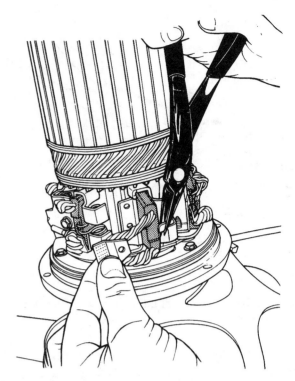

FIGURE 10–40 Installing brushes into brush holders. (Courtesy of General Motors Corporation.)

13. The brushes should be well seated to the commutator—that is, contacting the commutator over at least 60% of the brush contact area. If not, fit the brushes to the commutator with fine sandpaper until the desired seating area is obtained. Do only one brush at a time. Clean the brushes and commutator with compressed air. Be sure that no abrasive particles from the sandpaper are embedded in the contact area of the brushes **(Figure 10–41)**.

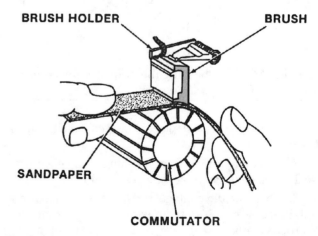

FIGURE 10–41 Seating the brushes to ensure full contact with the commutator. (Courtesy of Deere and Company.)

New brushes are not ground to fit the commutator; therefore, they must be seated as described.

Mark each brush and brush holder with which it was seated to permit assembly in the same position.

Check the tension of each brush spring at the point of contact with the brush. Use a spring scale hooked under the brush spring lip. Brush spring tension must be within specifications **(Figure 10–42)**.

14. Bendix Drive Check

Hold the drive in your hands with the fingers extending over the front end of the clutch and squeeze it. Doing this should compress the meshing spring inside the clutch housing. If the spring cannot be compressed, sufficient wear or damage has occurred to render the clutch inoperative, and the clutch assembly should be replaced.

NOTE: Compress the meshing spring before removing the drive from the armature shaft. It is possible for the meshing spring to fall out once the drive assembly is removed. When this happens, the check cannot be made.

15. Sprag Clutch Check

To check the overrunning clutch, clamp the drive pinion in a vise equipped with soft jaws. The drive should turn freely in one direction but should lock up in the other and show no evidence of slipping when torque is applied. Replace or renew a slipping clutch drive.

16. Assemble the starter in the reverse order of disassembly described earlier, making sure the marks made on the housing sections are aligned.

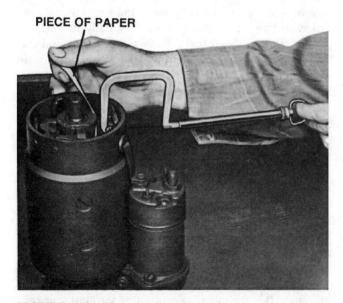

FIGURE 10–42 Using a spring scale to test brush spring tension. Read the scale when paper slips out from between brush and commutator. (Courtesy of Deere and Company.)

Drive Pinion Clearance Check

1. Energize the solenoid with the proper voltage battery. Connect the battery between the solenoid terminal and the ground terminal; the pinion should move forward. Adjust the clearance between pinion and pinion housing to specifications.

2. Disconnect the battery at the solenoid; the pinion must return to its normal position in one sharp movement.

To check the pinion clearance, connect a battery from the solenoid switch terminal to the motor frame. To prevent motoring, connect a heavy jumper lead from the solenoid motor terminal to the motor frame **(Figures 10–43 and 10–44)**.

With the solenoid energized and the clutch shifted toward the pinion, push the pinion back toward the commutator end as far as possible to take up any slack movement, then check the clearance between the pinion and housing **(Figure 10–45)**. Adjust the clearance by removing the plug on the lever housing and turning the nut on the plunger rod inside the housing. Turn the nut clockwise to decrease the clearance and counterclockwise to increase the clearance.

Install the starter and tighten the attaching bolts to specifications. Reconnect all the electrical connections to the starter and reconnect the battery ground cable.

AIR-STARTING SYSTEM

The air-starting motor consists of a five-vaned air motor with gear reduction, which drives the engine to be

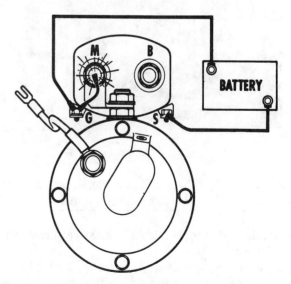

FIGURE 10–43 Hookup for checking pinion clearance on one type of sprag-type starter. (Courtesy of Mack Trucks Inc.)

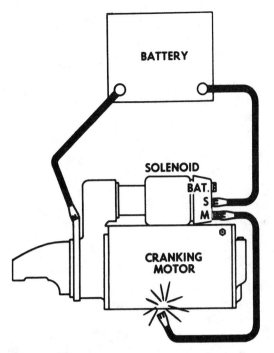

FIGURE 10-44 Hookup for another type of sprag-drive starter for checking pinion clearance. (Courtesy of Mack Trucks Inc.)

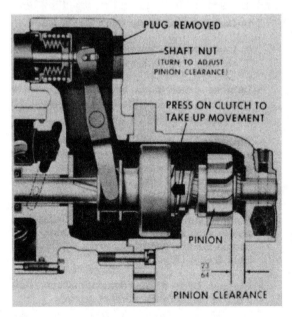

FIGURE 10-45 Checking sprag clutch pinion travel. (Courtesy of General Motors Corporation.)

started through a conventional Bendix-type drive (**Figures 10-46** and **10-47**).

An air-starting reservoir provides air for the cranking motor only. The connection to the cranking motor is through flexible hose with a quick-acting control valve to permit operation of the motor.

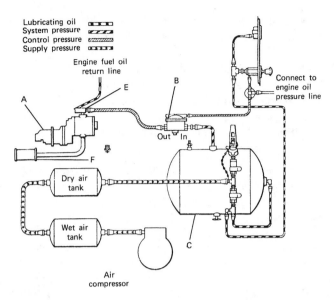

FIGURE 10-46 Air-starting system. (Courtesy of Stanadyne Diesel Systems.)

Air from the compressor, which is adjusted to deliver 95 to 120 psi (655 to 827 kPa) maximum pressure, flows through a check valve to the reservoir.

A trailer coupling connection or glad-hand is provided at the reservoir drain connection to permit charging of the reservoir from an external source, such as another vehicle or from the service shop air supply.

The energy source for an air-starting system is compressed air, which is usually stored in a separate receiver tank. The air starter has a rotor that is located eccentrically in a larger diameter bore and usually fitted with five vanes that can slide radially in and out within slots in the rotor. Between the starter and the compressed air tank is a control valve that holds the air in a ready condition. When the valve is opened, the compressor air is released, and the resultant force on the blades causes the rotation of the rotor—much like a paddle wheel.

In an inertia air-starting system, air is introduced into the air receiver tank by a one-way check valve from the air brake system. The compressed air within the receiver tank also serves as energy for the servo control. When the push button is activated, it sends a servo signal to the main relay valve, which then allows a high-volume air flow to pass on to the starter and crank the engine. Some starters are equipped with a device that is mounted on the inlet of the starter, which automatically injects a measured quantity of lubricant into the air stream so that the moving parts within the starter are adequately lubricated during operation. Also, to prevent an extremely loud discharging sound—made by exhaust air—from escaping into the environment, a muffler is

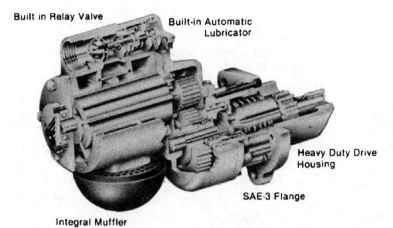

FIGURE 10–47 Cutaway view of Startmaster 250 air-starting motor. (Courtesy of Mack Trucks Inc.)

AIR STARTER DIAGNOSTIC CHART

Problem	Possible Cause	Correction
Starting motor not operating	1. Low air pressure	1. Check air pressure in system for 95 to 120 psi. Correct as required.
	2. Inoperative or defective starting control valve	2. Check operation of starting control valve. Repair if needed.
	3. Loose starter	3. Check mounting of starter to flywheel housing. Correct if needed.
	4. Loose or leaking air line	4. Check for air leaks between starting control valve and starter when starting control valve is operated. Correct as necessary.
	5. Seized starter	5. Remove starter from flywheel and check for rotation. Repair or replace as required.
Slow starting motor speed	1. Low air pressure	1. Check air pressure in system for 95 to 120 psi. Correct as needed.
	2. Defective starting control valve	2. Check operation of starting control valve. Repair as needed.
	3. Loose starter	3. Check mounting of starter to flywheel housing. Correct if necessary.
	4. Restricted or loose air line	4. Check for dented, kinked, restricted, or loose air line and connections. Repair as required.
	5. Dirty air cleaner (when used)	5. Remove and clean air cleaner.
	6. Improperly or overlubricated starter	6. Remove starter, disassemble, clean, lubricate, reassemble, and reinstall on flywheel housing.
Engine not turning over, starting motor operating	1. Defective motor drive	1. Check for broken or stripped drive pinion or flywheel ring gear or broken drive spring. Repair as necessary.

used on most starters to bring the overall sound level of the air-starting system down near the 80 db range.

Today most starting systems used on over-the-road vehicular applications employ a preengaged cranking motor. A preengaged air-starting system is almost identical to the inertia-type system, except that it utilizes a starter that receives a servo signal from the push button, whereupon an internal actuator engages the starter pinion with the engine ring gear. If, and only if, the starter pinion is meshed with the ring gear will the servo signal come out of the starter and onto the relay valve. The life of the engine ring gear has been appreciably extended by the use of preengaged-type starters.

HYDRAULIC STARTING SYSTEM

The hydrostarter system is a complete hydraulic system for cranking internal combustion engines. The system is automatically recharged after each engine start and can be manually recharged in an emergency. The starting potential does not deteriorate during long periods of inactivity; continuous exposure to hot or cold climates has no detrimental effect on the hydrostarter system. Also, the hydrostarter torque for a given pressure remains substantially the same regardless of the ambient temperature **(Figures 10–48** and **10–49).**

The hydrostarter system consists of a reservoir, an engine-driven charging pump, a manually operated pump, a piston-type accumulator, a starting motor, and connecting hoses and fittings.

Operation

Hydraulic fluid flows by gravity or slight vacuum from the reservoir to either the engine-driven pump inlet or hand pump inlet. The hand pump is used only to supply the initial charge or to recharge the system after servicing or overhaul. Fluid discharging from either pump outlet at high pressure flows into the accumulator and is stored at 3250 psi (22,408 kPa) under the pressure of compressed nitrogen gas. When the starter is engaged with the engine flywheel ring gear and the control valve is opened, high-pressure fluid is forced out of the accumulator (by the expanding nitrogen gas) and flows into the starting motor, which rapidly accelerates the engine to a high cranking speed. The used fluid returns from the starter directly to the reservoir.

The engine-driven hydrostarter charging pump runs continuously during engine operation, recharging the accumulator with fluid. When the proper amount of fluid has been returned to the accumulator, the pressure-operated unloading valve in the engine-driven pump opens and returns the pump discharge directly to the reservoir.

RESERVOIR

The reservoir is a cylindrical steel tank with a fine mesh screen at the outlet. The filler cap contains a filter to prevent dust and dirt from entering the reservoir.

ENGINE-DRIVEN CHARGING PUMP

The engine-driven charging pump is a single-piston positive displacement type and should run at approximately engine speed. It contains ball check valves and an unloading valve operated by the accumulator pressure. Its operation is entirely automatic and will operate in either direction of rotation.

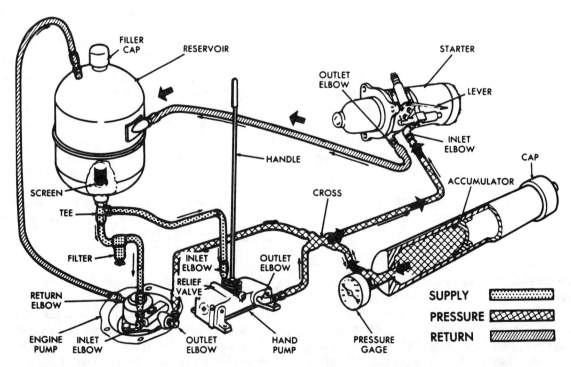

FIGURE 10–48 Hydrostarter system oil flow. (Courtesy of Detroit Diesel Corporation.)

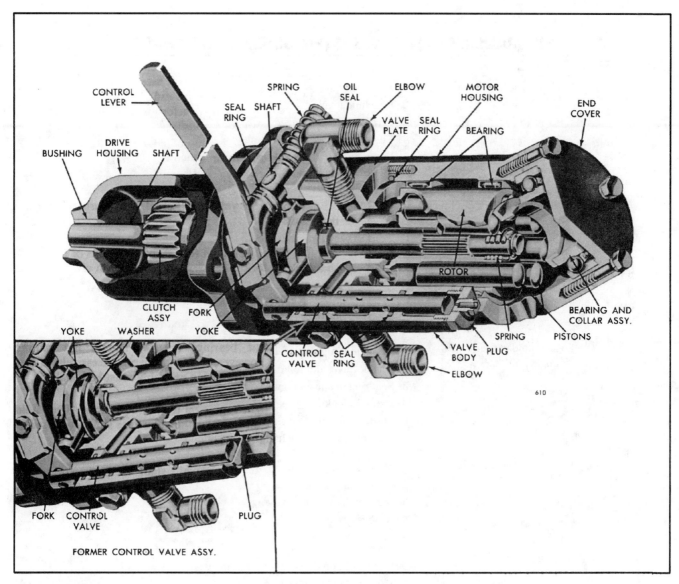

FIGURE 10–49 Cutaway view of hydrostarter. (Courtesy of Detroit Diesel Corporation.)

HAND PUMP

The hand pump is a single-piston, double-acting, positive displacement type. Flow through the pump is controlled by ball check valves. A manually operated relief valve is provided in this pump so that the accumulator pressure may be relieved when servicing of any components is required.

ACCUMULATOR

The piston-type accumulator is precharged with nitrogen through a small valve. A seal ring between the piston and the shell prevents the loss of gas into the hydraulic system. The accumulator is supplied with the proper precharge.

STARTER

The starter mounts on the flywheel housing and has a pinion gear with an overrunning clutch for engaging the flywheel ring gear. Movement of the starter control lever engages the pinion and opens the control valve in the proper sequence. The motor is a multipiston, swash plate type. Provision is made so that if pinion tooth abutment occurs, the motor rotates slowly until the pinion snaps into full engagement. When the control lever is released, the pinion is disengaged, and the valve is closed by spring action.

HYDRAULIC CRANKING SYSTEM DIAGNOSTIC CHART

Problem	Possible Cause
Low or no accumulator pressure	1. Air in system 2. Low fluid level 3. Screen or filter plugged 4. Check valves not functioning 5. Pump drive belt slipping (belt-driven pump) 6. Defective drive arm (direct-drive pump)
Cranking speed too slow	1. System fluid too heavy 2. Engine oil too heavy 3. Control valve not fully open
Loss of fluid from reservoir	1. External leaks 2. Starter shaft seal worn 3. Starter cover gasket defective 4. Worn shaft seal
Loss of fluid pressure with engine off	1. Ambient temperature decrease 2. Check valves not holding 3. Seal ring in starter control valve leaking 4. External leakage 5. Starter control valve shift fork bent 6. Loss of nitrogen precharge in accumulator
Hand pump does not discharge fluid	1. Manual relief valve open 2. Check valves leaking 3. Screen plugged in reservoir 4. Air in system 5. Piston seal rings leaking
Starter turns but engine does not	1. Starter pinion does not engage ring gear 2. Drive pinion clutch slipping 3. Overrunning clutch failure 4. Starter incorrectly assembled
Loss of accumulator precharge (nitrogen)	1. Piston seal ring failure 2. Air valve failure 3. Seal between shell and end cap damaged
Pressure in system too high	1. Defective gauge (incorrect pressure reading shown) 2. Engine-driven pump unloading valve defective
Fluid emerges from reservoir filler cap when starter is used	1. Filler cap filter plugged 2. Excess fluid in reservoir 3. Nitrogen returned to reservoir
Fluid leaks from starter control valve when starter is operated	1. Control valve seal ring defective 2. Bent shift fork causing valve to move past rear seal ring

REVIEW QUESTIONS

1. The starting system changes the _____ energy of the battery to _____ energy.
2. Magnetically, the starting motor is made up of the _____ and the _____ .
3. A strong _____ _____ pushes against the armature windings to cause rotation.
4. The commutator acts a _____ device.
5. In a shunt-wound starting motor, excess current is shunted to the ground. (T) (F)
6. A starter solenoid connects the starting motor to the _____ and shifts the _____ _____ into mesh with the _____ _____ _____ .
7. A starter relay is a _____ switch that connects the _____ to the _____ .
8. A rapid clicking or buzzing sound from a solenoid-operated starter is caused by a _____ battery or _____ or _____ battery _____ .
9. Very slow cranking is caused by nearly _____ _____, high starting motor _____ _____, or poor _____ _____ .
10. A starter current draw test reveals the condition of the _____ .
11. Voltage drop tests are performed to isolate _____ _____ defective _____ or a defective _____ .
12. Starter electrical parts should never be cleaned in solvent. (T) (F)
13. A starter drive should not be cleaned in solvent. (T) (F)
14. The starter armature should be tested for _____ _____ and _____ .
15. Name the five major components of the hydraulic cranking system.
16. Fluid under high pressure is stored in the _____ .
17. A continuously running (while engine is running) hydraulic pump is used to charge the accumulator. (T) (F)
18. What is the hand-operated hydraulic pump used for?
19. Name three causes of slow hydraulic cranking motor cranking speed.
20. Cranking air is exhausted to the atmosphere through a muffler. (T) (F)
21. List five causes of slow air cranking speed and the correction for each.

TEST QUESTIONS

1. The starter solenoid connects the
   a. battery to the relay
   b. hold-in winding to the bypass
   c. pull-in winding to the starting motor
   d. battery to the starting motor
2. Technician A says that the clutch interlock switch prevents starting when the clutch pedal is depressed. Technician B says that the neutral switch prevents starting when the transmission is in gear. Who is right?
   a. Technician A   c. both are right
   b. Technician B   d. both are wrong
3. Technician A says that the force that causes the starter armature to turn is mechanical. Technician B says that the armature is forced to rotate by invisible lines of force. Who is right?
   a. Technician A   c. both are right
   b. Technician B   d. both are wrong
4. "All starting motors have field windings." "Some starting motors are designed without field windings." Which of these statements is correct?
   a. the first   c. both are correct
   b. the second   d. both are incorrect
5. Technician A says that all starting systems provide a gear reduction. Technician B says that some starting systems provide two gear reductions. Who is correct?
   a. Technician A   c. both are right
   b. Technician B   d. both are wrong
6. A shunt winding is one that is connected to the main winding in
   a. series   c. series–parallel
   b. parallel   d. any of the above
7. In a series-wound cranking motor, current flows through the
   a. field windings and pole shoes
   b. pole shoes and armature
   c. starter frame and field windings
   d. armature and field windings

8. The starting motor should not be operated continuously for more than
   a. 30 seconds   c. 3 seconds
   b. 3 minutes   d. 30 minutes
9. An overrunning clutch is used in the
   a. Bendix starter drive
   b. inertia starter drive
   c. positive shift starter drive
   d. none of the above
10. A starter solenoid that clicks rapidly when the key is turned to the start position indicates a faulty
    a. battery
    b. solenoid switch
    c. cable connection
    d. any of the above
11. When diagnosing starting system problems, the procedure is first to test the
    a. field current draw
    b. battery and battery cables
    c. starting motor current flow
    d. solenoid hold-in winding
12. A single click sound when the key is turned to the start position indicates a faulty
    a. battery or cable connection
    b. solenoid switch
    c. starting motor
    d. any of the above
13. A growler is
    a. a buzzing solenoid switch
    b. used to test armatures
    c. a discontented technician
    d. caused by a slipping starter drive
14. To perform a starter current draw test set the voltmeter scale on
    a. 2 V   c. 200 V
    b. 20 V   d. 400 V
15. To perform a voltage drop test, connect the voltmeter in _____ across the portion of the circuit to be tested.
    a. parallel   c. series–parallel
    b. series   d. none of the above
16. The air-starting motor consists of
    a. a five-vaned air motor
    b. a special clutch drive
    c. a five-vaned air motor with gear reduction
    d. a one-way clutch
17. With a hydraulic starting system, the hand pump is used
    a. to supply hydraulic pressure for each start
    b. only to start the initial charge
    c. only when there is a hydraulic pressure failure
    d. all of the above
18. An engine-driven hydrostarter pump runs continuously during engine operation
    a. recharging the accumulator with fluid
    b. to keep the hydrostarter lubricated
    c. to keep external engine parts lubricated
    d. to lubricate the hydrostarter

CHAPTER 11

DIESEL FUEL AND FUEL SUPPLY SYSTEMS

INTRODUCTION

The diesel fuel injection system must provide the right amount of clean fuel to each cylinder at the correct time and must atomize the fuel adequately for good combustion. This must be achieved without entry of air into the fuel system. Injection of fuel into the cylinder must take place for a controlled period of time and must prevent leakage of fuel from the injector during noninjection time. Combustion must be controlled to limit exhaust emissions within emission standards. **(Figures 11–1 to 11–5.)**

There are four basic types of diesel fuel injection systems to consider: (1) the multiple pump system, (2) the pressure time system, (3) the unit injector system, and (4) the distributor pump system.

The multiple pump system is used on many four-stroke-cycle engines. The pressure time system is used only by the Cummins Engine Company, Inc. on its engines, which are used in many different manufacturers' products. The unit injector system is used by Detroit Diesel Corporation on its engines, both two stroke and four stroke, which are used in many different applications. Caterpillar uses the unit injector system on some of its engines and the multiplunger injection pump on others. The distributor pump system is used on various makes of four-stroke-cycle engines.

PERFORMANCE OBJECTIVES

After thorough study of this chapter and the appropriate service manuals and training models, you should be able to do the following:

1. List four types of diesel fuel injection systems.
2. Store and handle fuel in a clean and safe manner.
3. Describe the diesel engine fuel supply system.
4. Describe the basic operation of the following supply pump types: diaphragm, gear, vane, plunger, and priming pumps.
5. State the purpose of the fuel filtering system.
6. Define primary, secondary, final, series, and parallel fuel filtering.

TERMS YOU SHOULD KNOW

Look for these terms as you study this chapter, and learn what they mean.

cetane number
volatility
viscosity
cloud point
heat value
flash point
sulfur content
ash content

diesel fuel
classification 1D
   and 2D
alternative fuels
fuel tank
fuel supply (transfer) pump
diaphragm pump
plunger pump

203

204 Chapter 11 DIESEL FUEL AND FUEL SUPPLY SYSTEMS

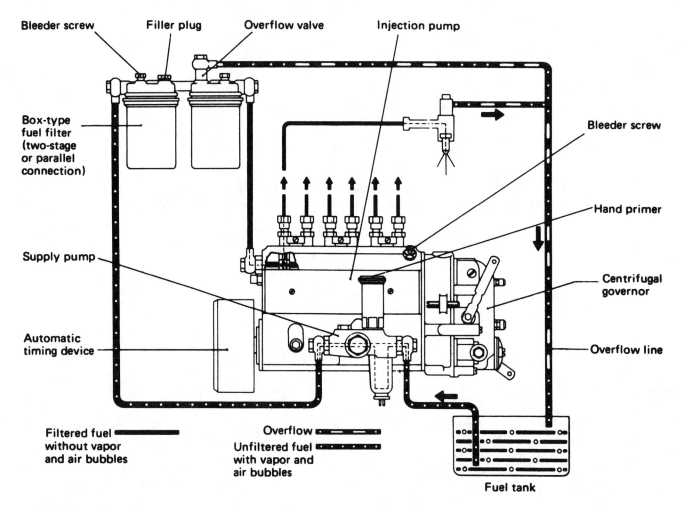

FIGURE 11-1 Fuel system with multiplunger injection pump. (Courtesy of Robert Bosch Canada Ltd.)

gear pump
vane pump
priming pump
primary filter
secondary filter
final filter
bleeding air from the fuel system

DIESEL FUEL QUALITY

Diesel fuels are obtained by the distillation of crude oil. The diesel fuel produced has certain characteristics that are modified or controlled during production. The fuel also contains some undesirable materials that are kept at a level that meets the requirements specified by different engine manufacturers. Engine performance and wear are affected by the following characteristics of diesel fuel.

1. *Cetane number:* a measure of the ignitability of the fuel.

2. *Volatility:* the boiling point of the fuel, which is a measure of its ability to vaporize.

3. *Viscosity:* a measure of the liquidity of the fuel, which affects its ability to vaporize.

4. *Cloud point:* the temperature at which wax crystals begin to form in the fuel and begin to plug the fuel filters.

5. *Heat value:* a measure of the amount of heat (and therefore power) a fuel is able to produce.

6. *Flash point:* the temperature of the fuel at which the vapors it produces ignite when a flame is passed over the surface of the liquid fuel.

7. *Sulfur content:* a nonmetallic element present in diesel fuel that can cause combustion chamber deposits and wear on pistons, rings, and cylinders; contributes to particulate emissions from the exhaust.

8. *Ash content:* a measure of the amount of ash produced by ash-forming materials contained in the fuel when a measured amount of fuel is burned.

Typical specifications for these fuel qualities are shown in **Figure 11-6.** Refer to the engine manufac-

DIESEL FUEL QUALITY

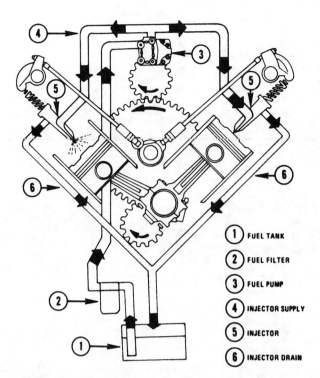

FIGURE 11-2 Typical Cummins pressure time (PT) fuel system. (Courtesy of Cummins Engine Company Inc.)

1. FUEL TANK
2. FUEL FILTER
3. FUEL PUMP
4. INJECTOR SUPPLY
5. INJECTOR
6. INJECTOR DRAIN

turer's service manual for actual specifications for any particular engine.

Viscosity

The viscosity of the fuel affects atomization and fuel delivery rate. The viscosity of diesel fuel is normally specified at 100°F (38°C). Fuels for medium-speed and high-speed engines generally lie in the range of 1.4 to 4.3 centistokes viscosity at 100°F (38°C). The lubricating properties of some low-viscosity, low-pour winter fuels can be improved by the addition of 1% crankcase oils or lubricity additives. This is important where injection pumps and injectors depend on fuel oil for lubrication, and the fuel oil viscosity is below 1.3 centistokes at 100°F (38°C).

Fuels with viscosities over 6 centistokes at 100°F (38°C) are limited to use in slow-speed engines and may require preheating for injection.

Cloud Point

As diesel fuel cools to a certain temperature, the fuel will become cloudy due to the formation of wax crystals. This is referred to as the *cloud point*. Wax crystals may begin to plug fuel filters when the fuel temperature drops to the cloud point. How critical this is in winter operation depends on the design of the fuel system with regard to fuel line bore, freedom from bends, size and location of filters, and degree of warm fuel recirculation, as well as the amount and kind of wax crystals.

Additives known as *flow improvers* are being used successfully to improve the fuel fluidity at low temperatures. Flow improvers modify wax crystal growth so that the wax, which forms at low temperatures, will pass through the fine (typically 10 to 20 microns) (a micron is 0.000001 m) fuel filter screens. The addition of 0.1% flow improver can result in satisfactory fuel flow at 15°F (−9°C) colder temperatures than possible with untreated fuel.

Fuel heaters are used on some trucks and other equipment to keep the temperature of the fuel above the cloud point. Different methods are used to provide the necessary heat depending on system design. Methods used include heat from warm engine coolant, from warm engine return fuel, or from an electric heating element **(Figure 11-7)**.

Heat Value

The heat value of fuel is a general indication of how much power the fuel will provide when burned. This is obtained by burning the oil in a special device known as a calorimeter. A measured quantity of fuel is burned, and the amount of heat is carefully measured in BTUs per pound of fuel. In the metric system the heat unit is the *joule*. To convert BTUs into joules, multiply by 1054.8 **(Figure 11-8)**.

Cetane Number and Ignition Delay Period

The cetane number is a measure of the autoignition quality of a diesel fuel. The shorter the interval between the time when the fuel is injected and the time when it begins to burn, called the *ignition delay period*, the higher the cetane number. It is a measure of the ease with which the fuel can be ignited and is most significant in low-temperature starting, warm-up, and smooth even combustion. Using a fuel with a cetane rating that is too low can cause the engine to knock. **(Figure 11-9)**.

Some hydrocarbons ignite more readily than others and are desirable because of this short ignition delay. The cetane number is measured in a single-cylinder test engine with a variable compression ratio. The refer-

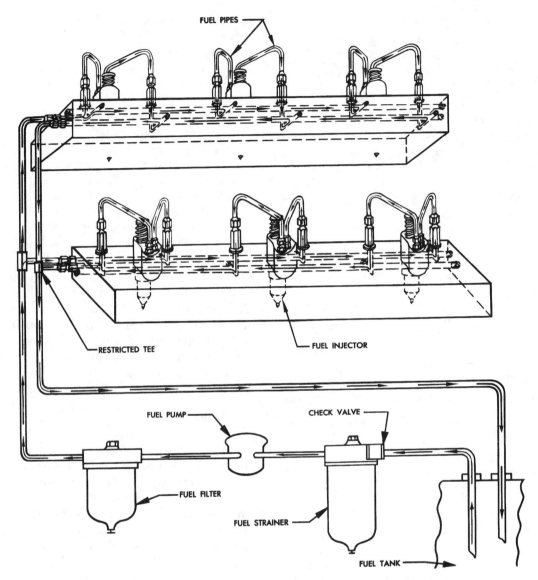

FIGURE 11-3 Fuel system for Detroit Diesel two-stroke-cycle V6 engines with unit injectors. (Courtesy of Detroit Diesel Corporation.)

ence fuels used are mixtures of cetane, which has a very short ignition delay, and alpha-methyl naphthalene, which has a long ignition delay. The percentage of cetane in the reference fuel that gives the same ignition delay as the test fuel is defined as the cetane number of the test fuel.

Diesel engines whose rated speeds are below 500 rpm are classed as slow-speed engines, from 500 to 1200 rpm as medium-speed, and over 1200 rpm as high-speed. Cetane numbers of fuels readily available range from 40 to 55, with values of 40 to 50 most common. These cetane values are satisfactory for medium- and high-speed engines, whereas low-speed engines may use fuels in the 25 to 35 cetane number range.

Volatility

The distillation characteristics of the fuel describe its volatility. A properly designed fuel has the optimum proportion of low-boiling components for easy cold starting and fast warm-up and heavier components, which provide power and fuel economy when the engine reaches operating temperature. Either too high or too low volatility may promote smoking, carbon deposits, and oil dilution due to the effect on fuel injection and vaporization in the combustion chamber. The 10, 50, and 90% distillation points and the final boiling point are the principal volatility controls. Diesel engines in automotive, agricultural, and construction ser-

ALTERNATIVE FUELS

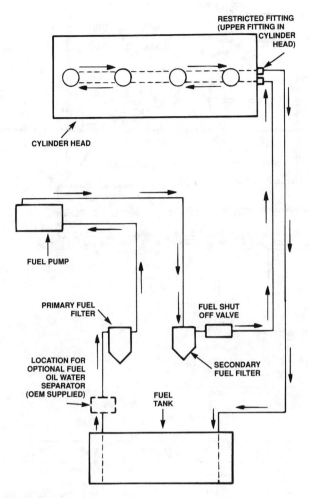

FIGURE 11-4 Fuel system for four-stroke-cycle engine with unit injectors. (Courtesy of Detroit Diesel Corporation.)

vice use fuels with a final boiling point approaching 700°F (371°C). Urban buses generally use a fuel with a lower final boiling point to minimize exhaust smoke and odor.

Flash Points

The flash point is determined by heating the fuel in a small enclosed chamber until the vapors ignite when a small flame is passed over the surface of the liquid. The temperature of the fuel at this point is the flash point. The flash point of a diesel fuel has no relation to its performance in an engine nor to its auto ignition qualities. It does provide a useful check on suspected contaminants such as gasoline, since as little as 0.5% of gasoline present can lower the flash point of the fuel very markedly.

Shipping, storage, and handling regulations are predicated on minimum flash point categories.

Sulfur Content

Sulfur in diesel fuel can cause combustion chamber deposits, exhaust system corrosion, and wear on pistons, rings, and cylinders—particularly at low water-jacket temperatures—and increased particulate emissions. Sulfur tolerance by an engine is dependent on the type of engine and the type of service. A fuel sulfur content above 0.4% is generally considered medium or high, whereas below 0.2% is considered low. Summer grades of commercially available diesel fuel are commonly in the 0.1 to 0.4% sulfur range. Winter grades often have less than 0.2% sulfur. Most diesel fuels are low in sulfur content, and by using the engine builder's recommended crankcase oil quality and oil change intervals, there should be no concern about the effects of fuel sulfur. As emissions regulations are tightened, sulfur content in diesel fuels has to be reduced.

Ash Content

Diesel fuel contains ash-forming materials that cause wear of injection equipment and engine cylinders, pistons, and rings. Ash from these materials contributes to the formation of deposits. The ash content of a fuel is determined by burning a specified amount of fuel in an open container until all the carbon is consumed. The weight of the remaining ash is expressed as a percentage of the weight of the fuel consumed in the test sample.

Diesel Fuel Classification

The American Society for Testing and Materials (ASTM) has set minimum quality standards for diesel fuel grades as a guide for engine operators. These are grades 1D, which specifies winter fuel, and 2D, which defines summer fuel. These definitions are very broad, and the diesel fuels marketed will meet one of these definitions. Quality requirements such as sulfur, ash, water, sediments, and corrosion rating are met by diesel fuel that is blended on a seasonal and geographical basis to satisfy anticipated temperature conditions (see **Figure 11-6**).

ALTERNATIVE FUELS

The U.S. National Energy Strategy (NES) provides tax incentives to encourage the use of alternative fuels to reduce pollution from engine exhausts. Alternative fuels include compressed natural gas (CNG), liquefied

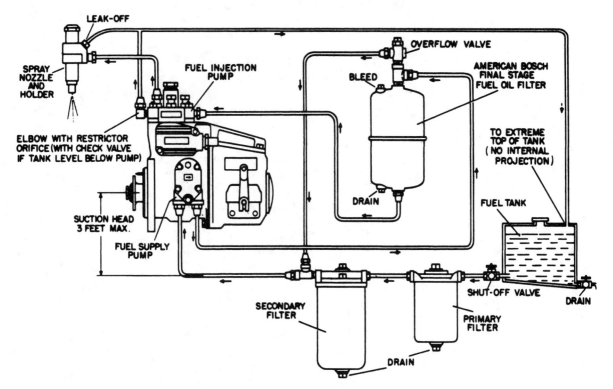

FIGURE 11-5 Fuel system with distributor injection pump. (Courtesy of AMBAC International Corporation.)

General Fuel Classification	ASTM Test	No. 1 ASTM 1D[a,b]	No. 2 ASTM 2D[c]
Gravity[d] (°API)	D287	40–44	33–37
Flash point, minimum [°F (°C)]	D93	100 (38)	125 (52)
Viscosity, kinematic at 100°F (40°C) (cSt)	D445	1.3–2.4	1.9–4.1
Cloud point[d] (°F)	D2500	e	e
Sulfur content, maximum (wt%)	D129	0.5	0.5
Carbon residue on 10% maximum, (wt%)	D524	0.15	0.35
Accelerated stability total insolubles, maximum[d] (mg/100 ml)	D2274	1.5	1.5
Ash, maximum (wt%)	D482	0.01	0.01
Cetane number, minimum[f]	D613	45	45
Distillation temperature [°F (°C)]	D86		
Initial boiling point, typical[d]		350 (177)	375 (191)
10%, typical[d]		385 (196)	430 (221)
50%, typical[d]		425 (218)	510 (256)
90%[f]		500 (260) maximum	625 (329) maximum
Endpoint[d]		550 (288) maximum	675 (357) maximum
Water and sediment, maximum (%)	D1796	0.05	0.05

[a] When prolonged idling periods or cold-weather conditions below 32°F (0°C) are encountered, the use of 1D fuel is recommended. No. 1 diesel fuel should also be considered when operating continuously at altitudes above 5000 ft.
[b] No. 1 diesel fuel is recommended for use in city coach engine models.
[c] No. 2 diesel fuel may be used in city coach engine models that have been certified to pass federal and California emission standards.
[d] Not specified in ASTM D975.
[e] The cloud point should be 10°F (6°C) below the lowest expected fuel temperature to prevent clogging of fuel filters by crystals.
[f] Differs from ASTM D975.

FIGURE 11-6 Example of diesel fuel specifications. (Courtesy of Detroit Diesel Corporation.)

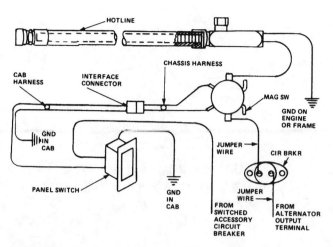

FIGURE 11-7 Fuel line heater electrical schematic. (Courtesy of Peterbilt Motors Company, a Division of Paccar.)

1D Diesel	137,000
2D Diesel	141,800
Gasoline	125,000
Butane	103,000
Propane	93,000

FIGURE 11-8 Heat value per gallon in BTUs.

natural gas (LNG), methanol, and ethanol. Of these, LNG is the most energy dense. LNG has 60% as much energy per gallon as No. 2 diesel fuel, compared with 55% for ethanol, 44% for methanol, and 22% for CNG. The quantity of fuel able to be stored on board and therefore the distance that can be traveled before refueling are considerably lower than for diesel fuel. A great deal of research and development is being expended on alternative fuels, with entire fleets being retrofitted and new alternative-fuel vehicles being added. More innovations, increased production, and an increasing number of refueling stations could result in lower costs and an increased adoption of these systems. At the same time, however, diesel fuel injection systems are much more efficient than they were previously and, with the very precise electronic controls on current systems, are able to meet the more stringent emissions regulations.

FUEL TANK DESIGN

The fuel supply tank is constructed of steel or aluminum and may be round or rectangular in shape. Many units have more than one fuel tank (**Figure 11-10**). In this case the operator is able to select which

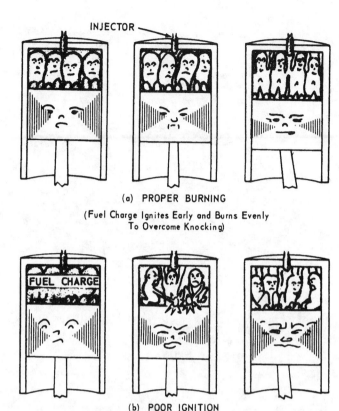

(a) PROPER BURNING
(Fuel Charge Ignites Early and Burns Evenly To Overcome Knocking)

(b) POOR IGNITION
(Ignition Of Fuel Charge Is Delayed, Followed By A Small Explosion)

FIGURE 11-9 In diesel engines, knock is due to the fuel igniting too slowly. It should start to burn almost as soon as it is injected (a). If there is any delay, a fuel buildup results, which burns with explosive force (b) and causes knocking.

tank is to supply the fuel by a fuel control valve. The fuel tank is provided with a vented filler cap, which allows the tank to "breathe" as the fuel expands and contracts due to temperature change and as fuel is consumed.

Also provided is a fuel supply line fitting to which the fuel line is connected. A second fuel line fitting is provided for fuel systems that use an injector "leak off" fuel return line. The fuel tank is usually equipped with a fuel shut-off valve at the fuel outlet fitting. Some fuel tanks have a water trap that is equipped with a drain tap. Water from condensation in the fuel tank collects in the water trap, from which it is periodically drained.

It is of the utmost importance that diesel fuel be kept absolutely clean. This means that dirt, moisture, and other foreign material must be kept out of the system when handling fuel or fuel system components.

The fill caps and fill nozzles should be cleaned before removal and refueling.

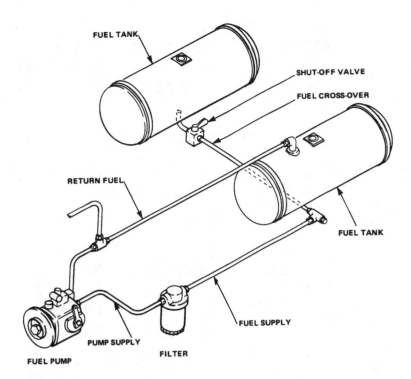

FIGURE 11-10 Heavy-duty truck fuel saddle tanks. (Courtesy of Peterbilt Motors Company, a Division of Paccar.)

FUEL SUPPLY PUMP DESIGN AND OPERATION

A fuel supply (or transfer) pump is needed in the fuel system to provide an adequate supply of fuel at specified pressure from the fuel tank to the fuel injection pump or the unit injectors via the necessary fuel filters. Several types of fuel supply pumps are being used. Camshaft-operated diaphragm pumps, gear pumps, plunger pumps, and hand-priming pumps are used, depending on the type of fuel system.

Diaphragm Pump

The mechanically operated fuel pump consists of a lever-operated spring-loaded diaphragm, an inlet valve, and an outlet valve. The lever is operated from an eccentric on the engine camshaft. A push rod may be used between the eccentric and the lever. The lever pivots on a pin to provide rocker arm action. The other end of the lever is connected to the diaphragm link with a sliding or slotted connection. This allows the diaphragm to be pulled by the lever (**Figure 11-11**).

When the diaphragm is lifted, it creates a low-pressure area in the pump. Atmospheric pressure on the fuel in the tank causes fuel to be forced into the low-pressure area in the pump past the one-way inlet valve. Spring pressure keeps the outlet valve closed during the pump intake stroke.

When the lever moves down, the diaphragm spring forces the diaphragm down and pressurizes the fuel. The pressure closes the inlet valve and opens the outlet valve.

Spring pressure forces the diaphragm down to supply the system with sufficient fuel for all operating conditions.

Fuel pressure is dependent on diaphragm spring pressure.

Some diaphragm pumps are equipped with a lever that is connected to the diaphragm so that the pump can be operated manually. This allows the operator to prime the fuel system manually and purge all air from the system before starting the engine after fuel system service.

Plunger Pump

The plunger type of supply pump is operated from a cam on the camshaft of the fuel injection pump. As the injection pump camshaft rotates, the plunger follows the camshaft profile by moving up and down in its bore. The plunger spring keeps the plunger roller against the cam and controls output pressure (**Figures 11-12 and 11-13**).

FUEL SUPPLY PUMP DESIGN AND OPERATION 211

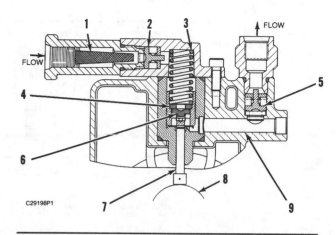

Fuel Transfer Pump
(1) Screen. (2) Inlet check valve. (3) Spring. (4) Piston assembly. (5) Outlet check valve. (6) Piston check valve. (7) Tappet assembly. (8) Cam. (9) Passage.

FIGURE 11-12 Cam-operated plunger pump components. (Courtesy of Caterpillar Inc.)

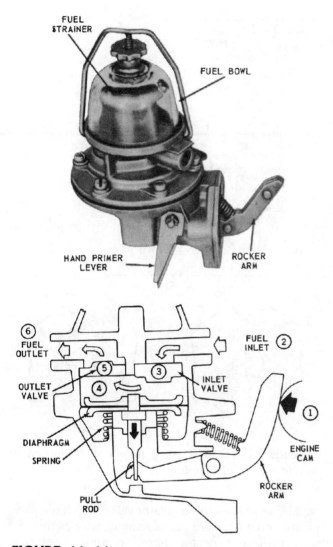

FIGURE 11-11 Typical diaphragm pump operation. (Courtesy of Deere and Company.)

As the plunger moves down, a low-pressure area is created in the pump body fuel chamber. This causes the outlet valve to close and the inlet valve to open. Atmospheric pressure in the fuel tank forces fuel to flow in past the inlet valve and fills the chamber above the plunger. At the same time fuel in the chamber below the plunger is forced out to the filter and then to the injection pump. As the injection pump cam continues to turn, the plunger moves up, creating pressure on the fuel above the plunger. This closes the inlet valve and opens the outlet valve, forcing fuel through the outlet valve to the chamber below the plunger. During this portion of plunger travel, fuel is simply transferred from above the plunger to the area below the plunger. This cycle of events is then repeated as the injector pump camshaft continues to rotate.

Gear Pump

The gear type of fuel supply pump is driven from the fuel injection pump camshaft, engine camshaft, or from the air blower drive on two-stroke-cycle engines (**Figures 11-14** and **11-15**).

Fuel enters the pump body on the inlet side of the gears when the rotation of the gears creates low pressure in this area. Atmospheric pressure in the fuel tank forces fuel out of the tank through the line to the fuel supply pump. Fuel trapped between the teeth of each gear is carried around to the outlet side of the housing. As the teeth of the two gears mesh, fuel is forced out from between the gear teeth and out of the fuel outlet to the filter.

A passage from the outlet is connected to the pressure-regulating valve. A spring-loaded valve is forced back against spring pressure as pressure builds up. When output pressure increases sufficiently, the relief valve is forced back far enough to open a connecting passage to the inlet side of the pump. Spring pressure on the pressure-regulating valve thereby controls pump output pressure.

Vane Pump

The vane-type fuel supply pump consists of a stationary liner or race, a set of spring-loaded vanes or blades, and a slotted drive rotor. The vanes fit into the slots of the rotor and are pushed away from center into contact

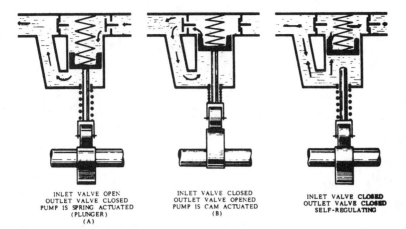

FIGURE 11–13 Cam-operated plunger pump operation. (Courtesy of AMBAC International Corporation.)

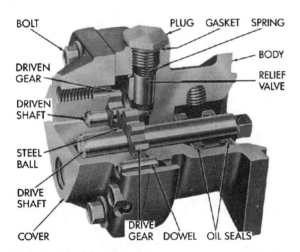

FIGURE 11–14 Cross section of gear-type fuel supply pump. (Courtesy of Detroit Diesel Corporation.)

FIGURE 11–15 Gear pump operation. (Courtesy of Stanadyne Diesel Systems.)

with the outer race or liner by the springs. Since the liner is offset or eccentric in relation to the rotor shaft, rotating the rotor causes the vanes to move in and out of their slots. This movement changes the volume between the vanes. **(Figure 11–16)**.

Turning the rotor causes an increase in volume between the vanes positioned over the fuel inlet passage. Increasing the volume creates a low-pressure area between the vanes, allowing atmospheric pressure in the fuel tank to force fuel into the pump low-pressure area. As the rotor continues to turn, the vanes pass the inlet passage. Continued rotation forces the vanes back into their respective slots in the rotor. This pressurizes the fuel trapped between the vanes. At this point, rotor rotation has positioned this area over the fuel outlet passage, and fuel is forced out of the pump.

This action takes place in turn between each pair of vanes as the rotor is forced to turn. The outlet passage connects to a pressure regulator valve, which regulates pump output pressure in the same manner as the gear pump pressure regulator. Output pressure is controlled by regulator valve–spring pressure.

The vane pump is usually driven by the fuel injection pump and may be part of the injection pump itself.

Priming Pumps

Some fuel systems are equipped with a hand-operated priming pump. This is a plunger type of pump designed to prime the system manually with fuel to eliminate air from the system before starting the engine after any air has been admitted **(Figure 11–17)**.

When the plunger is lifted, it creates a low-pressure area in the plunger barrel. Atmospheric pressure acting on the fuel in the fuel tank forces fuel into the plunger barrel by opening the check valve on the inlet side. The outlet check valve is closed by this action. As the plunger is forced down in the barrel, the pressure closes the inlet valve and opens the outlet valve, permitting pressurized fuel to flow to the fuel supply line of the injection system. This pumping action is re-

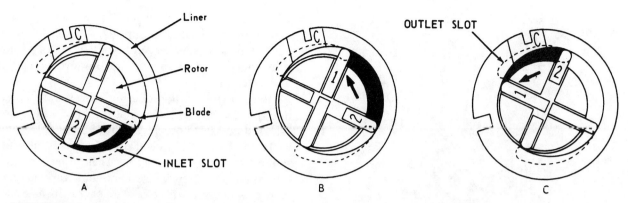

FIGURE 11-16 Vane type of fuel transfer pump, showing pump operation. (Courtesy of Deere and Company.)

peated until high pressure on the plunger is felt, indicating that the system is fully primed and all air is removed.

Most diaphragm-type supply pumps have a hand-operated lever provided to operate the diaphragm as a priming pump (see **Figure 11-11**).

FUEL FILTERS

Effective filtering of the fuel oil is most important for long, trouble-free fuel injection system operation. Dirt is the worst enemy of the fuel injection equipment. Even the most microscopic particles of abrasive foreign material can destroy the finely machined and lapped parts of the injection system. Clearances between moving parts of the injection system can be as low as from 0.000007 to 0.000008 in. (0.0001778 mm to 0.0002032 mm). Dirt of such dimensions that will affect such close tolerances is often suspended in the atmosphere. As the fuel tank "breathes" this air as a result of fuel level fluctuations, this type of dirt enters the fuel system even though very careful handling of the fuel may prevent entry of dirt through other means (**Figures 11-18 to 11-20**).

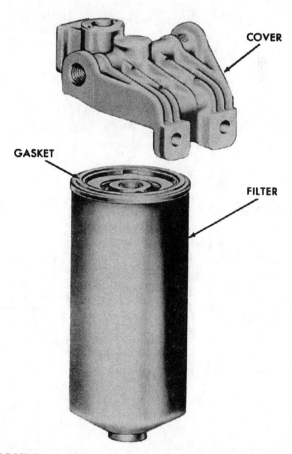

FIGURE 11-18 Spin-on fuel filter. (Courtesy of Detroit Diesel Corporation.)

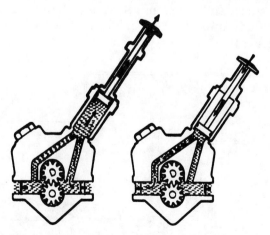

FIGURE 11-17 Hand-priming pump attachment for gear type of fuel transfer pump. (Left) plunger is raised to fill cylinder, (right) plunger is pushed down, discharging fuel past the outlet check valve. (Courtesy of Stanadyne Diesel Systems.)

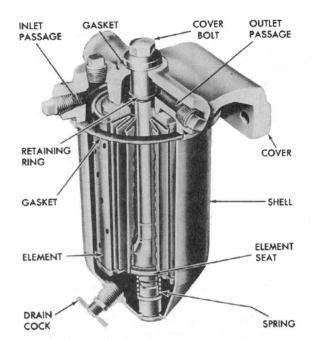

FIGURE 11-19 Fuel filter with replaceable element. (Courtesy of Detroit Diesel Corporation.)

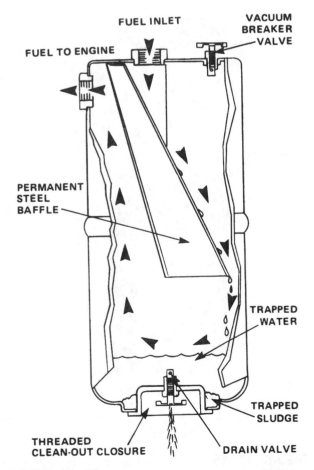

FIGURE 11-20 Fuel filter/water separator. (Courtesy of Peterbilt Motors Company, a Division of Paccar.)

For these reasons a very efficient filtering system that will remove these impurities and still allow free fuel flow is necessary. Several types of filters are used for this purpose: the strainer type, the surface type, and the depth type.

The porosity of the filtering material will determine the size of impurities that the filter will remove. This is stated in microns. One micron is 0.000039 in. or 0.000001 m. The clearance between an injector plunger and barrel may be as little as 2 microns. If the plunger is held in the hand and warmed, it will expand enough so that it cannot be inserted into the barrel until the temperature of the two units is equalized.

The strainer in the fuel system may be simply a finely woven screen or it may be a series of stacked disk-type screens. This screens out the larger impurities in the fuel.

The *surface type* of filter material is usually of the paper element accordion bellows design. Fuel oil passing through the paper element deposits impurities on the element surface.

The *depth type* of filter is a cotton-fiber-filled canister element. Impurities are collected at various depths in this type of filter as fuel oil passes through.

Filters are further classified as *primary* and *secondary* or *final* filters when they are connected in series (**Figure 11-21**). Series-connected filters cause fuel to pass through the primary filter first, then to the secondary and final filter. All the fuel must first pass through the primary filter before it can go to the secondary filter. In some cases, fuel filters are arranged in parallel. Fuel is able to flow through both filters at the same time. If one filter becomes plugged, the other can still function. Filters may be of the spin-on, throw-away type or of the replaceable element canister type.

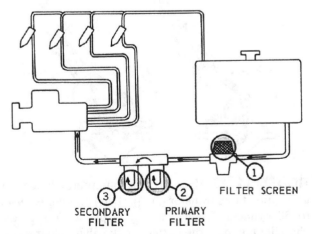

FIGURE 11-21 Three stages of fuel filtering. (Courtesy of Deere and Company.)

BLEEDING AIR FROM THE FUEL SYSTEM

Air is able to enter the fuel system any time it is opened by removing a fuel filter or disconnecting a fuel line. Air left in the system can form an air lock or aerate the fuel. Air in the system can prevent fuel from reaching the fuel injection pump or unit injectors or from going through them to the combustion chambers. This can prevent the engine from starting or cause misfiring and loss of power. Any time the fuel system has been opened for service, air trapped in the system must be removed.

The general procedure for bleeding the system is as follows.

1. Make sure there is an adequate supply of fuel in the fuel tank, filling it with the proper diesel fuel if necessary.

2. Make sure the fuel tank shut-off valve is fully open.

3. Loosen the bleed screw at the filter, strainer, or water separator nearest to the fuel tank in terms of fuel flow. Pump the priming pump. Operate the manual priming pump until a solid stream of fuel flows from the bleed screw opening. The fuel must be completely free of air bubbles.

NOTE: It may be necessary to turn the engine over to position the heel of the cam away from the priming pump. Tighten the bleed screw.

4. Repeat this procedure at each of the remaining filters in turn.

REVIEW QUESTIONS

1. Name the four basic types of diesel fuel injection systems.
2. Detroit Diesel engines use a _____ injector system.
3. Cummins Engine Company uses a _____ _____ injector system on some of its engines.
4. Cetane number is a measure of the _____ of diesel fuel.
5. Name eight characteristics of diesel fuel.
6. Diesel fuel may require _____ _____ to improve its fluidity at low temperatures.
7. Diesel engines with a rated speed of 1200 rpm are classified as _____ _____ engines.
8. Diesel fuel classification grade 1D specifies _____ fuel.
9. Diesel fuel tanks are constructed of _____ or _____ .
10. The fuel supply pump supplies fuel to the _____ _____ or the _____ _____ .
11. Name the five types of fuel supply pumps.
12. State the functions of the fuel filters.
13. Air in the fuel system can cause hard starting, _____ , and a loss of _____ .

TEST QUESTIONS

1. Mechanical fuel pumps
   a. are not reliable
   b. are easily repaired
   c. cannot be disassembled
   d. are all exactly alike
2. Technician A says fuel supply pumps should be tested for output pressure only. Technician B says fuel pumps should be tested for fuel output volume only. Who is correct?
   a. Technician A   c. both are right
   b. Technician B   d. both are wrong
3. Hydrocarbon emission results from
   a. engine misfiring
   b. light ends in the fuel
   c. high combustion temperature
   d. lean mixtures
4. The fuel system that uses a water separator is a
   a. gasoline system   c. natural gas system
   b. propane system   d. diesel system
5. A fuel transfer pump is used on a
   a. gasoline fuel system
   b. propane fuel system
   c. natural gas fuel system
   d. diesel fuel system
6. ASTM is the abbreviation for
   a. American Satellite Transmission Mode
   b. American Society of Transport Modes
   c. American Society of Tetraethyl Manufacturers
   d. American Society for Testing Materials

7. Fuel knock in a diesel engine is caused by
   a. fuel burning too slow
   b. fuel burning too fast
   c. fuel not reaching the cylinder
   d. a defective fuel pump

8. The water/fuel separator works on the following principles:
   a. relative weight difference and water filtering
   b. water filtering and trapping water
   c. water is heavier than fuel and penetrates fuel more easily
   d. all of the above

9. The clearance between an injector plunger and _____ may be as little as 2 _____.

10. Larger impurities are removed from the fuel by a finely _____ screen or a series of _____ disk-type screens.

11. The surface type fuel filter is usually of the paper _____ accordion _____ design.

12. Fuel filters are classified as being primary, _____, and _____ types.

◆ CHAPTER 12 ◆

Diesel Fuel Injection

INTRODUCTION

This chapter deals with the conditions required for proper fuel injection and efficient combustion to occur. Factors that contribute to good combustion include combustion chamber design and fuel injection nozzle design. A thorough understanding of these principles is required for accurate diagnosis and service of fuel injection nozzles. Mechanical and electronic unit injectors are discussed in other chapters of this book.

PERFORMANCE OBJECTIVES

After thorough study of this chapter and the appropriate training models and service manuals, you should be able to do the following:

1. List four requirements that a diesel fuel injection system must meet.
2. Define the ignition delay period.
3. List five combustion chamber designs and briefly describe each one.
4. Describe the operation of the hole type and the pintle type injection nozzles.
5. List the causes and effects of:
   a. too high a nozzle opening pressure
   b. too low a nozzle opening pressure
   c. incorrect nozzle spray pattern
   d. nozzle dribble or leakage
6. Remove, clean, test, adjust, and install fuel injection nozzles according to manufacturer's specifications.

TERMS YOU SHOULD KNOW

Look for these terms as you study this chapter, and learn what they mean.

- direct injection
- precombustion chamber
- turbulence (swirl) chamber
- M-type chamber
- energy cell
- injection nozzle
- nozzle holder
- hole-type nozzle
- pintle nozzle
- throttling pintle nozzle
- injection tubing
- nozzle opening pressure
- pressure adjusting screw
- pressure adjusting shims
- nozzle spray pattern
- nozzle dribble
- ultrasonic cleaning

FUEL INJECTION PRINCIPLES

Every diesel fuel injection system must have the ability to inject the fuel into the cylinders at sufficiently high pressures to assure good atomization and vaporization of the fuel in the cylinders. It must accurately meter the quantity of fuel injected to correspond with the engine's speed and load. It must do this at exactly the correct time in relation to piston position and crankshaft angle. It must inject this fuel over an accurately determined period of time to provide the correct rate of pressure rise in the cylinder and to maintain as much as possible a pressure level in the cylinder to provide the needed power and torque output. This requires a correct beginning, duration, and ending of injection related to engine speed.

Engine efficiency and the combustion process are most favorable when the amount of exhaust smoke is minimal. If combustion is incomplete or too much fuel is injected, the engine "smokes." This may result in overheating, with subsequent engine damage, such as piston and cylinder damage and oil dilution.

The preparation of the fuel in the cylinder for proper combustion takes a certain amount of time. The period from the time injection begins until the time when combustion begins is known as the *ignition delay period*. The injected amount of fuel should be small during this period, which is about 0.001 second. The objective is to obtain (as much as possible) constant pressure in the cylinder by means of a controlled rate of injection and combustion. This is not possible in high-speed diesel engines but can more nearly be achieved in large slow-speed engines. In direct injection diesel engines the injected amount of fuel should be generally constant per degree of crank angle for this reason.

Different manufacturers achieve these results in different ways. Some use an injection pump with injectors connected to each pumping element by thick-walled fuel injection lines of exactly equal inside diameter and length. In this system a separate cam-operated pumping element is used for each cylinder. The pumping elements may be contained in a single pump housing or in separate pump housings, one for each cylinder. The Bosch, AMBAC, Caterpillar, and CAV are examples of this system.

Other systems combine the high-pressure pumping element and the injector, one for each cylinder. Detroit Diesel and Caterpillar unit injectors use a similar pumping and metering element in each injector. The Cummins pressure time system uses a plunger and cup in each injector as the pumping element.

Other systems used include a plunger and barrel with a metering sleeve as the pumping and metering element. In the Caterpillar sleeve-metering system, each cylinder is provided with a pumping element with all the elements contained in one pump housing. In the AMBAC distributor system only one pumping element and metering sleeve are provided, with high-pressure fuel delivered to each cylinder by a distributing head and a rotating plunger.

COMBUSTION CHAMBER DESIGN

The principle of fuel injection is to inject a metered quantity of fuel into each cylinder at exactly the correct time in a properly atomized and definite spray pattern for the type of combustion chamber used, in a way that provides the most efficient means of combustion possible. Regardless of cylinder pressures, which can be quite high, fuel injection pressure must be considerably higher to inject fuel properly into the pressurized cylinder. Fuel injection pressures may be as high as 20,000 psi (137,900 kPa) or more.

To achieve this goal, different manufacturers have produced several different types of fuel systems and combustion chamber designs.

Combustion chamber designs include the open, precombustion, turbulence, energy cell, and M types.

In order to burn all the fuel injected into the cylinder, there must be adequate atomization and vaporization of the fuel. There must also be enough air around each fuel particle (mixing) to allow complete combustion.

The injectors spray fuel into the cylinders in a very fine mist under very high pressures. This provides the necessary vaporization; turbulence and mixing is provided by the combustion chamber design.

Direct Injection Combustion Chamber

In the direct injection design, fuel is injected directly into the combustion chamber. The cylinder head and valves provide a flat cover or top for the chamber. The design of the piston head forms the combustion chamber. Various piston head designs are used to accomplish this action.

Piston designs for the open combustion chamber include a flanged-domed design, a dished design with a raised center area (Mexican hat), and other irregular shaped designs. Advantages claimed for the open combustion direct injection design include good fuel economy and simplicity of design **(Figure 12–1)**.

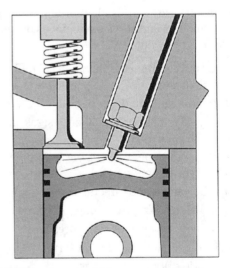

FIGURE 12–1 Direct injection system with combustion chamber in piston head. (Courtesy of Robert Bosch Canada Ltd.)

Precombustion Chamber

The precombustion chamber design consists of two interconnected chambers. The smaller chamber is located in the cylinder head or block close to the top of the cylinder. The fuel injector is mounted so that fuel is injected into the precombustion chamber, where combustion begins. The precombustion chamber is connected to the area above the piston, and as fuel injection and combustion continue, pressure is forced through the connecting passage to the top of the piston. The precombustion chamber design allows the use of a wider range of fuels and provides very smooth combustion **(Figure 12–2)**.

Turbulence Chamber (Swirl Chamber)

The turbulence chamber resembles the precombustion chamber in that a separate turbulence chamber is connected to the area above the piston. There is very little room above the piston when it is at TDC. As the piston moves up on the compression stroke, compressed air is forced into the turbulence chamber. The turbulent air promotes good mixing of fuel and air and provides good combustion **(Figure 12–3)**.

M-Type Chamber

The M type of combustion chamber consists of a spherical chamber in the head of the piston. This chamber has a small opening at the top. The fuel injector is positioned so that fuel will be injected into this chamber.

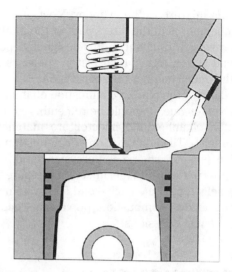

FIGURE 12–3 This swirl-chamber system uses a flat-head piston. (Courtesy of Robert Bosch Canada Ltd.)

This type of chamber has the advantage of eliminating the well-known diesel knock. It is also capable of using a variety of different fuels such as diesel fuels, kerosene, and gasoline **(Figure 12–4)**.

Energy Cell

The energy cell is a combination of the precombustion chamber and the turbulence chamber. It is also known as the *Lanova combustion chamber*. Combustion takes place mostly in the main figure 8 chamber. This design depends on a great deal of turbulence to provide the necessary mixing of air and fuel and the distribution of

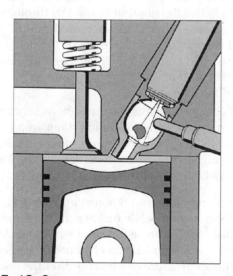

FIGURE 12–2 A dished piston head design is used with this prechamber combustion system. (Courtesy of Robert Bosch Canada Ltd.)

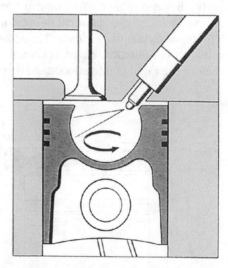

FIGURE 12–4 The M type of combustion system uses a piston with the combustion chamber in the piston head. (Courtesy of Robert Bosch Canada Ltd.)

the air–fuel mixture in the cylinder. Most of the combustion chamber is in the direct path of the intake and exhaust valves. Turbulence in this design is dependent on thermal action rather than piston speed as in the case of the open chamber **(Figure 12–5).**

Regardless of the type of combustion chamber used on any given engine, it should be remembered that the combustion chamber and injector are matched and work as a team. The combination of combustion chamber design (shape), engine compression ratio, fuel injection spray pattern, and fuel injection pressure is very carefully selected by the engine and fuel system manufacturers for good combustion, power, fuel economy, and low exhaust emissions.

COMBUSTION STAGES IN A DIESEL ENGINE

The combustion process begins as soon as the first part of the fuel charge enters the combustion chamber. Fuel injection continues for approximately 15° to 35° of crankshaft rotation. The diesel combustion process consists of the following four distinct periods or stages:

1. *Delay period*—first stage
2. *Rapid burn period*—second stage
3. *Constant pressure period*—third stage
4. *Afterburn period*—fourth stage

The first stage is the ignition delay period, which is the time from the start of injection until actual ignition of the fuel begins. The delay period includes the time required to atomize the fuel, mix it with air, and atomize it to produce the best possible air/fuel mixture for combustion. It also includes the chemical delay time, during which preflame oxidation occurs and localized temperatures in the combustion chamber reach 2000°F (1093°C) before the rapid combustion stage is reached.

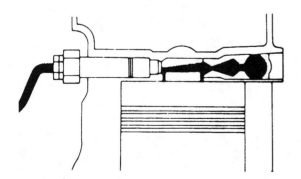

FIGURE 12–5 Lanova power cell (energy cell) type of combustion chamber.

Maximum combustion chamber temperature may reach 3000° to 4000°F (1649° to 2204°C).

The ignition delay period is a factor in the degree of roughness or combustion knock of an engine. The longer the delay period, the greater the amount of fuel that is injected into the cylinder before ignition occurs. The larger the amount of fuel injected before ignition begins, the greater the engine knock. When ignition finally occurs after the delay period, the large quantity of fuel in the cylinder burns very rapidly, resulting in extremely high cylinder pressures. The sudden violent combustion and consequent rapid rise in cylinder pressure causes the vibrations and noise known as combustion knock or diesel knock.

The following factors shorten the ignition delay period:

1. Greater atomization of the fuel being injected results in smaller fuel droplets, which speeds vaporization. Factors that reduce droplet size include higher injection pressures, higher fuel viscosity, smaller injection nozzle spray holes, and higher compression pressures.

2. Higher air charge temperatures shorten the delay period, since fuel ignition temperatures are reached earlier. Engine compression ratios, ambient air temperature and density, and cylinder wall and piston head temperature are factors that affect the temperature of the air charge at the time of injection.

3. Higher air charge pressures result in smaller droplets of fuel, faster vaporization, and a shorter delay period. Compression ratios, ambient air temperature and density, and whether the engine is naturally aspirated or turbocharged affect air charge pressures.

4. Greater turbulence of the compressed air charge results in better distribution of the fuel throughout the air charge, with a resultant increased rate of heating and fuel vaporization, which reduces the delay period. Intake port and combustion chamber design affect the direction of air flow and turbulence into and within the combustion chamber.

5. Higher cetane number fuels have shorter delay periods and lower auto-ignition temperatures. The higher the number of carbon molecules per molecule of fuel, the higher the cetane number of the fuel. The cetane number of fuel is discussed in Chapter 11.

The second stage of combustion, called the *rapid burn period*, occurs while fuel is still being injected. It begins just before the piston reaches TDC and results in a rapid rise in combustion chamber pressure. Maximum cylinder pressure occurs only after the piston has started moving downward in the cylinder. During the rapid burn stage fuel injection, atomization, vaporization, and combustion are occurring at the same time.

The third stage, called the *constant pressure stage*, is a period of combustion during which the remainder of the fuel charge is injected. Combustion continues and maintains a constant pressure on the downward-moving piston until injection ends.

The fourth stage, known as the *afterburn period*, occurs after fuel injection has ended. Any unburned fuel able to combine with oxygen is burned at this time.

INJECTION NOZZLE FUNCTION AND OPERATION

The fuel injection nozzle directs the metered quantity of fuel received from the injection pump into the combustion chamber of the engine. Fuel is discharged in a finely atomized pattern in such a manner that it is thoroughly mixed with compressed air in the combustion chamber.

These nozzles are held in a device known as a *nozzle holder*, which holds the nozzle in position in the cylinder head of the engine. In addition, the holder conveys the fuel from the injection tubing to the nozzle. The nozzle holder also contains the mechanism for controlling the nozzle opening pressure and leak-off ducting. A copper gasket is used between the injector and the cylinder head to seal compression and promote maximum heat transfer.

For maximum power and economy, it is essential that the fuel be consumed completely without excessive smoke or hydrocarbons, carbon monoxide, or oxides of nitrogen. This is accomplished by dispersing the fuel in a mist of infinitesimally fine droplets. In that way the surface area of each droplet is greatly reduced and maximum combustion can result.

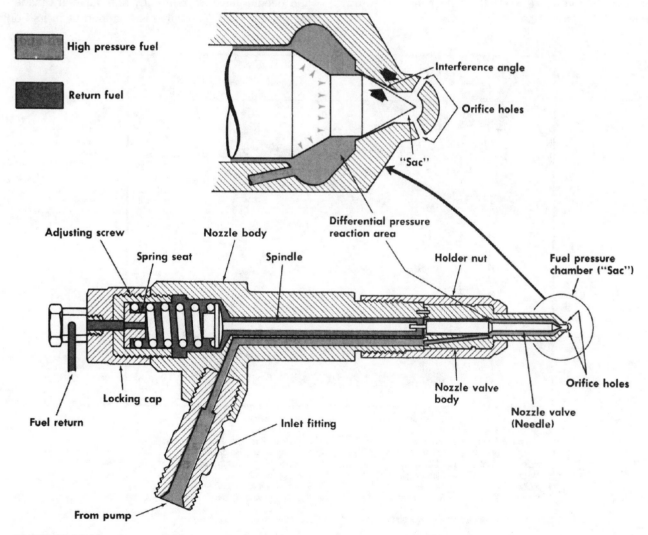

FIGURE 12–6 Typical nozzle operation and cross section of nozzle holder components. (Courtesy of Deere and Company.)

222   Chapter 12 DIESEL FUEL INJECTION

To obtain these characteristics, the design details of the nozzle are dependent largely on features of the engine, such as its speed, configuration of combustion chamber, size of engine, compression ratio, and intended use.

The nozzle is operated by fuel pressure. The pressure generated by the injection pump acts on the exposed annular area of the nozzle valve and lifts it from its seat as soon as the force of the pressure spring in the nozzle holder is exceeded. The fuel is then injected through the nozzle orifices into the combustion chamber **(Figures 12–6 to 12–9)**.

During injection, the fuel takes the following path: fuel injection tubing to the fuel passage of the nozzle holder (annular groove) inlet passages, pressure chamber, and spray orifice(s) of the nozzle to the combustion chamber. Fuel leaking past the valve stem is returned to the fuel tank via the leak-off connection on the nozzle holder and a return line.

After injection of the fuel delivered by the injection pump, the pressure spring again forces the valve back on its seat via the spindle and the valve stem. The injector is then closed for the next pressure stroke.

The following description of injection nozzle design and operation is typical of most diesel injection systems using an engine-driven injection pump with external high-pressure fuel injection lines used to deliver fuel to the injectors.

DESIGN AND APPLICATION OF INJECTION NOZZLES

(This section courtesy of Robert Bosch Canada Ltd.)

The Bosch injection nozzle consists of a nozzle body and needle valve. Nozzle body and valve are made of high-grade steel. The valve is a lapped fit in its body.

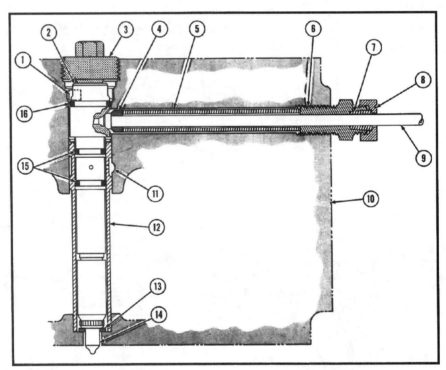

1. Nozzle Holder Locating Pin (Part of Nozzle Holder)
2. Nozzle Holder
3. Nozzle Holder Clamping Screw
*4. Nozzle Fuel Inlet Tube Collar
*5. Nozzle Fuel Inlet Tube Sleeve
*6. Nozzle Fuel Inlet Tube Clamping Screw
*7. Nozzle Fuel Inlet Tube Sleeve
*8. Nozzle Fuel Inlet Tube Sleeve Nut
9. Nozzle Fuel Inlet Tube
10. Cylinder Head
11. Fuel Drain Passage (Nozzle Leak Off)
12. Injection Nozzle Holder Insert
13. Injection Nozzle Holder Gasket
14. Injection Nozzle
15. Injection Nozzle Holder O-Ring Seal (1)
16. Injection Nozzle Holder O-Ring Seal (Upper)

*These parts are included with injection Nozzle Fuel Inlet Tube (Item 9) as an assembly only.

FIGURE 12–7 Cross section of injection nozzle and fuel inlet tube in cylinder head. (Courtesy of Mack Trucks Inc.)

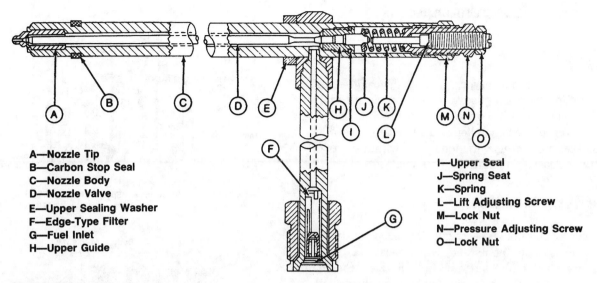

FIGURE 12-8 Cross section of fuel injection nozzle. (Courtesy of Deere and Company.)

A—Nozzle Tip
B—Carbon Stop Seal
C—Nozzle Body
D—Nozzle Valve
E—Upper Sealing Washer
F—Edge-Type Filter
G—Fuel Inlet
H—Upper Guide
I—Upper Seal
J—Spring Seat
K—Spring
L—Lift Adjusting Screw
M—Lock Nut
N—Pressure Adjusting Screw
O—Lock Nut

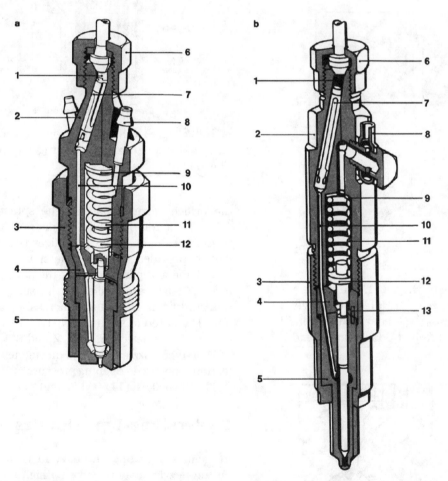

FIGURE 12-9 Bosch nozzle-and-holder assemblies. (a) With throttling pintle nozzle; (b) with hole-type nozzle. 1. Feed passage. 2. Holder body. 3. Nozzle-retaining nut. 4. Intermediate disk. 5. Nozzle. 6. Union nut with fuel injection tubing. 7. Edge filter. 8. Leak-off connection. 9. Pressure adjustment shims. 10. Pressure passage. 11. Pressure spring. 12. Pressure spindle. 13. Locating pins. (Courtesy of Robert Bosch Canada Ltd.)

These two parts are considered as a unit for replacement purposes.

There are two main types: *hole-type nozzles*—for open chamber engines (also known as direct injection engines)—and *pintle nozzles*—for precombustion chamber, turbulence chamber, and air-cell engines.

However, numerous variations exist within the two main types, based on the diversity of engines.

Hole-Type Nozzles

The valve of the hole-type nozzle (**Figures 12–10** to **12–12**) has a cone at its end, which serves as seat. There are single-hole and multihole nozzles. Single-hole nozzles have only one orifice, which may be drilled centrally or laterally. In the case of multihole nozzles,

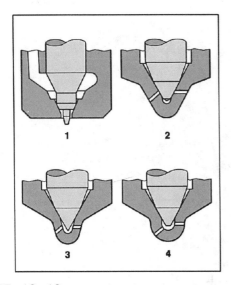

FIGURE 12–10 Nozzle shapes. 1. Throttling pintle nozzle. 2. Seat-hole nozzle. 3. Hole-type nozzle with conical blind hole. 4. Hole-type nozzle with cylindrical blind hole. (Courtesy of Robert Bosch Canada Ltd.)

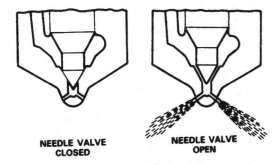

FIGURE 12–11 Hole-type nozzle operation. (Courtesy of Robert Bosch Canada Ltd.)

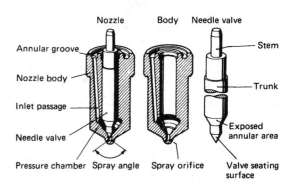

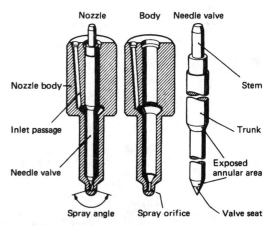

FIGURE 12–12 Standard (top) and long-stem (bottom) hole-type nozzles. (Courtesy of Robert Bosch Canada Ltd.)

the orifices form an angle—the spray angle (up to 180°). To obtain optimum fuel distribution in the combustion chamber, up to 12 orifices (usually symmetrical) are provided. Orifice diameter and orifice length affect shape and penetration of the spray. Conventional nozzle designs are available with spray orifice diameters starting at 0.2 mm and increasing in steps of 0.02 mm. The different sizes of hole-type nozzles are distinguished by the letters *S*, *T*, *U*, *V*, and *W*.

Hole-type nozzles are used on engines with direct injection. The nozzle opening pressure is usually between 2100 and 3600 psi (14,500 kPa and 24,800 kPa).

Cooled Hole-Type Nozzles

In engines operating on heavy fuels, the temperature on the nozzles may become so high that they require cooling.

The nozzle body of these nozzles is equipped with a passage for the fuel inlet, while the two others serve for coolant inlet and outlet. The lower end of the body has a double thread, which is closed to the outside by a cooling jacket. The coolant is discharged from the feed duct of the nozzle holder into the inlet passage of the nozzle, then through one thread of the double thread into the annulus; from there it is forced through the outlet passage into the outlet duct of the nozzle holder.

Oil as well as oil–water emulsions that do not attack steel can serve as coolants.

Pintle Nozzles

The valve of the pintle nozzle has a specially shaped pintle at its end, which projects into the spray hole of the nozzle body with a slight clearance. The spray pattern can be changed as needed by changing the dimensions and profile of the pintle. Furthermore, the pintle keeps the spray orifice free from carbon deposits **(Figure 12–13)**.

Pintle nozzles are used in precombustion chamber and turbulence chamber engines. In these engines, the fuel is prepared mainly by the air turbulence supported by a suitable shape of the spray. For pintle nozzles, the nozzle opening pressure is usually between 1100 psi and 1800 psi (7600 kPa to 12,400 kPa).

Throttling Pintle Nozzles

The throttling pintle nozzle is a pintle nozzle with special pintle dimensions.

A pilot injection is obtained by the shape of the pintle tip. The valve frees only a very narrow annular orifice during opening, allowing only a little fuel to pass (throttling effect). With further opening (produced by a pressure increase), the annular orifice area is increased and the main portion of the fuel is injected only near the end of the valve lift. Normally, combustion and engine operation become smoother with the throttling pintle nozzle because the pressure in the combustion chamber increases slowly. The desired throttling effect is obtained by the pintle configuration together with the characteristics of the pressure spring (in the nozzle holder) and the clearance in the throttling gap **(Figure 12–14)**.

DIESEL FUEL INJECTION TUBING

High-pressure fuel injection tubing is used to connect the injection pump to the fuel injectors. The tubing is of thick-walled, seamless design with precise inside-diameter dimensions to ensure equal fuel delivery to each engine cylinder. All the lines on an engine are of exactly the same length, for the same reason.

Injection tubing has a tensile strength of about 55,000 lb (25,000 kg) to prevent tubing expansion during injection. Inside diameters are accurately sized and free of any irregularities, to ensure even injection to all cylinders. This requires that outside dimensions also be precise and free of irregularities.

Injection tubing bends are formed with a mechanical tubing bender that prevents flattening and kinks. Internal dimensions must not be altered in bends. Tubing bends should have a radius of at least 1 in. (25.4 mm) or more to ensure that flow characteristics are maintained.

The most common injection tubing size is ¼ in. (6.4 mm) outside diameter. The tubing is available in several inside diameters to suit individual injection system requirements. Common inside dimensions and their exterior color code markings are as follows:

0.063 in. (1.6 mm)	red
0.067 in. (1.7 mm)	black
0.078 in. (2.0 mm)	yellow
0.084 in. (2.1 mm)	blue
0.093 in. (2.4 mm)	white

Color code markings are in the form of intermittent striping on the tubing exterior. Other sizes of injection tubing for larger diesel engines are also used and include outside diameters of 5/16 in. (7.9 mm), 3/8 in. (9.5 mm), 7/16 in. (11.1 mm), and ½ in. (12.7 mm).

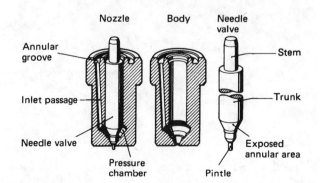

FIGURE 12–13 Pintle nozzle components. (Courtesy of Robert Bosch Canada Ltd.)

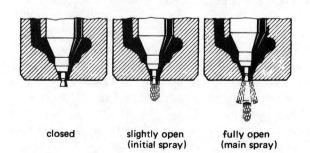

FIGURE 12–14 Throttling pintle nozzle operation. (Courtesy of Robert Bosch Canada Ltd.)

Several different types of injection tubing fittings are used to attach the tubing to the pump and injectors. These include the tapered seat type, the flared type, and the ferrule type.

INJECTOR PROBLEMS

Effects of Faulty Injector Operation

1. *Opening pressure too high:* tends to decrease the amount of fuel injected; tends to retard injection timing; loss of power and efficiency.
2. *Opening pressure too low:* decreases fuel atomization; may increase amount of fuel injected; can cause nozzle dribble, exhaust smoke, and loss of efficiency.
3. *Incorrect spray pattern:* exhaust smoke; carbon deposits; loss of power.
4. *Leaky nozzle, nozzle dribble:* exhaust smoke; detonation; loss of power; crankcase oil dilution; carbon deposits.

Causes of Faulty Injector Operation

1. *Opening pressure too high:* incorrect pressure setting; nozzle valve sticking; plugged nozzle spray holes.
2. *Opening pressure too low:* incorrect pressure setting; faulty pressure spring; nozzle valve sticking.
3. *Incorrect spray pattern:* nozzle valve eroded; plugged nozzle spray holes; nozzle valve sticking; dirt in nozzle.
4. *Leaky nozzle, nozzle dribble:* nozzle valve or seat damaged; dirt in nozzle; broken pressure spring or adjusting screw; nozzle valve sticking.

ISOLATING A FAULTY INJECTOR NOZZLE

A faulty injector nozzle may be identified by using the following procedure. With the engine at operating temperature and running at idle speed, loosen the injection line fittings in turn at each injector nozzle just enough to allow the fuel to escape and prevent it from reaching the engine cylinder. This prevents that cylinder from producing power. On cylinders with normally functioning nozzles the engine will run at a slightly reduced speed, and an engine miss condition will result. A faulty nozzle will have little or no effect on the engine during this procedure.

REMOVING INJECTION NOZZLES

1. Disconnect fuel injection and return lines from the injection nozzles (**Figure 12–15**).
2. Remove the nozzle retaining nuts and clamp.
3. Carefully remove the nozzle from the cylinder head.

CAUTION:

Do not damage the injector during removal. Use a roll-head prybar on injectors with mounting flanges. A special nozzle puller may be required in some cases (**Figures 12–16** and **12–17**).

INJECTION NOZZLE SERVICE

Injector service varies, depending on the particular make and model. All removal, disassembly, inspection, overhaul, reassembly, testing, adjusting, and installation procedures for any given make and model should follow the recommendations in the appropriate manufacturer's shop service manual.

It is critical to good engine performance that all injectors for any engine be of the same type and size. Never mix injectors or mated and lapped injector parts.

Some injectors are not repairable and must be replaced if defective. Others can be overhauled and adjusted.

Although many service shops are equipped to provide injector rebuilding, other shops rely on replacing injectors with rebuilt or new injectors.

FIGURE 12–15 Always use two wrenches when loosening or tightening fuel lines: (1) to hold the fitting adapter and prevent it from turning and (2) to turn the fitting nut. (Courtesy of Deere and Company.)

INJECTION NOZZLE SERVICE 227

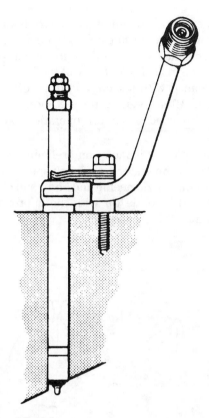

FIGURE 12–16 Pencil-type injector and injector mounting clamp. (Courtesy of Stanadyne Diesel Systems.)

FIGURE 12–17 Fuel inlet tube removal (top) and injector removal (bottom) on an E7 Mack Truck engine using an injector puller tool. (Courtesy of Mack Trucks Inc.)

The following service procedures apply in general to injectors. In addition, the appropriate service manual must be used for all special procedures and all service data. Unit injectors require additional service procedures as outlined in the appropriate chapters.

General Procedure

1. Clean the area around the injectors before removal to prevent dirt and other foreign matter from entering the fuel system. Any high-pressure washing system is good for this job.

CAUTION:

Never steam clean an injection pump while it is running. Pump parts may seize and be damaged.

2. Disconnect all fuel lines from injectors. Cap all openings.

3. Remove the injector-retaining device. Remove the injector. Some injectors must be removed using a special puller. Follow service manual procedures.

CAUTION:

Be careful not to damage injectors during removal. Pencil injectors are especially critical.

4. If the problem diagnosis procedure has established that an engine compression test is required, this should be done when the injectors are removed. Record the results of the compression test. If an engine compression problem is indicated, repair the engine as necessary.

5. Disassemble and clean the injectors according to the appropriate manufacturer's service manual. Do not mix parts from one injector with those of another injector.

6. Carefully clean and inspect all injector parts and compare wear and damage to service limits indicated in the appropriate manufacturer's service manual.

Testing Injection Nozzles

The injectors should be removed periodically from the engine in accordance with the engine manufacturer's instructions for routine testing and examination. Disassembly, cleaning, and reconditioning are carried out according to the condition of the injectors.

Each injector should be checked on a lever-operated test pump, before and after any cleaning or dismantling work is carried out, to determine the general condition of the injector when the injection pressure is applied. Injection setting pressures, together with spray hole and needle lift details, are given in manufacturer's service manuals for all types of injectors.

WARNING:

When the injectors are being tested, care must always be exercised not to allow the fuel spray to come into contact with the hands; otherwise the spray will penetrate the skin. Wear protective gloves and a face mask. Use a transparent spray containment device to contain fuel spray.

PRELIMINARY TESTING

1. Expel the air from the test pump, which has been filled with test fluid. Mount the injector on the test pump and check the pressure at which the needle valve opens. Compare with the specified setting pressure.

2. Examine the spray when pumping about 60 strokes per minute. The correct spray should appear as a fine mist without distortion and without visible streaks of unvaporized fuel (**Figure 12–18**).

3. Check the nozzle tip, which should remain dry on completion of the fuel injection. Nozzle dribble indicates seat leakage.

4. Check for leakage at the nozzle nut. Leakage may result if the joint faces of nozzle and nozzle holder do not register correctly to form a fuel-tight joint, or if there is dirt between the faces.

5. If the injector operates satisfactorily on preliminary testing, it should not be dismantled. The nozzle should be cleaned with a brass wire brush to remove any carbon deposit. The exterior of the injector should then be cleaned with solvent and dried off with compressed air. After cleaning the exterior, the injector should be checked again for operation on the test pump.

6. If the injector is faulty on preliminary testing, it must be disassembled for cleaning, inspection of parts, polishing of the nozzle needle and body, and cleaning of the holes in the nozzle tip with the appropriate size wires designed for the purpose.

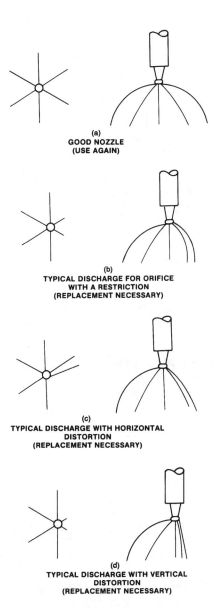

FIGURE 12–18 Spray pattern analysis for hole-type nozzle. (Courtesy of Caterpillar Inc.)

INJECTION NOZZLE SERVICE

CAUTION:
Never use anything other than tallow for polishing the nozzle parts. Never use anything other than the specified wires for cleaning the nozzle holes. See **Figures 12–19 to 12–21** for typical injector nozzle service tools.

Ultrasonic cleaning is a very effective method for cleaning injector parts. Ultrasonic cleaning uses soundwave vibrations that may reach 55,000 cycles per second. The cleaner uses a mixture of a special cleaning powder and water. The disassembled injector parts are placed on a cleaning tray that keeps parts for each injector separate. The tray of parts is immersed in the cleaning solution for about 15 to 30 minutes **(Figures 12–22 and 12–23)**. After ultrasound cleaning the nozzle spray holes are cleaned with the proper size cleaning tool.

Follow equipment manufacturer's instructions for proper use. Cleaning fluid must not be allowed to contact skin. Always use protective gloves and a face mask.

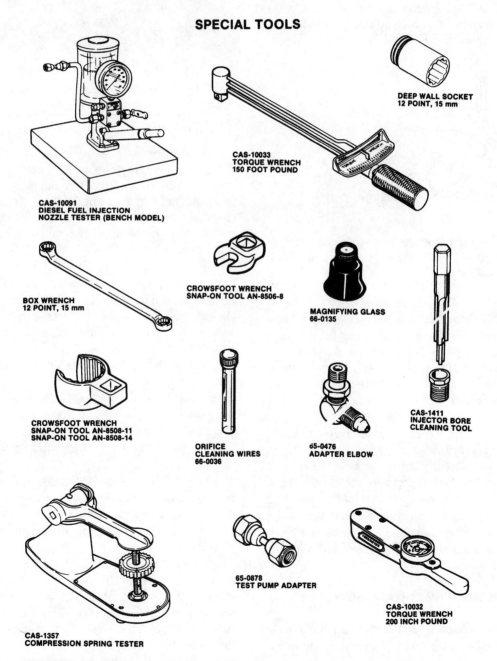

FIGURE 12–19 Typical nozzle service tools. (Courtesy of Case International.)

FIGURE 12-20 Cleaning the nozzle tip with a brass wire brush. (Courtesy of Stanadyne Diesel Systems.)

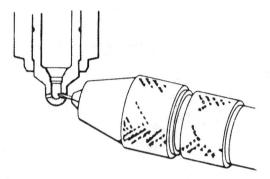

FIGURE 12-21 Cleaning injector nozzle orifices with special tool of specified diameter. (Courtesy of Stanadyne Diesel Systems.)

Inspection may require the use of a magnifying glass due to the extremely close tolerances and finely honed surfaces of injector parts. Some injector parts are not available separately. Mating parts such as nozzles must be purchased as an assembly due to the very close tolerances of these parts.

7. Assemble the injector according to the procedures given in the appropriate service manual. *Never apply undue pressure or force.* All parts must be at normal room temperature and be properly lubricated with the specified fuel oil. If parts are not at equal temperatures, it may be impossible to assemble them.

8. Mount the injector on the specified test stand and, using the specified test fluid, which should be at the proper testing temperature, proceed with injector testing and adjustment.

TESTING AFTER CLEANING AND REASSEMBLY

After thoroughly cleaning and inspecting the nozzle parts, assemble the nozzle and nozzle holder and retest as follows.

FIGURE 12-22 Ultrasonic cleaning equipment. (Courtesy of Mack Trucks Inc.)

1. *Specified nozzle opening pressure:* Operate the tester handle to expel any air, and thoroughly flush the nozzle. Adjust the opening pressure with the adjusting screw or shims depending on design. To assure balanced fuel delivery to all cylinders set all nozzles within the specified range (usually no more than 30 to 50 psi (210–345 kPa) difference between nozzles).

2. *Chatter test:* Isolate the tester gauge when performing the chatter test to avoid damage to the gauge. Operate the tester handle at 60 strokes per minute and listen for nozzle chatter, a buzzing noise resulting from the rapid opening and closing of the nozzle needle. Nozzle chatter indicates that the needle is moving freely.

3. *Nozzle drip or dribble:* With the tester handle apply pressure to the nozzle [200 psi (1380 kPa) below specified opening pressure] and observe the nozzle tip for any sign of leakage. A leaking nozzle must be replaced.

4. *Leak back:* With the nozzle under pressure as in step 4, note whether there is excessive leak back through the nozzle return passage (back to the tank). If leak back is excessive, the nozzle should be replaced.

5. *Spray pattern:* With the tester gauge isolated, operate the tester at 60 strokes per minute and note the shape of the spray pattern. The spray pattern should be uniform and well atomized with no offshoots of solid streams of fuel. The shape of the pattern should match that specified for the type of nozzle being tested. Dirty or eroded nozzle holes can cause irregular spray pat-

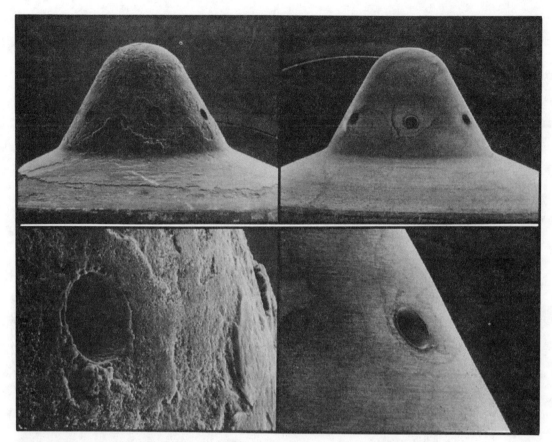

FIGURE 12-23 Nozzle tip before (left) and after (right) cleaning. (Courtesy of Mack Trucks Inc.)

terns. Clean or replace as required **(Figures 12-24 to 12-26).**

INSTALLING INJECTION NOZZLES

1. Thoroughly clean the nozzle recess in the cylinder head. Make sure the old sealing washer is removed from the recess. Use the special carbon-removing tool to clean the seating area and then wipe it clean. Absolute cleanliness is essential to forming a good seal and preventing combustion leakage **(Figure 12-27).**

2. Install a new sealing washer and carbon dam of the type specified on the nozzle **(Figure 12-28).** Insert the nozzle into the recess in the head and push it into place. Index the nozzle to the proper position in the head.

3. Install the retaining clamp (if used) and the retaining nuts finger tight. Tighten the nuts to specified torque **(Figure 12-29).**

4. Install the injection line.

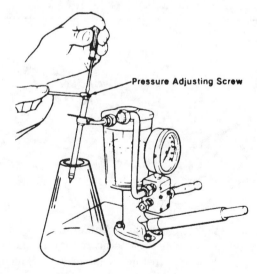

FIGURE 12-24 Adjusting nozzle opening pressure. (Courtesy of Stanadyne Diesel Systems.)

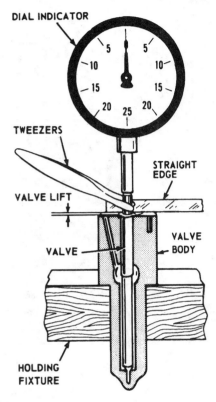

FIGURE 12-25 Using a dial indicator to measure nozzle valve lift. (Courtesy of Mack Trucks Inc.)

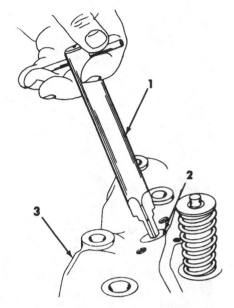

FIGURE 12-27 Cleaning the nozzle bore. 1. Bore cleaning tool. 2. Nozzle bore. 3. Cylinder head. (Courtesy of Allis-Chalmers.)

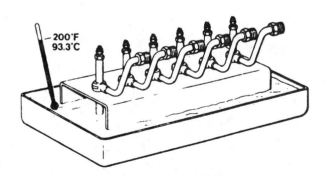

FIGURE 12-26 Controlled temperature of nozzles immersed in controlled temperature fluid is important for accurate nozzle testing. (Courtesy of Stanadyne Diesel Systems.)

FIGURE 12-28 Installing nozzle sealing washer. (Courtesy of Mack Trucks Inc.)

FIGURE 12-29 Installing the nozzle (top) and nozzle retainer (bottom). (Courtesy of Mack Trucks Inc.)

CAUTION:

Do not bend. Start the fitting nuts by hand to avoid cross threading. Leave the fitting nuts loose enough to allow air to be bled from the lines. Install any fuel return lines.

5. Install the remaining nozzles and lines in the same manner.

6. Place the fuel control (throttle) in the full fuel position and crank the engine until fuel is present at all nozzle fittings.

CAUTION:

Do not crank the engine for more than 30 seconds at a time to avoid starting motor damage. Allow enough time for starter cooldown between cranking perods.

7. Tighten the injection line fittings to specified torque and start the engine. Check for fuel leakage at the fittings. A leak at a fitting can be corrected by slightly loosening the fitting to allow fuel to wash any foreign particles away. Tighten the fitting and recheck for leaks.

8. Install all the injection line support brackets and clips and tighten to specifications.

REVIEW QUESTIONS

1. Explain the function of the diesel fuel injection system.
2. If combustion is incomplete or too much fuel is injected, the engine _____.
3. Name the four combustion chamber designs used in diesel engines.
4. Injector nozzles spray fuel into the cylinders in a _____ _____ or _____ under very high pressures.
5. The M type of combustion chamber consists of a _____ chamber in the piston head.
6. Explain the function of the injector nozzle.
7. The injector nozzle is operated by _____ _____.
8. The Bosch injection nozzle consists of a _____ _____ and _____ valve.
9. The valve in the hole-type nozzle has a cone at its end that serves as a _____.
10. Pintle nozzles are used in _____ chamber and _____ chamber engines.
11. All injection lines on an engine are of exactly the same _____.
12. List four causes of faulty injector operation.
13. Some injectors are not repairable and must be _____ if faulty.
14. Name the four steps in a nozzle test procedure.

TEST QUESTIONS

1. The injector assembly consists of two principal subassemblies:
   a. the nozzle and nozzle holder
   b. the nozzle holder and the pintle
   c. the pintle and the nozzle
   d. the pintle and the barrel

2. Positive cutoff of fuel injection is assured by the use of
   a. a governor
   b. a fuel shut-off valve
   c. injector nozzle size controls
   d. a drain-back valve

3. In a diesel fuel injection system fuel is injected at
   a. low pressure   c. variable pressures
   b. high pressure   d. the throttle body

4. The main purpose of the fuel injection nozzle is to
   a. deliver fuel to the intake manifold
   b. inject fuel into the throttle body
   c. direct atomized fuel into the combustion chamber
   d. direct fuel into the exhaust manifold

5. During injector service if injector parts are not at equal temperatures,
   a. it may be impossible to assemble them
   b. there will be too much clearance between parts
   c. severe parts damage may result
   d. injector opening pressure must be adjusted

6. Testing and adjusting injectors does not include
   a. obtaining proper injector chatter
   b. setting opening pressure
   c. testing for nozzle drip
   d. testing for nozzle closing pressure

7. Electronic control of injection timing is done through
   a. a rotary actuator and relay
   b. a transistor and rotary actuator
   c. a duty-cycle solenoid
   d. a conventional robot

8. The fuel delivery valve cuts off fuel
   a. from the fuel tank
   b. to the injector at the end of injection
   c. to the glow plugs
   d. to the altitude compensator

9. The diesel fuel injection system does not have
   a. high-pressure injectors
   b. a fuel pump
   c. fuel filters
   d. a throttle valve in the air intake

10. Injection lines are made
    a. from copper tubing
    b. to be equal in length
    c. as cheaply as possible
    d. from aluminum

◆ CHAPTER 13 ◆

GOVERNING FUEL DELIVERY

INTRODUCTION

The diesel engine governor controls the amount of fuel delivered to the engine at any given throttle setting. It changes the amount of fuel delivered as the engine speed and load change. Engine speed may change from low idle to high idle. Engine load may change from no load to full load. The governor must react quickly to these changes and increase or decrease fuel delivery accordingly. The governing of fuel delivery may be mechanical or electronic. This chapter deals with mechanical governors for Robert Bosch in-line injection pumps. Other governors and electronic controls are covered in other chapters.

PERFORMANCE OBJECTIVES

After thorough study of this chapter and the appropriate training models and service manuals, you should be able to do the following:

1. State the functions of the governor.
2. Name four types of governors.
3. State the function of the manifold-pressure compensator.
4. State the function of the altitude-pressure compensator.

TERMS YOU SHOULD KNOW

Look for these terms as you study this chapter, and learn what they mean.

governor
speed droop
hunting
stability
sensitivity
promptness
high-idle speed
full-load speed
low-idle speed
underrun

overrun
mechanical governor
hydraulic governor
pneumatic governor
isochronous governor
torque control
flyweights
governor springs
manifold-pressure compensator
altitude-pressure compensator

THE NEED FOR A GOVERNOR

In a diesel engine there is no fixed position of the control rod at which the engine will maintain its speed accurately without a governor. During idling, for example, without a governor the engine speed would either drop to zero or would increase continuously until the engine raced and ran completely out of control. The latter possibility results from the fact that the diesel engine operates with an excess of air; consequently, effective throttling of the cylinder charge does not take place as the speed increases (**Figures 13–1 to 13–3**).

If a cold engine is started, for example, by the starting motor, and if it is permitted to continue idling with a corresponding amount of fuel injected, the inherent friction in the engine, as well as the transmission resistance of parts driven by the engine (such as the generator, air compressor, and fuel injection pump) decreases after a certain length of time. As a result, if the position of the control rod were to remain

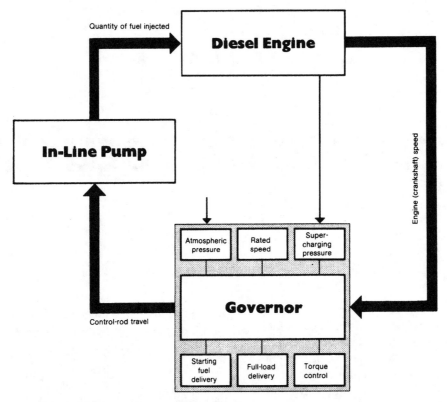

FIGURE 13–1 This diagram shows the factors that affect governor operation. (Courtesy of Robert Bosch Canada Ltd.)

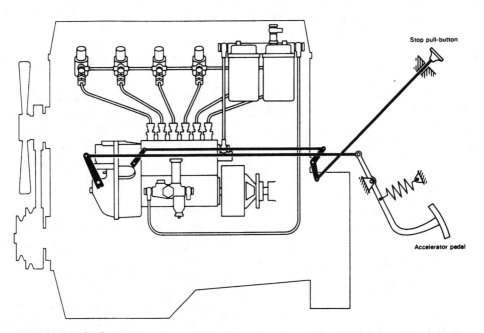

FIGURE 13–2 Operator controls and linkage to injection pump and governor. (Courtesy of Robert Bosch Canada Ltd.)

THE NEED FOR A GOVERNOR

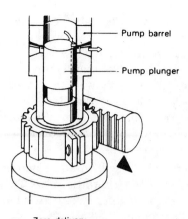
Zero delivery

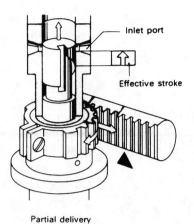
Partial delivery

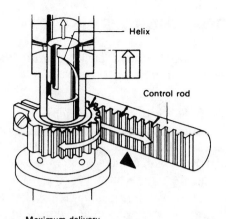
Maximum delivery

FIGURE 13-3 Metering of fuel delivery in a multiplunger pump is controlled by pump plunger position, which in turn is controlled by the control rod (rack). The control rod is connected to the governor, which moves the rack in response to engine speed and load as well as throttle position. (Courtesy of Robert Bosch Canada Ltd.)

unchanged without a governor, the engine speed would constantly increase and could rise to a level at which the engine would ultimately destroy itself. The delivery of fuel to the engine must be controlled to provide the following operating characteristics, depending on the type of service the engine is designed to perform:

- control of excess fuel for starting
- control of engine low-idle speed
- limiting of maximum engine speed
- maintenance of any desired engine speed regardless of changes in load
- delay of fuel delivery during acceleration on turbocharged engines
- control of fuel delivery to match air density at different altitudes
- fuel shut-off to stop engine
- torque control

The term *idle speed* of an engine is understood here to mean the lowest speed at which the engine will continue to operate reliably under no load; in this condition the engine is loaded only by its own internal friction and by the other equipment permanently connected to it such as the generator, fuel injection pump, and fan.

The following information on governor design and operation is provided courtesy of Robert Bosch. Other governors are covered in the appropriate chapters on fuel injection systems.

The governor is responsible for ensuring that the engine maintains a certain speed under various load conditions, that the engine speed does not exceed a certain speed as a protection against self-destruction, and that the engine does not stop during pauses in loaded operation, i.e., during idling. The governor accomplishes all this by controlling the amount of fuel injected into the engine.

Fuel injection must take place

- in an accurately metered quantity corresponding to the engine load,
- at the correct instant in time,
- for a precisely determined period of time, and
- in a manner suited to the particular combustion process.

Maintenance of these conditions is the function of the fuel injection pump and the governor. The quantity of fuel injected into the engine during each plunger lift is approximately proportional to the torque of the engine. This fuel delivery is adjusted by turning of the pump plungers, each of which has an inclined helix machined into it. As a plunger is turned, its effective stroke is varied. The plungers are turned by means of the control rod acting through either a set of gear teeth or some other transmission part. In a motor vehicle, the control rod is connected to the accelerator pedal through the governor and a linkage; when the accelerator pedal is pressed down, the pedal travel is converted to corresponding control-rod travel. Stationary engines can be operated with the control lever or by an electric speed-control device.

GOVERNOR TERMINOLOGY

Several terms that apply to governor operation should be learned in order to better understand governor action.

1. *Speed droop:* The difference in engine speed between no-load and full-load speed. Speed droop is expressed as a percentage of no-load speed and may be as high as 10% on some governors. Governors with speed droop are nonisochronous.

2. *Hunting:* An undesirable fluctuation of engine speed (alternately slowing down and speeding up). Hunting is the result of overcontrol by the governor, which requires adjustment or cleaning (since it may be sticking) to correct.

3. *Stability:* The ability of the governor to maintain the desired engine speed without hunting.

4. *Sensitivity:* The change in engine speed that occurs before the governor acts on the fuel control linkage.

5. *Promptness:* The time required for the governor to move the fuel control mechanism from a no-load position to a full-load position. Promptness is determined by governor design and its ability to overcome the resistance to movement of the control linkage.

6. *High-idle speed:* Maximum engine speed with the throttle control moved to the maximum position and no load applied to the engine. This speed can be adjusted to suit different applications within the range specified by the engine manufacturer.

7. *Rated full-load speed:* Engine speed at which maximum horsepower is produced as stated by the engine manufacturer.

8. *Low-idle speed:* Engine speed with the throttle control in the released or closed position.

9. *Underrun:* The inability of a governor to achieve and maintain proper low-idle speed as engine speed is dropped quickly from high speed.

10. *Overrun:* The inability of a governor to maintain proper high-idle speed when the engine is accelerated from idle. Idle speed may be exceeded by several hundred rpm.

GOVERNOR TYPES

Four types of governors are used on diesel engines:

1. *Mechanical governors:* Control fuel delivery by mechanical means only. All operate with centrifugal flyweights and springs that act on the fuel metering system through mechanical linkage only.

2. *Hydraulic governors:* Control fuel delivery through hydraulic pressure. These also operate with flyweights and springs, however, the flyweights and springs operate a control valve that controls the amount of hydraulic pressure allowed to act on a power piston. The power piston, in turn, acts on the fuel metering system to control fuel delivery. Engine lubricating oil pressure or pressure from a governor oil pump provides hydraulic pressure.

3. *Pneumatic governors:* Used on some automotive engines equipped with an air throttle valve. These governors operate on pressure differences between the atmosphere and engine vacuum.

4. *Isochronous governors:* Use compressed air from the vehicle storage tank to actuate a power piston to control fuel delivery. Isochronous governors incorporate a design feature that provides zero speed droop, holding the engine at constant speed regardless of engine load changes within the load range of the engine.

GOVERNOR FUNCTIONS

The basic function of every governor is to limit the high-idle speed; i.e., it must ensure that the speed of the diesel engine does not exceed the maximum value specified by the manufacturer. Depending on the type of governor, further functions can be the maintenance of certain specified speeds, e.g., the idle speed or speeds within a particular rotational-speed range or the entire range between low-idle speed and high-idle (maximum) speed.

- *Maximum-speed governors:* These governors are designed to limit the maximum speed only.
- *Minimum-maximum–speed governors:* These governors control the idle speed as well as the maximum speed.
- *Variable-speed governors:* These governors control the idle and maximum speeds as well as the speed range between them.

In addition to its basic function, the governor must also fulfill control functions, such as automatically providing or cutting off the starting fuel delivery (the increased fuel quantity that is required for starting) and varying the full-load delivery as a function of *speed* (torque control), *charge-air pressure*, or *atmospheric pressure*.

In order to carry out these functions, supplementary equipment is required in some cases, which will be described later.

Maximum-Speed Regulation

When the load is removed from the engine, the maximum full-load speed may rise no higher than rated (high-idle or no-load speed) in accordance with the permissible speed droop. The governor accomplishes this by drawing back the control rod in the shut-off direction.

The range is designated as the maximum-speed regulation.

The greater the speed droop, the greater is the increase in speed from the maximum full-load speed to the maximum noted no-load (high-idle) speed.

Intermediate-Speed Regulation

If required by the intended application of the governor (for example, in vehicles with an auxiliary drive), the governor can also maintain constant (within certain limits) various speeds between the idle and maximum speeds.

Depending on the load, therefore, the speed will fluctuate only between full load and no load within the performance range of the engine.

Low-Idle-Speed Control

The speed of a diesel engine can also be regulated in the lowest speed range. If the control rod returns from the starting position after a cold diesel engine is started, the frictional resistance of the engine is still relatively high. The amount of fuel required to keep the engine in operation is therefore somewhat larger, and the speed is somewhat slower, than would normally correspond to the idle-speed adjustment point.

After the friction during the warm-up period has diminished, the speed increases; the control rod moves back when the idle speed for the warm engine is reached.

The various demands made on governors have led to the development of the following different types.

Maximum-Speed Governors

Maximum-speed governors are designed for diesel engines that drive machines or machine systems—for example, engine-generator sets—at a fixed rated speed. Here the governor must ensure only that the maximum speed is maintained; there is no idle control or control of a particular starting fuel delivery. If the engine speed rises above the rated speed, as a result of a decreasing load on the engine, the governor shifts the control rod in the shut-off direction; i.e., the control-rod travel becomes smaller, and the fuel delivery decreases. The high-idle speed is reached when the entire load is removed from the engine.

Minimum-Maximum–Speed Governors

With diesel engines used in trucks, speed control in the range between idle and the maximum speed is often not required. In this rotational-speed range the driver operates the control rod in the fuel injection pump directly by means of the accelerator pedal and thus sets the required torque. The governor ensures that the engine does not stall in the idle-speed range, and it also controls the maximum speed.

The cold engine is started with the starting fuel delivery. The driver has pressed the accelerator pedal all the way down. If the driver releases the accelerator pedal, the control rod returns to the idle position. During the warm-up period, the idle speed fluctuates along the idle-speed-control curve and evens out.

When the engine has warmed up, the greatest starting fuel delivery is generally not required for a restart; many engines can even be started when the governor control lever is in the idle position.

If the driver presses the accelerator pedal all the way down with the engine operating, the control rod moves to the full-load delivery position. As a result, the engine speed increases, torque control of the fuel delivery starts, and the full-load delivery is slightly reduced.

The full-load delivery is injected into the engine with the accelerator pedal pressed all the way down until the maximum full-load speed is reached. Full-load speed regulation corresponding to the speed droop of the governor then starts. The engine speed rises somewhat, the control-rod travel is reduced, and, as a result, the fuel delivery is decreased. The high-idle speed is reached when the entire load is removed from the engine. During overrun (e.g., vehicle traveling downhill), the control-rod travel can become zero, and the engine speed can increase somewhat further.

Variable-Speed Governors

Vehicles with auxiliary drives (for example, for cistern pumps or for extending firefighting ladders) and agricultural tractors, which must maintain a certain operating speed, as well as boats and stationary assemblies of equipment, are equipped with variable-speed governors.

These governors control not only the idle and maximum speeds but also speeds between them independent of the engine load. The desired speed is set with the control lever. The governor operates as follows: starting of the engine with the starting fuel delivery, variation of the full-load regulation along the full-load characteristic curve, and extending the torque-control range until the onset of speed regulation at the maximum full-load speed.

Torque Control

Optimum exploitation of the engine torque can be achieved by means of torque control. Torque control is not an actual control process but is one of the regulation functions carried out by the governor. It is designed for the full-load delivery, i.e., the maximum amount of fuel delivered in the loadable range of the engine that can burn smoke-free (**Figures 13–4** and **13–5**).

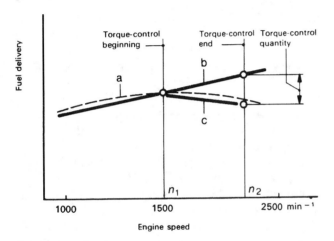

FIGURE 13–4 Fuel requirement and fuel-delivery characteristic with torque control. (a) Fuel requirement of engine, (b) full-load delivery without torque control, and (c) full-load delivery with torque control. (Courtesy of Robert Bosch Canada Ltd.)

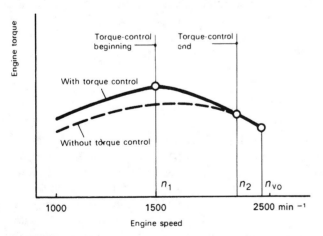

FIGURE 13–5 Torque characteristic of diesel engine, with and without torque control. (Courtesy of Robert Bosch Canada Ltd.)

The fuel requirement of the nonturbocharged diesel engine generally decreases as the speed increases (lower relative rate of air flow, thermal limiting conditions, changed mixture formation). At the same time the amount of fuel delivered by the injection pump increases within a certain range as the speed rises, as long as the control rod remains in the same position, because of the throttling effect at the control port in the pump plunger-and-barrel assembly. If too much fuel is injected into the engine, smoke will develop as the engine overheats.

The amount of fuel injected into the engine must be matched to the actual fuel requirement. As the speed increases, the fuel delivery decreases (positive torque control); as the speed drops, the fuel delivery increases.

Torque-control systems are arranged and designed according to the particular type of governor.

In engines with an exhaust turbocharger achieving a high measure of supercharging, the fuel requirement for full load in the lower speed range rises so sharply that the natural increase in fuel delivered by the fuel injection pump is no longer adequate. In such cases, torque control must be carried out as a function of speed or charge-air pressure; depending on the prevailing conditions, this can be accomplished with the governor, the manifold-pressure compensator, or both operating together.

This form of torque control is called *negative*. Negative torque control means an increase in the fuel delivery as the speed rises.

MECHANICAL GOVERNORS

Mechanical governors, i.e., governors employing the principle of centrifugal force, are used most widely with larger diesel engines.

The Bosch mechanical governor is mounted on the fuel injection pump. The injection-pump control rod is connected with the governor linkage through a flexible joint, and the connection to the accelerator pedal is made through the governor control lever.

Two different designs are used in mechanical governors:

- RQ, RQV, with the governor springs built into the flyweights (**Figure 13–6**).
- RS, RSV, with the centrifugal force acting through a system of levers on the governor spring located outside the two flyweights (**Figure 13–7**).

In mechanical governor types RQ and RQV, each of the two flyweights acts directly on a spring set, which is designed specifically for a given nominal speed.

MECHANICAL GOVERNORS 241

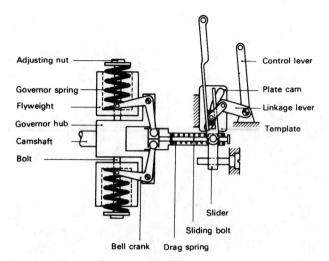

FIGURE 13-6 Components of RQ and RQV governor. (Courtesy of Robert Bosch Canada Ltd.)

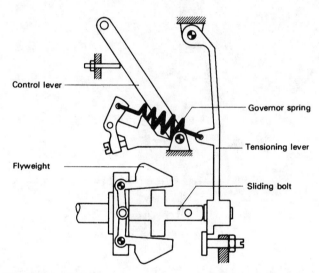

FIGURE 13-7 Components of RS and RSV governor. (Courtesy of Robert Bosch Canada Ltd.)

In mechanical governors types RS and RSV the action of the two flyweights presses the sliding bolt against the tensioning lever, which is drawn in the opposite direction by the governor spring. When the speed is set by the control lever, the governor spring is tensioned by an amount corresponding to the desired speed.

In both types of metering unit, the governor springs are so selected that at the desired speed the centrifugal force and the spring force are in equilibrium. If this speed is exceeded, the increasing centrifugal force of the flyweights acts through a system of levers to move the control rod, and the fuel delivery is decreased.

Minimum-Maximum-Speed Governor RQ

CONSTRUCTION

The fuel injection pump camshaft drives the governor hub through a vibration damper. The two flyweights with their bell cranks are supported in the governor hub, with one spring set built into each flyweight. By means of the bell cranks, the radial travel of the flyweights is converted to axial movements of the sliding bolt, which in turn transmits these movements to the slider. Movement of the slider is held in a straight line by the guide pin; the slider itself, operating through the fulcrum lever, forms the connection between the flyweight mechanism and the control rod. The lower end of the fulcrum lever is fastened to the slider (**Figures 13-8 and 13-9**).

In the fulcrum lever the movable guide block is guided radially by the linkage lever; this lever is connected with the control lever. The control lever is operated either manually or through a linkage system from the accelerator pedal.

When the control lever is moved, the guide block is shifted, and the control lever is tilted around the pivot at the slider; if the governor takes effect, the pivot for the fulcrum lever is at the guide block. The transmission ratio of the fulcrum lever changes. As a result, even in the idle range where the centrifugal forces are

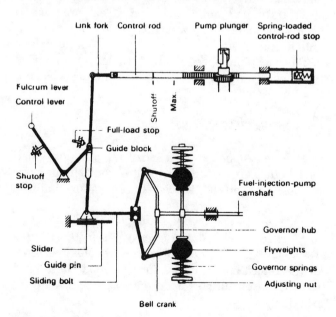

FIGURE 13-8 RQ minimum-maximum-speed governor in full shut-off position. (Courtesy of Robert Bosch Canada Ltd.)

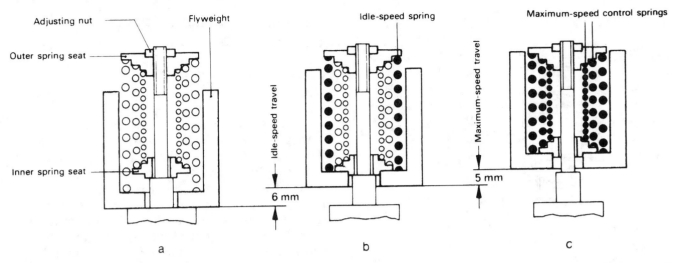

FIGURE 13-9 Flyweight travel and governor springs in the RQ governor. (Courtesy of Robert Bosch Canada Ltd.)

still low, there is more than sufficient adjustment force for the control rod.

The spring sets (governor springs) built into the flyweights consist generally of three cylindrical helical springs arranged as shown.

The outer spring is supported between the flyweight and the outer spring seat, while the two inner springs are positioned between the outer and inner spring seats. During low-idle-speed control, only the outer spring (the idle-speed spring) takes effect. As the speed increases, the flyweights, after surpassing the idle-speed travel path at the inner spring seat, are pressed against the inner spring seat and remain in this position until maximum-speed regulation begins. During control of the maximum speed, all springs act together. The two inner springs are designated maximum-speed control springs.

OPERATING CHARACTERISTICS

Engine Stopped. The control lever is positioned against the shut-off stop, the control rod is in the shut-off position, and the flyweights are positioned all the way in.

Start Position (Control lever at full load) **(Figure 13-10).** After the control rod overcomes the force of the return spring in the spring-loaded control-rod stop, it moves all the way to the starting-fuel-delivery position.

Idle Position **(Figure 13-11).** After the engine starts to operate and the control lever (accelerator pedal) is released, this lever returns to the idle position

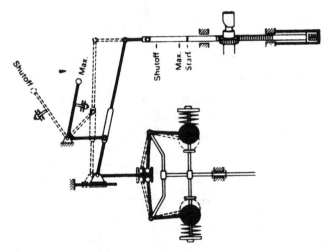

FIGURE 13-10 RQ governor in start (maximum-fuel) position. (Courtesy of Robert Bosch Canada Ltd.)

(a corresponding stop should be provided in the vehicle or on the engine). The control rod also returns to the idle position, which is determined by the now-operating governor.

Part-Load Position (when the engine is loaded (between no load and full load) **(Figure 13-12).** As the driver presses the accelerator pedal down somewhat, the engine speeds up. As a result, the flyweights are forced outward. In other words, the governor initially tends to prevent this increase in engine speed. However, after the speed exceeds the idle speed by only a slight amount, the flyweights are brought up against the spring seats, which are loaded by the maximum-speed control springs. They remain in this posi-

MECHANICAL GOVERNORS 243

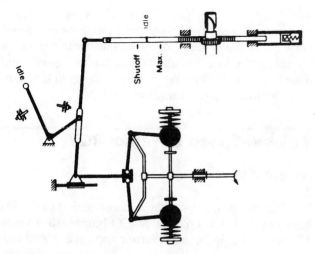

FIGURE 13-11 RQ governor in idle position. (Courtesy of Robert Bosch Canada Ltd.)

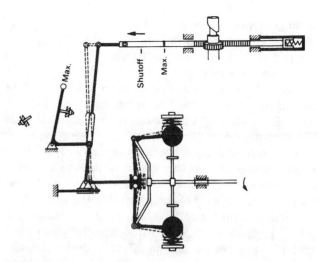

FIGURE 13-13 RQ governor maximum-speed regulation at full load. (Courtesy of Robert Bosch Canada Ltd.)

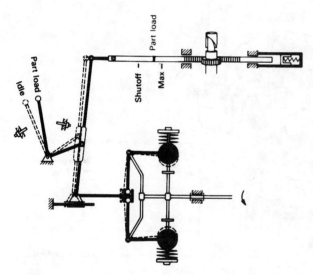

FIGURE 13-12 RQ governor in part-load position. Flyweight position remains unchanged (due to spring tension) until maximum engine speed is reached. (Courtesy of Robert Bosch Canada Ltd.)

tion until the maximum speed is reached, because the maximum-speed control springs yield to the centrifugal force only when the engine tends to exceed its nominal speed. For this reason, the governor has no effect between the idle speed and the maximum speed. In this intermediate range the position of the control rod, and thus the torque developed by the engine, is influenced only by the driver.

Maximum-Speed Regulation at Full Load **(Figure 13-13).** In the highest speed range of the engine (maximum-speed regulation), maximum-speed regulation begins when the engine exceeds the nominal speed, at full load or part load, depending on the position of the control lever. For this reason, as soon as maximum-speed regulation has started, the position of the control rod no longer depends solely on the driver but also on the governor. The maximum-speed regulation travel of the flyweights is 0.2 in. (5 mm). This results in a control-rod travel of about 0.6 in. (16 mm) (with a lever ratio of 1:3.23), which is sufficient to regulate fuel delivery from maximum delivery to shut-off.

TORQUE-CONTROL MECHANISM IN THE RQ GOVERNOR

In the RQ governor the torque-control mechanism is built into the flyweights, between the inner spring seat and the maximum-speed control springs **(Figure 13-14).** The torque-control spring is installed in a

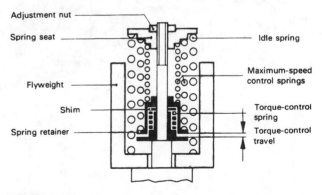

FIGURE 13-14 Torque-control mechanism in the RQ governor. (Courtesy of Robert Bosch Canada Ltd.)

spring retainer, on the outside of which the two maximum-speed control springs are supported. For operating purposes, therefore, the torque-control spring is connected in series with the maximum-speed control springs. The space between the inner spring seat and the spring retainer is the torque-control travel (0.3 to 1 mm). The width of this space can be adjusted with compensating shims.

The start of torque control depends on the fuel-requirement characteristic curve of the engine. At a point somewhat below the maximum speed, the torque-control spring is compressed to the extent that the inner spring seat and the spring retainer are pressed against each other. Without the torque-control spring, the governor has no effect between the idle speed and the maximum speed. However, since the torque-control springs yield, the flyweights can move outward by the distance of the torque-control travel in the range between the idle and maximum speeds and therefore shift the control rod accordingly in the shutoff direction (positive torque control).

Variable-Speed Governor RQV

CONSTRUCTION

The RQV governor is similar in construction to the RQ governor, but it is not identical (**Figures 13–15** and **13–16**). In the RQV, the governor springs are built into

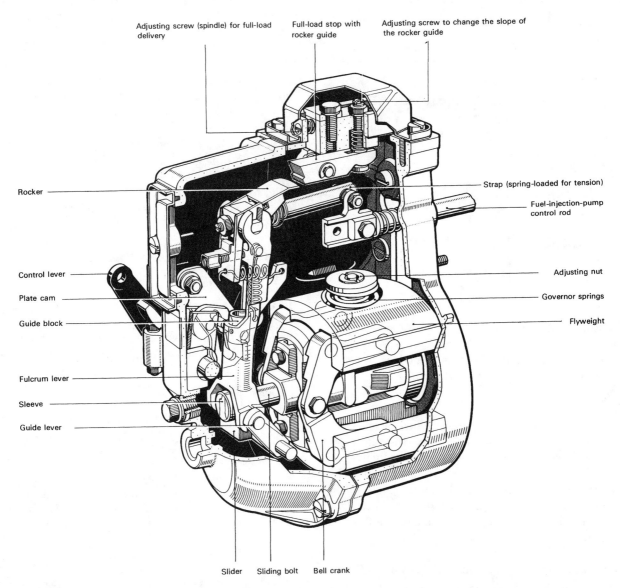

FIGURE 13–15 RQV governor. (Courtesy of Robert Bosch Canada Ltd.)

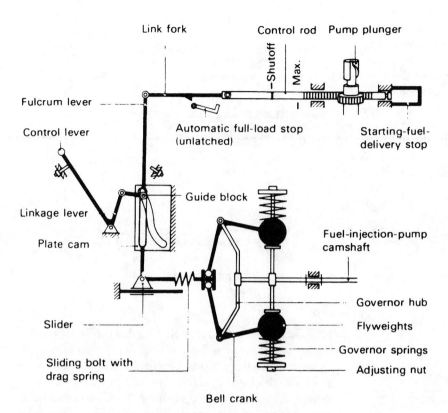

FIGURE 13-16 RQV variable-speed governor. (Courtesy of Robert Bosch Canada Ltd.)

the flyweights, but the flyweights move continuously outward within the specified adjustment range as the speed increases. A certain speed at which speed regulation begins is associated with each position of the control lever. Movements of the control lever are transmitted through the two-part linkage lever and the guide block to the fulcrum lever, and thus to the control rod. The pivot point of the fulcrum lever can be shifted. In addition, the pivot point is guided in a plate cam fastened on the governor housing so that the transmission ratio of the fulcrum lever changes in the range of 1:1.7 to 1:5.9.

The sliding bolt, being the connecting element between the flyweight assembly and the fulcrum lever, is spring-loaded for pressure and tension (drag spring).

As in the RQ governor, the spring sets built into the flyweights consist generally of three concentrically arranged cylindrical helical springs. The outer spring serves for low-idle-speed control and is supported between the flyweight and the adjusting nut for the spring preload. After moving across the short idle-speed travel path, the flyweight is positioned against the spring seat; the inner springs, which are installed between the spring seat and the adjusting nut, also take effect. From this point on, all the springs act together to control the speeds set by the control lever.

OPERATING CHARACTERISTICS

Engine Stopped. The control lever is positioned against the shut-off stop, and the control rod is in the shut-off position.

Starting Position **(Figure 13-17).** The control lever is positioned against the maximum-speed stop; the control rod moves to the starting-fuel-delivery stop.

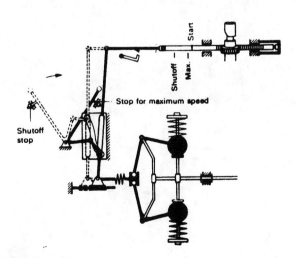

FIGURE 13-17 RQV governor in start (maximum-fuel) position. (Courtesy of Robert Bosch Canada Ltd.)

Idle Position (**Figure 13–18**). After the engine has started to operate and the control lever (accelerator pedal) has been released, this lever returns to the idle position. The control rod also returns to the idle position, which is determined by the now-operating governor.

Load on Engine (**Figures 13–19** to **13–28**). If the load on the engine at any speed set by the control lever (accelerator pedal) is increased or decreased, the

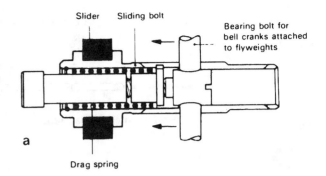

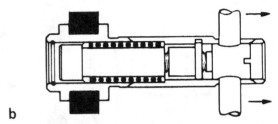

FIGURE 13–20 Sliding bolt with drag spring. (a) During acceleration or when the engine is overloaded, the flyweights have shifted the control rod to the full-load stop. The drag spring is tensioned. (b) Vehicle traveling downhill, engine driven by vehicle. The flyweights have shifted the control rod to the shut-off stop, and the drag spring is tensioned and absorbs further movement of the flyweights. (Robert Bosch Canada Ltd.)

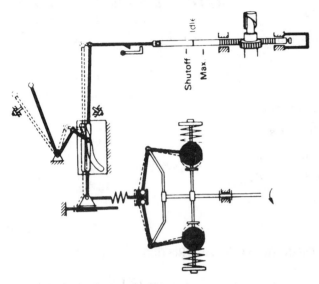

FIGURE 13–18 RQV governor in idle position. (Courtesy of Robert Bosch Canada Ltd.)

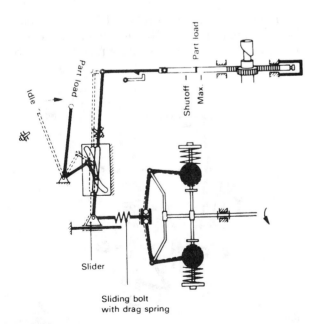

FIGURE 13–19 RQV governor in part-load position. (Courtesy of Robert Bosch Canada Ltd.)

variable-speed governor maintains the set speed by increasing or decreasing the amount of fuel delivered within the associated speed droop.

For example, the driver has moved the control lever from the idle position to a position intended to correspond to a desired vehicle speed. The movement of the control lever is transmitted through the linkage to the fulcrum lever. The transmission ratio of the fulcrum lever is variable and immediately above the idle-speed range becomes so large that even a relatively small part of the total control-lever or flyweight travel is adequate to shift the control rod to the set full-load stop. A control-rod stop (fixed or adjusted by hand; in no case spring-loaded) must therefore be available. Additional swiveling movement of the control lever results in tensioning of the drag spring. The control rod remains temporarily in the maximum-fuel-delivery position, and this results in a rapid increase in engine speed. This in turn forces the flyweights outward, but the control rod remains in the maximum-fuel-delivery position until the tension on the drag spring is released. Then the flyweights start to act on the fulcrum lever, and the control rod is thereby shifted in the shut-off direction. As a result, the amount of fuel delivered becomes smaller and the engine speed is restricted. This engine speed

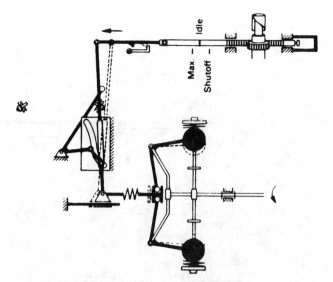

FIGURE 13-21 RQV governor maximum-speed regulation under full load. (Courtesy of Robert Bosch Canada Ltd.)

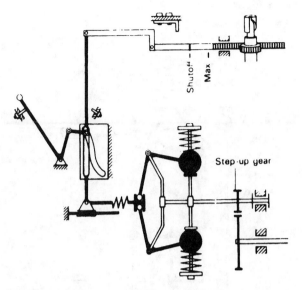

FIGURE 13-22 Schematic drawing of RQUV variable-speed governor. (Courtesy of Robert Bosch Canada Ltd.)

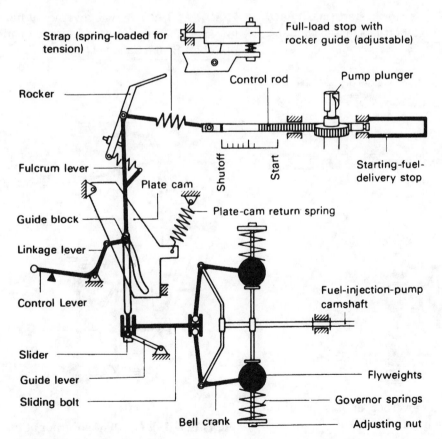

FIGURE 13-23 Schematic drawing of RQV-K variable-speed governor. (Courtesy of Robert Bosch Canada Ltd.)

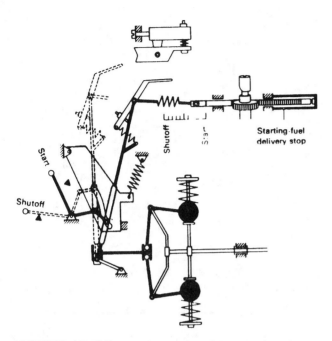

FIGURE 13-24 RQV-K governor in start (maximum-fuel) position. (Courtesy of Robert Bosch Canada Ltd.)

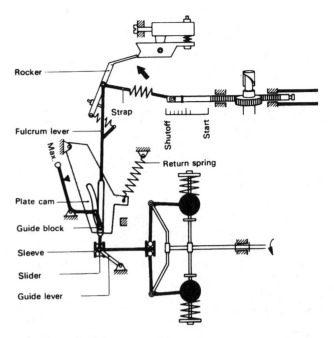

FIGURE 13-26 RQV-K governor at full-load delivery, low-speed, start of negative torque control. (Courtesy of Robert Bosch Canada Ltd.)

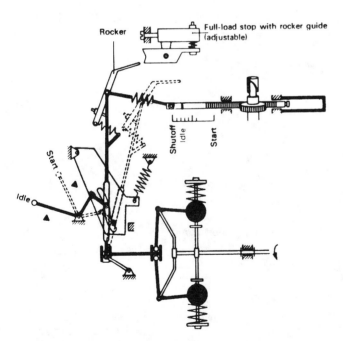

FIGURE 13-25 RQV-K governor in idle position. (Courtesy of Robert Bosch Canada Ltd.)

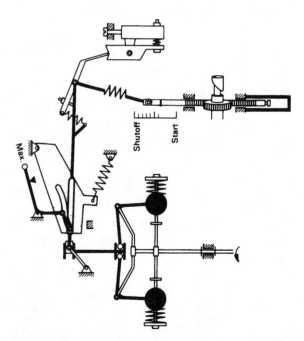

FIGURE 13-27 RQV-K governor at full-load delivery, medium-speed, reversal of torque control. (Courtesy of Robert Bosch Canada Ltd.)

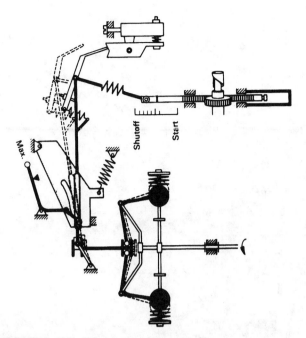

FIGURE 13–28 RQV-K governor at maximum full-load-speed end of positive torque control (dotted pattern: speed regulation). (Courtesy of Robert Bosch Canada Ltd.)

limit corresponds to the position of the control lever and to the flyweight stop.

During operation, therefore, one specific rotational-speed range is associated with every position of the control lever as long as the engine is not overloaded or driven by the vehicle when traveling downhill (overrun). If the engine loading becomes somewhat greater, for example when traveling uphill, the engine and governor speeds decrease. As a result, the flyweights move inward and shift the control rod to the maximum-delivery direction. This holds the engine speed constant at a level determined by the position of the control lever and by the speed droop. However, if the uphill slope (loading) is so great that, even though the control rod is shifted all the way to the maximum-fuel-delivery stop, the speed nevertheless still decreases, the flyweights are brought further inward in accordance with this speed, and they shift the sliding bolt to the left.

The flyweights therefore tend to shift the control rod farther in the maximum-fuel-delivery direction. However, since the control rod is already positioned against the full-load stop and cannot move farther in the maximum-fuel-delivery direction, the drag spring is tensioned. This means that the engine is overloaded. In this case the driver must shift to a lower gear.

When traveling downhill the opposite situation prevails. The engine is driven by the vehicle and its speed increases. As a result, the flyweights are forced outward and the control rod is shifted in the shut-off direction until it reaches the stop. If the engine speed then increases still further, the drag spring is tensioned in the opposite direction (control rod in shut-off position).

Such behavior of the governor applies basically for all positions of the control lever should for any reason the engine loading or speed change so greatly that the control rod is brought up against one of its terminal stops, i.e., maximum fuel delivery or shut-off.

Torque Control

In the RQV governor, the torque-control mechanism is built into the control-rod stop. For additional information on this point see control-rod stops described later in this section.

Full-Load Speed Regulation

If the engine exceeds its maximum speed, full-load speed regulation starts. During this process the flyweights move outward and the control rod shifts in the shut-off direction. If the entire load on the engine is removed, the high-idle speed is attained.

Variable-Speed Governor RQUV

The RQUV governor is used for regulating very low rotational speeds, for example, the speeds at which marine engines operate. It is a variant of the RQUV governor and affects two different speed-increasing ratios (about 1:2.2 or 1:3.76) between the driving element (i.e., the fuel-injection-pump camshaft) and the governor hub. Similar to the lever ratio in the RQV governor, the ratio of the fulcrum lever is variable here also (from 1:1.85 to 1:7). The RQUV governor can be used for PE. .P and PE. .ZW fuel-injection pumps. The operation and operating characteristics of this governor are similar to those of the RQV.

Variable-Speed Governor EP/RSV

CONSTRUCTION

Governor EP/RSV is constructed differently from the comparable type RQV (**Figures 13–29** and **13–30**). It has only one governor spring, and this spring can swivel. When the speed is set at the control lever, the position and tension of this spring change so that the effective torque at the tensioning lever is in equilibrium with the torque developed by the flyweights at the desired speed. All adjustments of the control lever and

250 Chapter 13 GOVERNING FUEL DELIVERY

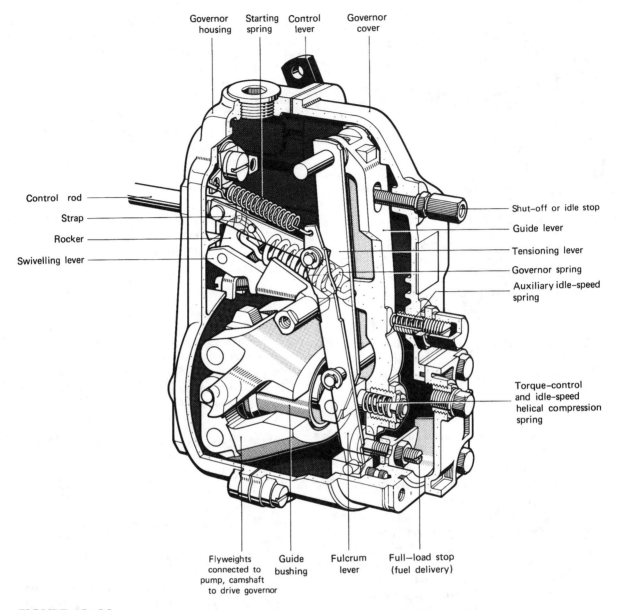

FIGURE 13-29 EP/RSV governor. (Courtesy of Robert Bosch Canada Ltd.)

the flyweight travel are transmitted through the governor linkage to the control rod.

The starting spring attached to the upper end of the fulcrum lever pulls the control rod to the starting position, automatically setting the starting fuel delivery.

A full-load stop and a torque-control mechanism are built into the governor. In order to stabilize the idle speed, an auxiliary idle-speed spring with an adjusting screw is built into the governor cover. The speed droop can be varied within certain limits by means of the governing spring adjusting screw. Lighter flyweights are required for higher speed ranges; when using springs that are under less tension, it is possible to set a smaller speed droop at lower speeds.

OPERATING CHARACTERISTICS

Starting the Engine (**Figure 13-31**). When the engine is stopped, the control rod is in the starting-fuel-delivery position so that the engine can be started with the control lever in the idle position.

Idle Position (**Figure 13-32**). The control lever is positioned against the idle stop. As a result, the governor spring is almost completely relaxed and stands almost vertical. It has a very weak effect, so that the flyweights swing outward at even a low speed. The sliding bolt, and with it the guide lever, therefore moves to the right. In turn, the guide lever swings the fulcrum lever

MECHANICAL GOVERNORS 251

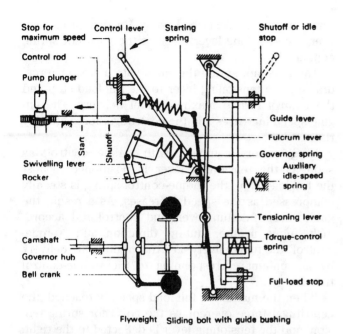

FIGURE 13-30 EP/RSV governor schematic. (Courtesy of Robert Bosch Canada Ltd.)

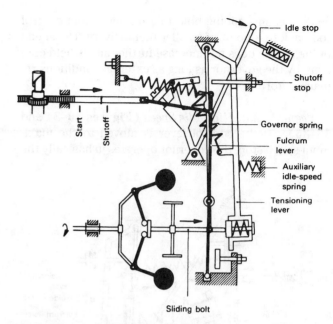

FIGURE 13-32 EP/RSV governor in low-idle position. (Courtesy of Robert Bosch Canada Ltd.)

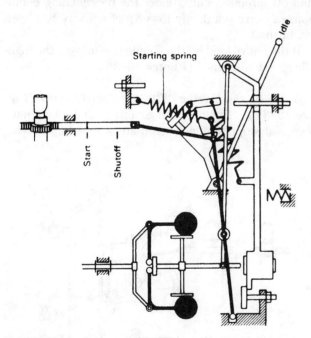

FIGURE 13-31 EP/RSV governor in start (maximum fuel) position. (Courtesy of Robert Bosch Canada Ltd.)

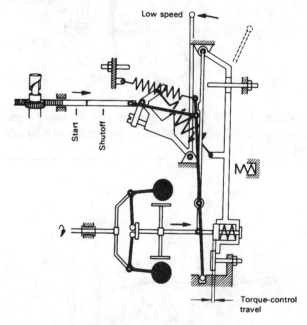

FIGURE 13-33 EP/RSV governor at full-load, low-speed, start of torque control position. (Courtesy of Robert Bosch Canada Ltd.)

to the right, so that the control rod is moved in the shut-off direction to the idle position. The tensioning lever is positioned against the auxiliary idle-speed spring, which assists the low-idle-speed control.

Regulation of Low Speeds **(Figure 13-33).** Even a relatively small shift of the control lever from the idle position is sufficient to move the control rod from its initial position to its full-load position. The fuel injection pump delivers the full-load fuel quantity into the engine cylinders, and the speed rises. As soon as the centrifugal force is greater than the tension of the governor spring corresponding to the position of the control lever, the flyweights swing outward and shift the guide bushing, sliding bolt, fulcrum lever, and control rod back to a point of smaller fuel delivery. The speed of the engine does not increase further and is held constant by the governor as long as external conditions remain uniform.

Regulation at High-Idle Speed (**Figures 13–34** and **13–35**). If the control lever is moved to the maximum-speed stop, the governor operates in basically the same manner as described above. In this case, however, the swiveling lever tensions the governor spring completely.

The governor spring thus acts with a greater force, drawing the tensioning lever to the full-load stop and the control rod to maximum fuel delivery. The engine speed increases and the centrifugal force steadily rises.

In governors equipped with torque control, as soon as the tensioning lever is positioned against the full-load stop, the torque-control spring is steadily compressed as the speed increases. As a result, the guide lever, fulcrum lever, and control rod accordingly move in the shut-off direction and "torque-control" the fuel delivery; i.e., they reduce the delivery by an amount corresponding to the torque-control travel.

When the maximum full-load speed is reached, the centrifugal force overcomes the governor spring tension, and the tensioning lever is deflected to the right. The sliding bolt with the guide lever and the control rod, coupled through the fulcrum lever, move in the shut-off direction until, under the new loading conditions, a correspondingly lower fuel delivery has been established.

If the entire load on the engine is removed, the high-idle speed is attained (**Figure 13–36**).

Stopping the Engine: With the Control Lever (**Figure 13–37**). Engines with governors that do not have a special stopping mechanism are stopped by moving

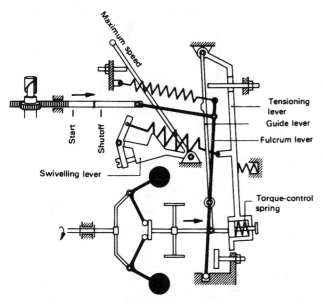

FIGURE 13–34 EP/RSV governor at no load, regulated from full load. (Courtesy of Robert Bosch Canada Ltd.)

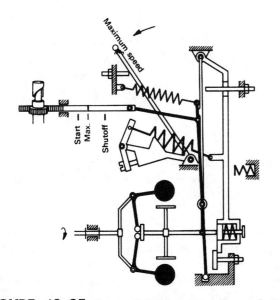

FIGURE 13–35 Start of full-load-speed regulation. (Courtesy of Robert Bosch Canada Ltd.)

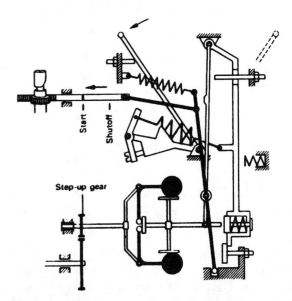

FIGURE 13–36 Schematic drawing of the variable-speed governor, EP/RSUV, maximum-speed position. (Courtesy of Robert Bosch Canada Ltd.)

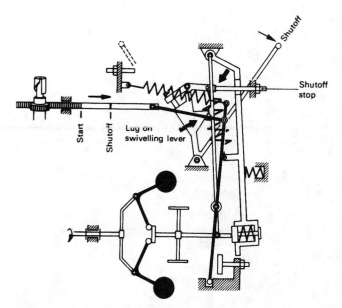

FIGURE 13-37 EP/RSV governor. Stopping the engine with the governor control lever. (Courtesy of Robert Bosch Canada Ltd.)

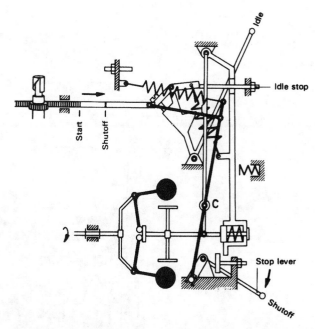

FIGURE 13-38 EP/RSV governor. Stopping the engine with the shut-off mechanism. (Courtesy of Robert Bosch Canada Ltd.)

the governor control lever to the shut-off position. As this is done, the lugs on the swiveling lever (inclined arrow) press on the guide lever. This lever swings to the right, taking the fulcrum lever, and thus the control rod as well, to the shut-off position with it. Since the tension exerted by the governor springs on the sliding bolt is released, the flyweights swing outward.

With the Stop Lever **(Figure 13–38).** In the case of governors fitted with a special shut-off mechanism, the control rod can be set to shut off if the stop lever is moved to the shut-off position.

When the stop lever is pressed to shut-off, the upper part of the fulcrum lever is swung to the right around the pivot point C in the guide lever. As a result, the control rod is drawn by the strap to shut-off. When the stop lever is released, a return spring not shown in the drawing brings it back into the initial position.

Variable-Speed Governor EP/RSUV

The RSUV governor is used to control very low speeds, for example, those at which low-speed marine engines operate **(Figures 13-39 to 13-45).** In terms of its construction, it differs essentially from the EP/RSV governor in its transmission gear for speed-increasing ratio (step-up gear), which is installed between the driving element (i.e., the fuel injection–pump camshaft), and the governor hub. The operation of this governor is basically the same as that of the EP/RSV. It is used with size P and Z fuel injection pumps.

Control-Rod Stops

In addition to the stops for shut-off and for full-load delivery or maximum speed (parts required in every governor to define the control-lever travel), a special stop is required for the control rod that limits the travel of this rod at full-load or starting fuel delivery. Depending on the particular purpose and use, there are various designs of control-rod stop: rigid and spring-loaded designs; stops for full-load delivery, with mechanical or electromagnetic unlocking for the starting fuel delivery; and stops with a built-in torque-control mechanism. In addition, there are full-load stops designed to carry out special compensation functions. Control-rod stops are produced for mounting on the fuel injection pump or on the governor.

RIGID EXCESS-FUEL STOP FOR STARTING

The rigid stop for starting-fuel delivery is used mainly in RQ governors with a low-idle speed. When the engine is running, the excess fuel for starting is withdrawn through the governor and cannot have a damaging effect (such as would be caused by the development of smoke).

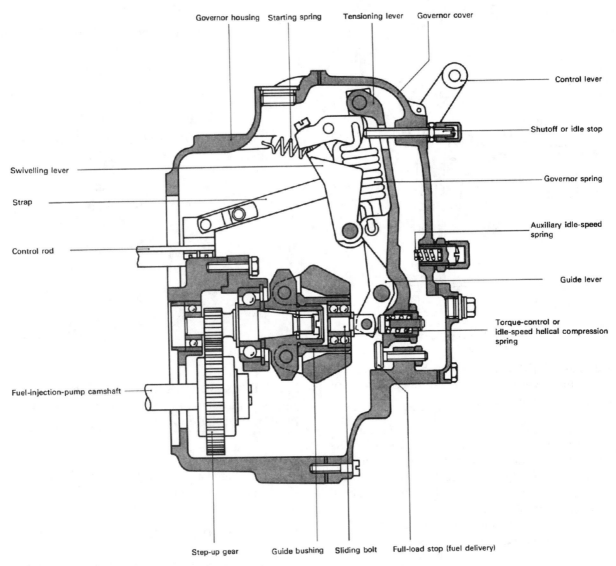

FIGURE 13-39 EP/RSUV governor. (Courtesy of Robert Bosch Canada Ltd.)

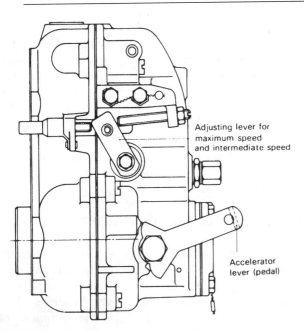

FIGURE 13-40 Minimum-maximum-speed EP/RS governor, external view. (Courtesy of Robert Bosch Canada Ltd.)

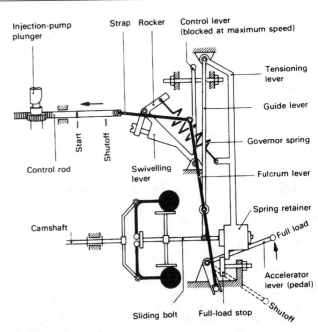

FIGURE 13-41 EP/RS governor in start position. (Courtesy of Robert Bosch Canada Ltd.)

MECHANICAL GOVERNORS 255

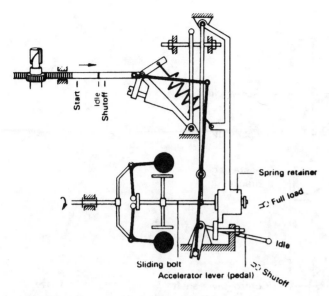

FIGURE 13–42 EP/RS governor in idle position. (Courtesy of Robert Bosch Canada Ltd.)

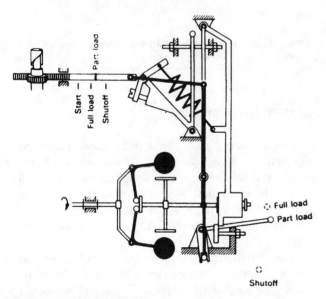

FIGURE 13–43 EP/RS governor in part-load position. (Courtesy of Robert Bosch Canada Ltd.)

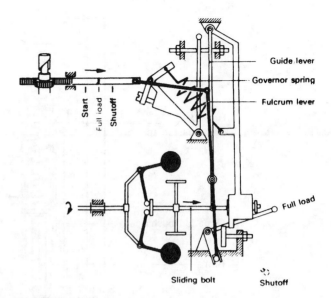

FIGURE 13–44 EP/RS governor. Start of speed regulation, at maximum speed and full load. (Courtesy of Robert Bosch Canada Ltd.)

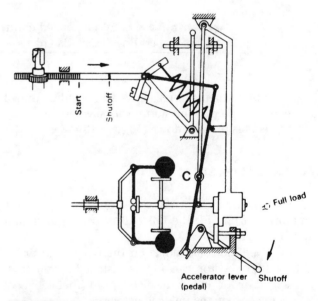

FIGURE 13–45 EP/RS governor in shut-off position. (Courtesy of Robert Bosch Canada Ltd.)

SPRING-LOADED CONTROL-ROD STOP FOR RQ GOVERNORS

If the accelerator pedal is pressed all the way down during the starting process, the stop bolt moves against the resistance of the spring to the set starting-fuel-delivery position. The spring built into the stop acts against the idle-speed spring and thus causes an early shift of the control rod back from the start position. This prevents a brief interim period of starting fuel delivery if the engine is accelerated rapidly from idling **(Figure 13–46)**.

AUTOMATIC FULL-LOAD CONTROL-ROD STOP

When the engine is at rest, the flyweight governor springs, acting through the sliding bolt, press on the rocker arm spring; as a result, the rocker arm forces the stop strap with the full-load stop lug downward.

256 Chapter 13 GOVERNING FUEL DELIVERY

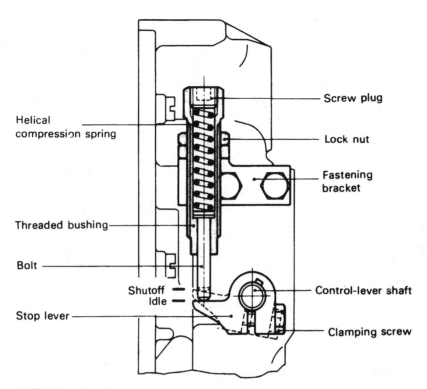

FIGURE 13-46 Spring-loaded control-rod stop, RQ and RQV governors. (Courtesy of Robert Bosch Canada Ltd.)

Therefore, when the engine is started, the control rod can be shifted to the start position when the accelerator pedal is pressed down **(Figure 13-47)**.

CONTROL-LEVER STOPS

On the governor cover are two stop screws, one for shut-off and one for the full-load delivery (maximum speed) **(Figure 13-48)**.

If desired, and depending on the type of governor (RQ or RQV), a stop can also be installed for the low-idle speed or for an intermediate speed (fuel delivery) that is lower than the full-load delivery.

SPRING-LOADED IDLE-SPEED STOP

The spring-loaded idle-speed stop consists of a sleeve with an external thread and from which a bolt under spring tension projects.

At the idle-speed fuel delivery, the stop lever is positioned against the spring-loaded bolt. In order to stop the engine, the governor control lever must be moved into the shut-off position against the force of the helical compression spring until the engine has come to a stop.

REDUCED-DELIVERY STOP

This stop serves as a fixed adjustment point for a fuel delivery lower than the full-load delivery or for an intermediate speed (depending on the type of governor). It is mounted on the governor cover and acts together with a short lever fastened on the control-lever shaft in such a way that it can be adjusted **(Figure 13-49)**.

The schematic cross-sectional drawing shows a spring-loaded bolt that can be shifted in the housing by means of a shaft with a furrow. In one end position, the lever strikes against the bolt and limits the travel of the control lever.

In the other end position, the bolt releases the lever, and the control lever can reach its end position.

MANIFOLD-PRESSURE COMPENSATOR

In turbocharged engines the full-load delivery is determined according to the charge-air pressure. In the lower speed range, however, the charge-air pressure is lower; therefore, the weight of the air charge in the engine cylinders is also lower. For this reason, the full-load delivery must be matched in a correspond-

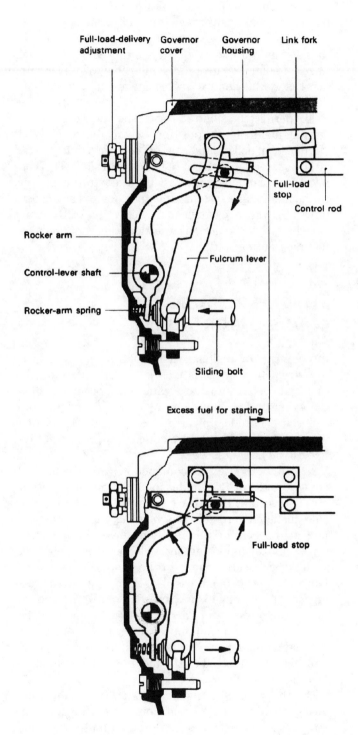

FIGURE 13–47 Automatic full-load control-rod stop for RQ governors. Release of starting fuel delivery (top) and limitation to full-load fuel delivery (bottom). (Courtesy of Robert Bosch Canada Ltd.)

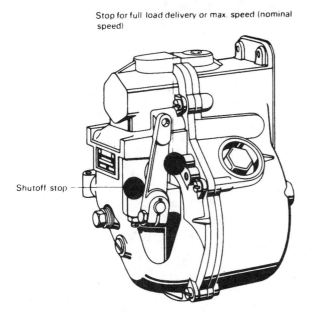

FIGURE 13-48 Shut-off and full-load stop screws location. (Courtesy of Robert Bosch Canada Ltd.)

ing ratio to the reduced weight of the air. This function is carried out by the manifold-pressure compensator (LDA), which reduces the full-load delivery in the lower speed range starting from a certain (selectable) charge-air pressure **(Figures 13–50 to 13–53)**. Various designs of the LDA are produced for mounting on fuel injection pumps and on either the side or the top of governors. The following description refers to an LDA designed for mounting on the RSV governor.

The construction of all these special control-rod stops is basically identical. Between the housing bolted onto the top of the governor and a suitable cover is a diaphragm that is tensioned and airtight. A connector fitting for the charge-air pressure is located in the cover. From below, a helical compression spring acts on the diaphragm; this spring is supported at its other end on a guide bushing attached to the housing by means of a thread. The initial tension of this spring can therefore be changed within certain limits.

A threaded pin is attached to the diaphragm through a plate washer and a guide washer; at the lower end of this pin, which projects out of the housing, a screw with a locknut is attached. The head of this screw is set to a certain distance from the housing surface and transmits the movement of the threaded pin through a bell crank to the control rod. This distance is preset, but after the LDA has been mounted, corrections can be made with the headless setscrew.

If charge-air pressure is applied to the diaphragm, the threaded pin moves against the force of the helical compression spring, traveling the greatest distance at the full charge-air pressure. Movement of the threaded pin is transmitted through the bell crank, which is supported in the governor housing on an axle so that it can turn, to the strap attached to the fuel injection–pump control rod. As the charge-air pressure decreases, the control rod is moved in the shut-off direction.

To permit the control rod to be moved into the starting-fuel-delivery position when the engine is started, the bell crank can be disengaged from the strap by lateral movement of the control shaft. This can be done manually either with a control cable or through a linkage system. Governor designs also exist, however, with electromagnetic activation of the control shaft; in these designs, the electromagnet takes effect only during starting.

ALTITUDE-PRESSURE COMPENSATOR (ADA)

In countries or regions where road traffic is subject to extremely wide variations in altitude, the amount of fuel injected into the engine must be matched to the worsening air charge in the engine cylinders from a certain altitude upward. This function is carried out by the altitude-pressure compensator (ADA) **(Figure 13–54)**.

The ADA is used in conjunction with mechanical governors RQ or RQV and is mounted on the governor cover. Basically, the ADA consists of an aneroid capsule, installed vertically in a housing; the capsule can be set to a certain altitude by means of an adjusting screw and an opposing spring-loaded threaded bolt. Within the effective range of the aneroid capsule, the length of the cell increases as the air pressure decreases. The spring-loaded threaded bolt at the bottom of the aneroid capsule and the fork attached to the threaded bolt transmit the changes in length to the swivel-mounted cam plate. This cam plate acts on the bolt that is connected to the fuel injection–pump control rod.

As the aneroid capsule expands, the cam plate swings downward. The bolt connected with the stop strap pulls the control rod in the shut-off direction, and the fuel delivery decreases; if the length of the aneroid capsule shortens, the amount of fuel delivered increases. To adjust the full-load delivery, the cam plate is adjustable in the horizontal plane by means of a screw.

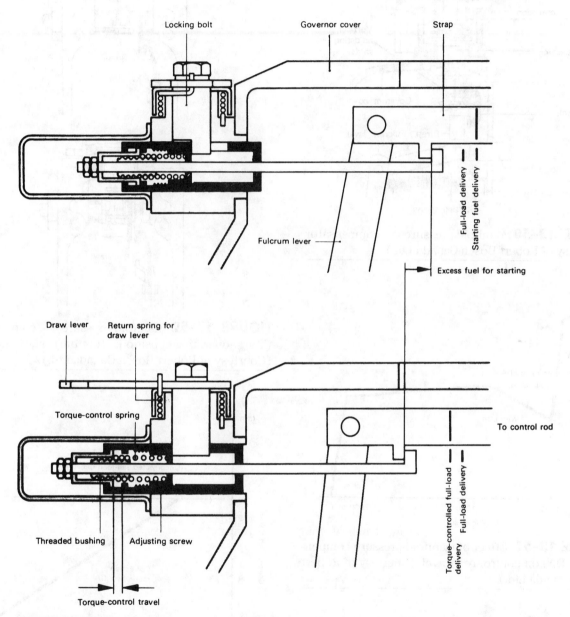

FIGURE 13-49 Control-rod stop for RQV governors with draw lever for starting fuel delivery and with torque-control mechanism. (Top) starting-fuel-delivery position; (bottom) full-load-delivery position with torque control. (Courtesy of Robert Bosch Canada Ltd.)

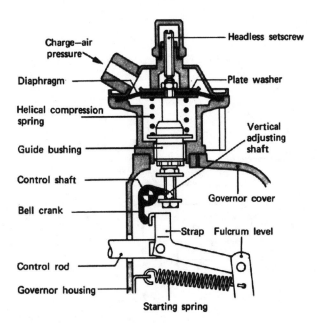

FIGURE 13-50 Manifold-pressure compensator. (Courtesy of Robert Bosch Canada Ltd.)

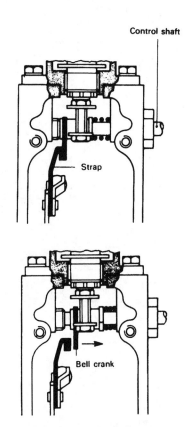

FIGURE 13-52 Manifold-pressure compensator. (Top) operating position; (bottom) start position. (Courtesy of Robert Bosch Canada Ltd.)

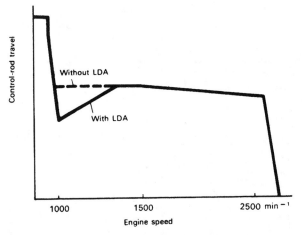

FIGURE 13-51 Effect of manifold-pressure compensator (LDA) on control-rod travel. (Courtesy of Robert Bosch Canada Ltd.)

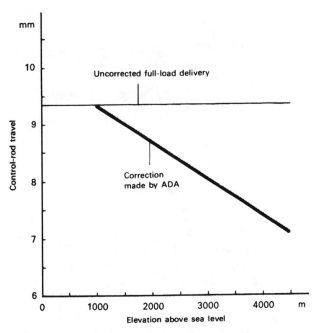

FIGURE 13-53 Example of correction of control-rod travel by the altitude-pressure compensator (ADA). (Courtesy of Robert Bosch Canada Ltd.)

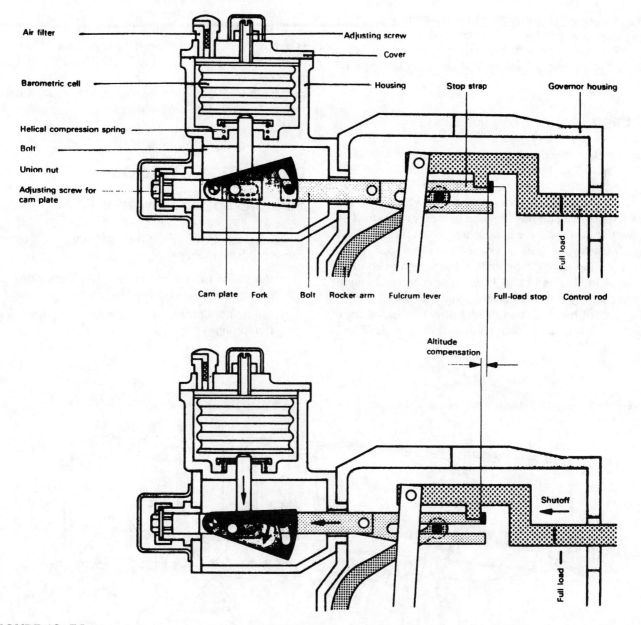

FIGURE 13–54 Altitude-pressure compensator (ADA). (Top) normal position; (bottom) operating position at low atmospheric pressure. (Courtesy of Robert Bosch Canada Ltd.)

ELECTRIC SPEED-CONTROL DEVICE

Use

Electric speed-control devices (**Figure 13-55**) are used for remote-controlled adjustment of the speeds of engines used with assemblies of equipment.

Various mounting parts are required to attach such a device to a fuel-injection-pump governor; the exact choice of mounting parts will depend on the type of governor concerned.

Design

The electric speed-control device is fitted with a base plate on which is mounted a 24 V motor. This motor is connected through a clutch with a threaded spindle, which can move a threaded nut back and forth in guide rails, depending on the direction of motor rotation. The governor control lever is attached to the threaded nut by a positive mechanical connection developed either by a torsion spring mounted on the governor-control-lever shaft (RQV and RQUV governors) or by a tension spring attached to the control lever. The spring forces the control lever in the full-load direction in the RQV and RQUV and in the shut-off direction in the RSV and RSUV governors. The motor and the adjustment spindle are joined by a releasable clutch fitted with overload protection. When the motor clutch has been released, the adjustable spindle can be turned by the handwheel.

The adjustment travel is restricted by electric limit switches in both directions.

To stop the diesel engine, the governor control lever is drawn to the shut-off position. The shut-off stop is adjusted in similar fashion to the full-load stop on the governor cover. Two models of the electric speed-control device are produced: one for left-hand mounting and one for right-hand mounting.

PNEUMATIC GOVERNOR

Variable-Speed Governor EP/M

Operating Principle. The pneumatic governor consists of two main parts:

- the venturi assembly fastened to the engine intake manifold on the inlet side
- the diaphragm block mounted on the fuel injection pump

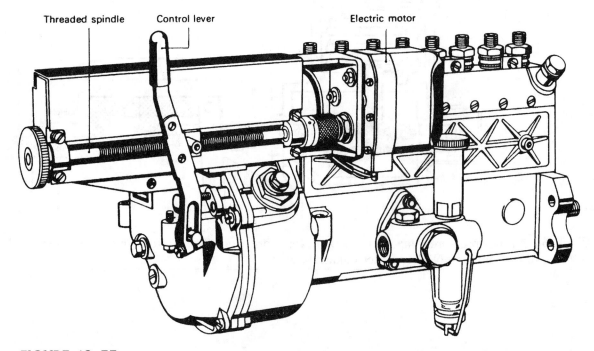

FIGURE 13-55 Electric speed-control device, left-hand mounting. (Courtesy of Robert Bosch Canada Ltd.)

Air drawn into the engine flows through the air filter and then through the venturi-shaped channel in the venturi assembly. At the narrowest point in this channel are a throttle valve and the connector fitting for the vacuum line leading to the diaphragm block. The throttle valve is connected with the accelerator pedal through the control lever and the linkage system (**Figures 13–56 to 13–58**).

Depending on the position of the throttle valve, the vacuum necessary for regulation is set (at full load

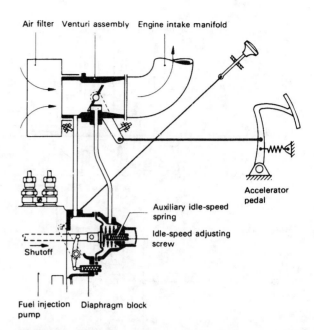

FIGURE 13–57 Pneumatic governor diaphragm block with auxiliary idle-speed spring in the idle-speed adjuster. (Courtesy of Robert Bosch Canada Ltd.)

about 400 mm water column) at a lower, medium, or higher speed. The diameter of the venturi channel must be so chosen that when the throttle valve is completely open, the required nominal speed can still be reached without bother. A stop screw serves to set the nominal speed accurately (limitation of the throttle valve opening). Pneumatic governors are used mainly in automobiles and agricultural tractors. The purpose of the auxiliary venturi at the point where the vacuum is taken is to ensure that if the engine should start to operate in the reverse direction, it does not run out of control; i.e., it can be stopped.

A diaphragm, connected with the fuel injection–pump control rod through a linkage system, divides the diaphragm block into two chambers:

- the vacuum chamber (connected by a hose or pipe with the venturi assembly) in which the governor spring tends to shift the diaphragm, and thus the control rod, in the maximum fuel direction
- the atmospheric chamber, which is connected with the outside air

When the engine is operating, the position of the diaphragm, and consequently the position of the control rod, depends on the magnitude of the difference between the pressures prevailing on the two sides of the diaphragm and developed by the engine loading at a particular time. This is because the engine speed rises or falls, respectively, if the load on the engine is in-

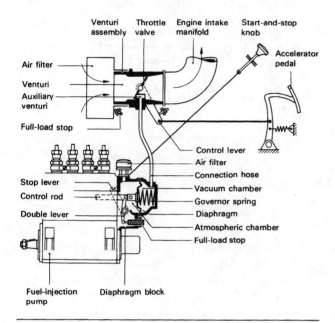

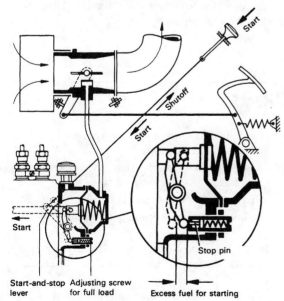

FIGURE 13–56 Pneumatic governor components and operation. Pneumatic governors are used on some light-duty trucks. (Courtesy of Robert Bosch Canada Ltd.)

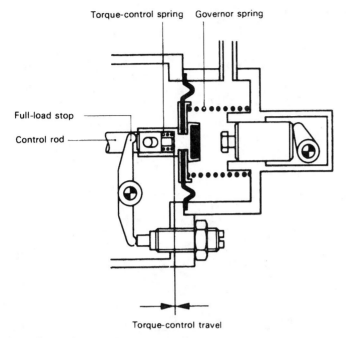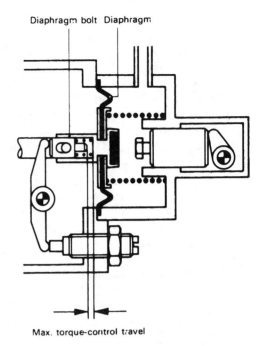

FIGURE 13-58 Pneumatic governor torque-control mechanism. (Courtesy of Robert Bosch Canada Ltd.)

creased or decreased at a certain position of the throttle valve. This results in the differing pressures in the vacuum chamber. If the initial tension of the governor spring is greater than the vacuum acting on the diaphragm, the control rod is shifted in the direction of increased fuel delivery. If the vacuum increases, the diaphragm is moved by the effect of the atmospheric air pressure against the spring pressure, and the control rod is shifted in the shut-off direction.

Speed regulation starts when the engine reaches that speed at which the vacuum is able to overcome the pressure exerted by the governor spring, or vice versa. The pneumatic governor is effective from the idle speed to the maximum speed.

REVIEW QUESTIONS

1. The diesel engine governor controls the amount of _____ delivered to the engine at any given _____ position.
2. Name eight duties of the governor.
3. The governor is responsible for ensuring the engine maintains a _____ _____ under various load conditions.
4. List 10 terms used in governor terminology.
5. Governors that operate on pressure differences between the atmosphere and engine vacuum are called _____ governors.
6. Minimum-maximum-speed governors control _____ and _____ speed.
7. The governor ensures that the engine does not _____ in the idle-speed range.
8. What is the function of governor torque control?
9. Mechanical governors employ the principle of _____ force.
10. In turbocharged engines, the full-load delivery is determined according to the _____ _____ pressure.
11. Why is a pressure compensator needed at high altitudes?

TEST QUESTIONS

1. Which of the following fuel delivery control functions is *not* performed by the governor?
   a. fuel viscosity
   b. fuel increase
   c. fuel decrease
   d. fuel quantity

2. A governor is needed on a diesel engine to
   a. control engine speed
   b. control the quantity of fuel delivered to the injectors
   c. reduce fuel delivery as the load on the engine is decreased
   d. all of the above

3. Governors with speed droop
   a. are isochronous
   b. are nonsynchronized
   c. allow engine speed to change as engine load is changed
   d. all of the above

4. A governor hunting condition can be corrected by
   a. employing a game warden
   b. cleaning or adjusting the governor
   c. stretching the governor springs
   d. changing the governor weights

5. Governors are classified as controlling
   a. minimum, maximum, and variable engine speeds
   b. low, medium, and high engine speeds
   c. fuel delivery and injection timing
   d. injection timing and duration

6. Speed droop as applied to governors is
   a. found only in variable-speed governors
   b. needed to ensure proper governor response
   c. the change in engine speed from high-idle to full-load speed.
   d. of very little concern, since it has little effect on engine performance

7. Variable-speed governors are
   a. used only on engines with power-generator sets
   b. used on engines in applications where engine speed must be adjusted to different operating speeds by the operator
   c. capable of holding the engine speed within 1% of high-idle speed when load is applied
   d. always of the hydraulic type

8. Which of the following is *not* a symptom of a worn governor?
   a. low engine power
   b. high-idle overrun
   c. low-idle underrun
   d. a rough-running engine

9. A weak governor spring may cause all of the following *except*
   a. low engine power
   b. slow governor response
   c. excess engine power
   d. low engine top speed

10. Erratic engine speed changes and severe hunting may be caused by all of the following *except*
    a. worn governor parts
    b. binding linkage
    c. worn metering valve
    d. improper high-idle adjustment

CHAPTER 14

ELECTRONIC CONTROL OF DIESEL FUEL INJECTION—OPERATING PRINCIPLES

INTRODUCTION

To meet the increased requirements of exhaust emission control legislation and to improve vehicle performance, manufacturers have developed electronic controls for a number of diesel fuel injection systems. The precise control of fuel injection timing and metering of fuel delivery result in reduced exhaust emissions, increased fuel economy, and better performance. This chapter discusses the general principles that apply to all electronically controlled diesel fuel injection systems. Learning about these principles will help you understand the different electronic control systems described in later chapters of this book. A typical electronic control system is shown in **Figure 14–1**. An electronic control module (ECM) is shown in **Figure 14–2**.

PERFORMANCE OBJECTIVES

After completing this chapter you should be able to:

1. State the three basic functions of an electronic control system.
2. Describe the basic sections of a diesel engine electronic control module (computer).
3. List the common diesel engine input sensors and state their function.
4. List the major output devices of an electronic diesel injection control system.
5. Describe the function of different types of computer memory.
6. Locate and name the components of different types of electronic control systems on different makes of vehicles.
7. Describe how to retrieve fault codes stored in computer memory.

TERMS YOU SHOULD KNOW

Look for these terms as you study this chapter, and learn what they mean.

computer	PROM
ECM	EEPROM
input	password
processing	reference voltage
output	limp-in mode
integrated circuit	variable resistance
digital	potentiometer
analog	pulse generator
memory	power transistor
RAM	power module
ROM	active sensor

ELECTRONIC CONTROL MODULE (ECM) OPERATION

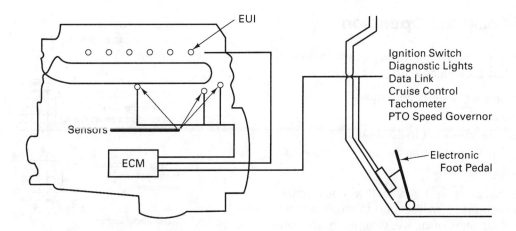

FIGURE 14–1 Detroit Diesel Electronic Control (DDEC) system. (Courtesy of Detroit Diesel Corporation.)

TPS
TBS
OTS
OPS
CTS
CLS
FTS
FTS
ESS
VSS
fault codes
passive sensor

CEL
pinpoint testing
engine power-down/shutdown
data extraction
progressive shifting
cruise control
vehicle speed limiting
automatic engine braking
engine fan braking
idle timer shutdown
isochronous governor
programmable low-idle rpm

ELECTRONIC CONTROL MODULE (ECM) OPERATION

The electronic control module is designed to operate in three steps: input, processing, and output **(Figure 14–3)**.

- *Input:* switches and sensors monitor conditions that they convert to electrical signals and send to the computer
- *Processing:* computer uses input data from sensors and computer memory to decide what the output actuators should do
- *Output:* computer produces output voltage to operate actuators

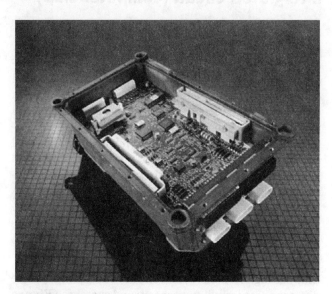

FIGURE 14–2 Internal view of DDEC electronic control module components. (Courtesy of Detroit Diesel Corporation.)

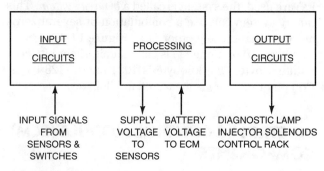

FIGURE 14–3 Basic computer (ECM) operation. (Courtesy of FT Enterprises.)

Speed of Computer Operation

Computer systems are extremely fast. How long does it take for the following sequence of events to occur?

- Sensor notes a change in condition.
- Sensor sends signal to computer.
- Computer decides what to do about it.
- Computer activates output device to compensate for change in condition.

All of this happens in a few milliseconds (thousandths of a second). A millisecond is much shorter and faster than the blink of an eye. Computers are constantly working at this speed to keep the systems they control at their most efficient and practical operating level.

Intergrated Circuit (Computer Chip)

A computer chip is an integrated circuit (IC) on a tiny piece of silicon. To *integrate* means to form or unite. An integrated circuit is the formation and interconnection of a number of semiconductor materials deposited on a silicon chip. These semiconductor deposits form the diodes, transistors, and resistors of the chip. All these components are microscopic in size and are produced under a microscope. Very fine high-quality metal conductors connect the components to each other and to the multipin input and output connectors **(Figure 14–4)**.

Digital System

Diesel engine computers are digital computers. That is, they use digits—zero (0) and one (1)—to distinguish the different data from input sensors and switches **(Figure 14–5)**. The computer can turn transistors on or off using digital data, where zero (0) represents OFF and one (1) represents ON. Because only two numbers (or digits) are used, the system is called a *binary system*. This binary system can use a combination of several zeros and ones to represent other data **(Figure 14–6)**.

A zero (0) or a one (1) is called a bit. Eight bits constitute a byte. One kilobyte is 1024 bits or 128 bytes. Computer memory is measured in kilobytes.

ELECTRONIC CONTROL MODULE (ECM) COMPONENTS

The computer (ECM) has a number of jobs it must perform. Each section of a computer is responsible for part of the action taking place. The following are the

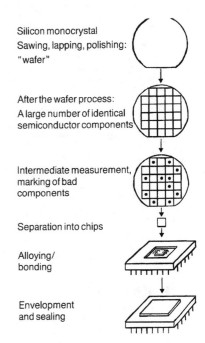

FIGURE 14–4 Steps in the production of a computer chip. (Courtesy of Robert Bosch Canada Ltd.)

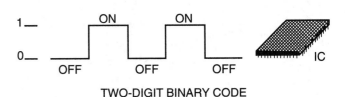

FIGURE 14–5 Binary digital voltage signals.

major components of a computer and their function **(Figure 14–7)**:

1. *Voltage regulator:* reduces input voltage to the computer and maintains it at a precise level. Voltage fluctuations in computer operating voltage cannot be tolerated.

2. *Clock:* pulse generator that produces steady pulses one bit in length. This constant pulse serves as a reference signal to which other signals are compared. This allows the computer to distinguish between different digital binary signals.

3. *Buffer:* serves as temporary storage space when data entry is too fast, then releases it as required.

4. *Analog converter:* converts analog voltage signals from sensors and switches to digital form that the microprocessor can handle (also called input interface).

5. *Digital converter:* converts digital computer output signals to analog voltage to operate actuators (also called an output interface).

6. *Microprocessor:* integrated circuit chip that analyzes data received from sensors, switches, and com-

ELECTRONIC CONTROL MODULE (ECM) COMPONENTS

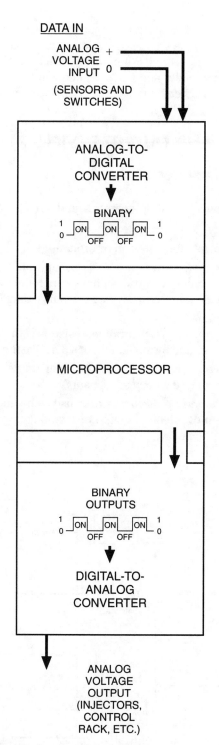

FIGURE 14–6 Electronic control module (ECM) operation showing analog and digital converter functions.

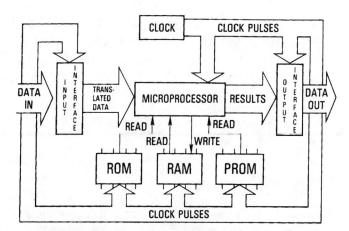

FIGURE 14–7 Computer memory functions. (Courtesy of General Motors Corporation.)

puter memory to produce output signals to operate actuators.

7. *Memory:* integrated circuit chip that stores data for use by microprocessor.

8. *Power transistors:* step up computer output voltage to operate actuators.

9. *Circuit board:* fiber panel with printed electrical circuitry that connects components mounted on board.

10. *Housing:* box that contains and protects computer components from damage and from outside electrical interference.

11. *Multiple connector:* multipin electrical connector that connects computer to wiring harness.

ECM Memory

The computer stores information or data in memory chips in the digital form just described. It uses this information and the information from switches and sensors to decide what the output devices should do. There are several kinds of memory in automotive computers: RAM, ROM, PROM, and EEPROM (**Figure 14–7**).

1. *Random access memory (RAM):* This system stores information or data on a temporary basis. This data comes from input sensors and switches. Information stored in the RAM unit can be lost when battery power is disconnected from the system.

2. *Read-only memory (ROM):* This information is permanent and is programmed into the computer at the factory based on the make and model of the vehicle. Disconnecting the battery does not affect the information stored in ROM.

3. *Programmed read-only memory (PROM):* This information is also permanent and is programmed into the computer at the factory. The data are specific to engine size, transmission type, fuel system, turbo or nonturbo, gear ratios, and a variety of other options. Although the PROM (or calibration unit) rarely fails, it can be replaced on many computers. Should the computer require replacing, the PROM is removed from the

old computer and installed in the new computer. Data stored in the PROM remain during computer replacement.

4. *Electrical erasable programmable read-only memory (EEPROM):* Unit in which programmed information can be erased electrically and new information programmed into the unit to tailor it to user-desired specifications.

Password

A password is a unique group of numbers or numbers and letters used to access the computer control system EEPROM in order to establish or change certain engine operating specifications. It may also be used to clear certain diagnostic trouble codes.

There are two kinds of passwords: *factory passwords* and *customer passwords*. The engine manufacturer uses the factory password to set the operating limits for which the engine was designed. This ensures that these limits will not be exceeded by the customer. The customer password is used by the end user of the vehicle or engine to set the engine operating specifications that best suit the customer's requirements. This ensures that these limits will not be exceeded by the vehicle or engine operator. Operating parameters that can be set include engine rpm limit, vehicle speed limit, low cruise control set limit, and the like.

Customer programming of the ECM is easily accomplished with the use of the appropriate electronic control analyzer as outlined in the operating manual provided by the analyzer manufacturer.

Reference Voltage

The ECM or computer puts out a reference voltage to sensors that do not generate their own voltage. This voltage is typically fixed at 5, 8, or 9 V depending on system design. Reference voltage sent to the sensor is modified by the sensor based on the conditions being sensed such as changes in pressure, temperature, speed, and the like. This modified voltage signal is sent back to the computer for processing.

Limp-in Mode

The computer control system has a limp-in mode designed to take over vehicle operation when any of its critical sensors fail to provide proper input signals. When the computer sees a problem with one of these signals, it operates on fixed values stored in the computer's permanent memory or it generates a replacement value by computing input from other input signals. This allows engine operation to continue until the problem is corrected. When the system is in the limp-in mode, driveability and performance will be below normal, and the check engine light (CEL) will be on.

INPUT AND OUTPUT DEVICES

Input Devices

Input devices include switches and sensors (**Figure 14–8**).

1. *Switch:* indicates on or off condition.
2. *Variable-resistance sensor:* internal resistance of sensor changes with a change in temperature or pressure. Sensor output voltage signals change accordingly.
3. *Potentiometer:* variable resistor with a sliding contact actuated by movement of a part like the throttle.
4. *Voltage-generating sensor:* generates its own voltage signal (e.g., oxygen sensor).
5. *Magnetic pulse generator:* uses a magnetic field and part movement to generate a voltage signal (e.g., crankshaft sensor).

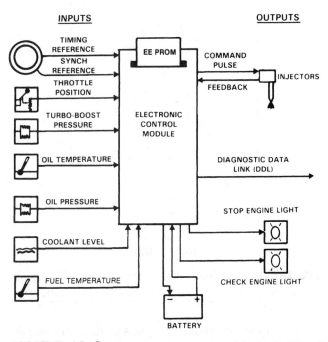

FIGURE 14–8 Inputs and outputs of Detroit Diesel Electronic Control system. Note diagnostic data link and EEPROM unit. (Courtesy of Detroit Diesel Corporation.)

Output Devices

Output device or actuator types include the following **(Figure 14–8)**:

1. *Relay:* Computer activates relay to control greater current.
2. *Solenoid:* Computer activates solenoid by energizing the solenoid windings. Solenoid acts on another device (e.g., electronically controlled injector).
3. *Transistor:* Computer energizes base current of transistor to control greater current: in effect, an electronic relay.
4. *Servo motor:* Computer sends current to motor to cause rotation.
5. *Lighted display:* Computer sends current to activate indicator lights in dash.
6. *Power module:* Contains power transistors to operate output devices based on signals received from the control module.

Active and Passive Sensors

Input devices or sensors are classified as *active* or *passive*. An active sensor generates its own voltage signal and does not require an external voltage source. Active sensors include magnetic pickup sensors. Passive sensors rely on an external voltage supplied by the ECM. The passive sensor's internal resistance changes with a change in sensed conditions to produce a variable-voltage output signal to the computer. Passive sensors include throttle position sensors, temperature sensors, and magnetic sensors.

INPUT SENSORS AND THEIR FUNCTION

The following sensors and their functions are typical of an electronically controlled diesel fuel injection system.

1. *Throttle position sensor (TPS):* The TPS is a potentiometer (variable resistor). It receives an input voltage (reference voltage) from the ECM. The TPS changes this voltage into an output voltage that varies with foot-pedal position. Output voltage is low at idle. This low-voltage signal is used by the ECM to maintain fuel delivery at a minimum level. As the pedal is pressed downward the voltage output signal increases, and the ECM causes fuel delivery to increase to meet engine demands. The actual quantity of fuel injected also depends on voltage signals sent to the ECM by a number of other sensors.

2. *Turbocharger boost sensor (TBS):* The TBS is mounted on the engine intake manifold. The output voltage of the TBS varies with boost pressures in the intake manifold. The ECM uses this signal to calculate the amount of air entering the engine at all times. It uses this information to increase or decrease the amount of fuel being injected in relation to the increase or decrease in boost pressure.

3. *Oil temperature sensor (OTS):* The OTS is screwed into the main oil gallery of the engine. It senses engine lubricating oil temperature and sends a variable-voltage signal to the ECM based on changes in oil temperature and sensor resistance. The ECM uses this information to improve cold-weather starting and to increase engine idle speed after a cold engine start for faster engine warm-up. If engine oil temperature exceeds preprogrammed limits for more than 2 seconds, the ECM will illuminate the CEL. When the engine oil temperature exceeds safe operating limits, the ECM will illuminate the SEL (stop engine light) and activate a warning buzzer. This will also cause the engine to reduce power, allowing the operator about 30 seconds to bring the vehicle to a stop in a safe area. If the vehicle is so equipped, an override switch can be activated by the operator to increase this time.

4. *Oil pressure sensor (OPS):* The OPS sends a signal to the ECM indicating engine oil pressure at all engine speeds. When oil pressure drops below the specified value for more than a few seconds, the ECM begins to activate the CEL and SEL functions.

5. *Coolant temperature sensor (CTS):* The CTS is located on the engine where it can sense the temperature of the engine coolant. Exact locations vary with the different engines. The CTS sends a variable-voltage signal to the ECM based on changes in coolant temperature and sensor resistance. The voltage signal is high when the coolant temperature is low, and low when it is hot. There are two types of temperature sensors. The negative temperature coefficient (NTC) sensor's resistance decreases as the temperature increases. The positive temperature coefficient (PTC) sensor's resistance increases as the temperature increases. The CTS receives a reference voltage from the ECM. At low coolant temperature the NTC-type sensor has a high resistance and provides a low-voltage output signal. If coolant temperature rises above a specified level, the ECM triggers the engine protection system.

6. *Coolant level sensor (CLS):* Located in the coolant reserve tank or the top radiator tank, the CLS sends a signal to the ECM indicating the engine coolant level. When the coolant level drops too low, the ECM begins to activate the CEL and SEL warning functions.

7. *Fuel temperature sensor (FTS):* The FTS tells the ECM the temperature of the fuel at the fuel inlet. The ECM uses this information to help calculate fuel consumption.

8. *Timing reference sensor (TRS):* The TRS sends a voltage signal to the ECM to indicate when each piston nears the TDC position. The ECM uses this information to help control fuel injection timing.

9. *Engine speed sensor (ESS):* The ESS tells the ECM at what speed (rpm) the engine is operating at all times. The ECM uses this information to help control injection timing and the quantity of fuel delivery.

10. *Vehicle speed sensor (VSS):* The VSS provides vehicle speed information to the ECM at all times. The ECM uses this information to govern maximum engine speed in different gear ratios, cruise control operation, and maximum vehicle speed.

COMPUTER-CONTROLLED ACTUATORS (OUTPUT DEVICES)

Actuators or output devices are designed to do the bidding of the electronic control module (computer). They do what they are told to do by the ECM. Power transistors in the output section of the computer provide the voltage required to operate the actuators when the computer energizes the transistor base circuit by grounding it. Actuators include solenoids, relays, servomotors, and light displays. An electronic unit injector (EUI) solenoid is an example. Other devices include governor control, cruise control, tachometer, data link, engine brake (retarder), and the like.

Check Engine Light (CEL)

The CEL turns on to warn the driver that an engine problem exists and that the vehicle should be serviced as soon as possible. The CEL also comes on when the ignition switch is first turned on to verify that the system is operational. The light goes off after the engine is started (**Figures 14–9 and 14–10**).

Engine Protection System

The engine protection system is designed to protect the engine from major damage should any of the following conditions occur:

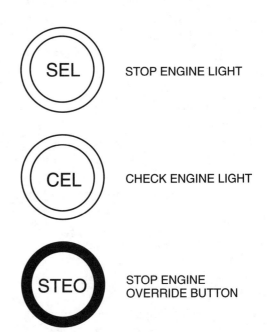

FIGURE 14–9 Engine protection system warning lights and override button.

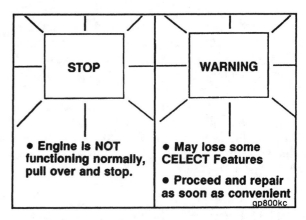

FIGURE 14–10 Warning lamp on Cummins Celect system. (Courtesy of Cummins Engine Company, Inc.)

- engine oil pressure falls below a specified value
- engine lubricating oil temperature exceeds the maximum safe value
- coolant level drops too low

If any of these conditions occurs, the SEL goes on, and the stop engine function is activated. Fuel delivery is reduced to reduce engine speed and power. If the problem persists, the engine will shut down 30 seconds after the SEL goes on. This normally gives the driver enough time to pull over and stop away from traffic. The stop engine override button can be held by the driver to pro-

FAULT CODES 273

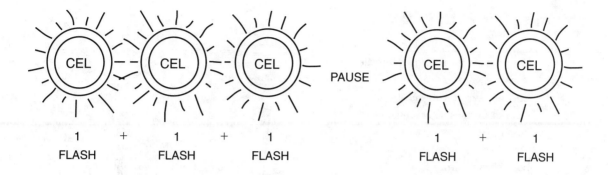

3 FLASHES + PAUSE + 2 FLASHES = CODE 32

FIGURE 14–11 Flashing-light fault code interpretation.

vide 30 seconds of override time should additional time be needed to park the vehicle safely.

FAULT CODES

The ECM or computer is designed to monitor the performance of input sensors, output devices, the wiring harness, and electrical connectors. When a problem occurs in any of these components, a numbered fault code is stored in the ECM memory. Manufacturers assign numbers to specific faults that may occur and program them into the computer memory. The technician can access these fault codes by using a dash-mounted switch to activate the CEL. The light flashes on and off several times depending on the fault code that is stored. The number of light flashes indicates the number of the fault code **(Figures 14–11 and 14–12)**. For example, if the light flashes once, pauses briefly, then flashes twice, the fault code is 12. Fault codes may also be retrieved with the use of a portable diagnostic tool that plugs into the vehicle wiring harness, such as Detroit Diesel's Diagnostic Data Reader (DDR), Caterpillar's Electronic Control Analyzer Programmer (ECAP), and Cummins' Compulink **(Figure 14–13)**. Fault code charts in the service manual are used to interpret fault code numbers. **Figure 14–14** is an example of one such chart. Always refer to the appropriate service manual for fault code interpretation for any specific vehicle.

Active and Inactive Codes

Active codes reflect problems that are ongoing and keep the CEL on. Inactive codes reflect problems that have occurred once or twice or are intermittent. The CEL goes on while the problem exists and goes off when the problem goes away. Inactive codes are stored in ECM memory. Inactive codes are also called *historic codes* or *logged codes*.

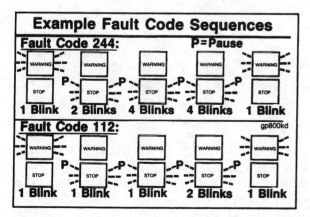

FIGURE 14–12 Example of Cummins fault code sequence interpretation. (Courtesy of Cummins Engine Company, Inc.)

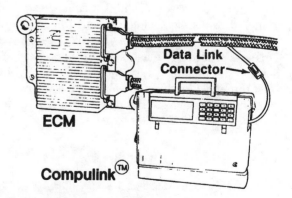

FIGURE 14–13 Cummins' Compulink diagnostic tool connected to data link in wiring harness. (Courtesy of Cummins Engine Company, Inc.)

3176 SYSTEM FAULT CODES

Fault Code	Description
01	Override of idle shutdown throttle
25	Boost pressure sensor fault
27	Coolant temperature sensor fault
28	Check throttle sensor adjustment
31	Loss of vehicle speed signal
32	Throttle position sensor fault
34	Loss of engine rpm signal
35	Engine overspeed warning
36	Vehicle speed signal out of range
37	Fuel pressure sensor fault
41	Vehicle overspeed warning
42	Check sensor calibrations
47	Idle shutdown timed out
51	Intermittent battery power to ECM
52	ECM or personality module fault
53	ECM fault
55	No detected faults
56	Check customer or system parameters
57	Parking brake switch fault
63	Low fuel pressure warning
72	Cylinder 1 or cylinder 2 fault
73	Cylinder 3 or cylinder 4 fault
74	Cylinder 5 or cylinder 6 fault

FIGURE 14–14 Caterpillar 3176 fault codes. (Courtesy of Caterpillar Inc.)

Pinpoint Testing

Pinpoint testing is normally done after completing self-diagnosis. Pinpoint tests involve testing individual components of the system identified as having a fault. This includes testing switches, sensors, actuators, wiring harness, and connectors (**Figure 14–15**). Service manual procedures must be followed precisely for pinpoint tests, including test equipment connections and test terminal identification. Making the wrong connections can damage testers and system components.

CAUTION:

Be very careful not to damage any electrical connectors when probing with tester leads. Damaged connectors must be replaced. Never pierce any electrical wiring insulation. Pierced insulation allows the entry of moisture and dirt that causes corrosion of wiring.

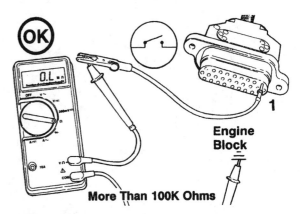

FIGURE 14–15 Example of pinpoint testing. (Courtesy of Cummins Engine Company, Inc.)

FEATURES AND ADVANTAGES OF ELECTRONIC CONTROLS

Electronic control systems provide very precise control over fuel metering and fuel injection timing. In addition, the system monitors engine operating factors and conditions. Fault codes and other operating data are stored for later retrieval. The following features and advantages are typical of a modern electronic control system. Not every vehicle is equipped with all the items listed. Features vary with the make and model of vehicle.

1. *Easier starting.* The ECM adjusts injection timing and fuel delivery to match precisely the start-up requirements, making cold-engine starts much easier.

2. *Improved fuel economy.* Fuel consumption is reduced by 5–10% due to precise control over injection timing and fuel delivery over all operating conditions. Operating costs are thereby reduced.

3. *Reduced maintenance.* Mechanical governor parts that are subject to wear and misadjustment are eliminated. Built-in logic in the ECM compensates for normal engine wear. This keeps the engine always in tune and reduces maintenance costs.

4. *Reduced exhaust emissions.* Very precise control of injection timing and fuel delivery are required to meet today's exhaust emission standards. Electronic controls decide exactly how much fuel to inject and exactly when to inject it in relation to piston position. Modern diesel engines do not belch out white exhaust smoke during engine warm-up, as older, mechanically controlled engines do.

5. *Easier troubleshooting.* Trouble diagnosis is made fast, easy, and accurate with either the on-board diagnostics or the portable electronic diagnostic tool that plugs into the system's wiring harness diagnostic connector.

6. *Engine protection system.* When the ECM receives a signal from any one of the engine's vital operating sensors that a condition is out of specifications, the system's power-down/shutdown function begins. Depending on the problem, the CEL goes on to warn the driver of an abnormal condition and/or the SEL goes on and engine shutdown begins. An override switch may be provided to allow the driver to maintain power longer and park the vehicle away from traffic.

7. *Data extraction.* Modern electronic control systems store vehicle operating data such as engine performance, engine operating hours, miles driven, fuel consumption, idle time, cruise time, and the like. These data can be extracted from the ECM memory for use in vehicle performance analysis, record keeping, and management. Operating data may also be displayed in a dash-mounted unit. Information displayed is obtained from that stored in the system (**Figures 14–16** to **14–18**).

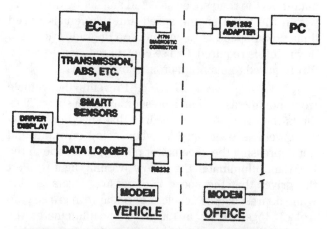

- WIRELESS DATA EXTRACTION ADDS AUTOMATION & UTILITY
- INTERFACES WITH AVAILABLE COMMUNICATIONS SYSTEMS
- ASSORTED TECHNOLOGIES MEET FLEET'S SPECIFIC NEEDS

FIGURE 14–17 Use of a personal computer and existing wireless communications systems allow data extraction from vehicle at a distance. (Courtesy of Detroit Diesel Corporation.)

The engine/vehicle information that is monitored and available on the Data Link includes the following:

Boost pressure
Cold start status
Coolant temperature
Cruise/PTO switch status
Cruise mode status
Cruise mode set speed
Customer specified parameters
Cylinder cutout/injection signal duration
Diagnostic messages
ECM identification
Engine identification
Engine speed—rpm
FRC fuel position
Fuel position
Fuel pressure
Idle shutdown timer status
Park brake switch status
PTO mode status
PTO mode set rpm
Throttle position in % (processed signal)
Throttle pulse width signal (raw data)
Rated fuel position
Resume switch status
Retarder enable status
Sensor calibration
Set switch status
System configuration parameters
Vehicle speed—mph

FIGURE 14–16 Data Link information available from the Caterpillar 3176 electronic control system. (Courtesy of Caterpillar Inc.)

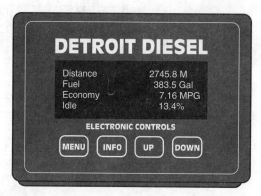

FIGURE 14–18 Detroit Diesel ProDriver dash-mounted display. (Courtesy of Detroit Diesel Corporation.)

8. *Progressive shifting.* Progressive shifting means shifting up quickly through the lower gears. Each upshift is made just above peak engine torque but just below rated engine speed. This avoids unnecessarily high engine speeds that waste fuel and increase the time needed for upshifts. The electronic control system can be programmed to require the driver to use progressive shifting techniques **(Figure 14–19)**.

9. *Cruise control.* Cruise control automatically maintains the road speed selected and set by the driver within the rated engine speed and the vehicle speed limit. Road speed is maintained on upgrades unless a downshift is required. Cruise control reduces driver effort required to maintain proper road speeds.

10. *Vehicle speed limiting.* Prevents the vehicle from moving faster than a preset maximum speed. This limits the driver from exceeding that speed on a level road. Vehicle overspeeding when going downhill and not depressing the accelerator pedal logs a code in the ECM and illuminates the yellow warning light to alert the driver that the speed is excessive. Overspeeding while depressing the accelerator pedal with cruise control off logs a code in the ECM and illuminates the warning light to alert the driver that the cruise control is off.

11. *Automatic engine braking.* The engine brake activates the low braking mode automatically when vehicle speed increases a few miles per hour above the cruise control speed setting. If vehicle speed continues to increase, engine braking increases progressively as required. When vehicle speed is no longer increasing, the engine brake stays off, and the vehicle returns to set cruise speed.

12. *Engine fan braking.* The engine fan clutch engages automatically when the engine brake is on HIGH. This adds from 20 to 45 HP to help retard the engine and vehicle speed.

13. *Idle timer shutdown.* Shuts down the engine after a preset amount of idle time (set by the driver) has expired. If engine idling is required for cab heating or cooling, the engine should be run at the specified high-idle speed. Excessive engine idling produces corrosive acids that eat into engine wear surfaces.

14. *Isochronous governor.* An isochronous governor provides zero or very near zero speed droop. This results in much better engine response to changes in engine load, no loss of engine speed when a load is applied, no overspeeding when the load is removed, and a constant-power take-off speed.

15. *Cold-idle mode.* After a cold-engine start, the cold-idle mode runs the engine at fast idle (approximately 900 to 1000 rpm) until the engine reaches a specified temperature. This provides faster engine warm-up. Should the driver start driving before the en-

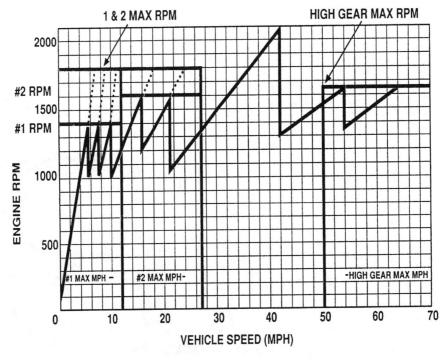

FIGURE 14–19 Progressive shifting chart. (Courtesy of Detroit Diesel Corporation.)

gine has reached the specified temperature, engine speed and power are reduced until the specified temperature is reached.

16. *Programmable low-idle rpm.* Low-idle rpm can be set by the driver to avoid certain idle speeds that can cause annoying vibrations. Mirrors can blur, and dash components may rattle at these idle speeds. The driver selects and sets the engine idle speed at which this does not occur.

17. *Data link standards.* The American Trucking Association (ATA) and the Society of Automotive Engineers (SAE) have agreed to certain data link standards for the industry. These standards, J1708 and J1587, make the retrieval of fault codes and other data possible using any electronic test equipment that meets these standards. This also allows this equipment to be used on any make or model of vehicle that meets these standards. Vehicle performance data such as fuel consumption, idle time, trip time, engine hours, and the like stored in computer memory can be extracted using a personal computer, data terminal connector, and printer. This information is used to help in fleet management **(Figure 14–20)**.

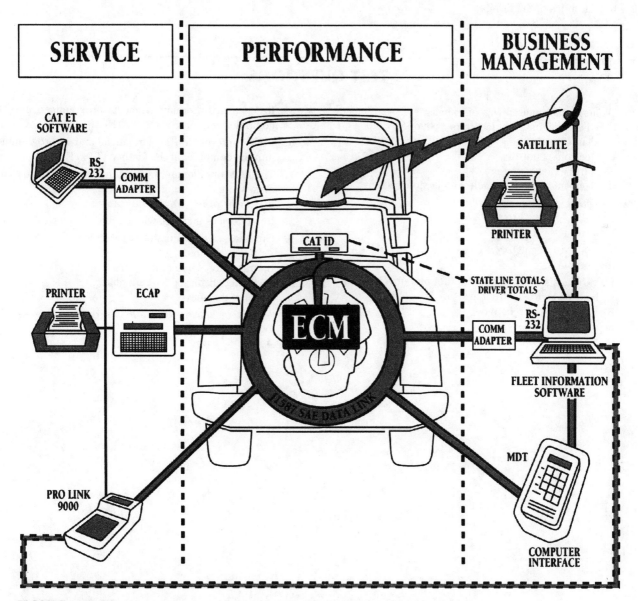

FIGURE 14–20 Electronic control system is used to maximize engine performance, ease troubleshooting and diagnostic procedures, and provide data for business management. (Courtesy of Caterpillar Inc.)

REVIEW QUESTIONS

1. To meet requirements of exhaust emission control and improved performance, vehicle manufacturers have developed _____ _____ for diesel fuel injection systems.
2. ECM is the abbreviation for _____ _____ _____.
3. Speed of computer operation is measured in _____.
4. A computer chip is an _____ _____ on a tiny piece of silicon.
5. Diesel engine computers are _____ computers.
6. List 11 duties of the ECM.
7. The kinds of memory in the ECM are _____, _____, _____ and _____.
8. What is the purpose of the limp-in mode?
9. Input devices or sensors are classified as _____ or _____.
10. A TPS sensor is a _____.
11. Name at least three duties of the engine protection system.
12. How are fault codes retrieved?
13. Name the advantages of electronic controls.

TEST QUESTIONS

1. To meet the requirements of exhaust emission legislation, diesel engine manufacturers developed
   a. a larger engine
   b. electronic controls
   c. better exhaust system
   d. electronic controls for diesel injection systems
2. A digital computer uses digits
   a. 1, 2, 3,   c. 1 to 10
   b. 0 and 1   d. 1 to 5
3. The voltage regulator reduces input _____ to the computer.
4. Power transistors step up computer output voltage to
   a. be used by the microprocessor
   b. be released or required
   c. operate actuators
   d. assist the microprocessor
5. RAM stores information and is programmed into the computer at the factory. (T) (F)
6. PROM information is permanent and is programmed into the computer at the factory. (T) (F)
7. The ECM or computer puts out reference voltage to the
   a. sensors   c. limp-in mode
   b. transistors   d. all of the above
8. Input devices include switches and sensors. (T) (F)
9. Output device or actuator types include
   a. relay, solenoid
   b. transistor, servo motor
   c. lighted display, power module
   d. all of the above
10. An active sensor generates its own _____ signal.

◆ CHAPTER 15 ◆

ROBERT BOSCH, AMBAC INTERNATIONAL, AND LUCAS CAV MULTIPLUNGER INJECTION PUMPS

INTRODUCTION

The most common diesel fuel injection pump design worldwide is the port and helix plunger-and-barrel type. Applications of this design include multiple plunger, in-line, and V-configuration pumps, and single element-types mounted separately on the engine. All these require one pumping element for each engine cylinder (**Figures 15–1 to 15–3**).

Injection pumps using the port and helix pumping element design are produced by various manufacturers. Robert Bosch pumps of this type include the in-line PE series and the flange-mounted single-element PF series. AMBAC International Corporation (formerly American Bosch and then United Technologies Diesel Systems) pumps of this type include the in-line and V-type APE series and the APF single-element flange-mounted series. These pumps may be mechanically or electronically controlled. Lucas CAV pumps of this type include the Majormec and Minimec series.

The information in this chapter is provided by and/or based on material provided by Robert Bosch Canada Ltd., AMBAC International Corporation, Lucas CAV, and Mack Trucks Inc.

PERFORMANCE OBJECTIVES

After thorough study of this chapter and the appropriate training models and service manuals and with the necessary tools and equipment, you should be able to do the following with respect to diesel engines equipped with port and helix type fuel injection pumps:

1. Describe the basic construction and operation of a port and helix fuel injection pump, including the following:
   a. method of fuel pumping and metering
   b. delivery valve action
   c. lubrication
   d. injection timing control
2. Diagnose basic fuel injection system problems and determine the required correction.
3. Clean, inspect, test, measure, repair, replace, and adjust fuel injection system components as needed to meet manufacturers' specifications.

TERMS YOU SHOULD KNOW

Look for these terms as you study this chapter, and learn what they mean.

port and helix
plunger and barrel
intake port
spill port
port closing
port opening
effective stroke
upper helix
lower helix
starting groove
delivery valve
constant-pressure valve
pump sizes
single-plunger unit pump
injection timing device
governor types
electronic diesel control
single-acting supply pump
double-acting supply pump
injection pump flushing
injection pump parts inspection
injection pump phasing
spill timing
pump calibration
injection pump timing

FIGURE 15-3 Bosch single-plunger pumps. (a) PFE1, (b) PFR1K, (c) PFR1W, (d) PF1D. (Courtesy of Robert Bosch Canada Ltd.)

FIGURE 15-1 Bosch injection pump. 1. Fuel injection pump, 2. Governor, 3. Fuel supply pump, 4. Timing device. (Courtesy of Robert Bosch Canada Ltd.)

FIGURE 15-2 AMBAC model APE 8 VBB injection pump. (Courtesy of AMBAC International Corp.)

INJECTION PUMP DESIGN AND OPERATION

The following discussion of injection pump design and operation applies to Robert Bosch pumps. Other port and helix plunger pumps have the same basic components and operate in a similar manner. **Figure 15-1** shows a PE injection pump with mechanical governor (2), fuel supply pump (3), and timing device (4). The supply pump transfers the fuel from the fuel tank and delivers it through the fuel filter into the suction gallery of the injection pump. The pump plunger driven by the injection pump camshaft delivers the fuel through the delivery valve and the fuel injection tubing to the injection nozzle. After completion of the plunger lift, the spring-loaded delivery valve closes the fuel injection tubing, and the plunger is returned to its starting position by the plunger return spring.

Pumping Element (Plunger-and-Barrel Assembly)

The number of pumping elements in the PE injection pump corresponds to the number of cylinders in the engine (**Figures 15-4 to 15-9**.)

Each pumping element consists of a plunger and a barrel. The plunger is fitted into the barrel so perfectly (clearance of $1/10,000$ in. or a few $1/1000$ mm) that it seals even at very high pressure and low speeds. Therefore, only complete pumping elements can be replaced, certainly not just the plunger or the barrel only.

The plunger is provided with a vertical groove as well as a helical cut. The edge formed on the upper sec-

INJECTION PUMP DESIGN AND OPERATION

tion of the plunger is known as the "helix." A single helix suffices for injection pressures up to 8700 psi (600 bar), but above this figure the plunger needs two diametrically opposite helixes. This measure serves to prevent plunger "seizure" because the plunger is no longer forced against the barrel wall by the injection pressure. The barrel is provided with one or two inlet ports for fuel inlet and end of delivery.

The barrel is usually provided with two radial ports opposite one another, through which the fuel flows from the suction gallery to the pressure chamber (intake and spill port) and through which it returns when the delivery ceases due to port opening.

However, there are also "single-port elements." Their barrels have only one lateral port (spill port). The plunger has an axial hole in place of the vertical groove and a slanted slot instead of a milled helix.

In the lowest plunger position, the pressure chamber above the plunger is filled with fuel that has entered from the suction gallery through the lateral barrel ports. As the plunger moves upward the barrel ports are closed and fuel is discharged through the delivery valve into the fuel injection tubing. Fuel delivery ceases when the helix registers with the spill port because from this moment on, the pressure chamber of the barrel is connected to the suction gallery—through the vertical and annular groove or the axial hole and slanted slot. Thus, the fuel is returned to the suction gallery.

Because the end of delivery is reached when the plunger helix opens the spill port, the effective stroke

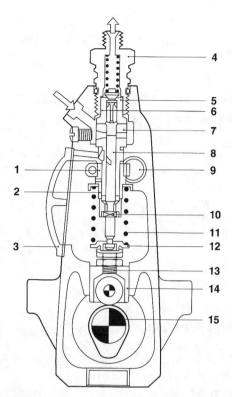

FIGURE 15-4 Injection pump cross section. 1. Control-sleeve gear, 2. Control sleeve, 3. Spring-chamber cover, 4. Delivery-valve holder, 5. Valve holder, 6. Delivery valve, 7. Pump barrel, 8. Pump plunger, 9. Control rack, 10. Plunger control arm, 11. Plunger return spring, 12. Spring seat, 13. Adjusting screw, 14. Roller tappet, 15. Camshaft. (Courtesy of Robert Bosch Canada Ltd.)

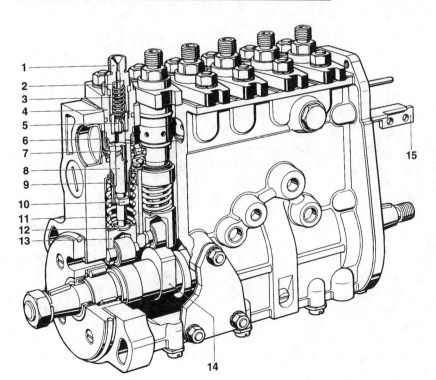

FIGURE 15-5 PES in-line fuel-injection pump. 1. Delivery-valve holder, 2. Filler piece, 3. Delivery-valve spring, 4. Pump barrel, 5. Delivery valve, 6. Inlet port and spill port, 7. Control helix, 8. Pump plunger, 9. Control sleeve, 10. Plunger control arm, 11. Plunger return spring, 12. Spring seat, 13. Roller tappet, 14. Cam, 15. Control rack. (Courtesy of Robert Bosch Canada Ltd.)

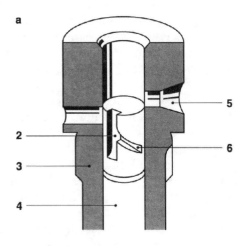

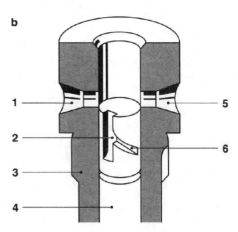

FIGURE 15–6 Plunger-and-barrel assemblies (pumping elements). (a) Single-port plunger-and-barrel assembly; (b) two-port plunger-and-barrel assembly. 1. Inlet port, 2. Vertical slot, 3. Barrel, 4. Plunger, 5. Spill port (inlet and return), 6. Control helix. (Courtesy of Robert Bosch Canada Ltd.)

can be varied by turning the pump plunger. If the plunger is turned until the vertical groove or the helix opens the spill port, the fuel in the pressure chamber is returned to the suction side during the plunger lift: no fuel is delivered.

In pump sizes PE (S) A, CW, Z and ZW, the control sleeve slipped over the pump plunger supports at its upper end a clamped-on gear segment (control sleeve gear); at the bottom it is provided with two vertical slots in which the plunger control arms slide. The teeth of the gear segment engage the teeth of the control rack (**Figure 15–10**). The pump plungers thus can be rotated by the control rack during operation so that the effective stroke and therefore the fuel delivery of the pump can be infinitely varied from zero to maximum delivery.

For pump sizes PE (S) M and P, the metering principle for fuel delivery is identical to that described above except that a link connected to the control rod is used for rotating the plunger in place of the rack and pinion.

Pumping Element with an Upper and a Lower Helix

There are pumping elements in which the plunger not only has a lower helix (for port opening) but also an upper one for control of the port closing. Rotating the plunger changes the start of delivery as a function of load (**Figure 15–11**).

These elements are used because noise reduction can be obtained when the speed-dependent variant in timing (by the timing device) is combined with a load-dependent variation in timing.

Starting Groove

Some diesel engines start more easily when fuel injection timing is retarded during the starting process relative to normal operation.

For this purpose, the plunger is provided with a starting groove at the top that results in a port closing with a 5–10° retard. As soon as the engine reaches operating speed, the governor pulls the control rod to the normal operating position (**Figure 15–12**).

Delivery Valve (Retraction Type)

As soon as the helix of the plunger opens the spill port, the pressure in the pump barrel drops. The higher pressure in the line to the injection nozzles and the valve spring force the delivery valve onto its seat. It closes the high-pressure delivery line against the pump barrel until fuel delivery starts again in the next delivery stroke (**Figure 15–13**).

The delivery valve furthermore has the task of "relieving" the high-pressure delivery line. Relief of the high-pressure line is necessary to obtain rapid closing of the nozzle valve and to prevent the dribbling of fuel into the combustion chamber. The "retraction volume" depends on the length of the high-pressure line and the quantity of fuel delivered. The delivery valve is guided with its stem in the valve body. During the delivery process it is lifted from its seat, so that the fuel can be discharged into the delivery valve holder through flutes terminating in an annulus. Above the annulus there is a short cylindrical shaft section (retraction piston); it fits tightly into the valve body and is followed by the conical valve.

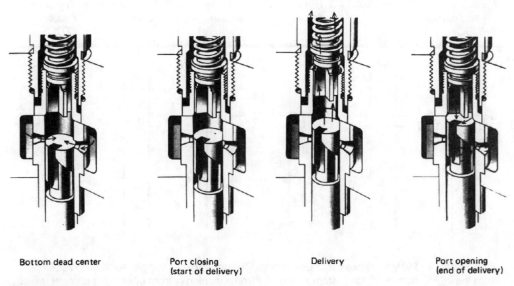

FIGURE 15–7 Operation of injection pump plunger. (Courtesy of Robert Bosch Canada Ltd.)

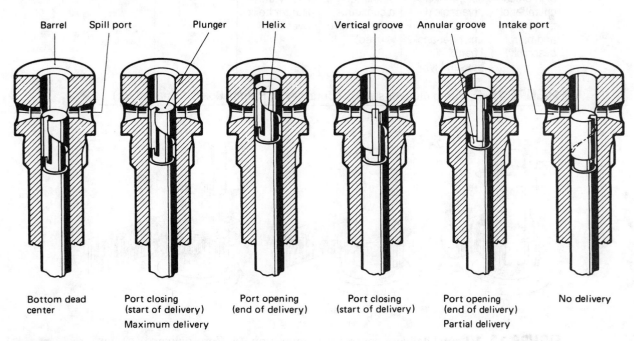

FIGURE 15–8 Pumping plunger action. Note vertical and rotational movement of plunger. (Courtesy of Robert Bosch Canada Ltd.)

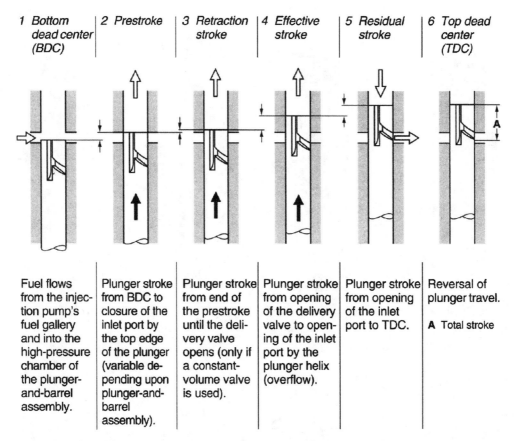

FIGURE 15–9 Pump plunger stroke. (Courtesy of Robert Bosch Canada Ltd.)

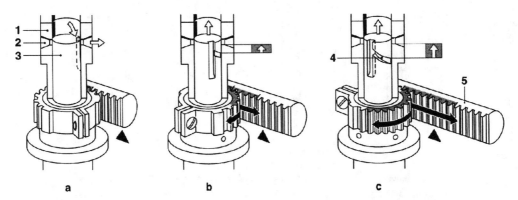

FIGURE 15–10 Fuel-delivery control-rack operation. (a) Zero delivery, (b) Partial delivery, (c) Maximum delivery. 1. Pump barrel, 2. Inlet port, 3. Pump plunger, 4. Helix, 5. Control rack. (Courtesy of Robert Bosch Canada Ltd.)

INJECTION PUMP DESIGN AND OPERATION 285

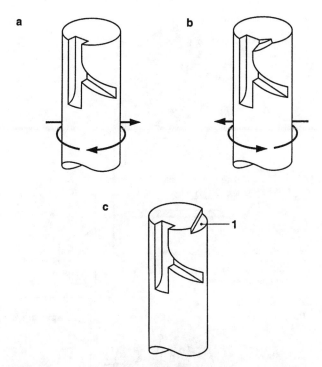

FIGURE 15–11 Pump plunger variations. (a) Lower helix, (b) upper and lower helix, (c) lower helix with starting groove (1). (Courtesy of Robert Bosch Canada Ltd.)

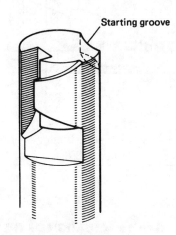

FIGURE 15–12 Pump plunger with starting groove. (Courtesy of Robert Bosch Canada Ltd.)

At the end of delivery, the retraction piston first enters the bore in the valve body and closes the high-pressure delivery line from the pressure chamber. Only then does the conical valve seat. The volume available for fuel in the high-pressure line is thereby increased by the volume of the retraction piston. The fuel in the high-pressure line can expand quickly, and the nozzle valve closes instantaneously.

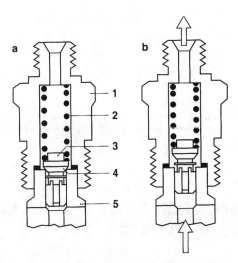

FIGURE 15–13 Delivery-valve holder with delivery valve. (a) Closed, (b) during fuel delivery. 1. Delivery-valve holder, 2. Delivery-valve spring, 3. Delivery valve, 4. Valve seat, 5. Valve holder. (Courtesy of Robert Bosch Canada Ltd.)

Constant-Pressure Type

The constant-pressure valve **(Figure 15–14)** is used with high-pressure fuel injection pumps that develop pressures above approximately 11,600 psi (800 bar), and small high-speed direct-injection (DI) engines. It comprises a forward-delivery valve in the delivery direction, and a pressure-holding valve in the

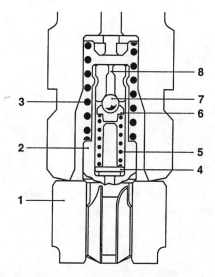

FIGURE 15–14 Constant-pressure valve. 1. Valve holder, 2. Valve element, 3. Valve spring, 4. Filler piece, 5. Compression spring, 6. Spring seat, 7. Ball, 8. Restriction passage. (Courtesy of Robert Bosch Canada Ltd.)

return-flow direction. Between injections, the latter maintains the static line pressure as constant as possible under all operating conditions. The advantage of the constant-pressure valve lies in the avoidance of cavitation and in improved hydraulic stability.

If the constant-pressure valve is to be employed efficiently, more precise adjustments and governor modifications are necessary.

CONTROL-ROD STOPS AND PRESSURE COMPENSATORS

The control-rod travel is limited by an adjustable full-load stop. Several different control-rod stop designs are used for different applications. These are discussed in detail in Chapter 13.

A manifold-pressure compensator controlled by boost pressure is used on turbocharged engines to reduce the full-load fuel delivery in the lower speed range in order to reduce exhaust smoke during acceleration. An altitude-pressure compensator is used on engines required to operate at wide variations in altitude. This device matches the amount of fuel injected to the air-charge density entering the engine at different altitudes (less fuel for a less dense air charge at higher altitudes). For a detailed description of control-rod stops, manifold-pressure compensators, and altitude-pressure compensators see Chapter 13 (See **Figures 15-15 to 15-17.**)

Actuation of the Excess-Fuel Device for Starting

During the starting process, the control lever on the governor (or the accelerator pedal) must be placed in full-load position. In this position, the stop pin attached to the control rod rests on the "full-load stop without charge-air pressure."

The starting cable is now actuated, i.e., the shaft is pulled out in the longitudinal direction by about 10 mm by a cable or a solenoid. The bell crank and stop screw follow this movement and permit the control rod to travel the distance of the starting quantity.

After completion of the starting process, the shaft together with the bell crank is forced back into the initial position by spring tension. See Chapter 13 for details.

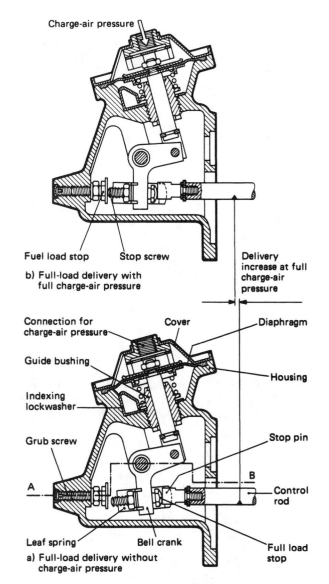

FIGURE 15-15 Bosch smoke limiter operation. (Courtesy of Robert Bosch Canada Ltd.)

PROTECTION AGAINST ACTUATION OF THE EXCESS-FUEL DEVICE DURING OPERATION

The control rod can reach the start position only when the control lever on the governor or the accelerator pedal has been placed at full load. If the starting cable is actuated while the control rod is not at full-load position, the leaf spring is placed in the path of the control rod and the latter cannot move to the starting position.

If the driver attempts to clamp the starting cable in the excess-fuel position, the control rod can no

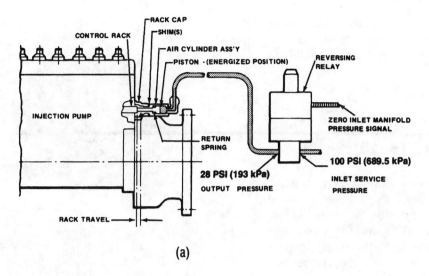

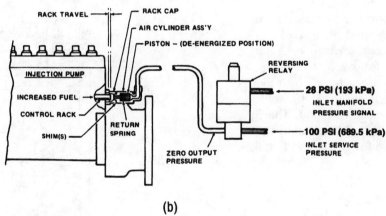

FIGURE 15–16 Mack Truck puff limiter operation. (a) Fuel-limiting position; (b) normal position. (Courtesy of Mack Trucks Inc.)

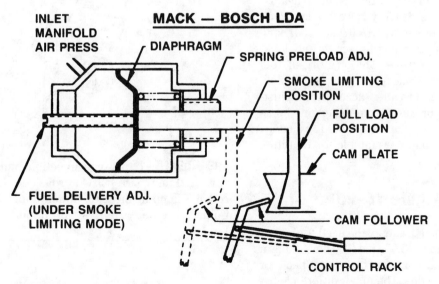

FIGURE 15–17 Improved puff limiter eliminates need for reversing relay. (Courtesy of Mack Trucks Inc.)

longer reach the full-load or starting positions after deceleration or shifting into another gear, and the engine thus receives even less fuel than the full-load quantity.

PUMP SIZES

Size M Pump (Figure 15–18)

The size M pump is the smallest of the PE-type pumps. It has a light metal housing that is flange-mounted to the engine. Maximum injection pressure is limited to 5800 psi (400 bar). A removable side cover and base plate provide access to the pump interior to allow adjustment of the plunger-and-barrel assemblies. Adjustment is made by repositioning the clamps on the control rod. Control-rod movement regulates the quantity of fuel delivered. A lever is attached to each pump barrel control sleeve. A pin riveted to its end engages a groove in the clamping piece. The pump plungers are actuated by roller tappets. Prestroke adjustment is achieved by selection of tappet rollers of the proper diameter. The M pump is lubricated from the engine lubrication system. Size M pumps are available for engines with 4, 5, or 6 cylinders.

Size A Pump (Figure 15–19)

The larger size A pump has a light metal housing that can be flange- or cradle-mounted to the engine. Pump barrels are inserted directly into the housing from above. The delivery valve is held against the pump housing by the delivery-valve holder. Peak injection pressure is limited to 8700 psi (600 bar). Each roller tappet has an adjusting screw and locknut for adjusting the prestroke. The quantity of fuel delivered is controlled by means of a control rack. A removable side cover provides access for adjustment. The A pump is lubricated from the engine lubrication system. Size A pumps are available for engines with up to 12 cylinders and are suitable for multifuel operation.

Size MW Pump (Figure 15–20)

The MW pump is designed to satisfy the need for higher injection pressures of up to 13,050 psi (900 bar). It has a light metal housing and can be cradle-, flat-bed-, or flange-mounted to the engine. The major difference between the MW pump and the A and M pumps is the plunger-and-barrel assembly, which comprises the

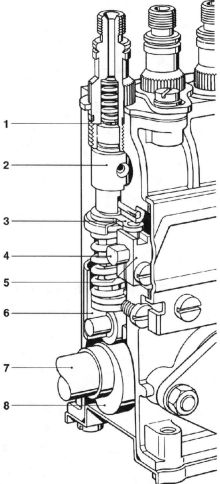

FIGURE 15–18 M-type injection pump. 1. Delivery valve, 2. Pump barrel, 3. Control-sleeve lever, 4. Control rod, 5. Clamping piece, 6. Roller tappet, 7. Camshaft, 8. Camshaft lobe. (Courtesy of Robert Bosch Canada Ltd.)

PUMP SIZES 289

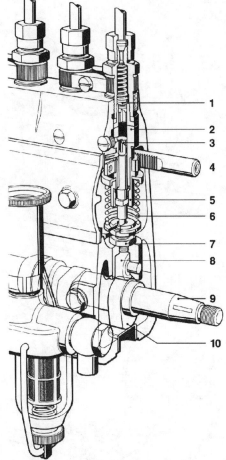

FIGURE 15–19 A-type injection pump. 1. Delivery valve, 2. Pump barrel, 3. Pump plunger, 4. Control rack, 5. Control sleeve, 6. Plunger return spring, 7. Adjusting screw, 8. Roller tappet, 9. Camshaft, 10. Cam lobe. (Courtesy of Robert Bosch Canada Ltd.)

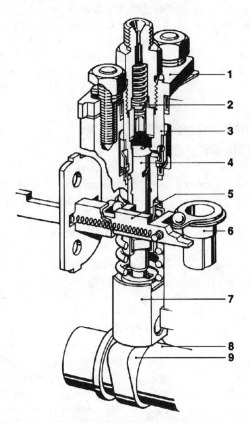

FIGURE 15–20 MW-type injection pump. 1. Fastening flange for the plunger-and-barrel assembly, 2. Delivery valve, 3. Pump barrel, 4. Pump plunger, 5. Control rack, 6. Control sleeve, 7. Roller tappet, 8. Camshaft, 9. Camshaft lobe. (Courtesy of Robert Bosch Canada Ltd.)

pump barrel, delivery valve, and delivery-valve holder. This assembly is inserted into the housing from above. Prestroke adjustment is made by the use of shims of varying thickness that are inserted between the housing and barrel assembly. Turning the barrel-and-valve

assemblies provides the means to adjust for uniform fuel delivery between individual barrels. The fastening flanges are slotted to provide for this adjustment. The MW pump is available for engines with up to 8 cylinders. Lubrication is provided from the engine lubrication system.

Size P Pump (Figure 15–21)

The size P pump is designed to provide peak injection pressures up to 16,675 psi (1150 bar). It is similar in design to the MW pump and may be base- or flange-mounted to the engine. Prestroke adjustment is accomplished in the same manner as with the MW pump. Fuel flow to the pump barrels is in parallel rather than in series (one after the other). In the series flow arrangement the temperature of the fuel may vary as much as 105° F (40° C) between the first and last barrels. The resulting difference in fuel density and volume causes a difference in the amount of fuel energy delivered to each engine cylinder. The parallel flow arrangement used in the size P pump eliminates this problem almost entirely. The P pump is available in versions for use with engines with up to 12 cylinders. See **Figures 15–22** and **15–23** for a comparison of pump sizes.

DESIGN AND OPERATION OF THE PF PUMP

The operation of the PF-type unit pumps is similar to that of the PE-type pumps **(Figures 15–3** and **15–24)**.

The PE pump does not have a built-in camshaft. PF injection pumps are usually operated by an additional cam on the engine camshaft. PF injection pumps are flange-mounted in varying forms. PF injection pumps usually are of the single-plunger type; however, pumps in sizes K, A, and B are also delivered in multiple-plunger models.

In the PF pump the port closing serves as the basis for timing the pump to the engine. In PF pumps with a timing window, the plunger is in port-closing position when the timing marks on the timing window and on the guide sleeve coincide. In PF pumps without a timing window, the value "a" is shown on the mounting flange or on the name plate. It refers to the distance between the underside of guide sleeve or roller tappet and the flange contact surface with the pump installed and pump drive in BDC position. An adjusting screw on the roller tappet or, in the PFR pump, shims placed under the pump flange serve for exact adjustment. The guide sleeve in the pump must be lubricated before installation. During operation, it receives oil for lubrication from the engine through the drive.

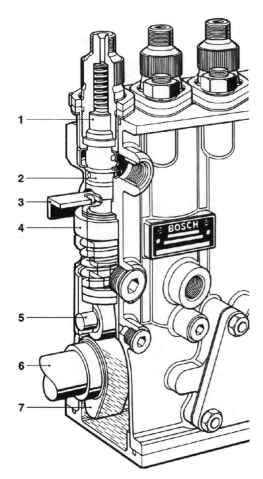

FIGURE 15–21 P-type injection pump. 1. Delivery valve, 2. Pump barrel, 3. Control rod, 4. Control sleeve, 5. Roller tappet, 6. Camshaft, 7. Cam lobe. (Courtesy of Robert Bosch Canada Ltd.)

INJECTION TIMING CONTROL DEVICE

The start-of-delivery (or port closing) time is selected according to injection delay and ignition delay. With the PE fuel injection pump, the speed-dependent adjustment of the start of delivery (port closing) is achieved by using a timing device. The timing device transfers the drive torque to the injection pump while at the same time performing its timing function.

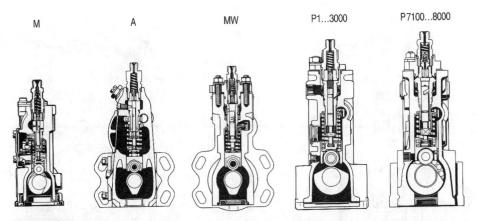

FIGURE 15-22 Size comparison of injection pumps. (Courtesy of Robert Bosch Canada Ltd.)

Pump model	M	A	MW	P1...3000	P7100...8000
Injection pressure (pump side)	400 bar 5800 psi	600 bar 8700 psi	900 bar 13,050 psi	800 bar 11,600 psi	1150 bar 16,675 psi
Application	4...6 cylinders	Light to medium commercial vehicles, tractors, industrial engines			Heavy commercial vehicles, industrial engines
Output per cylinder in kW/HP	10...15 13.4...20.1	25 33.5	35 46.9	60 80.5	70 93.8

FIGURE 15-23 Output and application of different models of injection pumps.

Design and Construction

The timing device for the in-line injection pump is mounted directly on the end of the pump's camshaft. There are two basic types, the *open type* and the *closed type*.

The closed timing device has its own lube oil reservoir, which makes it independent of the engine's lube oil circuit. The open design is connected directly to the engine's lube oil circuit. Its housing is screwed to a toothed gear, and the compensating and adjusting eccentrics are mounted in the housing so that they are free to pivot. The compensating and adjusting eccentrics are guided by a pin that is rigidly connected to the housing. The open type has the advantage of needing less room and of being more efficiently lubricated.

Operation (Figure 15-25)

The timing device is driven by a toothed gear in the engine timing case. The connection between the input and drive output (hub) is through interlocking pairs of eccentric elements. The largest of these, the adjusting eccentric elements (4), are located in holes in the backing disk (8), which in turn is bolted to the drive element (1). The compensating eccentric elements (5) fit in the adjusting eccentric elements (4) and are guided by them and the hub bolt (6). The hub bolt is directly connected with the hub (2). The flyweights (7) engage with the adjusting eccentric element (4) and are held in their starting positions by progressive springs.

GOVERNOR TYPES

There are a variety of different mechanical governor types in use:

- *Maximum-speed governor*, for limiting the maximum speed (high-idle speed).
- *Minimum-maximum-speed governor* (mainly for automotive applications) governs only at the upper and lower limits of the engine speed range, but not in between. The driver changes the injected fuel quantity by means of the accelerator pedal.
- *Variable-speed governor* governs throughout the complete speed range in addition to the maximum (high-idle) and idle speeds.

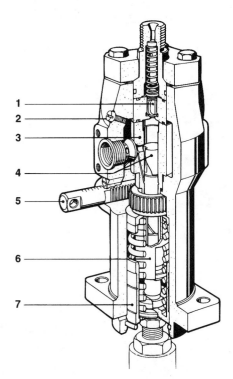

FIGURE 15–24 PF 1 D injection pump. 1. Delivery valve, 2. Vent screw, 3. Pump barrel, 4. Pump plunger, 5. Control rack, 6. Control sleeve, 7. Guide sleeve. (Courtesy of Robert Bosch Canada Ltd.)

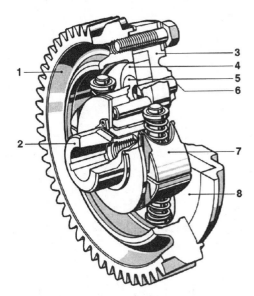

FIGURE 15–25 Timing device. 1. Drive element, 2. Hub, 3. Housing, 4. Adjusting eccentric element, 5. Compensating eccentric element, 6. Hub bolt, 7. Flyweights, 8. Backing disk. (Courtesy of Robert Bosch Canada Ltd.)

Governor components are shown in **Figures 15–26** and **15–27**. On in-line injection pumps, mechanical (flyweight) governors, or electronic diesel control (EDC), are used. Pneumatic governors are no longer used because they cannot comply with the severe requirements made on a modern diesel engine.

The above governors are described in Chapter 13.

ROBERT BOSCH ELECTRONIC DIESEL CONTROL (EDC) SYSTEM

The Bosch EDC system provides precise control of fuel injection timing and quantity of fuel injected. The system uses the Bosch PDE multiple-plunger injection pump with two solenoids and a control module (**Figure 15–28**). One solenoid controls the quantity of fuel delivered by regulating the position of the fuel control rack. The other controls injection timing advance by rotating the injection pump drive in the direction of pump shaft rotation. The transducers or solenoids are controlled by the electronic engine control unit (ECU) based on calculations made from programmed information in ECU memory and from signals received from a variety of input sensors and switches. System components include (**Figure 15–29**):

1. *Engine control unit (ECU)*
2. *Linear solenoid (fuel rack actuator)*
3. *Rack position sensor*
4. *Engine speed sensor*
5. *Auxiliary sensors.* Boost pressure, coolant temperature, intake air temperature, and fuel temperature sensors as provided by the engine and vehicle manufacturers provide additional information to the ECU for more precise control of fuel injection.

An example of the application of this electronically controlled injection pump is the V-MAC system used by Mack Trucks described in Chapter 16. The V-MAC system can use either a Robert Bosch pump or an AMBAC pump.

FUEL SUPPLY PUMPS

Single-acting or double-acting fuel supply pumps are used depending on the amount of fuel required by the engine (**Figure 15–30**).

Single-Acting Fuel Supply Pump Operation

Vacuum causes the suction valve to open, and the fuel enters the chamber between the suction valve and the pressure valve.

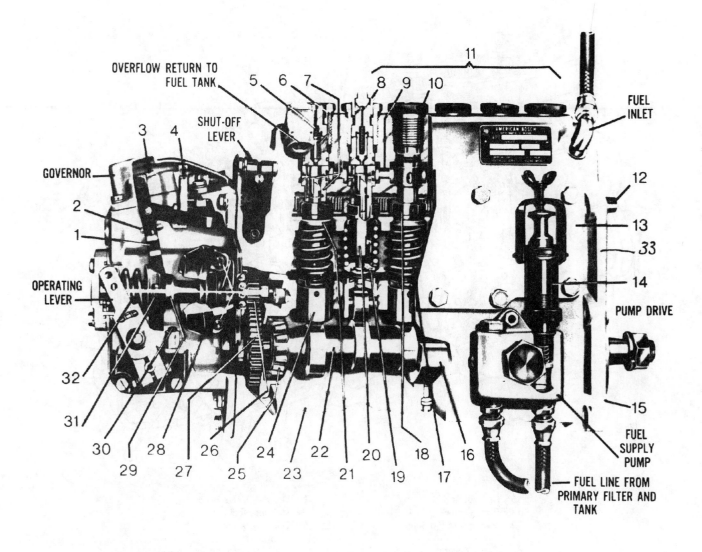

Nomenclature

1. Fulcrum Lever Assembly
2. Droop Screw
3. Torque Control Cam
4. Adjustable Stop Plate
5. Delivery Valve Spring and Holder
6. Delivery Valve Assembly
7. Plunger and Barrel Assembly
   Turned 90° from standard for illustration purposes
8. Tubing Union Nut
9. Fuel Sump
10. Retaining Nut
11. Fuel Discharge Outlets (6)
12. Control Rack
13. Inspection Cover
14. Hand Primer (Optional)
15. Lubricating Oil Outlet
16. Camshaft Center Bearing
17. Plunger Spring
18. Control Sleeve Gear Segment
19. Lower Spring Seat
20. Control Sleeve
21. Upper Spring Seat
22. Camshaft
23. Closing Plug
24. Tappet Assembly
25. Camshaft Bearing
26. Driven Gear, Governor
27. Drive Gear and Friction Clutch Assembly
28. Flyweight Assembly
29. Sleeve Assembly
30. Speed Adjusting Screw
31. Fulcrum Lever Pivot Pin
32. Inner/Outer Governor Springs
33. Pump Housing

FIGURE 15-26 AMBAC APE pump and governor components. (Courtesy of AMBAC International Corp.)

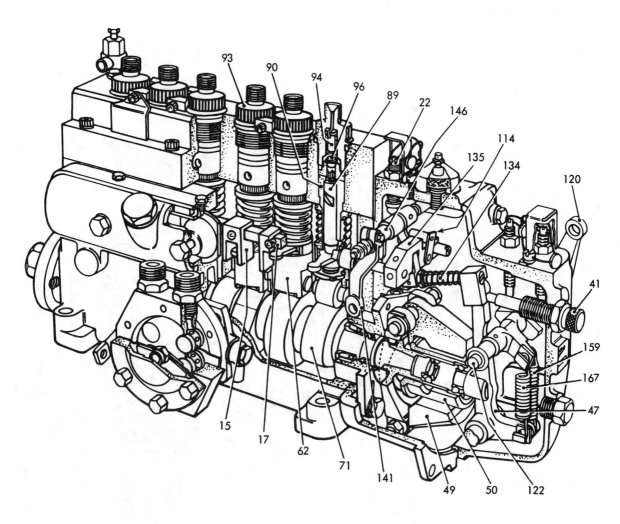

15	Control Fork	93	Delivery Valve Holder
17	Control Rod	94	Volume Reducer
22	Max. Fuel Stop Screw	96	Delivery Valve
41	Damper	114	Trip Lever
47	Crank Lever	120	Speed Control Lever
49	Governor Flyweight	122	Speed Lever Shaft
50	Governor Sleeve	134	Telescopic Link
62	Tappet Assy.	135	Bridge Link
71	Camshaft	141	Stop Control Lever
89	Plunger	146	Excess Fuel Device
90	Barrel	159	Governor Idling Spring
		167	Governor Main Spring

FIGURE 15-27 Components of Majormec pump and governor. (Courtesy of Lucas CAV Ltd.)

ROBERT BOSCH ELECTRONIC DIESEL CONTROL (EDC) SYSTEM

FIGURE 15-28 Electronically controlled in-line Bosch injection pump. (Courtesy of Robert Bosch Canada Ltd.)

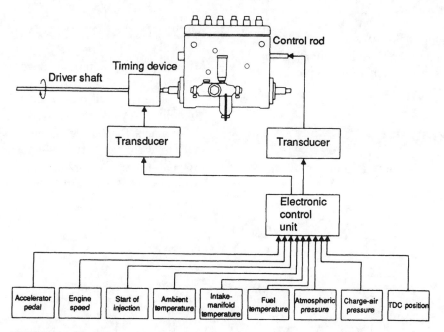

FIGURE 15-29 EDC system schematic diagram. (Courtesy of Robert Bosch Canada Ltd.)

When the camshaft rotates far enough, the spring forces the plunger back down again, the suction valve closes, and the pressure valve opens. Fuel is forced through the high-pressure line to the injection pump **(Figure 15-31)**.

Double-Acting Fuel Supply Pump Operation

Double-acting fuel supply pumps are used with injection pumps having large numbers of barrels and correspondingly higher delivery quantities.

Chapter 15 ROBERT BOSCH, AMBAC INTERNATIONAL, AND LUCAS CAV MULTIPLUNGER INJECTION PUMPS

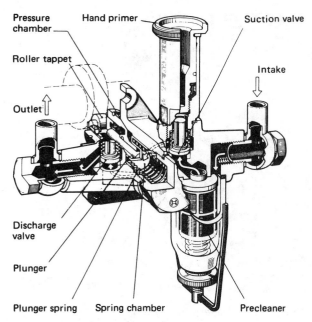

FIGURE 15–30 Fuel supply pump components. (Courtesy of Robert Bosch Canada Ltd.)

In contrast to the single-acting pump, the double-acting pump delivers fuel to the injection pump during both pump-plunger strokes, which is twice per camshaft revolution **(Figure 15–32)**.

Hand Pump Operation

The hand (primer) pump has the following functions:

- To fill the injection system's intake side before the system is put into operation for the first time
- To refill and bleed the system after repair or maintainance work
- To refill and bleed the system after the vehicle's fuel tank has run dry.

The hand pump is usually integrated in the fuel supply pump, although it may be installed in the line between the fuel tank and the supply pump in some cases.

TROUBLESHOOTING

For troubleshooting and problem diagnosis refer to **Figure 15–33**. This chart lists the symptoms, the probable causes, and the remedies for general problem diagnosis. For detailed diagnostic procedures refer to the appropriate service manual.

SERVICE PRECAUTIONS

- Be aware of the potential danger of disconnecting any pressurized systems such as hydraulics, fuel, air, or air conditioning, including refrigerant-handling precautions.
- Disconnect the battery ground cable at the battery to avoid electrical system damage.
- Be careful when handling hot cooling systems (the system is pressurized, and radiator cap re-

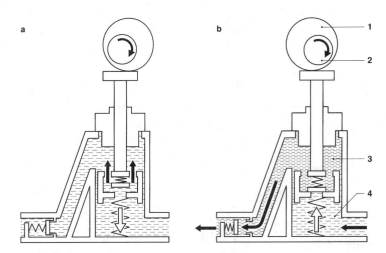

FIGURE 15–31 Single-acting fuel supply pump operation. (a) Cam stroke; (b) spring stroke. 1. Eccentric, 2. Camshaft, 3. Pressure chamber, 4. Suction chamber. (Courtesy of Robert Bosch Canada Ltd.)

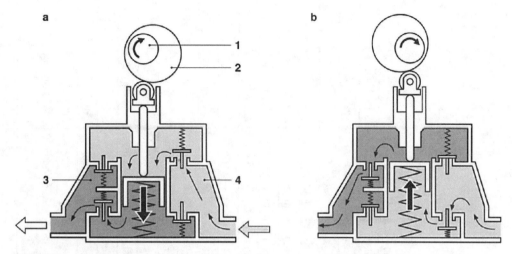

FIGURE 15–32 Double-acting fuel supply pump operation. (a) Cam stroke; (b) spring stroke. 1. Camshaft, 2. Eccentric, 3. Pressure chamber, 4. Suction chamber. (Courtesy of Robert Bosch Canada Ltd.)

moval can cause the system to boil and overflow, causing burns to the hands and face). Turn the cap to the first notch only to relieve system pressure (use a rag) and then remove the cap.
- Be careful of hot engine parts.
- Engine oil for pump lubrication may be hot enough to cause burns; handle with care to avoid injury.
- Use proper methods and equipment for handling parts to avoid personal injury and parts damage.
- Fuel system components have many precision machined surfaces; handle them with extreme care to avoid damage.
- Observe cleanliness of all parts and the entire area; this is essential during assembly. A very small piece of dirt can cause complete failure. Abrasives damage friction surfaces severely. Extremely close tolerances are destroyed by small particles of dirt or foreign material.
- Use only filtered compressed shop air for cleaning fuel system parts after washing.
- Use only filtered and temperature-controlled testing fluids for injector and injection pump service.
- Perform all injector and injection pump service in a controlled environment—a lab with controlled temperature and filtered air.

PREVENTIVE MAINTENANCE

The cost of operating a diesel engine can be greatly reduced by following a preventive maintenance program of careful inspection and checking. Many troubles are eliminated and anticipated before they can become costly problems. This applies to any diesel engine, regardless of where it may be operating—a truck, tractor, large earth-moving equipment, electric power generating equipment, a pleasure yacht, ocean-going ships, transcontinental bus, off-highway equipment, oil well drilling equipment, regardless of size and horsepower.

PUMP LUBRICATING OIL

Because the injection pump is lubricated by the engine oil, it is important that engine oil filters and engine oil be changed regularly to ensure that clean oil reaches the pump. The oil used in the engine should be the type recommended by the engine manufacturer.

FUEL TANK

Because moisture will condense and accumulate in the fuel supply tank, it is important that the tank be drained of such water and any sediment. In areas of high humidity, such draining must be done more frequently. In general, draining the fuel tank once each month is satisfactory.

SUPPLY HOSES, SUCTION HOSES, AND FITTINGS

Frequency of inspection will vary with the conditions under which the equipment is operated. Under severe operating conditions, as with heavy earth-moving equipment, once a week is desirable. Inspect for

Symptom												Cause	Remedy
Starting problem	Engine surges at idle	Rough idle at idle	Engine misses when engine is warm	Low power	Excessive fuel consumption	Engine cannot be shut off	Poor performance or black smoke or low power	Foglike exhaust in full-load range (white or blue)	Incorrect idle or maximum speed	Engine does not rev up	Injection pump runs hot		
•												Tank empty or tank vent blocked	Fill tank/bleed system, check tank vent
•	•						•					Air in fuel system	Bleed fuel system, eliminate air leaks
•						•						Shut off/start device defective	Repair or replace
•				•			•					Fuel filter blocked	Replace fuel filter
•				•			•	•				Injection lines blocked/restricted	Drill to nominal I.D. or replace
•		•		•			•					Fuel-supply lines blocked restricted	Test all fuel supply lines—flush or replace
•							•					Loose connections, injection lines leak or broken	Tighten the connection, eliminate the leak
•				•			•					Paraffin deposit in fuel filter	Replace filter, use winter fuel
•		•		•			•	•				Pump-to-engine timing incorrect	Readjust timing
		•	•	•			•	•				Injection nozzle defective	Repair or replace
				•			•					Engine air filter blocked	Replace air filter element
•												Preheating system defective	Test the glow plugs, replace as necessary
•		•		•			•	•				Injection sequence does not correspond to firing order	Install fuel injection lines in the correct order
	•								•			Low idle misadjusted	Readjust idle stop screw
									•	•		Maximum speed misadjusted	Readjust maximum speed screw
		•					•				•	Overflow valve defective or blocked	Clean the orifice or replace fitting
		•										Delivery valve leakage	Replace delivery valve (max. of 1 on 4 cyl., 2 on 6 cyl.)
	•											Bumper spring misadjusted (RS . . . governors)	Readjust bumper spring
				•			•	•				Timing device defective	Repair or replace timing device
		•	•	•			•	•				Low or uneven engine compression	Repair as necessary
	•	•		•			•	•	•	•		Governor misadjusted or defective	Readjust or repair
•	•	•	•	•			•	•	•	•	•	Fuel injection pump defective or cannot be adjusted	Remove pump and service

FIGURE 15–33 Troubleshooting guide for diesel fuel injection system with Bosch in-line fuel injection pumps. It is assumed that the engine is in good working order and properly tuned, and that the electrical system has been checked and repaired if necessary. (Courtesy of Robert Bosch Canada Ltd.)

dented, crimped or collapsed lines, which can restrict the flow of fuel. Check for abraded or cracked hoses and damaged or cracked fittings, which can cause suction air leaks. Make certain that all hose connections and fittings are tight.

PRIMARY FILTER(S)

The oil filter or diesel filters should be replaced in accordance with the engine manufacturer's recommendations. Remember, it costs far less to change a filter than it does to change an injection pump. Only filters approved by the engine manufacturer should be used.

SECONDARY FILTER

Follow the engine manufacturer's instructions regarding the servicing of this filter.

HAND PRIMING PUMP

It is necessary to bleed air from the low-pressure side of a system whenever a filter or filter element has been replaced or a line disconnected. The plunger retaining strap of the hand primer must always be kept in place.

HIGH-PRESSURE LINES

High-pressure fuel lines are made of seamless, cold-drawn, high–tensile strength steel. On each installation, the fuel lines are of equal length and inside diameter. Any variation will result in unequal quantities of fuel being delivered to the individual cylinders of the engine. Therefore, when replacement is necessary, the new lines must be of identical specifications in order to avoid impairment of injection characteristics.

Both ends of each fuel line must be inspected for possible crimping of the bore before an injection pump or nozzle holder is reinstalled. The inside diameter of the fuel line should be checked with a drill of the proper size as indicated by the manufacturer's specifications. Redrill or ream if necessary, and carefully flush the entire line.

The tubings must be securely fastened to prevent rubbing against each other or parts of the engine. Suitable clamps are provided to prevent such vibration and motion of the fuel lines.

INJECTION PUMP TIMING

Good preventive maintenance includes careful inspection of the pump timing to be sure it is correctly timed to the engine in accordance with the manufacturer's specifications.

INJECTION PUMP MOUNTING

It is important to make certain that injection pump mounting bolts are correctly torqued to the manufacturer's specifications. Loose bolts result in vibration and consequent leakage of fuel line connections and wear of pump parts.

THROTTLE AND STOP LINKAGE

Improperly adjusted throttle and stop linkage frequently results in complaints of low power. The throttle linkage must be adjusted so that the injection pump operating lever moves from the idle stop to the full-load stop (maximum-speed stop). "Breakaway" throttle levers deflect slightly at each end of their travel.

Worn, bent, or otherwise damaged throttle linkage and worn pivot points and joints must be replaced.

Stop linkage must also be checked periodically for proper operation. The stop linkage must be adjusted in accordance with the manufacturer's specifications. When making the adjustment or inspection, be sure the stop mechanism does not interfere with normal operation of the injection pump and governor.

INJECTION NOZZLES

To ensure maximum performance and economy of operation, the nozzle and nozzle holder assemblies should be removed from the engine on a scheduled basis and checked for proper opening pressure, chatter, and spray pattern. If a nozzle does not operate properly, it must be cleaned and repaired or replaced. If necessary, reset nozzle-opening pressures. Specialized equipment is required for making such checks. See Chapter 12 for injector nozzle service.

AIR INDUCTION SYSTEM

The ratio of the amount of air to the amount of fuel is seriously affected when the air intake is restricted. This, in turn, results in smoking exhaust, loss of power and reduced fuel efficiency. To overcome this condition, it is essential that the air cleaner be kept clean. The frequency of inspecting and cleaning the air cleaner is dependent largely on the conditions under which the diesel equipment operates. The manufacturer's instructions cover normal conditions. For dusty conditions, inspect and clean the air cleaner more frequently.

INJECTION PUMP REMOVAL

The following general procedure is typical for injection pump removal.

1. With the engine stopped, thoroughly clean the engine and injection pump exterior using the method approved in the service manual.
2. Position the No. 1 piston at TDC on the compression stroke.
3. Shut off the fuel supply at the fuel tank.
4. Disconnect the low-pressure fuel, lube, and air lines. Cap all lines and fittings to prevent contamination from dirt or condensation.
5. Disconnect the high-pressure fuel lines at the pump and injection nozzle holders. Cap all lines and fittings to prevent contamination.
6. Disconnect linkages that are connected to the pump.
7. Remove all accessories and any other items that will interfere with pump removal.
8. Remove the mounting bolts that fasten the pump to the engine.
9. Remove the pump drive gear cover and pump drive gear bolts as necessary (**Figures 15–34** and **15–35**).
10. Grasp the pump securely, slide it back, and lift it away from the engine. Use a lift sling if necessary (**Figure 15–36**).

INJECTION PUMP FLUSHING PROCEDURE

1. Mount the pump on an approved test stand (**Figure 15–37**).

NOTE: It may be necessary to replace the original pump drive hub and/or coupling with a coupling and/or hub that will fit the particular test stand being used.

2. Lubricate the pump as follows: APE type—if the pump is normally lubricated by engine oil, add approximately 1 pt (0.473 L) of SAE 30 lube oil to the camshaft compartment to lubricate the cam lobes, tappet rollers, etc.; if the pump is not engine oil–lubricated, add sufficient SAE 30 oil to the camshaft compartment to maintain oil plug level. PSB, PSJ & PSM type—use a pressurized lube oil system (system and ducted adapters and fixtures available from Bacharach Instrument Co., Pittsburgh, Pa.) filled with SAE 30 oil, or equivalent, to properly lubricate the pump. The lube oil pressure must be 20 psi (138 kPa) minimum or, if pump includes

FIGURE 15–34 Removing the pump drive gear cover. (Courtesy of Mack Trucks Inc.)

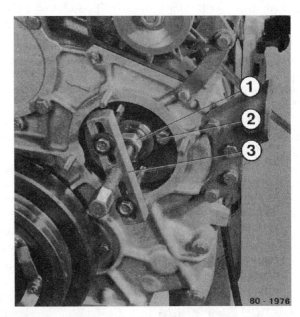

FIGURE 15–35 Removing injection pump drive gear screws. 1. Hub retaining nut, 2. Gear retaining screws, 3. Gear puller. (Courtesy of Mack Trucks Inc.)

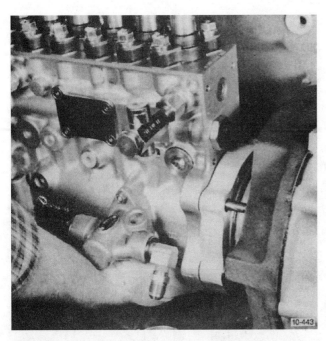

FIGURE 15–36 Removing the injection pump from the engine. (Courtesy of Mack Trucks Inc.)

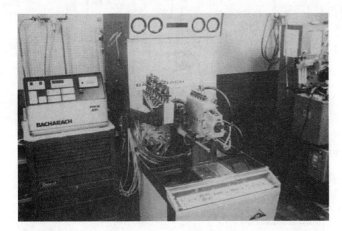

FIGURE 15–37 Injection pump mounted on test stand. (Courtesy of AMBAC International Corp.)

lube oil pressure adjusting devices, a pressure of 35–40 psi (241–276 kPa) is required.

NOTE: Do not splash-lubricate PSB, PSJ or PSM type pumps, as this provides insufficient lubrication to certain components.

IMPORTANT:

Do not use calibrating oil or fuel oil as a lubricant.

CAUTION:

Do not allow lube oil to mix with or contaminate calibrating or fuel oil.

---

3. Connect a set of high-pressure tubings to the pump outlets and direct the opposite ends of the tubings into a pail, thereby bypassing the nozzle and holder assemblies and test stand tank.

4. To flush the pump sump, proceed as follows: APE type—operate at about 600 rpm with the control rack in the midposition for approximately 5 minutes. PSB, PSJ and PSM types—operate at a speed of about 1000 rpm with the operating lever in the full-load position for approximately 3 minutes.

This procedure removes any fuel oil and foreign material from the pump sump and eliminates the possibility of contaminating the test oil or damaging the nozzles during the calibration check.

5. Stop the test stand and remove the high-pressure tubing.

INJECTION PUMP DISASSEMBLY

For disassembly and inspection, the normal procedure is to remove the inspection cover, pump body, delivery valves, pumping elements, tappets, drive coupling, governor cover, governor, and camshaft. The extent of disassembly thereafter depends on pump condition (**Figures 15–38A to 15–39**).

Prior to disassembly, the exterior of the pump should be cleaned with cleaning solvent and any lubricant drained. During disassembly it is critical that all parts pertaining to each pumping element be kept together and separate from other pumping element parts. Separate containers or a tray with separate compartments are recommended. Do not mix parts among pumping elements. These parts include the barrel and plunger, plunger spring, delivery valve, guide and holder, tappet, phasing spacers or adjusters, and the like.

The barrel and plunger of each element are selectively fitted or mated and must be kept together at all times during pump service. The delivery valves and guides are also selectively fitted. Phasing adjustment spacers are of selective thickness for each element.

Cleaning and Inspection of Parts

Clean all parts in cleaning solvent and blow dry with compressed air. Examine all parts visually for signs of

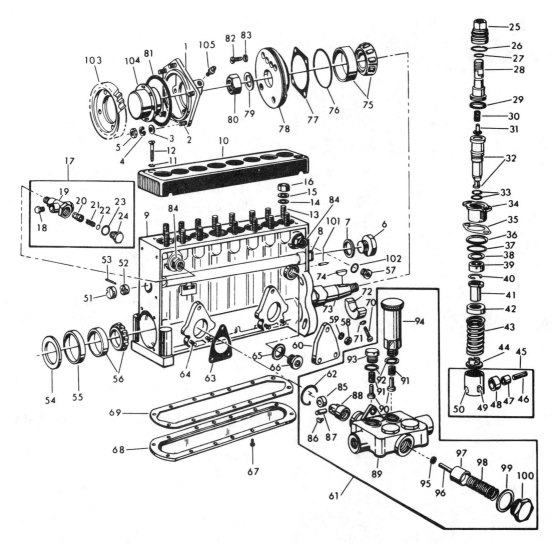

FIGURE 15-38A Injection pump disassembled. (Courtesy of Robert Bosch Canada Ltd.)

wear or damage and replace any that are excessively worn or damaged.

The upper parts of the barrel and plunger usually show the most wear. Fine scoring marks are normal after even a short period of operation, and the element should not be replaced because of them. If the scoring is deeper and the edge of the spill groove is damaged, the element should be replaced. A complete set of elements should be installed to ensure equal fuel delivery to all cylinders.

Inspect the delivery valves for pitting or damage. Replace if needed. Inspect the camshaft and camshaft bearings. Excessive wear of the cams will affect injection characteristics. Replace a worn camshaft and worn or damaged bearings. Inspect the governor weights, pins, ramp, rollers, collar, and spring for wear or damage. Examine all bushings for wear or damage. Replace any parts that are worn or damaged.

Reassembly

Assemble the pump in the reverse order of disassembly. Replace all gaskets, O-rings, seals, lock tab washers, lock nuts, and sealing washers. Cleanliness is extremely important during assembly. Lubricate all moving parts with clean fuel oil during assembly. This includes plungers and barrels, delivery valves and guides, camshafts and bearings, and governor components.

PUMP PHASING

Phasing is the procedure for checking and adjusting the phase angles between successive injections. For example, on a six-cylinder four-cycle engine there are 360° of

1. Adapter
2. Stud
3. Washer
4. Washer
5. Nut
6. Plug
7. Gasket
8. Rack
9. Housing
10. Cover
11. Washer
12. Screw
13. Stud
14. Washer
15. Washer
16. Nut
17. Check valve assembly
18. Plug
19. Body
20. Piston
21. Spring
22. Washer
23. Gasket
24. Plug
25. Cap
26. "O" ring
27. "O" ring
28. Holder
29. Gasket
30. Spring
31. Delivery valve
32. Plunger and barrel
33. "O" ring
34. Sleeve
35. Shim
36. Spacer
37. "O" ring
38. Washer
39. Cap
40. Snap ring
41. Sleeve
42. Upper spring seat
43. Spring
44. Lower spring seat
45. Tappet assembly
46. Pin
47. Bushing
48. Roller
49. Block
50. Tappet
51. Retainer
52. Rack guide
53. Rack stop pin
54. Shim
55. Rear bearing spacer
56. Bearing
57. Plug
58. Nut
59. Washer
60. Cover
61. Fuel supply pump
62. Snap ring
63. Gasket
64. Stud
65. Gasket
66. Plug
67. Screw
68. Cover
69. Gasket
70. Screw
71. Washer
72. Center support
73. Camshaft
74. Key
75. Bearing
76. "O" ring
77. Shim
78. Housing
79. Washer
80. Nut
81. "O" ring
82. Screw
83. Lock washer
84. Fitting
85. Roller
86. Pin
87. Lock pin
88. Tappet
89. Housing
90. Valve
91. Spring
92. Gasket
93. Plug
94. Hand pump
95. "O" ring
96. Spindle
97. Plunger
98. Spring
99. Gasket
100. Plug
101. Pointer
102. Gasket
103. Gear
104. Adjuster assembly
105. Plug

FIGURE 15–38B (Courtesy of Robert Bosch Canada Ltd.)

pump camshaft rotation for 720° of engine crankshaft rotation. Because all six cylinders fire in 360° of pump camshaft rotation, the phase angle is 360 ÷ 6, or 60° between the start of injection of all six elements.

Phasing is done manually or on a special test machine that uses a round degree plate graduated in degrees to measure the angular motion of the pump camshaft. Adjustment is provided by shims or adjusters for each pumping element. Phasing is necessary if any of the pumping element parts have been replaced. Calibration data is provided in the pump manufacturer's service manual data sheet.

Very precise handling and observation of the degree plate and adjustments is required to achieve accurate phasing and balanced fuel injection timing for all cylinders.

Spill Timing Method of Pump Timing and Phasing

Spill timing is a manual method of pump timing and phasing. Depending on pump plunger design, spill timing is done for port closing or for port opening. Spill timing for port closing is done on pumps with a lower helix plunger design. This design provides for a variable ending and a constant beginning of fuel delivery. Spill timing for port opening is done on pumps with an

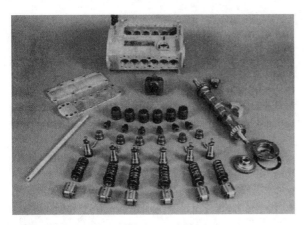

FIGURE 15–39 AMBAC APE6 pump disassembled. (Courtesy of AMBAC International Corp.)

FIGURE 15–41 Injection pump mounted on test stand with degree wheel in place in preparation for pump phasing. (Courtesy of AMBAC International Corp.)

upper helix plunger design, which provides for a variable beginning and a constant ending of fuel delivery.

In either case, the procedure requires that a spill pipe be installed in place of the number 1 delivery valve of the pump. This allows observation of the beginning or ending of fuel delivery during the procedure **(Figure 15–40)**. The injection pump is mounted to a suitable base plate, and the base plate is clamped securely in a vise. A degree wheel for the purpose is mounted to the drive end of the pump **(Figure 15–41)**. The degree wheel provides the means to turn the pump camshaft and indicates exact camshaft position. A clean gravity type of fuel supply is required with a shut-off valve in the supply line. The fuel supply is connected to the pump fuel inlet fitting.

Spill Timing Pump with Lower Helix

1. Move the fuel control rod through its travel range and place it in the stop position. This leaves the inlet port at the plunger open. Turn the camshaft in the proper drive direction to position No. 1 cam lobe at the BDC position. Open the fuel supply valve to allow fuel to flow until air-free fuel flows from the swan-neck tube. As solid fuel flows from the tube, slowly and carefully continue to turn the camshaft with the degree wheel. As fuel flow diminishes during port closing, continue to turn the camshaft very slowly until flow decreases to zero. Stop turning at that precise point. Repeat this procedure several times to ensure accuracy.

2. When the port-closing position has been established, adjust the degree wheel to index zero at the pointer. Check the timing line at the drive coupling to see if it coincides with the mark on the pump housing. If not, place a new timing mark on the coupling to reflect the newly established timing of port closing.

To complete the pump phasing, repeat the procedure above for each pumping element in the order of the engine injection sequence. Port closing for each succeeding element in the injection sequence should be exactly 45° later than the preceding element for an eight-cylinder engine. For a six-cylinder engine it is 60°, and for a four-cylinder it is 90°.

Adjustment is made by means of a tappet adjusting screw, shims, or spacers, depending on pump design. If the phase angle to the succeeding element is too great, the plunger is too low and must be raised. If the phase angle is too small, the plunger is too high and must be lowered **(Figures 15–42 to 15–47)**.

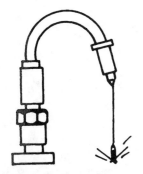 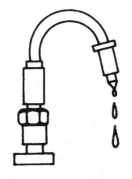

FIGURE 15–40 With the inlet port open a steady stream of fuel flows from the spill pipe (left). At port closing fuel flow is reduced to a drop (right). (Courtesy of Mack Trucks Inc.)

Spill Timing Pump with Upper Helix

The procedure for spill timing of port opening is very similar to that described above for port closing.

INJECTION PUMP CALIBRATION   305

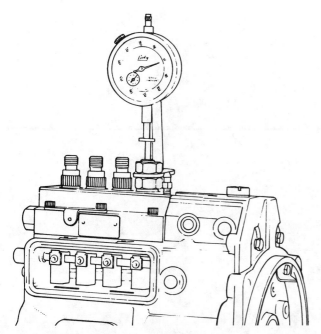

FIGURE 15-42 Measuring head clearance with dial indicator. Head clearance must be checked and adjusted to prevent the plunger from striking the delivery-valve body before pump timing can be done. Adjustment is by shim, spacer, or tappet screw. (Courtesy of Lucas CAV Ltd.)

FIGURE 15-44 Reading the phase angle on the degree wheel. (Courtesy of AMBAC International Corp.)

With the injection pump prepared as above for checking the No. 1 pumping element, open the fuel supply valve and turn the pump camshaft very slowly in the proper drive direction until fuel flow from the swanneck tube stops. Then very slowly continue to turn the camshaft until fuel just begins to flow from the tube. Repeat this procedure several times to establish accurately the exact point of port opening. When the point of port opening has been established, check the timing mark on the drive coupling as in step 2 above.

To complete the pump phasing, repeat the procedure for establishing port closing for each pumping element in the injection sequence to ensure that port closing for each succeeding element in the injection sequence will be exactly the number of degrees specified.

INJECTION PUMP CALIBRATION

CAUTION:

Never drive a fuel injection pump in a direction other than that shown on the pump, or serious damage will result.

Calibration is the procedure for adjusting the quantity of fuel delivery of each of the pumping elements so

FIGURE 15-43 Degree wheel pointer set at zero. (Courtesy of AMBAC International Corp.)

FIGURE 15-45 Checking phase angle adjusting shim. (Courtesy of AMBAC International Corp.)

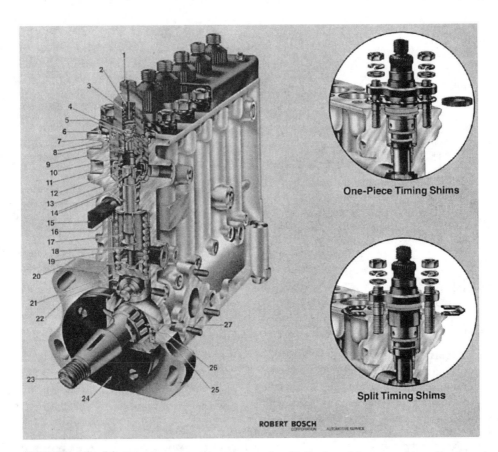

FIGURE 15-46 Bosch pump one-piece and split timing shims, used to adjust port closing. 1. Delivery-valve holder, 2. Fill piece, 3. Delivery-valve spring, 4. Delivery valve, 5. Delivery-valve gasket, 6. Timing shims, 7. Spacer, 8. O-rings, 9. Delivery-valve body, 10. Flange bushing, 11. Barrel, 12. Baffle ring, 13. Plunger, 14. O-rings, 15. Control rack, 16. Upper spring seat, 17. Control sleeve, 18. Plunger vane, 19. Plunger spring, 20. Lower spring seat, 21. Plunger foot, 22. Roller tappet, 23. Camshaft, 24. Bearing end plate, 25. End play shim, 26. O-ring, 27. Bearing. (Courtesy of Robert Bosch Canada Ltd.)

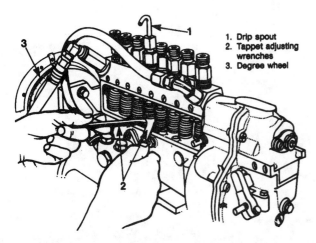

FIGURE 15-47 Tappet adjustment on Bosch model A pump. (Courtesy of Robert Bosch Canada Ltd.)

that each element will deliver precisely the same specified quantity of fuel. Adjustment is made by setting the control-sleeve gear on each element or by setting the control forks on the control rod, depending on pump design.

Mount the pump on the test machine and be sure that there is sufficient clearance at the drive coupling. Connect the fuel supply line to the pump inlet and the delivery pipes to the delivery valve holders. Connect the lubrication line from the tester to the pump or fill the pump sump with lubricating oil if pump is not engine oil–lubricated.

Set the specified (usually No. 6 or No. 1) element-sleeve gear or control fork to the dimension specified in the data sheet. Select the appropriate pump drive direction (cw, left hand or ccw, right hand).

Open the fuel supply valve to pressure flow. Loosen the pump bleed screw and start the pump drive. Run

FIGURE 15–48 Fuel-delivery calibration variation should not exceed ± 2% between pumping elements. (Courtesy of AMBAC International Corp.)

FIGURE 15–50 Pump calibration fuel-delivery adjustment on AMBAC pump. (Courtesy of AMBAC International Corp.)

the pump until all air has been expelled, then tighten the bleed screw.

Move the speed lever to the maximum-speed position. Maximum fuel delivery from the specified element is adjusted first to specifications. Run the pump at the specified speed and note the delivery from the specified element in its test tube for the specified number of shots (**Figure 15–48**). Adjust the maximum-fuel stop screw until the maximum delivery is as specified. Repeat several times to ensure accuracy.

Adjust the position of the other sleeve gears or forks (**Figures 15–49** and **15–50**) so that the maximum delivery from them is exactly the same as that from the specified element. Check the operation of the excess-fuel device and fuel stop control, and check fuel delivery at the excess-fuel position and at idle as specified.

After calibration install all plugs, covers, and the like with new gaskets and sealing washers as needed. Install tamper-proof locks and seals as specified.

INJECTION PUMP TO ENGINE TIMING

The injection pump is timed to the engine with the help of the reference marks for start of injection (port closing). These markings are on the engine as well as on the injection pump (**Figures 15–51** to **15–55**).

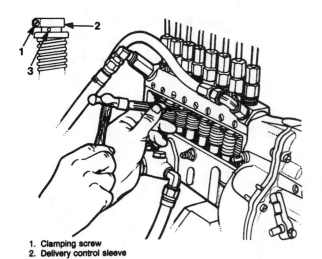

1. Clamping screw
2. Delivery control sleeve
3. Control sleeve indent

FIGURE 15–49 Adjusting pump plunger to achieve equal fuel delivery. (Courtesy of Robert Bosch Canada Ltd.)

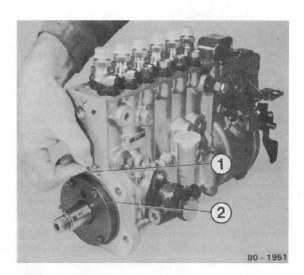

FIGURE 15–51 Injection pump adapter (1) and O-ring seal (2) ready for installation. (Courtesy of Mack Trucks Inc.)

FIGURE 15-52 Bosch PES pump timing marks. (Courtesy of Mack Trucks Inc.)

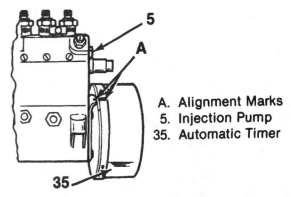

A. Alignment Marks
5. Injection Pump
35. Automatic Timer

FIGURE 15-53 Injection pump to drive coupling timing marks. (Courtesy of General Motors Corporation.)

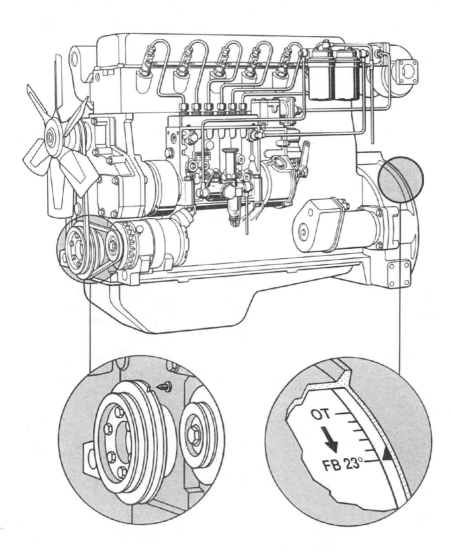

FIGURE 15-54 Diesel engine reference marks for timing the injection pump. (Courtesy of Robert Bosch Canada Ltd.)

308

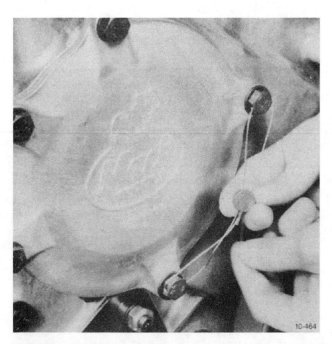

FIGURE 15–55 Installing a tamper-resistant wire and lead seal. (Courtesy of Mack Trucks Inc.)

The TDC position of piston no. 1 on the compression stroke is used as the basis for the timing adjustments. In most cases, the port-closing reference mark is on the engine's flywheel, on its V-belt pulley, or on its vibration damper. There are a number of methods for adjusting the injection pump to the correct port-closing (start-of-injection) setting.

1. The new or rebuilt fuel injection pump is delivered with its camshaft locked in a given position. After being bolted to the engine with the engine's crankshaft in the appropriate position, the pump camshaft is released.

This well-proven method is inexpensive and is gaining more and more in popularity.

2. The fuel injection pump is equipped with a port-closing indicator on the governor end that must be aligned with the reference marks when the injection pump is mounted.

3. There is a port-closing mark on the timing device or on the clutch that must be brought into alignment with a mark on the pump housing.

This method though is not as accurate as the preceding two methods.

4. After the injection pump has been mounted on the engine, the high-pressure overflow method is applied at one of the pump outlets to find the port-closing point (the instant in time when the pump plunger closes the inlet port).

This "wet" method is also being increasingly superseded by methods 1 and 2 described above.

Follow the vehicle or equipment manufacturer's service manual procedures and specifications to ensure accuracy.

BLEEDING THE INJECTION SYSTEM

Air bubbles in the fuel can impair injection pump operation, or even make it impossible. Therefore, installations that are to be taken into service for the first time, or that have been shut down temporarily, must be bled thoroughly.

If the fuel supply pump is equipped with a hand (primer) pump, this is used to fill the suction line, delivery line, fuel filter, and fuel injection pump with fuel. The vent screws on the filter cover and the fuel injection pump must remain open until the fuel flows out completely free of bubbles. The installation is always to be bled every time the filter has been changed or other work carried out on the system.

During actual operation, the installation automatically bleeds itself via the overflow valve on the Bosch fuel filter (permanent venting). A restriction is used instead if the pump is not equipped with an overflow valve.

REVIEW QUESTIONS

1. Robert Bosch injection pumps using the port and helix design include the in-line _____ series and the _____ series.
2. Lucas CAV pumps of the port and helix type include the _____ and _____ series.
3. The number of pumping elements in the PE injection pump corresponds to the _____ of _____ in an engine.
4. Each pumping element consists of a _____ and a _____.
5. As soon as the helix of the plunger opens the _____ _____, pressure in the pump barrel _____.
6. The constant-pressure type of delivery valve is used with _____ _____ fuel injection pumps.

7. A manifold-pressure compensator controlled by _____ _____ is used on turbocharged engines.
8. The size M injector pump is the _____ in the PE range of in-line pumps.
9. Lubrication of the M pump is through a _____ _____ _____ _____ with the engine.
10. The size A in-line injection pumps with their _____ _____ ranges follow directly after the size M.
11. The MW pump was developed to satisfy the need of higher _____ _____ .
12. The P pump is available in versions up to _____ _____ .
13. The timing device for the in-line injection is mounted directly on the end of the _____ .
14. The timing device is driven by a _____ _____ in the engine timing case.
15. The governor's main task is to limit an engine's _____ _____ .
16. ECU is the abbreviation for _____ .
17. Double-acting fuel supply pumps deliver fuel to the _____ pump during both pump _____ .
18. Hand pumps are used for _____ and _____ the system when required.
19. Perform all injector and injector pump service in a _____ _____ .
20. High-pressure fuel lines are made of high _____ _____ .
21. If the injector nozzle does not operate properly, it must be _____ and _____ or _____ .
22. If the air intake is restricted, it will cause _____ exhaust and _____ fuel consumption.
23. During disassembly of the injection pump, all parts for each pumping element must be kept _____ for reassembly.
24. Air bubbles in the fuel can _____ _____ _____ operation.

TEST QUESTIONS

1. The most common diesel fuel injection pump design is
   a. helix and in-line   c. port and plunger
   b. port and helix   d. PE and PF series
2. The Bosch fuel injection pump includes
   a. a governor, fuel supply pump, and timing device
   b. a governor, helix, and supply pump
   c. a fuel supply pump, timing device, and plunger
   d. all of the above
3. All diesel engines must have a governor to
   a. control minimum rpm
   b. control minimum and maximum rpm
   c. maintain proper fuel supply
   d. control maximum rpm
4. The end of fuel delivery is reached when the plunger helix
   a. closes the spill port
   b. is rotated
   c. opens the spill port
   d. is at the top
5. Some diesel engines start more easily when fuel injection timing is
   a. retarded for starting
   b. advanced for starting
   c. same as normal running
   d. a hand-held retard system
6. The pressure in the pump barrel drops as soon as the
   a. plunger reaches its maximum height
   b. helix of the plunger opens the spill port
   c. plunger starts to rotate
   d. engine reaches operating temperature
7. The constant-pressure type of valve is used with high-pressure fuel injection pumps that develop pressure of
   a. 20,000 psi   c. 11,600 psi
   b. 30,000 psi   d. 11,600 bar
8. An altitude compensator is used on engines required to operate
   a. in urban areas
   b. in the city (frequent stops)
   c. long, maximum-speed road driving
   d. at wide variations of altitude
9. The size M injection pump is the smallest pump in the
   a. PE range of in-line pumps
   b. A range of in-line pumps
   c. PF range of rotary pumps
   d. A range of rotary pumps

◆ CHAPTER 16 ◆

V-MAC ELECTRONIC CONTROL SYSTEMS—AMBAC AND ROBERT BOSCH

INTRODUCTION

The vehicle management and control (V-MAC) system is designed for use on Mack Trucks. It uses two control modules: the fuel injection control (FIC) module and the V-MAC module. Engine sensors and other vehicle sensors and switches send information on engine and vehicle operation to the control modules. The control modules use this information along with information stored in module memory to produce output signals. These output signals actuate the fuel control rack, injection timing device, and warning system as required. The injection pump may be AMBAC or Robert Bosch.

PERFORMANCE OBJECTIVES

After completing this chapter you should be able to:

1. List the performance features of the V-MAC system.
2. List the components of the V-MAC system and describe their function.
3. Diagnose the V-MAC system using the blink code method.
4. Perform diagnostic procedures when there are no active codes.

TERMS YOU SHOULD KNOW

Look for these terms as you study this chapter, and learn what they mean.

V-MAC
FIC
ECU
fuel rack actuator
fuel shut-off solenoid
econovance
engine position sensor
timing event marker
road speed sensor
coolant temperature sensor
intake air temperature sensor
oil pressure sensor
coolant level sensor
throttle position sensor
serial communications port
diagnostic blink codes
active faults
inactive faults
Pro Link
lap-top computer

V-MAC PERFORMANCE FEATURES

The V-MAC electronic control system provides the following performance features:

1. Improved exhaust emission control
2. Low-idle-speed adjustment
3. Idle shutdown
4. Road speed limiting

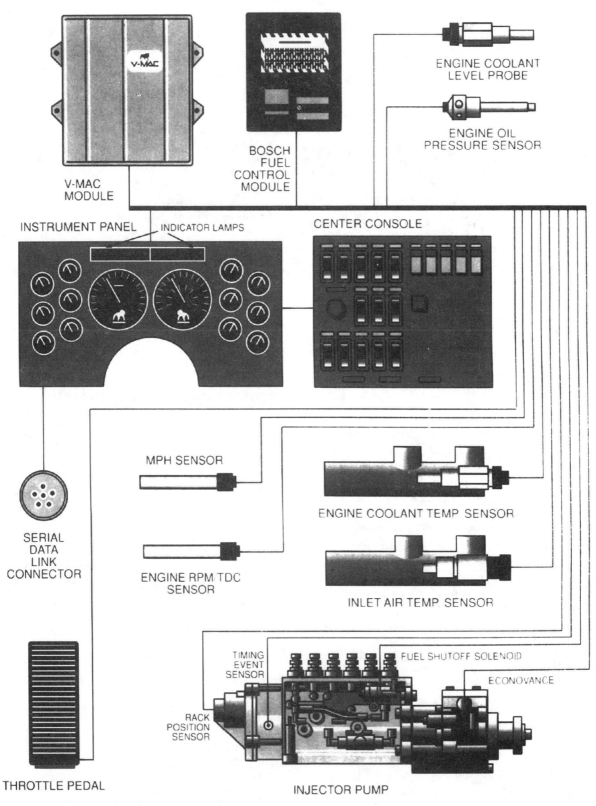

FIGURE 16-1 V-MAC system components. (Courtesy of Mack Trucks Inc.)

5. Cruise control
6. Engine braking control
7. PTO speed control
8. Engine warm-up (fast idle)
9. Self-diagnostics
10. Engine protection system
11. Communications data link

V-MAC SYSTEM COMPONENTS AND FUNCTION

The V-MAC system utilizes the following components (**Figures 16-1** and **16-2**).

1. *V-MAC module:* mounted under the dash in front of the passenger seat (**Figure 16-2**). A microprocessor-based electronic control unit (ECU) that receives input signals from various engine and vehicle sensors and switches. It processes this information in conjunction with data stored in module memory to produce output signals to operate various output devices as needed. It also logs faults and does password processing for both the V-MAC and FIC modules (**Figures 16-1** to **16-4**).

2. *Fuel injection control module (FIC):* mounted in the same area as the V-MAC module. Processes information from input sensors and switches to control fuel delivery and injection timing. It interfaces with the V-MAC module by sending certain data to it and receiving other data from it.

3. *Fuel injection pump:* AMBAC or Robert Bosch multiple-plunger injection pump injects fuel in the right amount under high pressure at the correct time for the most efficient engine operation.

4. *Fuel rack actuator:* mounted on the rear of the injection pump (**Figure 16-5**). An electrical solenoid acts on the fuel rack against return spring pressure. Increased current to the solenoid moves the rack to increase fuel delivery. With decreased current to the solenoid the return spring moves the rack to reduce fuel delivery. A rack position sensor provides feedback to the control module. An engine speed sensor is located in the actuator housing.

5. *Fuel shut-off solenoid:* shuts off fuel flow to the injection pump when the ignition switch is turned off and when the control module detects that the actual rack position does not correspond to the desired position (**Figure 16-6**).

6. *Econovance timing advance unit:* part of the fuel injection pump drive mechanism, it is hydraulically actuated and electronically controlled. It changes injection timing in response to commands from the control module (**Figure 16-7**).

7. *RPM/TDC engine position sensor:* senses engine speed (rpm) and piston position. Six notches in the face of the flywheel provide six pulses for every engine revolution on a six-cylinder engine (**Figure 16-8**). These notches are located at precise positions in relation to the TDC position of each piston. The engine rpm signal is used to determine the desired injection timing. The sensor also provides piston position information to the control module for comparison with the timing event marker.

8. *Timing event marker:* located in the rear outside of the injection pump, it senses passage of a timing bump on the injection pump speed sensor. It is used for injection pump calibration and to set initial injection pump timing (**Figure 16-9**).

9. *MPH (road speed) sensor:* senses transmission output shaft speed and sends this information to the control module for cruise control and road speed limiting (**Figure 16-10**).

10. *Coolant temperature sensor:* usually located in the rear of the water manifold, it sends information on coolant temperature to the control module to improve cold-engine starting and evaluates conditions that cause high coolant temperatures. Shuts down the engine if coolant temperature is too high (**Figure 16-11**).

FIGURE 16-2 V-MAC and FIC modules. (Courtesy of Mack Trucks Inc.)

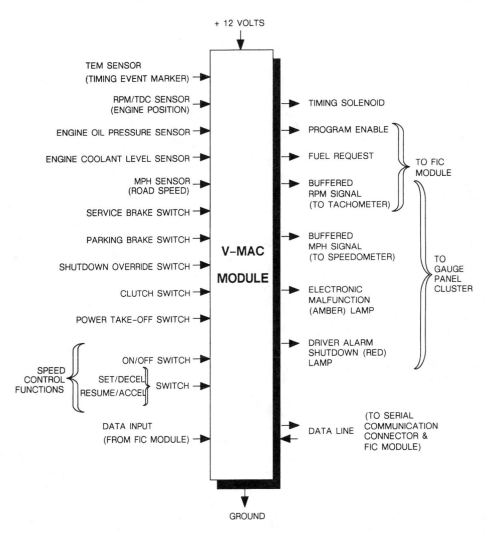

FIGURE 16–3 V-MAC module inputs and outputs. (Courtesy of Mack Trucks Inc.)

11. *Intake manifold air temperature sensor:* located in the intake manifold. Intake air temperature information is used by the control module to provide accurate air/fuel ratio control, injection timing control, and to control white exhaust smoke during engine warm-up.

12. *Engine oil pressure sensor:* This signal is used by the ECU to activate a warning alarm and engine shutdown after 30 seconds to avoid major engine damage **(Figure 16–12)**.

13. *Coolant level sensor:* located in the upper radiator tank **(Figure 16–13)**. When the coolant level drops too low, the engine shutdown process is activated.

14. *Throttle position sensor:* sends a variable electrical signal based on throttle position to the ECU to control fuel delivery. There is no mechanical throttle linkage; the TPS is connected to the ECU electrically **(Figure 16–14)**.

15. *Switches:* The following switches have input to the V-MAC module: Cruise ON/OFF, cruise SET/DECEL, cruise RESUME/ACCEL, shutdown override switch, service brake switch, parking brake switch, clutch switch, PTO switch, engine brake switch, and torque limiting switch **(Figure 16–15)**.

16. *Serial communications port:* located under the dash to the left of the steering wheel. Conforms to SAE 1708 Communications Link Standard.

DIAGNOSTIC AND TROUBLESHOOTING PROCEDURES
(Courtesy of Mack Trucks Inc.)

Before beginning any troubleshooting procedure, it is imperative to make certain that a problem actually exists. If possible, talk to the driver or the person who no-

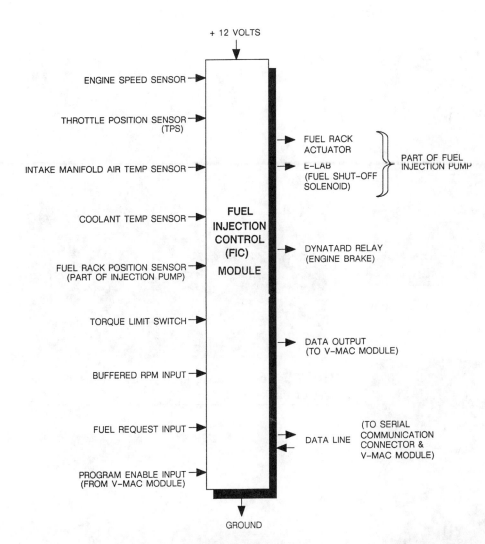

FIGURE 16–4 Fuel injection control (FIC) module inputs and outputs. (Courtesy of Mack Trucks Inc.)

FIGURE 16–5 Fuel rack actuator. (Courtesy of Mack Trucks Inc.)

FIGURE 16–6 Arrow indicates fuel shut-off solenoid (ELAB). On earlier models it was located on the fuel pump. (Courtesy of Mack Trucks Inc.)

FIGURE 16–7 Econovance timing advance. (Courtesy of Mack Trucks Inc.)

FIGURE 16–8 Engine speed (rpm) and TDC position sensor. (Courtesy of Mack Trucks Inc.)

FIGURE 16–9 Timing event marker sensor. (Courtesy of Mack Trucks Inc.)

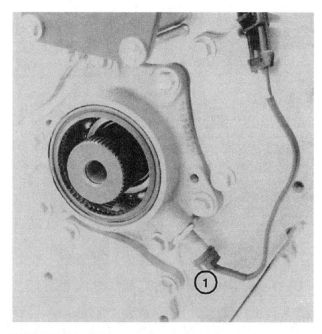

FIGURE 16–10 MPH (road speed) sensor. (Courtesy of Mack Trucks Inc.)

ticed the problem. Try to obtain as much information as possible. In some cases, there will only be a verbal complaint instead of a fault registered by V-MAC.

Diagnostic Blink Codes

The V-MAC module is capable of blinking a two-digit blink code for each of the detectable active faults in the V-MAC system. These codes are displayed on the electronic malfunction lamp located on the dashboard. The primary reason for the blink code is to allow for a quick diagnosis of an active fault in the system without requiring an expensive troubleshooting tool. The blink codes can be used for isolating and troubleshooting any active faults in the V-MAC system.

DIAGNOSTIC AND TROUBLESHOOTING PROCEDURES 317

FIGURE 16–11 Coolant temperature sensor (1) and intake manifold air temperature sensor (2). (Courtesy of Mack Trucks Inc.)

FIGURE 16–13 Coolant level sensor. (Courtesy of Mack Trucks Inc.)

FIGURE 16–12 Engine oil pressure sensor. (Courtesy of Mack Trucks Inc.)

FIGURE 16–14 Throttle position sensor and accelerator pedal. (Courtesy of Mack Trucks Inc.)

To properly activate and use the blink codes, follow the steps listed below.

1. Turn the key on and wait until the electronic fault lamp's 2-second power-up test is finished.

2. There must be an active fault that keeps the light on after the 2-second power-up test.

3. With the speed control ON/OFF switch in the OFF position, press and hold the SET/DECEL or the RESUME/ACCEL switch until the fault lamp goes off.

4. The fault lamp will remain off for approximately 1 second.

5. Immediately after the wait time, the V-MAC module will begin to flash a two-digit blink code. The

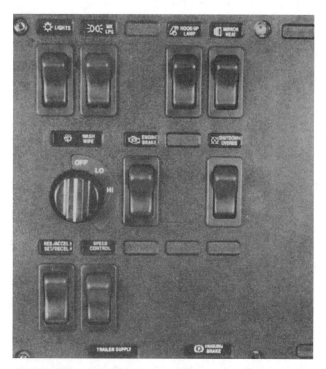

FIGURE 16–15 Switch panel on dashboard showing cruise control switches, parking brake switch, engine brake switch, and shutdown override switch. (Courtesy of Mack Trucks Inc.)

two digits of the code will be separated by a 1-second idle time (fault lamp off.)

6. Each digit of the blink code may consist of up to eight ON/OFF flashes. The ON and OFF time for each flash will be ¼ second.

7. Count the ON flashes of the fault lamp to determine the two-digit blink code.

8. Only one active fault will be blinked per request. There must be a separate request for each active fault where there are multiple active faults present in the system. To request another fault be displayed, hold in the SET/DECEL or the RESUME/ACCEL switch until the fault lamp goes off. The blinking sequence will begin again after a 1-second delay.

9. If the fault blinking request is repeated while V-MAC is in the process of blinking an active fault, that sequence will stop and the next active fault will be blinked.

10. If an active fault is cleared while V-MAC is blinking that fault, the procedure will not stop.

11. After every complete blinking sequence, the fault lamp will return to normal functions. It will remain on for active faults and off for inactive faults.

NOTE: If more than one active fault is present, continue the blink code sequence until the first active fault is redeployed to be sure all faults have been recovered.

The V-MAC blink code software will not provide codes for inactive faults. To access this data requires the use of a hand-held reader (Pro-Link 9000) or diagnostic computer.

A list of possible blink codes appears in **Figures 16–16A, B,** and **C.**

Once a fault has been determined, whether active or inactive, the tests for that particular fault must be completed in the order in which they are listed in the manual.

In performing electrical diagnostic tests, it is always a good practice to move (wiggle) the electrical wires and connectors while the test is being done. This will usually catch those instances where wires are making poor or intermittent contact. When working with the V-MAC system, this may cause a fault to be logged, which will help identify the problem.

NOTE: Before beginning the test sequence, check the condition of the wires and connectors. Check the batteries for correct voltage. Many electrical problems are caused by dirty, loose, or disconnected connectors. Examine the harness for places where sharp metal edges may have cut through the harnesses. Check for damaged connectors or wires that may have pulled out of connector terminals. Intermittent problems are usually an indication of a loose or poor connection, or marginal adjustment on a particular component. Inspect seals on the connector for looseness, cracking, missing, etc. Improper sealing can cause failures. There should be no grease of any kind on electrical connectors. If any is found, it should be cleaned off.

After determining the fault code from the diagnostic blink procedure explained earlier, turn to the section of the manual that explains the tests to follow for that fault. The test sequences are set up in a step-by-step method that will allow you to logically move to a solution to the problem. Follow the tests exactly as they are outlined. Start at the beginning of each sequence and perform each test in the order in which they are given. Do not skip steps or attempt to shortcut the test procedures. The result(s) from each test will tell which test should be performed next or will give possible causes for the problem. Not all tests are in sequence. For example: a particular test result may specify moving from test 1 to test 4, depending on the results of the previous test.

NOTE: The **least likely** components to fail are the V-MAC and FIC modules.

PROCEDURE FOR NO ACTIVE CODES, OR INTERMITTENT PROBLEMS

Function	Device	Inputs	Backup	Effects if Input Fails	Code
Fuel control	FIC	Accelerator pedal[a] —Signal line open, ground or short to 5 V —Reference line open or ground	None	Increased low-idle speed to 800 rpm 800 rpm isochronous governor *No* speed control	5–1 5–2
		Accelerator pedal[a] —Reference line short to 12 V	None	All the above at throttle greater than 0% Cruise control still functions at 0% throttle	5–1
		Engine speed sensor[a]	Buffered rpm	Lose rpm readout on PC or Pro-Link Usually a smooth transition to backup	3–3
		Buffered rpm[a] (signal from V-MAC)	Engine speed sensor	May or may not lose tachometer output, depending on the nature of the problem	3–1
	V-MAC	RPM/TDC[a] (buffered rpm)	Engine speed sensor	Engine timing readout goes to zero No tachometer readout Usually a smooth transition to backup If both speed sensors fail, then the engine stalls	3–1 3–2
		Rack position[a]	None	Rack actuator shutdown. Sensor is checked again at 900 rpm. If OK, normal operation resumes.	4–4 4–5
		Fuel request line	Accelerator pedal	No speed control functions: VSC, SSC, or cruise control	6–1
		Torque limit switch[a] shorted low/closed	None	Low power	
		Open	None	No torque limit in low hole	
		Inlet air temperature sensor	None	Fault on PC or Pro-Link display screen	2–3 2–4
		Coolant temperature sensor	None	Fault on PC or Pro-Link display screen	2–1 2–2

[a]Indicates that erratic engine speed or power can occur if failure of this input is intermittent.

FIGURE 16–16A (Courtesy of Mack Trucks Inc.)

NOTE: In the descriptions section of each test procedure, wires are referenced by pin designations. All pins at the V-MAC module and the FIC module are called out by numbers. All pins located at the sensor itself, or the harness that connects to it, are called out by letters. Each wire is identified by a description of what signal is transmitted on that wire and also by a pin letter or number and wire number.

CAUTION:

In any of the test procedures where the instructions are to disconnect the harness connector from the V-MAC module or the FIC module, the ignition key must be in the OFF position. Failure to follow this warning can cause internal electrical damage to the modules.

PROCEDURE FOR NO ACTIVE CODES, OR INTERMITTENT PROBLEMS
(Courtesy of Mack Trucks Inc.)

1. Verify that the electronic fault and engine shutdown lights work properly. When the key is turned on, both lamps should light for 2 seconds.

2. Check for inactive faults, or active faults if the electronic malfunction lamp was replaced. (The Pro-Link retains information from the previous vehicle. Reset the Pro-Link before looking for faults by activating the RESTART option on the Pro-Link menu.)

3. Review the "V-MAC System Diagnostics" section at the front of the V-MAC manual.

4. Start troubleshooting at the sensor, switch, or other device, then check the harness, then finally

Function	Device	Inputs	Backup	Effects if Input Fails	Code
Injection timing	V-MAC	Timing event marker	None	Full retard commanded	3–4
				Full retard assumed	
		RPM/TDC sensor (buffered rpm)	None	Full retard commanded	3–2
				Full retard assumed	
				Engine timing readout goes to zero	
				No tachometer readout	
		Proprietary data line	None	Full retard commanded	6–2
				Full retard assumed	
Speed control	V-MAC	Speed control ON/OFF SET/RESUME	None	Cannot set speed control	
				Cannot set or resume speed control	
		Accelerator pedal	None	See "Fuel control"	
		MPH sensor	None	Speed control functions (cruise, VSC, SSC) are canceled	4–1
				No speedometer output	4–2
				Reduced power	4–3
		RPM/TDC sensor	Engine speed sensor	VSC/SSC will drop out	3–1
				Engine timing readout goes to zero	3–2
				No tachometer readout	
		Clutch switch open	None	No speed control	
		Service brake switch shorted low/closed	None	No cruise control	
				No VSC if set with parking brake off	
		Open	MPH sensor	Normal state	
				A vehicle deceleration of 3 mph per second causes cruise control to drop out	
		Parking brake switch closed	None	Cannot set cruise control	7–2
		PTO switch open	None	Cannot set SSC	
	FIC	Fuel request line	None	No cruise control	6–1
		SAE/ATA serial line	None	No cruise control	6–3
				Reduced power throughout the speed range if the fuel request line has also failed	
				No PC or Pro-Link output	

FIGURE 16–16B (Courtesy of Mack Trucks Inc.)

check the ECU (V-MAC or FIC). Eliminate all other possible faults and repeat the procedure before replacing any ECU.

5. Check sensors, connectors and other components for broken pins, dirt, corrosion, loose terminals, excessive resistance, moisture, and bad grounds.

6. If a sensor or other component is suspected as being bad, disconnect, clean, and reconnect all terminals to that component. Retest to see if the problem has been resolved.

7. If a sensor or other component is suspected as being bad, try to isolate it from the system. Often, when a faulty component is removed from the system, the vehicle performance will change. If performance does not change when the component is disconnected, then that component may not be the problem.

8. When checking the harness refer to the wiring diagrams in the V-MAC manual. Check each wire that connects the component to an ECU for shorts to ground, power, or continuity with other wires if they should not exist. Check for continuity between both ends of the harness for each wire on the harness.

9. All "ground check" tests should be performed with a known good ground.

10. Continuity checks cannot be performed if there is any current flowing. Make sure the ignition key is off before performing any continuity checks.

Function	Device	Inputs	Backup	Effects if Input Fails	Code
Road speed limiting	V-MAC	MPH sensor	None	Reduced power	4–1
				No speedometer output	4–2
	FIC	Fuel request line	None	No cruise control	6–1
		SAE/ATA data line	None	No cruise control	6–3
				Reduced power throughout the speed range if the fuel request line has also failed	
				No PC or Pro-Link output	
Engine protection & shutdown	V-MAC	Oil pressure sensor	None	Protection/shutdown inactive for oil pressure	1–1
					1–2
		Coolant temperature sensor	None	Protection/shutdown inactive for coolant temperature	2–1
					2–2
		Coolant level sensor	None	Protection/shutdown inactive for coolant level	1–7
		Shutdown override	None	Override cannot be performed	7–4
	FIC	SAE/ATA data line	Fuel request line	Engine protection but no shutdown	
				Engine speed only drops to low idle	
Idle shutdown	V-MAC	Accelerator pedal	None	If vehicle is not moving, shutdown will occur even if vehicle is not idling unless PTO is engaged and PTO shutdown option is not enabled	5–1
					5–2
		Shutdown override	None	Override cannot be performed	7–4
		MPH sensor	None	Idle shutdown will occur if the accelerator pedal position is constant even if the vehicle is moving	4–1
					4–2
		Coolant temperature sensor	None	Shutdown occurs after the warmup and shutdown times have expired, but the engine may not reach the warm-up temperature	2–1
					2–2

FIGURE 16–16C (Courtesy of Mack Trucks Inc.)

11. The ignition key must be in the OFF position before disconnecting the harness to the V-MAC or FIC modules.

CAUTION:

Failure to have the key off may result in internal electrical damage to the modules.

12. Turning the key on with a sensor or the FIC module disconnected will result in an active fault.

13. If an ECU is suspected, verify that there is power from the circuit breaker to the ECU.

14. *Do not* open an ECU or take any measurements at the ECU pins. If a module is suspected, visually inspect the pins for any repairable problems.

15. As a final test before replacing an ECU, reconnect the suspect ECU and confirm that the fault still exists.

16. If an ECU is replaced, verify that the new module has fixed the problem.

CAUTION:

Do not program a replacement V-MAC module with MACK DATA until it is confirmed that the new V-MAC module has fixed the problem. MACK DATA reprogramming executes the password protection function, which assigns a new password to the vehicle. **This process is irreversible!!** If the old module is not the problem and it is reinstalled on the vehicle, it cannot be reprogrammed with MACK DATA because the password will no longer match.

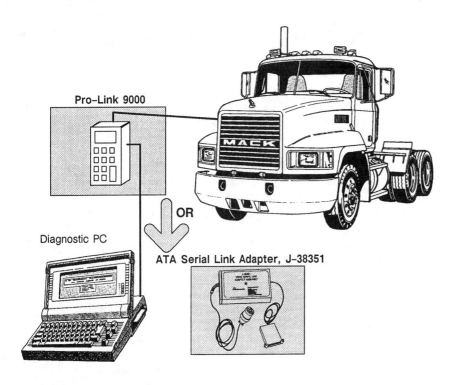

FIGURE 16–17 V-MAC diagnostic equipment. The Pro-Link 9000 or the ATA serial link adapter can be used to interface with a personal computer (PC). (Courtesy of Mack Trucks Inc.)

V-MAC DIAGNOSTIC TOOLS

In addition to the blink code method the V-MAC system can also be checked with the Pro-Link 9000 diagnostic tool and with an IBM-compatible lap-top computer **(Figures 16–17 and 16–18)**. The computer requires the use of the Pro-Link 9000 tool or the serial adapter link. You can use these tools to read inactive fault codes, to clear fault codes, or to program engine idle speed and customer-specified parameters. Follow the equipment manufacturer's instructions and the appropriate Mack Truck service manual for procedures and specifications. Repair or replace any faulty components according to instructions in the service manual.

Diagnostic Computer
(Courtesy of Mack Trucks Inc.)

To change certain proprietary data, such as engine horsepower, a computer is required. The computer also allows information specific to the vehicle to be entered and stored in the V-MAC module's internal memory, and it provides more flexibility in password selection. Any 100% IBM-compatible computer will work with the system. A serial link interface is required to connect a computer to the vehicle. The ATA Serial Link Adapter, J-38351, or the Pro-Link 9000, can be used as the interface device. In addition, the computer must be running the V-MAC Service Diag-

FIGURE 16–18 Pro-Link 9000 diagnostic tool. (Courtesy of Mack Trucks Inc.)

nostic Software. This software package is available from Mack Trucks Service Publications Department, through the normal Branch and Distributor Parts Ordering Network.

The software package contains complete instructions for installing and running the program. Follow the instructions completely.

REVIEW QUESTIONS

1. The vehicle management and control (V-MAC) is designed for use on _____ _____.
2. V-MAC uses two control modules, the _____ _____ and the _____ _____.
3. Name two performance features of the V-MAC electronic control system.
4. The fuel shut-off solenoid shuts off fuel flow to the _____ _____ when the ignition switch is turned _____.
5. The road speed sensor senses transmission _____ _____ _____ and sends this information to the control module for _____.
6. The TPS sensor send a variable _____ _____ based on throttle position to the ECU to control _____ _____.
7. The V-MAC module is capable of blinking a _____ _____ _____ code for each of the _____ detectable faults in the V-MAC system.
8. Before beginning a test sequence, check the condition of the _____ and _____.

TEST QUESTIONS

1. The V-MAC system has two control modules:
   a. FIC and V-MAC
   b. ECU and V-MAC
   c. TPS and ICU
   d. AMBAC and Bosch
2. The FIC module processes information from input sensors and switches to control
   a. injection timing
   b. fuel delivery
   c. fuel delivery and injection timing
   d. engine rpm
3. AMBAC and Robert Bosch fuel injection pumps are
   a. single-plunger type
   b. barrel type
   c. helix type
   d. multiplunger type
4. The injection pump fuel rack actuator is mounted on the
   a. rear of the pump
   b. front of the pump
   c. engine block with a linkage
   d. top right-hand side
5. The econovance timing advance unit changes injection timing in response to commands from the
   a. operator
   b. control module
   c. ECU
   d. TPS
6. If the engine coolant temperature is too high, the coolant temperature sensor
   a. opens the shutters
   b. decreases engine rpm
   c. shuts down the engine
   d. increases the rpm
7. The engine oil pressure sensor signal is used by the ECU to activate
   a. a warning alarm
   b. engine shut down
   c. a warning alarm and engine shutdown after 30 seconds
   d. a warning alarm to operate only
8. The V-MAC blink code will *not* provide codes for
   a. electrical problems
   b. wiggle tests
   c. all electrical problems
   d. inactive faults

CHAPTER 17

DETROIT DIESEL MECHANICAL UNIT INJECTOR FUEL SYSTEM

INTRODUCTION

The Detroit Diesel mechanical fuel injection system combines high-pressure pumping, fuel metering, and injection in one assembly called a *unit injector*. Each engine cylinder is provided with one injector. The helix-type plungers are controlled by the fuel control rack, which in turn is controlled by a foot- or hand-operated throttle (depending on engine application) and the governor. The unit injector plungers are actuated for pumping by a special lobe on the engine camshaft, a cam follower, push rod, and rocker arm (**Figures 17–1** and **17–2**). Fuel is injected into each cylinder once every camshaft (and crankshaft) revolution on two-stroke-cycle engines and once every two crankshaft revolutions on four-stroke-cycle engines.

PERFORMANCE OBJECTIVES

After thorough study of this chapter and the appropriate training models and service manuals, and with the necessary tools and equipment you should be able to do the following with respect to diesel engines equipped with the Detroit Diesel mechanical unit injector fuel system:

1. Describe the basic construction and operation of Detroit Diesel fuel system components, including the following:
    a. method of fuel pumping and metering
    b. injector lubrication
    c. injection timing control
    d. fuel control rack
    e. governor types

2. Diagnose basic fuel injection system problems and determine the correction required.

3. Remove, clean, inspect, test, measure, repair, replace, and adjust fuel injection system components as needed to meet manufacturer's specifications.

TERMS YOU SHOULD KNOW

Look for these terms as you study this chapter, and learn what they mean.

mechanical unit injector
injector plunger
injector barrel
injector bushing
helix
upper port
lower port
upper land
lower land
T hole
rack gear
control rack
injection timing
fuel metering
crown valve
needle valve
limiting-speed governor
variable-speed governor
fast-idle cylinder
fuel modulator
throttle delay mechanism
tailored torque governor
check fuel flow
control rack and plunger movement test
injector opening test
spray pattern test

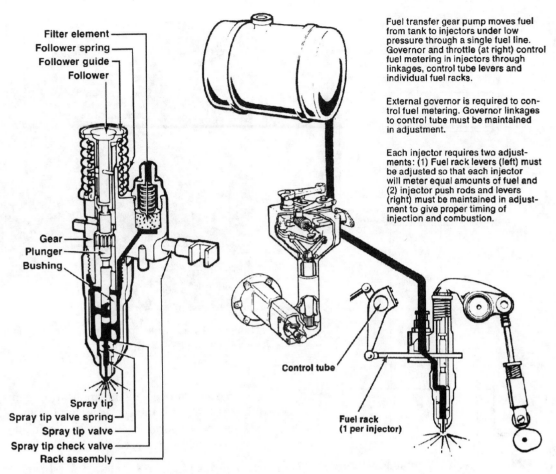

FIGURE 17-1 Detroit Diesel mechanical unit injector fuel system. (Courtesy of Detroit Diesel Corporation.)

spray tip test
high-pressure test
pressure holding test
fuel output test
spray tip concentricity test
set injector timing
adjust governor gap
adjust starting aid screw
position injector rack levers
adjust maximum no-load speed
adjust idle speed
adjust buffer screw
adjust throttle delay
adjust fuel modulator
adjust TT governor

UNIT INJECTOR FUNCTION

The fuel injector is a lightweight compact unit that enables quick, easy starting on diesel fuel and permits the use of a simple open-type combustion chamber. The simplicity of design and operation provides for simplified controls and easy adjustment. No high-pressure fuel lines are required.

The fuel injector performs four functions:

1. creates the high fuel pressure required for efficient injection
2. meters and injects the exact amount of fuel required to handle the load
3. atomizes the fuel for mixing with the air in the combustion chamber
4. permits continuous fuel flow

Combustion required for satisfactory engine operation is obtained by injecting, under pressure, a small quantity of accurately metered and finely atomized fuel oil into the cylinder.

The continuous fuel flow through the injector return serves to prevent air pockets in the fuel system, and to

326 Chapter 17 DETROIT DIESEL MECHANICAL UNIT INJECTOR FUEL SYSTEM

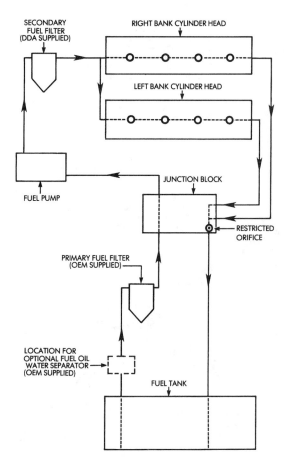

FIGURE 17–2 Fuel system schematic. (Courtesy of Detroit Diesel Corporation.)

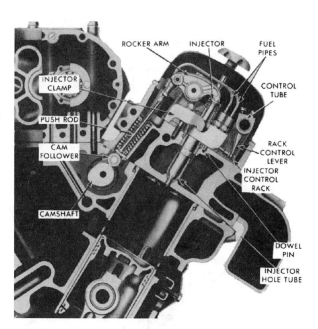

FIGURE 17–3 Fuel injector mounting. (Courtesy of Detroit Diesel Corporation.)

act as a coolant for those injector parts subjected to high combustion temperatures.

Each injector fits into an injector sleeve fitted into the cylinder head. A hold-down clamp and bolt keep the injector in place **(Figure 17–3).** Injector control racks are operated by a control rack tube that is connected to the governor by means of a fuel rod. Each control rack lever is independently adjustable on the control tube. This allows all injector racks to be adjusted in a uniform manner to ensure equal fuel delivery.

UNIT INJECTOR IDENTIFICATION

To vary the power output of the engine, injectors having different fuel output capacities are used. The fuel output of the various injectors is governed by the helix angle of the plunger and the type of spray tip used.

Because the helix angle on the plunger determines the output and operating characteristics of a particular type of injector, it is imperative that the correct injectors be used for each engine application. If injectors of different types are mixed, erratic operation will result and may cause serious damage to the engine or to the equipment that it powers.

NOTE: Do not intermix injectors with other types of injectors in an engine.

Each fuel injector has a circular disk pressed into a recess at the front side of the injector body for identification purposes. The identification tag indicates the nominal output of the injector in cubic millimeters **(Figure 17–4).**

PLUNGER, BARREL, AND HELIX CONSTRUCTION

The plunger and bushing assembly in the Detroit Diesel injector is an outstanding example of precision manufacturing. The mating surfaces of plunger and bushing are so smooth, uniform, and accurately fitted that you can hold the plunger, with the bushing assembled to it, spin the bushing—and it will continue to spin seemingly endlessly, just like a fine ball bearing that is almost frictionless. Yet, you can take this same assembly, place it in a test fixture, and test it for leakage with high-pressure air; the leakage may be almost immeasurable because the surfaces fit together so closely and uniformly.

The bushing (barrel) is bored and reamed to an accurate inside diameter, and the two ports are accu-

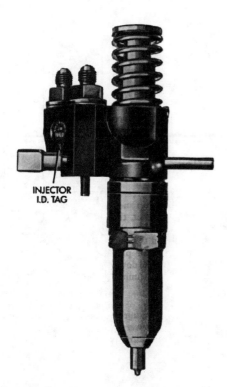

FIGURE 17-4 (Courtesy of Detroit Diesel Corporation.)

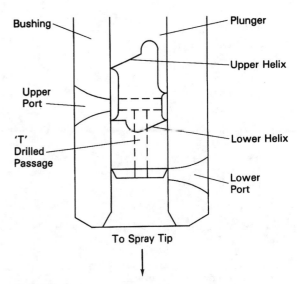

FIGURE 17-5 Injector plunger and bushing detail. (Courtesy of Detroit Diesel Corporation.)

rately located, drilled, and countersunk. The port is through-hardened. Then the sealing surface at the lower end is ground flat. The bushing is case-hardened to improve its strength, wear resistance, and corrosion resistance. Then a locating pin is inserted, the mating surface at the lower end is lapped, and the inside diameter is honed and lapped to make it smooth, round, and straight **(Figure 17-5)**.

Machine operations on the plunger start with an axial-drilled hole in each end. Then the circumferential groove with the helix edge is cut on an accurate cam-controlled milling machine, and a cross-hole is drilled to intersect the groove and the axial hole in the lower end. Two beveled undercuts are machined, one in the middle of the plunger and one at the upper end under the head, and the head is beveled. The plunger is through-hardened and then ground to approximately final diameter, with the "gear flat" on the upper section being ground to an accurate form. Then the plunger is lapped to the finished diameter.

The internal diameter of the bushings (barrels) is measured with an air gauge, and the bushings are segregated into groups to the nearest 12-millionths of an inch (0.0003048 mm). The plungers are measured and bushings are mated very closely to maintain a specified clearance that never exceeds 60-millionths of an inch (0.001524 mm). Each assembly is checked for leakage with compressed air before being sent on to final inspection.

UNIT INJECTOR OPERATION

The Detroit Diesel mechanical unit injector pump plunger and helix operate as follows. The fuel is injected into the cylinder by means of a plunger, which is forced downward by a rocker arm. As the plunger moves downward, it forces some fuel to spray into the cylinder combustion chamber in a fine, accurately timed spray. The plunger moves the same distance on each stroke, and a circumferential groove in the plunger determines the timing and amount of fuel injected. The upper edge of this groove is cut in the shape of a helix. The lower edge may be straight or a helix **(Figure 17-6)**. When the plunger is rotated by the fuel control rack, it changes the position of the helix, which changes the fuel output of the injector and the injection timing.

The plunger moves in a bushing inside the injector. There are two ports on opposite sides of the bushing, one higher than the other. Fuel is delivered to the injector under low pressure from a fuel delivery pump and may flow into or out of the bushing through either port. As long as either port is uncovered by the plunger, fuel is free to flow out of the bushing through either port; thus, no pressure can be developed by the plunger. As the plunger descends at the start of an injection stroke, the lower end of the plunger shuts off the lower port hole. As the plunger continues to descend, it reaches a point where the helix of the upper land closes off the upper port. With both ports shut off by the plunger, no

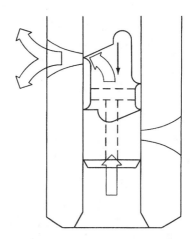 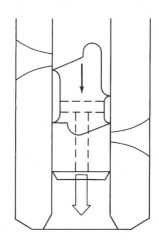 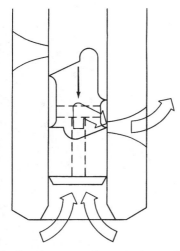

Injector Operation: The plunger descends, first closing off the lower port and then the upper. Before the upper port is shut off, fuel being displaced by the descending plunger may flow up through the "T" drilled hole in the plunger and escape through the upper port.

After the upper port has been shut off, fuel can no longer escape and is forced down by the plunger and sprays out the tip.

As the plunger continues to descend, it uncovers the lower port, so that fuel escapes and injection stops. Then the plunger returns to its original position and awaits the next injection cycle.

FIGURE 17–6 (Courtesy of Detroit Diesel Corporation.)

fuel may flow from the injector except to be forced out through the tip. Thus injection starts and continues under high pressure as the plunger moves downward until the lower land finally passes the lower port. At that point the groove is open to the lower port, and fuel may escape up through the drilled hole in the center of the plunger. At this point injection stops even though the plunger continues downward for a short distance before rising again to its starting position.

Thus, the start of injection depends on the point at which the upper land, with its helix edge, shuts off the upper port. The sooner this port is shut off, the sooner injection starts and the longer it takes; thus, more fuel is injected. If the plunger is turned so that the helix edge shuts off the port later, injection starts later, continues for a shorter length of time, and injects a smaller amount of fuel **(Figure 17–7)**.

The full-fuel position occurs when the lower part of the upper helix is positioned at the upper port so that this port is shut off; injection begins almost immediately after the lower port is shut off. The no-fuel position occurs when the plunger is positioned so that the upper helix is at its extreme upper position adjacent to the upper port; by the time the upper port is closed, the lower port has already been opened.

Rotation of the plunger between its no-fuel and full-fuel positions is accomplished by using a rack to rotate a gear on the plunger **(Figure 17–8)**. The gear has a flat on its internal diameter to match a flat surface on the upper part of the plunger. The plunger is free to slide up and down in the opening in the gear. A control tube assembly actuates the racks **(Figure 17–9)**.

The angle at which the upper and lower helixes are ground on Detroit Diesel unit injectors is controlled accurately to afford precise control of fuel injection. The relationship of the upper and lower helixes determines the amount of fuel injected for any given setting of the fuel control. If this engineered relationship is altered, the performance of the injector is changed. If such alteration is different from one plunger to another, engine operation at part-throttle settings will be rough and erratic.

INJECTOR TIMING

The upper helix controls the start-of-injection timing. At low power, injection starts late; at high power, it starts early. In the no-fuel position, the upper helix cuts off the upper port after the lower port is exposed, and thus no fuel is injected.

The lower helix controls the end-of-injection timing. There are three types of lower-land machining. One is flat (actually not a helix). It obviously always opens the lower port at the same point in the stroke regardless of fuel control position and thus always stops injection at the same time. This is known as the *constant ending* type of injector.

NEEDLE VALVE AND CROWN VALVE INJECTORS

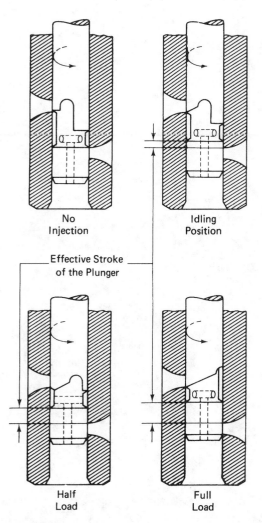

FIGURE 17-7 Fuel metering. The effective stroke of the plunger changes as it is rotated in the bushing. (Courtesy of Detroit Diesel Corporation.)

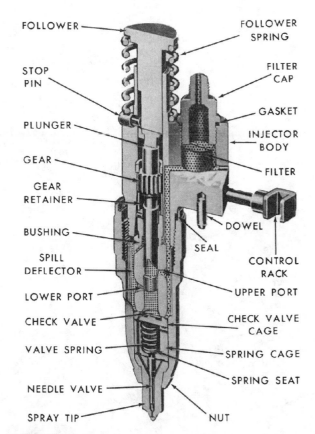

FIGURE 17-8 Cutaway view of injector showing control rack and gear. (Courtesy of Detroit Diesel Corporation.)

Detroit Diesel mechanical unit injector basic timing is achieved by proper alignment of the timing marks on the control racks and gears in relation to the piston and crankshaft position. Variation of injection timing in relation to engine speed is achieved through plunger helix design and position.

In the second type of injector, known as the *retarded* type, the lower helix is a right-hand spiral (as in the upper helix) but has a much shallower angle than the upper helix. Thus, as the upper helix advances the start-of-injection timing for greater fuel delivery, the lower helix also advances the shut-off timing but to a lesser extent.

In the third type of injector, known as the *50-50* type, the lower helix is a left-hand spiral (opposite of the upper helix), so that increased fuel delivery at the higher fuel settings is a combination of an earlier starting point and a later stopping point.

NEEDLE VALVE AND CROWN VALVE INJECTORS

Needle Valve Operation (Figure 17-10)

When sufficient pressure is built up, it opens the flat, nonreturn check valve. The fuel in the check valve cage, spring cage, tip passages, and tip fuel cavity is compressed until the pressure force acting upward on the needle valve is sufficient to open the valve against the downward force of the valve spring. As soon as the needle valve lifts off its seat, the fuel is forced through the small orifices in the spray tip and atomized into the combustion chamber.

The fuel injector outlet opening, through which the excess fuel oil returns to the fuel return manifold and then back to the fuel tank, is directly adjacent to the inlet opening.

As the plunger moves downward, under pressure of the injector rocker arm, a portion of that fuel trapped under the plunger is displaced into the supply chamber

330 Chapter 17 Detroit Diesel Mechanical Unit Injector Fuel System

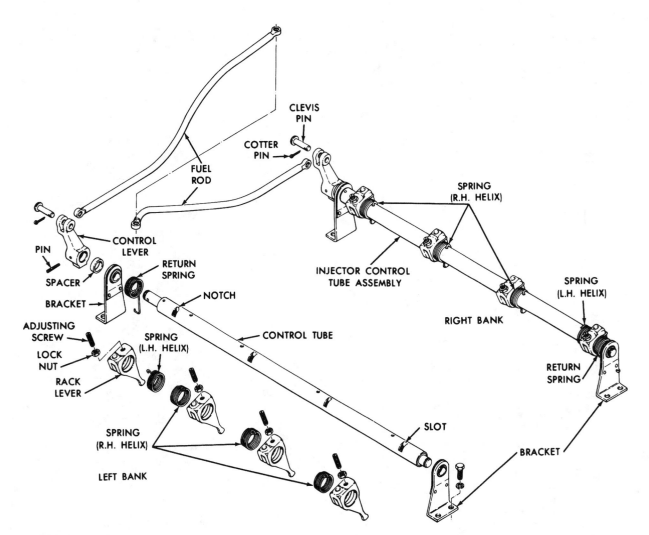

FIGURE 17–9 Injector control tube components. (Courtesy of Detroit Diesel Corporation.)

through the lower port until the port is closed off by the lower end of the plunger. A portion of the fuel trapped below the plunger is then forced up through a central passage in the plunger into the fuel metering recess and into the supply chamber through the upper port until that port is closed off by the upper ports, both closed off. The remaining fuel under the plunger is subjected to increased pressure by the continued downward movement of the plunger.

Changing the position of the helixes, by rotating the plunger, retards or advances the closing of the ports and the beginning and ending of the injection period. At the same time, it increases or decreases the amount of fuel injected into the cylinder. With the control rack pulled out all the way (no injection), the upper port is not closed by the helix until after the lower port is uncovered. Consequently, with the rack in this position, all the fuel is forced back into the supply chamber, and no injection of fuel takes place. With the control rack pushed all the way in (fuel injection), the upper port is closed shortly after the lower port has been covered, thus producing a maximum effective stroke and maximum injection. From this no-injection position to full-injection position (full rack movement), the contour of the upper helix advances the closing of the ports and the beginning of injection.

When the lower land of the plunger uncovers the lower port in the bushing, the fuel pressure below the plunger is relieved, and the valve spring closes the needle valve, ending injection.

A pressure relief passage has been provided in the spring cage to permit bleed-off of fuel leaking past the needle pilot in the tip assembly.

A check valve, directly below the bushing, prevents leakage from the combustion chamber into the fuel injector in case the valve is accidentally held open by a

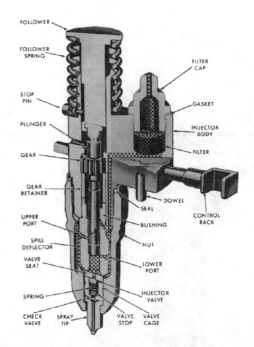

FIGURE 17–10 Crown valve injector. (Courtesy of Detroit Diesel Corporation.)

small particle of dirt. The injector plunger is then returned to its original position by the injector follower spring.

Crown Valve Operation

When sufficient pressure is built up, the injector valve is lifted off its seat, and the fuel is forced through small orifices in the spray tip and atomized into the combustion chamber.

A check valve, mounted in the spray tip, prevents air leakage from the combustion chamber into the fuel injector if the valve is accidentally held open by a small particle of dirt. The injector plunger is then returned to its original position by the injector follower spring. On the return upward movement of the plunger, the high-pressure cylinder within the bushing is again filled with fuel oil through the ports. The constant circulation of fresh cool fuel through the injector renews the fuel supply chamber, helps cool the injector, and also effectively removes all traces of air, which might otherwise accumulate in the system and interfere with accurate metering of the fuel (see **Figure 17–8**).

GOVERNOR TYPES

Horsepower requirements on an engine may vary due to fluctuating loads; therefore, some method must be provided to control the amount of fuel required to hold the engine speed reasonably constant during load fluctuations. To accomplish this control, a governor is introduced in the linkage between the throttle control and the fuel injectors. The governor is mounted on the front end of the blower and is driven by one of the blower rotors (**Figures 17–11** to **17–15**). The following types of mechanical governors are used:

1. limiting-speed mechanical governor
2. variable-speed mechanical governor

Engines requiring a minimum and maximum speed control, together with manually controlled intermediate speeds, are equipped with a limiting-speed mechanical governor.

Engines subjected to varying load conditions that require automatic fuel compensation to maintain a near-constant engine speed, which may be changed manually by the operator, are equipped with a variable-speed mechanical governor.

Each type of governor has an identification plate located on the control housing, containing the governor assembly number, type, idle-speed range, and drive ratio. The maximum engine speed, not shown on the identification plate, is stamped on the option plate attached to one of the valve rocker covers.

FAST-IDLE CYLINDER (FIGURE 17–16)

The limiting-speed governor equipped with a fast-idle air cylinder is used on vehicle engines where the engine powers both the vehicle and auxiliary equipment.

The fast-idle system consists of a fast-idle air cylinder installed in place of the buffer screw and a throttle-locking air cylinder mounted on a bracket fastened to the governor cover. An engine shutdown air cylinder, if used, is also mounted on the governor cover.

The fast-idle air cylinder and the throttle-locking air cylinder are actuated at the same time by air from a common air line. The engine shutdown air cylinder is connected to a separate air line.

The air supply for the fast-idle air cylinder is usually controlled by an air valve actuated by an electric solenoid. The fast-idle system should be installed so that it will function only when the parking brake system is in operation to make it tamper-proof.

The vehicle accelerator-to-governor throttle linkage is connected to a yield link, so the operator cannot overcome the force of the air cylinder holding the speed control lever in the idle position while the engine is operat-

332 Chapter 17 DETROIT DIESEL MECHANICAL UNIT INJECTOR FUEL SYSTEM

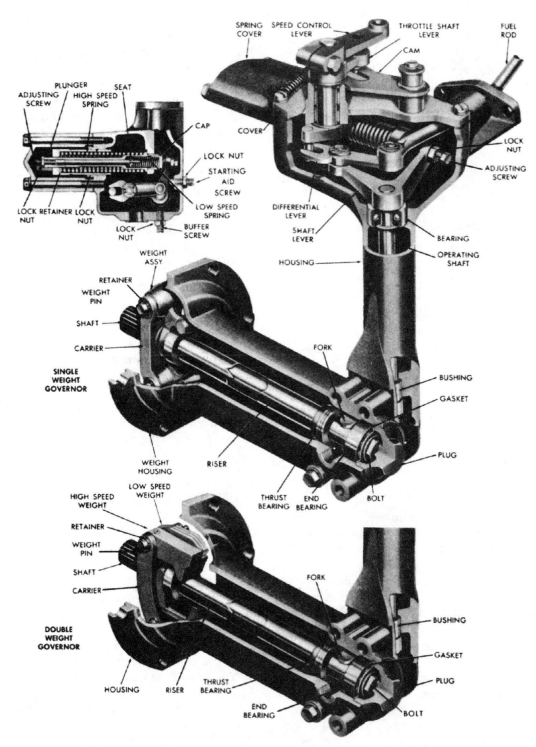

FIGURE 17–11 Detroit Diesel limiting-speed mechanical governor. (Courtesy of Detroit Diesel Corporation.)

ing at the single fixed high-idle speed. During highway operation, the governor functions as a limiting-speed governor.

For operation of auxiliary equipment, the vehicle is stopped and the parking brake set. Then, with the engine running, the low-speed switch is placed in the ON position. When the fast-idle air cylinder is actuated, the force of the dual idle spring is added to the force of the governor low-speed spring, thus increasing the engine idle speed.

FUEL MODULATOR

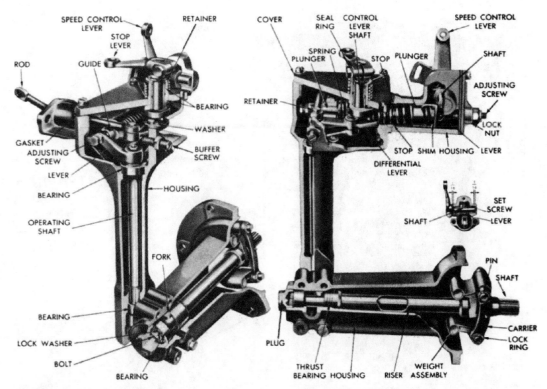

FIGURE 17-12 Detroit Diesel variable-speed mechanical governor. (Courtesy of Detroit Diesel Corporation.)

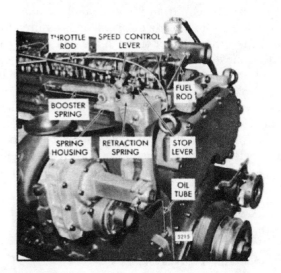

FIGURE 17-13 Variable-speed governor mounted on an in-line engine. (Courtesy of Detroit Diesel Corporation.)

The governor now functions as a constant-speed governor at the high-idle speed setting, maintaining a near-constant engine speed regardless of the load within the capacity of the engine. The fast-idle system provides a single fixed high-idle speed that is not adjustable, except by disassembling the fast-idle air cylinder and changing the dual idle spring. As with all mechanical governors, when load is applied, the engine speed is determined by the governor droop.

FUEL MODULATOR (FIGURE 17-17)

The fuel modulator, used on certain turbocharged aftercooled engines, maintains the proper fuel/air ratio in the lower speed ranges where the mechanical governor would normally act to provide maximum injector output. It operates in such a manner that although the engine throttle may be moved into the full-speed position, the injector racks cannot advance to the full-fuel position until the turbine speed is sufficient to provide proper combustion.

The fuel modulator reduces exhaust smoke and also helps improve fuel economy. The modulator mechanism is installed on the left bank between the No. 1 and No. 2 cylinders.

A fuel modulator consists of a cast housing containing a cylinder, piston, cam, and spring mounted on the cylinder head. A lever and roller that controls the injector rack is connected to the injector control tube. Tubes run from the air box to the housing to supply pressure to actuate the piston.

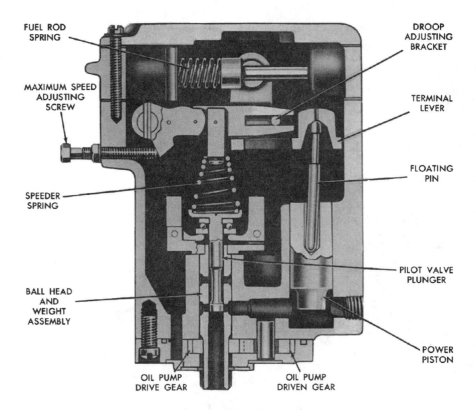

FIGURE 17-14 Hydraulic governor assembly. (Courtesy of Detroit Diesel Corporation.)

The modulator tells the fuel system how much fuel the engine can efficiently use based on air box pressure. Increased air box pressure forces the piston and cam out of the cylinder bore, allowing the rack to move toward full fuel.

Whenever the fuel injector rack control levers are adjusted, the fuel (air box) modulator lever and roller assembly must first be positioned free of cam contact. This is done by loosening the clamp screw.

THROTTLE DELAY MECHANISM

The throttle delay mechanism is used to retard full-fuel injection when the engine is accelerated. This reduces exhaust smoke and also helps improve fuel economy **(Figure 17–18)**.

The throttle delay mechanism is installed between the No. 1 and No. 2 cylinders on the right-bank cylinder head. It consists of a special rocker arm shaft bracket (which incorporates the throttle delay cylinder), a piston, throttle delay lever, connecting link, orifice plug, ball check valve, and U-bolt.

A yield link replaces the standard operating-lever connecting link in the governor.

Oil is supplied to a reservoir above the throttle delay cylinder through an oil supply fitting in the drilled oil passage in the rocker arm shaft bracket. As the injector racks are moved toward the no-fuel position, free movement of the throttle delay piston is assured by air drawn into the cylinder through the ball check valve. Further movement of the piston uncovers an opening that permits oil from the reservoir to enter the cylinder and displace the air. When the engine is accelerated, movement of the injector racks toward the full-fuel position is momentarily retarded while the piston expels the oil from the cylinder through an orifice. To permit full accelerator travel, regardless of the retarded injector rack position, a spring-loaded yield link replaces the standard operating-lever connecting link in the governor.

TAILORED TORQUE (TT) GOVERNOR OPERATION

The tailored torque (TT) engine is able to maintain reasonably constant horsepower over a wide speed range with a 6% torque rise per 100 rpm. These characteristics are achieved by the action of two Belleville springs (washers) in a limiting-speed governor.

NOTE: The horsepower for TT engines indicates a flat performance curve. However, during dynamometer testing an engine may exhibit horsepower readings slightly above or

CHECKING FUEL FLOW

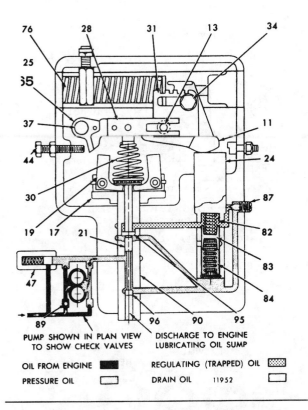

11. Lever—Terminal
13. Shaft—Terminal (long)
17. Ball Head Assy.
19. Flyweight
21. Plunger—Pilot Valve
24. Piston—Servo-Motor
25. Lever—Speed Adjusting
28. Lever—Floating
30. Spring—Speeder
31. Bracket—Droop Adjusting
34. Bolt—Droop Adjusting
37. Shaft—Speed Adjusting
44. Screw—Maximum Speed Adjusting
47. Valve—Relief
76. Spring—Terminal Lever Return
82. Spring—Buffer (upper)
83. Piston—Buffer
84. Spring—Buffer (lower)
87. Valve—Compensating Needle
89. Valve—Check
90. Bushing—Pilot Valve
95. Land—Receiving Compensating
96. Land—Pilot Valve Control

FIGURE 17-15 Position of hydraulic governor mechanism as load increases and engine speed tends to decrease. (Courtesy of Detroit Diesel Corporation.)

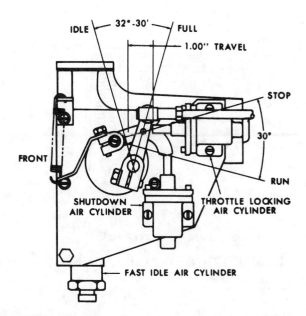

FIGURE 17-16 Governor with fast-idle cylinder. (Courtesy of Detroit Diesel Corporation.)

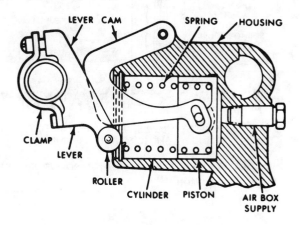

FIGURE 17-17 Typical fuel modulator. (Courtesy of Detroit Diesel Corporation.)

below the flat curve. A 5% horsepower variation from the flat published curve is acceptable.

The force provided by the Belleville springs works with the governor weights to pull the injector racks to reduced fuel as the engine speed is increased. Conversely, as the engine speed is reduced by increased load, the high-speed spring overcomes the force of the Belleville springs and moves the injector racks to an increased fuel position. The racks move progressively into more fuel to maintain the constant horsepower until the racks are in full fuel at a speed near the rpm specified for the engine (**Figure 17-19**).

UNIT INJECTOR SYSTEM DIAGNOSIS AND SERVICE

(Courtesy of Detroit Diesel Corporation)

CHECKING FUEL FLOW

A watch with a second hand and an appropriate size fuel container are required for this check. A fuel pres-

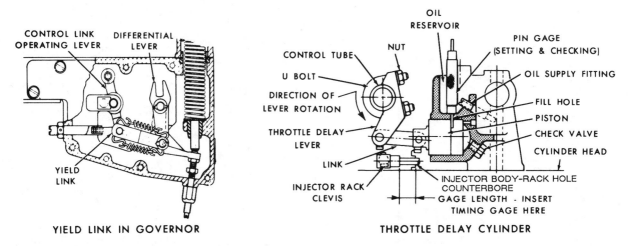

FIGURE 17-18 Yield link in governor, and throttle delay cylinder. (Courtesy of Detroit Diesel Corporation.)

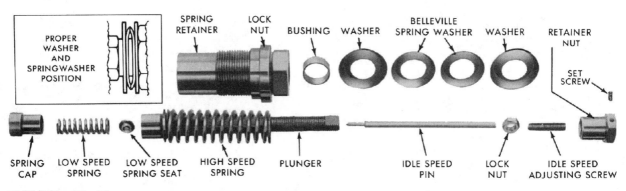

FIGURE 17-19 Tailored torque governor components. (Courtesy of Detroit Diesel Corporation.)

sure gauge and fitting tee are required to test fuel pump output pressure.

To check fuel flow the following procedure is typical **(Figure 17-20)**:

1. Disconnect the fuel return hose at a point where it is possible to allow fuel to run from it into the fuel container.

2. Start and run the engine at the specified speed (usually around 1200 to 1800 rpm) for 1 minute. Measure the amount of fuel delivered in that time and compare with specifications for that engine.

3. To check for air in the fuel, immerse the fuel return line well into the fuel in the container (well below the fuel surface) and watch for air bubbles rising to the surface. If air is present, it is being drawn into the fuel system at some point on the inlet side of the fuel pump. Inspect all the lines and fitting connections on the inlet side of the pump for air leaks, including the strainer, filter, and water separator. Correct as necessary.

4. If the fuel quantity delivered is below specifications, check the following and correct as necessary until the proper flow is achieved.

    a. Check all fuel lines for restriction or damage.

    b. Check to ensure that the engine has the proper size restriction fitting (restriction fitting size varies with engine size).

    c. Replace the fuel strainer element.

    d. Replace the fuel filter element.

    e. Check fuel pump output pressure. Replace the pump if required.

    f. Check for plugged injector filters as outlined in the service manual.

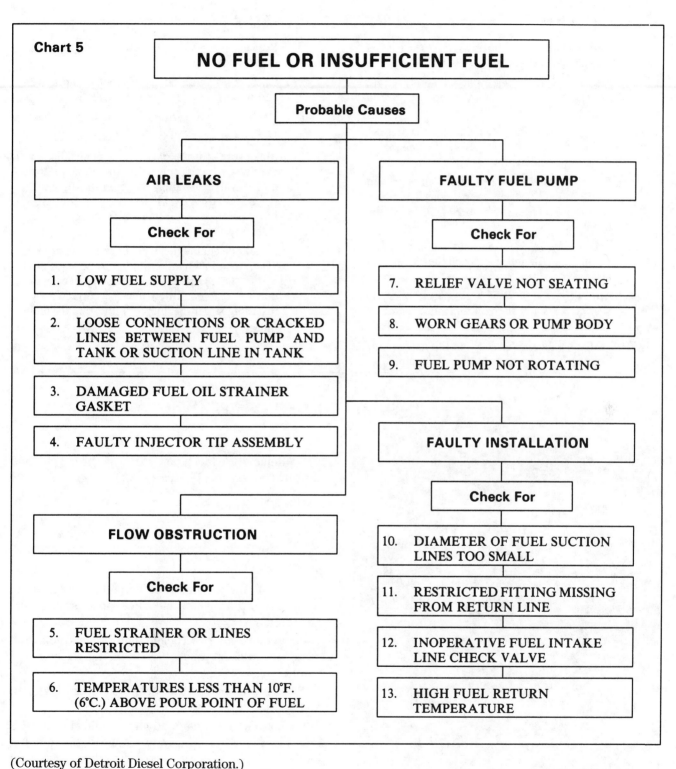

(Courtesy of Detroit Diesel Corporation.)

NO FUEL OR INSUFFICIENT FUEL

SUGGESTED REMEDY

1. The fuel tank should be filled above the level of the fuel suction tube.
2. Perform a *Fuel Flow Test* and, if air is present, tighten loose connections and replace cracked lines.
3. Perform a *Fuel Flow Test* and, if air is present, replace the fuel strainer gasket when changing the strainer element.
4. Perform a *Fuel Flow Test* and, if air is present with all fuel lines and connections assembled correctly, check for and replace faulty injectors.
5. Perform a *Fuel Flow Test* and replace the fuel strainer and filter elements and the fuel lines, if necessary.
6. Consult the *Fuel Specifications* in Section 13.3 for the recommended grade of fuel.
7. Perform a *Fuel Flow Test* and, if inadequate, clean and inspect the valve seat assembly.
8. Replace the gear and shaft assembly or the pump body.
9. Check the condition of the fuel pump drive and blower drive and replace defective parts.
10. Replace with larger tank-to-engine fuel lines.
11. Install a restricted fitting in the return line.
12. Make sure that the check valve is installed in the line correctly; the arrow should be on top of the valve assembly or pointing upward. Reposition the valve, if necessary. If the valve is inoperative, replace it with a new valve assembly.
13. Check the engine fuel spill-back temperature. The return fuel temperature must be less than 150°F (66°C) or a loss in horsepower will occur. This condition may be corrected by installing larger fuel lines or relocating the fuel tank to a cooler position.

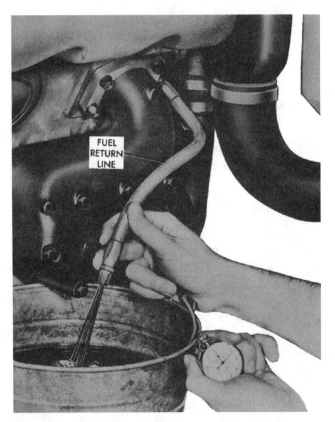

FIGURE 17-20 Checking fuel flow. (Courtesy of Detroit Diesel Corporation.)

INJECTOR REMOVAL

1. Clean and remove the valve rocker cover.
2. Remove the fuel pipes from both the injector and the fuel connectors (**Figure 17-21**).

NOTE: Immediately after removal of the fuel pipes from an injector, cover the filter caps with shipping caps to prevent dirt from entering the injector. Also protect the fuel pipes and fuel connectors against entry of dirt or foreign material.

3. Crank the engine to bring the outer ends of the push rods of the injector and valve rocker arms in line horizontally.
4. Remove the two rocker shaft bracket bolts and swing the rocker arms away from the injector and valves.
5. Remove the injector clamp bolt, special washer, and clamp.
6. Loosen the inner and outer adjusting screws (certain engines have only one adjusting screw and locknut) on the injector rack control lever, and slide the lever away from the injector.
7. Lift the injector from its seat in the cylinder head. Use a prybar if necessary (**Figure 17-22**).
8. Cover the injector hole in the cylinder head to keep foreign material out.

INSTALLING FUEL INJECTOR IN TESTER 339

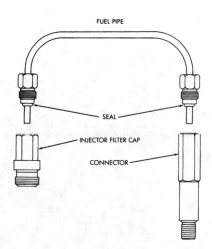

FIGURE 17-21 Fuel pipe removal. (Courtesy of Detroit Diesel Corporation.)

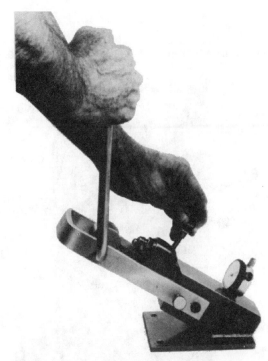

FIGURE 17-23 Checking rack freeness. (Courtesy of Detroit Diesel Corporation.)

FIGURE 17-22 Using a roll-head prybar to remove injector. (Courtesy of Detroit Diesel Corporation.)

9. Clean the exterior of the injector with clean fuel oil and dry it with compressed air.

CONTROL RACK AND PLUNGER MOVEMENT TEST

Place the injector in the injector fixture and rack freeness tester. Place the handle on top of the injector follower **(Figure 17-23)**.

If necessary, adjust the contact screw in the handle to ensure that the contact screw is at the center of the follower when the follower spring is compressed.

With the injector control rack held in the no-fuel position, push the handle down and depress the follower to the bottom of its stroke. Then very slowly release the pressure on the handle while moving the control rack up and down until the follower reaches the top of its travel. If the rack does not fall freely, loosen the injector nut, turn the tip, and retighten the nut. Loosen and retighten the nut a couple of times if necessary. Generally this will free the rack. Then, if the rack isn't free, change the injector nut. In some cases it may be necessary to disassemble the injector to eliminate the cause of the misaligned parts.

INSTALLING FUEL INJECTOR IN TESTER (FIGURES 17-24 AND 17-25)

1. Select the proper clamping head. Position it on the clamping post and tighten the thumb screw into the lower detent position.

2. Connect the test oil delivery piping into the clamping head.

3. Connect the test oil clear discharge tubing onto the pipe on the clamping head.

4. Locate the adapter plate on top of the support bracket by positioning the ⅜ in. diameter hole at the far right of the adapter plate onto the ⅜ in. diameter dowel pin. This allows the adapter plate to swing out for mounting the fuel injector.

Chapter 17 DETROIT DIESEL MECHANICAL UNIT INJECTOR FUEL SYSTEM

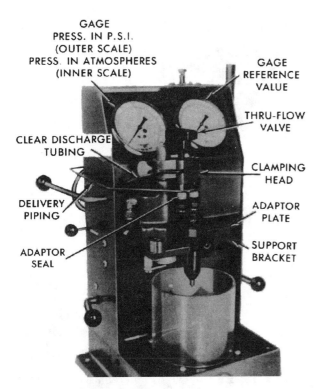

FIGURE 17-24 Injector in position on tester. (Courtesy of Detroit Diesel Corporation.)

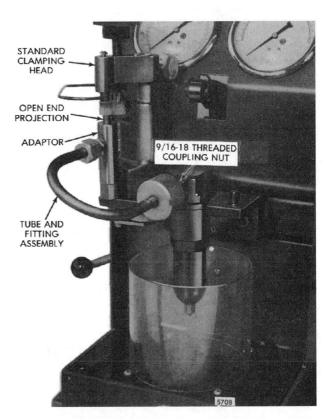

FIGURE 17-25 Injector fuel connection in tester. (Courtesy of Detroit Diesel Corporation.)

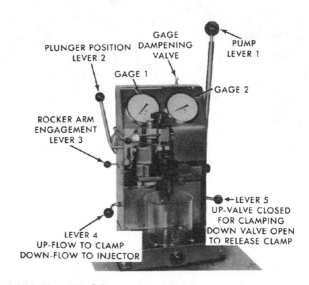

FIGURE 17-26 Injector tester operating levers. (Courtesy of Detroit Diesel Corporation.)

5. Mount the injector through the large hole and insert the injector pin in the proper locating pin hole.

6. Swing the mounted injector and adapter plate inward until they contact the stop pin at the rear of the support bracket.

VALVE OPENING AND SPRAY PATTERN TEST (FIGURE 17-26)

This test determines spray pattern uniformity and the relative pressure at which the injector valve opens and fuel injection begins.

1. Clamp the injector properly and purge the air from the system.

Move lever 4 down and operate pump lever 1 to produce a test oil flow through the injector. When air bubbles no longer pass through the clear discharge tubing, the system is free of air and is now ready for testing.

2. Move lever 4 *down*.
3. Position the injector rack in the full-fuel position.
4. Place pump lever 1 in the vertical position.
5. Move lever 3 to the forward detent position.
6. The injector follower should be depressed rapidly (at 40 to 80 strokes per minute) to simulate operation in the engine. Observe the spray pattern to see that all spray orifices are open and dispersing the test oil evenly. The beginning and ending of injection should be sharp, and the test oil should be finely atomized with no drops of test oil forming on the end of the tip.

The highest pressure reference number shown on gauge 2 will be reached just before injection ends. Use the reference values in the service manual to determine the relative acceptability of the injector.

SPRAY TIP TEST (FIGURE 17–26)

1. Move lever 4 *down* and operate pump lever 1 rapidly with smooth even strokes (40 strokes per minute) to simulate the action of the tip functioning in the engine.

2. Note the pressure at which the needle valve opens on gauge 1. The valve should open at specified pressure. The opening and closing action should be sharp and produce a normal, finely atomized spray pattern.

If the valve opening pressure is below specifications and/or atomization is poor, the cause is usually a weak valve spring or a poor needle valve seat.

If the valve opening pressure is within specifications, proceed to check for spray tip leakage as follows:

   a. Actuate pump lever 1 several times and hold the pressure at 1500 psi (10,335 kPa) for 15 seconds.
   b. Inspect the spray tip for leakage. There should be no fuel droplets, although a slight wetting at the spray tip is permissible.

INJECTOR HIGH-PRESSURE TEST (FIGURE 17–26)

This test checks for leaks at the filter cap gaskets, body plugs, and nut seal ring.

1. Clamp the injector properly and purge the air from the system.
2. Close the Thru-Flow valve, but do not overtighten.

NOTE: Make sure lever 4 is in the **down** position before operating pump lever 1.

3. Operate pump lever 1 to build up to 1600 to 2000 psi (11,024 to 13,780 pKa) on gauge 1. Check for leakage at the injector filter cap gaskets, body plugs, and injector nut seal ring.

PRESSURE HOLDING TEST (FIGURE 17–26)

This test determines if the body-to-bushing mating surfaces in the injector are sealing properly and indicates proper plunger-to-bushing fit.

1. Clamp the injector properly and purge the air from the system.
2. Close the Thru-Flow valve, but do not overtighten.
3. Move lever 2 to the rear, horizontal position.
4. Operate pump lever 1 until gauge 1 reads approximately 700 psi (4823 kPa).
5. Move lever 4 to the *up* position.
6. Time the pressure drop between 450 to 250 psi (3100 to 1723 kPa). If the pressure drop occurs in less than 15 seconds, leakage is excessive.

If the fuel injector passes all the preceding tests, proceed with the *Fuel Output Test*.

FUEL OUTPUT TEST

To check the fuel output, operate the injector in calibrator as follows (**Figure 17–27**).

NOTE: Place the cam shift index wheel and fuel flow lever in their respective positions. Turn on the test fuel oil heater switch and preheat the test oil to 95–105°F (35–40°C).

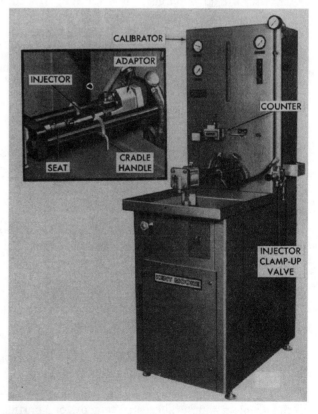

FIGURE 17–27 Injector fuel output tester. (Courtesy of Detroit Diesel Corporation.)

1. Place the proper injector adapter between the tie rods and engage it with the fuel block locating pin. Then slide the adapter forward and up against the fuel block face.

2. Place the injector seat into the permanent seat (cradle handle in vertical position). Clamp the injector into position by operating the air valve.

NOTE: Make sure the counter on the calibrator is preset at 1000 strokes. If for any reason this setting has been altered, reset the counter to 1000 strokes by twisting the cover release button to the left and hold the reset lever in the full up position while setting the numbered wheels. Close the cover. Refer to the calibrator instruction booklet for further information.

3. Pull the injector rack out to the no-fuel position.

4. Turn on the main power control circuit switch. Then start the calibrator by turning on the motor starter switch.

NOTE: The low oil pressure warning buzzer will sound briefly until the lubricating oil reaches the proper pressure.

5. After the calibrator has started, set the injector rack into the full-fuel position. Allow the injector to operate for approximately 30 seconds to purge the air that may be in the system.

6. After the air is purged, press the fuel flow start button (red). This will start the flow of fuel into the vial. The fuel flow to the vial will automatically stop after 1000 strokes.

7. Shut the calibrator off (the calibrator will stop in less time at full fuel).

8. Observe the vial reading and refer to specifications to determine whether the injector fuel output falls within the specified limits.

NOTE: The calibrator may be used to check and select a set of injectors that will inject the same amount of fuel in each cylinder at a given throttle setting, thus resulting in a smooth-running, well-balanced engine.

An injector that passes all the preceding tests may be put back into service. However, an injector that fails to pass one or more of the tests must be rebuilt and checked on the calibrator.

INJECTOR TROUBLESHOOTING

The following six troubleshooting charts cover both crown valve and needle valve injector problems revealed during testing. They indicate the probable cause of the problem and the suggested remedy for a particular engine. Refer to the appropriate service manual for any specific engine. These charts are reproduced here courtesy of Detroit Diesel Corporation (pp. 344–349).

INJECTOR DISASSEMBLY

If required, disassemble an injector as follows (**Figures 17–28 to 17–30**).

1. Support the injector upright in injector holding fixture and remove the filter caps, gaskets and filters.

NOTE: Whenever disassembling a fuel injector, discard the filters and gaskets and replace with new filters and gaskets. **In the offset injector, a filter is used in the inlet side only. No filter is required in the outlet side.**

2. Compress the follower spring. Then raise the spring above the stop pin with a screwdriver and withdraw the pin. Allow the spring to rise gradually.

3. Remove the plunger follower, plunger, and spring as an assembly.

4. Invert the fixture and, using a special socket, loosen the nut on the injector body.

5. Lift the injector nut straight up, being careful not to dislodge the spray tip and valve parts. Remove the spray tip and valve parts from the bushing and place them in a clean receptacle until ready for assembly.

FIGURE 17–28 Removing the filter cap with injector in holding fixture. (Courtesy of Detroit Diesel Corporation.)

CLEANING INJECTOR PARTS 343

FIGURE 17-29 Removing plunger follower and spring. (Courtesy of Detroit Diesel Corporation.)

When an injector has been in use for some time, the spray tip, even though clean on the outside, may not be pushed readily from the nut with the fingers. In this event, support the nut on a wood block and drive the tip down through the nut, using tool J 1291-02.

6. Remove the spill deflector. Then lift the bushing straight out of the injector body.

7. Remove the injector body from the holding fixture. Turn the body upside down and catch the gear retainer and gear in your hand as they fall out of the body.

8. Withdraw the injector control rack from the injector body. Also remove the seal ring from the body.

CLEANING INJECTOR PARTS

Because most injector difficulties are the result of dirt particles, it is essential that a clean area be provided on which to place the injector parts after cleaning and inspection.

Wash all the parts with clean fuel oil or a suitable cleaning solvent and dry them with clean, filtered compressed air. *Do not use waste or rags for cleaning purposes.* Clean out all the passages, drilled holes, and slots in all the injector parts.

Carbon on the inside of the spray tip may be loosened for easy removal by soaking for approximately 15 minutes in a suitable solution prior to the external

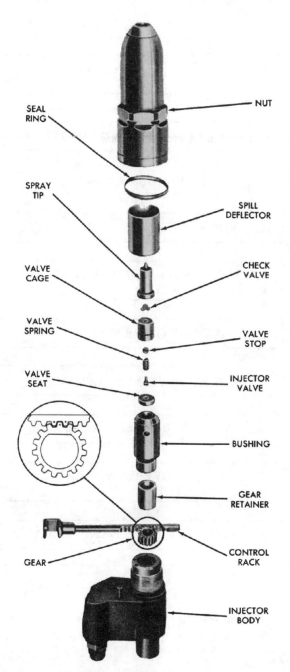

FIGURE 17-30 Disassembly of injector nut, body, and related parts. (Courtesy of Detroit Diesel Corporation.)

cleaning and buffing operation. Methyl Ethyl Ketone J 8257 solution is recommended for this purpose.

Clean the spray tip with the correct size tip cleaner.

NOTE: Care must be exercised when inserting the carbon remover in the spray tip to avoid contacting the needle valve seat in the tip.

CROWN VALVE INJECTORS

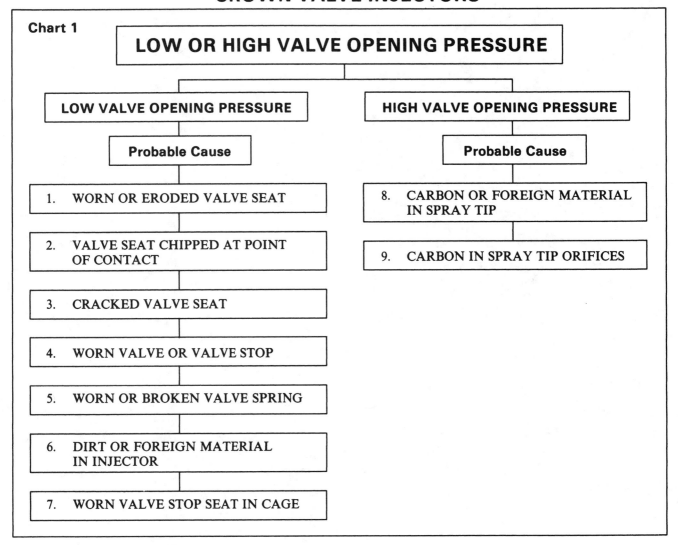

Chart 1

LOW OR HIGH VALVE OPENING PRESSURE

LOW VALVE OPENING PRESSURE

Probable Cause

1. WORN OR ERODED VALVE SEAT
2. VALVE SEAT CHIPPED AT POINT OF CONTACT
3. CRACKED VALVE SEAT
4. WORN VALVE OR VALVE STOP
5. WORN OR BROKEN VALVE SPRING
6. DIRT OR FOREIGN MATERIAL IN INJECTOR
7. WORN VALVE STOP SEAT IN CAGE

HIGH VALVE OPENING PRESSURE

Probable Cause

8. CARBON OR FOREIGN MATERIAL IN SPRAY TIP
9. CARBON IN SPRAY TIP ORIFICES

SUGGESTED REMEDY

1. A worn or eroded valve seat may be lapped, but not excessively as this would reduce thickness of the part causing a deviation from the valve stack–up dimension.

2. If the valve seat is chipped at the point of contact with the valve, lap the surface of the seat and the I.D. of the hole. Mount tool J 7174 in a drill motor and place the valve seat on the pilot of the tool, using a small amount of lapping compound on the lapping surface. Start the drill motor and apply enough pressure to bring the seat to the point of lap. Check the point of lap contact after a few seconds. If the edge of the hole appears sharp and clear, no further lapping is required. Excessive lapping at this point will increase the size of the hole and lower the injector valve opening pressure.

3. Replace the valve seat.

4. Replace the valve or valve stop.

5. Replace the spring. Check the valve cage and valve stop for wear; replace them, if necessary.

6. Disassemble and clean the injector.

7. Replace the valve cage.

8. Carbon in the tip should be removed with tip reamer J 1243 which is especially designed and ground for this purpose.

9. Check the size of the spray tip orifices. Then, using tool J 4298–1 with the proper size wire, clean the orifices.

CROWN VALVE INJECTORS

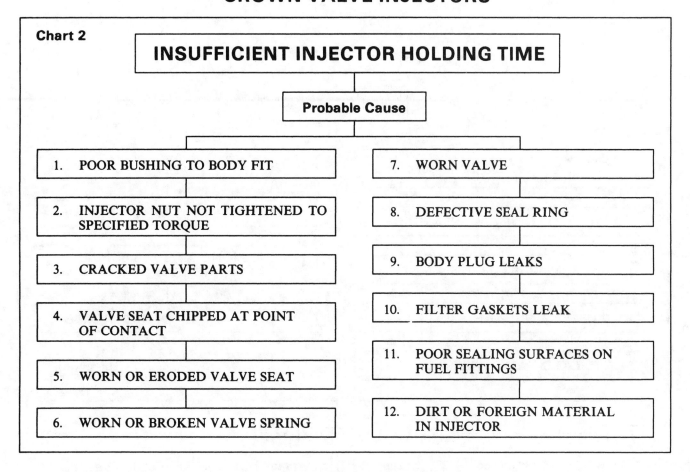

Chart 2 — INSUFFICIENT INJECTOR HOLDING TIME

Probable Cause:
1. POOR BUSHING TO BODY FIT
2. INJECTOR NUT NOT TIGHTENED TO SPECIFIED TORQUE
3. CRACKED VALVE PARTS
4. VALVE SEAT CHIPPED AT POINT OF CONTACT
5. WORN OR ERODED VALVE SEAT
6. WORN OR BROKEN VALVE SPRING
7. WORN VALVE
8. DEFECTIVE SEAL RING
9. BODY PLUG LEAKS
10. FILTER GASKETS LEAK
11. POOR SEALING SURFACES ON FUEL FITTINGS
12. DIRT OR FOREIGN MATERIAL IN INJECTOR

SUGGESTED REMEDY

1. Lap the injector body.

2. Tighten the nut to 55–65 lb–ft (75–88 N·m) torque. Do not exceed the specified torque.

3. Replace the valve parts.

4. If the valve seat is chipped at the point of contact with the valve, lap the surface of the seat and the I.D. of the hole. Mount tool J 7174 in a drill motor and place the valve seat on the pilot of the tool, using a small amount of lapping compound on the lapping surface. Start the drill motor and apply enough pressure to bring the seat to the point of lap. Check the point of lap contact after a few seconds. If the edge of the hole appears sharp and clear, no further lapping is required. Excessive lapping at this point will increase the size of the hole and lower the injector valve opening pressure.

5. A worn or eroded valve seat may be lapped, but not excessively as this would reduce the thickness of the part causing a deviation from the valve stack–up dimension.

6. Replace the spring. Check the valve cage and valve stop for wear; replace them, if necessary.

7. Replace the valve.

8. Replace the seal ring.

9. Install new body plugs.

10. Replace the filter gaskets and tighten the filter caps to 65–75 lb–ft (88–102 N·m) torque.

11. Clean up the sealing surfaces or replace the filter caps, if necessary.

12. Disassemble the injector and clean all of the parts.

CROWN VALVE INJECTORS

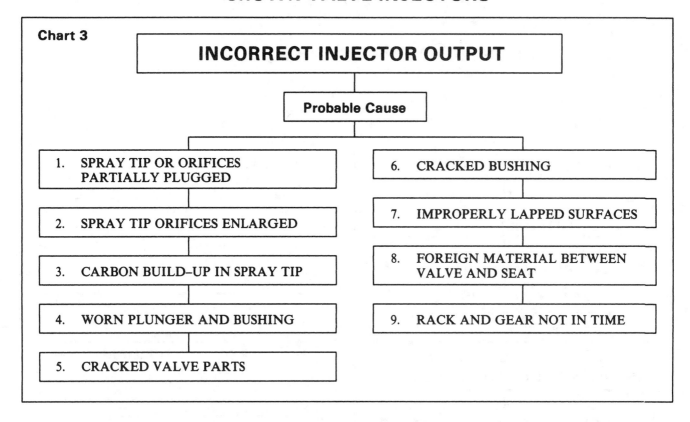

Chart 3 — INCORRECT INJECTOR OUTPUT

Probable Cause:
1. SPRAY TIP OR ORIFICES PARTIALLY PLUGGED
2. SPRAY TIP ORIFICES ENLARGED
3. CARBON BUILD-UP IN SPRAY TIP
4. WORN PLUNGER AND BUSHING
5. CRACKED VALVE PARTS
6. CRACKED BUSHING
7. IMPROPERLY LAPPED SURFACES
8. FOREIGN MATERIAL BETWEEN VALVE AND SEAT
9. RACK AND GEAR NOT IN TIME

SUGGESTED REMEDY

1. Clean the orifices with tool J 4298–1, using the proper size wire.
2. Replace the spray tip.
3. Clean the injector tip with tool J 1243.
4. After the possibility of an incorrect or faulty tip has been eliminated and the injector output still does not fall within its specific limits, replace the plunger and bushing with a new assembly.

   NOTICE: The fuel output of an injector varies with the use of different spray tips of the same size due to manufacturing tolerances in drilling the tips. If the fuel output does not fall within the specified limits of the *Fuel Output Check Chart,* try changing the spray tip. However, use only a tip specified for the injector being tested.

5. Replace the cracked parts.
6. Replace the plunger and bushing assembly.
7. Lap the sealing surfaces.
8. Disassemble the injector and clean all of the parts.
9. Assemble the gear with the drill spot mark on the tooth engaged between the two marked teeth of the rack.

NEEDLE VALVE INJECTORS

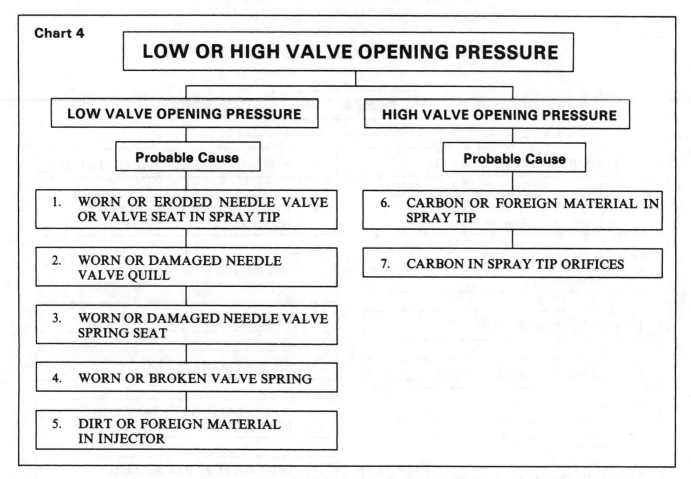

Chart 4

LOW OR HIGH VALVE OPENING PRESSURE

LOW VALVE OPENING PRESSURE

Probable Cause

1. WORN OR ERODED NEEDLE VALVE OR VALVE SEAT IN SPRAY TIP
2. WORN OR DAMAGED NEEDLE VALVE QUILL
3. WORN OR DAMAGED NEEDLE VALVE SPRING SEAT
4. WORN OR BROKEN VALVE SPRING
5. DIRT OR FOREIGN MATERIAL IN INJECTOR

HIGH VALVE OPENING PRESSURE

Probable Cause

6. CARBON OR FOREIGN MATERIAL IN SPRAY TIP
7. CARBON IN SPRAY TIP ORIFICES

SUGGESTED REMEDY

1. Replace the needle valve and spray tip assembly.
2. Replace the needle valve and spray tip assembly.
3. Replace the spring seat.
4. Replace the valve spring.
5. Disassemble the injector and clean all of the parts.

6. Remove the carbon in the spray tip with tip reamer J 24838 which is especially designed and ground for this purpose.
7. Check the size of the spray tip orifices. Then, using tool J 4298-1 with the proper size wire, clean the orifices.

NEEDLE VALVE INJECTORS

Chart 5

INSUFFICIENT INJECTOR HOLDING TIME

Probable Cause

1. POOR BUSHING TO BODY FIT
2. INJECTOR NUT NOT TIGHTENED TO SPECIFIED TORQUE
3. EXCESSIVE PLUNGER TO BUSHING CLEARANCE
4. CRACKED SPRAY TIP
5. WORN OR ERODED NEEDLE VALVE
6. WORN OR ERODED NEEDLE VALVE SEAT IN SPRAY TIP
7. WORN OR BROKEN NEEDLE VALVE QUILL
8. WORN OR BROKEN VALVE SPRING
9. WORN OR DAMAGED VALVE SPRING SEAT
10. DEFECTIVE SEAL RINGS
11. BODY PLUG LEAKS
12. FILTER GASKETS LEAK
13. POOR SEALING SURFACES ON FUEL FITTINGS
14. DIRT OR FOREIGN MATERIAL IN INJECTOR

SUGGESTED REMEDY

1. Lap the injector body.
2. Tighten the injector nut to 75–85 lb-ft (102–115 N·m) torque. Do not exceed the specified torque.
3. Replace the plunger and bushing.
4, 5, 6 and 7. Replace the needle valve and spray tip assembly.
8. Replace the valve spring.
9. Replace the valve spring seat.
10. Replace the seal rings.
11. Install new body plugs.
12. Replace the filter cap gaskets and tighten the filter caps to 65–75 lb-ft (88–102 N·m) torque.
13. Clean up the sealing surfaces or replace the filter caps, if necessary. Replace the filter if a cap is replaced.
14. Disassemble the injector and clean all of the parts.

NEEDLE VALVE INJECTORS

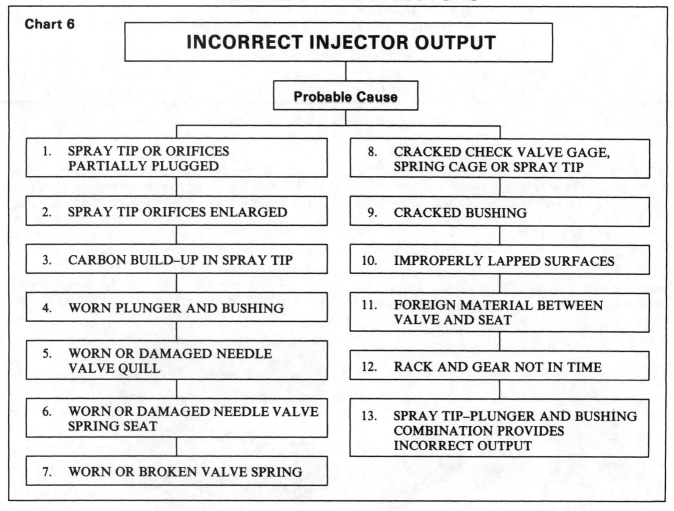

Chart 6 — INCORRECT INJECTOR OUTPUT

Probable Cause:

1. SPRAY TIP OR ORIFICES PARTIALLY PLUGGED
2. SPRAY TIP ORIFICES ENLARGED
3. CARBON BUILD-UP IN SPRAY TIP
4. WORN PLUNGER AND BUSHING
5. WORN OR DAMAGED NEEDLE VALVE QUILL
6. WORN OR DAMAGED NEEDLE VALVE SPRING SEAT
7. WORN OR BROKEN VALVE SPRING
8. CRACKED CHECK VALVE GAGE, SPRING CAGE OR SPRAY TIP
9. CRACKED BUSHING
10. IMPROPERLY LAPPED SURFACES
11. FOREIGN MATERIAL BETWEEN VALVE AND SEAT
12. RACK AND GEAR NOT IN TIME
13. SPRAY TIP-PLUNGER AND BUSHING COMBINATION PROVIDES INCORRECT OUTPUT

SUGGESTED REMEDY

1. Clean the spray tip as outlined under *Clean Injector Parts*.
2. Replace the needle valve and spray tip assembly.
3. Clean the spray tip with tool J 1243–01.
4. After the possibility of an incorrect or faulty spray tip has been eliminated and the injector output still does not fall within its specific limits, replace the plunger and bushing with a new assembly. The fuel output of an injector varies with the use of different spray tips of the same size due to manufacturing tolerances in drilling the tips. If the fuel output does not fall within the specified limits of the *Fuel Output Check Chart*, try changing the spray tip. However, use only a tip specified for the injector being tested.
5. Replace the needle valve and spray tip assembly.
6. Replace the spring seat.
7. Replace the valve spring.
8. Replace the cracked parts.
9. Replace the plunger and bushing assembly.
10. Lap the sealing surfaces.
11. Disassemble the injector and clean all of the parts.
12. Assemble the gear with the drill spot mark on the tooth engaged between the two marked teeth on the rack.
13. Replace the spray tip and the plunger and bushing assembly to provide the correct output.

Wash the tip in fuel oil and dry it with compressed air. Clean the spray tip orifices with a pin vise and the proper size spray tip cleaning wire **(Figures 17–31 and 17–32)**.

Before using the wire, hone the end until it is smooth and free of burrs and taper the end a distance of 1/16 in. with stone J 8170. Allow the wire to extend 1/8 in. from tool.

The exterior surface of an injector spray tip may be cleaned by using a brass wire buffing wheel. To obtain a good polishing effect and longer brush life, the buffing wheel should be installed on a motor that turns the wheel at approximately 3000 rpm. A convenient method of holding the spray tip while cleaning and polishing is to place the tip over the drill end of the spray tip cleaner tool and hold the body of the tip against the buffing wheel. In this way, the spray tip is rotated while being buffed.

NOTE: Do not buff excessively. **Do not use a steel wire buffing wheel, or the spray tip holes may be distorted.**

When the body of the spray tip is clean, lightly buff the tip end in the same manner. This cleans the spray tip orifice area and will not plug the orifices.

Wash the spray tip in clean fuel oil and dry it with compressed air.

Clean and brush all the passages in the injector body, using a fuel hole cleaning brush and a rack hole cleaning brush. Blow out the passages and dry them with compressed air. On needle valve injectors check the needle valve lift with the special tool. If lift is not within specifications, replace the valve assembly **(Figure 17–33)**.

Carefully insert a reamer in the injector body. Turn it clockwise a few turns, then remove the reamer and check the face of the ring for reamer contact over the entire face of the ring. If necessary, repeat the reaming procedure until the reamer makes contact with the entire face of the ring. Clean up the opposite side of the ring in the same manner **(Figure 17–34)**.

Carefully insert a 0.375 in. diameter straight fluted reamer inside the ring bore in the injector body. Turn

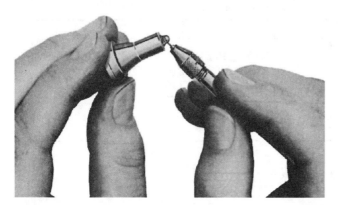

FIGURE 17–31 Cleaning spray tip orifices. (Courtesy of Detroit Diesel Corporation.)

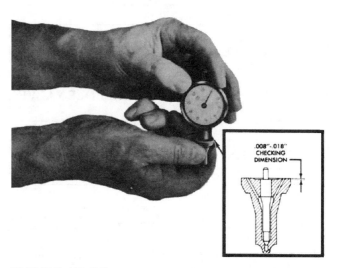

FIGURE 17–33 Checking needle valve lift. (Courtesy of Detroit Diesel Corporation.)

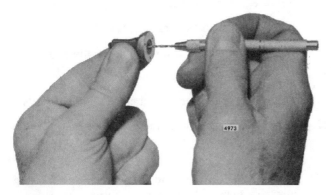

FIGURE 17–32 Cleaning injector spray tip. (Courtesy of Detroit Diesel Corporation.)

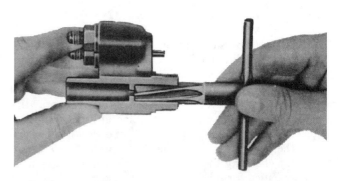

FIGURE 17–34 Cleaning the injector body. (Courtesy of Detroit Diesel Corporation.)

the reamer clockwise and remove any burrs inside the ring bore. Then wash the injector body in clean fuel oil and dry it with compressed air.

Remove the carbon deposits from the lower inside diameter taper of the injector nut with a carbon remover. Use care to minimize removing metal or setting up burrs on the spray tip seat. Remove only enough metal to produce a clean uniform seat to prevent leakage between the tip and the nut. Carefully insert the carbon remover in the injector nut. Turn it clockwise to remove the carbon deposits on the flat spray tip seat.

Wash the injector nut in clean fuel oil and dry it with compressed air. Carbon deposits on the spray tip seating surfaces of the injector nut will result in poor sealing and consequent fuel leakage around the spray tip.

When handling the injector plunger, do not touch the finished plunger surfaces with your fingers. Wash the plunger and bushing with clean fuel oil and dry them with compressed air. Be sure the high-pressure bleed hole in the side of the bushing is not plugged. If this hole is plugged, fuel leakage will occur at the upper end of the bushing, where it will drain out of the injector body vent and rack holes during engine operation, causing a serious oil dilution problem.

CAUTION:
Keep the plunger and bushing together, as they are mated parts.

After washing, submerge the parts in a clean receptacle containing clean fuel oil.

CAUTION:
Keep the parts of each injector assembly together.

Do not mix parts of one injector with those of another.

INSPECTING INJECTOR PARTS

Inspect the teeth on the control rack and the control rack gear for excessive wear or damage. Also check for excessive wear in the bore of the gear and inspect the gear retainer. Replace damaged or worn parts.

Inspect the injector follower and pin for wear.

Inspect both ends of the spill deflector for sharp edges or burrs that could create burrs on the injector body or injector nut and cause particles of metal to be introduced into the spray tip and valve parts. Remove burrs with a 500 grit stone.

Inspect the follower spring for visual defects. Then check the spring with a spring tester.

Check the seal ring area on the injector body for burrs or scratches. Also check the surface that contacts the injector bushing for scratches, scuff marks, or other damage. If necessary, lap this surface. A faulty sealing surface at this point will result in high fuel consumption and contamination of the lubricating oil. Replace any loose injector body plugs or a loose dowel pin. Install the proper number tag on a service replacement injector body.

Inspect the injector plunger and bushing for scoring, erosion, chipping, or wear. Check for sharp edges on that portion of the plunger that rides in the gear. Remove any sharp edges with a 500 grit stone. Wash the plunger after stoning it. Injector Bushing Inspectalite can be used to check the port holes in the inner diameter of the bushing for cracks or chipping. Slip the plunger into the bushing and check for free movement.

CAUTION:
Replace the plunger and bushing as an assembly if any of the above damage is noted, since they are mated parts.

Use new mated factory parts to ensure the best performance from the injector.

Injector plungers cannot be reworked to change the output. Grinding will destroy the hardened case at the helix and result in chipping and seizure or scoring of the plunger **(Figure 17–35)**.

Examine the spray tip seating surface of the injector nut and spray tip for nicks, burrs, erosion, or brinelling. Reseat the surface or replace the nut or tip if it is severely damaged.

The injector valve spring plays an important part in establishing the valve opening pressure of the injector assembly. Replace a worn or broken spring.

Inspect the sealing surfaces of the injector parts. Examine the sealing surfaces with a magnifying glass, for even the slightest imperfections will prevent the injector from operating properly. Check for burrs, nicks, erosion, cracks, chipping, and excessive wear. Also check for enlarged orifices in the spray tip.

Replace damaged or excessively worn parts. Check the minimum thickness of the lapped parts as noted in the chart in the service manual.

352 Chapter 17 DETROIT DIESEL MECHANICAL UNIT INJECTOR FUEL SYSTEM

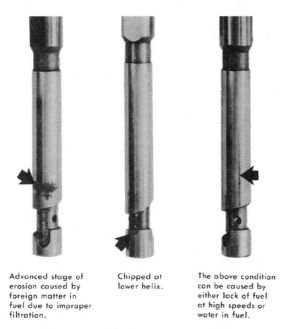

FIGURE 17–35 Unusable injector plungers. (Courtesy of Detroit Diesel Corporation.)

INJECTOR ASSEMBLY

Use an extremely clean bench to work on and to place the parts when assembling an injector. Also be sure all the injector parts, both new and used, are clean (**Figures 17–30 and 17–36**).

Assemble Injector Filters

Always use new filters and gaskets when reassembling an injector.

1. Insert a new filter, dimple end down, slotted end up, in each of the fuel cavities in the top of the injector body.

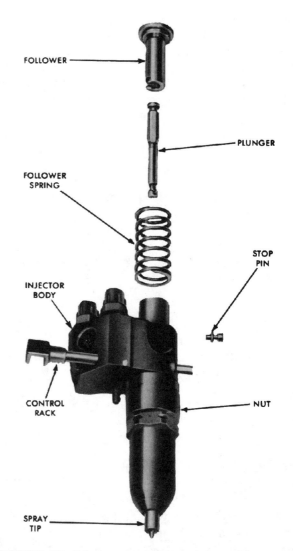

FIGURE 17–36 (Courtesy of Detroit Diesel Corporation.)

Examine the seating area of the needle valve for wear or damage. Also examine the needle quill and its contact point with the valve spring seat. Replace damaged or excessively worn parts.

Examine the needle valve seat area in the spray tip for foreign material. The smallest particle of such material can prevent the needle valve from seating properly. Polish the seat area with polishing stick J 22964. Coat only the tapered end of the stick with polishing compound J 23038 and insert it directly into the center of the spray tip until it bottoms. Rotate the stick 6 to 12 times, applying a light pressure with the thumb and forefinger.

NOTE: Be sure that no compound is accidentally placed on the lapped surfaces located higher up in the spray tip. The slightest lapping action on these surfaces can alter the near-perfect fit between the needle valve and tip.

Before reinstalling used injector parts, lap all the sealing surfaces. It is also good practice to lightly lap the sealing surfaces of new injector parts, which may become burred or nicked during handling.

NOTE: The sealing surface of current spray tips is precision lapped by a new process that leaves the surface with a dull, satinlike finish; the lapped surface on former spray tips was bright and shiny. It is not recommended to lap the surface of the **new** current spray tip.

NOTE: Install a new filter in the inlet side (located over the injector rack) in a fuel injector with an offset body. No filter is required in the outlet side of the offset body injector.

2. Place a new gasket on each filter cap. Lubricate the threads and install the filter caps. Tighten the filter caps to specified torque.

3. Purge the filters after installation by directing compressed air or fuel through the filter caps.

4. Install clean shipping caps on the filter caps to prevent dirt from entering the injector.

Assembling Rack and Gears

Note the drill spot marks on the control rack and gear. Then proceed as follows:

1. Hold the injector body, bottom end up, and slide the rack through the hole in the body. Look into the body bore and move the rack until you can see the drill marks. Hold the rack in this position.

2. Place the gear in the injector body so that the marked tooth is engaged between the two marked teeth on the rack.

3. Place the gear retainer on top of the gear.

4. Align the locating pin in the bushing with the slot in the injector body, then slide the end of the bushing into place.

Assembling Spray Tip, Spring Cage, and Check Valve Assemblies

1. Support the injector body, bottom end up, in injector holding fixture.

2. Place a new seal ring on the shoulder of the body.

NOTE: Wet the seal ring with test oil and install the ring all the way down past the threads and onto the shoulder of the injector body. This will prevent the seal from catching in the threads and becoming shredded. Use a seal ring protector if available.

3. Install the spill deflector over the barrel of the bushing.

4. Place the check valve (without the 0.010 in. hole) centrally on the top of the bushing. Then place the check valve cage over the check valve and against the bushing.

NOTE: The former and new check valve and check valve cage are not separately interchangeable in a former injector.

5. Insert the spring seat in the valve spring, then insert the assembly into the spring cage, spring seat first.

NOTE: Install a new spring seat in a former injector if a new design spray tip assembly is used.

6. Place the spring cage, spring seat, and valve spring assembly (valve spring down) on top of the check valve cage.

NOTE: When installing a new spray tip assembly in a former injector, a new valve spring seat must also be installed. The current needle valve has a shorter quill.

7. Insert the needle valve, tapered end down, inside the spray tip. Then place the spray tip and needle valve on top of the spring cage with the quill end of the needle valve in the hole in the spring cage.

8. Lubricate the threads in the injector nut and carefully thread the nut on the injector body by hand. Rotate the spray tip between your thumb and first finger while threading the nut on the injector body. Tighten the nut as tight as possible by hand. At this point there should be sufficient force on the spray tip to make it impossible to turn with your fingers.

9. Use a socket and a torque wrench to tighten the injector nut to specified torque.

NOTE: Do not exceed the specified torque. Otherwise, the nut may be stretched and result in improper sealing of the lapped surfaces in a subsequent injector overhaul.

10. After assembling a fuel injector, always check the area between the nut and the body. If the seal is still visible after the nut is assembled, try another nut that may allow assembly on the body without extruding the seal and forcing it out of the body-nut crevice.

Assembling Plunger and Follower

1. Slide the head of the plunger into the follower.

2. Invert the injector in the assembly fixture (filter cap end up) and push the rack all the way in. Then place the follower spring on the injector body.

3. Place the stop pin on the injector body so that the follower spring rests on the narrow flange of the stop pin. Then align the slot in the follower with the stop pin hole in the injector body. Next, align the flat side of the plunger with the slot in the follower. Then insert the free end of the plunger in the injector body. Press down on the follower and at the same time press

the stop pin into position. When in place, the spring will hold the stop pin in position.

CHECKING SPRAY TIP CONCENTRICITY

To assure correct alignment, check the concentricity of the spray tip as follows (**Figure 17-37**):

1. Place the injector in the concentricity gauge and adjust the dial indicator to zero.

2. Rotate the injector 360° and note the total runout as indicated on the dial.

3. If the total runout exceeds 0.008 in., remove the injector from the gauge. Loosen the injector nut, center the spray tip, and tighten the nut to specified torque. Recheck the spray tip concentricity. If after several attempts, the spray tip cannot be positioned satisfactorily, replace the injector nut.

TESTING RECONDITIONED INJECTOR

Before placing a reconditioned injector in service, perform all the tests (except the visual inspection of the plunger) previously outlined under *Test Injector*.

The injector is satisfactory if it passes these tests. Failure to pass any one of the tests indicates that defective or dirty parts have been assembled. In this case, disassemble, clean, inspect, reassemble, and test the injector again.

INJECTOR INSTALLATION

Before installing an injector in an engine, remove the carbon deposits from the beveled seat of the injector tube in the cylinder head. This will assure correct alignment of the injector and prevent any undue stresses from being exerted against the spray tip.

Use an injector tube bevel reamer to clean the carbon from the injector tube. Exercise care to remove *only* the carbon so that the proper clearance between the injector body and the cylinder head is maintained. Pack the flutes of the reamer with grease to retain the carbon removed from the tube.

Be sure the fuel injector is filled with fuel oil. If necessary, add clean fuel oil at the inlet filter cap until it runs out of the outlet filter cap.

Install the injector in the engine as follows:

1. Insert the injector into the injector tube with the dowel pin in the injector body registering with the locating hole in the cylinder head.

2. Slide the injector rack control lever over so that it registers with the injector rack.

3. Install the injector clamp, special washer (with curved side toward injector clamp), and bolt. Tighten the bolt to specified torque. Make sure that the clamp does not interfere with the injector follower spring or the exhaust valve springs.

NOTE: Check the injector control rack for free movement. Excess torque can cause the control rack to stick or bind.

4. Move the rocker arm assembly into position and secure the rocker arm brackets to the cylinder head by tightening the bolts to the torque specified.

NOTE: On four-valve cylinder heads, there is a possibility of damaging the exhaust valves if the exhaust valve bridge is not resting on the ends of the exhaust valves when the rocker shaft bracket bolts are tightened. Therefore, note the position of the exhaust valve bridge before, during, and after tightening the rocker shaft bolts.

FIGURE 17-37 Checking injector concentricity. (Courtesy of Detroit Diesel Corporation.)

5. Remove the shipping caps. Then install the fuel pipes and connect them to the injector and the fuel connectors. Use a socket to tighten the connections to specified torque.

IMPORTANT:

Do not bend the fuel pipes and do not exceed the specified torque. Excessive tightening will twist or fracture the flared end of the fuel line and result in leaks. Lubricating oil diluted by fuel oil can cause serious damage to the engine bearings.

ENGINE TUNE-UP

To tune-up an engine completely, perform all the adjustments in the applicable tune-up sequence given below after the engine has reached normal operating temperature. Since the adjustments are normally made while the engine is stopped, it may be necessary to run the engine between adjustments to maintain normal operating temperature.

NOTE: Before starting an engine after an engine speed control adjustment or after removal of the engine governor cover, the serviceperson must determine that the injector racks move to the no-fuel position when the governor stop lever is placed in the stop position. Engine overspeed will result if the injector racks cannot be positioned at no fuel with the governor stop lever.

CAUTION:

An overspeeding engine can result in engine damage that can cause personal injury.

Tune-up Sequence for Mechanical Governor

1. Adjust the exhaust valve clearance.
2. Time the fuel injectors.
3. Adjust the governor gap.
4. Position the injector rack control levers.
5. Adjust the maximum no-load speed.
6. Adjust the idle speed.
7. Adjust the Belleville spring for TT horsepower.
8. Adjust the buffer screw.
9. Adjust the throttle booster spring (variable-speed governor only).
10. Adjust the supplementary governing device, if used.

Tune-up Sequence for Hydraulic Governor

1. Adjust the exhaust valve clearance.
2. Time the fuel injectors.
3. Adjust the governor linkage.
4. Position the injector rack control levers.
5. Adjust the load limit screw.
6. Compensation adjustment (PSG governors only).
7. Adjust the speed droop.
8. Adjust the maximum no-load speed.

SETTING INJECTOR TIMING (TWO-STROKE-CYCLE ENGINE)

To time an injector properly, the injector follower must be adjusted to a definite height in relation to the injector body (**Figure 17–38**).

All the injectors can be timed in firing order sequence during one full revolution of the crankshaft. Refer to the service manual for the engine firing order and valve adjustment procedures.

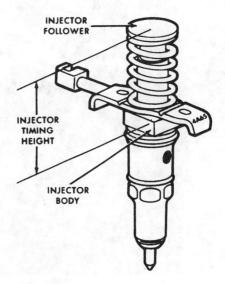

FIGURE 17–38 Injector timing height. (Courtesy of Detroit Diesel Corporation.)

Timing Fuel Injector

After the exhaust valve clearance has been adjusted, time the fuel injectors as follows:

1. Place the governor speed control lever in the *idle*-speed position. If a stop lever is provided, secure it in the *stop* position.

2. Rotate the crankshaft, manually or with the starting motor, until the exhaust valves are fully depressed on the particular cylinder to be timed.

NOTE: If a wrench is used on the crankshaft bolt at the front of the engine, do not turn the crankshaft in a left-hand direction, or the bolt may be loosened.

3. Place the small end of the injector timing gauge in the hole provided in the top of the injector body with the flat of the gauge toward the injector follower (**Figure 17–39**). Refer to the service manual for the correct timing gauge.

4. Loosen the injector rocker arm push rod locknut.

5. Turn the push rod and adjust the injector rocker arm until the extended part of the gauge will just pass over the top of the injector follower.

6. Hold the push rod and tighten the locknut. Check the adjustment and, if necessary, readjust the push rod.

7. Time the remaining injectors in the same manner.

A hole located in the injector body, on the side opposite the identification tag, may be used to visually determine if the injector rack and gear are correctly timed. When the rack is all the way in (full-fuel position), the flat side of the plunger will be visible in the hole, indicating that the injector is "in time." If the flat side of the plunger does not come into full view and appears in the "advanced" or "retarded" position, disassemble the injector and correct the rack-to-gear timing.

Setting Injector Timing Height (8.2 L Four-Stroke-Cycle Engine)

The 8.2 L engine may have different specifications for the timing height of each injector. This results from the precision timing method used at the factory or engine rebuilding shop to determine the precise dimension for each cylinder separately. This procedure ensures that fuel injection will begin at exactly the same time for each cylinder in relation to its piston and cam follower position. The dimension for each injector is provided for each injector on a label attached to the rocker cover.

The injector timing height checking and adjusting procedure requires the use of a dial indicator–type injector timing gauge and fixture and includes the following steps. Follow service manual instructions for details.

1. Calibrate the dial indicator using the master setting fixture (**Figure 17–40**).

2. Rotate the engine to position the No. 1 injector follower 0.20 in. (0.5 mm) below the top of its stroke.

3. Install the dial indicator tool on top of the No. 8 injector follower (**Figure 17–41**).

4. Compare the dial gauge reading with the dimension for No. 8 on the rocker cover.

5. If the dimension is incorrect, adjust the injector rocker arm adjusting screw until the dimension is correct. Tighten the locknut and recheck the dimension.

6. Repeat the procedure on cylinders 4, 3, and 6.

7. Rotate the engine to position the No. 6 follower 0.20 in. (0.5 mm) below the top of its stroke.

8. Check the dimension for cylinders 5, 7, 2, and 1 and adjust as required in the same manner.

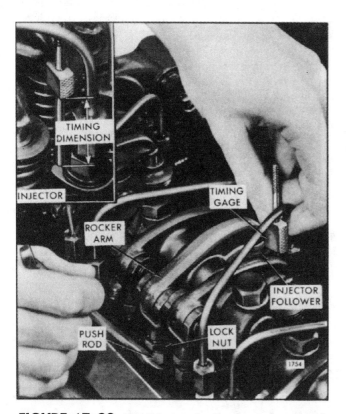

FIGURE 17–39 Checking and adjusting injector timing. (Courtesy of Detroit Diesel Allison.)

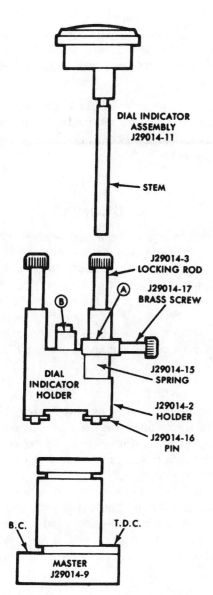

FIGURE 17-40 Calibrating fixture for dial indicator injector timing gauge. (Courtesy of Detroit Diesel Corporation.)

GOVERNOR GAP ADJUSTMENT

The governor gap provides for a gradual transition from low-idle speed to operator-actuated throttle linkage movement. The governor gap separates the travel of the governor weights between the low-speed and high-speed springs. Insufficient governor gap reduces governor weight travel in the low-speed range and may result in the engine stalling during deceleration. A governor gap that is too large reduces governor weight travel in the high-speed range and may cause the maximum no-load speed to exceed specifications.

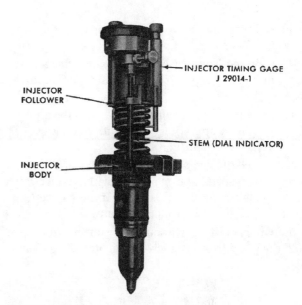

FIGURE 17-41 Dial indicator injector timing gauge in position over injector on 8.2 L four-stroke-cycle engine. (Courtesy of Detroit Diesel Corporation.)

The dimension of the governor gap varies with the type of mechanical governor used. The adjustment procedure also varies. The following procedure is typical for a 6V engine. Refer to the appropriate service manual for actual procedures and specifications.

After adjusting the exhaust valves and timing the fuel injectors, adjust the governor and position the injector rack control levers.

Before proceeding with the governor and injector rack adjustments, disconnect any supplementary governing device. On turbocharged engines, the fuel (air box) modulator lever and roller assembly must be positioned free from cam contact. After the adjustments are completed, reconnect and adjust the supplementary governing device.

With the engine stopped and at operating temperature, adjust the governor gap as follows:

1. Remove the high-speed spring retainer cover.
2. Back out the buffer screw or deenergize the fast-idle cylinder until it extends approximately ⅝ in. from the locknut.

NOTE: Do not back the buffer screw out beyond the limits given, or the control link lever may disengage the differential lever.

3. Start the engine and loosen the idle-speed adjusting screw locknut. Then adjust the idle screw to obtain the desired idle speed. Hold the screw and tighten the locknut to hold the adjustment. Governors used in turbocharged engines include a starting aid screw and

locknut threaded into the governor gap adjusting screw **(Figure 17–42)**.

4. Stop the engine. Clean and remove the governor cover and lever assembly and the valve rocker covers. Discard the gaskets.

5. Start and run the engine between 1100 and 1300 rpm by manual operation of the differential lever. *Do not overspeed the engine.*

6. Check the gap between the low-speed spring cap and the high-speed spring plunger with a feeler gauge. The gap should be 0.002–0.004 in. If the gap setting is incorrect, reset the gap adjusting screw.

7. On governors without the starting aid screw, hold the gap adjusting screw and tighten the locknut.

NOTE: Governors that include a starting aid screw threaded into the end of the gap adjusting screw do not require a locknut, because both screws incorporate a nylon patch in lieu of a locknut.

8. Recheck the gap with the engine operating between 1100 and 1300 rpm and readjust if necessary.

9. Stop the engine and, using a new gasket, install the governor cover.

CAUTION:

Before starting an engine after an engine speed control adjustment, or after removal of the engine governor cover and lever assembly, the technician must determine that the injector racks move to the **no-fuel** position when the governor stop lever is placed in the **stop** position. Engine overspeed will result if the injector racks cannot be positioned at no fuel with the governor stop lever. An overspeeding engine can result in engine damage that could cause personal injury.

An electric shutdown solenoid **(Figure 17–43)** is mounted on the engine governor. The solenoid must be removed from the governor prior to connecting the governor linkage to the rack control shaft, or engine overspeed may occur.

This precaution is necessary because the solenoid is designed to hold the fuel shutdown in the *stop* position at all times except when the solenoid is energized. If the injector racks are held in the *full-fuel* position and connected to the governor while the shutdown is in the

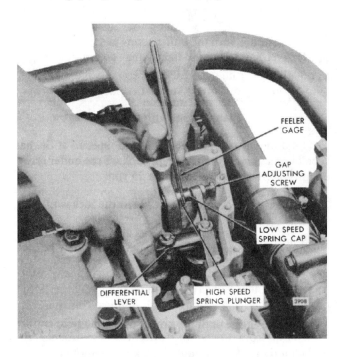

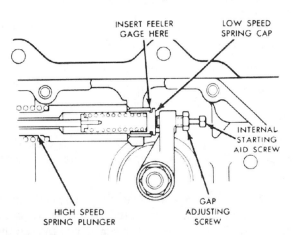

FIGURE 17–42 Governor gap adjustment. (Courtesy of Detroit Diesel Corporation.)

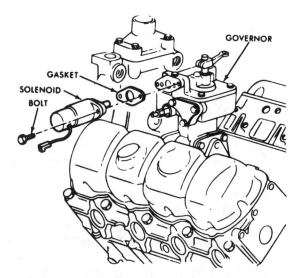

FIGURE 17–43 Shutdown solenoid on governor. (Courtesy of Detroit Diesel Corporation.)

stop position, the injector will be locked at full rack (maximum fuel).

SETTING THE GAP—DOUBLE-WEIGHT GOVERNOR

NOTE: Before adjusting the gap on TT governors, the Belleville spring retainer nut must be backed out until there is approximately 0.060 in. clearance between the washers and the retainer nut (see **Figure 17–44**).

1. Disconnect any supplementary governor devices.
2. Set the engine idle speed at 600 rpm and stop the engine.

CAUTION:

Disconnect the grounded battery cable(s) to prevent accidental engine cranking and possible personal injury while the gap is being checked or set.

3. Clean and remove the governor cover. Discard the gasket.
4. With the engine stopped, manually bar the engine over *in the direction of normal engine rotation* until the governor weights are in a horizontal position.
5. Insert governor weight wedge tool J35516 between the low-speed weight and the governor riser **(Figures 17–45 and 17–46)**. The tapered face of the

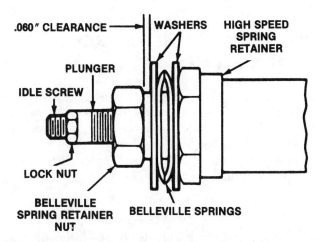

FIGURE 17–44 Belleville spring adjustment backed off to 0.060 in. clearance. (Courtesy of Detroit Diesel Corporation.)

FIGURE 17–45 Governor weight wedge tool. (Courtesy of Detroit Diesel Corporation.)

FIGURE 17–46 Wedge tool between riser and low-speed weight. (Courtesy of Detroit Diesel Corporation.)

wedge should be against the riser and positioned between the flanges on the ends of the riser.

6. Push the wedge as far to the bottom as it will go, forcing the weights against the maximum travel stop.

NOTE: Do not use a screwdriver to pry the weights, because damage to the weights, riser, or housing could result.

7. Use a feeler gauge to set the gap between the low-speed spring cap and the high-speed spring plunger at 0.008 in. Then tighten the governor gap adjusting screw locknut **(Figure 17–47)**.
8. Push down on the governor weight wedge tool to be sure it did not move while the gap was being set. Recheck the gap while holding the tool in this position. If the gap is incorrect, reset to 0.008 in., repeating the steps outlined above.
9. Remove the wedge.

NOTE: The buffer, idle-speed, no-load speed and starting aid screws, the injector racks, and supplemental governor

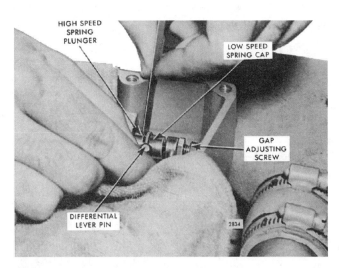

FIGURE 17–47 Adjusting governor gap. (Courtesy of Detroit Diesel Corporation.)

devices require adjustment whenever the governor gap is changed.

10. Reset the Belleville springs as follows.

Belleville Spring Adjustment

The tailored torque feature described earlier limits the fuel delivery at the maximum full-load rated speed. The Belleville spring adjustment must be made to achieve the specified horsepower for the engine within a 5% variation limit. Two methods are used to make the adjustment: the idle drop method and the power reduction method.

IDLE DROP METHOD

The idle drop method requires the use of an accurate tachometer (not the one in the vehicle) to ensure accurate results. A 2–3 HP error occurs with only a 1 rpm error.

The general procedure includes the following steps.

1. Obtain the idle drop setting specifications for the engine type, injector size, and governor type.

2. Run the engine until it reaches normal operating temperature.

3. Adjust the no-load speed to specifications.

4. Disconnect the throttle linkage at the governor speed control lever.

5. Adjust the idle speed to the specifications required for the idle drop method. This setting must be exact.

6. Turn the Belleville spring retainer nut clockwise until the exact specified idle-speed drop is achieved, then secure the setting with the locknut.

7. Adjust the idle speed to specifications using the idle-speed adjusting screw.

8. Adjust the buffer screw and starting aid screw to specifications.

POWER REDUCTION METHOD

The power reduction method of adjusting the TT governor setting requires the use of a dynamometer. The general procedure includes the following steps.

1. Obtain the power reduction factor for engine type, injector size, and governor type.

2. Adjust the Belleville spring retainer nut to achieve 0.060 in. (1.524 mm) clearance.

3. Run the engine until it reaches normal operating temperature.

4. Set the no-load speed to specifications.

5. Measure and record the full throttle horsepower at 100 rpm below rated engine speed using the dynamometer.

6. Multiply the power reduction factor by the horsepower reading obtained in step 5.

7. Turn the Belleville spring retainer nut clockwise until the horsepower is reduced to that recorded in step 5.

8. Adjust the idle speed to specifications using the idle-speed adjusting screw.

9. Adjust the buffer screw and starting aid screw to specifications.

POSITIONING INJECTOR RACK CONTROL LEVERS

The position of the injector racks must be correctly set in relation to the governor. Their position determines the amount of fuel injected into each cylinder and ensures equal distribution of the load.

Properly positioned injector rack control levers with the engine at full load will result in the following:

1. Speed control lever at the full-fuel position.

2. Governor low-speed gap closed.

3. High-speed spring plunger on the seat in the governor control housing.

4. Injector fuel control racks in the full-fuel position.

The letters R or L indicate the injector location in the right or left cylinder bank, viewed from the rear of the engine. Cylinders are numbered starting at the front of the engine on each cylinder bank. Adjust the No. 1L injector rack control lever first to establish a guide for adjusting the remaining injector rack control levers.

1. Disconnect the linkage attached to the speed-control lever.

2. Turn the idle-speed adjusting screw until about ½ in. of the threads (12–14 threads) project from the locknut when the nut is against the high-speed plunger. This adjustment lowers the tension of the low-speed spring so it can be compressed while closing the low-speed gap without bending the fuel rods.

NOTE: A false fuel rack setting may result if the idle-speed adjusting screw is not backed out as noted above.

3. If not already done, back out the buffer screw.

4. Remove the clevis pin from the fuel rod and the right cylinder bank injector control tube lever.

5. Loosen all the inner and outer injector rack control lever adjusting screws on both injector control tubes. Be sure all the injector rack control levers are free on the injector control tubes.

6. Move the speed control lever to the full-fuel position and hold it in that position with light finger pressure. Turn the inner adjusting screw of the No. 1L injector rack control lever down **(Figure 17–48)** until a slight movement of the control tube lever is observed or a step-up in effort to turn the screwdriver is noted.

This will place the No. 1L injector in the full-fuel position. Turn down the outer adjusting screw until it bottoms lightly on the injector control tube. Then alternately tighten both the inner and outer adjusting screws. This should result in placing the governor linkage and the control tube assembly in the same positions they will attain while the engine is running at full load as previously described.

NOTE: Overtightening of the injector rack control lever adjusting screws during installation or adjustment can result in damage to the injector control tube.

7. To be sure of the proper rack adjustment, hold the speed control lever in the full-fuel position and press down on the injector rack with a screwdriver or finger tip and note the "rotating" movement of the injector control rack when the speed control lever is in the full-fuel position **(Figure 17–49)**. Hold the speed control lever in the full-fuel position and, using a screwdriver, press downward on the injector control rack. The rack should tilt downward **(Figure 17–50)**, and when the pressure of the screwdriver is released, the control rack should "spring" back upward.

The setting is sufficiently tight if the rack returns to its original position. If the rack does not return to its original position, it is too loose. To correct this condition, back off the outer adjusting screw slightly and tighten the inner adjusting screw.

The setting is too tight if, when moving the speed control lever from the idle-speed to the maximum-speed position, the injector rack becomes tight before the speed control lever reaches the end of its travel (stop under the governor cover). This will result in a step-up in effort to move the speed control lever

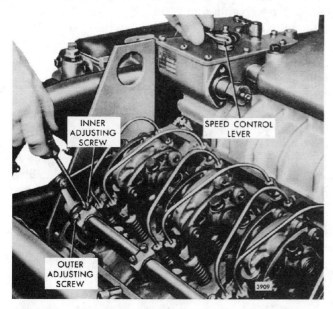

FIGURE 17–48 Positioning No. 1L rack control lever. (Courtesy of Detroit Diesel Corporation.)

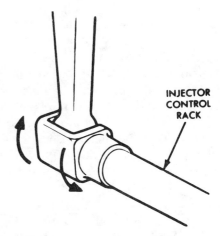

FIGURE 17–49 Checking rotating movement of injector control rack. (Courtesy of Detroit Diesel Corporation.)

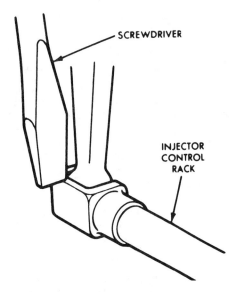

FIGURE 17–50 Checking control rack spring. (Courtesy of Detroit Diesel Corporation.)

to its maximum-speed position and a deflection in the fuel rod (fuel rod deflection can be seen at the bend). If the rack is too tight, back off the inner adjusting screw slightly and tighten the outer adjusting screw.

8. Remove the clevis pin from the fuel rod and the left bank injector control tube lever.

9. Insert the clevis pin in the fuel rod and the right cylinder bank injector control tube lever and position the No. 1R injector rack control lever as previously outlined in step 6 for the No. 1L injector rack control lever.

10. Insert the clevis pin in the fuel rod and the left cylinder bank injector control tube lever. Repeat the check on the No. 1L and No. 1R injector rack control levers as outlined in step 7. Check for and eliminate any deflection that occurs at the bend in the fuel rod where it enters the cylinder head.

11. Manually hold the No. 1L injector rack in the full-fuel position and turn down the inner adjusting screw of the No. 2L injector rack control lever until the injector rack of the No. 2L injector has moved into the full-fuel position. Turn the outer adjusting screw until it bottoms lightly on the injector control tube. Then alternately tighten both the inner and outer adjusting screws.

12. Recheck the No. 1L injector rack to be sure it has remained snug on the ball end of the rack control lever while positioning the No. 2L injector rack. If the rack of the No. 1L injector has become loose, back off the inner adjusting screw slightly on the No. 2L injector rack control lever. Tighten the outer adjusting screw.

When the settings are correct, the racks of both injectors must be snug on the ball end of their respective rack control levers.

13. Position the No. 3L and 4L injector rack control levers as outlined in steps 11 and 12.

14. Position the No. 2R, 3R, and 4R injector racks as outlined above for the left cylinder bank.

15. Turn the idle speed adjusting screw in until it projects 3/16 in. from the locknut to permit starting of the engine.

CAUTION:

Before starting an engine after an engine speed control adjustment, or after removal of the engine governor cover and lever assembly, the service technician must determine that the injector racks move to the no-fuel position when the governor stop lever is placed in the stop position. Engine overspeed will result if the injector racks cannot be positioned at no fuel with the governor stop lever. An overspeeding engine can result in engine damage that can cause personal injury.

---

16. Use new gaskets and reinstall the valve rocker covers.

STARTING AID SCREW ADJUSTMENT

The internal starting aid screw is threaded into the governor gap adjusting screw. This screw is adjusted to position the injector racks at less than full fuel when the governor speed control lever is in the idle position. The reduced fuel makes starting easier and reduces the amount of smoke on start-up.

NOTE: The effectiveness of the starting aid screw will be eliminated if the speed control lever is advanced to wide open throttle during starting.

After the normal governor running gap of 0.0015 in. has been set and the injector racks positioned, adjust the starting aid screw as follows:

1. With the engine stopped, place the governor stop lever in the run position and move the speed control lever to the idle position.

2. Hold the gap adjusting screw, to keep it from turning, and adjust the starting aid screw to obtain 0.330 in. to 0.360 in. clearance between the shoulder on the No. 1L injector rack clevis and the injector body, with the head of the starting aid screw against the governor wall.

With the engine stopped, this adjustment will provide a gap of 0.155 in. to 0.160 in. between the high-speed spring plunger and the low-speed spring cap **(Figure 17–51)**.

3. Move the stop lever to the stop position, with the speed control lever still in the idle position, and return it to the run position.

4. Recheck the injector rack clevis-to-body clearance. Movement of the governor stop lever is to take up clearances in the governor linkage. The clevis-to-body clearance can be increased by backing out the starting aid screw or reduced by turning it farther into the gap adjusting screw.

5. Start the engine and recheck the running gap (0.0015 in.) and, if necessary, reset it. Then stop the engine.

MAXIMUM NO-LOAD ENGINE SPEED ADJUSTMENT

All governors are properly adjusted before leaving the factory. However, if the governor has been reconditioned or replaced, and to ensure the engine speed will not exceed the recommended no-load speed as given on the engine name plate, set the maximum no-load speed as follows:

Type A Governor Springs (Figures 17–52 and 17–53)

1. Loosen the locknut with a spanner wrench and back off the high-speed spring retainer several turns. Then start the engine and increase the speed slowly. If the speed exceeds the required no-load speed before the speed control lever reaches the end of its travel, back off the spring retainer a few additional turns.

2. With the engine at operating temperature and no load on the engine, place the speed control lever in the maximum-speed position. Turn the high-speed spring retainer in until the engine is operating at the recommended no-load speed. Use an accurate hand tachometer to determine the engine speed. The maximum no-load speed varies with the full-load operating speed.

3. Hold the high-speed spring retainer and tighten the locknut.

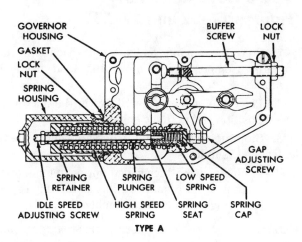
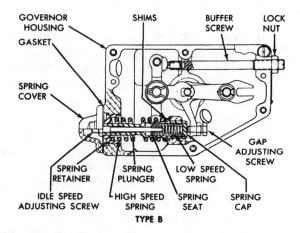

FIGURE 17–52 Type A and type B governor spring assemblies. (Courtesy of Detroit Diesel Corporation.)

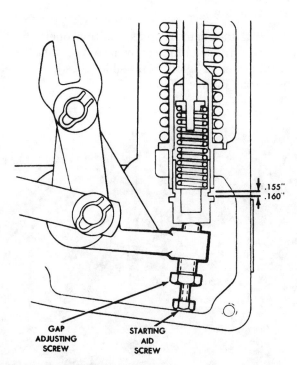

FIGURE 17–51 Starting aid screw adjustment. (Courtesy of Detroit Diesel Corporation.)

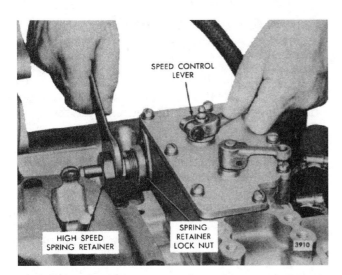

FIGURE 17–53 Adjusting maximum no-load speed. (Courtesy of Detroit Diesel Corporation.)

Type B Governor Springs (Figures 17–52 and 17–53)

1. Start the engine, and after it reaches normal operating temperature, remove the load from the engine.

2. Place the speed control lever in the maximum-speed position and note the engine speed.

3. Stop the engine, and if necessary, adjust the no-load speed as follows:

   a. Remove the high-speed spring retainer with tool J 5895 and withdraw the high-speed spring-and-plunger assembly.

NOTE: To prevent the low-speed spring and cap from dropping into the governor, be careful not to jar the assembly while it is being removed.

   b. Remove the high-speed spring from the high-speed spring plunger and add or remove shims as required to establish the desired engine no-load speed. For each 0.010 in. in shims added, the engine speed will be increased approximately 10 rpm.

   c. Install the high-speed spring on the plunger and install the spring assembly in the governor housing. Tighten the spring retainer securely. The maximum no-load speed varies with the full-load operating speed desired.

   If the full-load speed is to be 2800 rpm, then the no-load speed setting should be 2940 rpm (2800 rpm + 140 rpm) to ensure the governor will move the injector racks into the full-fuel position at the desired full-load speed.

   d. Start the engine and recheck the no-load speed. Repeat the procedure, as necessary, to establish the no-load speed required.

IDLE SPEED ADJUSTMENT

With the maximum no-load speed properly set, adjust the idle speed as follows:

1. With the engine running at normal operating temperature and with the buffer screw backed out to avoid contact with the differential lever, turn the idle-speed adjusting screw until the engine is operating at approximately 15 rpm below the recommended idle speed. The recommended idle speed is 500–600 rpm, but it may vary with the engine application **(Figure 17–54)**.

   If the engine has a tendency to stall during deceleration, install a new buffer screw. The current buffer screw uses a heavier spring and restricts the travel of the differential lever to the OFF *(no-fuel)* position.

2. Hold the idle screw and tighten the locknut.

3. Install the high-speed spring retainer cover and tighten the two bolts.

BUFFER SCREW ADJUSTMENT

With the idle speed properly set, adjust the buffer screw as follows:

1. With the engine running at normal operating temperature, turn the buffer screw in so it contacts the differential lever as lightly as possible and still eliminates engine roll **(Figure 17–55)**.

FIGURE 17–54 Adjusting engine idle speed. (Courtesy of Detroit Diesel Corporation.)

THROTTLE DELAY ADJUSTMENT

3. Hold the buffer screw and tighten the locknut.

After the governor adjustments are completed, adjust any supplementary governing device that may be used.

THROTTLE DELAY ADJUSTMENT

Whenever the injector rack control levers are adjusted, disconnect the throttle delay mechanism by loosening the U-bolt that clamps the lever to the injector control tube. After the injector rack control levers have been positioned, the throttle delay mechanism must be readjusted. With the engine stopped, proceed as follows:

1. Disconnect the throttle delay mechanism by loosening the U-bolt that clamps the lever to the injector control tube **(Figure 17–56)**.

2. To provide adequate lubrication of mechanical components, fill the throttle delay reservoir with clean engine oil. The oil reservoir does not have to remain full during the entire adjustment procedure.

3. Insert the appropriate throttle delay timing gauge on the rack between the injector body rack hole counterbore and the shoulder on the injector rack clevis. This is the No. 2 injector on 6V and 8V engines, the No. 5

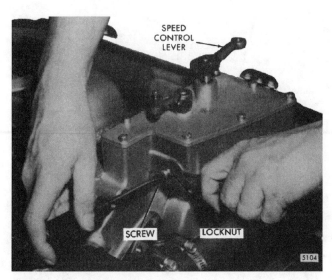

FIGURE 17–55 Adjusting the buffer screw. (Courtesy of Detroit Diesel Corporation.)

NOTE: Do not increase the engine idle speed more than 15 rpm with the buffer screw.

2. Recheck the maximum no-load speed. If it has increased more than 25 rpm, back off the buffer screw until the increase is less than 25 rpm.

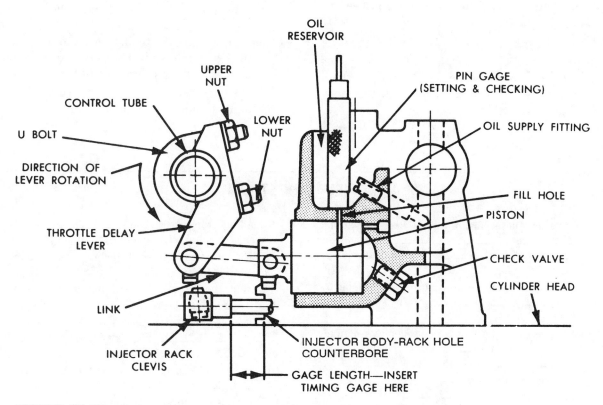

FIGURE 17–56 Throttle delay adjustment. (Courtesy of Detroit Diesel Corporation.)

injector on V-12 engines, and the No. 6 injector on 16V engines.

4. Hold the governor throttle lever in the maximum speed position. This should cause the injector rack to move toward the full-fuel position.

5. Insert pin gauge J 25558 with the "go" (green 0.069 in.) end in the cylinder fill hole. If the throttle delay housing has multiple holes, use the hole indicated in the service manual.

6. Rotate the throttle delay lever in the direction shown in **Figure 17–56** until further movement is limited by the piston contacting the pin gauge.

7. Tighten the U-bolt while exerting a slight pressure on the lever in the direction of rotation.

8. Check the setting as follows:
   a. Remove the pin gauge.
   b. Reinsert the "go" (green 0.069 in.) end of the gauge in the fill hole. If the gauge will not go past the piston without resistance, increase the torque on the lower U-bolt nut. Remove the gauge.

FUEL MODULATOR ADJUSTMENT (FIGURE 17–57)

After completing the injector rack control lever and governor adjustment, adjust the fuel modulator, as follows:

1. With the engine stopped, insert the proper gauge between the injector body and the shoulder on the injector rack of the No. 2 injector, which is adjacent to the modulator.

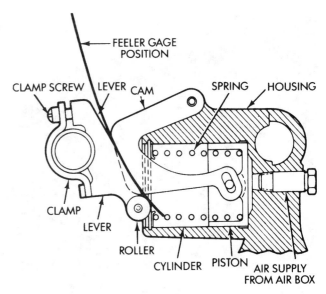

FIGURE 17–57 Fuel modulator adjustment. (Courtesy of Detroit Diesel Corporation.)

2. Position the governor speed control lever in the maximum-speed position and the governor run stop lever in the run position.

3. Rotate the air box modulator lever assembly and clamp on the injector control tube until the lever roller contacts the modulator cam with sufficient force to take up the pin clearance.

4. Check to make sure only the roller contacts the cam and not the lever stamping. Tighten the lever and clamp screw. After tightening, check to make sure that the gauge is still in contact with the injector body and the shoulder on the injector rack of the No. 2 injector.

5. Remove the gauge from between the injector body and the shoulder on the injector rack.

REVIEW QUESTIONS

1. The Detroit Diesel fuel injection system combines _____ _____ _____, fuel metering and _____ in one assembly called a mechanical unit injector.

2. The unit injector plungers are actuated for pumping by a _____ _____ on the engine camshaft.

3. Fuel output of the injector is controlled by rotation of the plunger by the _____ _____ and the position of the _____.

4. The lower helix controls the _____ _____ timing.

5. All injectors must begin _____ at exactly the same point in relation to injector timing.

6. The internal diameter of the bushings (barrels) is measured with an _____ _____.

7. Name the four functions the unit fuel injector performs.

8. Explain unit injector operation.

9. The governor is mounted on the front end of the _____ and is driven by one of the _____ _____.

10. Two types of mechanical governors are _____ _____ and _____ _____.

11. What is the purpose of the fuel modulator?

12. The throttle delay mechanism is used to delay full fuel injection when the engine is _____.
13. Explain the injector fuel output test.
14. Do *not* _____ _____ of one injector with those of another.
15. Explain how to reassemble an injector after cleaning.
16. Identify the steps in reinstalling a reconditioned injector.
17. Identify the 10 steps in tuning-up a mechanical governor.
18. To time an injector properly, the injector _____ must be adjusted to a _____ _____ in relation to the injector body.
19. The dimension of the governor gap varies with the _____ _____ _____ governor used.
20. The idle drop method requires the use of an _____ _____ to ensure accurate results.
21. The power reduction method of setting the TT governor requires the use of a _____.
22. Identify the eight steps in adjusting throttle delay.

TEST QUESTIONS

1. Helix-type plungers are controlled by
   a. the fuel pressure
   b. the fuel control rack
   c. the supply pump
   d. all of the above
2. On two-stroke-cycle engines fuel is injected into the cylinder every
   a. two crankshaft revolutions
   b. two camshaft revolutions
   c. camshaft revolution
   d. two camshaft revolutions
3. The Detroit Diesel mechanical unit injector
   a. develops its own high pressure
   b. has a low-pressure supply pump
   c. has a plunger and helix
   d. all of the above
4. All injectors must begin injection at exactly the same point in relation to
   a. piston travel
   b. engine rpm
   c. camshaft rotation
   d. throttle setting
5. The internal diameter of the bushings of an injector is measured with a(n)
   a. air gauge
   b. feeler gauge
   c. dial gauge
   d. micrometer
6. The unit injector requires
   a. steel high-pressure steel lines
   b. steel high-pressure lines of exactly the same length
   c. no high-pressure fuel lines
   d. less fuel
7. A continuous flow of fuel through the injector
   a. prevents air pockets
   b. acts as a coolant
   c. assists in lubrication
   d. all of the above
8. When sufficient pressure is built up, the injector valve is lifted off its seat and fuel is
   a. forced through small orifices in the spray tip
   b. atomized into the combustion chamber
   c. forced through the spray tip at the exact correct time
   d. all of the above
9. Two types of mechanical governors used are
   a. limited- and variable-speed
   b. minimum-speed
   c. maximum-speed
   d. idle-speed control
10. Fuel modulators are on certain
    a. types of load conditions
    b. turbocharged aftercooled engines
    c. stationary engines
    d. high rpm engines
11. Throttle delay mechanisms are used to retard fuel injection when the engine is
    a. accelerated
    b. overheated
    c. used for braking
    d. overloaded
12. TT governors are used to
    a. keep the vehicle at constant speed
    b. maintain maximum rpm
    c. maintain constant horsepower
    d. avoid excess fuel supply
13. To remove the injector for cleaning and reconditioning you must remove
    a. the rocker arm and cover
    b. the fuel lines
    c. two rocker arm shaft brackets
    d. all of the above
14. Incorrect injector output could be caused by
    a. partially plugged spray tip orifices
    b. enlarged spray tip orifices
    c. a worn plunger and bushing
    d. all of the above

CHAPTER 18

DETROIT DIESEL ELECTRONIC CONTROL SYSTEMS

INTRODUCTION

The Detroit Diesel Electronic Control (DDECI) system was the first such system offered by a U.S. heavy-duty diesel engine manufacturer. It was introduced in 1985. This system was replaced by the improved DDEC II and later by the DDEC III system.

The DDEC system uses electronically controlled unit injectors (EUI). Input signals from various input switches and sensors are sent to the electronic control module (ECM). The ECM processes this information in consultation with stored programmed information and sends electrical command pulses to the power module section of the ECM. This high-current switching unit provides the voltage to operate the EUI solenoids. The ECM can limit engine power or shut the engine down completely (depending on selected options) should engine lubrication pressure drop too low, engine oil temperature be too high, or the engine coolant level drop too low. In the DDEC I the main control module (ECM) and electronic distributor unit (EDU) are separate units. The DDEC II and DDEC III combine both units in a single ECM.

DDEC systems control injection timing and fuel delivery electronically. DDEC systems are used on the four-stroke-cycle series 60 engines and on the two-stroke-cycle series 71 and 92 engines. Operation and service are the same on different engines, although the locations of some components differ.

PERFORMANCE OBJECTIVES

After thorough study of this chapter and the appropriate training models, service manuals, and equipment, you should be able to:

1. List the advantages of the DDEC system over nonelectronic systems.

2. List the programmable options available with DDEC.

3. List the DDEC III system components and state their function.

4. Describe DDEC system fuel flow.

5. Diagnose system faults by accessing trouble codes with the check engine light (CEL) and the digital diagnostic reader (DDR).

6. Remove and install a DDEC unit injector.

TERMS YOU SHOULD KNOW

Look for these terms as you study this chapter, and learn what they mean.

DDEC	SRS
EUI	TRS
EDU	CTS
ECM	CLS
EEPROM	OPS
TPS	OTS
TBS	FPS

DDEC FUNCTION AND DEVELOPMENT

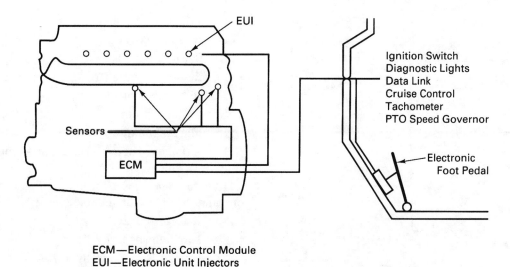

FIGURE 18–1 Detroit Diesel Electronic Control (DDEC) system. (Courtesy of Detroit Diesel Corporation.)

FTS
ATS
engine protection system
data communications link
DDR
Pro-Link
fault codes

CEL
SEL
STEO
hold-down crab
crab washer
injector height

DDEC FUNCTION AND DEVELOPMENT

The Detroit Diesel Electronic Control (DDEC) system electronically controls the timing of fuel injection and the amount of fuel injected. It monitors several engine functions with sensors that send electrical signals to the ECM **(Figure 18–1)**. The DDEC offers an engine protection system that displays warnings or provides engine shutdown should damaging engine conditions occur. There are three generations of the DDEC system.

1. *DDEC I:* The electronic control module (ECM) and electronic distributor unit (EDU) are separate units. The EDU is fuel-cooled **(Figure 18–2)**.

2. *DDEC II:* Has an engine-mounted ECM containing the EDU components of the earlier DDEC I and an electronic erasable programmable read-only memory (EEPROM) unit. Some units are fuel-cooled depending on application **(Figure 18–3)**.

3. *DDEC III:* This version of the DDEC ECM is 50% smaller, eight times faster, has seven times more memory, and twice the number of features in comparison

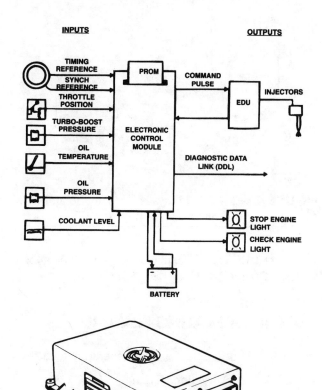

FIGURE 18–2 DDEC I system schematic and ECM. (Courtesy of Detroit Diesel Corporation.)

370 Chapter 18 DETROIT DIESEL ELECTRONIC CONTROL SYSTEMS

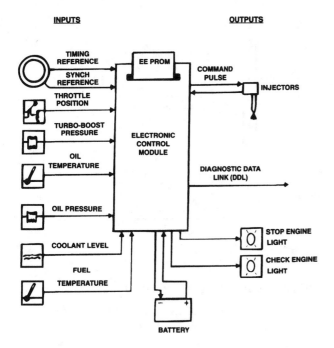

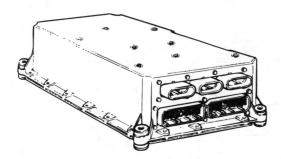

FIGURE 18–3 DDEC II system schematic and ECM. (Courtesy of Detroit Diesel Corporation.)

10. Vehicle ID number
11. Pressure governor
12. Starter lockout
13. Remote throttle PTO control
14. High-idle control
15. Idle adjustment
16. Idle timer shutdown
17. Auxiliary engine protection—vehicle sensors shut down engine
18. Customer password: four-digit password, alphanumeric digits, 1.6 million combinations
19. Rating security
20. Maximum security
21. Low DDEC voltage light
22. Low coolant light
23. Deacceleration light
24. 12 V or 24 V ECM
25. Communications links; SAE J1587, J1922, J1939

with the DDEC I and II, and includes the self-diagnostic system **(Figure 18–4)**.

DDEC III OPERATING FEATURES

The DDEC III offers the following operating features:

1. Engine protection system
2. Cruise control
3. Cruise control automatic resume with double clutching
4. Progressive engine braking in cruise control
5. Fan controls
6. Engine fan braking
7. Progressive shifting
8. Vehicle speed limiting
9. Vehicle overspeed diagnostics

DDEC III COMPONENT DESCRIPTIONS

1. *Electronic control module (ECM):* receives electrical signals from engine and vehicle sensors and switches. It processes this information in consultation with programmed information in ECM memory and provides electrical power to operate the EUI. The ECM runs tests on all inputs and outputs and stores any malfunctions in ECM memory for diagnostic purposes **(Figures 18–2 to 18–4)**.

2. *Electronic erasable programmable read-only memory (EEPROM):* that part of the ECM in which programmed information can be erased electrically and new information programmed into it to user-desired specifications.

3. *Throttle position sensor (TPS):* sends a voltage signal to the ECM that varies with the amount the throttle pedal is depressed. Output voltage is low at idle and increases as the pedal is pressed farther downward. An increasing voltage signal to the ECM results in increased fuel delivery (longer injector ON time) **(Figure 18–5)**.

4. *Turbo boost sensor (TBS):* monitors turbocharger output air pressure. The ECM uses this information to adjust fuel delivery to control exhaust smoke during acceleration **(Figure 18–6)**.

5. *Synchronous reference sensor (SRS):* sends a signal to the ECM once per cylinder. A raised metal pin on the bull gear of the engine gear train generates the signal. It is generated as the No. 1 piston reaches ap-

DDEC III COMPONENT DESCRIPTIONS 371

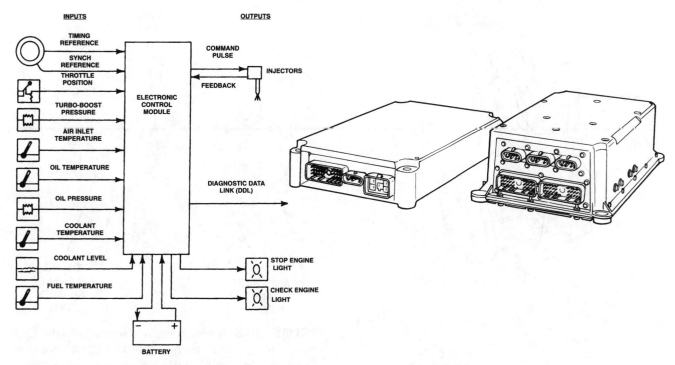

FIGURE 18-4 DDEC III system schematic and ECM. (Courtesy of Detroit Diesel Corporation.)

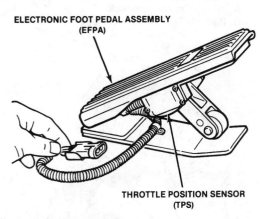

FIGURE 18-5 Throttle position sensor provides information on throttle position to the ECM. (Courtesy of Detroit Diesel Corporation.)

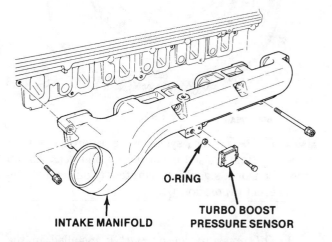

FIGURE 18-6 Turbo boost pressure sensor sends a voltage signal to the ECM to control boost pressure. (Courtesy of Detroit Diesel Corporation.)

proximately 45° BTDC. The ECM uses this signal to determine engine speed **(Figure 18-7)**.

6. *Timing reference sensor (TRS):* mounted on the engine gear case, it extends into the gear case near the teeth of the timing wheel. A series of evenly spaced teeth on the timing wheel generate the signal, which is sent to the ECM. A tooth passes the TRS each time a piston reaches the 10° BTDC position. The ECM uses these signals to determine injector solenoid operating times **(Figure 18-8)**.

7. *Coolant temperature sensor (CTS):* sends a variable-voltage signal to the ECM based on coolant temperature and sensor resistance. If coolant temperature rises above the specified level, the ECM activates the stop engine or warning function.

8. *Coolant level sensor (CLS):* sends a signal to the ECM when coolant level drops too low. A low coolant level causes the ECM to activate the stop engine or warning system **(Figure 18-9)**.

372 Chapter 18 DETROIT DIESEL ELECTRONIC CONTROL SYSTEMS

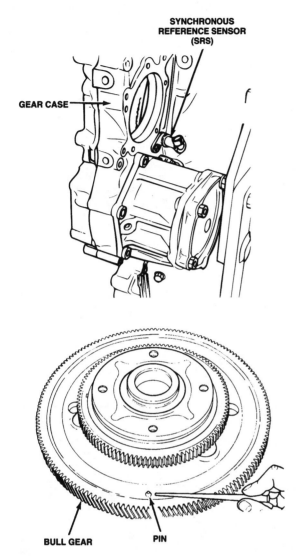

FIGURE 18–7 (Top) synchronous reference sensor (SRS), (bottom) raised metal pin on bull gear. As pin passes sensor a signal is generated. (Courtesy of Detroit Diesel Corporation.)

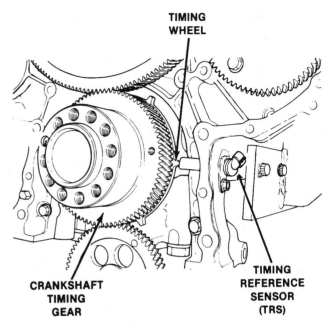

FIGURE 18–8 Timing reference sensor tells the ECM when each piston reaches the 10° BTDC position. (Courtesy of Detroit Diesel Corporation.)

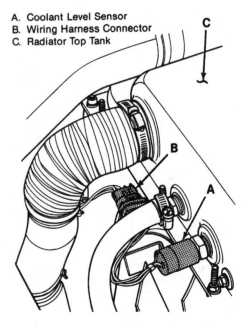

FIGURE 18–9 Coolant level sensor in top radiator tank. (Courtesy of Detroit Diesel Corporation.)

9. *Oil pressure sensor (OPS):* installed in the engine's main oil gallery, it provides information on engine lubricating oil pressure to the ECM. If low oil pressure exceeds 7 seconds, the ECM begins the stop engine or warning function **(Figure 18–10).**

10. *Oil temperature sensor (OTS):* installed in the engine's main oil gallery, the OTS provides information on oil temperature to the ECM. The ECM uses this information to help cold-engine starting. If oil temperature exceeds specifications for 2 seconds or more, the ECM begins the stop engine or warning function **(Figure 18–10).**

11. *Fuel pressure sensor (FPS):* installed at the secondary fuel filter, it provides fuel pressure information to the ECM **(Figure 18–11).**

12. *Fuel temperature sensor (FTS):* installed at the secondary fuel filter, it provides fuel temperature infor-

DDEC III COMPONENT DESCRIPTIONS 373

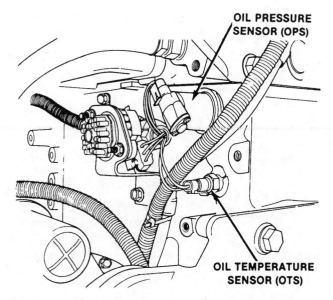

FIGURE 18–10 Oil pressure sensor (OPS) and oil temperature sensor (OTS). (Courtesy of Detroit Diesel Corporation.)

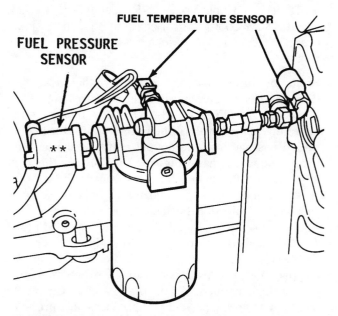

FIGURE 18–11 Fuel temperature sensor (FTS) and fuel pressure sensor (FPS). (Courtesy of Detroit Diesel Corporation.)

mation to the ECM. The ECM uses this information to control fuel delivery based on changes in fuel density caused by changes in fuel temperature **(Figure 18–11)**.

13. *Air temperature sensor (ATS):* located in the intake air manifold, it provides information on intake air temperature entering the engine to the ECM. The ECM uses this information to adjust injection timing to improve cold starting and reduce white exhaust smoke.

14. *Electronic unit injector (EUI):* mechanically operated, electronically controlled injector that injects atomized fuel into the combustion chamber. The ECM operates a solenoid-actuated poppet valve that controls fuel flow in the injector. This allows the ECM to control injection timing and fuel delivery **(Figure 18–12)**.

15. *Wiring harness:* connects electronic control system components to each other and to vehicle wiring.

DDEC Engine Protection System

The engine protection system is designed to protect the engine from major damage should any of the following conditions occur:

- Engine oil pressure falls below specified value.
- Engine lubricating oil temperature exceeds maximum safe value.
- Coolant level drops too low.

If any of these conditions occurs, the stop engine light (SEL) goes on, and the stop engine function is activated **(Figure 18–13)**. Fuel delivery is reduced to reduce engine speed and power. If the problem persists, the engine will shut down 30 seconds after the SEL goes on. This normally gives the driver enough time to pull over and stop away from traffic. The stop engine override

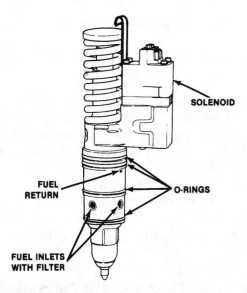

FIGURE 18–12 Detroit Diesel electronic unit injector (EUI). (Courtesy of Detroit Diesel Corporation.)

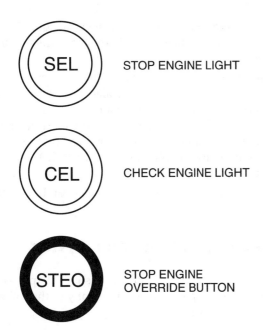

FIGURE 18–13 (Courtesy of FT Enterprises.)

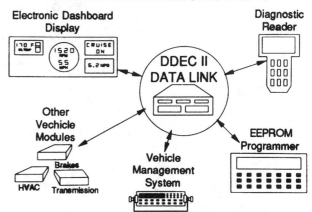

FIGURE 18–14 Series 60 engine DDEC II data links. (Courtesy of Detroit Diesel Corporation.)

button can be held by the driver to provide 30 seconds of override time should additional time be needed to park the vehicle safely.

The check engine light (CEL) turns on to warn the driver that an engine problem exists and that the vehicle should be serviced as soon as possible. The CEL also comes on when the ignition switch is first turned on to verify that the system is operational. The light goes off after the engine is started.

Data Communication Links

Communication links provide connections for the following (Figure 18–14):

1. *Electronic dashboard display:* provides operating information to the driver.
2. *Diagnostic reader:* test equipment used to access trouble codes and diagnose system problems.
3. *Other vehicle control modules:* interface with modules governing brakes, transmission, and HVAC (heating, ventilation, and air conditioning).
4. *Data extraction:* allows data extraction concerning vehicle performance for purposes of vehicle management.

DDEC Fuel Flow and EUI Operation

An engine-driven gear pump draws fuel from the tank through the primary fuel filter (Figure 18–15). The pump pushes fuel under low pressure through the secondary filter via an external line to the drilled supply passage in the cylinder head.

NOTE: Prior to unit No. 6R-8950 fuel flowed through the EDU/ECM cooler plate before entering the cylinder head. The cooler plate was removed from later-model engines.

From the supply passage in the cylinder head fuel flows into each unit injector through two fuel filter inlet screens (Figure 18–16). With the solenoid-operated poppet valve open, fuel flows through a drilled passage into the poppet control valve and injector plunger areas (Figure 18–17). As the cam lobe acts on the cam follower it causes the injector rocker arm to push the injector plunger downward. Just before injection begins, the ECM energizes the injector solenoid, closing the poppet valve (Figure 18–18) and trapping fuel below the plunger. As the plunger continues downward the pressure on the trapped fuel increases very rapidly. When the pressure is high enough, it lifts the needle valve off its seat against spring pressure. This causes high-pressure fuel to be sprayed into the combustion chamber. Because the fuel is forced through tiny holes in the injector tip, the fuel is finely atomized. At the end of the injector pulse width time, the ECM turns off current to the injector solenoid, allowing the spring to open the poppet valve. The resulting pressure drop in the injector causes the spring to seat the needle valve, stopping injection. Fuel then flows through the return and back to the fuel tank. Continuous fuel flow through the injector and return provides component cooling and prevents air pockets from forming in the fuel.

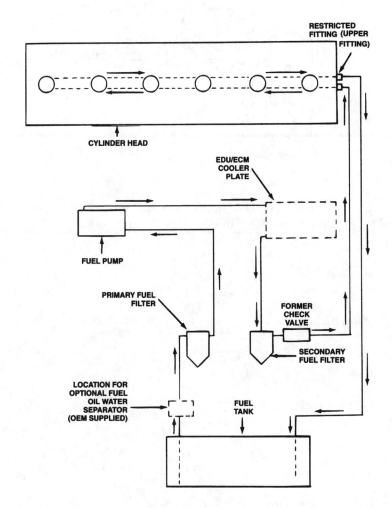

FIGURE 18–15 DDEC system fuel flow schematic. (Courtesy of Detroit Diesel Corporation.)

DDEC DIAGNOSIS AND TESTING

Either of two troubleshooting methods may be used by the technician. One method is to activate the CEL to flash out any stored trouble codes. The other method is to use the diagnostic data reader (DDR). The DDR is also used to program certain customer-specified parameters and to clear trouble codes from the ECM.

The following examples of these procedures are provided here as general information only. Always refer to the appropriate Detroit Diesel service manual and the test equipment instructions for specific procedures and specifications on how to replace and calibrate faulty system components.

Diagnostic Procedure Using the CEL

If the vehicle has a diagnostic switch, hold the switch in the ON position with the ignition switch on and the engine off. The CEL will begin flashing any active codes **(Figure 18–19)**. Two flashes followed by a pause, then five more flashes indicate a code 25. A code 25 means no trouble codes have been logged since the last system check. The flashing codes will repeat until the diagnostic switch is turned off.

If the vehicle is not equipped with a diagnostic switch, locate the diagnostic data link (DDL) under the dash. With the ignition key off, disconnect the data link. Connect a jumper wire from pin A to pin M and turn on the ignition switch (engine off) **(Figure 18–20)**. Any active codes will be flashed by the CEL. The code will repeat until the switch is turned off. Active codes are those keeping the CEL on.

Diagnostic Procedure Using the DDL Reader (DDR)

The DDL reader or DDR is used to access the system through the vehicle's diagnostic data link connector. It monitors sensor outputs, displays diagnostic codes, and can be used to program certain engine calibration factors **(Figure 18–21)**. The Pro-Link model has a touch-sensitive keypad with 0 to 9, up, down, lateral, and enter keys and a display screen. Use the keypad as

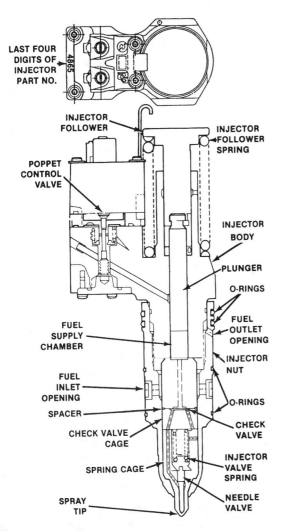

FIGURE 18-16 EUI cross section and part number location. (Courtesy of Detroit Diesel Corporation.)

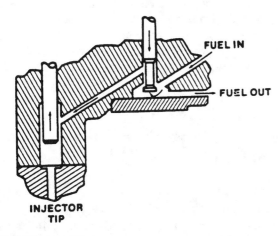

FIGURE 18-17 Fuel flowing into injector with poppet valve open. (Courtesy of Detroit Diesel Corporation.)

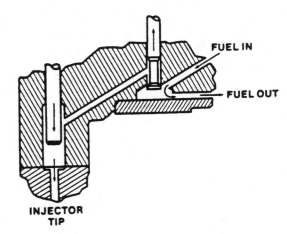

FIGURE 18-18 Poppet valve closed—start of injection. (Courtesy of Detroit Diesel Corporation.)

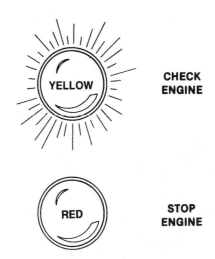

FIGURE 18-19 Check engine light (CEL) and stop engine light (SEL). (Courtesy of Detroit Diesel Corporation.)

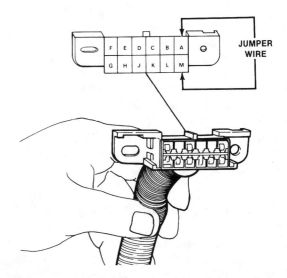

FIGURE 18-20 Diagnostic data link with jumper wire connecting pin A to pin M. (Courtesy of Detroit Diesel Corporation.)

FIGURE 18–21 Diagnostic data reader and printer. (Courtesy of Detroit Diesel Corporation.)

outlined in the instructions provided with the tester and in the Detroit Diesel service manual. An available printer interfaces with the Pro-Link to provide hard copy data printouts. Calibration changes that can be made include:

1. Automatic idle shutdown timer (3 to 100 minutes)
2. Engine droop
3. Engine shutdown protection
4. Cruise control
5. Road speed governing
6. Vehicle speed sensor

DDEC Diagnostic Codes and Their Meaning

CODE NO.	MEANING
11	Power take off sensor—voltage too low
12	Power take off sensor—voltage too high
13	Coolant sensor—voltage too low
14	Coolant or oil temperature sensor—voltage too low
15	Coolant or oil temperature sensor—voltage too high
16	Coolant level sensor voltage
21	Throttle position sensor—voltage too high
22	Throttle position sensor—voltage too low
23	Fuel temperature sensor—voltage too high
24	Fuel temperature sensor—voltage too low
25	No codes—no faults have been detected
26	Power control switch enabled—CEL and SEL lights turn on, which powers down and eventually shuts down the engine
31	Fault on auxiliary output—One of the following conditions has been detected: open in CEL or SEL circuit, short or ground to the engine, stop engine, cruise light circuits, crank position, engine brake, auxiliary drive circuits (if so equipped).
32	ECM backup system failure
33	Turbo boost sensor—voltage too high
34	Turbo boost sensor—voltage too low
35	Oil pressure sensor—voltage too high
36	Oil pressure sensor—voltage too low
37	Fuel pressure sensor—voltage too high
38	Fuel pressure sensor—voltage too low
41	Timing reference sensor—pulses incorrect or missing
42	Synchronous reference sensor—pulse not received for every firing of No. 1 cylinder
43	Low coolant level for 7 seconds—causes CEL and SEL to go on, powers down and eventually shuts down engine (if so equipped)
44	Oil or coolant over temperature for more than 2 seconds—causes CEL and SEL to go on, powers down and eventually shuts down engine (if so equipped)
45	Low oil pressure for 7 seconds below specifications—causes CEL and SEL to go on, powers down and eventually shuts down engine (if so equipped)
46	Low battery voltage—if battery voltage is below specifications for more than 30 seconds
47	High fuel pressure—fuel pressure too high
48	Low fuel pressure—fuel pressure too low

51	EEPROM error—fault in EEPROM
52	ECM failure—unable to convert sensor inputs into numbers for computer use
53	EEPROM memory failure
54	Vehicle speed sensor failure
55	Proprietary communication link failure
56	ECM failure—unable to correctly convert sensor signals to numbers
61 to 68	Injector response time—longer than maximum limit
71 to 78	Injector response time—shorter than minimum limit
85	Engine overspeed—for at least 2 seconds

If a detected problem goes away, the CEL will go off, but the code is stored in ECM memory as a historical (or logged) code.

Clearing Trouble Codes

On DDEC I systems trouble codes can be cleared by disconnecting the battery cables. On DDEC II and DDEC III systems trouble codes are cleared using the DDR tool. Follow the DDR instructions to clear codes. Always be sure that you do not clear trouble codes that you may still need.

Checking Electrical Connectors

All DDEC system connectors are protected from environmental damage and corrosion. Never puncture or pierce the insulation on any wiring, because this will allow moisture to enter and cause corrosion.

Disconnect any suspected connector and inspect for damaged locking tabs or screws, dirty, bent, or broken pins or sockets. Use the recommended solvent and tools to clean and straighten faulty items. Replace any badly damaged wiring or connectors. Reconnect the wiring and lock connectors. Make sure proper connections are made, and recheck system operation.

ELECTRONIC UNIT INJECTOR REPLACEMENT

The EUI can be replaced as an assembly, however the solenoid can be replaced without injector removal. The general procedure is as follows.

Injector Assembly Removal

1. Thoroughly clean the rocker arm cover exterior and remove the cover.

2. Drain the fuel gallery in the cylinder head for at least 5 minutes before removing the injector. Collect the draining fuel in a suitable container.

3. Disconnect the fuel line inlet and outlet fittings at the rear of the cylinder head. Use low-pressure shop air to blow into the inlet fitting until all remaining fuel is removed from the cylinder head **(Figure 18–22)**.

CAUTION:

If any fuel is left in the cylinder head fuel passages and an injector is removed, fuel will drain into the cylinder and collect in the piston dome recess. It cannot drain from the dome, and if not removed, can cause a hydrostatic lock that can bend the connecting rod.

---

4. Loosen the adjusting screws in the valve and injector rocker arms and back off the adjustment two turns.

5. Remove the rocker shaft through bolts and nut from the assembly being removed (front or rear).

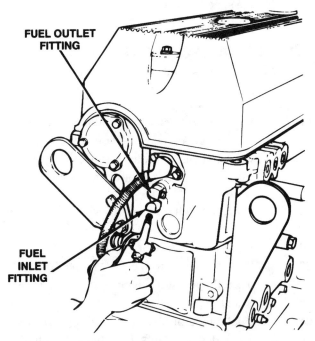

FIGURE 18–22 Blowing fuel from cylinder head internal passages. (Courtesy of Detroit Diesel Corporation.)

6. Use the proper rocker arm shaft removal tool and remove the assembly as a unit **(Figure 18–23)**.

7. Loosen the injector electrical wire terminal screws exactly two turns and slip the terminals over the screw heads **(Figure 18–24)**.

CAUTION:

Turning the screws too far will damage the threads in the solenoid housing.

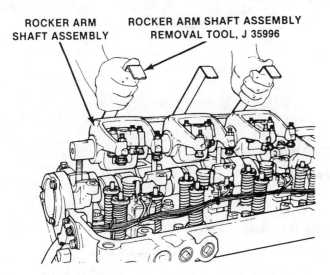

FIGURE 18–23 Rocker arm shaft assembly removal using special tool. (Courtesy of Detroit Diesel Corporation.)

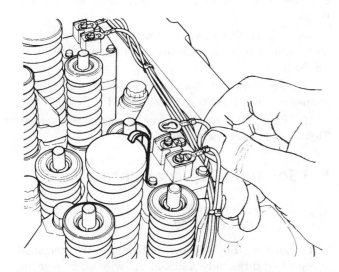

FIGURE 18–24 EUI electrical connections. Note keyhole-type openings in wire ends that slip over screw head. (Courtesy of Detroit Diesel Corporation.)

8. Remove the injector hold-down bolt and clamp.

9. Pry the injector upward using a roll-head prybar and lift out the injector **(Figure 18–25)**.

CAUTION:

Use extreme care when handling an electronic unit injector. Mishandling or dropping an electronic injector can cause damage that cannot be repaired, requiring costly replacement.

EUI Installation

1. Inspect the injector tube bore for any damage. Remove any carbon from the injector bore. Replace the tube if badly damaged.

2. Remove the old O-rings from the injector and wipe the injector clean **(Figure 18–26)**.

3. Install new O-rings on the injector, making sure they are not twisted and are seated properly. Use an O-ring pick and slide it around under the O-ring to allow it to untwist.

4. Lubricate the O-rings with clean engine oil and insert the injector into its bore.

FIGURE 18–25 Using a roll-head pry bar to remove the unit injector. (Courtesy of Detroit Diesel Corporation.)

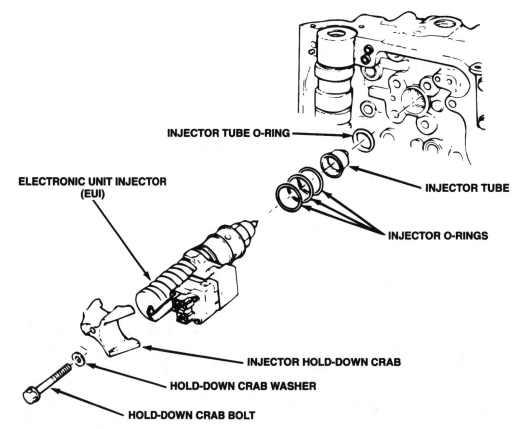

FIGURE 18–26 EUI and related mounting parts. (Courtesy of Detroit Diesel Corporation.)

5. Rotate the injector to center it between the valve springs. (There is no locating dowel on the underside of the injector.)

6. Use the heel of your hand to press the injector down into place until it is seated in its bore.

7. Install the hold-down crab, crab washer (**Figure 18–27**) (flat side against the bolt head), and the bolt. Tighten the hold-down bolt to specifications.

8. Install the wires onto the EUI solenoid screws. Slip the large part of the opening in the wire end over

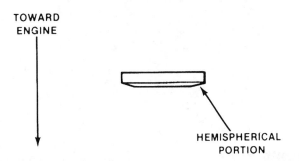

FIGURE 18–27 Hold-down crab washer installation. (Courtesy of Detroit Diesel Corporation.)

the screw head and pull the terminal into place. Tighten the screws to specifications.

9. Install the rocker shaft assembly, start the bolts and nut by hand, then tighten evenly to specifications.

10. Adjust the valve clearances, injector height, and clearance to specifications (**Figures 18–28 to 18–31**).

11. Reconnect the fuel lines to the inlet and outlet fittings in the back of the cylinder head.

12. Install the rocker cover using a new gasket. Tighten the fasteners to specifications.

EUI Solenoid Replacement

The EUI solenoid can be replaced without removing the injector (**Figure 18–32**).

1. Loosen the injector solenoid wire terminal screws two turns only, and slip the wire ends over the screw heads to remove.

2. Remove the four hex head screws and the solenoid.

ELECTRONIC UNIT INJECTOR REPLACEMENT 381

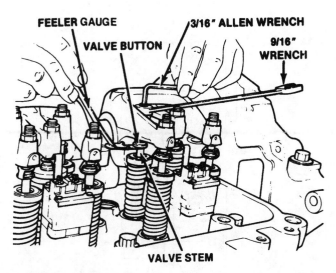

FIGURE 18-28 DDC 60 series engine valve clearance adjustment. (Courtesy of Detroit Diesel Corporation.)

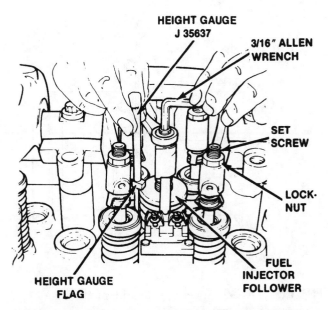

FIGURE 18-30 DDC 60 series engine injector height adjustment. (Courtesy of Detroit Diesel Corporation.)

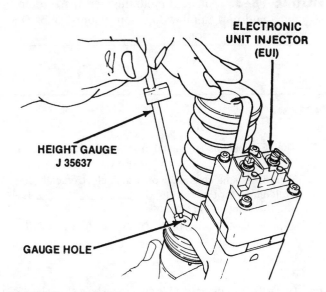

FIGURE 18-29 Using the injector timing height gauge. (Courtesy of Detroit Diesel Corporation.)

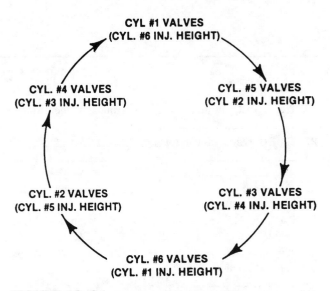

FIGURE 18-31 Sequence of valve adjustment and injector height adjustment. (Courtesy of Detroit Diesel Corporation.)

3. Discard the solenoid, load plate, follower retainer, and the screws. (New screws must be used during assembly.)

4. Remove the spacer and seals from the injector body. Keep the spacer (it is a matched component with the injector and must remain with it) and discard the seals.

5. Install a new seal in the spacer groove and position the spacer on the injector body with the seal facing the injector.

6. Install a new seal in the solenoid groove and position the solenoid on the spacer.

7. Install new screws through the load plate, follower retainer, solenoid, and spacer. Start the screws by hand. Tighten the screws evenly until the heads contact the retainer and load plate (less than 5 lb-in. or 0.6 N·m) in the sequence shown in **Figure 18-33**. Tighten the screws in the same sequence to specified torque.

8. Use a metal scribe marking tool to etch the last four digits of the injector part number on the load plate.

9. Pressurize the fuel system and check for leaks.

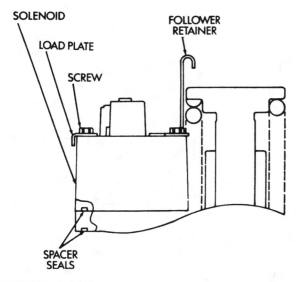

FIGURE 18–32 Injector solenoid mounting detail. (Courtesy of Detroit Diesel Corporation.)

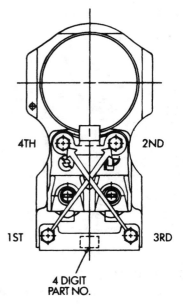

FIGURE 18–33 Solenoid mounting screw tightening sequence and part number location. (Courtesy of Detroit Diesel Corporation.)

REVIEW QUESTIONS

1. The DDEC system uses electronically controlled _____ _____ .
2. The DDEC systems control _____ _____ and _____ _____ electronically.
3. There are _____ generations of the DDEC system.
4. The ECM receives electrical signals from engine and vehicle _____ and _____ .
5. The TPS sends a varying _____ _____ to the ECM.
6. The TBS monitors _____ output air pressures.
7. The CLS sends a signal to the ECM when the _____ level drops too low.
8. The EUI is a mechanically operated, electronically controlled injector that injects _____ fuel into the combustion chamber.
9. Explain the operation of the DDEC fuel flow and EUI.
10. The DDL reader is used to access the system through the vehicle's diagnostic _____ .
11. On DDEC I systems trouble codes can be cleared by disconnecting the _____ _____ .
12. The EUI solenoid can be replaced without removing the injector. (T) (F)

TEST QUESTIONS

1. The Detroit Diesel Electronic Control system uses
   a. mechanically controlled injectors
   b. electronically controlled unit injectors
   c. an electric high-pressure pump
   d. a low-pressure supply pump

2. DDEC systems control
   a. injection timing and fuel delivery electronically
   b. fuel supply electronically
   c. ignition timing electronically
   d. oil pressure electronically

3. The DDEC III offers the following operating feature(s):
   a. engine protection system
   b. cruise control
   c. fan control
   d. all of the above
4. The TPS sends a variable-voltage signal to the ECM that varies with the amount
   a. of battery voltage
   b. of engine rpm
   c. that the throttle pedal is depressed
   d. of road speed
5. DDEC engine protection is designed to protect the engine from major damage should any of the following occur:
   a. engine oil pressure falls below a specified value
   b. engine lubricating oil temperature exceeds maximum safe value
   c. coolant level drops too low
   d. all of the above
6. An engine-driven gear pump draws fuel from the tank through the
   a. secondary filter
   b. primary fuel filter
   c. injectors
   d. bypass fuel line
7. On DDEC I systems trouble codes can be cleared by
   a. disconnecting the battery cables
   b. use of DDR tool
   c. switching the starting key off and on
   d. waiting 5 minutes with the key off
8. The EUI can be replaced
   a. without injector removal
   b. but injector must be removed first
   c. as an assembly
   d. but not without the solenoid
9. The EUI solenoid can be replaced without removing the
   a. cylinder head
   b. injector
   c. fuel lines
   d. valve cover
10. After the injector is reinstalled and all the lines are connected, be sure to
    a. pressurize the system and check for leaks
    b. remove any excessive fuel
    c. reconnect the battery
    d. start the engine and check for leaks

CHAPTER 19

CUMMINS PT FUEL INJECTION SYSTEM

INTRODUCTION

The Cummins PT fuel injection system is used exclusively on Cummins engines. A low-pressure positive-displacement pump supplies fuel to the injectors at a pressure that varies with engine speed. The pump also contains the governor assembly, which controls fuel pressure, engine idle speed, maximum speed, and torque. A single low-pressure fuel line provides fuel for all injectors. Cummins diesel engines are four-stroke-cycle engines, and the injectors are operated from special lobes on the engine camshaft through cam followers, push rods, and rocker arms. The camshaft turns at one-half crankshaft speed. The injectors meter the fuel and inject it into the cylinder at high pressure **(Figures 19–1 and 19–2).** To meet the increasingly stringent emission control legislation, electronic fuel control systems were developed. These include the PACE, PACER, and CELECT systems. These are discussed in Chapter 26.

PERFORMANCE OBJECTIVES

After thorough study of this chapter and the appropriate training models and service manuals, and with the necessary tools and equipment, you should be able to do the following with respect to diesel engines equipped with the Cummins PT fuel injection system:

1. Describe the basic construction and operation of a Cummins PT fuel injection system including the following:
    a. method of fuel pumping and metering in the injector
    b. fuel pump operation
    c. injector and pump lubrication
    d. injection timing control
    e. governor types
2. Diagnose basic fuel injection system problems and determine the correction required.
3. Remove, clean, inspect, measure, repair, replace, and adjust fuel injection system components as needed to meet manufacturer's specifications.

TERMS YOU SHOULD KNOW

Look for these terms as you study this chapter, and learn what they mean.

PT fuel system
gear pump
governor
throttle
shutdown valve
pressure regulation
variable-speed governor
mechanical governor
hydraulic governor
AFC
PT-D injector
PT-D top-stop injector
injector pumping and metering
injector plunger
injector cup

FUEL PUMP DRIVE AND MAJOR COMPONENTS

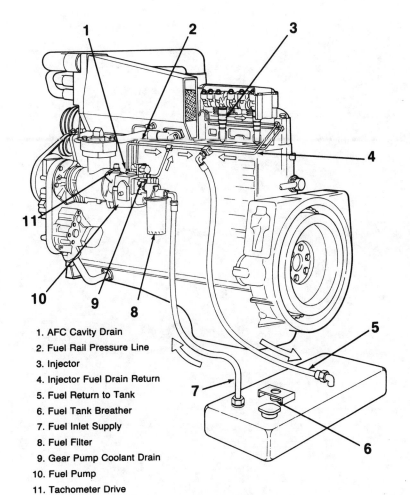

1. AFC Cavity Drain
2. Fuel Rail Pressure Line
3. Injector
4. Injector Fuel Drain Return
5. Fuel Return to Tank
6. Fuel Tank Breather
7. Fuel Inlet Supply
8. Fuel Filter
9. Gear Pump Coolant Drain
10. Fuel Pump
11. Tachometer Drive

FIGURE 19–1 Cummins PT fuel system. (Courtesy of Cummins Engine Company, Inc.)

HVT
STC
injector adjustment
injection timing
injector calibration
fuel system restriction checks
air leakage checks
fuel rail pressure check
fuel rate adjustment
throttle linkage check
low-idle-speed adjustment
high-idle-speed adjustment
no-air valve adjustment

FUEL PUMP DRIVE AND MAJOR COMPONENTS

The fuel pump is coupled to the compressor or fuel pump drive, which is driven from the engine gear train. The fuel pump main shaft turns at engine crankshaft speed and in turn drives the gear pump, governor, and tachometer shaft.

The fuel pump assembly is made up of three main units (**Figures 19–3 to 19–5**):

1. The *gear pump*, which draws fuel from the supply tank through a fitting on top of the fuel pump housing or an inlet fitting in the gear pump and forces it through the pump filter screen to the governor,

2. The *governor*, which controls flow of fuel from the gear pump, as well as maximum and idle engine speeds, and

3. The *throttle*, which provides a manual control of fuel flow to injectors under all conditions in the operating range.

Gear Pump and Pulsation Damper

The gear pump and pulsation damper are located at the rear of the fuel pump (rear being end farthest from drive coupling).

The gear pump is driven by the pump main shaft and contains a single set of gears to pick up and deliver fuel throughout the fuel system. A pulsation damper

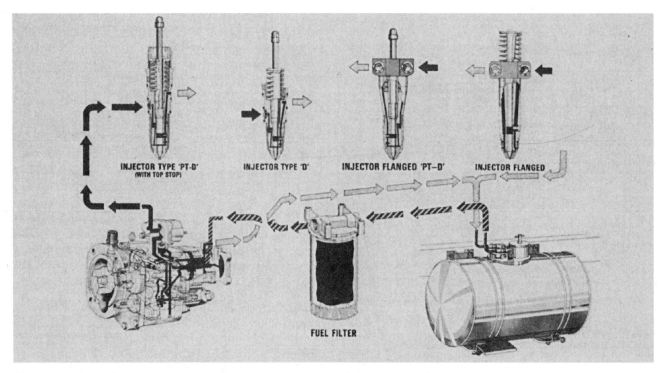

FIGURE 19-2 Fuel flow in Cummins PT fuel system. Note different injector types. (Courtesy of Cummins Engine Company, Inc.)

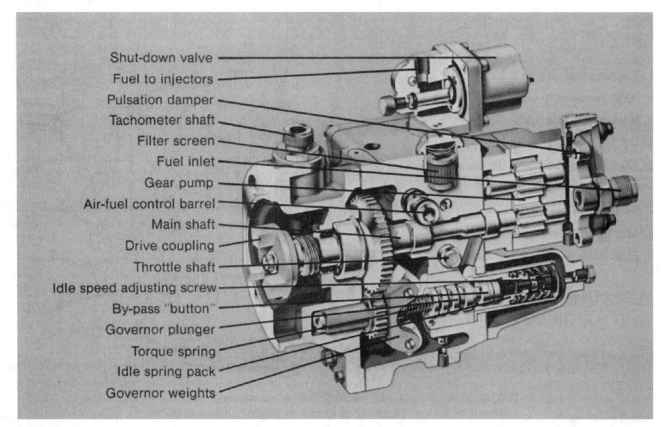

FIGURE 19-3 PT-G AFC fuel pump components. (Courtesy of Cummins Engine Company, Inc.)

FUEL PUMP DRIVE AND MAJOR COMPONENTS

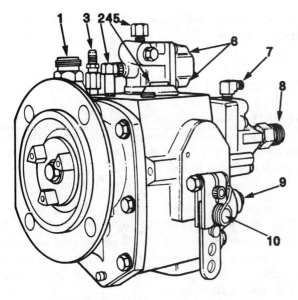

1. Tachometer Drive
2. AFC Fuel Return
3. AFC Air Supply
4. Priming Plug
5. Fuel to the Injector
6. Shutoff Valve Electric Connection
7. Gear Pump Fuel Return
8. Fuel Inlet Connection
9. Idle Speed Screw Location
10. Fuel Rate (Pressure) Screw

FIGURE 19-4 PT-G AFC fuel pump connections and location of adjustments. (Courtesy of Cummins Engine Company, Inc.)

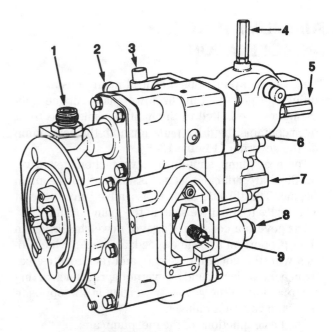

1. Tachometer Drive
2. AFC Air Supply
3. Fuel to the Injectors
4. VS High Speed Screw
5. VS Low (Idle) Speed Screw
6. Gear Pump Fuel Return
7. Fuel Inlet Connection
8. Idle Speed Screw Location
9. Fuel Rate (Pressure) Screw

FIGURE 19-5 PT-G AFC-VS fuel pump connections and location of adjustments. (Courtesy of Cummins Engine Company, Inc.)

mounted to the gear pump contains a steel diaphragm that absorbs pulsations and smoothes fuel flow through the fuel system. From the gear pump, fuel flows through the filter screen and to the governor assembly.

Throttle

The throttle provides a means for the operator to manually control engine speed above idle as required by varying operating conditions of speed and load.

In the PT (type G) fuel pump, fuel flows through the governor to the throttle shaft. At idle speed, fuel flows through the idle port in the governor barrel, around the throttle shaft sleeve. Above idle speed, fuel flows through the main governor barrel port to the throttling hole in the throttle shaft and onward to the injectors.

Shutdown Valve

Either a manual or an electric shutdown valve is used on Cummins fuel pumps.

With a manual valve, the control lever must be fully clockwise or open to permit fuel flow through the valve.

With the electric valve, the manual control knob must be fully counterclockwise to permit the solenoid to open the valve when the "switch key" is turned on. For emergency operation in case of electrical failure, turn the manual knob clockwise to permit fuel to flow through the valve.

FUEL PUMP FUNCTION AND OPERATION

The PT-G fuel pump has a central main housing containing the governor in the lower half. All the subassemblies are bolted to the main housing. The front cover contains the drive parts for both the gear pump and the governor **(Figure 19–6)**.

The gear pump, shutdown valve, tachometer drive, and filter are attached as subassemblies to the main housing. The throttle shaft is assembled in a bore at right angles to the drive shaft in the main housing.

The governor consists of a set of flyweights driven in direct proportion to engine speed; a plunger, which rotates with the governor weights and reciprocates axially in a fixed sleeve; and a governor spring pack acting in opposition to the governor weights at the opposite end of the governor plunger.

The basic functions of the fuel pump are:

1. to provide the fuel pressure to the injectors as required to produce the desired full-load torque curve;

2. to provide a means for part-throttle operation;

3. to limit the maximum speed of the engine with stability and consistent regulation;

4. to control the idle speed with closed throttle and varying load.

Pressure Regulation

Figure 19–7 shows a simplified version of the pressure-regulating system in the fuel pump with which the governor weight controls the maximum fuel pressure as a function of engine speed.

The basic parts of the flow-regulating system are:

1. a source of flow and pressure, which in the PT-G fuel pump is a positive-displacement gear pump;

2. a governor and plunger with porting and a relieved groove on the plunger that enables the fuel to flow through the governor when the port and grooves are appropriately indexed;

3. a throttle valve between the governor and the fuel pump discharge to reduce the pressure in the fuel manifold and therefore provide part-load control;

4. the fuel conduit to the injectors;

5. a fuel bypass circuit from the governor plunger fuel groove through an axial drilling out of the end of the plunger and back to the inlet of the fuel supply pump.

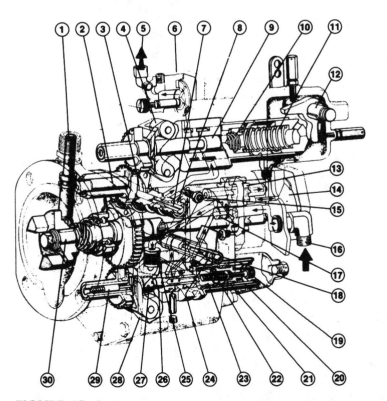

1. Tachometer Drive
2. Idler Gear and Shaft
3. AFC Piston
4. VS Governor Weights
5. Fuel to Injectors
6. Shutoff Valve
7. AFC Plunger
8. AFC Fuel Barrel
9. VS Governor Plunger
10. VS Idle Spring
11. VS High Speed Spring
12. VS Throttle Shaft
13. Gear Pump
14. Pulsation Damper
15. AFC Needle Valve
16. Fuel From Filter
17. Pressure Regulator Valve
18. Throttle Shaft
19. Idle Adjustment Screw
20. Spring Spacer
21. High Speed Spring
22. Idle Spring
23. Idle Spring Plunger
24. Fuel Adjustment Screw
25. Filter Screen
26. Governor Plunger
27. Torque Spring
28. Governor Weights
29. Governor Assist Plunger
30. Main Shaft

FIGURE 19–6 Cummins PT-G fuel pump with AFC (air fuel control) and variable-speed governor. Fuel flow is indicated by arrows. (Courtesy of Cummins Engine Company, Inc.)

FUEL SUPPLY PUMP OPERATION

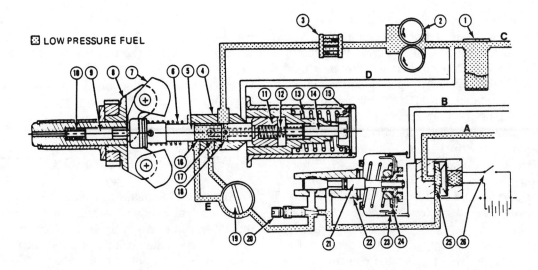

FIGURE 19-7 Schematic of PT-G AFC fuel pump components. (Courtesy of Cummins Engine Company, Inc.)

1	PRIMARY FUEL FILTER	17	MAIN GOVERNOR PORT
2	GEAR PUMP	18	GOVERNOR DUMP PORTS
3	FILTER SCREEN	19	THROTTLE
4	GOVERNOR SLEEVE	20	AFC NEEDLE VALVE
5	GOVERNOR PLUNGER	21	AFC CONTROL PLUNGER
6	TORQUE CONTROL SPRING	22	AFC BARREL
7	GOVERNOR WEIGHTS	23	DIAPHRAGM (BELLOWS)
8	GOVERNOR WEIGHT CARRIER	24	AFC SPRING
9	WEIGHT ASSIST PLUNGER	25	SOLENOID VALVE
10	WEIGHT ASSIST SPRING	26	IGNITION SWITCH
11	IDLE SPRING PLUNGER	A	FUEL TO INJECTORS
12	IDLE SPEED SPRING	B	AIR FROM INTAKE MANIFO
13	MAXIMUM SPEED GOVERNOR SPRING	C	FUEL FROM TANK
14	IDLE SPEED ADJUSTING SCREW	D	BY-PASSED FUEL
15	MAXIMUM SPEED GOVERNOR SHIMS	E	IDLE FUEL PASSAGE
16	IDLE SPEED GOVERNOR PORT		

The pressure delivered by the fuel pump is determined by the forces acting to close the gap between the ends of the plunger and the pressure control button. This pressure within the fuel pump is independent of throttle position but is directly related to engine speed. All fuel that is not delivered to the injectors is discharged out of the gap formed by the end of the governor plunger and the adjacent pressure control button. Some fuel is bypassed under all operating conditions.

The pressure force that is acting at the bypass gap is opposed by the centrifugal force of the governor weights applied axially to the driven end of the plunger.

The fuel pressure required to separate the pressure control button from the governor plunger is determined by the cross-sectional area of the recess in the pressure control button. The equilibrium condition is reached when the forces generated by the fuel pressure in the recess area in the pressure button are equal to the governor weight force on the governor plunger.

At a given engine speed, the pressure delivered by the pump is a function of the governor weight force. Since the governor weight varies directly as the square of the engine speed, the pressure delivered by the pump also varies in this manner. This characteristic produces approximately constant engine torque. This flat torque curve provides a basis for obtaining almost *any desired torque characteristic by means of minor adjustments*, which is another important and unique feature of the PT-G fuel pump.

FUEL SUPPLY PUMP OPERATION

The supply pump for the PT fuel system is designed to deliver several times more fuel than required to operate the engine at maximum power. The excess fuel is bypassed and returned to the pump inlet. Hence, several different engine models can use the same size supply pump.

390 Chapter 19 CUMMINS PT FUEL INJECTION SYSTEM

Different engine models such as six- or eight-cylinder engines of the same basic design may require the same fuel pressure to the injector but different volumes of fuel flow. For this variation, no change in pump parts is required, since the bypassed fuel quantity simply changes to accommodate the different flows. It also follows that variations in pump capacity caused by tolerances and pump wear have no effect on the pressure delivered by the fuel pump. *The ability of the fuel pump to accommodate the changes in supply pump delivery throughout the life of the fuel pump is an important and unusual feature, providing unchanged engine characteristics throughout the life of the engine.*

If change of injector supply pressure is desired, this change can be made by replacing the control button with one that has a different size recess.

Engine operation at part load is accomplished simply by throttling the fuel pressure between the fuel pump delivery and the injector. The throttle consists of a shaft with a cross-drilled fuel passage, which is indexed with holes in the bore of the pump housing. It should be noted that the pressure within the fuel pump is unaffected by throttle position; the throttle affects only the downstream pressure.

GOVERNOR TYPES AND OPERATION

The idling and high-speed mechanical governor, often called an "automotive governor," is actuated by a system of springs and weights and has two functions. First, the governor maintains sufficient fuel for idling with the throttle control in the idle position; second, it cuts off fuel to injectors above maximum rated rpm. The idle springs in the governor spring pack position the governor plunger so the idle fuel port is opened

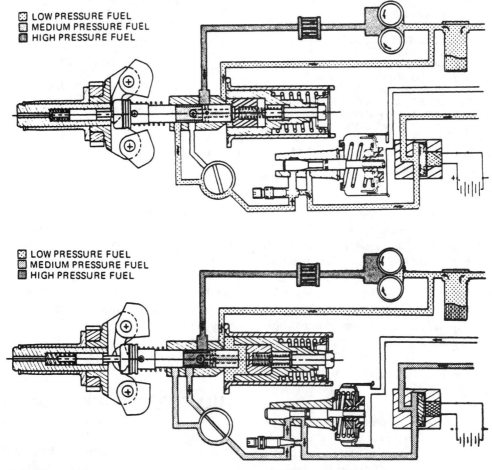

FIGURE 19–8 PT-G AFC fuel flow during starting and idle (top) and during normal driving (bottom). (Courtesy of Cummins Engine Company, Inc.)

enough to permit passage of fuel to maintain engine idle speed.

During operation between idle and maximum speeds, fuel flows through the governor to injectors in accordance with engine requirements as controlled by throttle and limited by size of idle spring plunger counterbore on PT-G fuel pumps. When the engine reaches governed speed, governor weights move the governor plunger, and fuel passages to throttle and injectors are shut off. At the same time another passage opens and dumps the fuel into the main pump body. In this manner engine speed is controlled and limited by the governor regardless of throttle position (**Figures 19–8** to **19–11**). Fuel leaving the governor flows through the throttle, shutdown valve, and inlet supply lines into internal drillings in the cylinder heads and on to the injectors.

Mechanical Variable-Speed (MVS) Governor

This governor supplements the "standard or automotive governor" to meet the requirements of ma-

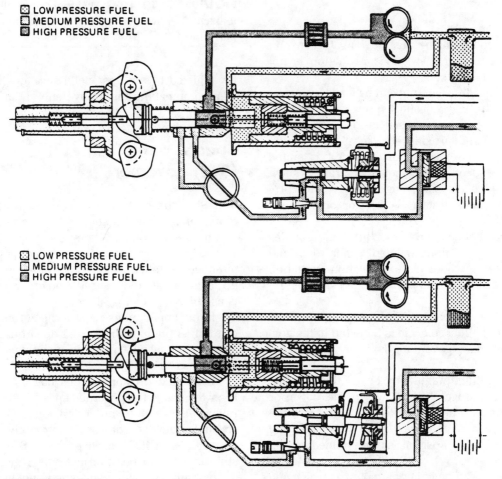

FIGURE 19–9 PT-G AFC fuel flow during beginning of high-speed governing (top) and full high-speed governing (bottom). (Courtesy of Cummins Engine Company, Inc.)

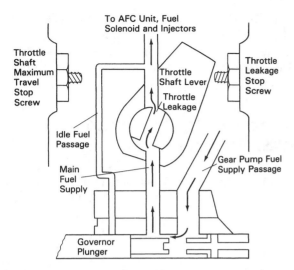

FIGURE 19–10 Fuel flow through throttle shaft. Note throttle stop screws. (Courtesy of Cummins Engine Company, Inc.)

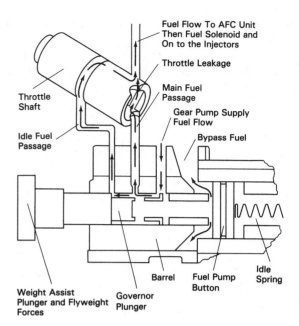

FIGURE 19–11 PT-G AFC fuel pump idle and throttle leakage fuel flow. (Courtesy of Cummins Engine Company, Inc.)

chinery on which the engine must operate at a constant speed, but where extremely close regulation is not necessary.

Adjustment for different rpm can be made by means of a lever control or adjusting screw. At full rated speed, this governor has a speed droop between full load and no load of approximately 8%.

As a variable-speed governor, this unit is suited to the varying speed requirements of cranes, shovels, etc., in which the same engine is used for propelling the unit and driving a pump or other fixed-speed machine.

As a constant-speed governor, this unit provides control for pumps, nonparalleled generators, and other applications where close regulation (variation between no-load and full-load speeds) is not required.

The (MVS) governor assembly mounts on top of the fuel pump, and the fuel solenoid is mounted to the governor housing. The governor also may be remote mounted.

Fuel from the fuel pump body enters the variable-speed governor housing and flows to the governor barrel and plunger. Fuel flows past the plunger to the shutdown valve and on into the injector according to the governor lever position, as determined by the operator.

The variable-speed governor cannot produce engine speeds in excess of the automotive governor setting. The governor can produce idle speeds below the automotive pump idle-speed setting but should not be adjusted below this speed setting when operating as a combination automotive and variable-speed governor.

Special Variable-Speed (SVS) Governor

The SVS governor provides many of the same operational features as the MVS governor but is limited in application. An overspeed stop should be used with SVS governors in unattended applications; in attended installations, a positive shutdown throttle arrangement should be used if no other overspeed stop is used.

Marine applications require the automotive throttle of the fuel pump to be locked open during operation, and engine speed control is maintained through the SVS governor lever.

Power take-off applications use the SVS governor lever to change the governed speed of the engine from full rated speed to an intermediate power take-off speed. During operation as an automotive unit, the SVS governor is in high-speed position. See operation instructions for further information.

Hydraulic governor applications, not having variable-speed setting provisions, use the SVS governor to bring engine speed down from rated speed for warm-up at or slightly above 1000 rpm.

Hydraulic Governor

Hydraulic governors are used on stationary power applications where it is desirable to maintain a constant speed with varying loads.

The Woodward SG Hydraulic Governor uses lubricating oil, under pressure, as an energy medium.

The governor acts through oil pressure to increase fuel delivery. An opposing spring in the governor control linkage acts to decrease fuel delivery.

Speed droop is introduced into the governing system to make its operation stable. Speed droop means decreasing speed with increasing load. The desired magnitude of this speed droop varies with engine applications and may easily be adjusted to cover a range of approximately 0.5–7%.

Assume a certain amount of load is applied to the engine. The speed will drop, and the flyballs will be forced inward, lowering the pilot valve plunger. This will admit oil pressure underneath the power piston, which will rise. The movement of the power piston is transmitted to the terminal shaft by the terminal lever. Rotation of the terminal shaft through the linkage to the fuel pump causes the fuel setting of the engine to be increased.

AIR/FUEL CONTROL (AFC)

The air/fuel control is an acceleration exhaust smoke control device built internally in the fuel pump. It restricts fuel proportionally to the engine air intake manifold pressure during engine acceleration (**Figures 19–12** and **19–13**).

During acceleration, turbocharger speed (intake manifold air pressure) lags behind the almost instantaneous fuel-delivering capability of the fuel pump and injectors, thus supplying an overrich fuel/air mixture, which causes excessive exhaust smoke.

The air/fuel control provides a more completely combustible fuel/air mixture by continuously monitoring turbocharger air pressure and proportionally responding to load or acceleration changes.

FUEL INJECTOR OPERATION

PT (Type D) Injectors

The injector provides a means of introducing fuel into each combustion chamber. It combines the acts of metering, timing, and injection. Principles of operation are

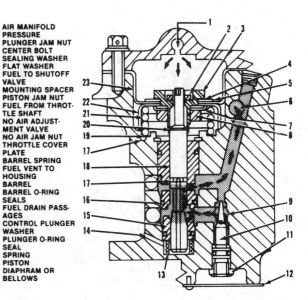

FIGURE 19–12 AFC unit components. (Courtesy of Cummins Engine Company, Inc.)

the same for in-line and V engines, but injector size and internal design differ slightly.

Fuel supply and drain flow are accomplished through internal drillings in the cylinder heads. A radial groove around each injector mates with the drilled passages in the cylinder head and admits fuel through an adjustable (adjustable by burnishing to size at test stand) orifice plug in the injector body. A fine mesh screen at each inlet groove provides final fuel filtration.

The fuel grooves around the injectors are separated by O-rings that seal against the cylinder head injector bore. This forms a leakproof passage between the injectors and the cylinder head injector bore surface.

Fuel flows from a connection atop the fuel pump shutdown valve through a supply line into the lower drilled passage in the cylinder head. A second drilling in the head is aligned with the upper injector radial groove to drain away excess fuel. A fuel drain allows return of the unused fuel to the fuel tank.

The injector contains a ball check valve. As the injector plunger moves downward to cover the feed opening, an impulse pressure wave seats the ball and at the same time traps a positive amount of fuel in the injector cup for injection. As the continuing downward plunger movement injects fuel into the combustion chamber, it also uncovers the drain opening and the ball rises from its seat. This allows free flow through the injector and out the drain for cooling

Chapter 19 CUMMINS PT FUEL INJECTION SYSTEM

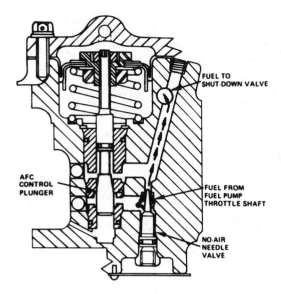

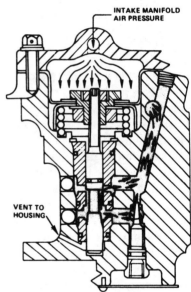

FIGURE 19-13 AFC unit fuel flow without turbocharger boost pressure (top) and with boost pressure (bottom). (Courtesy of Cummins Engine Company, Inc.)

purposes and purging gases from the cup **(Figures 19-14 to 19-16).**

PT (Type D) Injector Top-Stop

The top-stop injector functions like the standard PT (type D) injector except the upward travel of the injector plunger is limited by an adjustable stop. The stop is set before the injector is installed in the engine.

When the injector is installed and properly adjusted in the engine, plunger spring load is carried against the stop, which allows engine lubricating oil to better lubricate the sockets, reducing wear in the injection train. Consequently, the injector train remains in adjustment longer.

PT-D Injector Pumping and Metering

Figure 19-17 shows the cycle of events in the injector. (See also **Figure 19-15.**) The plunger is retracted during about one-half of each engine cycle. The fuel, from the "common rail" or fuel passage, enters into the injector through an orifice plug. This adjustable orifice provides a means for calibrating the injector for a desired identical fuel delivery. The PT-D injector contains a ball valve. When the plunger is retracted and the metering or feed orifice is open, the fuel pressure opens the check ball valve and the fuel enters into the cup. *The amount of fuel entering into the cup is—as stated before—a function of the fuel pressure and the time that the metering orifice is open.* As the injector plunger moves downward, a pressure wave closes the check ball valve; shortly after the metering edge of the plunger closes the metering orifice, the crossover groove on the plunger opens the cross-holes, and the fuel circulation or purging begins.

The metering portion of the cycle is followed by the injection, in which the plunger moves farther down, and the cam-operated positive-displacement plunger then generates whatever injection pressure is required to discharge the fuel through the spray holes into the combustion chamber. During injection and after completion of injection, the purging cycle continues, until the retracted plunger once again closes the crossover holes. It is important to note that no purging of fuel takes place during the metering cycle. The fuel circulation within the injector starts only after metering of the fuel is completed, thereby eliminating the effect of drain-flow fluctuation or drain restriction during metering. This is the so-called closed-drain metering cycle.

The PT-D injector cup is made of two pieces: the cup itself and the cup retainer. There is ample clearance between the cup and the retainer to allow the cup to center on the injector plunger. Because of this self-centering feature, the seating of the plunger in the cup is assured, preventing leakage between the cup and the plunger.

The PT-D injector plunger has a relieved diameter, which operates in the cup with a small clearance, and the inside diameter of the plunger remains engaged in the cup, so the metered fuel enters the volume between the cup and the metering edge on the plunger. There is always a fuel film in the annulus below the metering chamber, which works as a *liquid seal*, sealing off the metering portion of the injector, to prevent carboning.

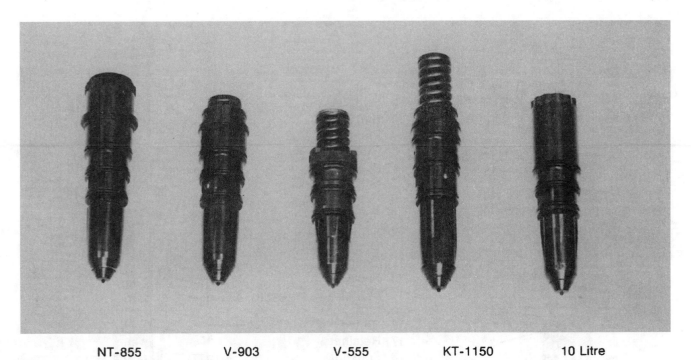

FIGURE 19–14 PT (type D) injectors and engine application. (Courtesy of Cummins Engine Company, Inc.)

Start upstroke (fuel circulates)

Fuel at low pressure enters the injector at (A) and flows through the inlet orifice (B), internal drillings, around the annular-groove in the injector cup and up passage (D) to return to the fuel tank. The amount of fuel flowing through the injector is determined by the fuel pressure before the inlet orifice (B). Fuel pressure in turn is determined by engine speed, governor and throttle.

Upstroke complete (fuel enters injector cup)

As the injector plunger moves upward, metering orifice (C) is uncovered and fuel enters the injector cup. The amount is determined by the fuel pressure. Passage (D) is blocked, momentarily stopping circulation of fuel and isolating the metering orifice from pressure pulsations.

Downstroke (fuel injection)

As the plunger moves down and closes the metering orifice, fuel entry into the cup is cut off. As the plunger continues down, it forces fuel out of the cup through tiny holes at high pressure as a fine spray. This assures complete combustion of fuel in the cylinder. When fuel passage (D) is uncovered by the plunger undercut, fuel again begins to flow through return passage (E) to the fuel tank.

Downstroke complete (fuel circulates)

After injection, the plunger remains seated until the next metering and injection cycle. Although no fuel is reaching the injector cup, it does flow freely through the injector and is returned to the fuel tank through passage (E). This provides cooling of the injector and also warms the fuel in the tank.

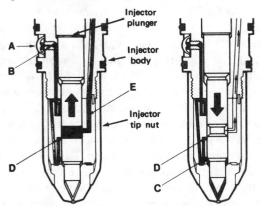
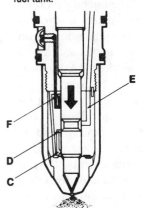
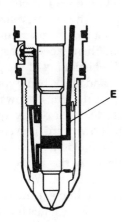

FIGURE 19–15 PT-D injector operation and fuel flow. (Courtesy of Cummins Engine Company, Inc.)

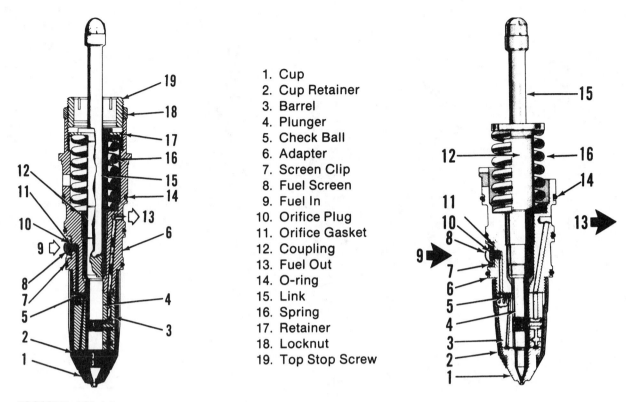

FIGURE 19–16 PT-D top-stop injector (left) and standard PT-D injector (right). Note the difference in components. (Courtesy of Cummins Engine Company, Inc.)

While the metering orifice is open-metered, fuel is delivered into the injector cup, but this fuel charge never completely fills the cavity. The remaining portion is filled with compressed and heated air, which enters the cup through the spray holes from the engine cylinder. Thus, a mixture of heated air and fuel is prepared in the cup before injection.

Because of the vaporous mixture within the cup, actual injection does not commence when the camshaft starts the downward movement of the injector plunger; injection cannot begin until the pressure in the cup equals and then exceeds that in the combustion chamber. At the beginning of the plunger movement, a mixture of air and fuel is being compressed, so that the pressure rise in the cup follows the law of compressible fluids, producing a relatively slow discharge from the cup. The downward movement of the plunger compresses the fuel to a substantially "solid mass," and the rate of pressure rise in the cup increases sharply to inject the main charge. With all other things equal, the rate of the injector plunger travel during the injection of the main charge determines the degree of atomization, penetration, and distribution of the fuel emerging into the combustion chamber. Since the small pilot charge is displaced and arrives in the combustion chamber slightly ahead of the main charge, it passes through the usual ignition delay but has already started the combustion when the main charge arrives. Thus, the condition for minimum ignition delay and fast, smooth combustion of the main charge is created by the initial or pilot injection characteristic of the injector.

The beginning of the injection is a variable depending on the quantity of fuel in the injector cup cavity. At light load, the charge is relatively small, and the injector plunger reaches the solid fuel level relatively late, i.e., near the TDC of the piston travel, as desired. As the load increases, the injected quantity increases and the solid fuel level is met earlier, BTDC. The pilot injection, during the compressible phase, always precedes the main charge injection. Thus, the *actual injection timing varies automatically* in the proper direction to suit varying engine load conditions; this is an inherent characteristic of the injector. At constant injected fuel quantity, however, the beginning of the injection remains the same regardless of engine speed. This characteristic is a significant advantage over other conventional fuel systems where variable timing devices are needed to compensate for the pressure wave propagation or injection lag due to high-pressure fuel lines between the fuel pump and injectors.

As with most injector systems, if good combustion is to be obtained, the rate of fuel injection must be main-

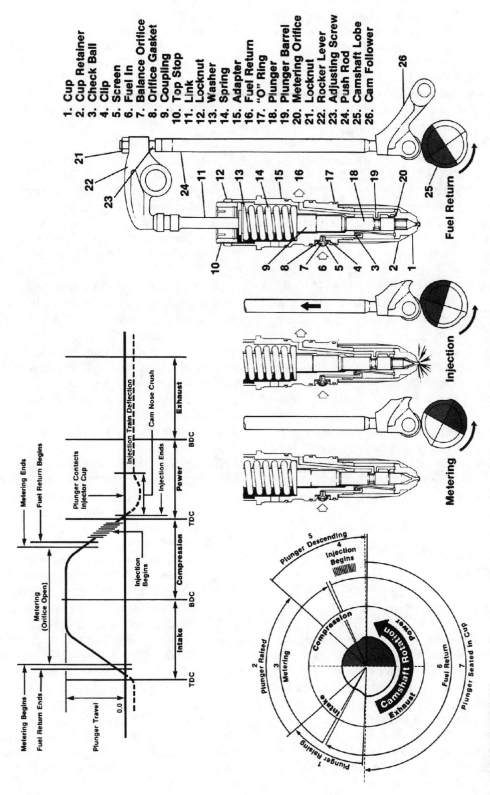

FIGURE 19–17 PT-D top-stop injection cycle. (Courtesy of Cummins Engine Company, Inc.)

tained right up to the end of the injection process. This is accomplished in the design of the injector cam. The injection cam incorporates a "nose," which theoretically projects the travel of the injector plunger beyond the cup seat by the amount below the dwell level. This nose therefore compensates for the elasticity of the injection mechanism during actual injection and ensures holding the plunger tight against its seat while relatively high pressures exist in the engine cylinder. This nose does not increase the maximum load on the injector mechanism but prevents the maximum load, or the injection pressure, from decreasing near the end of injection.

The fuel quantity delivered through the metering orifice is regulated by changing the fuel pressure at the entrance of the metering orifice in the injector.

The second metering variable is time. Since the injector plunger is controlled by the camshaft, the metering time increment per cycle decreases as engine speed increases. For example, doubling the engine speed reduces the absolute metering time per cycle by half. For constant pressure, the fuel per cycle decreases with increasing engine speed, but the total engine fuel rate remains unchanged.

In combining both variables of metering pressure and metering time, it is evident that constant engine torque throughout the engine speed range requires the fuel-metering pressure to vary as the square of the engine speed. *The important feature of the PT-G fuel pump is its inherent ability to produce a fuel-metering pressure that varies as the square of the engine speed.* This is achieved by taking unique advantage of the fact that a rotating mechanical governor flyweight also has a square-law characteristic.

CUMMINS INJECTION TIMING CONTROL

Cummins injection timing is controlled by the amount of fuel metered into the injector cup. With less fuel in the cup (low engine speed) injection begins later in relation to piston travel than it does when more fuel is metered into the injector cup (higher engine speeds). In other words, less fuel means less injection timing advance, which means lower engine speeds; more fuel means more injection timing advance, which means higher engine speeds. Precise adjustment of the injector operating mechanism is also required to obtain accurate and even injection timing between all cylinders.

Cummins also uses automatic injection timing advance mechanisms to improve timing control. One system, introduced in 1980, requires the engine to be equipped with an air compressor. This system is known as *mechanical variable timing*. Two later systems, known as *hydraulic variable timing*, and *step timing control* (STC), do not require air pressure to operate. They use engine lubricating oil instead. Currently, electronic control systems are used to control injection timing and fuel metering. These include the PACE, PACER, and CELECT systems discussed in Chapter 20.

Mechanical Variable Timing

Mechanical variable timing varies the injection timing by changing the location of the injector cam followers relative to the camshaft. This is accomplished by mounting an eccentric rigidly to the cam follower shaft. The shaft changes the location of the injector cam followers on the camshaft when the cam follower shaft is rotated. To rotate the cam follower shaft, an air cylinder with a rack-and-pinion gear arrangement is used **(Figure 19–18)**.

The MVT actuator is an air-operated cylinder. The cylinder shaft has a rack of gear teeth, which mesh with the pinion gear mounted to the cam follower shaft. Raising and lowering the cylinder piston cause the pinion to rotate. The cylinder piston is held in the down position by a large spring. When air pressure is applied to the cylinder, the piston rises against the spring pressure. An oversized first tooth on the rack provides positive alignment with the pinion.

Mounted to the cam follower shaft are eccentrics. These eccentrics and the cam followers are held to the shaft with a set screw. Rotation of the cam follower shaft provides the in-and-out movement of the followers.

Air flow to and from the cylinder is controlled by a three-way solenoid control valve mounted to the actuator. The solenoid valve is activated by an electrical pressure switch located in the discharge side of the fuel pump pressure line. The air supply to the actuator is plumbed from the dry storage tank of the truck.

When the engine is first started, there usually is no truck system air pressure. The spring holds the piston down in the cylinder, and the engine starts with retarded injection timing. When air pressure builds up in the truck system, this pressure enters the cylinder and raises the piston. This causes the injection timing to advance.

Air is directed to the actuator as long as the fuel pressure is below the preset pressure point or when the engine is from 0 to 25% of load. Under these conditions the electrical pressure switch in the fuel pump remains activated, supplying current to the three-way solenoid valve. The solenoid valve remains activated, allowing air pressure to enter the actuator.

When the driver puts the truck under load, fuel pressure increases. When fuel pressure reaches the preset

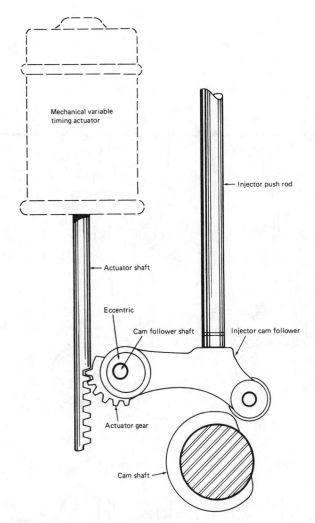

FIGURE 19–18 (Courtesy of Cummins Engine Company, Inc.)

point or approximately 25% of load, the electrical pressure switch deactivates. The three-way solenoid valve shifts, closing off the port that supplies air to the actuator and allows trapped air in the cylinder to be exhausted.

With the exhaust port of the solenoid valve open, the internal spring in the actuator forces the piston down, discharging the trapped air. So that the injection timing does not go from advance to retard too rapidly, an orifice is located in the discharging line. The air is gradually discharged from the cylinder. Thus, the change in timing is very gradual.

All three cam follower housings and the actuator are splined together on a shaft. The whole assembly weighs around 80 lb and is mounted to the engine completely assembled. A one-piece cam follower gasket of uniform thickness is used.

A slow retraction injector cam is used. This camshaft limits the amount of fuel that can enter into the injector cup. This is accomplished by decreasing the metering time in the advance mode.

Hydraulic Variable Timing

Hydraulic variable timing (HVT), a step-timing system, is relatively simple. It is made up of a special PT (type D) top-stop injector with a hydraulic tappet between the injector lever and the plunger **(Figure 19–19)**.

At any engine speed the system allows the engine to operate at *advanced* injection timing under *light-load* conditions, and at *retarded* timing during *high-load* conditions.

The HVT

- improves cold-weather idling characteristics
- reduces white smoke in cold climates
- reduces injector carboning

HVT OIL FLOW (ADVANCED TIMING) (FIGURE 19–20)

1. When engine fuel pressure (5) is less than 32 psi, the fuel pressure switch (6) is *closed;* the HVT oil control valve (3) is *open.*

2. Engine oil, under pressure, flows from the oil filter head (2) to the open oil control valve (3).

3. The pressure relief valve (4) returns a small amount of oil to the oil pan.

4. Oil flows from the control valve to the oil manifold (7).

5. From the manifold oil flows through the oil transfer connection (9) and into the HVT tappet assembly (8).

6. The tappet fills with oil, causing injection timing to be advanced.

7. At the end of the injection cycle, the increased oil pressure in the tappet moves the load-cell check ball from its seat. The oil then drains from the tappet through the drain holes in the injector adapter.

HVT OIL FLOW (RETARDED TIMING)

1. When engine fuel pressure (5) is more than 32 psi, the fuel pressure switch (6) is *open;* the oil control valve (3) is *closed* **(Figure 19–20)**.

2. Engine oil, under pressure, flows from the oil filter head (2) to the closed oil control valve (3).

3. The pressure relief valve (4) prevents oil pressure in the oil manifold from exceeding 6 psi. This oil pressure must be 10 psi or more before it can flow into the tappet. Injection timing is retarded when there is no oil in the tappet.

Advanced Timing

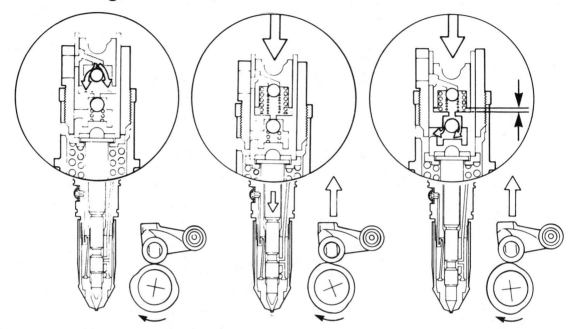

Retarded Timing

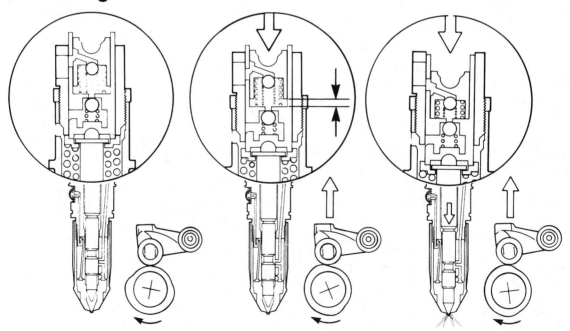

FIGURE 19–19 Oil flow—advanced and retarded timing. (Courtesy of Cummins Engine Company, Inc.)

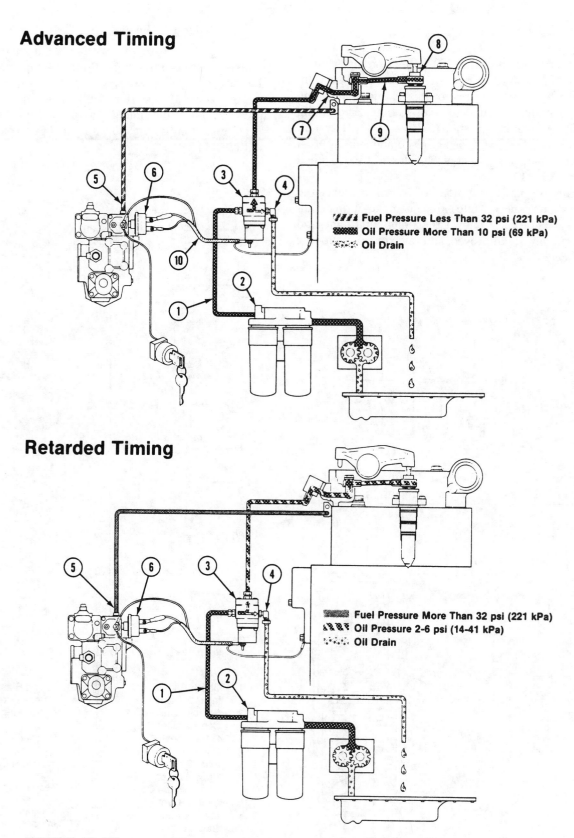

FIGURE 19–20 (Courtesy of Cummins Engine Company, Inc.)

Step-Timing Control System

The step-timing control (STC) system provides injection timing advance during start-up and light-load engine operation and returns to normal injection timing during medium- to high-load operation. Step-timing control is similar in operation to the earlier HVT system but does not have any electrical components. A fuel pressure–controlled oil control valve is used instead **(Figure 19-21)**.

During the advanced mode of injection timing the STC system reduces white exhaust smoke during cold-weather operation. Cold-weather idling and light-load fuel economy are improved, and injector tip carboning is reduced.

STC components include:

1. *Top-stop injector* with a hydraulic tappet mounted on top. Pumping oil into the tappet changes the effective length of the tappet, thereby changing the start of injection timing. Oil in the tappet advances injection timing. No oil in the tappet returns injection timing to normal **(Figures 19-22 and 19-23)**.

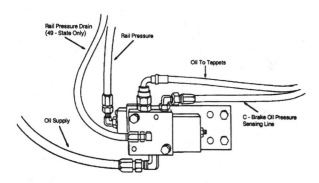

FIGURE 19-21 STC control valve connections. (Courtesy of Cummins Engine Company, Inc.)

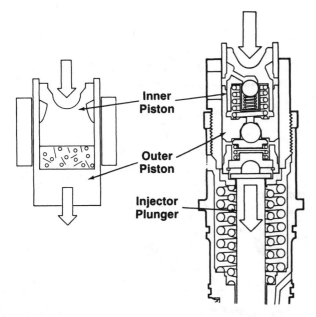

FIGURE 19-22 STC top-stop injector cross section. (Courtesy of Cummins Engine Company, Inc.)

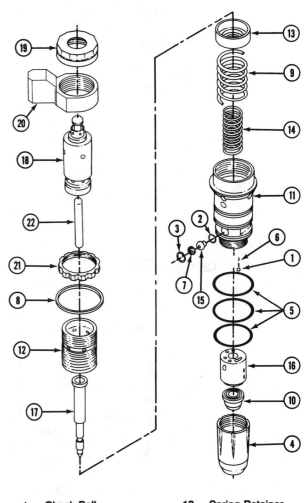

1	Check Ball	13	Spring Retainer
2	Injector Gasket	14	Compression Spring
3	Screen Retainer	15	Orifice Plug
4	Cup Retainer	16	Barrel
5	O-Ring Seal	17	Plunger
6	Roll Pin	18	STC Tappet
7	Filter Screen	19	Tappet Top Stop Cap
8	Plain Washer		
9	Compression Spring	20	Tappet Top Stop Locknut
10	Injector Cup		
11	Injector Adapter	21	Locknut
12	Stop Screw	22	Injector Link

FIGURE 19-23 STC injector components. (Courtesy of Cummins Engine Company, Inc.)

2. The *STC control valve* controls the oil supply to the injector tappet (**Figures 19–24** and **19–25**). The control valve is actuated by fuel pressure on one side and spring pressure on the other. During normal operation the injection tappet is collapsed, there is no oil in the tappet, and rocker arm action on the downstroke is transmitted to the injector by mechanical contact through the tappet.

When injection timing advance is required, the STC control valve shifts to allow oil pressure to be supplied to the injector tappet. Oil between the inner and outer pistons increases the effective length of the tappet, thereby starting injection earlier.

When fuel pressure is below 10 psi (69 kPa), as during start-up and light load, the oil control valve is open, allowing engine lube oil to flow into the oil manifold, which supplies oil to the injector tappets. When fuel pressure exceeds 10 psi (69 kPa), the upper check ball in the tappet unseats, allowing oil to enter the tappet between the two pistons. As the rocker arm force acts on the tappet, the inner piston moves down, and the oil is trapped below the piston, forcing the outer piston and injector plunger to move down. At the end of the injection cycle the lower check ball unseats and allows tappet oil to drain.

When fuel pressure exceeds the predetermined value, the control valve closes and prevents engine oil from entering the system, and injection timing returns to normal.

FUEL SYSTEM DIAGNOSIS

(Courtesy of Cummins Engine Company, Inc.)

The following diagnostic charts (**Figures 19–26** and **19–27**) and remedies on the Cummins PTG-AFC air fuel control fuel system are provided courtesy of the Cummins Engine Company, Inc. For appropriate overhaul and calibration procedures and service data, refer to the appropriate manufacturer's shop service manual.

INJECTOR SERVICE

Locating a Faulty Injector

A faulty or misfiring injector can be located as follows.

1. Run the engine at the speed where the problem occurs.

2. Hold one injector at a time down on its seat by pushing down on the rocker arm at the injector end.

3. Listen for a change in the sound of the engine. A good injector held down will noticeably change the

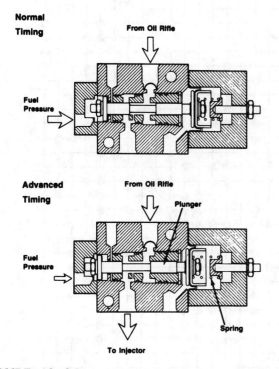

FIGURE 19–24 STC control valve operation during normal timing (top) and advanced timing (bottom). (Courtesy of Cummins Engine Company, Inc.)

Model	Part Number	Shifts into Normal Timing As Rail psi Increases [psi (kPa)]	Shifts into Advanced Timing As Rail psi Decreases [psi (kPa)]	Rail psi Drain Line	Plunger Spring Part Number
NTC	3056564	53 (370)	25 (170)	Yes	3041069
NTCC	3056565	25 (170)	20 (140)	No	3042420

FIGURE 19–25 Relationship between fuel rail pressure and timing shift. (Courtesy of Cummins Engine Company, Inc.)

PT (type G) FUEL SYSTEM TROUBLESHOOTING

POSSIBLE CAUSE	Acceleration Slow	Air Leaks	Carboned Valves, Injector Cups	Deceleration Slow	Failure To Pick Up Fuel	Fuel Consumption Excessive	Fuel Pressure High	Fuel Pressure Low	Governed Speed High	Governed Speed Low	High Speed Surge MVS	Idle Speed Too High	Idle Surge MVS	Idle Undershoot MVS	Low Power	Rough Operation	Smoke Black, Low Speed	Throttle Leakage Excessive	Wear Rate High
Air Signal Attenuator Filter Plugged	●														●				
Air Leaks		●			●			●							●				●
Aneroid Not Opening					●										●		●		
Aneroid Stuck Open	●					●													
Aneroid Valve Stuck Open	●					●													
Cooling Line By-Passing						●													
Cranking Speeds Slow					●														
Filter Suction Restricted	●							●							●				
Fuel Dirty			●					●									●	●	●
Fuel With Water								●									●	●	●
Fuel, Wrong Type	●					●									●				
Gear Pump Worn					●		●	●							●				
Governor Plunger Chamfer, Inadequate									●		●		●						
Governor Plunger Scored									●	●			●	●					
Governor Plunger, Wrong/Worn/Sticking	●								●	●		●	●	●					
Governor Plunger, Worn/Scored	●									●				●					
Governor Spring Shims Low										●									
Governor Spring Shims High									●										
Governor Weights Incorrect (Heavy)	●									●				●	●				
Governor Weight, Pin Wear									●				●						●
High Speed Spring Shimming Wrong									●	●					●				
Idle Plunger (Button) Wrong					●								●						
Idle Plunger Spring Weak					●								●						
Idle Spring Wrong													●	●					
Injector Adjustment Loose															●	●	●		
Injector Cup Cracked, Wrong, Damaged				●		●									●		●		
Injector Flow High			●			●											●		
Incorrect Injector	●														●		●		
Injector Orifice Size Wrong						●									●		●		
Injector Plunger Worn	●																		
Pressure Valve Failure	●	●																	
Reversed Rotation, Drive Failure					●														
Screw Adjust, Incorrect									●	●		●	●	●					
Shutoff Valve Restriction								●							●				
Speed Settings, Unmatched auto/MVS or VS									●	●	●			●					
Spring Fatigue									●	●									
Throttle Leakage Excessive				●		●						●						●	
Throttle Linkage				●								●							
Throttle Shaft Restricted				●															
Throttle Shims Excessive						●						●						●	
Throttle Shims Insufficient	●														●				
Torque Spring Wrong						●									●		●		
Weight Assist Setting High						●			●								●		
Weight Assist Set Wrong						●	●	●	●	●					●	●			
Torque Limiting Valve Coil Not Energized	●														●				
Torque Limiting System (Air Valve) Switch Closed	●														●				

FIGURE 19–26 (Courtesy of Cummins Engine Company, Inc.)

DRIVEABILITY: DRIVEABILITY IS CLASSIFIED INTO TWO GENERAL AREAS, EACH WITH SEPARATE COMPLAINTS

1. Starting and/or sluggish acceleration 2. Low power and/or no acceleration

Complaint	Possible Cause	Checking Procedure	Correction
Startability and/or sluggish acceleration			
Hard starting or engine not starting	No-air needle valve shut off or set too lean	Apply 25 psi air pressure to AFC unit.	If engine starts, reset no-air needle valve.
		Caution: Excessive pressure will rupture bellows.	If engine does not start, AFC is not at fault.
	AFC plunger too short	Length should be 3.340/3.360 in.	Replace plunger.
	AFC cover center die diameter too deep	Check if face of cover to bottom of die hole 0.190/0.200 in. deep.	Replace cover.
	Barrel meter hole too deep	Check distance to hole, 1.615/1.635 in.	Replace barrel.
No-air valve not set to limits	Set before AFC plunger adjustment not after	Reset.	Make AFC adjustment, then no-air valve.
	Pressure too high	Reset.	Loosen jam nut; turn valve (cw) until pressure correct; tighten jam nut.
	Pressure too low	Reset.	Loosen jam nut; turn valve (ccw) until pressure correct; tighten jam nut.

Note: After any adjustment to pump, readjust AFC and no-air valve settings.

Complaint	Possible Cause	Checking Procedure	Correction
AFC pressure not set to limits	No-air needle valve not bottomed during adjustment	Reset. Check thread depth so valve seats.	Loosen jam nut on valve and bottom valve against seat; check for no flow before adjusting AFC. Bottom tap threads 5/16–24 UNF.
	Pressure set too high	Reset.	Loosen jam nut; turn AFC plunger (ccw) until pressure is correct; tighten jam nut.
	Plunger sticking	Differential pressure should not exceed 15 psi.	See section on AFC doesn't repeat.
Scored plunger and barrel	Vent screw filter media loose	Check if media loose in screw.	Install new vent screw, barrel, and plunger seals.
AFC not repeating within recheck limits; AFC recheck +3 psi; no-air recheck +2 psi	High plunger-to-barrel seal movement resistance	Set AFC to limits, increase air to 25 psi and decrease to setting; record value. Must not exceed 15 psi difference in values.	Work AFC plunger in and out several times. Check barrel surface finish and replace glyd ring.
Sluggish acceleration	Plugged vent capscrew	Blow through freely.	Clean or replace each time removed, and at maintenance check.
	Loose AFC barrel	Add parts noted, as necessary.	Correct installation or add 139585 spring to bottom of barrel and add S-16240 snap ring.
Low power	Air pressure required to actuate AFC higher than specifications; bellows	Apply 25 psi air pressure to AFC unit; large air leakage will be noted at vent capscrew.	Replace bellows. Bellows action forces air out. A small air leak is expected.
	Wrong AFC spring or setting	Check AFC spring setting and/or color code.	Adjust AFC setting or change spring.

FIGURE 19–27 (Courtesy of Cummins Engine Company, Inc.)

DRIVEABILITY: DRIVEABILITY IS CLASSIFIED INTO TWO GENERAL AREAS, EACH WITH SEPARATE COMPLAINTS—CONT'D

1. Starting and/or sluggish acceleration 2. Low power and/or no acceleration

Complaint	Possible Cause	Checking Procedure	Correction
Low power –cont'd	AFC plunger stuck in starting position	Apply 25 psi pressure to AFC unit and check for rise in fuel pressure as throttle is advanced.	Remove and check plunger and barrel and clean. Check seals.
	Ruptured bellows.	Check for large air leak.	Install new bellows. Watch; small air leak may be bellows action.
	AFC restricting early; power falling off rapid near torque peak	Check AFC setting.	Reset. AFC control should occur at least 100 rpm below torque peak
	Plugged filter vent cap-screw	Blow through freely.	Clean or replace each time removed.
Bellows leak	Improper use of AFC service tool or over-tightened cover	Check for frayed edges, elongated holes, or stretched holes. Torque capscrew.	Index tool sockets over nuts and tighten to 30/55 in.-lb. only.
	Bellows upside down; not aligned with bolt pattern.	Visually check.	Change or replace.
	Bubbles, cracks, separated bellows	Visually check.	Replace bellows.
Fall-off in power	AFC barrel movement	See sluggish acceleration section.	
Engine not accelerating properly	Air line to AFC leaking; plugs not connected	Check for loose line, connections, or plugged line.	Tighten connection lines or replace line.
	Ruptured bellows	Apply 25 psi air pressure to AFC air leakage at vent screw.	Replace bellows.
	Plugged vent capscrew.	Remove screw and check AFC operation.	Clean or replace.
	No-air needle valve set low	Reset.	Reset.
	Sticking AFC plunger	Remove cover and check plunger movement. Also see AFC doesn't repeat section.	Remove and check plunger and barrel and clean. Check seals.
	Wrong AFC setting, delay too much	Check plunger setting.	Reset to specifications.
Air leakage at AFC vent screw	Bellows damaged	Large air leakage from vent screw.	Replace bellows.
	AFC seal gasket (cork) not compressed, damaged, or missing	Small air leakage from vent screw.	Tighten nut on piston assembly to 30/40 in.-lb. Center bolt in piston so gasket seats. Replace gasket.
	Teflon tape seal on AFC plunger threads	Remove plunger nut and check for tape or sealant.	Retape plunger thread.
Excessive smoke	Fuel pump flow exceeding calibration specifications	Connect ST-435 pressure gauge to pump shutdown valve. Apply 25 psi air pressure to AFC unit. Take a snap pressure reading and check pressure against calibration specifications.	If fuel pressure is within specification, fuel rate is satisfactory. If above specification, adjust fuel pump pressure.

FIGURE 19–27 (Continued)

Complaint	Possible Cause	Checking Procedure	Correction
Excessive smoke –cont'd	Wrong AFC spring, plunger setting, or no-air setting	Check plunger and no-air settings on test stand. Also check: injectors, turbocharger, engine timing.	Change spring; adjust plunger or no-air needle valve.
Low engine fuel pressure	AFC not wide open	With gauge at shut-off valve, idle engine, then snap throttle wide open. Compare psi reached to no-air snap rail check in calibration data.	Reset AFC. May use for other pump problems as troubleshooting aid on engine.
Fuel leakage, no no-air needle valve	No-air needle valve O-ring	Check O-ring for damage and valve bore for burrs.	Replace O-ring. Remove burrs as necessary using appropriate tools.

Notes:
1. When checking AFC barrel inside diameter, a polished appearance is normal, but deep scratches will cause leakage even with a new seal; replace barrel.
2. AFC plungers will show a polished appearance. Light scratches are normal, but severe scratches require plunger replacement.
3. When any fuel pump component is replaced, special attention should be given to cleanliness and lubrication. With the AFC section be sure to use glyd ring installation, forming tools, and torque wrenches and make sure the hysteresis check described under AFC doesn't repeat within recheck limits section.

FIGURE 19–27 (Continued)

sound of the engine. A bad injector will have little or no effect on the engine when held down.

Injector Removal

1. Loosen the injector by adjusting the locknut on the rocker arm and backing out the adjusting screw.
2. Push the rocker arm down against the injector and remove the push rod.
3. Remove the injector link and hold-down bolts **(Figure 19–28)**.
4. Use the appropriate injector puller and remove the injector from the head **(Figure 19–29)**.

Testing Injectors

Cummins injectors are tested on a special test stand designed for this purpose. Procedures vary somewhat among different makes of testers, therefore instructions given for any particular tester must be followed for the tests to be valid. Follow the equipment manufacturer's instructions for testing injectors.

The following tests are performed on Cummins injectors.

1. Check for a sticking plunger **(Figure 19–30)**.
2. Check the spray pattern. **(Figure 19–31)**.

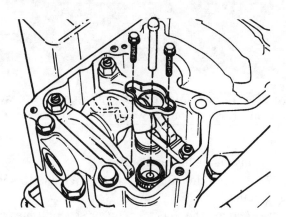

FIGURE 19–28 One-piece injector hold-down clamp removal. (Courtesy of Cummins Engine Company, Inc.)

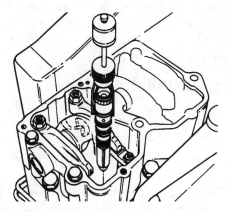

FIGURE 19–29 Removing an injector with a slide hammer puller. (Courtesy of Cummins Engine Company, Inc.)

408 Chapter 19 CUMMINS PT FUEL INJECTION SYSTEM

FIGURE 19–30 Injector plunger sticking tool. (Courtesy of Cummins Engine Company, Inc.)

FIGURE 19–31 Injector spray tester. (Courtesy of Cummins Engine Company, Inc.)

3. Check for injector leakage at
   a. plunger to barrel
   b. plunger to cup
   c. check ball **(Figure 19–32)**.
4. Calibration and flow testing **(Figure 19–33)**.

Injectors that do not pass these tests must be disassembled, cleaned, inspected, repaired, tested, and calibrated according to Cummins service manual procedures and specifications **(Figures 19–34 to 19–39)**.

Injector Installation

1. Install new O-rings on the injector.
2. Lubricate the O-rings with clean oil.
3. Install and align the injector in the cylinder head.
4. Use an injector installing tool to seat injector in sleeve **(Figure 19–40)**.
5. Install the retaining ring, clamp, and capscrews. Tighten the capscrews to specified torque.

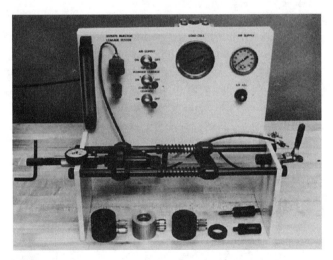

FIGURE 19–32 Injector leakage tester. (Courtesy of Cummins Engine Company, Inc.)

6. Install the link and push rod.
7. Tighten the adjusting screw locknuts to specified torque.
8. Actuate the injector plunger several times as a check of the adjustment.

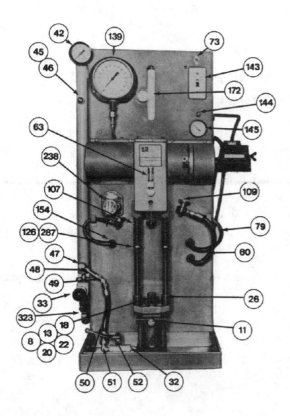

ST-790-8	Orifice Screw (.013)
ST-790-11	Exhaust Muffler
ST-790-13	Seat Block
ST-790-18	O-ring
ST-790-20	Orifice Screw (.011)
ST-790-22	Check Valve Spring
ST-790-26	Tie Rod Spacer
ST-790-32	Hex Nipple
ST-790-33	Air Valve
ST-790-42	Air Gauge
ST-790-45	O-ring
ST-790-46	Hyd. Oil Level Sight Gauge
ST-790-47	Elbow
ST-790-48	Hyd. Fluid Sight Glass
ST-790-49	Elbow
ST-790-50	Hose Assembly
ST-790-51	Valve Elbow
ST-790-52	Hydraulic Valve
ST-790-60	Inlet Hose
ST-790-63	Sight Glass
ST-790-73	Vial Light Switch
ST-790-79	Inlet Hose
ST-790-107	Tie Rod Spring
ST-790-108	Cover Assembly
ST-790-109	Inlet Connection
ST-790-126	Tie Rod Bushing
ST-790-139	Fuel Pressure Gauge
ST-790-143	Start & Stop Switch
ST-790-144	Flow Start Switch
ST-790-145	Temperature Gauge
ST-790-154	Hose Grommets
ST-790-172	Vial Graduate
ST-790-238	Pressure Regulator
ST-790-287	Tie Rod
ST-790-323	Air Valve
ST-790-335	Orifice (.020)
ST-790-338	Counter
ST-790-363	Orifice (.026)

FIGURE 19–33 ST-790 injector test stand. (Courtesy of Cummins Engine Company, Inc.)

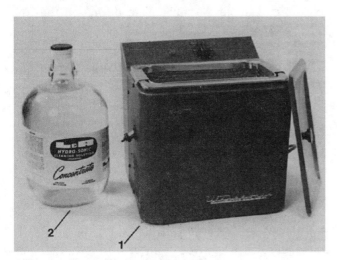

FIGURE 19–34 Ultrasonic injector parts cleaner. (Courtesy of Cummins Engine Company, Inc.)

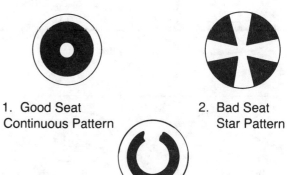

1. Good Seat Continuous Pattern
2. Bad Seat Star Pattern
3. Bad Seat Broken Pattern

FIGURE 19–35 Seat patterns in plunger cup. (Courtesy of Cummins Engine Company, Inc.)

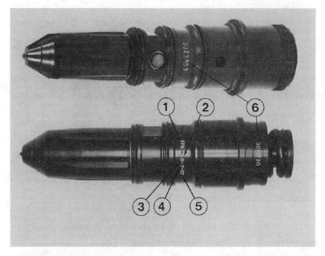

1. 178—Injector Flow
2. A—80% Flow
3. 8—Number of Holes
4. 7—Size of Holes (.007)
5. 17—Degree of Holes
6. Assembly Number

FIGURE 19–36 Assembly number of PT-D injector. (Courtesy of Cummins Engine Company, Inc.)

Chapter 19 CUMMINS PT FUEL INJECTION SYSTEM

NEW — Laser Marked
1. Cup Part Number
2. Year Quarter Made
   A.—First Quarter
   B.—Second Quarter
   C.—Third Quarter
   D.—Fourth Quarter
3. Year

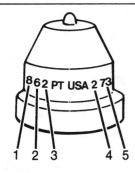

OLD — Stamped
1. Number of Holes
2. Size of Holes
3. Degree of Holes
4. Month
5. Year

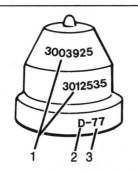

Old — Acid Etched
1. Cup Part Number
2. Year Quarter Made
   A.—First Quarter
   B.—Second Quarter
   C.—Third Quarter
   D.—Fourth Quarter
3. Year

FIGURE 19–37 Size marking on PT injector cups. (Courtesy of Cummins Engine Company, Inc.)

NOTE: The same engine position used for setting injectors is used to adjust the intake and exhaust valves on the same cylinder.

9. Adjust the valve clearance by putting the correct feeler gauge between the rocker lever and crosshead contact pads. Turn the adjusting screw down until the rocker lever touches the feeler gauge. See the service manual for valve clearance values.

CAUTION:

Make sure the valve tappet rollers are against the lobe on the camshaft before adjusting the valves.

10. Continue through the firing order until all the injectors and valves have been adjusted.

INJECTOR ADJUSTMENT (TYPICAL FOR L10 SERIES ENGINE)

This procedure describes the valve and injector adjustment procedures for fixed-time and STC engines.

All valves and injector adjustments *must* be made when the engine is cold and coolant temperature is stabilized at 140° F (60° C) or below.

The valve set marks are located on the accessory drive pulley. The marks align with a pointer on the gear cover **(Figure 19–41)**.

Use the accessory drive shaft to rotate the crankshaft.

The crankshaft rotation is *clockwise* when viewed from the front of the engine.

The cylinders are numbered from the front gear housing end of the engine.

The engine firing order is 1-5-3-6-2-4.

Each cylinder has three rocker levers **(Figure 19–42)**:

- The long rocker lever (E) is the exhaust lever.
- The center rocker lever is the injector lever.
- The short rocker lever (I) is the intake lever.

Refer to the accompanying chart for valve rocker lever locations.

On fixed-time and STC engines, the valves and injectors on the same cylinders are *not* adjusted at the same index mark on the accessory drive pulley.

One pair of valves and one injector are adjusted at each pulley index mark before rotating the accessory drive to the next index mark.

Two crankshaft revolutions are required to adjust all the valves and injectors.

Adjustment can begin on any valve set mark. In the following example the adjustment begins on the "A" valve set mark with cylinder No. 5 valves closed and cylinder No. 3 injector ready for adjustment.

1. Rotate the accessory drive *clockwise* until the "A" valve set mark on the accessory drive pulley is aligned with the pointer on the gear cover.

2. When the "A" mark is aligned with the pointer, the intake and exhaust valves for cylinder number five *must* be closed. If these conditions are *not* correct, cylinder No. 4 injector and cylinder No. 2 valves *must* be ready to set. According to the sequence chart **(Figure 19–43)**, set the injector on the cylinder that corresponds with the cylinder on which both the intake and exhaust valve rocker lever arms are loose and can be moved from side to side by hand **(Figure 19–44)**.

1. Fascia panel with instruments, controls and operating instructions
2. Dial indicator — injector output display
3. Fuel arm
4. Pressure select valve
5. Rotary control valve
6. Guard
7. Anti-splash flap
8. Cambox
9. Top tray
10. Stowage box — cams, box spanner, T bar etc.
11. Stowage box — adaptors and links
12. Work tray
13. Base with front access door

FIGURE 19-38 HA290 PT injector test stand. (Courtesy of Cummins Engine Company, Inc.)

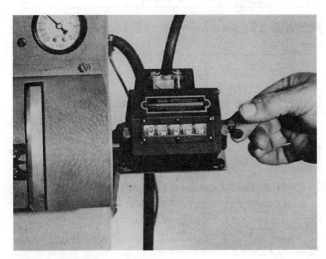

FIGURE 19-39 Setting the stroke counter at zero for flow testing an injector. (Courtesy of Cummins Engine Company, Inc.)

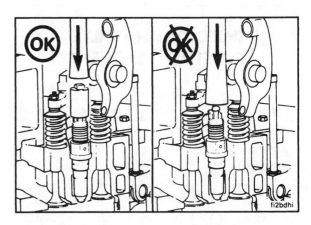

FIGURE 19-40 STC injector installation. (Courtesy of Cummins Engine Company, Inc.)

412 Chapter 19 CUMMINS PT FUEL INJECTION SYSTEM

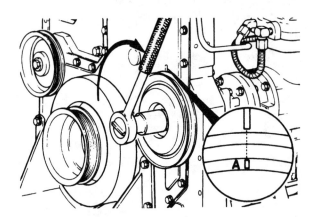

FIGURE 19–41 Valve set marks on accessory drive pulley. Accessory drive shaft must be used to rotate engine. (Courtesy of Cummins Engine Company, Inc.)

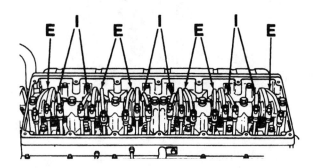

FIGURE 19–42 (Courtesy of Cummins Engine Company, Inc.)

INJECTOR AND VALVE ADJUSTMENT SEQUENCE

Bar Engine in Direction of Rotation	Pulley Position	Set Cylinder Injector	Set Cylinder Valve
Start	A	3	5
Advance to	B	6	3
Advance to	C	2	6
Advance to	A	4	2
Advance to	B	1	4
Advance to	C	5	1

Firing order: 1-5-3-6-2-4

FIGURE 19–43 (Courtesy of Cummins Engine Company, Inc.)

3. Loosen the locknut on the injector adjusting screw on cylinder No. 3 **(Figure 19–45)**. Tighten the adjusting screw until all the clearance is removed from the injector train.

4. Tighten the adjusting screw one additional turn to correctly seat the link **(Figure 19–46)**.

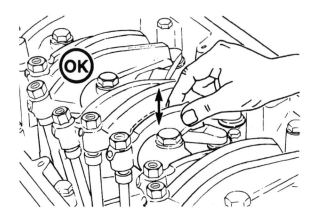

FIGURE 19–44 (Courtesy of Cummins Engine Company, Inc.)

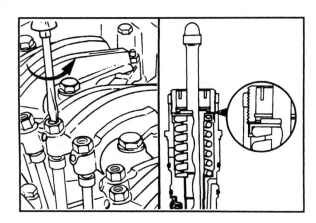

FIGURE 19–45 (Courtesy of Cummins Engine Company, Inc.)

Fixed-Time Injector Adjustment

1. Loosen the injector adjusting screw until the injector spring retainer washer touches the top stop screw.

2. Hold the adjusting screw in this position. The adjusting screw *must not* turn when the locknut is tightened. Tighten the locknut. The torque values are given with and without the torque wrench adapter, Part No. ST-669 **(Figure 19–47)**.

TORQUE VALUE:
- Without torque wrench adapter: 45 ft-lb (61 N·m)
- With torque wrench adapter: (1) 35 ft-lb (47 N·m)

Use a torque wrench and screwdriver socket to tighten the adjusting screw.

TORQUE VALUE: 5 to 6 in.-lb (0.6 to 0.7 N·m)

INJECTOR ADJUSTMENT (TYPICAL FOR L10 SERIES ENGINE)

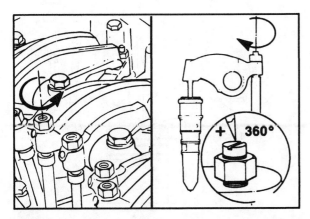

FIGURE 19–46 (Courtesy of Cummins Engine Company, Inc.)

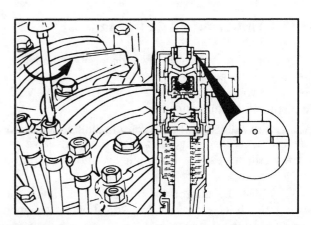

FIGURE 19–48 (Courtesy of Cummins Engine Company, Inc.)

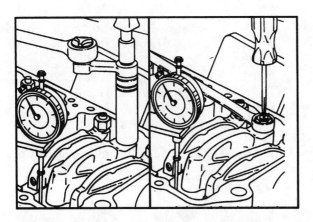

FIGURE 19–47 Tightening the injector adjusting screw with an ST-669 tool or screwdriver and box-end wrench. (Courtesy of Cummins Engine Company, Inc.)

CAUTION:

An overtightened setting on the injector adjusting screw will produce increased stress on the injector train and the camshaft injector lobe, which can result in engine damage.

STC Injector Adjustment

1. Loosen the injector adjusting screw until the STC tappet touches the top cap of the injector on STC engines **(Figure 19–48)**.

2. Place the STC tappet adjusting tool, part no. 3823348, on the upper surface of the STC injector top cap. Rotate the tool around the tappet until the tool's locating pin is inserted into one of the four holes in the top of the tappet.

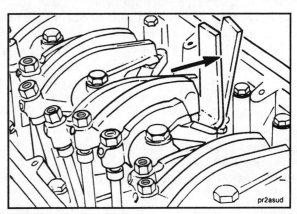

FIGURE 19–49 (Courtesy of Cummins Engine Company, Inc.)

Apply thumb pressure to the tool handle to hold the tappet in the maximum upward position **(Figure 19–49)**.

NOTE: Do **not** apply more force than needed to hold the tappet in the maximum upward position. The tool can break if excess force is used.

Fixed-Time and STC Engines

1. Adjust the valves on the appropriate cylinder according to the sequence chart before rotating the accessory drive to the next valve set mark. Refer to "Valve Adjustment" in this procedure.

2. After adjusting the valves on cylinder No. 5, rotate the accessory drive and align the next valve set mark on the accessory drive pulley with the pointer on the gear cover.

3. Adjust the appropriate injector and valves following the Injector and Valve Adjustment Sequence Chart **(Figure 19–43)**.

4. Repeat the process to adjust all injectors and valves.

INJECTION TIMING (TYPICAL FOR L10 SERIES ENGINE)

The injection timing *must* be correct to achieve the best performance, fuel economy, and lowest emissions. The injection timing *must* be checked when the engine is rebuilt and when a component of the gear train is replaced or removed and reinstalled.

Injection timing refers to injecting fuel in the combustion chamber at the correct time during the compression stroke. The timing *must* be checked when the piston is on the compression stroke at 0.2032 in. (5.160 mm) BTDC. When the piston is at this position, measure the amount of travel left in the injector push rod with the Part No. 3823451 Injector Timing Tool. To verify the correct injection timing for a particular engine, check the control parts list (CPL) number on the engine dataplate, then refer to CPL Bulletin No. 3379133. Timing codes are listed as two-letter alpha characters, for example, an "HM" code indicates a nominal setting of 0.199 in. (5.08 mm). Refer to the accompanying chart **(Figures 19–50 and 19–51)**.

L10 injection timing can be adjusted by removing the camshaft gear and changing the camshaft key. The camshaft key controls the position of the camshaft lobes during the operating cycles of the engine.

If an offset camshaft key is installed with the arrow marked on the top of the key pointing toward the engine (1), the timing will be *retarded*. If the offset key is installed with the arrow pointing away from the engine (2), the timing will be *advanced*. **(Figure 19–52)**.

Retarded timing (1) begins the fuel injection process *later*, and advanced timing (2) begins the fuel injection process *earlier* relative to the TDC position of the piston.

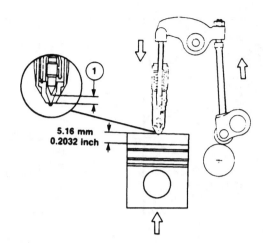

FIGURE 19–50 (Courtesy of Cummins Engine Company, Inc.)

Timing Code	Push Rod Travel @ 0.2032 in. (5.161 mm) BTDC Piston Travel Position
HM	0.197 to 0.201 in. (5.00 to 5.16 mm)
HN	0.144 to 0.148 in. (3.66 to 3.76 mm)

FIGURE 19–51 (Courtesy of Cummins Engine Company, Inc.)

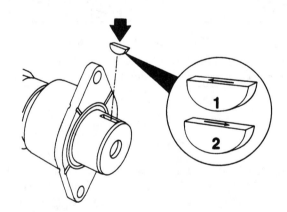

FIGURE 19–52 (Courtesy of Cummins Engine Company, Inc.)

Refer to the chart for a list of offset keys by part number and degree of offset. (Refer to **Figure 19–57**).

Injection Timing Check

1. Install the piston plunger rod of the Part No. 3823451 Injection Timing Tool into the injector bore of the No. 1 cylinder.

INJECTION TIMING (TYPICAL FOR L10 SERIES ENGINE)

2. Align the swivel bracket with the injector hold-down cap screw hole.

Install the cap screw through the swivel bracket. The cap screw is included in the timing tool kit.

NOTE: If the cap screw is tightened too tight, it will restrict the piston rod travel.

3. Tighten the cap screw enough to hold the timing fixture rigid.

4. Position the timing tool push tube plunger bracket (4) on the back side of the center bracket (5) **(Figure 19–53)**.

5. Use the alignment tool (6) to align the push rod plunger rod (7).

6. Tighten the clamp handle (8) after the plunger rod is aligned, and remove the alignment tool.

7. Install the injector push rod (9) between the injector camshaft follower and the plunger rod **(Figure 19–54)**.

NOTE: The push tube **must** be aligned properly to attain a correct reading.

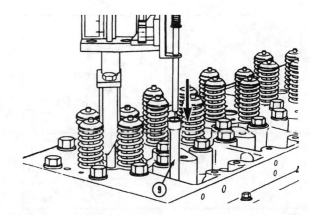

FIGURE 19–54 (Courtesy of Cummins Engine Company, Inc.)

CAUTION:

Always use the accessory drive shaft to rotate (bar) the crankshaft for injection timing. Using any other method will cause an error in the injection timing or can damage the engine.

8. Determine the piston TDC on the compression stroke by rotating the accessory drive shaft in the direction of engine rotation *(clockwise)*.

The piston is on the compression stroke when both plungers move in an upward direction at the same time. TDC is indicated by the maximum *clockwise* indicator position of the piston travel indicator pointer.

CAUTION:

Both indicators must have a travel range of at least 0.250 in. (6.35 mm) or the indicators will be damaged.

9. Position the gauge contact tip in the center of the plunger rod and lower the gauge to within 0.025 in. (0.63 mm) of the fully compressed position.

10. Set the dial indicator over the piston plunger rod to 0 (zero) when the piston plunger rod reaches maximum upward movement TDC.

11. Rotate the accessory drive shaft back and forth (before and after) the 0 (zero) indicator reading for approximately 3° to be sure the piston is at TDC.

Rotate the accessory drive shaft in the direction of engine rotation *(clockwise)* to 90° ATDC **(Figure 19–55)**.

The piston plunger will be at the "L10 90°" mark on the timing fixture.

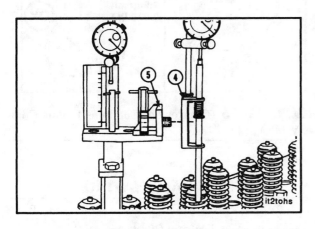

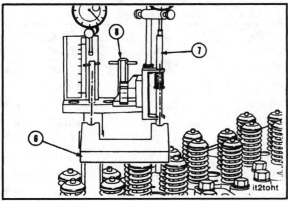

FIGURE 19–53 (Courtesy of Cummins Engine Company, Inc.)

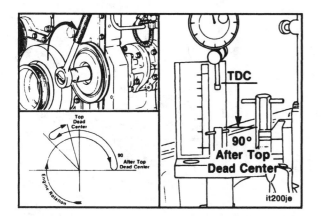

FIGURE 19-55 (Courtesy of Cummins Engine Company, Inc.)

12. Position the push rod dial indicator contact tip in the center of the plunger rod and lower the gauge to within 0.025 in. (0.63 mm) of the fully compressed position.

13. Set the push rod dial indicator to 0 (zero).

14. Rotate the accessory drive shaft in the opposite direction of crankshaft rotation *(counterclockwise)* to TDC.

Continue to rotate the accessory drive shaft *counterclockwise* until the crankshaft is at 45° BTDC.

NOTE: This step is necessary to remove gear backlash in the engine. When rotating the crankshaft **counterclockwise** near the TDC position, the push rod indicator will move from 0 (zero) to 0.005 to 0.015 in. (0.13 to 0.38 mm) in a **clockwise** direction.

15. Rotate the accessory drive shaft *clockwise* slowly until the piston travel gauge is at 0.2032 in. (5.160 mm) BTDC.

NOTE: If the crankshaft is rotated beyond the 0.2032 in. (5.160 mm) BTDC position, the crankshaft **must** be rotated **counterclockwise** back to the 45° BTDC mark.

16. Read the push rod travel gauge *counterclockwise* from 0 (zero). This travel represents the injection timing value. In the example shown, the value is 0.078 in. (1.98 mm).

To verify the correct timing for a particular engine, check the CPL number on the engine dataplate, then refer to the Control Parts List, Bulletin No. 3379133. Timing codes are listed as two-letter alpha characters. An example is a "HM" code, which indicates a nominal setting of 0.199 in. (5.08 mm). Refer to the accompanying chart.

If the indicator reading is lower than the specification, the timing is *advanced*.

If the indicator reading is higher than the specification, the timing is *retarded*.

Injection timing can be changed by removing the camshaft gear and installing an offset key **(Figures 19-56 and 19-57)**.

NT Engine Timing Adjustment

On NT engines the injection timing can be changed by changing the thickness of the cam follower housing gasket. Minimum and maximum thicknesses allowable are shown in **Figure 19-58**. Refer to the service manual for information for the effect of changing the gasket pack thickness on injection timing and push tube travel.

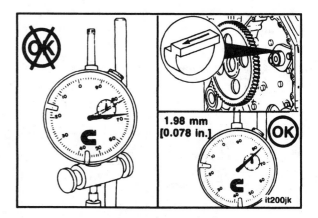

FIGURE 19-56 (Courtesy of Cummins Engine Company, Inc.)

OFFSET CAMSHAFT KEY

Key Part No.	Degree of Offset (To the Camshaft)	Change in Push Rod Travel	
		in.	mm
3030893	0.25	0.0020	0.051
3009948	0.50	0.0040	0.102
3030894	0.75	0.0060	0.152
3009949	1.00	0.0080	0.203
3030895	1.25	0.0100	0.254
3009950	1.50	0.0120	0.305
3030896	1.75	0.0140	0.356
3009951	2.00	0.0160	0.406
3030897	2.25	0.0180	0.457
3030898	2.50	0.0200	0.508

FIGURE 19-57 (Courtesy of Cummins Engine Company, Inc.)

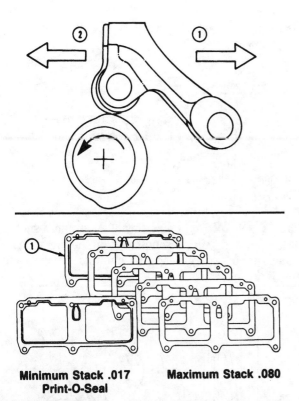

FIGURE 19-58 Moving the cam follower position in relation to the camshaft changes the push tube lift dimension and the injector timing on a fixed-timing NT engine. This can be done by changing the thickness of the cam follower housing gasket. (Courtesy of Cummins Engine Company, Inc.)

FUEL SYSTEM CHECKS AND ADJUSTMENTS

Before making the fuel system checks or adjustments on the engine, make sure of the following:

1. The engine must be at the operating temperature and the fuel temperature must not be above 110° F (43° C).
2. The engine parts are the same as those in the Control Parts List and are in good condition. The timing, valves, and injectors are correctly adjusted.
3. The instruments (gauges and tachometers) must have high accuracy.
4. The control linkage of the vehicle throttle is adjusted for full throttle travel. When released, the throttle is stopped by the throttle adjustment screw (throttle leakage adjustment screw).

NOTE: The control linkage of the vehicle throttle must have a maximum throttle stop. When the fuel pump is in the full throttle position, there must not be any pressure on the throttle shaft.

5. When the fuel pump is correctly calibrated, very little adjustment is required after the installation on the engine. A small adjustment of the idle setting and the fuel rail pressure is acceptable.
6. Fuel system problems may occur due to restrictions in the fuel supply line, air in the fuel, or air leaks. To diagnose the system for possible problems of this type several checks and tests as described here can be performed to isolate the cause. Refer to the service manual for exact procedures and specifications.

Fuel Supply Line Restriction Check

This check determines whether fuel flow from the tank to the pump is restricted. A vacuum gauge or mercury manometer may be used. The connection is made close to the pump or at the primary fuel filter. During the check the gauge must be held at the level of the gear pump to ensure accuracy. The check is preferably made at rated engine speed under full load. If this is not possible, it may be made at the high-idle speed. Restriction should not exceed 4 to 8 in. (100 to 200 mm) Hg **(Figure 19-59)**.

Fuel Drain Line Restriction Check

This check is made with a vacuum gauge connected to the drain line as shown in **Figure 19-60.** Restriction is checked at rated engine speed under full load. Restriction should not exceed 2.5 in. (62.6 mm) Hg on systems

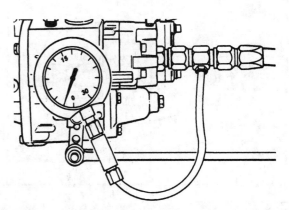

FIGURE 19-59 Using a vacuum gauge to check PT fuel system restriction. (Courtesy of Cummins Engine Company, Inc.)

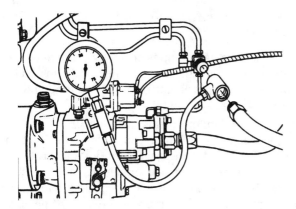

FIGURE 19-60 PT fuel pump drain line restriction check. (Courtesy of Cummins Engine Company, Inc.)

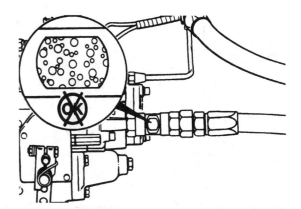

FIGURE 19-62 Sight glass installed in pump suction line is used to detect presence of air in the fuel with the engine running. (Courtesy of Cummins Engine Company, Inc.)

without a check valve or 6.5 in. (162.5 mm) Hg on systems with a check valve.

Suction Side Air Leak

Several different methods may be used to check for air leakage into the suction side of the pump. These include (a) installing a temporary fuel return line submerged in a beaker partially filled with fuel, (b) installing a clear plastic line between the pump and the filter, (c) installing a sight glass in the line between the primary filter and the pump, or (d) pressurizing the fuel system between the fuel tank and the gear pump. Specified pressure must not be exceeded during this check **(Figures 19-61 to 19-63)**.

Any of these procedures will reveal suction side air leaks.

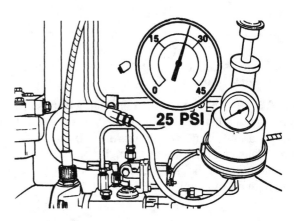

FIGURE 19-63 Hand pump and pressure gauge connected to AFC fuel line. (Courtesy of Cummins Engine Company, Inc.)

Fuel Pump Air Leak

This check involves loosening the PT fuel pump outlet line above the fuel solenoid to vent any trapped air. If air continues to vent, the pump must be removed and repaired **(Figure 19-64)**.

Fuel Rail Pressure

Low fuel rail pressure will reduce engine power. To check fuel rail pressure, remove the access plug at the fuel shut-off valve and connect a pressure gauge (Cummins ST435) to the opening. Take the pressure reading at the full-load rated engine speed and compare it with specifications for the engine **(Figure 19-65)**.

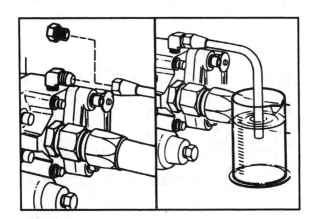

FIGURE 19-61 Using a temporary test line connected to the fuel pump suction fitting to check for presence of air in the fuel. (Courtesy of Cummins Engine Company, Inc.)

FUEL SYSTEM CHECKS AND ADJUSTMENTS 419

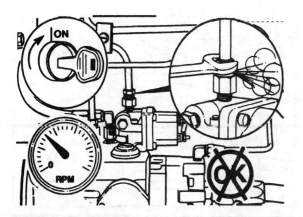

FIGURE 19-64 Venting air from fuel pump outlet fitting. (Courtesy of Cummins Engine Company, Inc.)

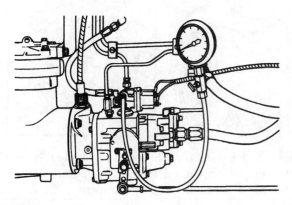

FIGURE 19-65 Checking fuel system pressure at the fuel shut-off valve. (Courtesy of Cummins Engine Company, Inc.)

PTG-AFC Pump Fuel Rate

The amount of fuel consumed per horsepower-hour can be determined on an engine or chassis dynamometer using the appropriate fuel flow rate equipment. The test is performed at full-load rated speed. Another method of determining fuel consumption is to record the amount of fuel consumed during operation in the engine's actual operating environment over a specified time period **(Figure 19-66)**.

PTG-AFC Fuel Rate Adjustment

The fuel rate can be adjusted by turning the throttle shaft adjusting screw. The screw may have a hex socket or a slot for adjustment. Access to the screw requires removal of the tamper-resistant ball at the end of the throttle shaft. Turning the screw clockwise reduces the fuel rate, and turning it counterclockwise increases

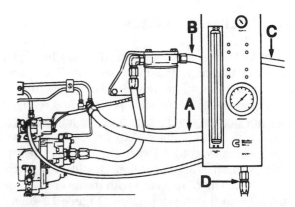

FIGURE 19-66 Checking the fuel rate. (Courtesy of Cummins Engine Company, Inc.)

the fuel rate. The screw must never be turned out to less than 0.250 in. (6 mm) from the end of the shaft. Doing so may force the plunger out of the throttle shaft **(Figure 19-67)**.

At idle speed, fuel must be able to flow through the opening to keep the engine running. If the opening is too restricted, the engine will stall. At full-load rated speed, the size of the fuel pump button in the pump regulates the maximum fuel pressure, the flow rate, and therefore the maximum horsepower **(Figure 19-68)**.

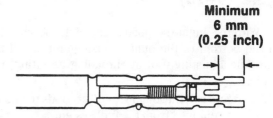

FIGURE 19-67 Throttle shaft adjusting screw must not be backed out beyond this dimension. (Courtesy of Cummins Engine Company, Inc.)

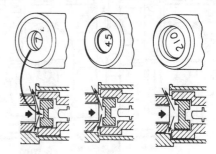

FIGURE 19-68 The engine's power rating can be changed by changing to a different size idle spring plunger. Several sizes are available. (Courtesy of Cummins Engine Company, Inc.)

Throttle Linkage Check

The throttle linkage must be checked for any binding or interference that may inhibit its operation. The linkage must be able to move the throttle lever freely from its low-idle position to its high-idle position and return. If it does not do this, the linkage must be adjusted. The linkage must also be adjusted to provide the correct amount of breakover travel when the lever is in the full-throttle position. This dimension should be between 0.125 and 0.250 in. (3 to 6 mm) with the lever stop contacting the rear throttle stop screw **(Figure 19–69)**.

Engine Speed Adjustment

To adjust the engine low- or high-idle speed:

1. Obtain the correct low- and high-idle speed specifications for the engine from the data plate or the service manual.
2. Bring the engine to normal operating temperature.
3. Connect an accurate tachometer to the pump tach drive, or use a digital or optical tachometer.

PTG-AFC Low-Idle-Speed Adjustment (Figure 19–70)

1. With the engine stopped, remove the access plug from the bottom of the pump housing and install the special idle-adjusting tool by threading its fitting into the hole.
2. Start and run the engine for 30 seconds to ensure that all air is removed from the fuel system.
3. Return engine speed to low idle.

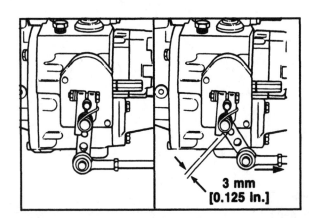

FIGURE 19–69 Throttle lever breakover dimension. (Courtesy of Cummins Engine Company, Inc.)

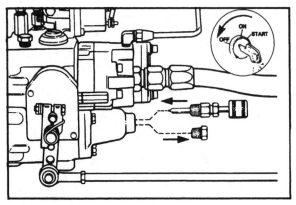

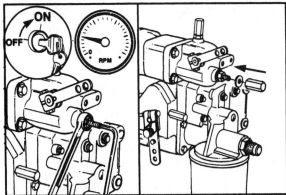

FIGURE 19–70 Low-idle-speed adjustment on PTG-AFC pump (top) and PTG-AFC-VS pump (bottom). (Courtesy of Cummins Engine Company, Inc.)

4. Turn the adjusting screw to increase or decrease idle speed until the specified idle speed is obtained.
5. Stop the engine, remove the idle-adjusting tool, install the access plug, and tighten it to specifications.

PTG-AFC-VS Low-Idle-Speed Adjustment (Figure 19–71)

1. With the engine stopped, remove the lock- and jam nuts from the idle-speed screw. Discard the old copper washers.
2. Install a new copper washer and the jam nut on the idle-speed screw. Do not tighten.
3. Start the engine and hold the variable-speed lever in the idle position.
4. Use a hex wrench to adjust the idle-speed screw. Increase or decrease the engine speed to obtain the specified idle speed.
5. Hold the hex wrench firmly to avoid any change in the adjustment while tightening the jam nut to specifications. Recheck the idle speed. Install a new copper

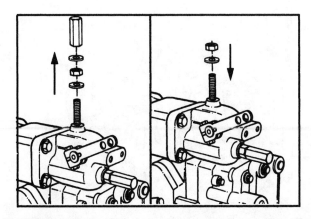

FIGURE 19–71 PTG-AFC-VS high-idle-speed adjustment. (Courtesy of Cummins Engine Company, Inc.)

washer and the locknut on the adjusting screw and tighten the locknut to specifications.

PTG-AFC-VS High-Idle-Speed Adjustment

1. With the engine stopped, remove the lock- and jam nuts from the top screw on the VS governor cover. Discard the old copper washers.

2. Install a new copper washer and the jam nut on the adjusting screw. Do not tighten at this point.

3. Start the engine and slowly rotate the VS lever clockwise until it reaches the maximum high-idle position and hold it there.

4. Use a hex wrench and increase or decrease the engine high-idle speed to obtain the specified speed.

5. Hold the hex wrench firmly to prevent any change in the adjustment and tighten the jam nut. Recheck the high-idle speed.

6. Install a new copper washer and the locknut on the adjusting screw and tighten the locknut to specifications.

PTG-AFC No-Air Valve Adjustment

During low-idle or high-idle operation when there is very little or no turbocharger boost pressure, the AFC cuts off fuel flow through the fuel solenoid and the fuel rail. The only fuel being fed to the injectors flows from the throttle shaft to the AFC and the no-air needle valve. The no-air valve must be set to provide this fuel flow.

The no-air valve adjustment is made with the engine under load at the specified rpm (usually 1600 rpm). The engine load may be applied with a chassis dynamometer, during a road test using the vehicle brakes, or during a stall test on torque converter–equipped vehicles or equipment. Refer to the service manual for specifications and procedures.

FUEL PUMP REMOVAL, INSTALLATION, AND TESTING (FIGURES 19–72 TO 19–74)

The fuel pump should be removed only after it has definitely been established through testing that it is at fault. Fuel pump removal and installation is a simple procedure, since no pump timing is required.

To remove the fuel pump, first clean the pump and surrounding area thoroughly. Disconnect the fuel lines and cap all openings to prevent the entry of dirt. Disconnect the throttle linkage. Remove the pump mounting bolts and then the pump. To install the pump, simply follow the above procedures in reverse order, making sure that all connections and mounting bolts are tightened to specifications and that there are no fuel leaks.

A faulty fuel pump must be disassembled, cleaned, inspected, repaired, and reassembled. After assembly the pump is installed on a test stand for testing and calibration. The procedures and specifications of the test equipment manufacturer and the appropriate Cummins service manual must be applied during pump overhaul, testing, and calibration.

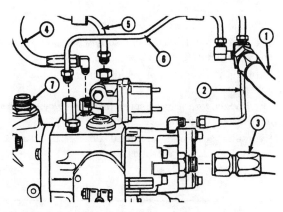

FIGURE 19–72 Removing connections from fuel pump prior to pump removal. 1. Fuel drain line from cylinder head, 2. Gear pump cooling drain line, 3. Gear pump suction line, 4. AFC fuel drain line, 5. Fuel supply line to injectors, 6. AFC air supply hose to intake manifold, 7. Tachometer drive cable. (Courtesy of Cummins Engine Company, Inc.)

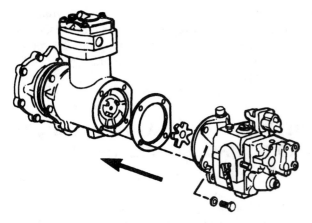

FIGURE 19–73 Removing PT pump from air compressor housing. (Courtesy of Cummins Engine Company, Inc.)

FIGURE 19–74 Removing PT pump from accessory drive. (Courtesy of Cummins Engine Company, Inc.)

REVIEW QUESTIONS

1. The Cummins PT injection system is used exclusively on _____ engines.
2. A _____ _____ positive-displacement pump supplies fuel to the injectors.
3. The pump also contains the governor assembly, which controls _____ pressure, engine _____ maximum speed, and _____.
4. The injectors meter the fuel and inject it into the cylinder at _____ _____.
5. The fuel pump assembly comprises three main units: _____ _____, _____, and _____.
6. State the basic functions of the fuel pump.
7. The excess fuel from the fuel supply pump is bypassed and returned to the _____ _____.
8. Explain the operation of the idle-speed and high-speed governor.
9. The MVS governor has a speed droop between full load and no load of about _____.
10. Hydraulic governors are used on _____ _____ applications.
11. PT (type D) injectors combine the acts of _____, _____, and _____.
12. Cummins injection timing is controlled by the amount of fuel metered into the _____ _____.
13. Hydraulic variable timing (HVT), a _____-_____ system, is relatively simple.
14. During the advanced mode of injection timing, the STC system reduces the _____ _____ _____ during cold-weather operation.
15. Explain proper injector removal procedure.
16. The injector timing must be correct to achieve the best performance _____ _____ and _____ _____.
17. Fuel system problems may occur due to restrictions in the fuel _____ _____, _____ in the fuel, or _____ _____.
18. Fuel pump removal and installation is a simple procedure, since _____ _____ _____ is required.

TEST QUESTIONS

1. The Cummins PT pump contains the governor assembly, which controls
   a. fuel pressure
   b. engine idle speed
   c. maximum speed and torque
   d. all of the above
2. Electronic fuel control systems were developed to
   a. meet emission control legislation
   b. give better performance
   c. have better rpm control
   d. improve fuel economy

3. The fuel pump assembly is made up of three main units:
   a. vane pump, governor, and throttle
   b. gear pump, governor, and throttle
   c. vane pump, throttle, and rack
   d. gear pump, governor, and springs
4. A gear pump is driven by the
   a. camshaft
   b. crankshaft
   c. pump main shaft
   d. all of the above
5. The excess fuel supplied by the PT system supply pump is bypassed and returned to the
   a. fuel tank
   b. injectors
   c. pump and inlet
   d. pump and outlet
6. The idling and high-speed mechanical governor is actuated by
   a. a system of springs and weights
   b. hydraulic pressure
   c. weights and fuel pressure
   d. engine oil pressure
7. A hydraulic governor is used on stationary power applications where it must maintain
   a. a minimum speed
   b. a minimum and maximum speed
   c. constant speed with varying loads
   d. constant speed
8. The PT injector provides a means of introducing fuel into the combustion chamber. It combines the acts of
   a. metering and injection
   b. metering, timing, and injection
   c. injection and timing
   d. vaporizing the fuel and mixing it with the air
9. Cummins injection timing is controlled by the amount of
   a. fuel metered
   b. fuel from the supply pump
   c. injection pressure
   d. fuel metered into the injector cup
10. Air flow to and from the cylinder is controlled by a
    a. mechanical valve
    b. damper in the air intake
    c. three-way solenoid control valve
    d. solenoid control valve
11. Fuel rate can be adjusted by turning the
    a. fuel pump actuator
    b. throttle shaft slightly
    c. governor spring nut
    d. throttle shaft adjusting screw
12. Fuel pump removal and installation is a simple procedure, since
    a. no pump timing is required
    b. the pump can easily be retimed
    c. pump *cannot* be installed wrong
    d. timing marks are quite visible

CHAPTER 20

CUMMINS ELECTRONIC CONTROL SYSTEMS

INTRODUCTION

The PACE fuel control system introduced in 1988 is an electronic control system added to the existing PT fuel system described earlier that replaced its mechanical controls. The later PT PACER system is an improved version of PACE. The Cummins electronically controlled injection (ECI) system called CELECT became available in 1990. It uses a gear pump to supply fuel to the mechanically operated electronic injectors. Fuel metering and injection timing are controlled electronically. Both systems use the dash-mounted flashing-light fault code retrieval method and the Compulink diagnostic and programming tool.

PERFORMANCE OBJECTIVES

After thorough study of this chapter and the appropriate training models, service manuals, and equipment, you should be able to:

1. List the PT PACER system components and describe their basic function.
2. Describe the PACER system fuel flow.
3. Diagnose PACER system faults using the flashing-light codes and the Compulink tool.
4. List the CELECT system components and describe their function.
5. Describe CELECT system fuel flow.
6. Describe CELECT system electronic injector operation.
7. Diagnose CELECT system faults using the flashing-light codes and the Compulink tool.

TERMS YOU SHOULD KNOW

Look for these terms as you study this chapter, and learn what they mean.

PACE	CLS
PT PACER	OPS
PTCM	OTS
PDM	APS
EPS	IATS
VSS	BPS
EFC	TPS
TBV	VSS
TPS	gear pump
brake switch	ECUI
clutch switch	warning lights
cruise control	idle-speed adjust switch
warning light	compression brake control switch
fault codes	cruise control switches
CELECT	cooling fan clutch control
ECI	gear-down protection
ECM	progressive shifting
PDM	idle shutdown
EPS	engine protection system
CTS	road relay

sensor plus Echek
data link Compulink

PACE AND PT PACER ELECTRONIC CONTROL SYSTEM COMPONENTS AND FUNCTION

The PACE and PT Pacer fuel control systems are an addition to the Cummins PT fuel system **(Figure 20–1)**. They replace the mechanical fuel controls. System components include those of the PT system and the following electronic components:

1. *Pressure time control module (PTCM)*: receives operating information from all other PACE/PACER components **(Figure 20–1)**. It uses this information along with information stored in module memory to control fuel system and engine operation. The earlier PTCM did not contain the power distribution module (PDM). Later PTC modules incorporated the two components into a single unit. The PTCM is mounted on the side of the engine with a cooling plate in between. Fuel

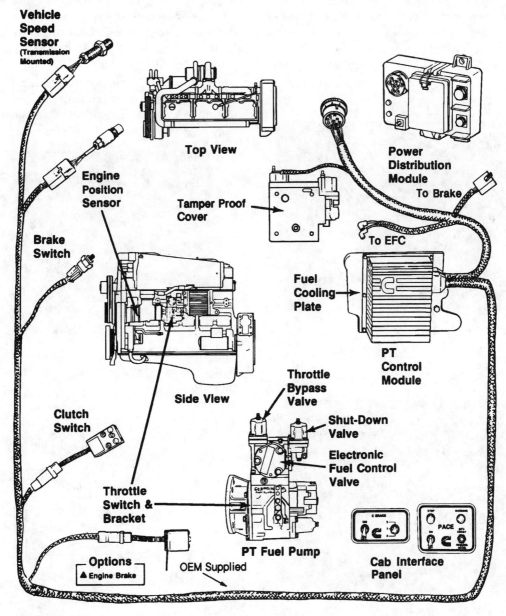

FIGURE 20–1 PACE fuel control system. (Courtesy of Cummins Engine Company, Inc.)

flow through the cooling plate cools the module electronics. The PTCM (and PDM) provides electrical power for system operation. Battery voltage is supplied to the PTCM.

2. *Power distribution module (PDM):* receives its power from the vehicle battery. The PDM provides power to the PTCM and protects it from polarity reversal and power surges **(Figure 20–1)**. On later models the PTCM and PDM were combined into a single unit.

3. *Engine position sensor (EPS):* detects the passage of notches in the back face of the camshaft gear, which provides the PTCM with information on engine speed and piston position **(Figure 20–2)**.

4. *Vehicle speed sensor (VSS):* senses transmission output shaft speed and sends this information to the PTCM for use in governing road speed **(Figure 20–3)**.

5. *Electrical fuel control actuator (EFC):* receives signals from the EPS, PTCM, and VSS to control vehicle road speed and engine maximum speed **(Figures 20–4 and 20–5)**.

6. *Throttle bypass valve (TBV):* located in the electronic governor housing, the TBV bypasses fuel from the throttle to allow the EFC to control fuel flow to the engine **(Figure 20–6)**.

7. *Throttle position switch (TPS):* located in the PT fuel pump housing, the TPS is closed when the throttle is at idle position. When the driver depresses the throttle pedal, the TPS opens to deactivate the engine compression brake and override the cruise control.

8. *Brake switch:* located in the service brake line, it opens when pressure approaches 4 psi (28 kPa). With the switch open the PACER system cannot operate. With the brakes off the switch closes, allowing the PACER system to function.

9. *Clutch switch:* located at the clutch pedal or clutch linkage, it is closed with the pedal in the UP position. This allows cruise control and engine brake acti-

FIGURE 20–2 Engine position sensor. (Courtesy of Cummins Engine Company, Inc.)

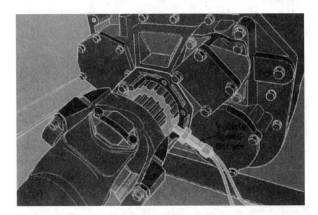

FIGURE 20–3 Vehicle speed sensor senses transmission output shaft speed. (Courtesy of Cummins Engine Company, Inc.)

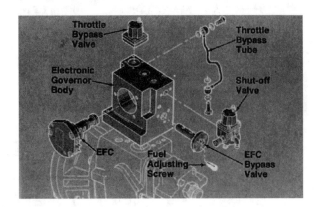

FIGURE 20–4 PACE pump-mounted module output components. (Courtesy of Cummins Engine Company, Inc.)

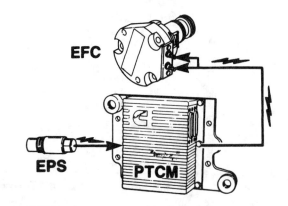

FIGURE 20–5 Electrical fuel control actuator. (Courtesy of Cummins Engine Company, Inc.)

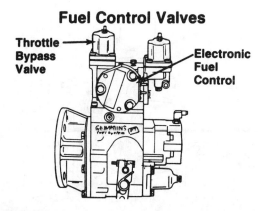

FIGURE 20–6 (Courtesy of Cummins Engine Company, Inc.)

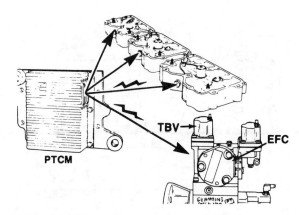

FIGURE 20–8 PTCM outputs. (Courtesy of Cummins Engine Company, Inc.)

vation to occur. Depressing the clutch pedal opens the switch and deactivates engine braking and cruise control.

10. *Compression brake switch:* a dash-mounted ON/OFF switch that controls activation of the engine compression brake. A second switch with three positions allows the brake to be activated on two, four, or six cylinders.

11. *Limp-home mode:* In the event of system failure a fuel bypass valve can be manually rotated to bypass the EFC (electronic fuel control) valve. This allows the vehicle to be moved to safe parking or to a service facility. Engine power and speed are reduced in this mode.

Input and output components are shown in **Figures 20–7** and **20–8.**

Cruise Control Operation. The cruise control can be activated when vehicle speed is greater than 30 mph (50 km/h). The cruise control is deactivated by applying the tractor or trailer brakes, depressing the clutch pedal, activating the engine brake, or switching the cruise control off.

Power Take-off Operation. The PACER system can be used to control PTO speed. To utilize the PTO mode the vehicle must be stopped. Several PTO speeds may be selected. The cruise control switches are used to set the PTO speed.

PT PACER SYSTEM OPERATION

Electrical power is supplied to the system by the battery through the PDM and the PTCM. Input switches and sensors provide operating information to the PTCM. The PTCM controls engine operation based on this information and programmed information in PTCM memory. Components controlled by the PTCM include the TBV and the EFC **(Figure 20–1).**

FUEL FLOW

Fuel flows from the fuel filter to the PTCM cooling plate to the PT pump. Fuel flow for the pump is the same as for the nonelectronic PT pump described in the previous chapter. The electronic governor body mounted on top of the pump contains the throttle bypass valve, throttle bypass tube, EFC valve, EFC bypass valve, fuel adjusting screw, fuel shut-off valve, and the fuel passages that connect these components to the pump **(Figure 20–9).**

With the cruise control and PTO off, fuel flows from the fuel pump to the restriction screw, EFC bypass valve, EFC valve, shutdown valve, and through drilled passages in the cylinder head to the injectors **(Figure**

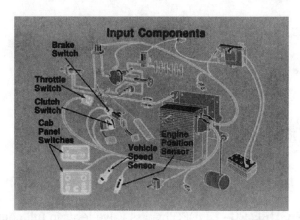

FIGURE 20–7 PACE system inputs. (Courtesy of Cummins Engine Company, Inc.)

428 Chapter 20 CUMMINS ELECTRONIC CONTROL SYSTEMS

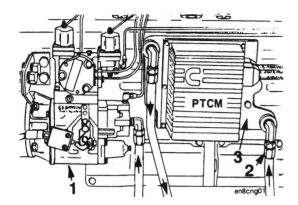

FIGURE 20–9 Fuel flow through cooler plate. (Courtesy of Cummins Engine Company, Inc.)

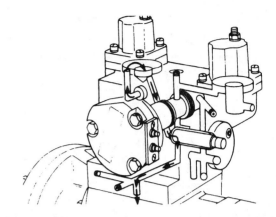

FIGURE 20–11 Fuel flow in cruise control mode bypasses the closed throttle shaft. (Courtesy of Cummins Engine Company, Inc.)

20–10). The PTCM signals the EFC valve to remain wide open, and fuel delivery is controlled by the accelerator pedal. On signals from the vehicle speed sensor, the PTCM limits fuel delivery through the EFC valve to limit top road speed **(Figure 20–11)**.

With cruise control or PTO in operation, fuel bypasses the throttle valve, flows to the AFC, EFC valve, and fuel shut-off valve. The EFC valve controls fuel flow to control engine speed and road speed.

Should the system fail, it can be bypassed by turning the EFC bypass valve lever to the full clockwise position **(Figure 20–12)**. During normal operation the EFC bypass lever must be in the full counterclockwise position **(Figure 20–13)**. During bypass operation fuel flow and engine speed and power are reduced.

FAULT CODES

Fault codes stored in the PTCM can be read by activating the dash-mounted warning and stop lights. The

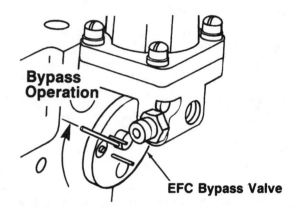

FIGURE 20–12 EFC bypass valve lever in bypass position. (Courtesy of Cummins Engine Company, Inc.)

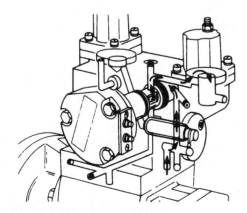

FIGURE 20–10 Fuel flow through the AFC, EFC valve, and fuel shut-off valve. (Courtesy of Cummins Engine Company, Inc.)

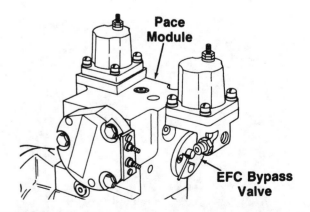

FIGURE 20–13 EFC bypass valve lever in full counterclockwise position for normal operation. (Courtesy of Cummins Engine Company, Inc.)

warning light is yellow, and the stop light is red. If the yellow warning light comes on during vehicle operation, a fault has been detected by the PACER system; however, the fault is not serious enough to warrant engine shutdown. The fault should be corrected as soon as possible. If the red warning light comes on, the vehicle should be moved over to the side of the road and the engine stopped as soon as possible to avoid major engine damage. The problem must be corrected before proceeding with the vehicle. Fault codes are stored in the PTCM for later retrieval by the technician.

To access the fault codes turn on the ignition key (engine stopped) and turn on the diagnostic switch on the dash panel. The warning light will then flash any stored fault codes. The yellow light will flash first, then after a short pause the red light will flash the code **(Figure 20–14)**. The code will repeat until the system is asked to perform another function or is switched off. To access other fault codes, push the cruise control switch to the SET/COAST position and release it. To clear fault codes, the 9-pin connector must be disconnected from the PDM for about 30 minutes.

PACER System Fault Codes and Their Meaning

CODE	MEANING
115	Engine position sensor
144	TBV circuit
152	Vehicle out of fuel or poor PACER system ground or short in actuator harness
153	EPS and VSS circuit
154	EFC valve circuit
155	EFC bypass valve circuit
212	PTCM and VSS circuit
214	Cylinder 1 and 2 C brake problem
215	Cylinder 3 and 4 C brake problem
221	Cylinder 5 and 6 C brake problem
225	PDM circuit
234	PTCM and cruise control circuit
241	PTCM system check required
242	PDM relay and PTCM circuit
243	PTCM and harness

The Compulink test equipment can be used on both the PACER and CELECT systems and is described later in this chapter.

CELECT ECI (ELECTRONICALLY CONTROLLED INJECTION) SYSTEM COMPONENTS AND FUNCTION

The CELECT electronic control system uses a gear pump to supply fuel to the injectors. The injectors are mechanically operated by the engine camshaft, whereas fuel metering and injection timing are controlled electronically. The system is designed to reduce harmful exhaust emissions and improve fuel economy **(Figures 20–15 to 20–19)**.

1. *Electronic control module (ECM):* receives information about operating conditions from various input sensors and switches. It processes this information in conjunction with programmed information stored in ECM memory to produce output signals that control fuel metering and injection timing and operate other output devices as required. It logs faults for later retrieval for diagnosis and it can be programmed to meet certain customer-specified operating parameters **(Figures 20–15 to 20–19)**.

2. *Wiring harness:* connects the various switches, sensors, and output devices to the ECM. It includes the sensor harness, actuator harness, and vehicle OEM harness, which are all plugged into the ECM **(Figures 20–16 and 20–17)**.

3. *Engine position sensor (EPS):* detects the passage of notches in the back face of the camshaft gear to provide the ECM with engine speed and piston position information **(Figure 20–2)**.

4. *Coolant temperature sensor (CTS):* monitors engine coolant temperature and provides this information to the ECM **(Figure 20–20)**.

5. *Coolant level sensor (CLS):* informs the ECM when the coolant level drops below the specified level **(Figure 20–21)**.

FIGURE 20–14 (Courtesy of Cummins Engine Company, Inc.)

430 Chapter 20 CUMMINS ELECTRONIC CONTROL SYSTEMS

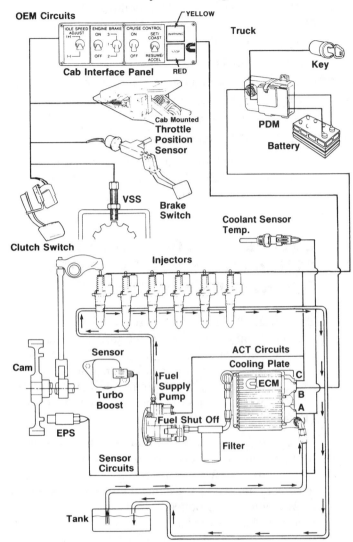

FIGURE 20–15 CELECT ECI system schematic. (Courtesy of Cummins Engine Company, Inc.)

6. *Oil pressure sensor (OPS):* provides oil pressure signal to the ECM. If oil pressure drops to a specified level, the warning light goes on, and engine power is reduced. If oil pressure drops too low, the engine shutdown function is activated **(Figure 20–22)**.

7. *Oil temperature sensor (OTS):* used by the ECM to determine cold-engine speed (fast idle when the engine is cold). If oil temperature rises too high, engine speed and power are reduced by the ECM **(Figure 20–22)**.

8. *Ambient air pressure sensor (APS):* used by the ECM to help control fuel delivery at different altitudes (less fuel at higher altitudes because air is less dense) **(Figure 20–23)**.

9. *Intake air temperature sensor (IATS):* The ECM uses turbo boost output air temperature to adjust the fuel rate in accordance with changes in air temperature and density. Hotter air is less dense and requires less fuel **(Figure 20–24)**.

10. *Boost pressure sensor (BPS):* measures boost pressure in the intake manifold. The ECM uses this information to match fuel delivery to the amount of air entering the engine **(Figure 20–25)**.

11. *Throttle position sensor (TPS):* The accelerator pedal–mounted potentiometer produces a variable output signal in relation to pedal position. The ECM uses this information to vary fuel delivery according to driver demands. The TPS includes an idle verification switch that tells the ECM when the pedal is in the idle position **(Figure 20–26)**.

12. *Vehicle speed sensor (VSS):* tells the ECM the road speed of the vehicle. The ECM uses this information for cruise control operation and to control the vehicle maximum speed **(Figure 20–27)**.

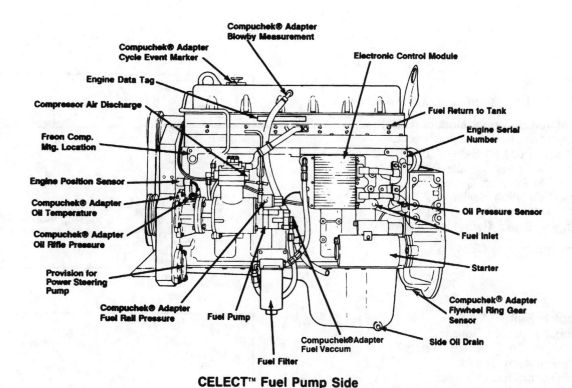

CELECT™ Fuel Pump Side

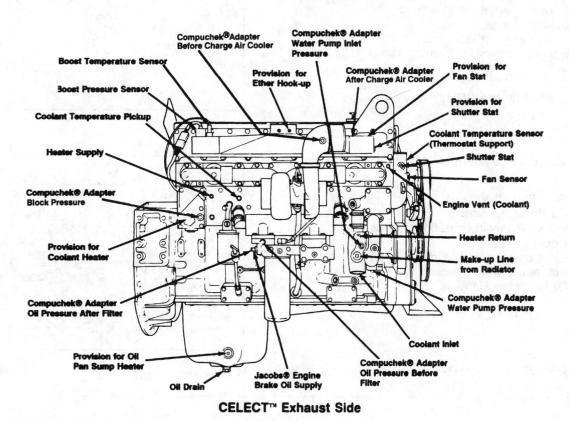

CELECT™ Exhaust Side

FIGURE 20–16 (Courtesy of Cummins Engine Company, Inc.)

Components

The CELECT™ system contains the following parts:

1. Electronic Control Module (ECM)
2. Electronically Controlled Injectors
3. Engine Position Sensor (EPS)
4. Vehicle Speed Sensor (VSS)**
5. Throttle Position Sensor**
6. Idle Validation Switch**
7. Brake Switch**
   Engine Brake Relay** (not shown)

**Provided by the OEM

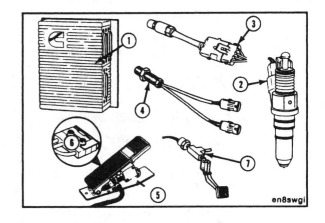

8. Clutch Switch**
9. Boost Pressure Sensor
10. Ambient Air Pressure Sensor*
11. Oil Pressure Sensor
12. Oil Temperature Sensor
13. Intake Manifold Temperature Sensor
14. Coolant Temperature Sensor
15. Coolant Level Sensor**
    Fault Lamps (Electronics)** (not shown)
    Engine Protection Warning Lamp** (not shown)

* **Not** used on all engine ratings.

**Provided by the OEM

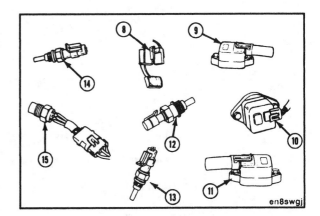

The CELECT™ system on engine components consists of:

1. Fuel Shutoff Valve
2. Oil Pressure Sensor
3. Intake Manifold Pressure Sensor
4. Cooling Plate
5. Electronic Control Module (ECM)
6. Actuator Wiring Harness
7. OEM Wiring Harness

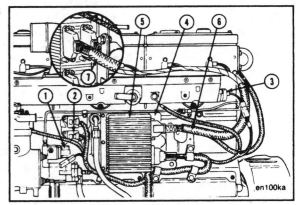

FIGURE 20–17 CELECT system components. (Courtesy of Cummins Engine Company, Inc.)

CELECT ECI (ELECTRONICALLY CONTROLLED INJECTION) SYSTEM COMPONENTS AND FUNCTION

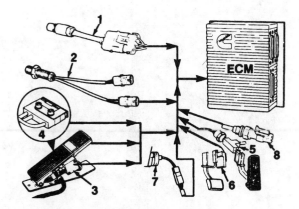

1. Engine Position Sensor (EPS)
2. Vehicle Speed Sensor (VSS)
3. Throttle Position Sensor
4. Idle Validation Switch
5. Service Brake Pedal Switch
6. Clutch Pedal Switch
7. Turbocharger Boost Pressure Sensor
8. Coolant Temperature Sensor
9. Cab Panel Switches (Not Shown)

FIGURE 20–18 ECM inputs. (Courtesy of Cummins Engine Company, Inc.)

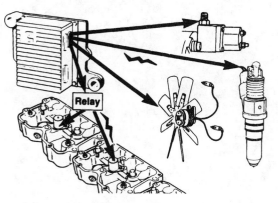

FIGURE 20–19 CELECT ECM outputs. (Courtesy of Cummins Engine Company, Inc.)

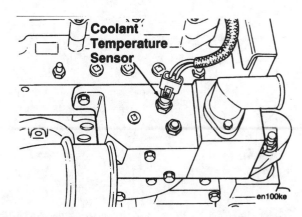

FIGURE 20–20 (Courtesy of Cummins Engine Company, Inc.)

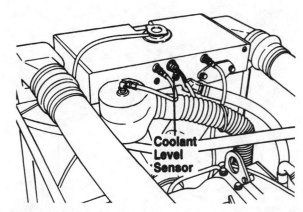

FIGURE 20–21 (Courtesy of Cummins Engine Company, Inc.)

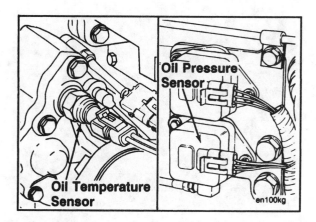

FIGURE 20–22 Oil temperature and oil pressure sensors. (Courtesy of Cummins Engine Company, Inc.)

13. *Clutch switch:* pushing the clutch pedal down opens the switch and deactivates the engine braking system **(Figure 20–28)**.

14. *Brake switch:* located in the service brake air line, it is normally closed. Applying the brakes deactivates the cruise control and the PTO **(Figure 20–28)**.

15. *Gear pump and pressure regulator:* supplies fuel to the ECM cooling plate and the injectors. Pressure is regulated at 150 psi maximum. The pump is equipped with a fuel filter and ECM-controlled shutdown valve **(Figure 20–29)**.

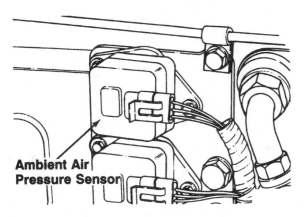

FIGURE 20–23 Ambient air pressure sensor. (Courtesy of Cummins Engine Company, Inc.)

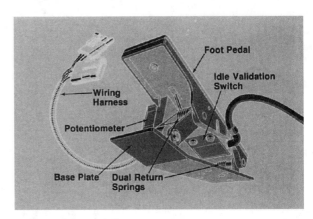

FIGURE 20–26 Throttle pedal components. (Courtesy of Cummins Engine Company, Inc.)

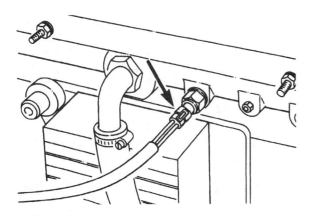

FIGURE 20–24 Intake air temperature sensor. (Courtesy of Cummins Engine Company, Inc.)

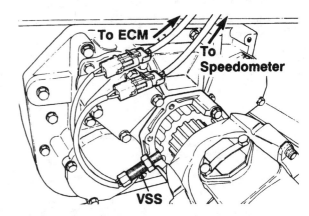

FIGURE 20–27 Vehicle speed sensor. (Courtesy of Cummins Engine Company, Inc.)

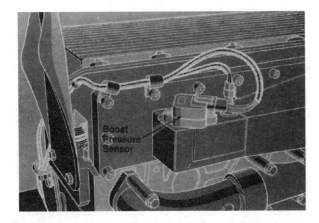

FIGURE 20–25 Turbo boost pressure sensor. (Courtesy of Cummins Engine Company, Inc.)

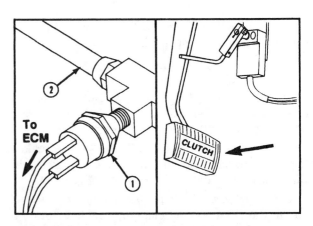

FIGURE 20–28 Brake switch (left) and clutch switch (right). (Courtesy of Cummins Engine Company, Inc.)

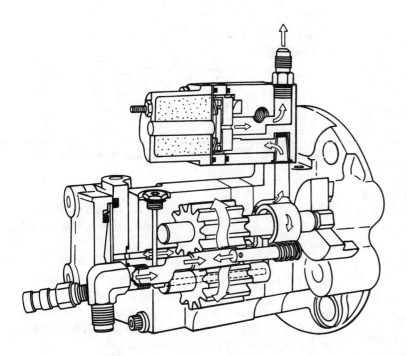

FIGURE 20–29 Gear-type fuel supply pump showing fuel flow and direction of gear rotation. (Courtesy of Cummins Engine Company, Inc.)

16. *Electronically controlled unit injectors (ECUI):* inject the correct amount of atomized fuel into the combustion chamber at the correct time in relation to piston position **(Figure 20–30)**.

17. *Warning lights:* The ECM uses a dash-mounted yellow check engine light and a red stop engine light to tell the driver when a malfunction occurs (yellow) that requires attention or that requires stopping the vehicle and engine for immediate attention (red) **(Figures 20–31 and 20–32)**.

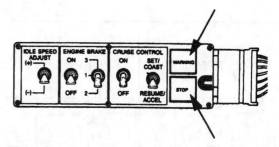

FIGURE 20–31 Warning and stop lamps. Also shown are the idle-speed adjust switch, engine brake switches, and cruise control switches. (Courtesy of Cummins Engine Company, Inc.)

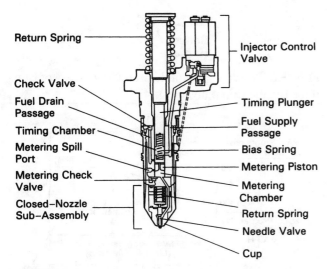

FIGURE 20–30 Electronically controlled injector (ECI) components. (Courtesy of Cummins Engine Company, Inc.)

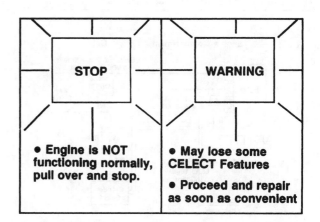

FIGURE 20–32 Stop (red) and warning (yellow) lights. (Courtesy of Cummins Engine Company, Inc.)

18. *Idle-speed adjust switch:* allows the driver to adjust the engine idle speed between 650 and 800 rpm in 25 rpm increments **(Figure 20–33)**.

19. *Compression brake control switch:* an ON/OFF switch that controls activation of the engine compression brake. A second three-position switch allows the brake to be activated on two, four, or six cylinders as desired **(Figure 20–34)**.

20. *Cruise control switches:* two toggle switches; one is an ON/OFF switch, while the other has two positions. They may be either SET/ACCEL and RESUME/COAST or SET/COAST and RESUME/ACCEL depending on OEM equipment **(Figure 20–35)**.

Other Electronic Control Features

The CELECT system offers electronic features that include the following.

1. *Cooling fan clutch control:* The ECM controls the cooling fan output based on coolant temperature

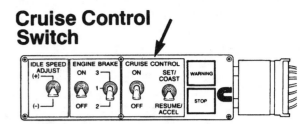

FIGURE 20–35 (Courtesy of Cummins Engine Company, Inc.)

and intake manifold air temperature **(Figure 20–36)**.

2. *Cruise control:* provides "foot off the pedal" cruise operation.

3. *Power take off:* controls the engine at a constant rpm in the PTO mode as selected by the customer.

4. *Gear-down protection:* limits vehicle speed in lower gears. For example, 65 mph in top gear and 62 mph (100 km/h) in lower gear encourages driving in top gear to reduce fuel consumption.

5. *Progressive shifting:* Limits the rate of acceleration and engine rpm in lower gears to promote upshifting and reduce fuel consumption **(Figure 20–37)**.

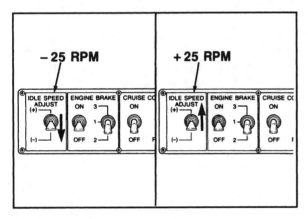

FIGURE 20–33 Idle-speed adjust switch. (Courtesy of Cummins Engine Company, Inc.)

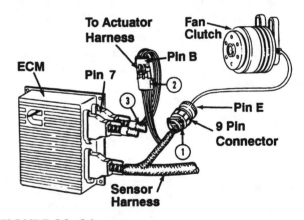

FIGURE 20–36 Fan clutch circuit. (Courtesy of Cummins Engine Company, Inc.)

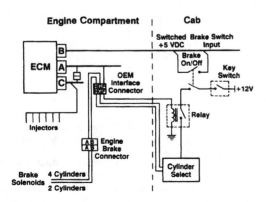

FIGURE 20–34 Engine brake enable circuit. (Courtesy of Cummins Engine Company, Inc.)

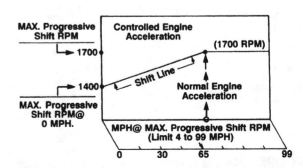

FIGURE 20–37 Progressive shift operating limits. (Courtesy of Cummins Engine Company, Inc.)

6. *Governor options:* The automotive governor performs the same function as the standard PT governor. Throttle pedal position and engine load determine engine speed. The VS all-speed governor performs at any engine speed. Throttle pedal position changes engine speed. Engine load changes do not affect engine speed when throttle position remains steady.

7. *Idle shutdown:* automatically shuts engine down after a period of idling. The shutdown delay period can be set between 3 and 30 minutes.

8. *Engine protection system:* monitors coolant temperature, coolant level, oil temperature, oil pressure, and intake manifold air temperature. A diagnostic fault is recorded by the ECM when any of these register conditions over or under specified norms. Engine speed and power are reduced, or engine shutdown may occur in accordance with the severity of the over or under condition **(Figure 20–38)**.

9. *Road relay and sensor plus:* dash-mounted units that provide the driver with vehicle running information. Include miles per gallon, overspeed and excess idle information, elapsed trip time, and miles driven. After-trip information available includes average miles per gallon, time spent over customer specified miles per hour, PTO idle hours and fuel consumption, and accumulated trip mileage. Sensor plus has the added features of a two-way radio network with the home office, theft prevention, and pinpointing of vehicle location, and it allows reconfiguration from the home office **(Figure 20–39)**.

10. *Data link:* conforms to the American Trucking Association (ATA) and Society of Automotive Engineers (SAE) standards J1708 and J1587, making it compatible with industry standards. The data link communicates with service tools and a personal computer database.

CELECT SYSTEM BASIC FUEL FLOW

A gear-type fuel supply pump draws fuel from the fuel tank. Fuel flows from the tank to the ECM cooling

Engine Protection System Out-of-Range Fault Codes
Coolant temperature
Coolant level
Oil temperature
Oil pressure
Intake manifold temperature

FIGURE 20–38 (Courtesy of Cummins Engine Company, Inc.)

FIGURE 20–39 CELECT system RoadRelay monitor and display. (Courtesy of Cummins Engine Company, Inc.)

plate, fuel filter, pressure regulator valve, fuel shutdown valve, and fuel line to the fuel galleries in the cylinder head and to each injector through a filter screen **(Figure 20–40)**. The ECM controls the start and ending of injection. The start of injection (injection timing) varies with engine operating conditions. The duration (pulse width) of injection varies with throttle pedal position, engine load, and other sensor inputs. With longer pulse widths more fuel is injected. Shorter pulse widths inject less fuel. This is called *pulse width modulation*.

CELECT ELECTRONIC INJECTOR OPERATION

Fuel metering starts with the injector metering plunger and timing plunger at the bottom of their travel **(Figures 20–41 and 20–42)**. As the camshaft rotates, the

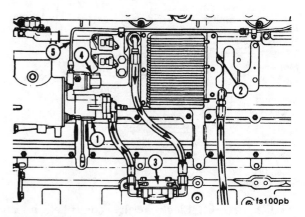

FIGURE 20–40 CELECT system fuel flow. (Courtesy of Cummins Engine Company, Inc.)

438 Chapter 20 CUMMINS ELECTRONIC CONTROL SYSTEMS

timing plunger begins its upward movement. The plunger return spring keeps the plunger up against the injector rocker arm. With the injector control valve closed (on ECM command) fuel flows into the metering chamber, forcing the metering plunger up against the timing plunger **(Figure 20–43)**. Fuel continues to flow into the metering chamber until the injector control valve opens on signal from the ECM. This equalizes pressure above and below the metering plunger, stopping movement of the metering piston. Continuing upward movement of the timing plunger allows fuel to fill the timing chamber. Downward movement of the timing plunger forces fuel out of the injector control valve to the fuel inlet. Injection begins (injection timing) when the ECM determines closing of the injector control valve. The timing plunger acting on the fuel trapped below it starts to move the metering plunger downward. Because the metering check ball is seated and the injector control valve is closed, fuel pressure rises until the needle valve is lifted off its seat

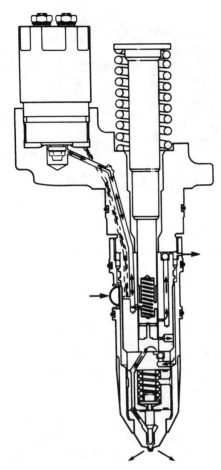

FIGURE 20–41 Electronically controlled injector (ECI) cross section showing fuel flow. (Courtesy of Cummins Engine Company, Inc.)

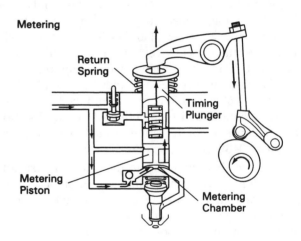

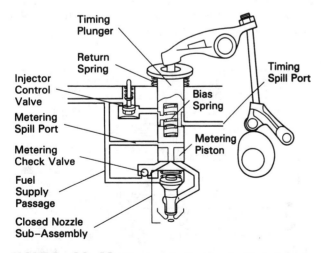

FIGURE 20–42 Electronically controlled injector (ECI) schematic diagram. (Courtesy of Cummins Engine Company, Inc.)

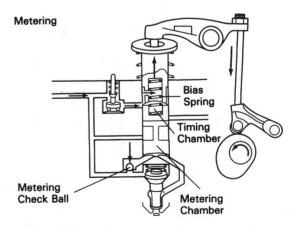

FIGURE 20–43 Start (top) and end (bottom) of fuel metering. (Courtesy of Cummins Engine Company, Inc.)

and fuel is forced out of the injector tip orifices into the combustion chamber. This occurs when fuel pressure reaches about 5000 psi (34,500 kPa). Injection continues until the spill port in the injector and the metering piston spill port are indexed **(Figure 20–44)**. Fuel pressure drops, the needle valve is seated by the return spring, and injection ends. Fuel in the timing chamber spills back into the fuel drain, ending the injection cycle.

CELECT SYSTEM DIAGNOSIS AND SERVICE

(Courtesy of Cummins Engine Co. Inc.)

FAULT CODE DIAGNOSIS

The CELECT system can display and record certain problems. The problems are displayed as fault codes, which makes troubleshooting easier. The fault codes are retained in the ECM.

There are two types of fault codes: engine electronic fuel system fault codes and engine protection system fault codes.

All fault codes recorded are either active (fault code is currently active on the engine) or inactive (fault code was active at some time, but is *not* currently active).

The active electronic fault codes can be read using the two fault lamps in the cab panel or on the Compulink screen. Inactive fault codes can be seen only on the Compulink screen.

The stop fault light is red. The warning light is yellow or red, depending on the OEM's preference. When the vehicle key switch is turned on, and the diagnostic switch is off, both fault lamps illuminate to check their operation. The fault lamps go off after about 2 seconds **(Figure 20–45)**.

1. With the vehicle key switch off, turn on the diagnostic switch (or plug in the shorting plug if so equipped) **(Figure 20–46)**.

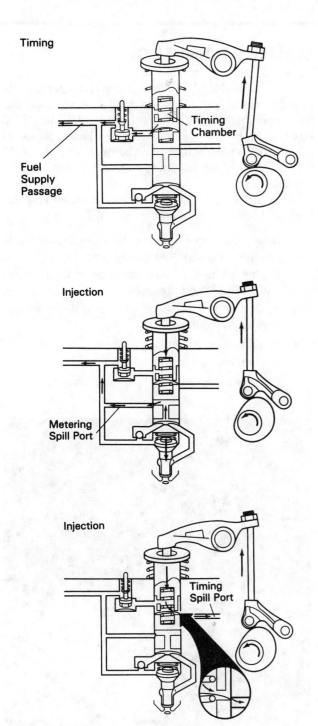

FIGURE 20–44 Injection action. (Courtesy of Cummins Engine Company, Inc.)

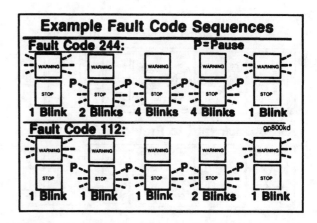

FIGURE 20–45 Fault code flashing-light sequence. (Courtesy of Cummins Engine Company, Inc.)

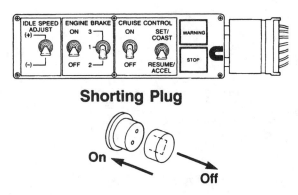

FIGURE 20–46 A shorting plug is used on some vehicles instead of a diagnostic switch. (Courtesy of Cummins Engine Company, Inc.)

2. Turn on the vehicle key switch. If no active fault codes are recorded, both lights will come on and stay on. If *active* fault codes are recorded, both lights will come on momentarily. Then the lights will begin to flash the code of the recorded fault.

The fault code will flash in the following sequence. First, a yellow light will flash, followed by a short 1- or 2-second pause. Then the number of the recorded fault code will flash in red. There will be a 1- or 2-second pause between numbers. When the number is done flashing in red, a yellow light will appear again. The number will repeat in the same sequence.

The lights will continue to flash the same fault code until the system is told to do something else. To go to the second fault code, move the idle-speed adjust switch to "+", then release it. You can also go back to the previous fault code by moving the switch to "−", then releasing it. To check the third or fourth fault code, move the switch to "+", then release it when all active fault codes have been viewed. Moving the switch to "+" will go back to the first fault code **(Figure 20–47)**.

FIGURE 20–47 Stepping forward or backward in the fault code sequence. (Courtesy of Cummins Engine Company, Inc.)

The explanation and correction of all of the fault codes is in the troubleshooting charts, Section T, in the front of the CELECT system troubleshooting service manual.

To stop the trouble code sequence, turn off the diagnostic switch (or remove the shorting plug) and vehicle key switch.

CUMMINS ECHEK TOOL

Echek is a hand-held management and diagnostic tool that delivers instant information on how a vehicle has been operated. Echek can be used with CELECT-PT, PACER-, and PACE-equipped engines. It also supports competitive engines that have incorporated the TMC (Truck Manufacturers Council)/SAE standards **(Figure 20–48)**.

Working with CELECT, Echek provides the following:

- Diagnostic capabilities such as fault code tracking and single-cylinder cutout tests
- Engine protection data such as critical engine temperatures and pressures
- Audit trail that includes when and if parameters have been modified
- Display of customer parameters
- A complete fuel log
- Outputs to a printer and personal computer

FIGURE 20–48 Echek tool and printer. (Courtesy of Cummins Engine Company, Inc.)

From Echek's fault monitor screen, you can quickly read active or inactive fault codes, as well as check all switches and sensor input. All faults are recorded when they first happen, then displayed whenever you choose. When a fault first occurs, Echek records critical-status items such as miles per hour, rpm, and percent throttle. The CELECT control module stores the data in memory so it can be recalled later to assist when troubleshooting the engine.

Once a fault is detected, the pocket-size CELECT Fault Code Manual leads you through a step-by-step diagnosis of the problem. Fault Code Manuals are also available for PT PACER and PACE.

Echek also displays engine protection data on how a vehicle has actually been operated on the road. Echek can maintain a historical record of the ECM time and how long critical engine temperatures and pressures have been outside their specified range. This allows permanent service records to be maintained.

On competitive engines that have incorporated TMC/SAE standards within their control modules, you can use Echek to monitor information like operating parameters and switch/sensor input, request and receive fault information, and clear inactive faults.

USING THE COMPULINK TOOL

The Compulink tool can be used to troubleshoot system problems while driving the vehicle or while it is parked with the key on (engine off). Connect the Compulink to the data link in the wiring harness. Press the NEXT key on the keyboard (**Figure 20–49**).

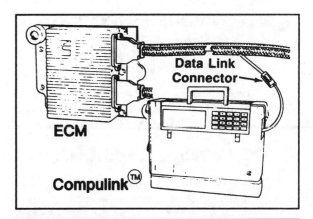

FIGURE 20–49 Compulink connection (top) and menu selections (bottom). (Courtesy of Cummins Engine Company, Inc.)

WARNING:

Two people are required when using the Compulink while road testing the vehicle: one to operate the vehicle and the other to monitor the system.

Select the troubleshooting option. Press the "1" key on the keyboard. This will display the troubleshooting menu on the Compulink screen. Select the desired option on the screen from the four that are offered: fault code information, monitor, fault tree, and special function. The monitor option is used to monitor engine functions and switch control functions. Press the NEXT key to monitor switch functions. Press the NEXT key to return to monitoring engine functions (**Figures 20–50** and **20–51**). After completing diagnosis, disconnect the Compulink from the data link.

PINPOINT TESTING

After identifying which circuit is at fault, further tests may be required to determine which component is at fault. It could be wiring, wiring connectors, low battery voltage, input switches or sensors, or output devices. The wiring and wiring connectors should be carefully inspected for damage or corrosion. A pull test may be required to determine if pins and sockets are making good contact. If connectors disconnect with almost no effort, there may not be a satisfactory electrical connection. Pinpoint tests of wiring and components with a digital VOM of proper capacity can identify problem components (**Figures 20–52** and **20–53**). Refer to the appropriate Cummins service manual for testing procedures, replacement procedures, and specifications.

CYLINDER MISFIRE TEST

NOTE: If the Compulink is available, use it to perform the "Cylinder Cutout" test instead of this procedure.

SELECT TROUBLESHOOTING MENU OPTION:

1. Fault Code Information
2. *Monitor*
3. Fault Tree
4. Special Function

MONITORING FUNCTIONS:

% Fuel = XXX RPM = XXXX MPH = XXX
% Throttle = XXX Degrees Adv. = XX
Cool Temp. = XXXF
Boost Pressure XX.XX In. Hg.
Oil Temp. = XXXF Oil Pressure XXX PSI
Intake Air = XXXF Fuel Rate XX.X G/Hr
Ambient Air Pressure = XXXX In. Hg.
Active Keys: *NEXT*, CNCL

FIGURE 20–50 Compulink troubleshooting menu options (top) and monitoring functions (bottom). (Courtesy of Cummins Engine Company, Inc.)

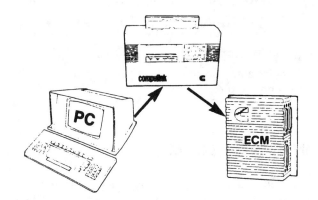

FIGURE 20–51 Compulink and a personal computer are used to program owner-specified parameters into the ECM. (Courtesy of Cummins Engine Company, Inc.)

1. Remove the rocker lever cover and crankcase breather tube. Remove the rocker housing cover gaskets.

NOTE: The rocker housing cover gaskets can be reused if they are **not** damaged. Do **not** discard.

2. Operate the engine until the coolant temperature is 160° F (70° C).

WARNING:

Do not touch or remove the injector electric leads while the engine is running. This can cause an electric shock.

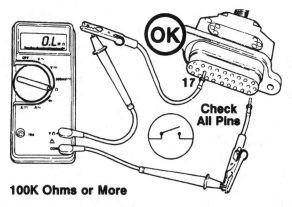

FIGURE 20–52 Using a VOM to perform pinpoint tests. (Courtesy of Cummins Engine Company, Inc.)

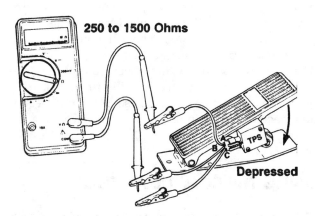

FIGURE 20–53 Performing a full-fuel (pedal depressed) resistance test on a throttle position sensor. (Courtesy of Cummins Engine Company, Inc.)

3. Remove the C-brakes, if applicable.

Disconnect the C-brake harness wire from the electrical connector on each C-brake housing.

Remove the cap screws and washers from each C-brake housing.

Remove the C-brake housing.

Remove the C-brake housing gasket.

Install a plug, part no. 3051329, into the C-brake oil supply drilling.

NOTE: The cylinder misfire test used on engines with PT fuel systems will **not** work on CELECT engines because the CELECT system automatically compensates to maintain idle speed when a cylinder is cut out. To use this procedure, the engine speed **must** be raised to 800 to 1000 rpm using the throttle pedal. Do **not** use the idle or PTO controls to raise the engine speed.

4. Use the throttle pedal to set the engine speed to 800 to 1000 rpm. Install the rocker lever actuator, part no. 3823609, or a wrench on an injector rocker lever.

CELECT INJECTOR REPLACEMENT

5. Hold the injector plunger down while the engine is running. This will stop the fuel flow to that injector.

If the engine rpm decreases when an injector plunger is held down, the injector is good **(Figure 20–54)**.

NOTE: This procedure can also be used to detect smoking problems caused by individual cylinders. If the exhaust smoke disappears when a cylinder is cut out, this cylinder is causing the smoke problem.

If the engine rpm does *not* decrease, replace the defective injector.

NOTE: Also be aware of any change in how the engine sounds. If no change in sound is heard, it is a good indication the injector is defective.

CELECT INJECTOR REPLACEMENT

Injector Removal

1. Disconnect the battery cables before removing or installing the injectors.
2. Remove the crankcase breather tube.
3. Remove the 10 cap screws and washers from each rocker lever cover.
4. Remove the rocker housing cover gaskets.

NOTE: The rocker lever cover gasket can be used again if it is **not** damaged. Do **not** discard it.

NOTE: Do **not** use solvent to clean the rocker housing cover gasket. Solvent will damage the O-ring material and cause it to swell.

5. Remove the C-brakes, if applicable.

Disconnect the C-brake harness wire from the electrical connector on each C-brake housing.

Remove the cap screws and washers from each C-brake housing.

Remove the C-brake housing.

Remove the C-brake housing gasket.

Loosen the valve and injector locknuts and adjusting screws.

Remove the rocker lever shaft cap screws.

Remove the rocker lever shaft assembly.

NOTE: Hold the shaft at both ends so that the rocker levers do **not** slide off.

6. Remove the crossheads. Be sure and mark them appropriately so they can be installed in the same location and orientation during the installation procedure.

NOTE: Excessive crosshead wear can result if the crossheads are **not** installed in their original locations. The larger hole on the underside of the crosshead **must** be oriented toward the exhaust side of the engine.

7. Disconnect the injector solenoid leads from the pass-through connector in the rocker housing.
8. Remove the injector hold-down clamp cap screw.
9. Remove the injector and hold-down clamp **(Figure 20–55)**.

Use an injector puller, part no. 3823579, to remove CELECT injectors. Insert the pin of the tool into the hole provided in the body of the injector. The hole faces the exhaust side of the engine.

If the designated injector puller is *not* available, carefully use a pry bar. Pry upward on the injector against the cylinder head.

Injector Installation

CAUTION:

Do not use anything metal to scrape the copper injector sleeves.

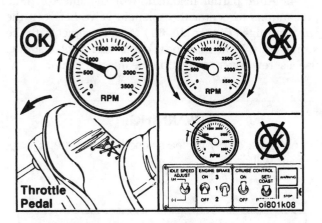
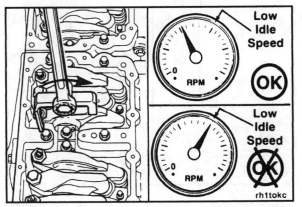

FIGURE 20–54 (Courtesy of Cummins Engine Company, Inc.)

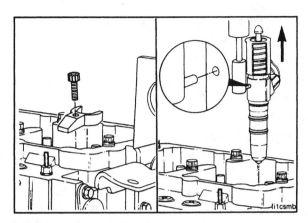

FIGURE 20–55 Injector removal. (Courtesy of Cummins Engine Company, Inc.)

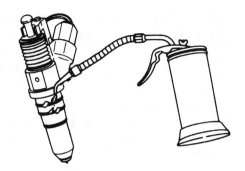

FIGURE 20–57 (Courtesy of Cummins Engine Company, Inc.)

CAUTION:

Do not strike the top-stop spring cage when installing CELECT injectors **(Figure 20–58)**.

1. Use a clean wooden stick with a clean cloth wrapped around the end to remove all the carbon from the copper injector sleeves in the cylinder head **(Figure 20–56)**.

NOTE: Use the chip removing unit, part no. 3823461, to remove the carbon from the top of the piston.

NOTE: When installing injectors for reuse, new O-rings **must** be installed on the injector.

NOTE: The CELECT injectors require three different injector O-rings. The O-rings are color coded as follows:

Top O-ring—black
Center O-ring—brown
Bottom O-ring—black with a white identification dot

2. Lubricate the O-rings with lubricating oil just before installation **(Figure 20–57)**.

3. Using the CELECT injector puller/installer, part no. 3823579, install the injector into the cylinder head injector bore with the injector solenoid valve facing the intake side of the engine **(Figure 20–59)**.

4. After partial installation of the injector, take precautions to center the solenoid valve between the valve springs **(Figure 20–60)**. Avoid contact with the spring coils. If the injector is contacting a valve spring, use a screwdriver to position the injector again.

WARNING:

Do not strike or pry on the solenoid. Otherwise, injector damage will occur.

FIGURE 20–56 (Courtesy of Cummins Engine Company, Inc.)

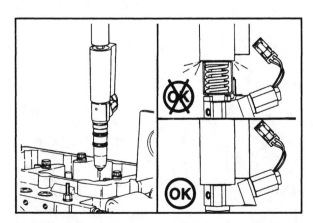

FIGURE 20–58 (Courtesy of Cummins Engine Company, Inc.)

CELECT INJECTOR REPLACEMENT 445

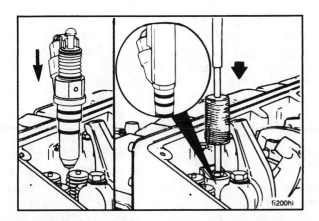

FIGURE 20-59 (Courtesy of Cummins Engine Company, Inc.)

FIGURE 20-61 (Courtesy of Cummins Engine Company, Inc.)

6. Install the hold-down clamp capscrew **(Figure 20-62)**.

Torque value: 35 ft-lb (47 N·m)

7. Take care to route the injector solenoid leads to avoid contact with the valve and injector push tubes **(Figure 20-63)**.

8. Connect the injector solenoid leads again to the pass-through connector in the rocker housing. Push the lead into the connector until a "snap" is heard **(Figure 20-64)**.

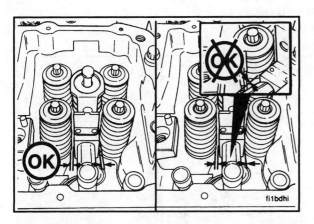

FIGURE 20-60 (Courtesy of Cummins Engine Company, Inc.)

WARNING:

Loosen the valve and injector rocker lever adjusting screws before installing the rocker lever shaft assembly. Otherwise, engine damage will occur.

5. Continue driving the injector into the bore using the puller/installer.

If an injector puller is *not* available, a screwdriver can be used to install the injector by putting the screwdriver on the injector body at the base of the injector solenoid and striking the screwdriver with a soft mallet. Do *not* strike on the injector solenoid or on the top-stop spring cage **(Figure 20-61)**.

WARNING:

The injector must be fully seated before installing the hold-down clamp. The hold-down clamp cannot pull the injector into the bore. Engine damage can occur if the injector is not fully seated.

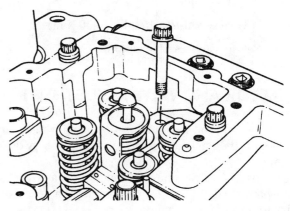

FIGURE 20-62 (Courtesy of Cummins Engine Company, Inc.)

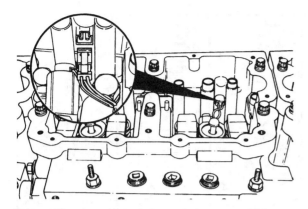

FIGURE 20-63 (Courtesy of Cummins Engine Company, Inc.)

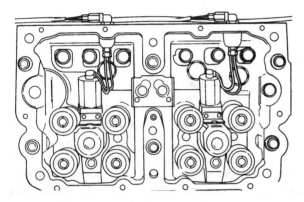

FIGURE 20-64 (Courtesy of Cummins Engine Company, Inc.)

9. Install the rocker lever shaft assembly. Make sure both ends of the rocker lever shaft are properly located on the dowel alignment pins in the rocker housing.

10. Install the two rocker lever shaft cap screws.

11. Tighten the cap screws alternately and evenly to the specified torque.

   Torque value: 115 ft-lb   (156 N·m)

12. Set the valves and injectors.

INJECTOR AND VALVE ADJUSTMENT

The valves and the injectors on the same cylinder are adjusted at the same index mark on the accessory drive pulley.

One pair of valves and one injector are adjusted at each pulley index mark *before* rotating the accessory drive to the next index mark.

Two crankshaft revolutions are required to adjust all the valves and the injectors.

1. Rotate the accessory drive in the direction of engine rotation. The accessory drive will rotate *clockwise* on a right-hand engine when looking at the front of the engine **(Figure 20-65)**. Align the "A" or "1-6 VS" mark on the accessory drive pulley with the pointer on the gear cover **(Figure 20-66)**. Check the valve rocker levers on cylinder No. 1 to see if both valves are closed.

NOTE: Both valves are closed when both rocker levers are loose and can be moved from side to side. If both valves are **not** closed, rotate the accessory drive one complete revolution, and align the "A" mark with the pointer again.

2. If the valve rocker lever adjusting screws have been loosened and *not* yet adjusted, watch the valve push tubes as the engine rolls upon the "A" mark. Both valve push tubes will have moved to the downward (valve closed) position if the engine is on the correct stroke.

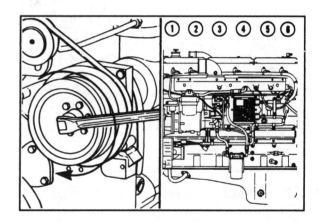

FIGURE 20-65 (Courtesy of Cummins Engine Company, Inc.)

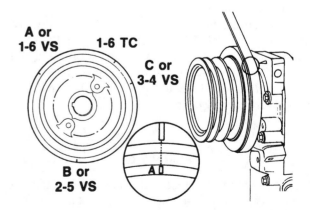

FIGURE 20-66 (Courtesy of Cummins Engine Company, Inc.)

INJECTOR AND VALVE ADJUSTMENT   447

3. Loosen the injector adjusting screw locknut on cylinder No. 1. Bottom the injector plunger by tightening and loosening the adjusting screw 25 in.-lb (2.8 N·m) three or four times to remove the fuel **(Figure 20–67)**.

NOTE: Do **not** bottom the plunger any tighter than 25 in.-lb (2.8 N·m) when removing the excess fuel.

4. Tighten the adjusting screw on the injector rocker lever.

   Torque value: 25 in.-lb   (2.8 N·m)

CAUTION:

After preloading the CELECT injector to 25 in.-lb (2.8 N·m), make sure to back out the adjusting screw two flats (120°), or damage to the injector will result.

---

5. Back out the adjusting screw on the injector rocker lever two flats (120°). **(Figure 20–68)**.

NOTE: Two flats will provide 0.025 in. (0.63 mm) lash. The specification is 0.020 to 0.029 in. (0.50 to 0.74 mm) lash.

6. Hold the adjusting screw and tighten the lock nut.

   Torque value: 50 ft-lb   (68 N·m)

7. After setting the injector on a given cylinder, set the valves on the same cylinder.

8. With the "A" set mark aligned with the pointer on the gear cover and both valves closed on cylinder No. 1, loosen the locknuts on the intake and the exhaust valve adjusting screws.

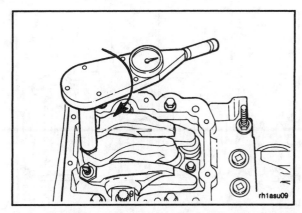
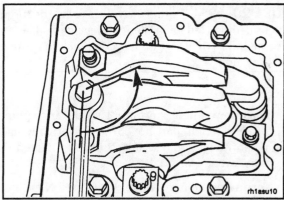
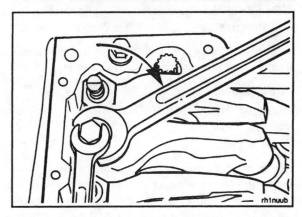

FIGURE 20–68 (Courtesy of Cummins Engine Company, Inc.)

9. Select a feeler gauge for the correct valve lash specification.

Intake: 0.014 in. (0.35 mm)   Exhaust: 0.027 in. (0.68 mm)

10. Insert the feeler gauge between the top of the crosshead and the rocker lever pad.

Two different methods for establishing valve lash clearance are described below. Either method can be used; however, the torque wrench method has proved to be the most consistent **(Figure 10–69)**.

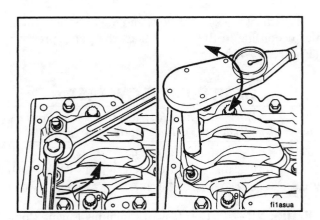

FIGURE 20–67 (Courtesy of Cummins Engine Company, Inc.)

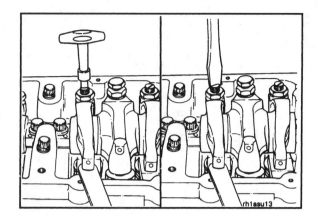

FIGURE 20–69 Adjusting the valves using the torque wrench method (left) and the feeler gauge and screwdriver method (right). (Courtesy of Cummins Engine Company, Inc.)

- **Torque wrench method:** Use the inch-pound torque wrench, part no. 3376592 (normally used to set preload on STC injectors), and tighten the adjusting screw.

  Torque value 5 to 6 in.-lb (0.6 to 0.7 N·m)

- **Feel method:** Tighten the adjusting screw until a slight drag is felt on the feeler gauge.

Hold the adjusting screw in this position. The adjusting screw *must not* turn when the locknut is tightened.

Torque values:

With torque wrench adapter, part no. ST-669	40 ft-lb	(54 N·m)
Without adapter	50 ft-lb	(88 N·m)

11. After tightening the locknut to the correct torque value, check to make sure the feeler gauge will slide backward and forward between the crosshead and the rocker lever with only a slight drag.

12. If using the feel method, attempt to insert a feeler gauge that is 0.001 in. (0.03 mm) thicker between the crosshead and the rocker lever pad. The valve lash is *not* correct when a thicker feeler gauge will fit.

13. After adjusting the injector on cylinder No. 1 and the valves on cylinder No. 1, rotate the accessory drive; and align the next valve set mark with the pointer.

14. Adjust the appropriate injector and valves following the Injector and Valve Adjustment Sequence Chart **(Figure 20–70)**.

15. Repeat the process to adjust all injectors and valves correctly.

CELECT ENGINE INJECTOR AND VALVE ADJUSTMENT SEQUENCE

Bar Engine in Direction of Rotation	Pulley Position	Set Cylinder	
		Injector	Valve
Start	A	1	1
Advance to	B	5	5
Advance to	C	3	3
Advance to	A	6	6
Advance to	B	2	2
Advance to	C	4	4

Firing Order: 1-5-3-6-2-4

FIGURE 20–70 (Courtesy of Cummins Engine Company, Inc.)

SUPPLY PUMP AND SHUT-OFF VALVE TESTS

Installation

1. Install the supply pump assembly on the test stand with four cap screws.

2. The plumbing of the test stand *must* be as shown. The minimum size of the supply line is No. 10 with a maximum length of 70 in. The minimum size of the drain is No. 5.

3. Connect the correct voltage to operate the shut-off valve.

Gear Pump Test

1. Operate the system with the inlet and outlet valves open to remove the air from the pump and the system.

2. Adjust the speeds, flow, and inlet restriction values according to the table. Check the pump pressure **(Figure 20–71)**.

GEAR PUMP FLOW TEST

rpm	Flow lb/hr (kg/hr)	Inlet Restriction in. Hg (mm Hg)	Outlet Pressure psi (kPa)
600		2.0 (51)	
1200	520 (236)	5.0 (127)	19–23 (130–160)
2100	650 (295)	5.0 (127)	22–26 (150–180)

FIGURE 20–71 (Courtesy of Cummins Engine Company, Inc.)

NOTE: The fuel supply **must** follow the minimum and maximum pressure guidelines for correct injector performance.

Shut-off Valve Test

1. Remove the air from the pump and system.
2. Open the inlet restriction and flow control valve.
3. Adjust the pump speed to 2400 rpm.
4. Disconnect the voltage from the shutoff valve.
5. Check the outlet for leakage. No leakage *must* be evident **(Figure 20–72)**.

NOTE: Do **not** operate the pump at zero flow for more than 1 minute.

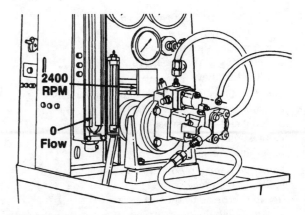

FIGURE 20–72 (Courtesy of Cummins Engine Company, Inc.)

REVIEW QUESTIONS

1. The PACE fuel control system added to the existing PT fuel system and replaced its _____.
2. The PACE system uses a _____ _____ to supply fuel to the mechanically operated injectors.
3. The PDM receives its power from _____ _____.
4. The VSS senses transmission output _____ _____.
5. The TPS is located in the _____ _____ housing.
6. Fuel flows from the fuel filter to the PTCM _____ _____ to the PT pump.
7. Fault codes stored in the PCM can be read by activating the _____ _____ _____ and stop lights.
8. The CELECT electronic control system use a _____ _____ to supply the fuel injectors.
9. The injectors are mechanically operated by the engine camshaft, whereas _____ _____ and _____ _____ are controlled electronically.
10. The APS signal is used by the ECM to help control fuel delivery at _____ _____.
11. Name the CELECT system electronic features.
12. The ECM controls the start and _____ of injection.
13. The CELECT can display and _____ certain problems.
14. The problems are displayed as _____ _____, which makes troubleshooting easier.
15. The Cummins Echek tool is a hand-held management and _____ tool.
16. When injectors are installed for reuse, new _____ must be installed on the injector.
17. One pair of valves and one _____ are adjusted at each pulley index.
18. The fuel supply must follow the _____ _____ _____ pressure guidelines for correct injector performance.

TEST QUESTIONS

1. The CELECT fuel system uses a gear pump to supply fuel to the
   a. mechanically operated electronic injectors
   b. mechanically operated injectors
   c. electronic injectors
   d. gear-operated injector
2. The PDM receives its power from the
   a. generator   c. PCM
   b. vehicle battery d. PACE
3. The TPS located in the PT pump housing is closed when the throttle
   a. is shut down   c. is in cruise control
   b. is at full load d. is at idle position

4. The cruise control is deactivated by
   a. applying the brakes
   b. depressing the clutch pedal
   c. activating the engine brake
   d. all of the above
5. The PTCM receives operating information from
   a. input switches and sensors
   b. input and output sensors
   c. fuel pump sensor
   d. air intake sensor
6. Fault codes are stored in the PTCM for
   a. driver reference
   b. later retrieval by the technician
   c. accountability of driver speed
   d. reference data
7. The CELECT system is designed to
   a. reduce emissions
   b. improve engine performance
   c. reduce emissions and improve fuel economy
   d. obtain more power at high rpm
8. The BPS measures boost pressure
   a. in the intake manifold
   b. at the air cleaner
   c. in the exhaust manifold
   d. in the combustion chamber
9. There are two types of fault codes:
   a. orange and red
   b. stationary and operating
   c. electronic fuel system and engine protection
   d. mechanical and engine protection
10. When installing the injector be sure to install
    a. gaskets
    b. a new fuel line seal
    c. new O-rings
    d. new electrical wire

◆ CHAPTER 21 ◆

CATERPILLAR FUEL INJECTION SYSTEMS

INTRODUCTION

Current Caterpillar diesel engines are equipped with one of the following fuel injection systems, depending on engine model, size, and application.

1. New scroll injection pump system, 3306 and 3406 engines.
2. Sleeve-metering injection pump system, 3208 engine.
3. Mechanical unit injector system, 3114 and 3116 engines.
4. Programmable electronic engine control (PEEC) system, 3406B engine.
5. Electronic unit injector (EUI) system, 3176B and 3406E engines.
6. Hydraulic electronic unit injector (HEUI) system, 3100 engine.

The first three systems are discussed in this chapter. Electronic control systems (items 4 and 5) are discussed in Chapter 22. The HEUI system is discussed in Chapter 23. Although the earlier compact housing pump is not discussed, it is similar to the new scroll injection pump. The older flanged body pump is not discussed.

This chapter is based on information and material provided through the courtesy of Caterpillar, Inc.

PERFORMANCE OBJECTIVES

After thorough study of this chapter and the appropriate training models and service manuals, and with the necessary tools and equipment, you should be able to do the following with respect to diesel engines equipped with Caterpillar fuel injection systems:

1. Describe the basic construction and operation of Caterpillar fuel injection systems, including the new scroll, sleeve-metering, and unit injection systems and including the following for each system:

    a. method of fuel pumping and metering
    b. delivery valve action
    c. lubrication
    d. injection timing control
    e. governor types

2. Diagnose basic fuel injection system problems and determine the needed correction.

3. Remove, clean, inspect, test, measure, repair, replace, and adjust fuel injection system components as needed to meet manufacturer's specifications.

TERMS YOU SHOULD KNOW

Look for these terms as you study this chapter, and learn what they mean.

new scroll injection pump
sleeve-metering injection pump
mechanical unit injector
fuel transfer pump
priming pump
injection pump camshaft
injection pump barrel
injection pump plunger
spill port
bypass port
fuel rack
rack gear
scroll
check valve
governor flyweights
governor servo
governor dashpot
stop bar
torque spring
fuel ratio control
automatic timing advance
constant-bleed orifice
constant-bleed valve
metering sleeve
sleeve control shaft
static timing
dynamic timing
timing pin
fuel setting adjustment
fuel ratio control setting
low-idle-speed setting
crossover lever adjustment
set point (balance point) adjustment
fuel pump calibration
unit injector
rack control linkage
injector synchronization
unit injector fuel setting
unit injector fuel timing

NEW SCROLL INJECTION PUMP FUEL SYSTEM—COMPONENTS AND FUEL FLOW (FIGURE 21-1)

Fuel is drawn from the fuel tank (9) through the primary fuel filter by the fuel transfer pump (8). It is then pushed through the secondary fuel filter (11) to the fuel manifold in the injection pump housing (5). The fuel transfer pump spring determines fuel pressure in the manifold. A constant-bleed orifice in the return line elbow (4) allows a constant fuel flow back to the tank through the return line (3). This helps cool the fuel and keep it free from air. There is one injection pump for each engine cylinder. The injection pumps push fuel under very high pressure through the fuel lines (2) to the injection nozzles (1). The fuel transfer pump **(Figure 21-2)** is a piston pump operated by an eccentric on the injection pump camshaft.

INJECTION PUMP OPERATION (FIGURE 21-3)

The fuel injection pump plungers are actuated by cams (10) on the pump camshaft. The cam raises the lifter (9) and plunger (5) until a full stroke is made. Further camshaft rotation allows the spring (6) to return the plunger to the bottom of the stroke. At this point fuel flows through the spill port (1) and bypass port (4), filling the barrel (3) in the area above the plunger **(Figures 21-4 and 21-5)**.

During the upstroke, fuel is pushed out through the bypass port (4) until the top of the plunger closes the port. As plunger travel continues, fuel pressure is increased to about 100 psi (690 kPa). At this point the check valve (2) opens to allow fuel to flow into the fuel injection line. As the plunger continues upward travel, the scroll (14) uncovers the spill port (1). Fuel above the plunger goes through the slot (15) along the edge of the scroll and out the spill port back to the fuel manifold. Uncovering the spill port ends the effective stroke of the plunger. When the spill port is opened, the spring (13) closes the check valve (2) as pressure above the plunger drops below 100 psi (690 kPa). The reverse-flow check valve (11) prevents rough idle by stopping secondary injection of fuel between injection strokes. This valve is effective only below 1200 psi (8250 kPa). When the pump plunger moves downward and uncovers the bypass port (4), fuel again fills the area above the plunger, making it ready for the next stroke.

Rotating the pump plunger changes the amount of fuel sent to the injection nozzle. The fuel rack teeth (7) are in mesh with teeth on the plunger gear (8) **(Figure 21-3)**. The governor moves the fuel rack and turns the plungers to provide the correct amount of fuel for the engine's needs. Turning the plungers changes their effective stroke by changing the position of the scroll (14), thereby changing the point where the scroll uncovers the spill port **(Figure 21-6)**. To stop the engine, the plungers are rotated until the slot (15) on the plunger is in line with the spill port. Fuel now flows out the spill port instead of to the injection nozzles. The engine lubrication system provides oil for injection pump lubrication **(Figure 21-7)**. Lubricating oil is also directed to the governor servo (1), fuel ratio control, and the dashpot (6). The fuel injection nozzle is shown in **Figure 21-8**.

GOVERNOR OPERATION

The governor controls the amount of fuel needed by the engine to keep it at the desired rpm. Governor components are shown in **Figure 21-9**. The governor moves the fuel rack to vary the amount of fuel injected. The injection pump camshaft drives the governor flyweights (8). The flyweights and governor spring (1) act on the riser (12). A lever (7) connects the riser to the sleeve (2) and the valve (3). The valve is part of the governor servo (5), which moves the piston (4) and fuel rack (6). The governor spring always pushes toward

GOVERNOR OPERATION 453

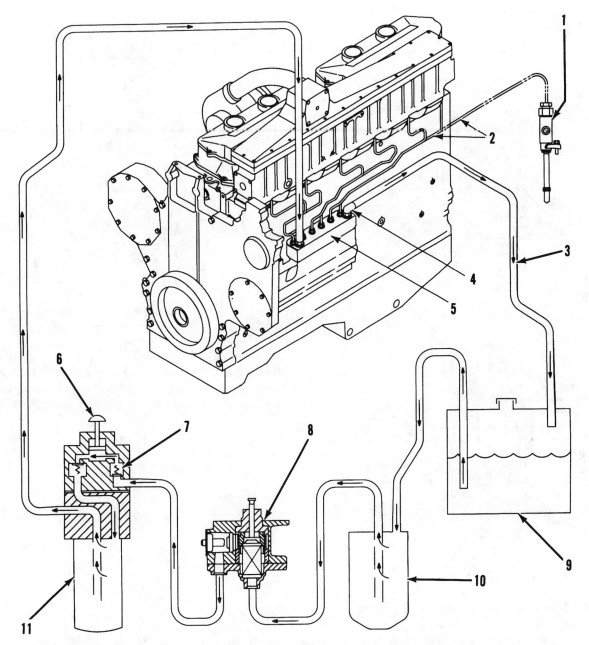

FIGURE 21–1 New scroll fuel system components and fuel flow on 3406B engine. 1. Fuel injection nozzle. 2. Fuel injection lines. 3. Fuel return line. 4. Constant-bleed orifice (part of elbow). 5. Fuel injection pump housing. 6. Fuel priming pump. 7. Check valves. 8. Fuel transfer pump. 9. Fuel tank. 10. Primary fuel filter. 11. Secondary fuel filter. (Courtesy of Caterpillar, Inc.)

the increased fuel position, while the weights push toward the reduced fuel position. With these two forces in balance, the engine runs at a constant rpm.

When the engine is cranked, an over fueling spring (9) moves the riser to provide extra fuel for starting. When the engine starts and runs, the flyweight force increases and overcomes the spring force, reducing fuel delivery to the amount required for low-idle speed.

Moving the control lever toward the high-idle position compresses the governor spring (1) and moves the riser (12) toward the flyweights. The sleeve (2) moves the valve (3) through the link spring, stopping oil flow through the governor servo (5). Oil pressure moves the piston (4) and the fuel rack to the increased fuel direction. With increasing engine speed the flyweight force increases, moving the riser toward the governor spring.

454 Chapter 21 CATERPILLAR FUEL INJECTION SYSTEMS

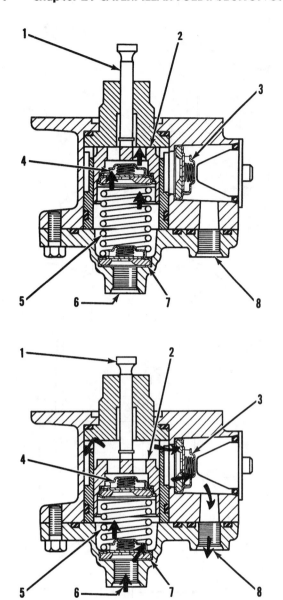

FIGURE 21-2 New scroll fuel transfer pump operation showing start of downstroke (top) and start of upstroke (bottom). Arrows show direction of fuel flow. 1. Push rod. 2. Piston. 3. Outlet check valve. 4. Pumping check valve. 5. Pump spring. 6. Inlet port. 7. Inlet check valve. 8. Outlet port. (Courtesy of Caterpillar, Inc.)

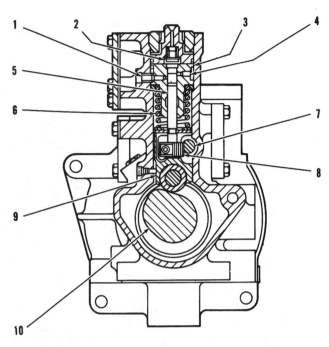

FIGURE 21-3 Fuel injection pump cross section. 1. Spill port. 2. Check valve. 3. Pump barrel. 4. Bypass port. 5. Pump plunger. 6. Spring. 7. Fuel rack. 8. Gear. 9. Lifter. 10. Cam. (Courtesy of Caterpillar, Inc.)

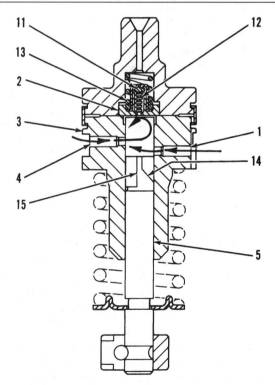

FIGURE 21-4 New scroll injection pump barrel-and-plunger assembly. Spill and bypass ports are both open. 1. Spill port. 2. Check valve. 3. Barrel. 4. Bypass port. 5. Plunger. 11. Orificed reverse-flow check valve. 12. Spring. 13. Spring. 14. Scroll. 15. Slot. (Courtesy of Caterpillar, Inc.)

The lever (7), sleeve (2), and valve (3) are moved forward. Oil pressure is directed to the rear of the piston (4), moving the piston and fuel rack forward in the decreased fuel direction. When the governor spring force and the flyweight become equal, the engine runs at the high-idle speed. A high-idle screw is provided to allow adjustment of the high-idle rpm. The screw limits governor spring compression.

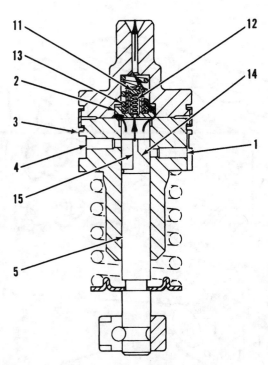

FIGURE 21–5 Pump barrel and plunger assembly during injection. 1. Spill port. 2. Check valve. 3. Pump barrel. 4. Bypass port. 5. Pump plunger. 11. Orificed reverse-flow check valve. 12. Spring. 13. Spring. 14. Scroll. 15. Slot. (Courtesy of Caterpillar, Inc.)

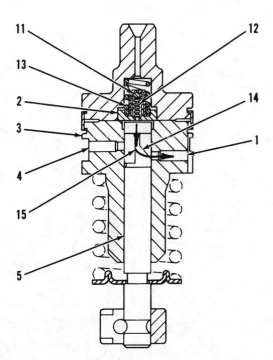

FIGURE 21–6 Pump barrel and plunger assembly with check valve closed. 1. Spill port. 2. Check valve. 3. Pump barrel. 4. Bypass port. 5. Pump plunger. 11. Orificed reverse-flow check valve. 12. Spring. 13. Spring. 14. Scroll. 15. Slot. (Courtesy of Caterpillar, Inc.)

When the load is increased (with the engine at high idle), engine speed decreases, and the governor moves the fuel rack toward the increased fuel direction. As the load is increased further the governor spring pushes the riser farther forward, and the spring seat pushes on the stop bolt (15). The other end of the stop bolt has a stop collar (18) equipped with a fuel setting screw (17) and a torque rise setting screw (14). The torque rise screw limits maximum fuel rack travel. The fuel setting screw controls the engine horsepower at full-load speed. As the stop bolt moves forward the fuel setting screw makes full contact with the torque spring (16) at the full-load speed of the engine. The torque spring now controls the fuel rack movement.

Torque Control

If the load is increased further with the engine in the maximum horsepower output mode, a lug condition is developed. This causes engine speed to decrease, moving the stop bolt and fuel setting screw forward. This bends the torque spring (16), allowing the fuel rack (6) to move farther in the increased fuel direction. This movement is limited by the torque rise setting screw (14) when it contacts the stop bar (11) **(Figures 21–9 and 21–10)**. This is the maximum fuel setting position. Adjustment of the torque rise setting screw (14) controls the additional amount of fuel rack travel below full-load engine speed as the peak torque speed of the engine is reached. Should the governor become stuck in the FUEL ON position, the engine can be stopped by the fuel shut-off solenoid or by moving the manual shut-off lever to the OFF position.

Governor Servo Operation

The governor servo provides hydraulic assistance to the governor to move the fuel rack **(Figure 21–11)**. When the governor moves in the increased fuel direction, the valve (1) moves to the left, opening oil outlet (B) and closing oil passage (D). Oil from the inlet (A) pushes the piston (2) and fuel rack (5) to the left to increase fuel delivery. Oil from behind the piston flows through oil passage (C) and outlet (B). When the governor is in balance, engine speed is constant and valve (1) stops moving. Oil pressure pushes the piston (2), opening oil passages (C and D). With no oil pressure acting on the piston, it stops moving. When the governor moves to the increased fuel direction, the valve (1) moves to the right, closing oil outlet (B) and opening oil passage (D). Oil pressure

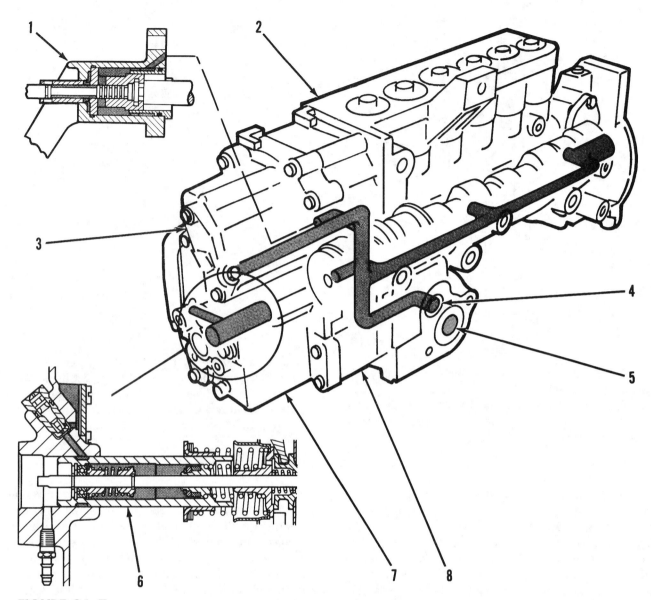

FIGURE 21-7 Fuel injection pump and governor components. 1. Servo. 2. Fuel injection pump housing. 3. Cover. 4. Oil supply from cylinder block. 5. Oil drain into cylinder block. 6. Dashpot. 7. Governor rear housing. 8. Governor center housing. (Courtesy of Caterpillar, Inc.)

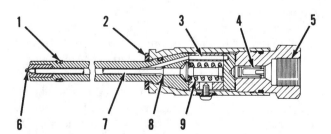

FIGURE 21-8 Fuel injection nozzle. 1. Carbon dam. 2. Seal. 3. Passage. 4. Filter screen. 5. Inlet passage. 6. Orifice. 7. Valve. 8. Diameter. 9. Spring. (Courtesy of Caterpillar, Inc.)

from inlet (A) is now on both sides of the piston. Because the reaction area of the piston is greater on the left side than on the right side, the piston and fuel rack move to the right to reduce fuel delivery.

Dashpot Operation

The dashpot helps the governor provide better speed control when there are sudden load and speed changes **(Figure 21-12)**. When the load changes, the spring seat (6), dashpot spring (5), and piston (4) move. As the piston moves it causes oil to be moved in or out of the

GOVERNOR OPERATION   457

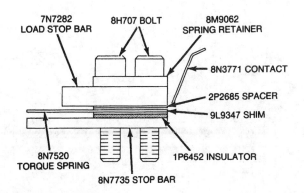

FIGURE 21–10 Torque spring components. (Courtesy of Caterpillar, Inc.)

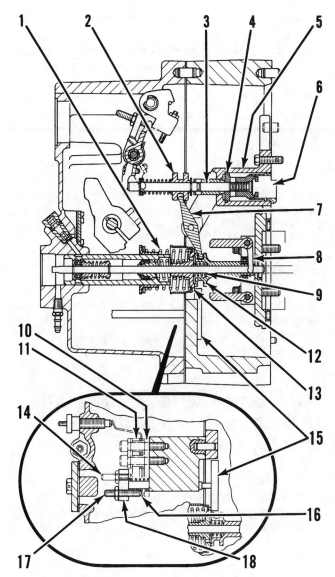

FIGURE 21–9 Governor components. 1. Governor spring. 2. Sleeve. 3. Valve. 4. Piston. 5. Governor servo. 6. Fuel rack. 7. Lever. 8. Flyweights. 9. Over fueling spring. 10. Load stop bar. 11. Stop bar. 12. Riser. 13. Spring seat. 14. Torque rise setting screw. 15. Stop bolt. 16. Torque spring. 17. Fuel setting screw. 18. Stop collar. (Courtesy of Caterpillar, Inc.)

cylinder through the needle valve (1) and oil reservoir (2). The needle valve provides a restriction to oil flow that dampens the movement of the piston and spring seat. The faster the governor tries to move the greater the dampening effect.

Fuel Ratio Control Operation

The function of the fuel ratio control is to restrict the amount of fuel delivered to the engine until sufficient

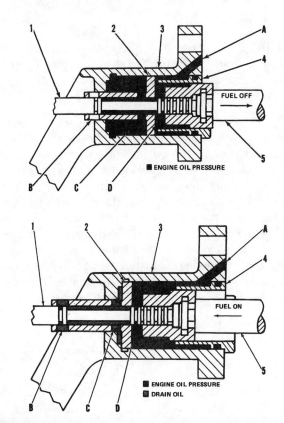

FIGURE 21–11 Governor servo in FUEL OFF (top) and FUEL ON position (bottom). 1. Valve. 2. Piston. 3. Cylinder. 4. Cylinder sleeve. 5. Fuel rack. A. Oil inlet. B. Oil outlet. C. and D. Oil passages. (Courtesy of Caterpillar, Inc.)

boost (inlet manifold pressure) is achieved to ensure complete combustion on turbocharged engines **(Figures 21–13 to 21–15)**.

The fuel ratio control limits the amount of fuel delivery to the engine when the accelerator pedal is pressed down, thereby reducing exhaust smoke. When the stem (6) moves the lever (11), fuel rack movement in the FUEL ON direction is restricted with the engine run-

458 Chapter 21 CATERPILLAR FUEL INJECTION SYSTEMS

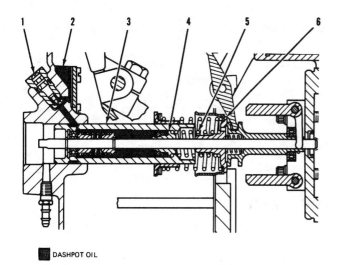

FIGURE 21–12 Dashpot. 1. Needle valve. 2. Oil reservoir. 3. Cylinder. 4. Piston. 5. Dashpot spring. 6. Spring seat. (Courtesy of Caterpillar, Inc.)

ning. With the engine stopped there is no oil pressure and fuel rack movement is not restricted, allowing maximum fuel to the engine for easier starting. The stem (6) does not move until inlet manifold pressure increases enough to move the internal valve (3). A line connects the air chamber (1) to the inlet manifold. With the increased manifold pressure the diaphragm assembly (2) and the internal valve (3) move to the right, closing off the oil passage (9). Rising oil pressure in the chamber (10) moves the piston (8) and stem (6) to the left into operating position until the engine is stopped or manifold pressure increases with increased load on the engine. The stem (6) limits movement of the lever (11) in the FUEL ON direction when the governor control is moved to increase fuel delivery. As inlet manifold pressure increases the diaphragm assembly and internal valve move to the right, opening the oil passage (9), allowing oil in the chamber (10) to drain through the oil passage (4). This allows the spring to move the piston and stem to the right until the oil passage (9) is closed by the internal valve (3). The lever (11) can now move to allow the fuel rack to move to the full-fuel position. The fuel ratio control restricts fuel delivery until air pressure in the inlet manifold is high enough for complete combustion. This reduces exhaust smoke caused by an air–fuel ratio that is too rich.

Sealed Unit Fuel Ratio Control

The sealed unit fuel ratio control is serviceable only as a complete unit. Individual parts are not replaceable. No static lever setting is required with this sealed unit **(Figure 21–16)**.

AUTOMATIC TIMING ADVANCE UNIT OPERATION (4MG3600-UP, 5KJ1-UP, 3ZJ1-UP)

The timing advance unit connects the injection pump camshaft to the timing gear in the engine gear train. Engine oil pressure is used to control a double-acting servo. The servo directs oil pressure to either side of

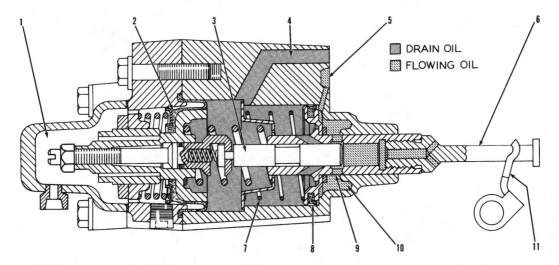

FIGURE 21–13 Fuel ratio control with the stem (6) in fully extended position, allowing extra fuel for starting. 1. Inlet air chamber. 2. Diaphragm assembly. 3. Internal valve. 4. Oil drain passage. 5. Oil inlet. 6. Stem. 7. Spring. 8. Piston. 9. Oil passage. 10. Oil chamber. 11. Lever. (Courtesy of Caterpillar, Inc.)

AUTOMATIC TIMING ADVANCE UNIT OPERATION (4MG3600-UP, 5KJ1-UP, 3ZJ1-UP)

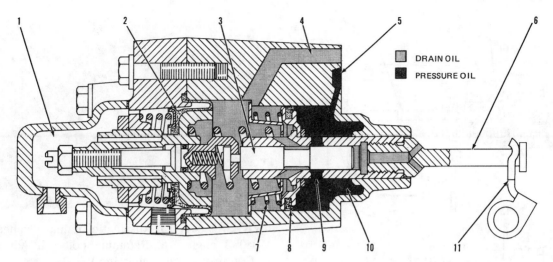

FIGURE 21–14 Fuel ratio control during normal idle operation. Oil pressure has increased in the chamber (10), and the piston (8) has moved to the left. 1. Inlet air chamber. 2. Diaphragm assembly. 3. Internal valve. 4. Oil drain passage. 5. Oil inlet. 6. Stem. 7. Spring. 8. Piston. 9. Oil passage. 10. Oil chamber. 11. Lever. (Courtesy of Caterpillar, Inc.)

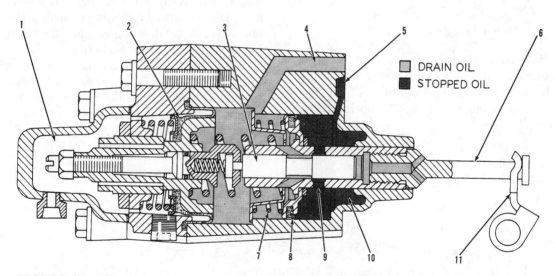

FIGURE 21–15 Fuel ratio control during acceleration. 1. Inlet air chamber. 2. Diaphragm assembly. 3. Internal valve. 4. Oil drain passage. 5. Oil inlet. 6. Stem. 7. Spring. 8. Piston. 9. Oil passage. 10. Oil chamber. 11. Lever. Oil pressure in the chamber (10) resists movement of the stem (6). When boost pressure increases, the diaphragm (2) and valve (3) move to the right, and oil drains from the chamber (10) through the passage (4). (Courtesy of Caterpillar, Inc.)

the drive carrier to advance or retard timing within a range of 12 crankshaft degrees **(Figure 21–17)**. Components of the unit are shown in **Figure 21–18** before timing advance begins.

When the engine starts to run, the flyweights (5) in **Figure 21–19** move out and pull the valve spool (4) to the left. This closes the oil drain passage. Oil pressure now pushes the body assembly (10) and the carrier (8) to the left. The carrier slides between the splines on the ring (6), causing the helical splines on the carrier to turn the pump camshaft in relation to the gear (7) to advance injection timing. Timing continues to advance against the pressure of the spring (3) until the spring pressure and flyweight force are balanced, and the

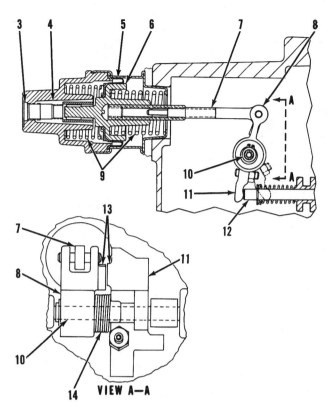

FIGURE 21-17 Automatic timing advance unit on 3406B engine. 3. Retaining bolts. 4. Advance unit. (Courtesy of Caterpillar, Inc.)

FIGURE 21-16 Fuel ratio control and governor. 3. Port. 4. Chamber. 5. Diaphragm. 6. Retainer assembly. 7. Rod. 8. Lever assembly. 9. Springs. 10. Shaft. 11. Lever assembly. 12. Servo valve. 13. Tangs. 14. Spring. (Courtesy of Caterpillar, Inc.)

valve spool is held in position. Oil pressure now moves the carrier to the right, and oil pressure is directed to the left side of the carrier. At this point the carrier moves slightly from left to right to maintain the required timing advance. The adjustment of the screw (1) and setscrew (2) determines when the timing advance begins and ends **(Figure 21-19)**.

When engine speed is reduced, the flyweights move in, and the force of the spring (3) in **Figure 21-20** moves the valve spool (4) to the right. Oil pressure builds to move the body assembly (10) and carrier (8) to the right to retard fuel injection timing.

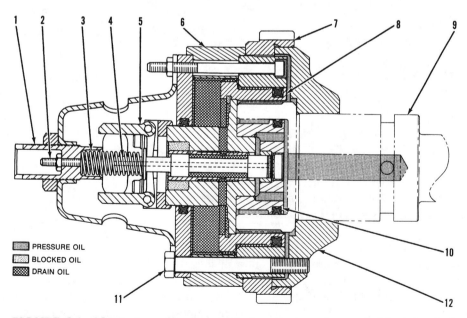

FIGURE 21-18 Timing advance unit (before timing advance begins) (typical example). 1. Screw. 2. Setscrew. 3. Spring. 4. Valve spool. 5. Flyweights. 6. Ring. 7. Gear. 8. Carrier. 9. Fuel injection pump camshaft. 10. Body assembly. 11. Bolt. 12. Ring. (Courtesy of Caterpillar, Inc.)

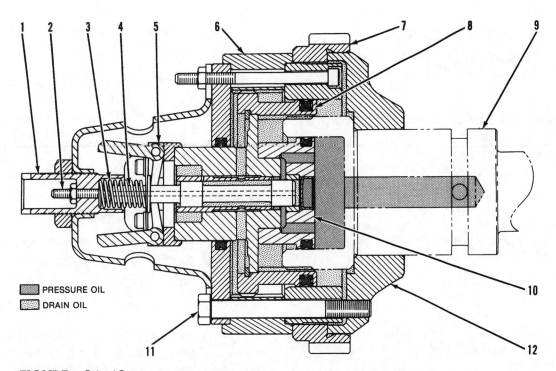

FIGURE 21–19 Maximum timing advance (typical example). 1. Screw. 2. Setscrew. 3. Spring. 4. Valve spool. 5. Flyweights. 6. Ring. 7. Gear. 8. Carrier. 9. Fuel injection pump camshaft. 10. Body assembly. 11. Bolt. 12. Ring. (Courtesy of Caterpillar, Inc.)

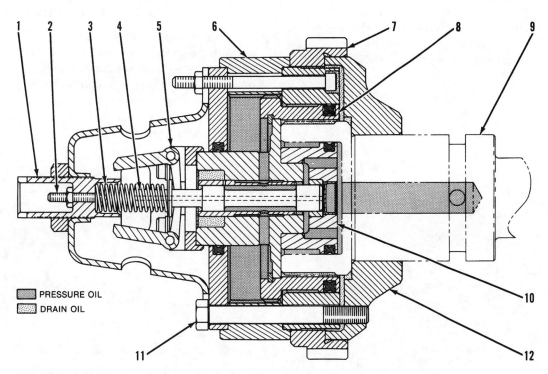

FIGURE 21–20 Timing fully retarded (typical example). 1. Screw. 2. Setscrew. 3. Spring. 4. Valve spool. 5. Flyweights. 6. Ring. 7. Gear. 8. Carrier. 9. Fuel injection pump camshaft. 10. Body assembly. 11. Bolt. 12. Ring. (Courtesy of Caterpillar, Inc.)

NEW SCROLL FUEL SYSTEM DIAGNOSIS AND SERVICE

(Courtesy of Caterpillar, Inc.)

REMOVAL OF FUEL INJECTION PUMPS

NOTE: Before any parts are removed from the fuel injection pump housing, thoroughly clean all dirt from the housing. Dirt that gets inside the pump housing will cause much damage.

NOTE: The fuel rack must be in the zero (center) position before the fuel injection pumps can be removed or installed. Follow Steps 1 through 5.

1. Remove the plug and cover from the fuel injection pump housing.

2. Install the 6V4186 Timing Pin (3) in the top of the fuel injection pump housing. Make sure the timing pin (3) engages in the slot of the fuel rack as shown **(Figure 21-21)**.

3. Install the bracket assembly on the fuel injection pump housing. Make sure the lever of the bracket assembly is engaged in the slot of the fuel rack.

4. Install the 1U5426 Compressor Assembly all the way into the 8T9198 Bracket Assembly to compress the spring.

5. Tighten the collet on the bracket assembly to hold the compressor assembly. Spring force now holds the fuel rack against the timing pin (3) in the zero position.

6. Remove the fuel injection line from the pump to be removed and also the fuel injection lines on each side of the pump to be removed.

7. Use the 8T5287 Wrench (6) to loosen the bushing (8) one-quarter turn. Do not remove the bushing at this time.

8. Install the 6V7050 Compressor Group (7) on the pump housing over the 8T5287 Wrench (6). Lower the screw in the compressor ram to the fuel line seat before the nut is tightened to hold the compressor group in position. This centers the compressor group **(Figure 21-22)**.

WARNING

There is spring force on the fuel injection pump plunger and barrel assembly. Removal of the retainer bushing (8) without the 6V7050 Compressor Group correctly installed can cause bodily injury.

---

9. Use the 8T5287 Wrench (6) to loosen the retainer bushing (8) until it is out of the threads. Slowly raise the compressor tool handle to release the spring force.

10. Remove the 6V7050 Compressor Group and the 8T5287 Wrench. Install the 8S2244 Extractor on the injection pump threads. Carefully pull the pump straight up and out of the pump housing bore **(Figure 21-23)**. Remove the spacer from the pump housing bore.

Be careful when disassembling an injection pump not to damage the surface on the plunger. The plunger and barrel are made as a set. Do not put the plunger of one pump in the barrel of another pump. If one part is worn, install a complete new pump assembly. Be careful when putting the plunger into the bore of the barrel. When removing injection pumps from the fuel injection

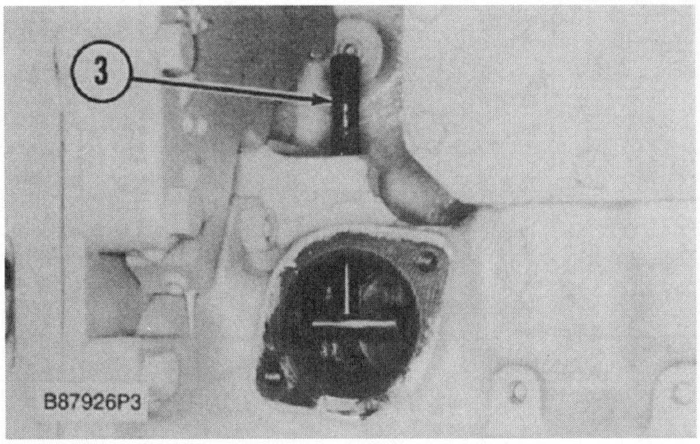

Fuel Rack Against Timing Pin in the Zero Position
(3) 6V4186 Timing Pin.

FIGURE 21-21 (Courtesy of Caterpillar, Inc.)

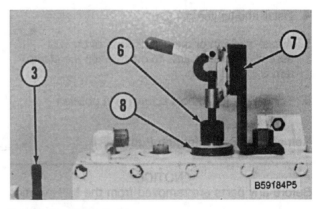

Fuel Injection Pump Housing
(3) 6V4186 Timing Pin. (6) 8T5287 Wrench. (7) 6V7050 Compressor Group. (8) Retainer bushing.

FIGURE 21-22 (Courtesy of Caterpillar, Inc.)

INSTALLATION OF FUEL INJECTION PUMPS 463

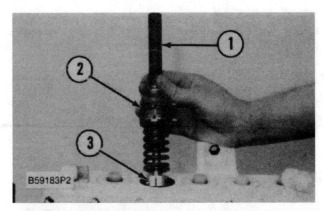

Fuel Injection Pump Installation
(1) 8S2244 Extractor. (2) Pump barrel. (3) Gear segment.

FIGURE 21–23 (Courtesy of Caterpillar, Inc.)

pump housing, keep the parts together so they can be installed in the same location in the housing.

Checking the Plunger and Lifter of an Injection Pump

NOTE: There are no different size spacers available to adjust the timing dimension of a fuel injection pump. If the pump plunger or the lifter is worn, it must be replaced. Because no adjustment to the timing dimension is possible, there is **no** off-engine lifter setting procedure.

When there is too much wear on the fuel injection pump plunger, the lifter may also be worn, and there will not be good contact between the two parts **(Figure 21–24)**. To stop fast wear on the end of a new plunger, install new lifters in place of the worn lifters.

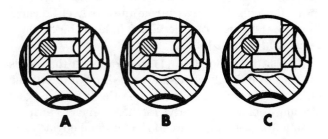

Wear Between Lifter and Plunger
Fig. A. Illustrates the contact surfaces of a new pump plunger and a new lifter. In Fig. B the pump plunger and lifter have worn considerably. Fig. C shows how the flat end of a new plunger makes poor contact with a worn lifter, resulting in rapid wear to both parts.

FIGURE 21–24 (Courtesy of Caterpillar, Inc.)

An injection pump can have a good fuel flow coming from it but not be a good pump because of slow timing that is caused by wear on the bottom end of the plunger. When making a test on a pump that has been used for a long time, use a micrometer and measure the length of the plunger. If the length of the plunger is shorter than the minimum length (worn) dimension given in the chart, install a new pump.

FUEL PUMP PLUNGER

Length (new)	3.1120 ± .0005 in. (79.044 ± 0.013 mm)
Minimum length (worn)	3.1115 in. (79.031 mm)

INSTALLATION OF FUEL INJECTION PUMPS (FIGURE 21–25)

NOTE: The fuel rack **must be in the center position** before the correct installation of an injection pump is possible.

1. Put the fuel rack in the center position.
2. Put the 8S2244 Extractor (9) on the threads of the fuel injection pump.
3. Make sure the lifter for the pump to be installed is at the bottom of its travel (cam lobe is at its lowest point).
4. Align the groove in the barrel (2) with the slot (groove) in the gear segment (3).
5. Be sure the spacer is in position in the pump housing bore.
6. Carefully install the pump straight down into the pump housing bore.

NOTE: The slot (groove) in the gear segment (3) must be in alignment with the pin (4) in the side of the lifter, and the groove in the barrel (2) must be in alignment with the dowel (5) in the housing bore.

7. Remove the 8S2244 Extractor. Put the O-ring seal, the retainer bushing (9), and the 8T5287 Wrench (7) in position on top of the injection pump. Install the 6V7050 Compressor Group (8).
8. Slowly move the handle of the 6V7050 Compressor Group down to push the injection pump into the bore.

NOTE: The handle of the 6V7050 Compressor Group must move smoothly down to the lock position. Do not force the handle if it stops. If the handle does not move smoothly down to the lock position, raise the handle, remove the 6V7050 Compressor Group, and repeat steps 3 through 8.

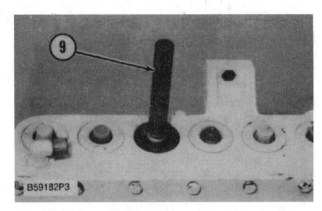

Fuel Injection Pump Housing
(9) 8S2244 Extractor.

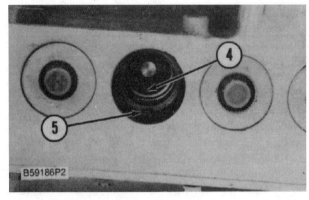

Fuel Injection Pump Housing (Top View)
(4) Pin. (5) Dowel.

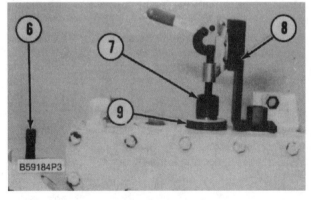

Fuel Injection Pump Housing
(6) 6V4186 Timing Pin. (7) 8T5287 Wrench. (8) 6V7050 Compressor Group. (9) Retainer bushing.

FIGURE 21–25 (Courtesy of Caterpillar, Inc.)

9. Put the O-ring seal in position in the pump housing bore. Use the 8T5287 Wrench to install the retainer bushing.

10. Remove the 6V7050 Compressor Group. Tighten the retainer bushing to 170 ± 11 lb-ft (230 ± 15 N·m).

NOTE: The bushing must be tightened to the correct torque. Damage to the housing will result if the bushing is too tight. If the bushing is not tight enough, the pump will leak.

11. Install the fuel injection lines to the pump and tighten to 30 ± 5 lb-ft (40 ± 7 N·m). See Fuel Injection Lines in this section for more information.

WARNING

Be sure the fuel injection line clamps are installed in the correct locations. Incorrectly installed clamps may allow the fuel injection lines to vibrate and become damaged. The damaged lines may leak and cause a fire.

FINDING THE TOP CENTER COMPRESSION POSITION FOR THE NO. 1 PISTON

TOOLS NEEDED
9S9082 Turning Tool 1

The No. 1 piston at top center (TC) on the compression stroke is the starting point of all timing procedures.

NOTE: On some engines there are two threaded holes in the flywheel. These holes are in alignment with the holes with plugs in the left and right front of the flywheel housing. The two holes in the flywheel are at a different distance from the center of the flywheel, so the timing bolt cannot be put in the wrong hole.

1. The timing bolt (1) is kept in storage at location (3) and can be installed in either the left side of the engine at location (2) or in the right side of the engine. Remove the bolts and cover from the flywheel housing. Remove the plug from the timing hole in the flywheel housing **(Figure 21–26)**.

2. Put the timing bolt (1) (the long bolt that holds the cover on the flywheel housing) through the timing hole in the flywheel housing. Use the 9S9082 Engine Turning Tool and ½ in. drive ratchet wrench to turn the engine flywheel in the direction of normal engine rotation until the timing bolt engages with the threaded hole in the flywheel.

NOTE: If the flywheel must be turned opposite normal engine rotation approximately 45°, then turn the flywheel in the direction of normal rotation until the timing bolt engages with the threaded hole. The reason for this procedure is to make sure the play is removed from the gears when the No. 1 piston is put on TC.

CHECKING ENGINE TIMING WITH THE 8T5300 TIMING INDICATOR GROUP AND THE 8T5301 DIESEL TIMING ADAPTER GROUP

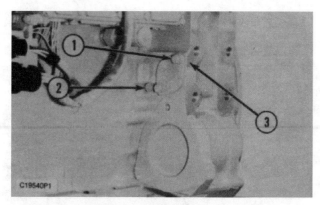

Locating Top Center (Left Side of Engine)
(1) Timing bolt. (2) Timing bolt location. (3) Storage location.

Using 9S9082 Engine Turning Tool
(1) Timing bolt. (5) 9S9082 Engine Turning Tool.

FIGURE 21–26 (Courtesy of Caterpillar, Inc.)

1. See the engine information plate for the performance specification number and refer to the Caterpillar Fuel Setting And Related Information Fiche for the correct timing specifications to use.

2. Refer to the operating instructions inside the lid of the 8T5300 Timing Indicator or Special Instruction Form No. SEHS8580 for complete instructions and calibration.

WARNING

The engine must be stopped before the timing indicator group is installed. A high-pressure fuel line must be disconnected and a probe must be installed against the flywheel.

3. Loosen all fuel line clamps that hold No. 1 fuel injection line, and disconnect the fuel injection line (13) for the No. 1 cylinder at the fuel injection pump. Slide the nut up and out of the way. Put the 5P7436 Adapter in its place and turn it onto the fuel pump bonnet until the top of the bonnet threads are approximately even with the bottom of the "window" in the adapter.

4. Put the 5P7435 Tee Adapter (11) on the injection transducer (10), and put the end of the 5P7435 Tee Adapter in the "window" of the 5P7436 Adapter.

5. Put the fuel injection line (13) on top of the 5P7435 Tee Adapter. Install the 5P7437 Adapter and tighten it to 30 lb-ft (40 N·m) **(Figure 21–27)**.

6. Remove the plug from the flywheel housing. Install the transducer adapter (5) into the hole from which the plug was removed. Tighten only a small amount.

7. Push the magnetic transducer (3) into the pipe adapter (5) until it makes contact with the flywheel. Pull it back out 0.06 in. (1.5 mm) and lightly tighten the knurled locknut.

8. Connect the cables from the transducers to the Engine Timing Indicator. Calibrate and make adjustments. For the calibration procedure, refer to Special Instruction Form No. SEHS8580.

9. Start the engine, and let it reach operating temperature. Then run the engine at approximately half throttle for 8 to 10 minutes before measuring the timing.

10. Run the engine at the speeds required, and record the timing indicator readings. If the engine timing is not correct, refer to Checking Engine Timing By The Timing Pin Method for static adjustment of the fuel injection pump drive. If the timing advance is not correct, perform the following steps to make an adjustment.

3. Remove the front valve cover from the engine.

4. The intake and exhaust valves for the No. 1 cylinder are closed if No. 1 piston is on the compression stroke and the rocker arms can be moved by hand. If the rocker arms cannot be moved and the valves are slightly open, the flywheel must be turned again. Remove the timing bolt and turn the flywheel in the direction of normal engine rotation 360° until the timing bolt can be installed. The No. 1 piston is now in the top center compression position.

CHECKING ENGINE TIMING WITH THE 8T5300 TIMING INDICATOR GROUP AND THE 8T5301 DIESEL TIMING ADAPTER GROUP

The 8T5300 Timing Indicator Group must be used with the 8T5301 Diesel Timing Adapter Group.

466 Chapter 21 CATERPILLAR FUEL INJECTION SYSTEMS

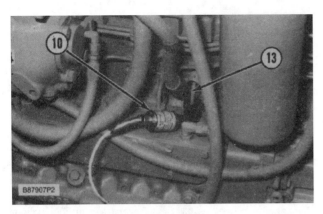

Transducer In Position (Typical Example)
(10) Injection transducer. (13) Fuel injection line for No. 6 cylinder.

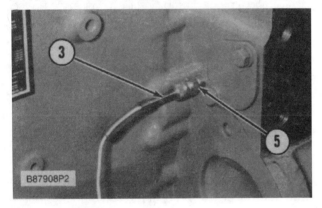

Transducer In Position
(3) TDC magnetic transducer. (5) Transducer adapter.

FIGURE 21-27 (Courtesy of Caterpillar, Inc.)

Automatic Timing Advance Unit
(Governor And Fuel Pump Drive Group)
(15) Locknut. (16) Screw. (17) Locknut. (18) Setscrew.

FIGURE 21-28 (Courtesy of Caterpillar, Inc.)

11. Stop the engine and remove the timing advance cover.

Steps 12 through 15 are for earlier timing advance units 7FB1-Up and 4MG1-4MG3599 **(Figure 21-28).**

12. To adjust the speed where the fuel injection timing starts to advance, loosen the locknut (15) and turn the screw (16). Turn the screw (16) clockwise to increase the speed where the timing advance starts. One turn of the screw (16) will change the START speed approximately 50 rpm. Torque the locknut (15) to 50 ± 11 lb-ft (70 ± 15 N·m).

NOTE: If the speed where timing advance starts is adjusted, the speed where the automatic timing advance stops should also be adjusted.

13. To adjust the speed where the fuel injection automatic timing advance stops, loosen the locknut (17) and turn the setscrew clockwise to decrease the stop speed. One turn of the setscrew (18) will change the STOP speed approximately 30 rpm. Tighten the locknut (17) to a torque of 20 ± 2 lb-in. (2.25 ± 0.25 N·m). The 6V2106 Tool Group (19) (part of the 6V6070 Governor Adjusting Tool Group) can be used to make this adjustment.

14. After each adjustment, install the timing advance cover and recheck the automatic timing advance with the 8T5300 Timing Indicator Group.

15. If the automatic timing advance unit cannot be adjusted to operate within the correct range, or the operation of the unit is not smooth, replace the automatic timing advance unit (governor and fuel pump drive group).

Steps 16 through 20 are for later timing advance units 4MG3600-Up, 5KJ1-Up, and 3ZJ1-Up **(Figure 21-29).**

Later Automatic Timing Advance Unit
(15) Locknut. (16) Screw. (18) Setscrew-located inside of screw (11).

FIGURE 21-29 (Courtesy of Caterpillar, Inc.)

16. If the adjustments are being made because of an engine horsepower rating change for replacement of parts, start with an initial setting as follows:
    a. Adjustment screw (16) must extend out from the retainer approximately 1.10 in. (27.9 mm).
    b. The setscrew (18) must be below the end of the screw approximately 17.78 mm (.700 in.) on an engine with an 11° timing advance and 0.750 in. (19.05 mm) on an engine with a 10° timing advance.

17. To adjust the speed where the fuel injection timing starts to advance, loosen the locknut (15) and turn screw (16) clockwise to increase the speed where the timing advance starts. Tighten the locknut (15) to a torque of 50 ± 11 lb-ft (70 ± 15 N · m).

NOTE: If the speed where timing advance starts is adjusted, the speed where the automatic timing advance should also be adjusted.

18. To adjust the speed where the fuel injection automatic timing advance stops, loosen the locknut and turn the setscrew (18) clockwise to decrease the STOP speed. Tighten the locknut to a torque of 20 ± 2 lb-in. (2.25 ± 0.25 N · m). The 6V2106 Tool Group (19) (part of the 6V6070 Governor Adjusting Tool Group) can be used to make this adjustment.

19. After each adjustment, install the cover and recheck the automatic timing advance with the 8T5300 Timing Indicator Group.

20. If the automatic timing advance unit cannot be adjusted to operate within the correct range, or the operation of the unit is not smooth, repair or replace the automatic timing advance unit (governor and fuel pump drive group).

ENGINE TIMING BY THE TIMING PIN METHOD 4MG3600-UP, 5KJ1-UP, 3ZJ1-UP

TOOLS NEEDED		
9S9082	Engine Turning Tool	1
6V4186	Timing Pin	1
1U8271	Timing Advance Holding Tool	1
8T5300	Engine Timing Indicator Group	1
8T5301	Diesel Timing Adapter Group	1

1. Put the No. 1 piston at top center on the compression stroke. Remove the timing bolt from the flywheel and use the 9S9082 Engine Turning Tool to rotate the crankshaft clockwise 45° as seen from the flywheel end of the engine.

NOTE: The crankshaft can be turned from the front of the engine by using a wrench on the vibration damper bolts, if necessary.

2. Remove the timing advance cover.

3. Loosen the nuts (3) and remove the retainer (2) and the flyweight spring from the timing advance unit. Make sure that the flyweight spring does not fall out and get lost **(Figure 21–30)**.

NOTE: Do not loosen the locknuts and adjustment screws in the end of the retainer (2). If the adjustment screws are moved from their original settings, the dynamic engine timing must be set.

4. Loosen the bolts (4) that hold the timing advance unit together.

5. Tighten the bolts (4) to a torque of 20 ± 1 lb-in. (2.2 ± 0.1 N · m). This puts a slight clamping force on the fuel pump drive gear to hold it in position. Also, the fuel pump camshaft can be turned or held in position separate from the engine crankshaft. The drive gear is allowed to slip.

6. Remove the timing pin plug from the fuel injection pump housing.

7. With the No. 1 piston 45° before top center, slowly rotate the crankshaft counterclockwise (as seen from the flywheel end of the engine) until the timing pin (6) goes into the slot in the fuel pump camshaft **(Figure 21–31)** and the timing bolt can be installed in the timing hole in the flywheel.

8. Install the 1U8271 Holding Tool (7) and push the timing advance unit piston back as follows **(Figure 21–32)**:
    a. Turn the knurled nuts on the holding tool out until each stud is 0.25 in. (6.4 mm) below the surface of the nut.

Timing Advance Unit
(2) Retainer. (3) Nuts. (4) Bolts.

FIGURE 21–30 (Courtesy of Caterpillar, Inc.)

468 Chapter 21 CATERPILLAR FUEL INJECTION SYSTEMS

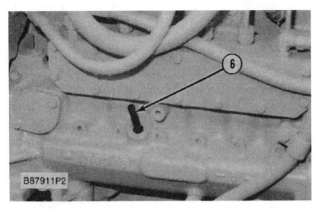

Timing Pin Installed
(6) 6V4186 Timing Pin.

FIGURE 21–31 (Courtesy of Caterpillar, Inc.)

Install Holding Tool
(3) Nut. (7) 1U8271 Holding Tool.

FIGURE 21–32 (Courtesy of Caterpillar, Inc.)

b. Put the holding tool (7) in position on the cap screws that hold the retainer (2). Install and tighten the nuts (3) finger tight. Make sure that the four tangs on the loose inner ring of the holder tool are positioned at the corners of the four flyweights, and the flyweights are free to move.

c. Tighten the four large knurled nuts evenly by hand until a positive stop is felt. No external component contact can be seen. The positive stop is the timing advance piston making contact at the bottom of its travel. This step makes sure that the timing advance unit is in its most retarded timing position.

9. Tighten the four drive bolts to a torque of 41 ± 5 lb-ft (55 ± 7 N·m).

10. Remove the holding tool from the automatic timing advance unit.

11. Remove the timing pin bolt from the flywheel and the timing pin from the fuel injection pump housing.

12. Install the flyweight spring and retainer on the timing advance unit. Make sure the spring is in its correct position, and tighten the four nuts to hold the retainer in position.

13. Install the cover on the timing gear housing and plug in the fuel injection pump housing.

FUEL RATIO CONTROL AND GOVERNOR CHECK (LATER ENGINES)

	TOOLS NEEDED	
2W9161	Manual Shutoff Group	1
6V4186	Timing Pin	1
6V6070	Governor Adjusting Tool	1

NOTE: The governor seals **do not** have to be cut or removed for the procedure that follows:

1. Remove the plug and cover from the fuel injection pump housing.

2. Install the rack position indicator as follows **(Figures 21–33 and 21–34)**:

  a. Install the 5P4814 Collet (7) on the 8T9198 Bracket Assembly (4).

  b. Position the indicator arm in approximately the middle of its travel to make sure that it will engage in the slot (9) in the rack **(Figure 21–35)**. Put the 8T9198 Bracket Assembly (4) in position on the fuel injection pump housing.

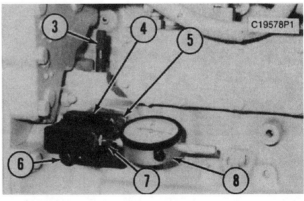

Dial Indicator And Centering Pin Installed
(3) 6V4186 Timing Pin. (4) 8T9198 Bracket Assembly. (5) 2A0762 Bolt. (6) 8H9178 Ground Body Bolt. (7) 5P4814 Collet. (8) 6V6106 Dial Indicator.

FIGURE 21–33 (Courtesy of Caterpillar, Inc.)

FUEL RATIO CONTROL AND GOVERNOR CHECK (LATER ENGINES) 469

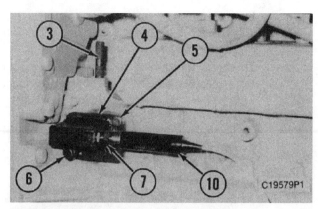

Electronic Indicator And Centering Pin Installed
(3) 6V4186 Timing Pin. (4) 8T9198 Bracket Assembly. (5) 2A0762 Bolt. (6) 8H9178 Ground Body Bolt. (7) 5P4814 Collet. (10) 8T1000 Electronic Position Indicator.

FIGURE 21–34 (Courtesy of Caterpillar, Inc.)

Slot In Fuel Injection Pump Rack
(9) Slot.

FIGURE 21–35 (Courtesy of Caterpillar, Inc.)

4. Move the governor control linkage to the full FUEL ON position and hold or fasten it in this position.

5. Install the 6V4186 Timing Pin (3) in the rack zeroing hole near the front of the fuel injection pump housing.

6. With the governor control lever in the full FUEL ON position, use a 1N9954 Lever and move the manual shut-off shaft slowly to the FUEL OFF position (counterclockwise). Watch and make sure the timing pin drops and engages with the slot in the fuel rack.

7. Release the manual shut-off shaft and zero the dial or electronic indicator. If using the dial indicator, move the indicator assembly in the collet and tighten the collet when all three needles are on zero. If using the electronic indicator, press the zero button.

8. Remove the 6V4186 Timing Pin (3) and note the change in indicator reading. The indicator should show movement in the FUEL ON direction. If no movement occurs, repeat steps 5, 6, and 7 to zero the indicator.

9. Release the governor control shaft and linkage.

10. Remove the boost air line (11) from the engine. Put plugs over the openings to keep dirt out of the system **(Figure 21–36)**.

11. Start the engine and operate it for a minimum of 5 minutes to get the governor and engine up to normal operating temperatures.

12. Check the leak-down rate of the fuel ratio control (with the engine operating at low idle) as follows:

    a. Connect a pressure gauge, a shut-off valve, a pressure regulator, and an air supply to the fitting (12) **(Figure 21–36)**.

    b. Apply 10 psi (70 kPa) air pressure to the fuel ratio control.

    c. Install the 8H9178 Ground Body Bolt (6) first. Then install the 2A0762 Bolt.

    d. Be sure the indicator arm moves freely.

    e. Put the dial indicator (8) of the 8T1000 Electronic Position Indicator (10) in position in the collet (7).

    f. Put the 9S8903 Contact Point on the 6V2030 Extension and install it on the indicator (8) or (10).

NOTE: The 9S8903 Contact Point will not go through the collet and must be assembled after the indicator stem has passed through the collet.

    g. Tighten the collet (7) just enough to hold the dial or electronic indicator.

3. Turn the engine start key ON to activate the shut-off solenoid. Do not start the engine at this time.

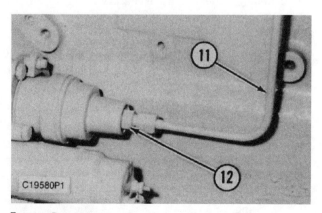

Remove Boost Line
(11) Air Line. (12) Fitting.

FIGURE 21–36 (Courtesy of Caterpillar, Inc.)

c. Turn the shut-off valve off and check the leak-down rate. Leakage of 3 psi (20 kPa) in 30 seconds is acceptable.

d. If leakage is more than 3 psi (20 kPa) in 30 seconds, the fuel ratio control must be replaced before continuing to the next step.

e. Keep 10 psi (70 kPa) air pressure on the fuel ratio control for step 13.

13. From low idle, rapidly move the governor control shaft to the full FUEL ON position and note the reading on the indicator. If using the dial indicator, read the indicator carefully because this reading will be a maximum for only a moment. The electronic indicator will hold and display the maximum reading. Record the maximum indicator reading.

The maximum indicator reading is the *dynamic full torque fuel setting* of the engine. This setting is 0.02 in. (0.5 mm) greater than the static full torque setting given on the engine information plate of later engines or in the Fuel Setting And Related Information Fiche. If dynamic full torque setting is not reached, increase the air pressure to make sure there is full fuel ratio control movement. Do not exceed 15 psi (100 kPa).

14. Release all air pressure from the fuel ratio control. Start at 900 rpm and rapidly move the governor control shaft to the full FUEL ON position and note the reading on the indicator. If using the dial indicator, read the indicator carefully because this reading will be a maximum for only a moment. The electronic indicator will hold and display the maximum reading. Record the maximum indicator reading. This is the *dynamic fuel ratio control setting* for the engine.

15. If the dynamic fuel ratio control setting is *within* ± 0.010 in. (0.25 mm) of the specification given on the Engine Information And Related Information Fiche, an adjustment is not necessary.

16. For adjustment of the control see *Fuel Ratio Control Adjustment.*

17. Check the boost pressure that gives full torque rack travel, as follows:

a. Connect a pressure gauge, a pressure regulator, and an air supply to the fitting (12).

b. Apply 4 psi (25 kPa) air pressure to the fuel ratio control.

c. Start at 900 rpm and rapidly move the governor control shaft to the full FUEL ON position and record the maximum indicator reading.

d. Repeat this procedure several times, each time increasing the air pressure 0.5 psi (5 kPa).

e. Record the first air pressure setting that gives the full torque fuel setting. The full torque fuel setting was measured in step 13.

f. This is the boost pressure that moves the fuel ratio control out of the rack control position. This pressure permits dynamic full torque fuel setting. If the pressure recorded in step *e* is less than rated boost, the fuel ratio control cannot be a cause of low power complaint.

LOW-IDLE-SPEED ADJUSTMENT

Start the engine and run until normal operating temperature is reached. Check low-idle rpm with no load on the engine. If an adjustment is necessary, use the following procedure:

To adjust the low-idle rpm, start the engine and run it with the governor control lever in the low-idle position. Loosen the locknut for the low-idle screw (1). Turn the low-idle screw to get the correct low-idle rpm. Increase engine speed and return to low idle and recheck the low-idle speed. Tighten the locknut **(Figure 21–37)**.

CHECKING THE SET POINT (BALANCE POINT)

The engine set point is an adjusted specification and is important to the correct operation of the engine. High-idle rpm is *not* an adjusted specification. The set point (formerly balance point) is full load rpm plus an additional 20 rpm. The set point is the rpm at which the fuel setting adjustment screw and stop or first torque spring just start to make contact. At this rpm, the fuel setting adjustment screw and stop or first torque spring still

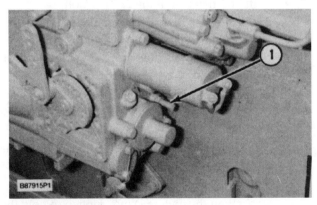

Low Idle Adjustment
(1) Low idle stop screw.

FIGURE 21–37 (Courtesy of Caterpillar, Inc.)

have movement between them. When additional load is put on the engine, the fuel setting adjustment screw and stop or first torque spring will become stable against each other. The set point is controlled by the fuel setting and the high-idle adjustment screw.

There is a new and more accurate method for checking the set point of the engine. If the tools for the new method are not available, there is an alternative method for checking the set point.

TOOLS NEEDED
6V4060 Engine Set Point Indicator Group 1

The 6V4060 Engine Set Point Indicator Group with the 6V2100 Multitach can be used to check the set point. Special Instruction, *Form No. SEHS7931* gives instructions for installation and use of this tool group.

Alternative Method

TOOLS NEEDED
8T0500 Circuit Tester 1
6V3121 Multitach Group 1

NOTE: Do not use the vehicle tachometer unless its accuracy is known to be within ± 1 rpm.

If the set point is correct and the high-idle speed is within specifications, the fuel system operation of the engine is correct. The set point for the engine is

- at 20 rpm greater than full-load speed.
- the rpm where the fuel setting adjustment screw stop or first torque spring just make contact

Use the following procedure to check the set point. Refer to Techniques for Loading Engines in Special Instruction *Form No. SEHS7050*.

1. Connect a tachometer that has good accuracy to the tachometer drive.

2. Connect the clip end of the 8T0500 Circuit Tester to the brass terminal screw (2) on the governor housing. Connect the other end of the tester to a place on the fuel system that is a good ground connection.

3. Start the engine.

4. With the engine at normal operating conditions, run the engine at high idle.

5. Record the speed of the engine at high idle.

6. Add load on the engine slowly until the circuit tester light just comes on (minimum light output). This is the set point.

7. Record the speed (rpm) at the set point.

8. Repeat step 6 several times to make sure that the reading is correct.

9. Stop the engine. Compare the records from steps 6 and 7 with full-load speed from the engine information plate. If the engine information plate is not available, see the Fuel Setting And Related Information Fiche. The tolerance for the set point is ± 10 rpm. The tolerance for the high-idle rpm is ± 50 rpm in chassis and ± 30 rpm on a bare engine. If the readings from steps 5 and 7 are within the tolerance, no adjustment is needed.

NOTE: Later-model engines have the actual dyno high idle stamped on the engine information plate. It is possible in some applications that the high-idle rpm will be less than the actual lower limit. This can be caused by high parasitic loads such as hydraulic pumps or compressors.

Adjusting the Set Point (Balance Point) (Figure 21–38)

1. If the set point and the high-idle rpm are within tolerance, no adjustment is to be made.

2. If the set point rpm is not correct, remove the cover (3).

This adjustment controls the amount of restriction to oil flow into and out of the dashpot chamber. Too much oil flow will cause the governor to hunt, and too little oil flow will cause a slow governor action.

1. Turn the needle valve (1) in (clockwise) until it stops **(Figure 21–38)**. Now, open the needle valve (1) two full turns (counterclockwise). The exact point of adjustment is where the governor gives the best performance.

NOTE: Do not keep the needle valve (1) fully closed. This can cause excessive overshoot on start-up or load rejection.

2. Check governor operation.

With the engine running at medium (mid) speed, load the engine (at least one-quarter load) to find the stability of the setting. Quickly remove the load. A slight overshoot of speed is desired, as it reduces response time. The engine speed should return to smooth, steady operation. If it does not have a slight overshoot and return to a smooth, steady operation, adjust the needle valve and repeat the above procedure.

3. Loosen the locknut (4) and turn the adjustment screw (5) to adjust the set point to the midpoint of the tolerance.

4. When the set point is correct, check the high-idle rpm. The high-idle rpm must not be more than the high limit of the tolerance.

472 Chapter 21 CATERPILLAR FUEL INJECTION SYSTEMS

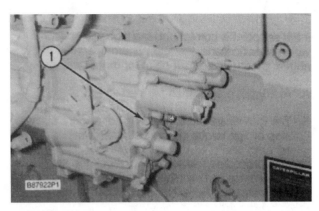

Adjustment Of Dashpot
(1) Needle Valve.

Terminal Location
(2) Brass terminal screw.

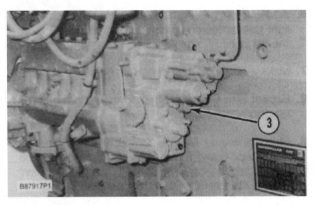

Remove Cover
(3) Cover.

Set Point Adjustment
(4) Locknut. (5) Adjustment screw.

FIGURE 21–38 (Courtesy of Caterpillar, Inc.)

If the high-idle rpm is more than the high limit of the tolerance, check the governor spring and flyweights. If the high-idle rpm is less than the low limit of the tolerance, check for excess parasitic loads and then the governor spring and flyweights.

FUEL RATIO CONTROL ADJUSTMENT (LATER ENGINES)

TOOLS NEEDED
2W9161	Manual Shutoff Group	1
6V4186	Timing Pin	1
6V6070	Governor Adjusting Tool Group	1

NOTE: Before the governor seals are cut or removed, see Fuel Ratio Control and Governor Check to make sure an adjustment is needed.

1. See the engine information plate or the Fuel Setting And Related Information Fiche for the correct dynamic fuel ratio control setting specification. Before an adjustment is made, record the difference between the correct specification and the dynamic fuel ratio control reading recorded in step 14 of Fuel Ratio Control and Governor Check.

2. Install and zero the rack position indicator group **(Figure 21–39).**

WARNING

To help prevent an accident caused by rotating parts, work carefully around an engine that has been started.

3. Start the engine and operate it for a minimum of 5 minutes to get the governor and engine up to normal operating temperature.

4. Loosen the bolts on the clamp plate. The mating surfaces of the clamp plate and fuel ratio control are serrated or "notched." A movement of one "notch" will change the dynamic fuel ratio control setting by 0.0008

in. (0.02 mm). Refer to the difference recorded in step 1. Turn the fuel ratio control (4) clockwise to get a more negative (less fuel) setting. Turn the control counterclockwise to get a more positive (more fuel) setting.

Fuel Ratio Governor Adjustment Check
(17) 6V2017 Governor Adjusting Tool. (18) 6V2106 Rack Adjusting Tool (outer part).

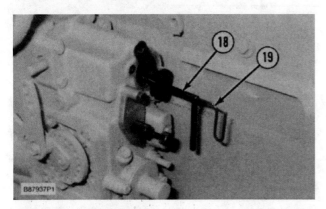

Tools For Linkage Adjustment
(18) 6V2106 Rack Adjusting Tool (outer part). (19) 6V2104 Hex Wrench.

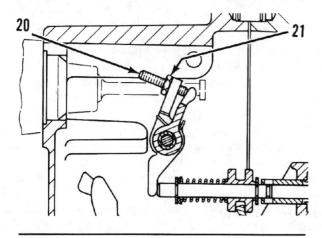

Location Of Linkage Adjustment
(20) Adjustment screw. (21) Locknut.

FIGURE 21–39 (Courtesy of Caterpillar, Inc.)

5. After an adjustment is made, tighten the clamp bolts and recheck the dynamic fuel ratio control setting. Start at 900 rpm and rapidly move the governor control shaft to the full FUEL ON position and note the reading on the indicator. If using the dial indicator, read the indicator carefully because this reading will be a maximum for only a moment. The electronic indicator will hold and display the maximum reading. Record the maximum indicator reading. Compare this reading with the correct dynamic fuel ratio control setting specification. Repeat the adjustment procedure as required until the correct setting is achieved.

6. Stop the engine.
7. Install the wire and seal on the fuel ratio control.
8. Install the air line on the engine.
9. Remove the rack position indicator tooling.

SLEEVE-METERING FUEL INJECTION SYSTEM

The sleeve-metering fuel injection pump has a pumping plunger for each cylinder of the engine. The name is derived from the method used to meter the amount of fuel sent to each engine cylinder. There is a gear-type fuel transfer pump at the front of the injection pump and a governor at the rear of the injection pump housing. The pump camshaft is supported in the housing by a bearing at each end. The pump is driven by the timing gears at the front of the engine. The pump housing and governor housing are full of fuel at transfer pump pressure for lubrication.

SLEEVE-METERING FUEL SYSTEM FUEL FLOW (FIGURE 21–40)

The transfer pump (11) draws fuel from the fuel tank (7) through the water separator (F) (if so equipped), through the fuel filter (9) to the injection pump housing (14) at the top and through a passage to the fuel transfer pump. Fuel under transfer pump pressure fills the housing. The bypass valve (12) controls fuel supply pressure at 30 psi (±5 psi) (205 ± 35 kPa). Excess fuel from the bypass valve is returned to the transfer pump intake **(Figure 21–40)**. A hand-operated priming pump is provided to prime the system with fuel and eliminate any air that may be present. The injection pump housing must be full of fuel before the pump camshaft is turned. The constant-

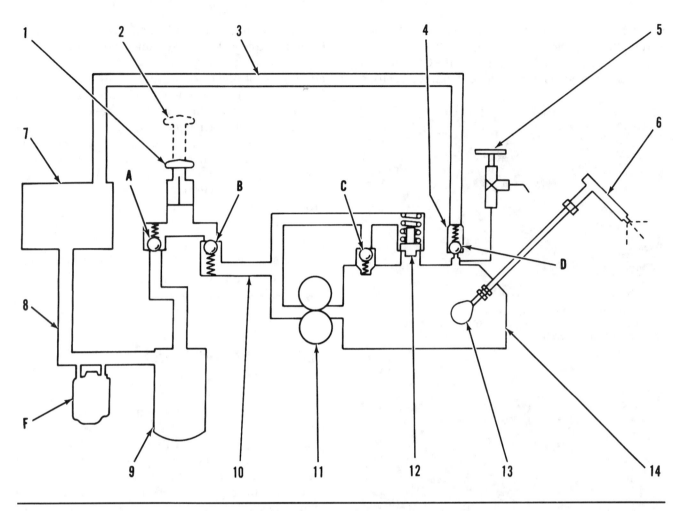

Schematic Of Fuel System
(1) Fuel priming pump (closed position). (2) Fuel priming pump (open position). (3) Return line for constant bleed valve. (4) Constant bleed valve. (5) Manual bleed valve. (6) Fuel injection nozzle. (7) Fuel tank. (8) Fuel inlet line. (9) Fuel filter. (10) Fuel line to injection pump. (11) Fuel transfer pump. (12) Fuel bypass valve. (13) Camshaft. (14) Housing for fuel injection pumps. (A) Check valve. (B) Check valve. (C) Check valve. (D) Check valve. (F) Water Separator.

FIGURE 21–40 (Courtesy of Caterpillar, Inc.)

bleed valve (4) lets approximately 9 gallons (35 liters) of fuel per hour go back to the fuel tank to remove any air and aid in cooling the injection pump **(Figure 27–41)**.

SLEEVE-METERING INJECTION PUMP OPERATION (FIGURES 21–42 AND 21–43)

With the engine running, fuel under transfer pump pressure enters the center of each pumping plunger (B) through the fuel inlet (C) during the downstroke of the plunger. The fuel outlet (E) is closed off by the sleeve (D) (see position 1 in **Figure 21–42**). In position 2 fuel injection starts when the plunger (B) is raised by cam action. Fuel above the plunger is now injected into the engine cylinder under increased pressure. Injection ends (position 3) when the fuel outlet (E) is raised above the top edge of the sleeve (D) by the camshaft (4). Fuel that is above and in the plunger now flows through the outlet (E) back to the pump housing. Raising the sleeve (D) keeps the fuel outlet (E) closed for a longer time, causing more fuel to be delivered to the engine cylinders. Lowering the sleeve causes the outlet to be uncovered sooner, thereby delivering less fuel to the engine **(Figures 21–44 and 21–45)**. Delivery valves are used to help cut off fuel injection and prevent secondary injection. The different types of de-

GOVERNOR OPERATION 475

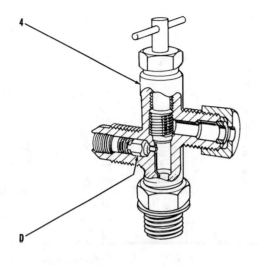

Constant Bleed Valve
(4) Constant bleed valve. (D) Check valve.

FIGURE 21–41 (Courtesy of Caterpillar, Inc.)

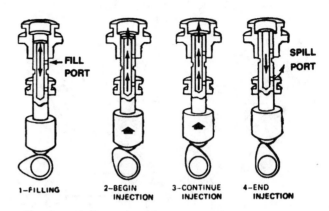

FIGURE 21–43 Injection pump action. (Courtesy of Caterpillar, Inc.)

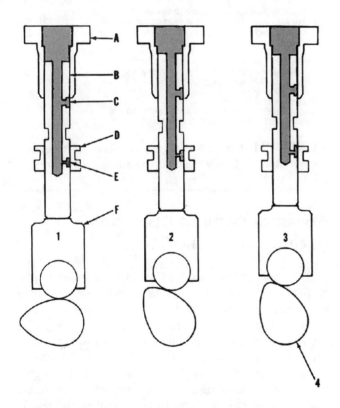

Fuel Injection Sequence
(1), (2), (3) Injection stroke (positions) of a fuel injection pump.
(4) Injection pump camshaft. (A) Barrel. (B) Plunger. (C) Fuel inlet.
(D) Sleeve. (E) Fuel outlet. (F) Lifter.

FIGURE 21–42 (Courtesy of Caterpillar, Inc.)

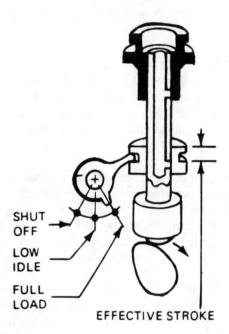

FIGURE 21–44 The effective stroke of the plunger is determined by vertical positioning of the sleeve. The lower the sleeve, the less fuel is injected, since the spill port is uncovered sooner. (Courtesy of Caterpillar, Inc.)

livery valves used and their application are shown in **Figure 21–46.**

GOVERNOR OPERATION

Governor components are shown in **Figure 21–47.** A cross-sectional view of the injection pump and governor components showing their relative position is shown in **Figure 21–48.** With reference to this illustra-

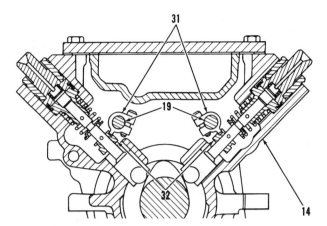

FIGURE 21–45 Cross section of V pump. 14. Injection pump housing. 19. Sleeve control shafts. 31. Sleeve levers. 32. Sleeves. (Courtesy of Caterpillar, Inc.)

tion and **Figure 21–49,** when the lever (15) moves to increase fuel delivery, lever (22) pressure is exerted against the governor springs (25), which causes the thrust collar (28) to move forward. This causes the sleeve control shafts (19) to turn and the levers (31) to raise the sleeves **(Figure 21–50)** to increase fuel delivery to the engine.

When the engine is started, the overfueling spring (27) pushes the thrust collar (28) to the full-fuel position. At around 400 rpm the force of the governor weights (30) compresses the spring (27), causing the thrust collar (28) and spring seat (26) to come into contact. From this point on the governor controls the engine speed.

When the lever (15) is moved to the maximum fuel position, the governor springs are put under compression, and the spring seat (26) contacts the load stop lever (29). As the load stop lever is rotated the load stop pin (17) moves up until it contacts the stop bar or stop screw. This stops movement of the thrust collar, levers, and sleeve control shafts. In this position fuel delivery is at maximum.

As engine speed increases, the force of the governor weights increases, causing the thrust collar to move toward the governor springs. When the force of the governor springs and governor weights comes into balance, the injection pump sleeves are held in position to maintain engine speed. When the governor control lever is moved toward the reduced fuel direction (FUEL OFF), the control shafts move the sleeves down, and less fuel is delivered to the engine. To stop the engine the ignition switch must be turned to the OFF position. This causes the fuel shut-off solenoid to move the sleeves down to the no-fuel position. With no fuel reaching the engine cylinders, the engine stops **(Figure 21–51).**

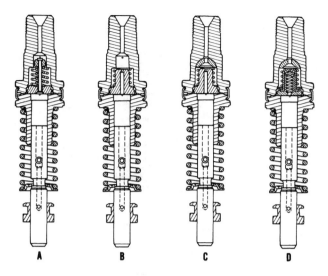

A. 9N5857 Fuel Injection Pump uses a reverse flow check valve (RFC) and is used in engines with serial nos. 2Z1 to 2Z13749.

B. 4W5868 Fuel Injection Pump uses an orificed delivery valve (ODV) and is used in engines with serial nos. 2Z13750 to 2Z34220.

C. 7W8040 Fuel Injection Pump also has an orificed delivery valve (ODV). This fuel injection pump is used in engines with serial nos. 2Z24754 to 2Z34220*.

D. 7C8927 Fuel Injection Pump uses an orificed reverse flow check. Valve (ORFC) is used in engines with serial nos. 2Z34221-Up.

*The engines rated at 240 hp at 2200 rpm and the air to air aftercooled (ATAAC) engines above 250 hp that have 12.5° BTC timing use this fuel system.

FIGURE 21–46 Different applications of injection pumps and delivery valves. (Courtesy of Caterpillar, Inc.)

Governor Dashpot Operation (Figures 21–52 and 21–53)

The dashpot is designed to steady the engine rpm. A piston (4) operates inside a cylinder (3) filled with fuel. Movement of the piston moves fuel into or out of the cylinder through an orifice (E). The restriction of the orifice dampens governor action to prevent engine speed from oscillating as the engine load changes. When the load is increased, the engine runs slower, resulting in less governor weight force. The governor spring starts to push against the spring seat (6) to increase fuel delivery and puts the spring (5) in tension. This spring is connected to the dashpot piston (4), which now starts to move, creating a low-pressure area in the cylinder. This draws fuel into the cylinder through the orifice restriction, slowing piston movement until the forces are again in balance.

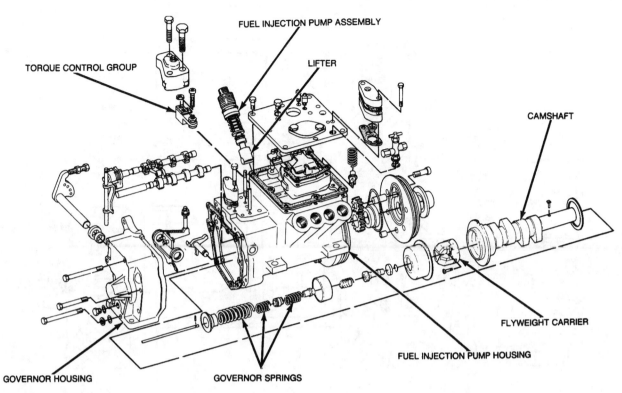

FIGURE 21-47 Sleeve-metering pump and governor components. (Courtesy of Caterpillar, Inc.)

When the load on the engine is decreased, the engine starts to run faster, resulting in increased governor weight force. This causes the spring (5) to be put under compression and to push against the servo piston (4). The pressure of the piston against the fuel oil causes fuel to be pushed out of the cylinder through the orifice. The orifice-restricted fuel flow makes piston movement more gradual, thereby dampening governor action.

Automatic Timing Advance Operation (Figure 21-54)

The automatic timing advance unit drives the gear on the fuel injection pump camshaft. The weights (4) are driven by two slides (6) that fit into angled notches in the weights (4). During rotation, centrifugal force moves the weights outward against the force of the springs (5). Movement of the notches in the weights changes the angle between the timing advance gear and the two drive dowels in the drive gear for the engine camshaft, thereby changing the fuel injection timing. Timing advance begins at low-idle speed and stops at rated engine speed. Two different timing advance units are used on the 3208 truck engine, one with 5° of advance and the other with 3.5°. No adjustment is possible on these advance units. Lubrication is provided from drilled holes that connect with the front bearing for the engine camshaft.

Fuel Ratio Control Operation

The fuel ratio control limits the amount of fuel delivery to the engine during acceleration to reduce exhaust smoke **(Figure 21-55).** During acceleration the bolt (3) will not allow the fuel control shaft to go to the full-fuel position. When boost pressure is high enough for good combustion in the cylinders, inlet manifold pressure goes through a line into the air chamber (1), where it pushes against the diaphragm (2), which moves the bolt (3) down to allow the fuel control shaft to go to the full-fuel position.

A hydraulic override was added to allow maximum fuel delivery for easier starting. Oil from the rear of the right cylinder head is used to actuate a diaphragm and plunger in the override unit **(Figure 21-56).** Oil flow from the cylinder head to the override unit is allowed when the starter switch is activated and stopped when the switch is released. A solenoid connected to the starter switch controls oil flow.

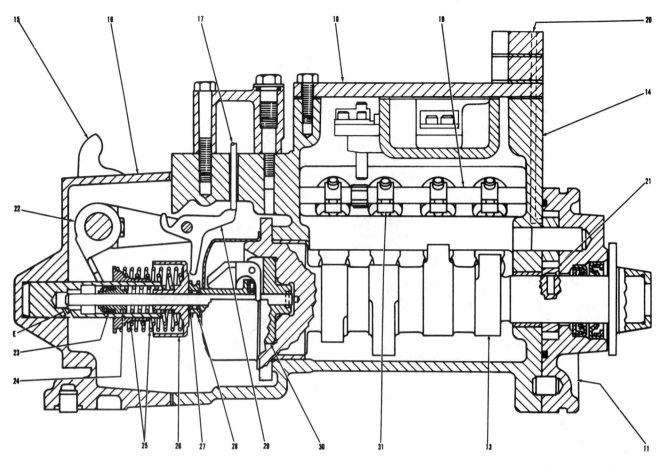

11. Fuel transfer pump. 13. Camshaft. 14. Housing for fuel injection pumps. 15. Lever. 16. Governor housing. 17. Load stop pin. 18. Cover. 19. Sleeve control shafts (two). 20. Inside fuel passage. 21. Drive gear for fuel transfer pump. 22. Lever on governor shaft. 23. Piston for dashpot governor. 24. Spring for dashpot governor. 25. Governor springs (inner spring is for low idle; outer spring is for high idle). 26. Spring seat. 27. Over fueling spring. 28. Thrust collar. 29. Load stop lever. 30. Carrier and governor weights. 31. Sleeve levers. E. Orifice for dashpot.

FIGURE 21-48 Cross section of fuel system with dashpot governor. (Courtesy of Caterpillar, Inc.)

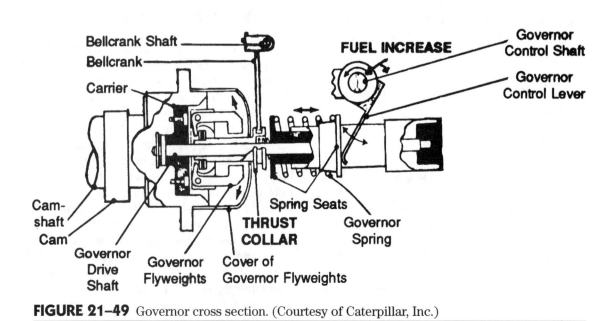

FIGURE 21-49 Governor cross section. (Courtesy of Caterpillar, Inc.)

FUEL INJECTION NOZZLES

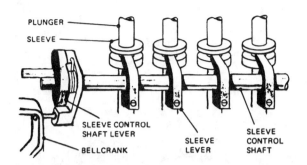

FIGURE 21–50 Bellcrank to sleeve control shaft connections. (Courtesy of Caterpillar, Inc.)

FIGURE 21–51 1. Fuel shut-off solenoid. 2. Top cover. (Courtesy of Caterpillar, Inc.)

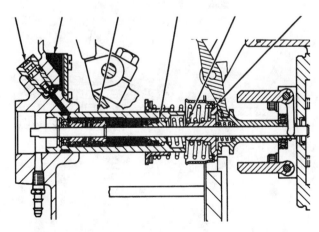

FIGURE 21–52 Dashpot governor components. 1. Needle valve. 2. Oil reservoir. 3. Cylinder. 4. Piston. 5. Dashpot. 6. Spring seat. (Courtesy of Caterpillar, Inc.)

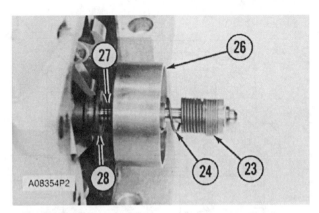

Governor Parts
(23) Piston for dashpot governor. (24) Spring for dashpot governor. (26) Spring seat. (27) Over fueling spring. (28) Thrust collar.

FIGURE 21–53 (Courtesy of Caterpillar, Inc.)

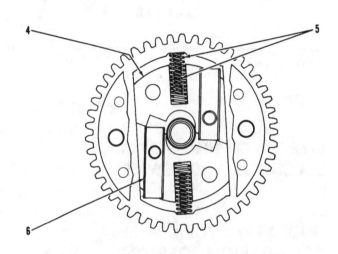

Automatic Timing Advance Unit
(4) Weights. (5) Springs. (6) Slides.

FIGURE 21–54 (Courtesy of Caterpillar, Inc.)

FUEL INJECTION NOZZLES

The fuel injection nozzles go through the cylinder head into the combustion chambers. High-pressure fuel opens the nozzle valve against spring pressure, and a fine spray of fuel is injected for good combustion. A compression seal is fitted between the nozzle body and the cylinder head. A carbon dam keeps carbon out of the nozzle bore in the cylinder head. Two different nozzle styles are used with the sleeve-metering fuel system, the 7000 series and the adjustable nozzle models 9N3979 and 1W5829 **(Figures 21–57 and 21–58)**.

480 Chapter 21 CATERPILLAR FUEL INJECTION SYSTEMS

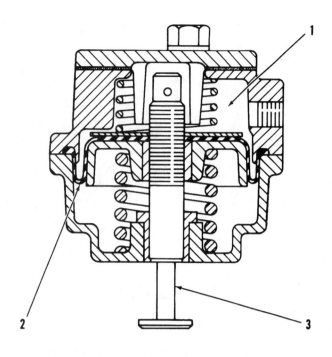

Fuel Ratio Control
(1) Air Chamber. (2) Diaphragm. (3) Bolt.

FIGURE 21–55 (Courtesy of Caterpillar, Inc.)

SLEEVE-METERING SYSTEM DIAGNOSIS AND SERVICE
(Courtesy of Caterpillar Inc.)

FINDING THE TOP CENTER COMPRESSION POSITION FOR THE NO. 1 PISTON

The No. 1 piston at top center (TC) on the compression stroke is the starting point for all timing procedures.

1. Remove the fitting from the timing hole in the front cover. Put a bolt in the timing hole.

2. Turn the crankshaft *counterclockwise* (as seen from rear of engine) until the bolt will go into the hole in the drive gear for the camshaft.

3. Remove the valve cover on the right side of the engine (as seen from rear of engine). The two valves at the right front of the engine are the intake and exhaust valves for the No. 1 cylinder.

4. The intake and exhaust valves for the No. 1 cylinder must now be closed, and the timing pointer is in alignment with the TDC-1 on the damper assembly. The No. 1 piston is now at top center on the compression stroke.

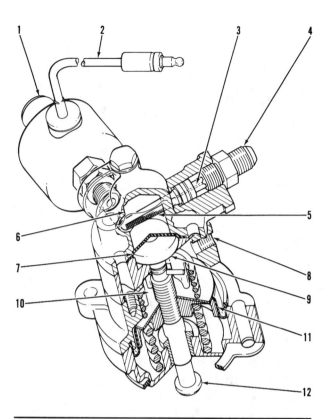

Fuel Ratio Control
(1) Solenoid. (2) Wire. (3) Orifice. (4) Fitting (oil outlet). (5) Screen. (6) Oil chamber. (7) Diaphragm. (8) Air inlet. (9) Plunger. (10) Air chamber. (11) Diaphragm. (12) Bolt.

FIGURE 21–56 (Courtesy of Caterpillar, Inc.)

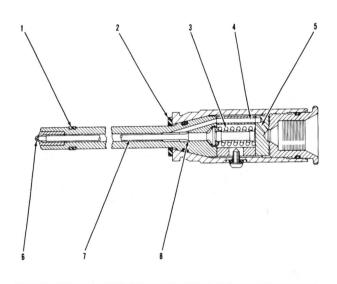

Fuel Injection Nozzle
(1) Carbon dam. (2) Seal. (3) Spring. (4) Passage. (5) Inlet passage. (6) Orifice. (7) Valve. (8) Diameter.

FIGURE 21–57 (Courtesy of Caterpillar, Inc.)

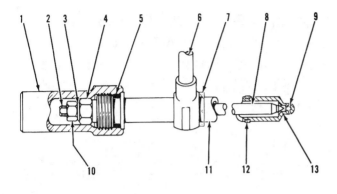

Fuel Injection Nozzle
(1) Cap. (2) Lift adjustment screw. (3) Pressure adjustment screw. (4) Locknut for pressure adjustment screw. (5) O-ring seal. (6) Fuel inlet. (7) Compression seal. (8) Valve. (9) Orifices (four). (10) Locknut for lift adjustment screw. (11) Nozzle body. (12) Carbon dam. (13) Nozzle tip.

FIGURE 21–58 (Courtesy of Caterpillar, Inc.)

CHECKING FUEL INJECTION PUMP TIMING ON THE ENGINE

The timing of the fuel injection pump can be checked and changed if necessary, to compensate for movement in the taper sleeve drive or worn timing gears. The timing can be checked and changed using the following method.

Checking Timing by the Timing Pin Method

TOOLS NEEDED
6V4069   Puller   1
3P1544   Timing Pin   1

1. Remove the bolt from the timing pin hole.
2. Turn the crankshaft *counterclockwise* (as seen from rear of engine) until the timing pin (2) goes into the notch in the camshaft for the fuel injection pumps **(Figure 21–59)**.
3. Remove the fitting from the timing hole (4) in the front cover. Put the bolt (3) through the front cover and into the hole with threads in the timing gear. The bolt from the hole (5) can be used **(Figure 21–60)**.
4. If the timing pin is in the notch in the camshaft for the fuel injection pumps, and the bolt (3) goes into the hole in the timing gear through the timing hole (4), the timing of the fuel injection pump is correct.

NOTE: If the bolt (3) does not go in the hole in the timing gear with the timing pin (2) in the notch in the camshaft, use the following procedure.

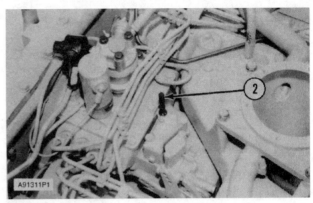

Timing Pin Installed
(2) 3P1544 Timing Pin.

FIGURE 21–59 (Courtesy of Caterpillar, Inc.)

a. Remove the nuts and the cover for the tachometer drive assembly.
b. Remove the tachometer drive shaft (9) and washer (8) from the camshaft for the fuel injection pumps **(Figure 21–61)**.

NOTE: The tachometer drive shaft (9) and washer (8) are removed as an assembly.

c. Put the 6V4069 Puller (10) on the camshaft for the fuel injection pumps. Tighten the bolts (11) until the drive gear on the camshaft for the fuel injection pumps comes loose **(Figure 21–61)**.
d. Remove the 6V4069 Puller.
e. Turn the crankshaft *counterclockwise* (as seen from rear of engine) until the bolt goes

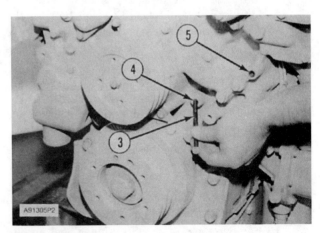

Installing Bolt
(3) 1D4539 Bolt, 5/16 in-18 NC, 63.5 mm (2.5 in) long. (4) Timing hole. (5) Hole.

FIGURE 21–60 (Courtesy of Caterpillar, Inc.)

482 Chapter 21 CATERPILLAR FUEL INJECTION SYSTEMS

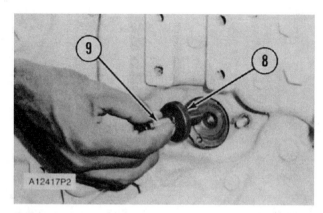

Location Of Bolt
(8) Washer. (9) Tachometer drive shaft.

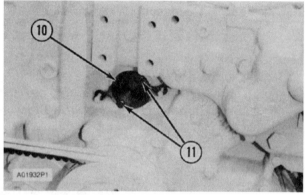

Loosening Drive Gear
(10) 6V4069 Puller. (11) Bolts.

FIGURE 21–61 (Courtesy of Caterpillar, Inc.)

Checking Engine Timing and the Automatic Timing Advance Unit with the 8T5300 Timing Indicator Group and 8T5301 Diesel Timing Adapter Group

	TOOLS NEEDED	
8T5300	Timing Indicator Group	1
8T5301	Diesel Timing Adapter Group	1

WARNING

A high-pressure fuel line must be disconnected. To avoid personal injury or fire from fuel spray, the engine must be stopped before the fuel line is disconnected.

When checking the dynamic timing on an engine that has a mechanical advance, Caterpillar recommends that the serviceperson calculate and plot the dynamic timing specifications first on a worksheet like Form No. SEHS8140. These worksheets are available in pads of 50 sheets. See Special Instruction, Form No. SEHS8580 for information required to calculate the timing curve. For the correct timing specifications to use, see the engine information plate for the performance specification number and refer to the Fuel Setting And Related Information Fiche.

After the timing values are calculated and plotted, the dynamic timing should be checked with the 8T5300 Timing Indicator Group and the 8T5301 Diesel Timing Adapter Group. To do this, operate the engine from 1000 rpm (base rpm) to high idle and from high idle to 1000 rpm (base rpm). Unstable readings are often obtained below 1000 rpm. Record the dynamic timing at each 100 rpm and at the specified speeds during both acceleration and deceleration. Plot the results on the worksheet.

Inspection of the plotted values will show if the fuel injection timing is within specification and if it is advancing correctly.

1. Refer to Special Instruction, Form No. SEHS8580 for complete service information and use of the 8T5300 Timing Indicator Group.

2. Loosen all fuel line clamps that hold the No. 1 fuel injection line, and disconnect the fuel injection line for the No. 1 cylinder at the fuel injection pump. Slide the nut up and out of the way. Put the 5P7436 Adapter in its place and turn the adapter onto the fuel pump bonnet until the top of the bonnet threads are approxi-

into the hole in the timing gear. With the timing pin in the notch in the camshaft for the fuel injection pumps, and the bolt in the hole in the timing gear, the timing for the engine is correct.

f. Install the washer and tachometer drive shaft. Tighten the tachometer drive shaft to 110 ± 10 lb-ft (149 ± 14 N·m). Remove the timing pin (2).

g. Turn the crankshaft two complete revolutions *counterclockwise* (as seen from rear of engine) and put the timing pin and bolt in again. If the timing pin and bolt cannot be installed, do steps *a* through *f* again.

h. Remove the bolt from the timing gear and install in hole. Install the plug in timing hole. Remove the timing pin and install the bolt. Install the cover for the tachometer drive assembly.

mately even with the bottom of the "window" in the adapter.

3. Put the 5P7435 Adapter on the 6V7910 Transducer and put the end of the 5P7435 Adapter in the "window" of the 5P7436 Adapter.

4. Put the fuel injection line (13) on top of the 5P7435 Adapter. Install the 5P7437 Adapter and tighten to 30 lb-ft (40 N·m) **(Figure 21–62)**.

5. Remove the fitting (14) from the front housing. Install the transducer adapter into the hole from which the fitting was removed. Tighten only a small amount.

6. Push the TDC magnetic transducer (3) into the transducer adapter until it makes contact with the camshaft gear. Pull it back out 0.06 in. (1.6 mm) and lightly tighten the knurled locknut **(Figure 21–63)**.

7. Connect the cables from the transducers to the engine timing indicator. Make a calibration check of the indicator. For the calibration procedure, refer to Special Instruction, Form No. SEHS8580.

8. Start the engine and let it reach operating temperature. Then run the engine at approximately half throttle for 8 to 10 minutes before measuring the timing.

9. Run the engine at the speeds required to check low idle, timing advance, and high idle. Record the engine timing indicator readings. If the engine timing is not correct, refer to Fuel System Adjustments: On Engine, Fuel Injection Pump Timing (Timing Pin Method) for static adjustment of the fuel injection pump drive.

10. If the timing advance is still not correct, or if the operation of the advance is not smooth, repair or replace the automatic advance unit. There is no adjustment to the unit.

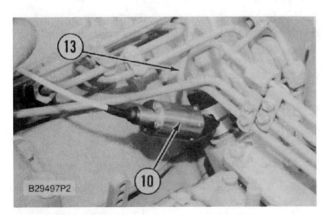

Transducer In Position
(10) Injection transducer. (13) Fuel injection line for No. 1 cylinder.

FIGURE 21–62 (Courtesy of Caterpillar, Inc.)

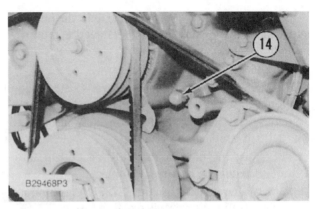

Location Of Fitting
(14) Fitting.

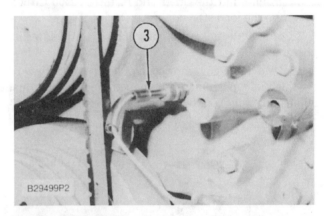

Transducer In Position
(3) TDC magnetic transducer.

FIGURE 21–63 (Courtesy of Caterpillar, Inc.)

FUEL SETTING

TOOLS NEEDED
5P4203   Field Service Tool Group   1

The following procedure for fuel setting can be done with the housing for the fuel injection pumps either on or off the engine.

NOTE: If the fuel injection pump group is equipped with a fuel ratio control, the control must be removed before the fuel setting is checked or adjusted.

1. Remove the shut-off solenoid and cover.

2. Put the 5P0298 Zero Set Pin (5) in the pump housing **(Figure 21–64)**.

3. Put the adapter (3) and spring (4) over the zero set pin (5). Use a 1D4533 Bolt and a 1D4538 Bolt to fasten the adapter (3) to the housing for the fuel injection pumps.

4. Put the screw (6) in the hole over the pin (5) and spring.

484 Chapter 21 CATERPILLAR FUEL INJECTION SYSTEMS

5. Turn the screw (6) clockwise until the pin (5) is held against the housing for the fuel injection pump. *Do not* tighten the screw (6) too tight.

6. Put the clamp in the adapter.

7. Move the governor control lever to the full-load position.

8. Put the 5P6531 Point (9) on the dial indicator (8). Clamp the indicator assembly.

9. Adjust the dial indicator (8) so both pointers (10) are on 0 (zero) **(Figure 21-65)**.

10. Use a wrench to turn the 8S7271 Screw counterclockwise. Turn the screw six or more turns.

11. Put the clip end of the 8T0500 Circuit Tester to a good ground. Put the other end of the 8T0500 Circuit Tester on the load stop contact.

12. Move the governor control lever to the low-idle position.

13. Move the governor control lever slowly toward the high-idle position until the continuity light just comes on. Make a note of the reading on the dial indi-

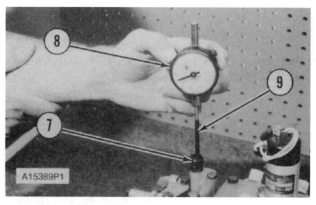

Installation Of Dial Indicator
(7) 3P1565 Collet Clamp. (8) 3P1567 Dial Indicator. (9) 5P6531 Contact Point, 57.2 mm (2.25 in) long.

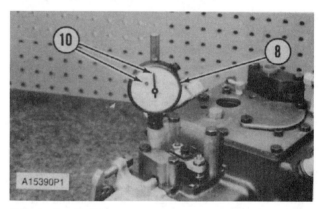

Indicator Set On Zero
(8) 3P1567 Dial Indicator. (10) Pointers.

FIGURE 21-65 (Courtesy of Caterpillar, Inc.)

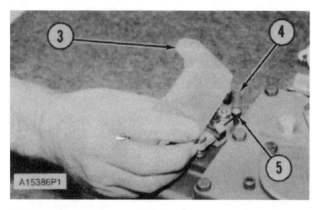

Installation Of Cover
(3) 5P6602 Adapter. (4) 3J6956 Spring. (5) 5P0298 Zero Set Pin, with 17.8507 on it.

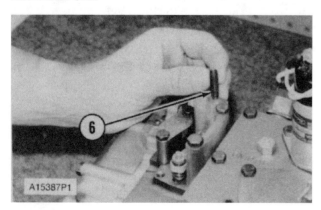

Installation Of Screw
(6) 8S7271 Screw.

FIGURE 21-64 (Courtesy of Caterpillar, Inc.)

cator (8). Do this step several times to make sure the reading is correct.

14. Compare this reading and the fuel setting in the Fuel Setting And Related Information Fiche.

15. If the reading on the dial indicator (8) is not correct, make the following load stop adjustment.

LOAD STOP ADJUSTMENT

Stop Bar Torque Control (Figures 21-66 and 21-67)

a. Use a wrench (18) and loosen the locknut.

b. Use a screwdriver (16) to turn the adjustment screw (17) until the reading on the dial indicator is the same as the dimension given in the Fuel Setting And Related Information Fiche.

c. When the adjustment is correct, tighten the locknut. Check the adjustment again by repeating steps 11 through 15.

FUEL RATIO CONTROL ADJUSTMENT   485

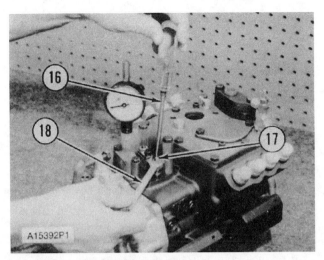

Adjustment Of Fuel Setting
(16) Screwdriver. (17) Adjustment screw. (18) Wrench.

FIGURE 21–66 (Courtesy of Caterpillar, Inc.)

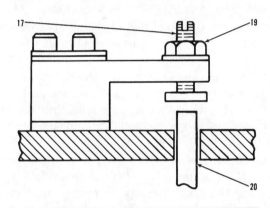

Adjustment Screw For Fuel Setting
(17) Adjustment screw. (19) Locknut. (20) Load stop pin.

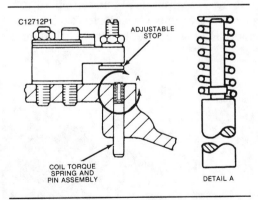

FIGURE 21–67 (Courtesy of Caterpillar, Inc.)

A coil-type spring control group has been added to 3208 truck engines rated at 215 hp (160 kW) at 2200 rpm. The new 4W9007 Pin Assembly (detail A) is a direct replacement for the load stop pin 4N0443 effective with Engine Serial No. 2Z34221. This new pin assembly can be adapted to 3208 truck engines 2Z13750 through 2Z34220, which have the above engine rating specifications.

The coil-type torque control group improves the torque rise as the engine lugs below full load rpm. The pin-and-spring assembly are matched and must be kept together.

Leaf-Type Torque Spring (Figure 21–68)

a. Write down the dimension that is on the dial indicator.

b. Write down the dimension given in the Fuel Setting And Related Information Fiche.

c. Remove the test tools (adapter, spring, and dial indicator) from the housing for the fuel injection pumps.

d. Install or remove shims at location (12) to get the correct dimension as given in the Fuel Setting And Related Information Fiche. The difference between the dimensions in (a) and (b) is the thickness and amount of shims to remove or install to get the correct setting.

e. Install the correct number of shims (12), the torque spring (14), and the stop bar (13) on the housing for the fuel injection pumps.

f. Install the test tools and repeat the test procedure. Do this until the dimension on the dial indicator is the same as the dimension given in the Fuel Setting And Related Information Fiche. After the fuel setting is correct, remove the test tools. Install the cover and shutoff solenoid.

FUEL RATIO CONTROL ADJUSTMENT (FIGURE 21–69)

NOTE: The same tools are needed for the fuel ratio control adjustment that were used for the fuel setting.

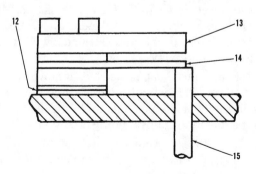

Leaf Type Torque Spring
(12) Location of shims. (13) Stop bar. (14) Leaf type torque spring. (15) Load stop pin.

FIGURE 21–68 (Courtesy of Caterpillar, Inc.)

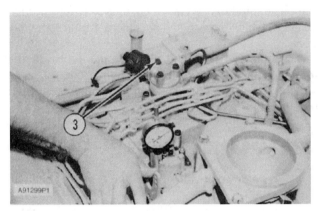

Checking Fuel Ratio Control Setting
(3) Bolts.

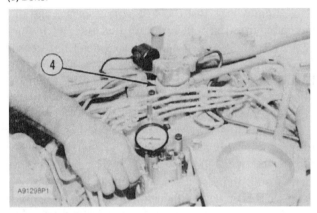

Adjustment Of Fuel Ratio Control Setting
(4) Flange.

FIGURE 21–69 (Courtesy of Caterpillar, Inc.)

NOTE: The fuel setting must be correct before an adjustment is made to the fuel ratio control.

1. Remove the shut-off solenoid (1) and cover (2). Install the tools, and zero the dial indicator as in Fuel Setting steps 2 through 10.

2. To check the fuel ratio control setting, move the governor lever slowly to the high-idle position. Make a record of the reading on the dial indicator. Compare the reading with the specification given in the Fuel Setting And Related Information Fiche.

3. If an adjustment is needed, remove the three bolts (3) from the fuel ratio control. Hold the governor lever in the high-idle position, and turn the flange (4) until the fuel ratio control setting is correct.

4. Move the governor lever to low idle and again move the lever slowly to high idle to check the fuel ratio control setting.

5. Install the bolts (3). The flange (4) can be turned a small amount to align the bolts (3).

6. Remove the tools and install the cover and shut-off solenoid.

ADJUSTMENT OF CROSSOVER LEVERS

1. Remove the fuel shut-off solenoid top cover of the fuel pump housing and the cover over the torque control group.

2. Remove the fuel that is in the injection pump housing and the governor housing.

3. Put the 3P1546 Calibration Pin (3) in the calibration hole **(Figure 21–70)**.

4. Install the 5P6602 Adapter (7). Fasten it in position with a 1D4533 Bolt (5) and a 1D4538 Bolt (6).

5. Put the 8S7271 Screw (4) (setscrew) in the hole over the calibration pin (3). Tighten the setscrew (4) to 20 to 25 lb-in. (2.3 to 2.8 N·m) with the 2P8264 Socket.

6. Adjust the low-idle screw (8) to position the lever (9) to 0.35 ± 0.04 in. (8.9 ± 1.0 mm) from the governor housing boss **(Figure 21–71)**.

7. Loosen the bolts that hold the sleeve levers (A) and slide the levers (A) out of the way **(Figure 21–72)**.

8. Loosen the bolts that hold the crossover lever (10) and (11), and move the lever (10) off the dowel pin (12).

9. Put the gauge (13) on the shafts (14) and (15), put the crossover lever (11) in a position so that the dowel pin (12) will fit in the gauge hole. Hold the gauge (13) down, and torque the bolt that holds the crossover lever (11) to 24 ± 2 lb-in. (2.8 ± 0.2 N·m).

10. Check the adjustment again with the 5P4209 Gauge (13). Put the gauge (13) on the shafts (14) and (15), slide the gauge toward the crossover lever (11) to engage the dowel pin (12) into the hole in the gauge (13).

11. If the dowel pin (12) must be lifted to go into the gauge, the lever must be adjusted again. If the gauge (13) is lifted, a maximum of 0.008 in. (0.20 mm) clear-

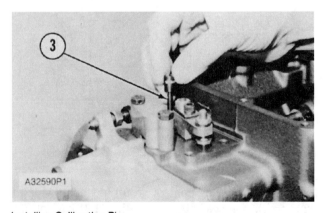

Installing Calibration Pin
(3) 3P1546 Calibration Pin with 15.9410 on it.

FIGURE 21–70 (Courtesy of Caterpillar, Inc.)

CHECKING THE SET POINT (BALANCE POINT) 487

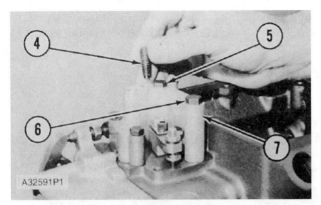

Installing 5P6602 Adapter And 8S7271 Screw
(4) Screw. (5) 1D4533 Bolt. (6) 1D4538 Bolt. (7) 5P6602 Adapter.

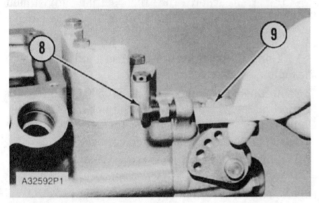

Adjustment Of Low Idle Screw
(8) Low idle screw. (9) Lever.

FIGURE 21–71 (Courtesy of Caterpillar, Inc.)

Crossover Levers
(10) Crossover lever. (11) Crossover lever. (12) Dowel pin.
(A) Sleeve levers.

Installing 5P4209 Gauge
(13) 5P4209 Gauge. (14) Shaft. (15) Shaft.

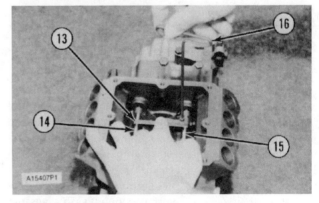

Adjustment Of Crossover Levers
(13) Gauge. (14) Shaft. (15) Shaft. (16) 5P4206 Wrench.

FIGURE 21–72 (Courtesy of Caterpillar, Inc.)

ance is acceptable under one side of the gauge (13). Use a feeler gauge to check clearance.

12. Slide the crossover lever (10) onto the dowel pin (12). Torque the bolt that holds the crossover lever (10) to 24 ± 2 lb-in. (2.8 ± 0.2 N·m).

13. Check the adjustment again with the 5P4209 Gauge.

NOTE: After the adjustment of the crossover levers is completed, all the fuel injection pumps must be calibrated. See Fuel Pump Calibration.

LOW-IDLE ADJUSTMENT (FIGURE 21–73)

To adjust the low-idle rpm, start the engine and run it in the low-idle position. Loosen the locknut (2) for the low-idle screw (1). Turn the low-idle screw to get the correct low-idle rpm. Increase engine speed and return to low idle. Check the low-idle speed again. Tighten the locknut.

CHECKING THE SET POINT (BALANCE POINT)

The engine set point is an adjusted specification and is important to the correct operation of the engine. High-idle rpm is *not* an adjusted specification. The set point

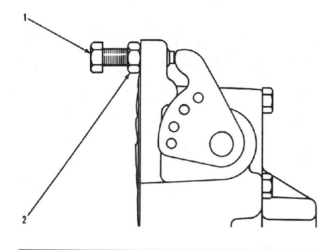

Adjustment Of Low Idle RPM
(1) Adjustment bolt for low idle. (2) Locknut.

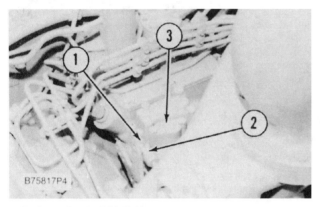

Adjustment Of Low Idle RPM
(1) Adjustment bolt for low idle. (2) Locknut. (3) Cover.

FIGURE 21-73 (Courtesy of Caterpillar, Inc.)

(formerly balance point) is full load rpm plus an additional 20 rpm. The set point is the rpm at which the fuel setting adjustment screw and stop or first torque spring just start to make contact. At this rpm the fuel setting adjustment screw and stop or first torque spring still have movement between them. When additional load is put on the engine, the fuel setting adjustment screw and stop or first torque spring become stable against each other. The set point is controlled by the fuel setting and the high-idle adjustment screw.

There is a new and more accurate method for checking the set point of the engine. If the tools for the new method are not available, there is an alternative method for checking the set point.

TOOLS NEEDED
6V4060   Engine Set Point Indicator Group   1

The 6V4060 Engine Set Point Indicator Group with the 6V2100 Multitach can be used to check the set point. Special Instruction, Form No. SEHS7931 gives instructions for installation and use of this tool group.

Alternative Method

TOOLS NEEDED
8T0500   Circuit Tester   1
6V3121   Multitach Group   1

If the set point is correct and the high-idle speed is within specifications, the fuel system operation of the engine is correct. The set point for the engine is

- at 20 rpm greater than full load speed
- the rpm where the fuel setting adjustment screw and stop or first torque spring just make contact

1. Connect a tachometer that has good accuracy to the tachometer drive.

2. Connect the clip end of the 8T0500 Circuit Tester to the brass terminal screw on the governor housing. Connect the other end of the tester to a place on the fuel system that is a good ground connection.

3. Start the engine.

4. With the engine at normal operating conditions, run the engine at high idle.

5. Record the speed of the engine at high idle.

6. Add load on the engine slowly until the circuit tester light just comes on (minimum light output). This is the set point.

7. Record the speed (rpm) at the set point.

8. Repeat step 6 several times to make sure that the reading is correct.

9. Stop the engine. Compare the records from steps 5 and 7 with the information from the engine information plate. If the engine information plate is not available, see the Fuel Setting And Related Information Fiche. The tolerance for the set point is ± 10 rpm. The tolerance for the high-idle rpm is ± 50 rpm. If the readings from steps 5 and 7 are within the tolerance, no adjustment is needed.

NOTE: It is possible in some applications that the high-idle rpm will be less than the lower limit. This can be caused by high parasitic loads such as hydraulic pumps or compressors.

Adjusting the Set Point (Balance Point) (Figure 21-74)

1. If the set point and the high-idle rpm are within tolerance, no adjustment is to be made.

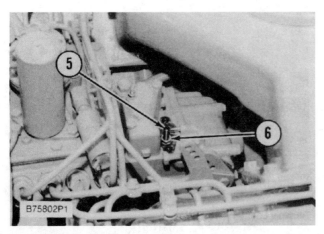

Adjustment Of Set Point
(5) Adjustment screw. (6) Locknut.

FIGURE 21-74 (Courtesy of Caterpillar, Inc.)

2. If the set point rpm is not correct, remove the cover and loosen the locknut (6). Turn the adjustment screw (5) to adjust the set point to the midpoint of the tolerance.

3. When the set point is correct, check the high-idle rpm. The high-idle rpm must not be more than the high limit of the tolerance.

If the high-idle rpm is more than the high limit of the tolerance, check the governor spring and flyweights. If the high-idle rpm is less than the low limit of the tolerance, check for excess parasitic loads and then the governor spring and flyweights.

FUEL INJECTION PUMP SERVICE

When removing injection pumps, sleeves, and lifters from the injection pump housing, keep the parts of each pump together so they can be installed back in their original location.

Be careful when disassembling injection pumps. Do not damage the surface on the plunger. The plunger, sleeve, and barrel for each pump are made as a set. Do not put the plunger of one pump in the barrel or sleeve of another pump. If one part is worn, install a complete new pump assembly. Be careful when putting the plunger in the bore of the barrel or sleeve.

When an injection pump is installed correctly, the plunger is through the sleeve, and the adjustment lever is engaged with the groove on the sleeve. The bushing that holds the injection pump in the pump housing must be kept tight. Tighten the bushing to 60 ± 5 lb-ft (80 ± 7 N·m). Damage to the housing will result if the bushing is too tight. If the bushing is not tight enough, the pump will leak.

NOTE: If the sleeves on one or more of the fuel injection pumps have been installed wrong, damage to the engine is possible if precautions are not taken at first starting. When the fuel injection pumps have been removed and installed with the fuel injection pump housing on the engine, take the following precautions when first starting the engine.

a. Remove the air cleaner, leaving the air inlet pipe open as shown.

b. Set the governor control at low idle.

WARNING

Be careful when putting the plate against the air inlet opening. Due to excessive suction, the plate can be pulled quickly against the air inlet pipe. To avoid crushed fingers, do not put fingers between the plate and the air inlet pipe.

c. Start the engine, and if the engine starts to overspeed (run out of control), put a steel plate over the air inlet as shown to stop the engine.

CHECKING FUEL PUMP CALIBRATION

The following procedure for fuel pump calibration can be done with the housing for the fuel injection pumps on or off the engine.

	TOOLS NEEDED	
5P4203	Tool Group	1
8S2243	Wrench	1
5P6602	Adapter	1
5P4205	Wrench	1
1D4533	Bolt	1
1D4538	Bolt	1
8S7271	Screw	1
5P7253	Socket Assembly	1
5P4206	Wrench	1
6V0190	Clamp	1
3P2200	Sleeve Metering Calibration Tool Group	1

(See **Figure 21-75**).

NOTE: The 3P1540 Calibration pump must have the 5P6557 Spring installed instead of the 1P7377 Spring.

1. Remove the fuel shut-off solenoid, the top cover of the fuel pump housing, and the cover over the torque control group.

490   Chapter 21 CATERPILLAR FUEL INJECTION SYSTEMS

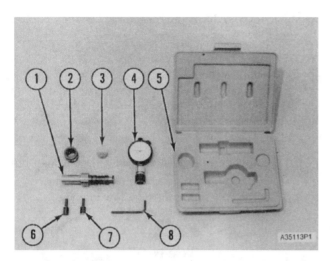

3P2200 Tool Group
(1) 3P1540 Calibration Pump. (2) 4N218 Bushing. (3) 1P7379 Microgauge. (4) 3P1568 Dial Indicator with 3P2226 Collet. (5) 5P6510 Box. (6) 3P1545 Calibration Pin with 17.3734 on it, (in-line engines). (7) 3P1546 Calibration Pin with 15.9410 on it. (Vee engines). (8) 1S9836 Wrench.

FIGURE 21–75 Fuel pump calibration tool group. (Courtesy of Caterpillar, Inc.)

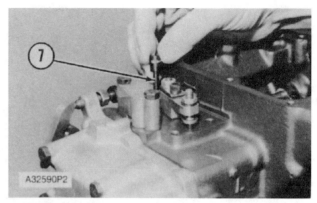

Installing Calibration Pin
(7) 3P1546 Calibration Pin with 15.9410 on it.

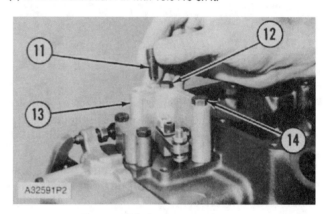

Installing 5P6602 Adapter And 8S7271 Screw
(11) Screw. (12) 1D4533 Bolt. (13) 5P6602 Adapter. (14) 1D4538 Bolt.

2. Remove the fuel that is in the injection pump housing and the governor housing.

3. Install the 3P1546 Calibration Pin (7) in the calibration hole.

4. Install the 5P6602 Adapter (13) as shown. Fasten it in position on the injection pump housing with a 1D4533 Bolt (12) and a 1D4538 Bolt (14).

5. Put the 8S7271 Screw (11) in the hole over the calibration pin (7). Tighten the screw (11) to 20 to 25 lb·in. (2.3 to 2.8 N·m).

6. Adjust the low-idle screw (15) to position the lever (16) to 0.35 ± 0.04 in. (8.9 ± 1.0 mm) from the governor housing boss.

7. Use the 8S2243 Wrench and remove the fuel injection pumps to be checked **(Figure 21–76)**.

NOTE: If the pump is removed carefully, the sleeve will remain on the plunger. If the sleeve falls off the pump plunger during removal, find it immediately and replace it on the pump plunger before removing another pump. The original sleeve must remain with the same pump plunger.

NOTE: When the sleeve is installed on the pump plunger, the narrower of the two lands on the sleeve must be toward the top of pump (nearest the pump spring).

8. Clean the barrel and plunger of the calibration pump (1). Put clean diesel fuel on the calibration pump for lubrication.

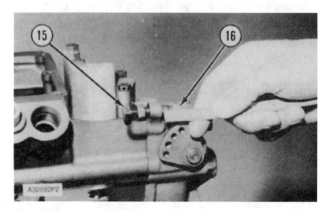

Adjustment Of Low Idle Screw
(15) Low idle screw. (16) Lever.

FIGURE 21–76 (Courtesy of Caterpillar, Inc.)

NOTE: Be sure that the spring on the calibration pump (1) is the 5P6557 Spring instead of the 1P7377 Spring that was installed on earlier calibration pumps.

9. Put the calibration pump (1) in place of the pump to be checked with the flat place (20) on the

CHECKING FUEL PUMP CALIBRATION 491

plunger toward the tang (17) on the lever (18). When the calibration pump (1) is all the way in the bore, turn it 180° either clockwise or counterclockwise. The tang (17) on the lever (18) is now in the groove (19) of the calibration pump (1). Then install the 4N218 Bushing (2). Use the 8S2243 Wrench and a torque wrench to tighten the bushing to 60 ± 5 lb-ft (80 ± 7 N·m) **(Figure 21–77)**.

NOTE: Turning the calibration pump (1) 180° gives the same reference point for all measurements.

NOTE: Use the 4N0218 Bushing and calibration pump together. The contact surfaces of the standard bushing, fuel injection pump, and the housing for the fuel injection pumps are sealing surfaces. Keep them clean and free of scratches to prevent leaks.

10. Put the dial indicator on the microgauge, and hold them together tightly. Loosen the lockscrew, and turn the face of the dial indicator to put the pointer at 0. Tighten the lockscrew.

Remove the dial indicator from the microgauge. Look at the face of the dial indicator and put the dial indicator on the microgauge again. The pointer must move through one to one and a half revolutions before stopping at exactly 0. If the number of revolutions is not correct, loosen the locknut on the 3P2226 Collet (23), and adjust the position of the dial indicator until the adjustment is correct.

NOTE: If the locknut on the 3P2226 Collet is too tight, it can interfere with the operation of the dial indicator.

11. Put the 6V0190 Clamp (25) in the position shown **(Figure 21–78)**, next to the transfer pump end. The clamp (25) pushes the shaft down against the bottom of its bearing. The other end of the shaft is held down against its bearing by the 3P1546 Calibration Pin.

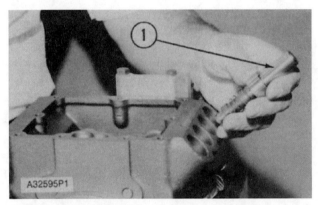

Installing Calibration Pump
(1) 3P1540 Calibration Pump.

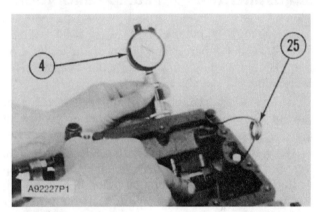

Dial Indicator Position
(4) 3P1568 Dial Indicator with 3P2226 Collet. (25) Clamp.

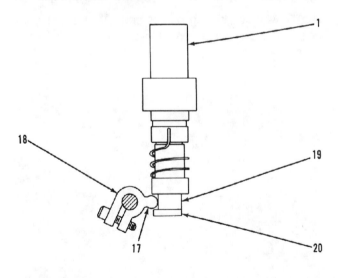

Calibration Pump Installed
(1) 3P1540 Calibration Pump. (17) Tang on lever. (18) Lever. (19) Groove of calibration pump. (20) Flat on plunger.

FIGURE 21–77 (Courtesy of Caterpillar, Inc.)

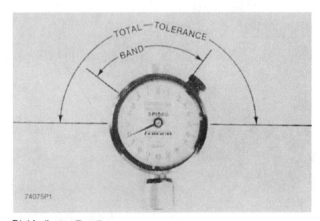

Dial Indicator Reading
Desired reading for all pumps is "0.000".

FIGURE 21–78 (Courtesy of Caterpillar, Inc.)

The combination of forces from the clamp (25) and the calibration pin is necessary to hold the shaft in its normal operating position against the lifting force from the spring in the calibration pump.

NOTE: When checking pumps on the "slave" side (side opposite the governor control lever), clamp (25) both ends of the sleeve shaft.

12. Put the dial indicator (4) on the calibration pump (1) as shown. Hold it tightly in place. Move the shaft toward the governor end to remove end play. To remove any clearance in the linkage, lift the crossover lever dowel and rapidly let it go. Do this several times. Then look at the reading on the dial indicator (4).

13. If the dial indicator (4) reading is more than ±0.050 mm from 0.000 (outside the total tolerance), repeat steps 17 through 20, Adjusting Fuel Pump Calibration.

Adjustment of Fuel Pump Calibration

14. Remove all pumps.

15. Clean the barrel and pump of the calibration pump. Put clean diesel fuel on the calibration pump for lubrication.

16. Install the calibration pump in the place of one of the pumps according to the procedure in step 9.

17. Loosen the bolt (26) with a 5P4206 Wrench. Turn the lever (18) on the shaft enough to move the top of the plunger of the calibration pump below the top surface (27) of the calibration pump. Tighten the bolt (26) just enough for the lever (18) to hold the plunger (28) stationary **(Figure 21–79)**.

NOTE: When the bolt (26) has the correct torque, pushing with a small amount of force on the lever (18) through the wrench moves the plunger in the calibration pump.

18. Move the shaft toward the governor to remove end play. Then push down on the lever (18) through the wrench until the top of the plunger (28) is almost even with the top surface (27) of the calibration pump (1), as shown **(Figure 21–80)**.

19. Check the dial indicator according to step 10. Then put the dial indicator in place over the center of the calibration pump (1) and hold it there tightly. Now move the plunger (28) of the calibration pump (1) by pushing on the lever through the wrench. Stop moving the plunger when the dial indicator is at approximately 0.009 mm past 0.000. Tighten the bolt (26) to 24 ± 2 lb-in. (2.8 ± 0.2 N·m).

NOTE: When moving the plunger (28), make sure that the last direction of plunger (28) movement is in the up direction. If the plunger (28) goes up too far, move the plunger (28) down to a position below that desired. Then move the plunger (28) up to the desired position.

NOTE: The action of tightening the bolt (26) usually changes the reading on the dial indicator (4) by approximately (0.010 mm) in the minus direction.

20. Move the shaft toward shut-off several times to remove clearance in the linkage. The dial indicator reading must be 0.000 ± 0.010 mm as shown **(Figure 21–81)**.

When the pump calibration is correct, record it and then repeat the procedure for all the other pumps.

NOTE: When calibrating pumps on the "slave" side (side opposite the governor control lever), clamp both ends of the sleeve shaft.

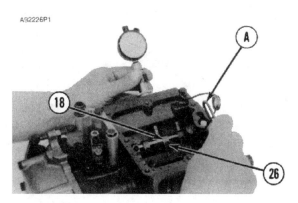

5P4206 Wrench
(18) Lever. (26) Bolt. (A) 5P4206 Wrench.

FIGURE 21–79 (Courtesy of Caterpillar, Inc.)

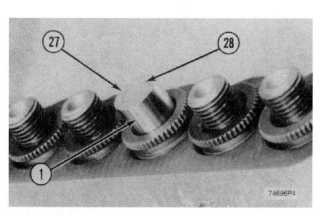

Plunger Position
(1) Calibration pump. (27) Top surface of calibration pump. (28) Plunger.

FIGURE 21–80 (Courtesy of Caterpillar, Inc.)

UNIT INJECTOR OPERATION 493

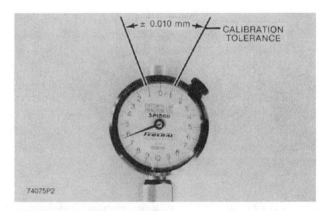

± 0.010 mm Calibration Tolerance

FIGURE 21-81 (Courtesy of Caterpillar, Inc.)

CATERPILLAR MECHANICAL UNIT INJECTOR FUEL SYSTEM (FIGURE 21-82)

Fuel is drawn from the fuel tank through the screen (1) by the fuel transfer pump (3), which sends it through the fuel filter (5) and from there to a drilled passage in the cylinder head (6). The drilled passage connects to a gallery around each unit injector to provide fuel for injection. Excess fuel flows through a pressure-regulating orifice (7) and check valve (8) and back to the fuel tank (10). The pressure-regulating orifice ensures adequate fuel pressure during low-idle speeds. The check valve prevents fuel bleed-off during engine shutdown. A fuel priming pump (A) is used on some models to prime the system with fuel before engine startup when required after servicing the system.

FUEL TRANSFER PUMP OPERATION

The fuel transfer pump (**Figure 21-83**) is integral with the governor and is located in the front housing of the governor. A cam (8) on the governor drive gear shaft actuates the tappet (7) and pumping piston (4). A return spring (3) keeps the tappet in contact with the cam. Fuel flows through the inlet screen (1) and inlet check valve (2) on the downstroke of the piston. On the upstroke of the piston the inlet check valve (2) and the outlet check valve (5) close. Closure of the check valve (5) prevents fuel from being drawn back from the outlet into the pump. With increased pressure above the piston (4), the check valve (6) opens to allow fuel to fill the passage (9). On the downstroke, the piston check valve (6) closes and the outlet check valve (5) opens due to increased fuel pressure in the passage (9), pushing fuel to the main fuel filter and engine. Springs hold the check valves closed when the engine is not running.

UNIT INJECTOR OPERATION

The unit injector (supplied by LUCAS CAV Ltd.) is held in place by a hold-down bracket and bolt (1) (**Figure 21-84**). Two O-rings (2) provide the seal between the

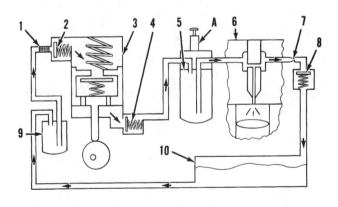

Fuel System Schematic
(1) Screen. (2) Inlet check valve. (3) Fuel transfer pump (integral with governor). (4) Outlet check valve. (5) Fuel filter. (6) Cylinder head. (7) Pressure regulating orifice. (8) Check Valve. (9) Primary fuel filter (if equipped). (10) Fuel tank. (A) Fuel priming pump (if equipped).

FIGURE 21-82 (Courtesy of Caterpillar, Inc.)

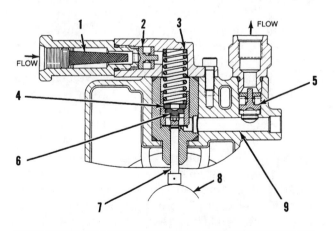

Fuel Transfer Pump
(1) Screen. (2) Inlet check valve. (3) Spring. (4) Piston assembly. (5) Outlet check valve. (6) Piston check valve. (7) Tappet assembly. (8) Cam. (9) Passage.

FIGURE 21-83 (Courtesy of Caterpillar, Inc.)

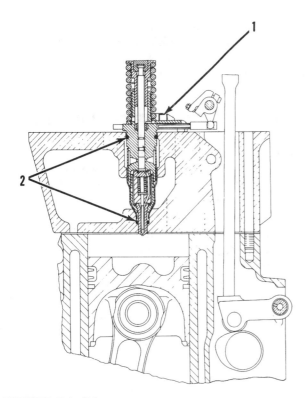

FIGURE 21–84 (Courtesy of Caterpillar, Inc.)

injector and the injector sleeve. The injector sleeve (11) **(Figure 21–85)** isolates the injector from the coolant passages in the cylinder head. Each injector is surrounded by a fuel gallery (10). Injection timing is controlled by the location of the cam (14) and the vertical position of the plunger (6). Plunger location is adjusted by a setscrew (2) in the injector rocker arm. Cam action is transferred by the lifter (13), push rod (5), and rocker arm (10). When the plunger (6) **(Figure 21–85)** is at the top of its stroke, fuel fills the area below the plunger (6). As the plunger moves downward, fuel is displaced into the gallery through the two ports in the barrel (9) **(Figure 21–86)**. As the bottom edge of the plunger closes the upper port (19), fuel continues to flow through the lower port (18) until it is closed by plunger movement. At this point fuel is trapped and becomes pressurized, and the effective stroke begins. As plunger movement continues, fuel pressure increases and opens the check valve (needle valve) (21), and fuel under high pressure is forced out of the nozzle openings into the combustion chamber. This continues until the upper port (19) is uncovered by the helix (17) on the plunger. At this point the effective stroke ends and injection stops. Fuel spills through the upper port and into the gallery, and the check valve (19) is closed by spring pressure.

The plunger continues downward movement until the lifter reaches the nose of the cam. As the cam retreats, the spring (4) returns the plunger upward, and the cavity below the plunger again fills with fuel. The plungers can be rotated by the control rack (7), which is in mesh with a gear (15) located on the plunger. Rotating the gear alters the position of the helix (17) in relation to the upper port (19), thereby changing the effective length of the stroke and the amount of fuel injected.

CONTROL RACK OPERATION (FIGURES 21–87 AND 21–88)

Governor action is transmitted to each fuel injector (11) **(Figure 21–87)** by the rack control linkage. The governor output is linked to the link (4) on the control shaft (1). When the governor link moves toward the reduced fuel direction, the shaft (1) and clamps (3) rotate in the FUEL OFF direction. The torsion spring (2) causes the lever (7) to rotate clockwise, which pushes the injector rack (10) in the shut-off direction. The torsion spring at each lever allows the rack control to move toward shut-off even when an injector rack is stuck open. The power setting for the No. 1 cylinder is made with the screw (5) in the clamp (9). Adjusting this screw changes the rotational position of the shaft (1) in relation to the link (4) and lever (6). The remaining injectors are synchronized to No. 1 by adjusting the screws (8).

CAUTION:

Do not loosen the screws holding the clamps (3) or clamp (9) to shaft (1). These screws are set at the factory and must not be loosened. These screws have socket heads which are filled with sealant. Loosening the screws will cause poor engine performance and can cause engine damage.

GOVERNOR OPERATION (FIGURES 21–89 TO 21–92)

Engine speed is determined by throttle position and the governor. When the throttle is moved, the governor output shaft moves immediately. As throttle movement changes, engine speed is stabilized by the governor corresponding to throttle position.

GOVERNOR OPERATION 495

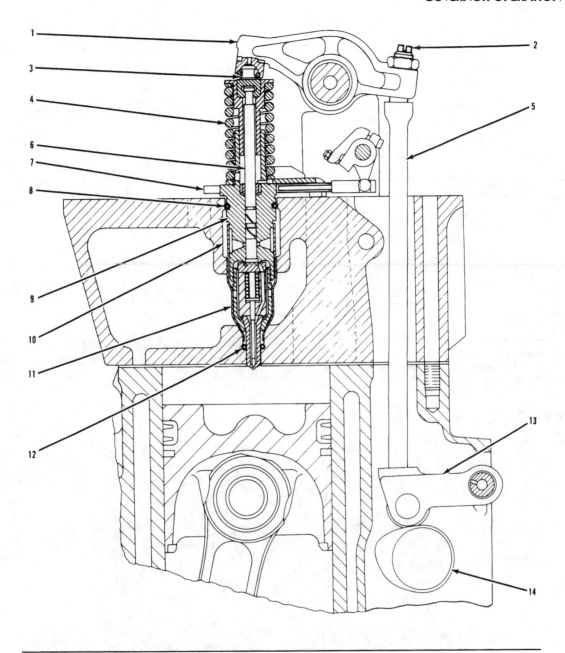

Fuel Injection System
(1) Rocker arm. (2) Setscrew. (3) Follower. (4) Tappet spring. (5) Push rod. (6) Plunger. (7) Rack. (8) O-ring. (9) Barrel. (10) Fuel gallery. (11) Sleeve. (12) O-ring. (13) Lifter. (14) Cam.

FIGURE 21-85 (Courtesy of Caterpillar, Inc.)

The governor is engine-driven through the governor drive gear (1) **(Figure 21-89),** which drives the shaft (2), flyweight carrier (3), and flyweights (4). Action between the centrifugal force of the rotating flyweights and the low- and high-idle springs (6) and (7) determines whether engine speed will increase or decrease. This action is similar to that described earlier for other Caterpillar engines and is illustrated in **Figures 21-91** and **21-92.** When a lug condition (rated load) is reached, the lever (14) is against the limit lever set screw (10) and the output shaft (9) is in the maximum-fuel position. If more load is applied to the engine while the torque lever (15) is in contact with the torque cam (16), engine speed will decrease. The design profile of the torque cam determines the torque characteristics of the engine.

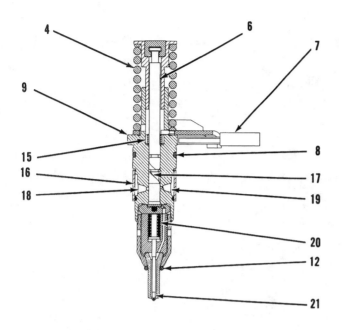

Fuel Injector Pump (Unit Injector)
(4) Tappet spring. (6) Plunger. (7) Rack. (9) Barrel. (12) O-ring. (15) Gear. (16) Sleeve filter. (17) Helix. (18) Lower port. (19) Upper port. (20) Spring. (21) Check (needle valve).

FIGURE 21–86 (Courtesy of Caterpillar, Inc.)

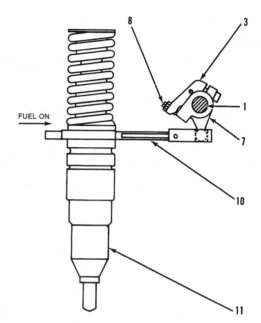

FIGURE 21–88 Rack control linkage. 1. Shaft. 3. Clamp. 7. Lever assembly. 8. Synchronization screw. 10. Rack. 11. Injector. (Courtesy of Caterpillar, Inc.)

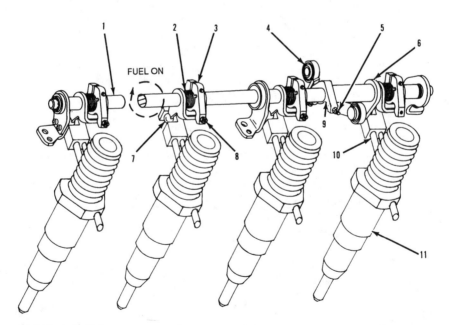

FIGURE 21–87 Injectors and rack control linkage. 1. Shaft. 2. Spring. 3. Clamp. 4. Link. 5. Fuel setting screw. 6. Lever assembly. 7. Lever assembly. 8. Synchronization screw. 9. Clamp assembly. 10. Rack. 11. Injector. (Courtesy of Caterpillar, Inc.)

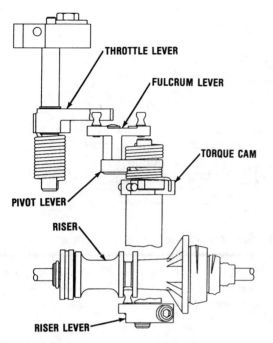

FIGURE 21-89 Major components of governor for 3116 engine. (Courtesy of Caterpillar, Inc.)

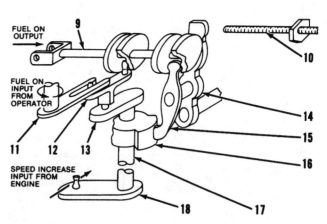

FIGURE 21-91 Governor linkage components. 9. Governor output shaft. 10. Limit lever setscrew. 11. Throttle lever. 12. Fulcrum lever. 13. Pivot lever. 14. Limit lever. 15. Torque lever. 16. Torque cam. 17. Pivot shaft. 18. Riser lever. (Courtesy of Caterpillar, Inc.)

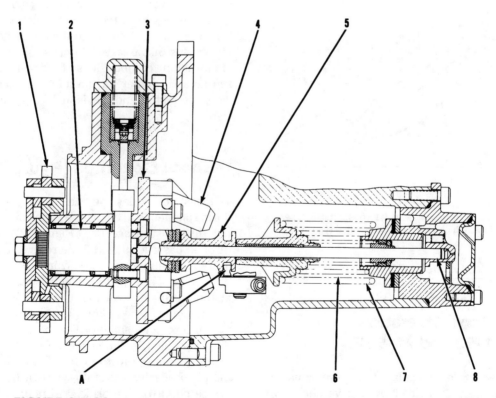

FIGURE 21-90 Mechanical governor cross section. 1. Drive gear. 2. Shaft. 3. Flyweight carrier. 4. Flyweights. 5. Riser. 6. Low-idle spring. 7. High-idle spring. 8. Shaft. A. Pin. (Courtesy of Caterpillar, Inc.)

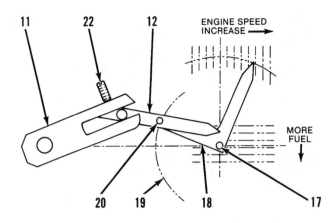

Example C
Engine speed becomes stable at high idle resulting in a new riser position which decreases the amount of fuel.
(11) Throttle lever. (12) Fulcrum lever. (17) Pivot shaft. (18) Riser lever. (19) Path of fulcrum pin. (20) Fulcrum pin. (22) High idle stop.

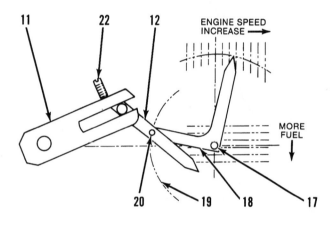

Example D
As the engine is loaded and the speed is decreasing, the riser is seeking a new position causing the fuel to increase.
(11) Throttle lever. (12) Fulcrum lever. (17) Pivot shaft. (18) Riser lever. (19) Path of fulcrum pin. (20) Fulcrum pin. (22) High idle stop.

FIGURE 21-92 (Courtesy of Caterpillar, Inc.)

Governor Servo Operation (Figures 21-93 and 21-94)

On engines with close regulation requirements hydraulic assistance is provided by a governor servo. When the governor calls for more fuel, the valve (1) moves to the left, the outlet (B) opens, and oil passage (D) closes. Engine lubricating oil entering through the inlet (A) pushes the piston (2) and clevis (5) to the left

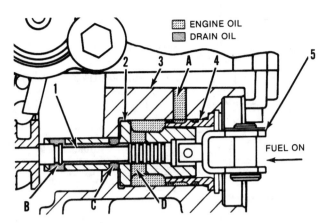

FIGURE 21-93 Governor servo in FUEL ON direction. 1. Valve. 2. Piston. 3. Cylinder. 4. Cylinder sleeve. 5. Clevis for fuel injector rack control linkage. A. Oil inlet (shown out of position). B. Oil outlet. C. Oil passage. D. Oil passage. (Courtesy of Caterpillar, Inc.)

in the FUEL ON direction until oil passage (D) is opened. Oil from behind the piston now flows through outlet (B), relieving pressure on the piston, which stops piston and clevis movement (balanced position). When the governor moves in the decreased fuel direction, the valve (1) moves to the right, the oil outlet (B) closes, and oil passage (D) opens. Oil from the inlet (A) is now acting on both sides of the piston. Since the piston area is greater on the left side than on the right side, the piston and clevis are forced to move to the right.

Dashpot Operation (Figure 21-95)

The dashpot is designed to improve engine stability during sudden load changes. Engine oil enters the dashpot through passage (7) **(Figure 27-95)** and flows through passage (9) to fill the reservoir (8). Excess oil flows through passage (5). When the governor moves the spring seat (1), the dashpot spring (2) moves the piston (3) in the seat (4). During a load decrease or speed increase the piston (3) moves to the right, forcing oil through a small orifice in the plug (6) and into the reservoir (8). When engine speed decreases, the piston (3) moves to the left, creating low pressure in the piston cavity. Oil is now drawn back through the orifice plug (6). The orifice restricts oil flow into and out of the piston cavity, thereby slowing piston (3) and seat (1) movement. The faster the governor tries to move the seat, the greater the resistance to seat movement.

GOVERNOR OPERATION   499

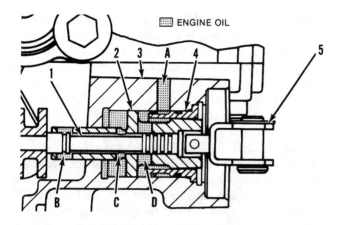

Governor Servo
(Balanced Position)
(1) Valve. (2) Piston. (3) Cylinder. (4) Cylinder sleeve. (5) Clevis for fuel injector rack control linkage. (A) Oil inlet (shown out of position). (B) Oil outlet. (C) Oil passage. (D) Oil passage.

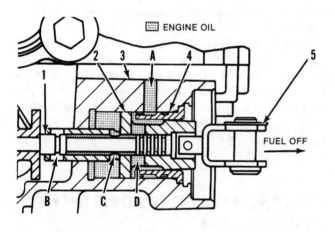

Governor Servo
(Fuel Off Direction)
(1) Valve. (2) Piston. (3) Cylinder. (4) Cylinder sleeve. (5) Clevis for fuel injector rack control linkage. (A) Oil inlet (shown out of position). (B) Oil outlet. (C) Oil passage. (D) Oil passage.

FIGURE 21–94 (Courtesy of Caterpillar, Inc.)

Fuel Ratio Control Operation (Figures 21–96 and 21–97)

The fuel ratio control (FRC) is used on turbocharged engines to reduce exhaust smoke during acceleration at low boost pressures. The amount of fuel delivered to the combustion chambers is restricted until adequate boost pressure is achieved.

A tube connects the engine inlet manifold to an inlet port on the FRC. At low boost pressure, shaft (4) is held stationary by springs. When the driver accelerates, the

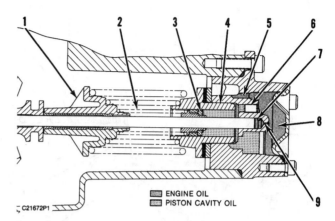

FIGURE 21–95 Governor dashpot cross section. 1. Spring seat. 2. Dashpot spring. 3. Piston. 4. Seat. 5. Overflow passage. 6. Orifice plug. 7. Oil passage. 8. Reservoir. 9. Passage. (Courtesy of Caterpillar, Inc.)

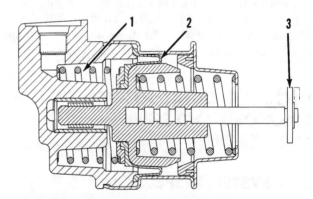

FIGURE 21–96 Fuel ratio control cross section. 1. Air cavity. 2. Diaphragm. 3. Washer. (Courtesy of Caterpillar, Inc.)

governor output shaft moves in the FUEL ON direction until the lever (8) contacts the setscrew (7) of the FRC lever (5). With the lever (5) unable to rotate in the FUEL ON direction, further movement of the governor output shaft is stopped, thereby preventing overfueling of the engine. As engine power and exhaust flow increase, boost pressure also increases. When boost pressure is high enough, spring force in the FRC is overcome and the shaft (4) moves to the right. This allows the FRC lever (5) and the limit lever (8) to rotate clockwise and the governor output shaft to move in the FUEL ON direction until the limit lever (8) contacts the setscrew (6). When boost pressure decreases, springs inside the FRC (3) return the retainer shaft (4) to the normal position of restricting governor output shaft movement in the FUEL ON direction.

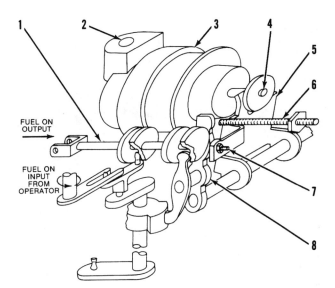

FIGURE 21–97 Governor linkage and fuel ratio control schematic. 1. Governor output shaft. 2. Inlet port. 3. Fuel ratio control. 4. Retainer shaft. 5. FRC lever. 6. Limit lever setscrew. 7. FRC lever setscrew. 8. Limit lever. (Courtesy of Caterpillar, Inc.)

MECHANICAL UNIT INJECTOR SYSTEM DIAGNOSIS AND SERVICE

(Courtesy of Caterpillar, Inc.)

FUEL SYSTEM INSPECTION

A problem with the components that send fuel to the engine can cause low fuel pressure. This can decrease engine performance.

1. Check the fuel level in the fuel tank. Look at the cap for the fuel tank to make sure the vent is not filled with dirt.
2. Check the fuel lines for air or fuel leakage. Be sure the fuel supply line does not have a restriction or a bad bend.
3. Install a new secondary fuel filter. Clean the screen located in the inlet fitting of the fuel transfer pump.
4. Inspect the orifice in the tube assembly to see that there is no restriction.

CHECKING ENGINE CYLINDERS SEPARATELY

The temperature of an exhaust manifold port, when the engine is running at low-idle speed, can be an indication of the condition of a fuel injector. Low temperature at an exhaust manifold port is an indication of no fuel to the cylinder. This can be an indication of an inoperative injector. Extra-high temperature at an exhaust manifold port can be an indication of too much fuel to the cylinder, caused by a malfunctioning injector. The difference between cylinders should be no more than 158°F (70°C).

With the valve cover removed and the engine idling, the control shaft levers allow each injector to be actuated individually to the FUEL ON position for a few seconds. This causes excess fuel to be injected into that particular cylinder, causing a loud combustion "knock." If actuating an injector in this fashion briefly does not result in a loud combustion knock, there may be a problem with the injector, the fuel supply to the injector, or the seal between the injector and the brass sleeve.

Use the 1U8865 Infrared Thermometer to check exhaust temperature. The Operator's Manual, Form No. NEHS0510, for the 1U8865 Infrared Thermometer gives complete operating and maintenance instructions for this tool.

Testing of the injectors must be done off of the engine. Use the 1U6661 Pop (Injector) Tester (or 6V4022 Injector Tester with 1U6667 Conversion Group) and a 1U6662 Injector Holding Block to test the injectors. For the 6V4022 Injector Tester, refer to Special Instruction, Form No. SEHS8858 for the test procedure. For the 1U6661 Pop (Injector) Tester, refer to Special Instruction, Form No. SEHS8867.

GOVERNOR LOW-IDLE ADJUSTMENT (FIGURE 21–98)

The only on-engine adjustment to the governor is the low-idle setting. All other governor adjustments (including the high-idle/throttle stop setting and fuel ratio

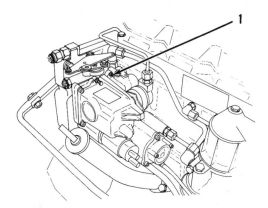

FIGURE 21–98 1. Low-idle speed screw. (Courtesy of Caterpillar, Inc.)

control) are to be done off engine on the governor calibration bench.

NOTE: Refer to Special Instruction, Form No. SEHS8868 for procedures and tooling required to bench test the governor. If governor disassembly and assembly is required, refer to the Disassembly And Assembly Section of the service manual.

The low-idle screw (1) is initially set during the bench test procedure. However, if the low-idle setting is not satisfactory, use the 1U6672 Wrench to loosen the locknut and adjust the screw (1) to the specified low-idle speed. Tighten the locknut and recheck the low-idle setting.

FUEL PRESSURE CHECK (FIGURE 21–99)

The fuel pressure to the cylinder head fuel gallery should be 29 to 58 psi (200 to 400 kPa) at rated rpm and load condition. At low idle, the pressure-regulating orifice in the tube assembly maintains a minimum fuel pressure of 7.2 psi (50 kPa) to the injectors. The check valve (4) prevents fuel from leaving the cylinder head fuel gallery during shutdown.

To check the unfiltered fuel pressure, remove the plug from the fuel pressure tap (7). Install the adapter, seal, and 1U5470 Engine Pressure Group to the fuel pressure tap (7). Operate the engine. To check the filtered fuel pressure, remove the plug from the fuel pressure tap (9). Install the adapter, seal, and 1U5470 Engine Pressure Group to the fuel pressure tap (9). Operate the engine.

NOTE: Make sure the fuel filter is clean before checking fuel pressure. A restricted fuel filter will cause lower fuel pressure at tap (9) than at tap (7).

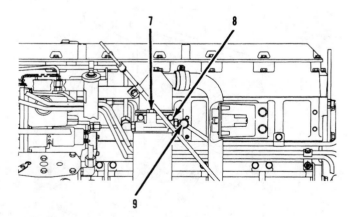

FIGURE 21–99 Fuel pressure test. 7. Unfiltered fuel pressure tap. 8. Fuel filter base. 9. Filtered fuel pressure tap. (Courtesy of Caterpillar, Inc.)

FINDING THE TOP CENTER POSITION FOR THE NO. 1 PISTON (FIGURE 21–100)

TOOLS NEEDED
8T0292 Bolt (Part Of 1U6675 Injector Spring 1
Compressor used in 1U6680 Tool Group)

NOTE: Depending on engine application, the timing hole (2) is located at either the left front face or right front face of the flywheel housing.

1. Remove the plug from the timing hole (2) on the front of the flywheel housing.

NOTE: Turn the engine with the four large bolts on the front of the crankshaft. **Do not** use the eight small bolts on the front of the crankshaft pulley.

2. Put the 8T0292 Bolt (1) in the hole. Now turn the engine flywheel counterclockwise until the timing bolt engages with the threaded hole in the flywheel.

NOTE: If the flywheel is turned beyond the point that the timing bolt engages in the threaded hole, the flywheel must be turned clockwise approximately 30°. Then turn the flywheel counterclockwise until the timing bolt engages with the threaded hole. The reason for this procedure is to make sure the play is removed from the gears when the No. 1 piston is put on top center.

3. Remove the valve cover.

4. The intake and exhaust valves for the No. 1 cylinder are fully closed if the No. 1 piston is on the *compression stroke* and the rocker arms can be moved by hand. If the rocker arms cannot be moved and the

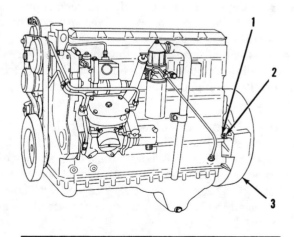

Timing Bolt Location
(1) Timing bolt. (2) Timing hole. (3) Flywheel housing.

FIGURE 21–100 (Courtesy of Caterpillar, Inc.)

valves are slightly open, the No. 1 piston is on the *exhaust stroke*. Refer to charts for Crankshaft Positions For Fuel Timing And Valve Clearance Setting to find the correct cylinder(s) to be checked/adjusted for the stroke position of the crankshaft when the timing bolt has been installed in the flywheel.

NOTE: When the actual stroke position is identified, and the other stroke position is needed, it is necessary to remove the timing bolt from the flywheel, turn the flywheel counterclockwise 360°, and reinstall the timing bolt.

INJECTOR SYNCHRONIZATION (RACK ADJUSTMENT)

TOOLS NEEDED

1U6680	Governor And Fuel System Adjusting Tool Group	1
1U6675	Injector Spring Compressor	1
1U6679	Indicator Group	1
1U6673	Wrench	1
6V6106	Dial Indicator	1
4C4734	Spanner (Solenoid) Wrench	1

Injector synchronization is the setting of all injector racks to a reference position so that each injector gives the same amount of fuel to each cylinder. This is done by setting each injector rack to the same position while the control linkage is in a fixed position (called the synchronizing position).

The control linkage is at the synchronizing position when the injector of the No. 1 cylinder is at fuel shut-off. Since the No. 1 injector is the reference point for the other injectors, *no synchronizing adjustment* is made to the No. 1 injector.

Always synchronize an injector when it has been removed and reinstalled or replaced. If the No. 1 injector is reinstalled or replaced, all the injectors must be synchronized.

NOTE: This procedure includes steps to remove the rocker arm assemblies for the No. 1 injector and for the injector to be checked. Removing the rocker arm assemblies provides greater access to the control linkage and injector rack. Checking valve clearance and fuel timing is recommended after the rocker arm assemblies are reinstalled.

1. With the engine stopped and the engine's electrical system shut off, remove the fuel shut-off solenoid (2) **(Figure 21–101)** to allow free movement of the injector rack control linkage during injector synchronization. The 4C4734 Spanner Wrench can be used to remove the fuel shut-off solenoid.

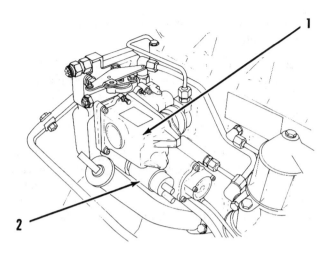

Fuel Shutoff Solenoid (Energize To Run)
(1) Governor. (2) Fuel shutoff solenoid.

FIGURE 21–101 (Courtesy of Caterpillar, Inc.)

2. Remove the valve cover and No. 1 cylinder rocker arm assembly.

NOTE: Hold the rocker arm assembly level when removing it from the engine to prevent accidental disassembly.

3. Apply a small amount of clean engine oil to the top of the injector and install the 1U6675 Injector Spring Compressor (6) **(Figure 21–102)** to the cylinder head and injector as shown. Tighten the bolt (part of the tool (6)) that is threaded into the cylinder head hole for the rocker arm assembly. This will compress the injector.

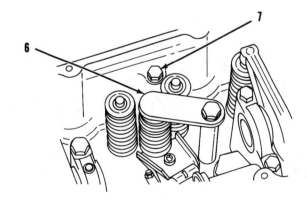

Injector Spring Compressor Installed On Injector To Be Checked
(6) 1U6675 Injector Spring Compressor. (7) Bolt.

FIGURE 21–102 (Courtesy of Caterpillar, Inc.)

INJECTOR SYNCHRONIZATION (RACK ADJUSTMENT)

4. Tap lightly with a soft hammer on the spring compressor directly above the injector to ensure free movement of the injector rack bar.

NOTE: The injector spring must be slightly compressed to allow free movement of the injector rack. Either the rocker arm assembly or the 1U6675 Injector Spring Compressor **must** be installed on all injectors to prevent internal damage to the injectors.

5. Remove the bolt (7) nearest to the injector to be checked.

6. Install the 6V6106 Dial Indicator (8) in the 1U6679 Indicator Group (9). Do not tighten the indicator yet **(Figure 21-103)**.

7. Make sure the end face of the rack bar (12) is clean. Install the assembled 1U6679 Indicator Group (9) and the 6V6106 Dial Indicator (8) where the bolt (7) was removed. The lever (10) must contact the end face of the rack bar (12).

8. Firmly push the rack head (11) of the injector to be checked, by hand, toward the injector until the rack stop (14) touches the injector base (13). The rack is now in shut-off position.

9. While holding the rack head (11) in the shut-off position, adjust dial indicator (8) until all three dials read zero. Then tighten the dial indicator. Release the rack head (11) **(Figure 21-103)**.

10. Push down on the clamp assembly (15) **(Figure 21-104)** to rotate the rack control linkage in the FUEL ON direction. Now quickly release the clamp assembly (15). This ensures that the springs and bearings of the control linkage are in their "normal" positions.

11. Firmly push the injector rack head (11) of the No. 1 injector, by hand, toward the injector until the rack stop (14) touches the injector base (13). The No. 1 injector is now at FUEL SHUTOFF. Hold the rack head (11) in this position for steps 12 and 13.

12. Push down, then quickly release, the lever (16) **(Figure 21-105)** of the injector being checked. "Flip" the lever in this manner to make sure there is smooth movement of the injector rack.

13. The dial indicator must indicate +0.01 to +0.05 mm. Repeat steps 10 through 13 two or three times to confirm the reading.

14. If the reading is correct, go to step 21.

15. If the dial indicator does not indicate +0.01 to +0.05 mm, go to step 16.

16. Use a 1U6673 Wrench to loosen the locknut and back out (turn counterclockwise) the setscrew (17).

NOTE: Do not loosen the screws holding the clamps (18) to the control shaft. (The screws can be identified as those with socket heads filled with sealant.) The clamps are fac-

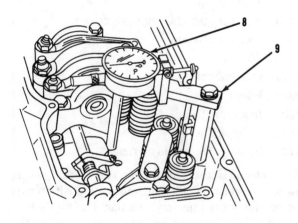

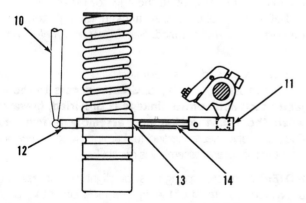

FIGURE 21-103 Injector synchronization. 8. 6V6106 Dial Indicator. 9. 1U6679 Indicator Group. 10. Lever (part of 1U6679 Indicator Group). 11. Rack head. 12. Rack bar. 13. Injector base. 14. Rack stop. (Courtesy of Caterpillar, Inc.)

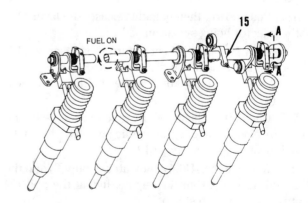

Injector Rack Control Linkage
(15) Clamp assembly.

FIGURE 21-104 (Courtesy of Caterpillar, Inc.)

504 Chapter 21 CATERPILLAR FUEL INJECTION SYSTEMS

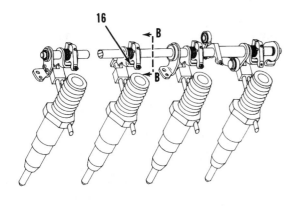

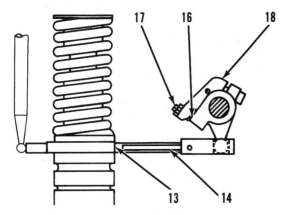

FIGURE 21–105 Rack control linkage (top) and closeup of view B-B (bottom). 16. Lever being checked. 13. Injector base. 14. Rack stop. 16. Lever. 17. Setscrew. 18. Clamp. (Courtesy of Caterpillar, Inc.)

tory set to the shaft. Loosening the clamps will cause poor engine performance and may damage the engine.

17. Hold the rack head of the No. 1 injector in the FUEL SHUTOFF position (as described in step 11) for steps 18 and 19.

18. Push down, then quickly release, the lever (16) of the injector being synchronized.

19. With a 1U6673 Wrench, turn the setscrew (17) clockwise until the dial indicator reads +0.01 to +0.05 mm. Tighten the locknut while holding the setscrew (17).

20. Check the adjustment by repeating steps 10 and 11. If the indicator does not indicate +0.01 to +0.05 mm, repeat steps 16 through 19.

21. Remove the 1U6679 Indicator Group (9) and the dial indicator (8) from the engine. Install the bolt (7) that was removed in step 5.

22. Remove the 1U6675 Injector Spring Compressor (6) from the injector that was checked. Install the rocker arm assembly. Make sure the push rods are properly seated in the rocker arms and lifters.

23. Synchronize the other injectors as necessary.

24. If the fuel setting is to be checked, refer to Fuel Setting.

25. If the fuel setting is known to be correct, remove the 1U6675 Injector Spring Compressor from the No. 1 injector. Install the rocker arm assembly. Make sure the push rods are properly seated in the rocker arms and lifters.

26. Check valve clearance and fuel timing for those cylinders that had rocker arms removed. See Valve Clearance and Fuel Timing.

27. With the engine stopped and the engine's electrical system shut off, install the fuel shut-off solenoid (2). Install the valve cover.

FUEL SETTING

	TOOLS NEEDED	
1U6680	Governor And Fuel System Adjusting Tool Group	1
1U6675	Injector Spring Compressor	1
1U6679	Indicator Group	1
1U6673	Wrench	1
6V6106	Dial Indicator	1
6V0006	Pliers	1
1U7305	Insertion Tool	1
1U6681	Holding Tool	1
4C4734	Spanner (Solenoid) Wrench	1

Fuel setting is the adjustment of the fuel setting screw to provide a specified injector rack position measured on the No. 1 injector rack bar. The fuel setting screw limits the power output of the engine by setting the maximum travel of all the injector racks.

Before the fuel setting is checked, the injectors must be correctly synchronized. See *Injector Synchronization*.

NOTE: This procedure is illustrated with the No. 1 cylinder rocker arm assembly installed. **Greater access to the injector rack and control linkage is provided, however, when the No. 1 rocker arm assembly is removed.** Checking valve clearance and fuel timing is recommended after the rocker arm assembly is reinstalled.

NOTE: The injector spring must be slightly compressed to allow free movement of the injector rack. Either the rocker arm assembly or the 1U6675 Injector Spring Compressor **must** be installed on all injectors to prevent internal damage to the injectors.

1. With the engine stopped and the engine's electrical system shut off, remove the fuel control solenoid

FUEL SETTING 505

to allow free movement of the injector rack control linkage during fuel setting.

2. Remove the appropriate bolt from the inlet manifold.

3. Install the 6V6106 Dial Indicator (2) in the 1U6679 Indicator Group (3). Do not tighten the indicator yet **(Figure 21–106)**.

4. Make sure the end of the rack bar (6) is clean. Install the assembled 1U6679 Indicator Group (3) and 6V6106 Dial Indicator (2) as shown where the bolt (1) was removed. Position the dial indicator group so that the dial indicator stem is parallel to the rack bar of the No. 1 injector (perpendicular to the rocker arm shafts). The lever (4) must contact the end face of the rack bar (6) **(Figure 21–107)**.

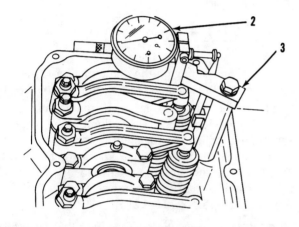

Fuel Setting Indicator Installed
(2) 6V6106 Dial Indicator. (3) 1U6679 Indicator Group.

FIGURE 21–106 (Courtesy of Caterpillar, Inc.)

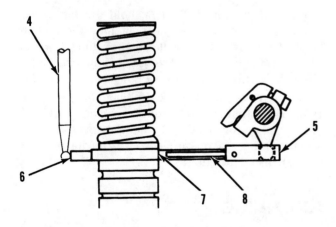

No. 1 Injector
(4) Lever (part of 1U6679 Indicator Group). (5) Rack head. (6) Rack bar. (7) Injector base. (8) Rack stop.

FIGURE 21–107 (Courtesy of Caterpillar, Inc.)

5. Firmly push the rack head (5) of the No. 1 injector, by hand, toward the injector until the rack stop (8) touches the injector base (7). The No. 1 injector is now at fuel shut-off **(Figure 21–107)**.

NOTE: If the rocker arm assembly is installed, rotate the control shaft in the FUEL OFF direction (against spring pressure) until all the injector levers are in the shut-off position.

6. While holding the rack head in the shut-off position, adjust the dial indicator until all three dials read zero. Then tighten the dial indicator. Release the rack head.

7. Remove the clip (9) that keeps the sleeve (11) in position between the governor (10) and the inlet manifold (12) **(Figure 21–108)**.

NOTE: Do not use hard-jawed pliers or a screwdriver to move the sleeve (11). Damage may result to the sleeve that will damage the wiper seal in the inlet manifold when the sleeve is installed in the inlet manifold.

8. With 6V0006 Pliers, slide the sleeve from the governor toward the inlet manifold.

9. Install the 1U7305 Insertion Tool (14) into the link pin (13) of the governor output shaft. When properly installed, equal lengths of the small diameter of the tool (14) will extend from both ends of the link pin (13) **(Figure 21–109)**.

10. Install the 1U6681 Holding Tool (15) between the sleeve (11) and the small diameter (B) of the 1U7305 Insertion Tool (14). Push the holding tool (15) down until the small diameter (B) of the tool (14) contacts the face (A) of the governor (10). This is the governor calibration point **(Figure 21–110)**.

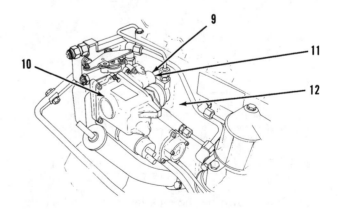

Governor
(9) Clip. (10) Governor. (11) Sleeve. (12) Inlet manifold.

FIGURE 21–108 (Courtesy of Caterpillar, Inc.)

506 Chapter 21 CATERPILLAR FUEL INJECTION SYSTEMS

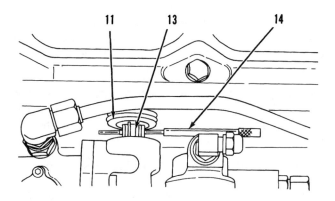

Install Governor Pin
(11) Sleeve. (13) Link pin. (14) 1U7305 Insertion Tool.

FIGURE 21-109 (Courtesy of Caterpillar, Inc.)

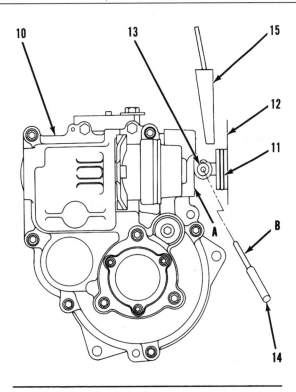

Governor Viewed From Rear
(10) Governor. (11) Sleeve. (12) Inlet manifold. (13) Link pin.
(14) 1U7305 Insertion Tool. (15) 1U6681 Holding Tool. (A) Face of governor. (B) Small diameter of tool.

FIGURE 21-110 (Courtesy of Caterpillar, Inc.)

11. Push down on the rack lever (17) and quickly release it. Flip the lever in this manner to make sure there is smooth movement of the injector rack.

12. Refer to the engine information plate or Fuel Setting And Related Information Fiche for the correct fuel setting.

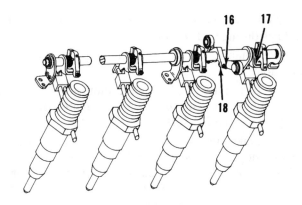

Rack Control Linkage
(16) Fuel setting screw. (17) Rack lever for No. 1 injector.
(18) Clamp assembly.

FIGURE 21-111 (Courtesy of Caterpillar, Inc.)

13. If the reading on the dial indicator (2) is within ±0.25 mm of the specified fuel setting, go to step 16.

14. If the fuel setting needs adjustment, use a 1U6673 Wrench to loosen the locknut of the fuel setting screw (16). Adjust the fuel setting screw (16) until the indicator reading is the same as the correct fuel setting. Turn the screw counterclockwise for more fuel (greater fuel setting) or clockwise for less fuel (lower fuel setting) **(Figure 21-111)**.

NOTE: Do not loosen the screw holding the clamp assembly (18) to the control shaft. (This screw can be identified by sealant in the socket head). The clamp assembly is factory set to the shaft. Loosening the clamp will cause poor engine performance and may damage the engine.

15. With a 1U6673 Wrench, hold the fuel setting screw (16) in position while tightening the locknut. Recheck the fuel setting by flipping (pushing down and releasing) the lever (17). Read the dial indicator again. If the fuel setting is not correct, repeat steps 14 and 15.

16. Remove the 1U6681 Holding Tool (15) and 1U7305 Insertion Tool (14) from the engine. Use 6V0006 Pliers to slide the sleeve (11) into the governor (lubricate the O-ring seal on the sleeve with engine oil if necessary). Install the clip (9).

FUEL TIMING

TOOLS NEEDED

1U6680	Governor and Fuel System Adjusting Tool Group	1
4C4716	Injector Timing Fixture	1

1U8702	Injector Timing Block	1
1U6678	Calibration Fixture Group	1
6V6106	Dial Indicator	1
9S8883	Contact Point	1
	or	1
1U8869	Digital Dial Indicator	1
3S3269	Contact Point	1

CRANKSHAFT POSITIONS FOR FUEL TIMING SETTING

SAE Standard (Counterclockwise) Rotation Engines As Viewed from the Flywheel End

Check/adjust with No. 1 piston on TC compression stroke	Injectors 3-5-6[a]
Check/adjust with No. 1 piston on TC exhaust stroke	Injectors 1-2-4[a]

[a]Note that the No. 1 injector is not checked or adjusted with the No. 1 piston on TC compression stroke. Be certain that the No. 1 piston is TC on the correct stroke (compression or exhaust) to check or adjust each injector. For more information refer to the table **(Figure 21–112)** Crankshaft Positions for Fuel Timing.

NOTE: Always turn the engine with the four large bolts on the front of the crankshaft. **Do not** use the eight small bolts on the front of the crankshaft pulley.

NOTE: If the rocker arm assemblies are removed and installed prior to setting the fuel timing dimension, rotate the crankshaft two complete revolutions to allow the rocker arms to properly seat on the injectors.

1. Put the No. 1 piston at TC and identify the correct stroke (compression or exhaust).

See the table Crankshaft Positions for Fuel Timing **(Figure 21–112)**. With the two crankshaft positions given in the chart, all the injectors can be checked or adjusted. This will make sure the push rod lifters are off the lobes and on the base circles of the camshaft.

NOTE: Two dial indicators, each with its own procedure, are available for this procedure. If the 1U8869 Digital Dial Indicator is to be used, use Procedure B. If the 6V6106 Dial Indicator is to be used, use Procedure A as outlined in the service manual.

Procedure with the 1U8869 Dial Indicator

Before a check or an adjustment of the fuel timing dimension can be made, the tooling must be calibrated as follows:

CRANKSHAFT POSITIONS FOR FUEL TIMING AND VALVE CLEARANCE SETTING

SAE Standard (Counterclockwise) Rotation Engines As Viewed from the Flywheel End

Check/adjust with No. 1 piston on	TC compression stroke[a]
Injectors	3-5-6
Intake valves	1-2-4
Exhaust valves	1-3-5
Check/adjust with No. 1 piston on	TC exhaust stroke[a]
Injectors	1-2-4
Intake valves	3-5-6
Exhaust valves	2-4-6
Firing order	1-5-3-6-2-4

[a]Put No. 1 piston at top center (TC) position and identify the correct stroke. Refer to Finding Top Center Position for the No. 1 Piston. After finding the TC position for a particular stroke and making adjustments for the correct cylinders, remove the timing bolt and turn the flywheel counterclockwise 360°. This will put the No. 1 piston at the TC position on the other stroke. Install the timing bolt in the flywheel and complete the adjustments for the remaining cylinders.

FIGURE 21–112 (Courtesy of Caterpillar, Inc.)

2. Program the 1U8869 Digital Dial Indicator to read the actual timing dimension. Since the gauge block in the timing and fuel setting group is 62.00 mm, the digital dial indicator is to be programmed for 62.00 mm.

   a. Turn the indicator on by pushing the ON/OFF button.
   b. Push the "in/mm" button so the display shows mm.
   c. A negative sign (−) should be in the display window under REV. If that space is blank, push the "±" button so the display shows (−). When this is done, plunger movement into the indicator will show on the display as negative movement, and plunger movement out of the indicator as positive movement.
   d. Push and hold the "preset" button down until there is a flashing (P) in the upper right corner of the display, and then release the button.
   e. Push and hold the "preset" button down until the (P) stops flashing, and a flashing indicator bar is seen in the lower left corner of the display, and then release. Momentarily pushing the "preset" button will cause a minus sign to appear or disappear above the flashing indica-

tor. Use the "preset" button to make this position blank.

f. Push and hold the "preset" button down until the flashing indicator begins to flash under the first number position (fourth position to the left of the decimal), then release. Momentarily pushing the "preset" button will cause the display number in that position to change. Use the "preset" button to make the position show zero (0).

g. Use the "preset" button to move the flashing indicator and change the display numbers until the display shows 0062.00 mm.

h. Push and hold the "preset" button until the flashing (P) is shown in the upper right corner of the display, and then release. Momentarily push the "preset" button so the flashing (P) and the zeros to the left of 62.00 mm disappear.

i. The indicator can now be turned off if desired. The indicator will retain the preset number in memory (only one preset number is retained). To recall the preset number, repeat steps 2(a) through 2(d). Then momentarily push the "preset" button so the flashing (P) and the zeros to the left of 62.00 mm disappear.

j. Turn the indicator off. Make sure the indicator plunger is extended fully. Repeat steps 2(a) through 2(d). Then momentarily push the "preset" button so the flashing (P) and the zeros to the left of 62.00 mm disappear.

3. Install the 3S3269 Contact Point [25.45 mm (1.0 in.) excluding threads] on the dial indicator stem.

4. Install the 1U8869 Digital Dial Indicator in the collet of the 4C4716 Timing Fixture. Leave the collet loose.

5. Put the dial indicator and timing fixture on the 1U8702 Injector Timing Block (3) and 1U6678 Calibration Fixture (4) as shown. Make sure the locating pin (C) in the left side of the timing fixture (2) engages the hole in the calibration fixture (4). Install the bolt (B) on the right side and tighten to secure the timing fixture (2) firmly to the calibration fixture (4). The long pin of the timing fixture (2) must be in the hole located in the center of the injector timing block (3). The ends of the injector timing block and surface of the calibration fixture must be clean (**Figure 21-113**).

6. Position the 1U8869 Digital Dial Indicator in the bracket so the indicator display shows 55.40 mm ± 0.40 mm and tighten the collet. Turn off the indicator. This step is to ensure that the indicator plunger has adequate travel in both directions.

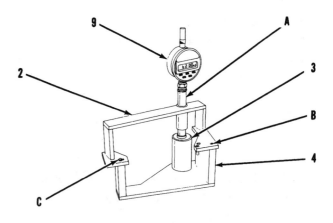

Calibrate Injector (Fuel) Timing Fixture
(2) 4C4716 Timing Fixture. (3) 1U8702 Injector Timing Block. (4) 1U6678 Calibration Fixture. (9) 1U8869 Digital Dial Indicator. (A) Collet sleeve. (B) Bolt. (C) Locating pin.

FIGURE 21-113 (Courtesy of Caterpillar, Inc.)

7. Repeat steps 2(a) through 2(d).

8. Momentarily push the "preset" button. The display should show 62.00 mm. Lift up on the injector timing block slightly. The display should increase in the positive direction. If it does not, repeat steps 2 through 7.

9. Make sure the top surfaces of injector tappet and shoulder are clean and dry (**Figure 21-114**).

NOTE: After the timing fixture (2) has been installed on the inlet manifold, **do not** rotate the engine. The timing fixture may be damaged.

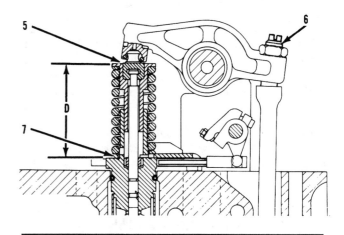

Fuel Timing Dimension
(5) Tappet. (6) Adjustment screw. (7) Shoulder. (D) Fuel Timing Dimension.

FIGURE 21-114 (Courtesy of Caterpillar, Inc.)

10. Remove the bolt (B) from the timing fixture. While holding up the collet sleeve (A), gently put the 1U8869 Digital Dial Indicator (9) and the 4C4716 Timing Fixture (2) in position on the inlet manifold over the injector to be checked.

When properly positioned, the locating pin (C) and the bolt (B) on the timing fixture (2) will engage holes in the top face of the inlet manifold (8). Tighten the bolt (B) to secure the timing fixture (2) firmly to the inlet manifold (8).

Slowly lower the collet sleeve (A) until the long pin of the timing fixture (2) contacts the shoulder (7) of the injector.

NOTE: The sliding locating pin (C) and the two hole positions for the bolt (B) are provided in the timing fixture because of a different valve cover bolt hole position on the rear cylinder.

11. The digital dial indicator will display the fuel timing dimension directly, for example, 63.00 mm will be displayed as 63.00 mm, and 63.50 mm will be displayed as 63.50 mm **(Figure 21–115)**.

Refer to the engine information plate or the Fuel Setting And Related Information Fiche for the correct fuel timing dimension to use.

The digital dial indicator must show the correct fuel timing dimension within ± 0.20 mm. Each injector must be checked separately and adjusted if necessary.

12. If the digital dial indicator displays the correct dimension or is within the ±0.20 mm checking tolerance, no adjustment is necessary. Proceed to step 15.

13. If the dial indicator points *do not* indicate 0.00 ± 0.20 mm, loosen the locknut on the push rod adjustment screw (6) for the injector to be adjusted.

14. Turn the adjustment screw (6) until the digital dial indicator displays the correct fuel timing dimension. Tighten the locknut on the adjustment screw (6)

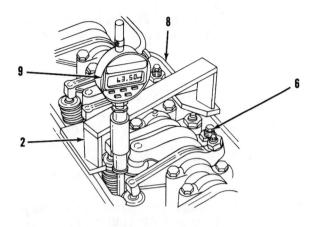

4C4716 Timing Fixture Installed
(2) 4C4716 Timing Fixture. (6) Adjustment screw. (8) Inlet manifold. (9) 1U8869 Digital Dial Indicator.

FIGURE 21–115 (Courtesy of Caterpillar, Inc.)

to a torque of 18 ± 5 lb-ft (25 ± 7 N·m) and check the adjustment again. If necessary, repeat this procedure until the adjustment is correct.

NOTE: After the timing fixture (2) has been installed on the inlet manifold, **do not** rotate the engine. The timing fixture may be damaged.

15. The fuel timing dimension can be checked or adjusted on half of the cylinders with the timing bolt installed, then the timing bolt and the timing fixture must be removed. Only then may the flywheel be rotated counterclockwise 360° (one revolution) and the timing bolt and the timing fixture reinstalled to check the fuel timing dimension on the remaining cylinders. Refer to the table Crankshaft Positions for Fuel Timing **(Figure 21–112)**.

16. Remove the timing bolt from the flywheel when the fuel timing check is completed.

REVIEW QUESTIONS

1. The fuel transfer pump is a _____ pump that is moved by a cam (eccentric) on the camshaft for the fuel injection pump.
2. The fuel injection pump _____ _____ _____ of the fuel and sends an exact amount of fuel to the fuel injection nozzle.
3. The fuel injection nozzle is installed in an _____ in the cylinder head and extended into the _____ _____.
4. The governor controls the amount of _____ needed by the engine to maintain a desired _____.
5. The governor servo gives hydraulic assistance to the _____ governor force to move the fuel rack.
6. The dashpot helps give the governor better speed control when there are sudden _____ _____.

Chapter 21 CATERPILLAR FUEL INJECTION SYSTEMS

7. The fuel ratio control (FRC) limits the amount of _____ to the cylinders during an increase of engine speed (acceleration) to reduce _____ .

8. The FRC restricts the amount of _____ to the _____ until sufficient boost has been achieved.

9. The timing advance unit uses engine oil pressure to change the fuel _____ _____ .

10. The No. 1 piston at TC on the _____ _____ is the starting point of all timing procedures.

11. To adjust the low-idle rpm, start the engine and run it with the governor control lever in the _____ _____ .

12. The sleeve-metering fuel system has an injection pump for each _____ of the engine.

13. The constant-bleed valve lets approximately 9 gallons of fuel per hour go back to the _____ .

14. The automatic timing advance is installed on the front of the _____ for the engine.

15. The _____ , _____ , and _____ for each pump are made as a set.

16. The fuel transfer pump is integral with the _____ .

17. The fuel injection pump (unit injector) allows a small _____ of fuel to be injected at the proper time into the _____ _____ .

18. The _____ _____ _____ connects the governor output to the fuel injector at each cylinder.

19. The governor transfers the operator's requirements to the fuel injector rack _____ _____ .

20. The governor _____ gives hydraulic assistance to the mechanical governor.

21. The dashpot improves engine stability during sudden _____ _____ .

22. A turbocharged engine uses a fuel ratio control to control _____ during acceleration at low boost levels.

23. Low temperature at an exhaust manifold port is an indication of _____ _____ .

24. The only on-engine adjustment to the governor is the _____ _____ _____ .

25. Before the fuel setting is checked, the injectors must be correctly _____ .

TEST QUESTIONS

1. Fuel is drawn from the fuel tank
   a. by gravity
   b. by the transfer pump
   c. by vacuum
   d. all of the above

2. There is one fuel injection pump for each cylinder. (T) (F)

3. The fuel injector pump increases the pressure of the fuel and sends
   a. high-pressure fuel to the injector
   b. it through high-pressure steel lines
   c. an exact amount of fuel to the injector nozzle
   d. fuel for adequate power

4. The fuel injection pump sends fuel with high pressure to the fuel injector nozzle, where fuel is
   a. made into a fine spray
   b. atomized
   c. ignited by high temperatures
   d. vaporized

5. When the engine is cranked to start, the governor is at
   a. high-idle position
   b. off-idle position
   c. high rpm
   d. low-idle position

6. The governor servo gives hydraulic assistance to the mechanical governor force to
   a. control rpm
   b. move the fuel rack
   c. operate the cruise control
   d. adjust according to altitude

7. The fuel ratio control restricts the amount of fuel to the combustion chamber
   a. until sufficient boost has been achieved
   b. to prevent black exhaust smoke
   c. to give better economy
   d. to prevent engine damage

8. All timing procedures are started with the No. 1 piston at
   a. TDC
   b. TDC on the compression stroke
   c. TDC on valve overlap
   d. BDC with the companion piston on TDC

9. To adjust the low-idle rpm, start and run the engine with the governor control lever in the
   a. fast-idle position
   b. neutral position
   c. cruise position
   d. low-idle position

10. The sleeve-metering fuel system has an injection pump for
    a. all cylinders
    b. each pair of cylinders
    c. each cylinder of the engine
    d. each cylinder bank

11. The automatic advance timing unit drives the gear on the camshaft for the
    a. governor
    b. fuel injection pump
    c. supply pump
    d. timing advance

12. Technician A says the plunger, barrel, and sleeve of each pump are made as a set. Technician B says they are interchangeable as long as they are cleaned. Who is correct?
    a. Technician A
    b. Technician B
    c. both are correct
    d. neither is correct

13. The governor achieves the desired engine speed by the position of the
    a. throttle
    b. air control valve
    c. governor weights
    d. all of the above

14. No fuel in a cylinder can be detected by low temperature at
    a. an intake manifold port
    b. an exhaust manifold port
    c. the exhaust tail pipe
    d. catalytic converter

◆ CHAPTER 22 ◆

CATERPILLAR ELECTRONIC CONTROL SYSTEMS

INTRODUCTION

Caterpillar has three types of electronic fuel injection control systems. The programmable electronic engine control (PEEC) system uses a multiplunger fuel injection pump, electronic control module (ECM), and a number of input sensors and switches. The 3176B and 3406E electronic control system utilizes an ECM, input sensors and switches, and electronically controlled unit injectors (EUI). Both of these systems are discussed in this chapter. The HEUI (hydraulic electronic unit injector) system is described in Chapter 23.

PERFORMANCE OBJECTIVES

After completing this chapter you should be able to:

1. List 12 advantages of the PEEC system over nonelectronic systems.
2. List the components of the PEEC system and state their function.
3. Access PEEC system fault codes using the cruise control switches.
4. Connect the ECAP tool to the PEEC system and check the stored fault codes.
5. List the 3176 system components that differ from PEEC system components and describe their function.
6. Describe the fuel flow in a 3176B and 3406E system electronic unit injector.
7. Connect the ECAP tool to the 3176B and 3406E system and check the stored faults.
8. Remove and install a 3176 system electronic unit injector.

TERMS YOU SHOULD KNOW

Look for these terms as you study this chapter, and learn what they mean.

PEEC
rack actuator
engine speed sensor
timing advance solenoid
throttle position sensor
rack position sensor
vehicle speed buffer
diagnostic lamp
check engine light (CEL)
data link
personality module
limp-home system
pressure transducer module
boost pressure sensor
inlet air pressure sensor
DDT tool
ECAP tool
3176B and 3406E electronic control system
electronic unit injector (EUI)
fuel manifold
fuel pressure sensor
engine protection system
pulse width
three-cylinder cutout
engine derate
engine shutdown
active code
logged code
TPS adjustment
injection timing calibration
speed/timing sensor calibration

PROGRAMMABLE ELECTRONIC ENGINE CONTROL (PEEC) SYSTEM

Caterpillar's Programmable Electronic Engine Control (PEEC) system was provided as an option on the 3406B engines. Fuel delivery and injection timing are controlled electronically. A number of sensors and switches provide input to the electronic control module (ECM). The ECM consults programmed information and processes input signals to produce output signals to operate the rack actuator, timing advance unit, warning lights, and warning buzzer.

ADVANTAGES AND FEATURES OF THE PEEC SYSTEM

The PEEC system provides the following features and advantages over nonelectronic systems. See Chapter 5 for a description of these features.

1. Easier starting
2. Better fuel economy
3. Lower exhaust emissions
4. Cold-idle mode
5. Vehicle speed limiting
6. Progressive shifting
7. Engine protection system
8. Isochronous governor
9. Constant PTO speeds
10. Idle shutdown timer
11. Cruise control
12. Programmable low-idle rpm
13. Diagnostic system

PEEC SYSTEM COMPONENT DESCRIPTIONS (FIGURES 22-1 TO 22-3)

1. *Injection pump:* a Caterpillar scroll pump that has no mechanical governor.
2. *Rack actuator:* a solenoid that is mounted on the side of the housing that is bolted to the rear of the injection pump. The solenoid is a brushless torque motor (BTM) that is controlled by the ECM and operates the fuel control rack in the pump. Located in the housing are the rack position sensor, engine speed sensor, fuel shut-off solenoid, and manual override.

3. *Engine speed sensor:* a magnetic pickup that senses engine speed from milled grooves in the injection pump drive shaft and sends this information to the ECM.

4. *Transducer module:* mounted on the rear of the injection pump, it contains the boost pressure sensor, inlet air pressure sensor, and engine oil pressure sensor that provide signals to the ECM.

5. *Timing advance solenoid:* A brushless torque motor (BTM)–type solenoid actuates a valve that controls hydraulic pressure to a piston that changes injection pump timing. As engine speed, load, boost pressure, and throttle position change, the ECM causes the timing advance unit to change injection timing.

6. *Throttle position sensor:* sends information about throttle position to the ECM. The sensor is operated by a linkage from the accelerator pedal and produces a pulse width modulated signal rather than an analog voltage signal.

7. *Timing position sensor:* provides information on the position of the timing advance unit to the ECM to aid in controlling injection timing.

8. *Rack position sensor:* provides information on the position of the fuel control rack to the ECM to help control fuel delivery.

9. *Vehicle speed sensor:* a magnetic sensor that senses vehicle speed from a toothed wheel on the transmission output shaft. This information is sent to the ECM for progressive shifting, cruise control, and road speed governing.

10. *Vehicle speed buffer:* amplifies vehicle speed sensor output signals and provides the signal for the vehicle speedometer.

11. *Diagnostic lamp (check engine light):* When the ECM detects a fault, the lamp goes on and stays on for 5 seconds, blinks off for 5 seconds, on for 5 seconds, off for 5 seconds, and on for 5 seconds. After that, fault codes (if any) are flashed out by the lamp.

12. *Cruise control switches:* used to operate the cruise control and to retrieve any stored fault codes.

13. *Exhaust brake enable switch:* prevents the application of the Jacob's brake when the throttle is in any position other than idle.

14. *Data link:* A multiple-pin wiring harness connector used to communicate engine operating information to other vehicle electronic devices and to interface with Caterpillar's ECAP (electronic control analyzer and programmer) and DDT (digital diagnostic tool) service tools.

15. *ECM (electronic control module):* The ECM is mounted on the engine block through rubber bushings and is cooled by fuel from the transfer pump as it circu-

514 Chapter 22 CATERPILLAR ELECTRONIC CONTROL SYSTEMS

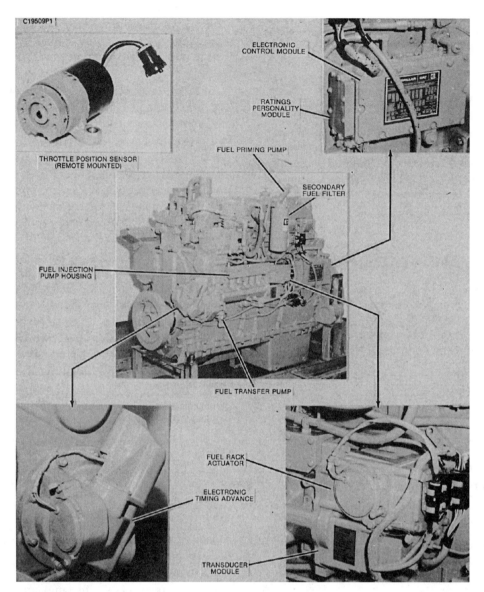

FIGURE 22–1 Caterpillar 3406B engine electronic control module, throttle position sensor, transducer module, and electronic timing advance unit. (Courtesy of Caterpillar, Inc.)

lates through passages in the control module. The ECM receives information from various input sensors and switches and processes this information in conjunction with programmed information in ECM memory, and provides electrical power to engine sensors and output devices. The ECM tests continuous runs on all inputs and outputs and stores any malfunction in ECM fault code memory.

16. *Ratings personality module:* that part of the ECM that contains certain customer-specified operating parameters that can be programmed into the module within the range that applies to a certain family group of engines. This may include road speed limit, fuel/air ratio, injection timing, and owner-entry passwords that prevent unauthorized alteration of programmed settings.

17. *Wiring harness:* provides electrical connections between PEEC system components.

PEEC system fuel flow is shown in **Figure 22–4.**

PEEC Limp-Home System

Should certain PEEC system components fail, a backup system allows the vehicle to be driven home or to a repair facility. In the event of such failure, the CEL

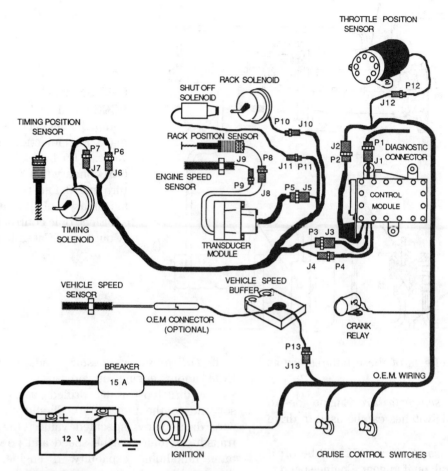

FIGURE 22–2 PEEC system wiring harness and components. (Courtesy of Caterpillar, Inc.)

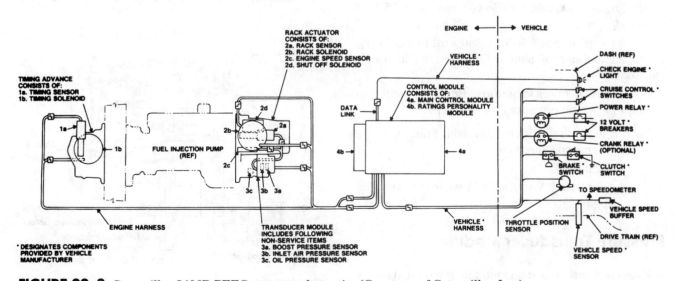

FIGURE 22–3 Caterpillar 3406B PEEC system schematic. (Courtesy of Caterpillar, Inc.)

516   Chapter 22 CATERPILLAR ELECTRONIC CONTROL SYSTEMS

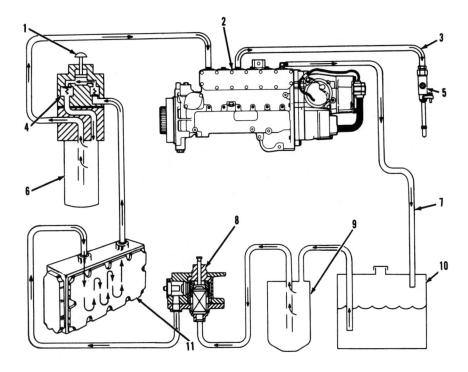

FIGURE 22-4 Caterpillar 3406B PEEC fuel flow. 1. Fuel priming pump. 2. Fuel injection pump housing. 3. Fuel injection line. 4. Check valve. 5. Fuel injection nozzle. 6. Secondary fuel line. 7. Fuel return line. 8. Fuel transfer pump. 9. Primary fuel filter. 10. Fuel tank. 11. Electronic control module. (Courtesy of Caterpillar, Inc.)

will be activated. The effects of these failures are as follows:

1. *Throttle position sensor failure:* Engine goes to low idle. Cruise control switches can be used to drive the vehicle.

2. *Vehicle speed sensor failure:* CEL will be off if transmission is in neutral; on if in gear. Engine speed is limited in any gear to VSL speed only.

3. *Rack position sensor, boost pressure sensor, timing sensor, timing BTM, and sluggish rack BTM:* Engine is limited to lower rpm and 75% of power.

4. *Shut-off solenoid failure:* Shuts off fuel delivery. Rotate manual shut-off shaft to ON position with a suitable tool.

5. *Oil pressure sensor failure:* Engine is limited to 1350 rpm when oil pressure is below 20 psi.

Towing is required with the following failures:
- Engine speed sensor
- Rack BTM
- Less than 9 V to the ECM while cranking the engine

Pressure Transducer Module

A sealed pressure transducer module is mounted below the injection pump rack actuator. Three sensors are contained in the module **(Figure 22-5)**:

1. *Oil pressure sensor:* sends a signal to the ECM about engine oil pressure. Engine oil pressure is directed through drilled passages to the sensor and to the rack servo. Should engine oil pressure drop below specified values, the oil pressure transducer limits engine speed and power but maintains limp-home capability. If engine oil pressure drops too low, engine shutdown takes place in 30 seconds.

2. *Turbocharger boost pressure sensor:* Air pressure upstream of turbo air is directed to the boost pres-

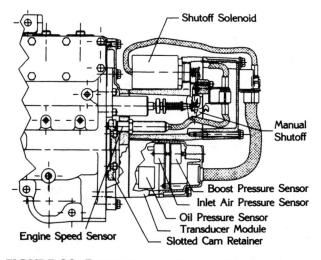

FIGURE 22-5 PEEC shut-off solenoid, rack actuator, and transducer module. (Courtesy of Caterpillar, Inc.)

sure sensor. This sensor is used to control engine exhaust smoke during acceleration.

3. *Inlet air pressure sensor:* This sensor is used to protect the engine during operation at high altitude or with a restricted air cleaner.

Rack Actuator and Transducer Operation

An electronic actuator and transducer are used to operate the injection pump control rack on PEEC-equipped engines. Engine oil pressure is supplied to the rack actuator servo **(Figures 22–6 and 22–7)**. A rotary solenoid (BTM) and spool valve control hydraulic pressures that regulate the position and movement of the rack. The BTM is spring loaded so that its force is applied in a direction toward the fuel shut-off, engine stop position. Output voltage from the ECM, based on sensor inputs and programmed data, moves the rack to increase fuel delivery, reduce fuel delivery, or maintain steady fuel delivery.

Timing Advance Unit Operation (Figures 22–8 to 22–10)

The PEEC electronic advance unit provides injection timing advance in a manner similar to that of the mechanical speed-sensitive variable timing unit. A helically splined drive carrier has replaced the earlier straight splined carrier to increase timing advance 10° to 15° for a total range of from 9 to 25 crankshaft degrees. Instead of flyweights actuating the control valve, an electronically controlled BTM is used to control a

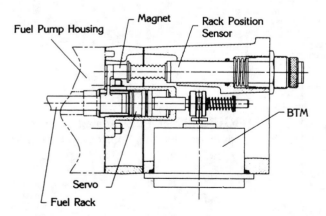

FIGURE 22–6 PEEC system rack servo. (Courtesy of Caterpillar, Inc.)

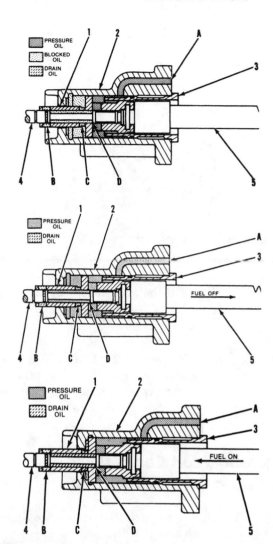

FIGURE 22–7 Various rack positions. (Top) Stationary, (middle) high-idle, no-load position, (bottom) full-load position. (Courtesy of Caterpillar, Inc.)

double-acting hydraulic servo. A timing advance feedback sensor and linear potentiometer provide feedback control.

PEEC SYSTEM TESTS

When a problem occurs, the check engine light (CEL) stays on until the problem is corrected. If the problem is only momentary, the CEL will go off after 5 seconds. Fault codes are stored in the ECM memory. The technician can access these fault codes by using the cruise control switches or by using special tools like the DDT or the ECAP. The DDT is a handheld tool, while the ECAP is a larger portable tool

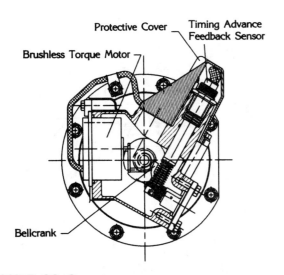

FIGURE 22-8 PEEC timing advance unit and feedback sensor. (Courtesy of Caterpillar, Inc.)

with a keyboard (**Figures 22-11** to **22-13**). The following procedures are given here as examples only. For specific tests, procedures, and specifications always refer to the appropriate Caterpillar service manual.

Accessing Fault Codes Using the Cruise Control Switches

1. Have the vehicle switch on or the engine running and the cruise control switch off.

2. Flip the resume switch to ON and hold it there until the CEL begins to flash, then release the switch.

3. Note any flashing codes and refer to the fault code list in the service manual to identify the faulty circuit. A code 55 indicates there are no fault codes.

4. Test the components in the faulty circuit and repair or replace the faulty item. Tests can be performed with a digital multimeter of the correct capacity. Voltage and/or resistance are normally checked. The most likely problem is usually an electrical connection. Check every connection, pin, and wire carefully.

Testing the System Using the ECAP Tool

The ECAP tool can be used to access fault codes even with the engine running. It can be used to simulate normal operation of system sensors and actuators. The general procedure is as follows. Refer to the appropriate Caterpillar service manual for specific procedures and specifications.

1. Connect the ECAP tool into the vehicle wiring harness diagnostic connector using the connector cable.

2. Turn the ignition switch ON to power up the ECAP.

3. Observe the ECAP display—title screen and main menu.

4. Select the Display Diagnostic Messages Menu.

5. With the ECAP cursor on 1, DISPLAY ACTIVE DIAGNOSTIC CODES, press RETURN on the keyboard. The ECAP will display any stored diagnostic messages and code numbers.

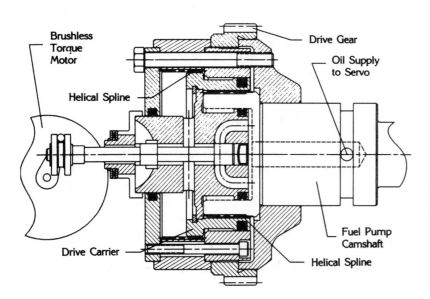

FIGURE 22-9 PEEC timing advance unit cross section. (Courtesy of Caterpillar, Inc.)

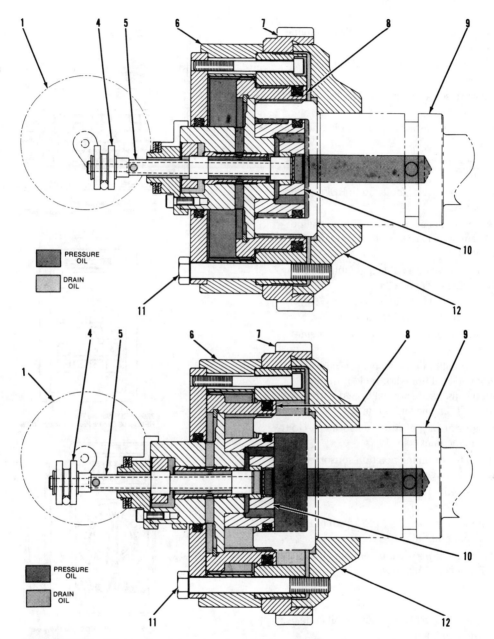

FIGURE 22-10 Timing fully retarded (top) and fully advanced (bottom). 1. Timing solenoid (BTM). 4. Sleeve. 5. Valve spool. 6. Ring. 7. Gear. 8. Carrier. 9. Fuel injection pump camshaft. 10. Body assembly. 11. Bolt. 12. Ring. (Courtesy of Caterpillar, Inc.)

Code	Meaning
21	Sensor supply voltage fault. Go to PEEC engine fault code 21.
22	Rack position sensor fault. Go to PEEC engine fault code 22.
23	Timing position sensor fault. Go to PEEC engine fault code 23.
24	Oil pressure sensor fault. Go to PEEC engine fault code 24.
25	Boost pressure sensor fault. Go to PEEC engine fault code 25.
26	Inlet air pressure sensor fault. Go to PEEC engine fault code 26.
31	Loss of vehicle speed signal. Go to PEEC engine fault code 31.
32	Throttle position sensor fault. Go to PEEC engine fault code 32.
33	Engine speed signal out of range. Faulty engine speed sensor signal. Go to PEEC engine fault code 33.
34	Engine speed sensor loss of signal fault. Go to PEEC engine fault code 34.
35	Engine overspeed. Faulty engine speed sensor signal. Go to PEEC engine fault code 35.
36	Vehicle speed sensor fault. Go to PEEC engine fault code 36.
42	Check sensor calibrations. Boost pressure sensor has not been calibrated or rack or timing position sensors are not calibrated correctly. Go to PEEC engine fault code 42.
43	Rack actuator or rack position sensor fault. Go to PEEC engine fault code 43.
44	Timing actuator or timing position sensor fault. Go to PEEC engine fault code 44.
45	Shut down solenoid fault. Go to PEEC engine fault code 45.
46	Low oil pressure detected by PEEC. Go to PEEC engine fault code 46.
52	Control module or personality module fault. Go to PEEC engine fault code 52.
53	Personality module fault. Go to PEEC engine fault code 53.
54	Data link fault. Go to PEEC engine fault code 54.
55	No PEEC-detected faults.
56	Check customer-specified parameters. Parameters not programmed in or programmed incorrectly. Go to PEEC engine fault code 56.

FIGURE 22–11 PEEC system fault code chart directs technician to fault code diagnosis in the service manual. (Courtesy of Caterpillar, Inc.)

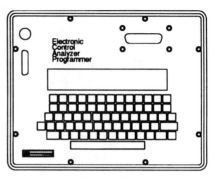

8T8697 ECAP Service Tool
NEXG4522 Service Program Module (SPM)

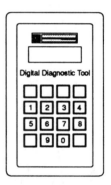

7X6400 DDT Service Tool
NEXG4508 Service Program Module (SPM)

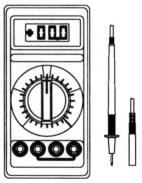

6V7800 or 6V7070 Digital Multimeter

FIGURE 22–12 Tools used to troubleshoot PEEC and 3176 electronic control systems. (Courtesy of Caterpillar, Inc.)

3406E/3176B ELECTRONIC CONTROL SYSTEM 521

9. Disconnect the ECAP from the diagnostic connector.

Component Replacement and Calibration

For accurate information to be sent to the ECM the calibration of the following components must be checked and adjusted to meet specifications. Use the ECAP tool to check and adjust the timing position sensor, rack position sensor, and throttle position sensor. The engine speed sensor must also be checked and adjusted as required. Refer to the appropriate Caterpillar service manual for component replacement procedures and adjustment specifications.

3406E/3176B ELECTRONIC CONTROL SYSTEM (FIGURES 22-14 TO 22-17)

The 3406E/3176B electronic control system uses mechanically operated, electronically controlled unit injectors (EUI) instead of a multiplunger injection pump, high-pressure fuel lines, and injection nozzles. Injectors are actuated by an overhead cam or by push rods and rocker arms **(Figure 22-18)**. The 3406E/3176B control system consists of the following components:

1. Electronic control module
2. Engine speed/timing sensor
3. Coolant temperature sensor
4. Coolant level sensor
5. Intake manifold air temperature sensor
6. Boost pressure sensor
7. Atmospheric pressure sensor
8. Oil pressure sensor
9. Fuel temperature sensor
10. Throttle position sensor
11. Vehicle speed sensor
12. Clutch switch
13. Service brake switch
14. Parking brake switch
15. Cruise control switches
16. Air conditioning pressure switch
17. Electronic unit injectors
18. Retarder solenoids
19. Retarder solenoid switches
20. Check engine light
21. Warning light and alarm
22. Speedometer

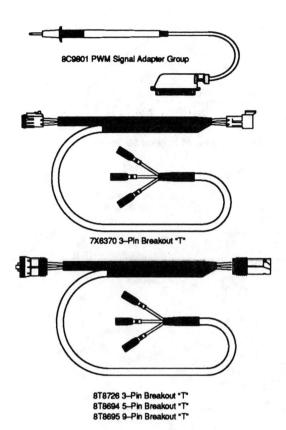

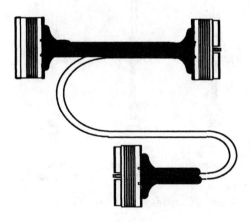

FIGURE 22-13 Adapter cables and harness connectors for DDT and ECAP service tools. (Courtesy of Caterpillar, Inc.)

6. Inspect and repair or replace any faulty wiring or component.

7. After all repairs have been made, press RETURN to clear the fault codes from the ECM memory. The ECAP screen will ask whether you wish to clear the fault. This feature prevents accidental clearing of faults.

8. To clear the fault, press Y and RETURN.

522 Chapter 22 CATERPILLAR ELECTRONIC CONTROL SYSTEMS

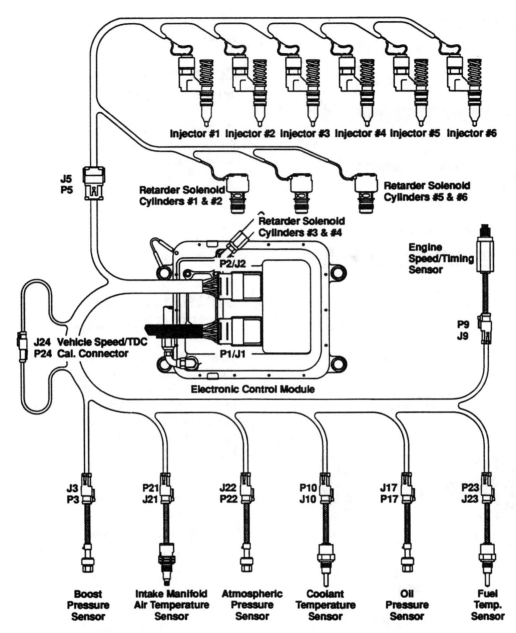

FIGURE 22-14A 3406E/3176B component diagram. (Courtesy of Caterpillar, Inc.)

23. Tachometer
24. Cab ATA data link connector
25. SAE J1922 data link connector
26. Wiring harness

COMPONENT DESCRIPTIONS

1. *Electronic control module (ECM):* The ECM is mounted on the engine and is cooled by fuel from the transfer pump as it circulates through passages in the cooling plate. The ECM receives information from the various input switches and sensors, processes this information in consultation with programmed information in ECM memory, provides voltage to some input sensors, and power to operate the output devices. On earlier models the transducer module and speed buffer were separate units. On current models these units are integrated with the ECM. The current ECM is also installed on Caterpillar 3500 and 3600 series engines.

2. *Engine speed/timing sensor:* a magnetic pickup that sends engine speed and piston position information to the ECM.

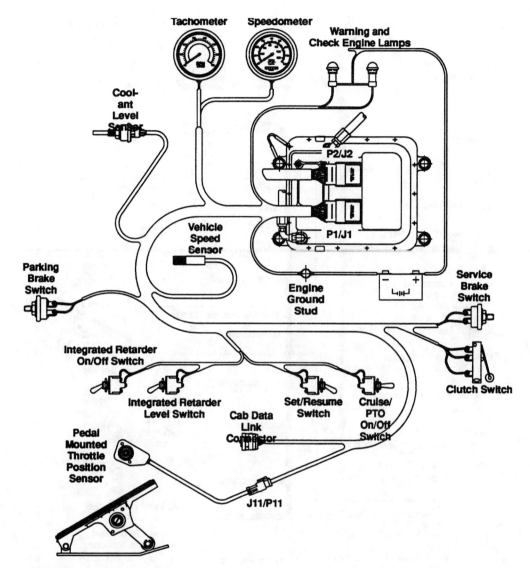

FIGURE 22–14B 3406E/3176B component diagram continued. (Courtesy of Caterpillar, Inc.)

3. *Coolant temperature sensor:* sends coolant temperature information to the ECM to help control radiator fan clutch operation.

4. *Coolant level sensor:* When coolant level drops below a specified level, the ECM begins to activate the engine warning functions.

5. *Intake manifold air temperature sensor:* provides information on intake air temperature to the ECM. The ECM adjusts injection timing and fuel quantity accordingly.

6. *Boost pressure sensor:* The ECM adjusts fuel injection and injection timing in relation to changes in boost pressure.

7. *Atmospheric pressure sensor:* The ECM adjusts fuel quantity injected in relation to the quantity of air entering the engine.

8. *Oil pressure sensor:* When oil pressure drops below a specified level, a signal from the OPS tells the ECM to activate the engine warning system.

9. *Fuel temperature sensor:* The ECM uses the signal from the FTS to help calculate fuel consumption.

10. *Throttle position sensor:* produces an output signal that varies with foot-pedal position. The ECM uses this signal to vary the amount of fuel injected in relation to throttle pedal position.

11. *Vehicle speed sensor:* The ECM uses this signal to govern maximum engine speed in different gear ratios, cruise control operation, and maximum vehicle speed.

12. *Clutch switch:* Pushing the clutch pedal down opens the switch and deactivates the engine braking system.

524 Chapter 22 CATERPILLAR ELECTRONIC CONTROL SYSTEMS

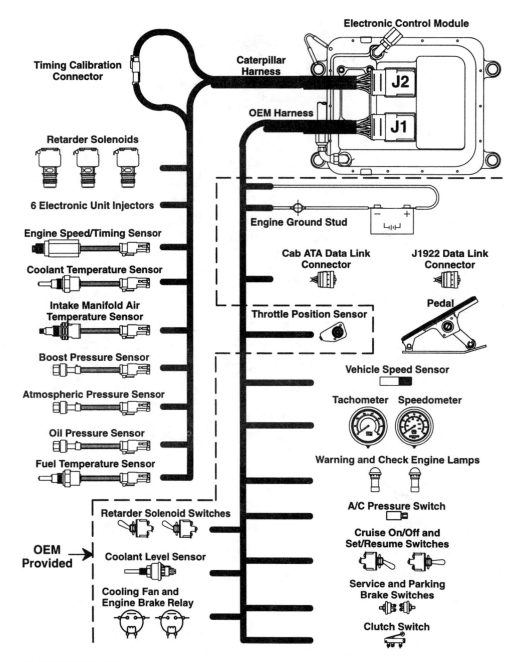

FIGURE 22-15 3406E electronic control system diagram. (Courtesy of Caterpillar, Inc.)

13. *Service brake switch:* Depressing the brake pedal opens the switch and deactivates the cruise control.

14. *Integrated retarder ON/OFF switch and integrated retarder level switch:* control operation of the engine retarder (brake) system on 2, 4, or 6 cylinders.

15. *Cruise control switches:* Cruise/PTO ON/OFF switch and SET/RESUME switch control operation of the cruise control and PTO.

16. *Air conditioning pressure switch:* This signal helps the ECM control the operation of the radiator clutch and fan.

17. *Electronic unit injectors (EUI):* The electronically controlled unit injectors are mechanically operated. They inject the correct amount of atomized fuel into the combustion chambers at the correct time in relation to piston position and engine operating conditions.

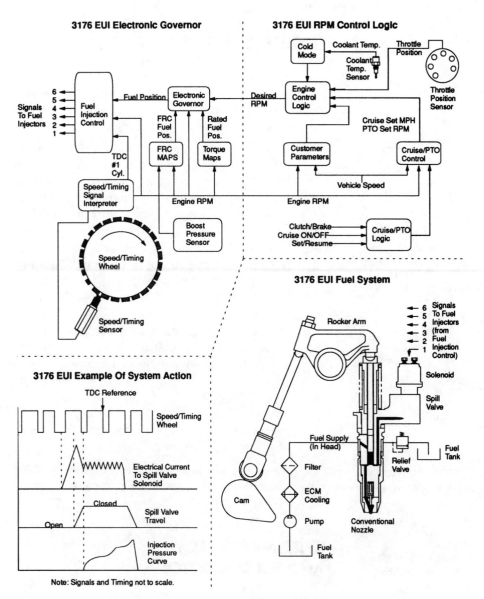

FIGURE 22–16 3176B EUI electronic system. (Courtesy of Caterpillar, Inc.)

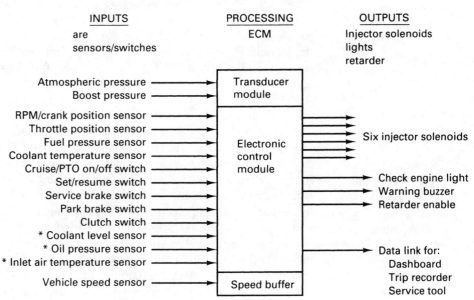

FIGURE 22–17 Inputs and outputs of Caterpillar electronic control system for the 3176 engine. (Courtesy of Caterpillar, Inc.)

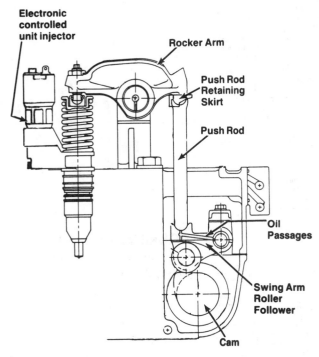

FIGURE 22-18 (Courtesy of Caterpillar, Inc.)

18. *Retarder solenoids:* control exhaust valve operation to convert the engine into an air compressor to retard vehicle speed.

19. *Retarder solenoid switches:* control retarder solenoids at ECM command.

20. *Check engine light:* Dash-mounted warning light tells the driver when a malfunction occurs that requires attention as soon as possible.

21. *Warning light (alarm):* tells the driver that a malfunction has occurred that requires immediate attention.

22. *Speedometer:* indicates vehicle speed.

23. *Tachometer:* indicates engine speed in rpm.

24. *Cab data link connector:* American Trucking Association (ATA) standard data link connector for in-cab components.

25. *SAE data link connector:* Society of Automotive Engineers (SAE) J1587 standard data link connector that allows industrywide use of standard diagnostic and test equipment that connects to the wiring harness connector.

26. *Wiring harness:* includes engine manufacturer (OEM) harness and vehicle manufacturer–supplied harness.

ENGINE PROTECTION SYSTEM

The engine protection system monitors engine oil pressure, coolant temperature, and coolant level. The system can be programmed to function in any of four different modes as follows:

1. *Off:* The control system disables those engine protection system diagnostics associated with low oil pressure, very low oil pressure, oil pressure sensor open/short circuit, high coolant temperature, low coolant level, very low coolant level, and coolant level sensor fault.

2. *Warning:* The system will warn the operator of a potential problem only by turning on the warning lamp to light continuously and the diagnostic lamp to indicate there is an active diagnostic code.

3. *Derate:* The warning lamp will first go on continuously, and if the problem continues, the system will go into the derate mode. The warning lamp will begin to flash, and engine power and speed will be reduced.

4. *Shutdown:* If conditions persist through the derate mode, the engine will shut down 30 seconds after entering the derate mode.

The engine protection system is enabled or disabled through a customer-programmable parameter.

SYSTEM FUEL FLOW AND EUI OPERATION

The engine-driven gear-type fuel transfer pump draws fuel from the fuel tank. It pumps fuel through the ECM cooling passages, fuel filters, fuel supply passage in the fuel manifold, and to the electronic unit injectors (**Figure 22-19**). When the EUI spill port is open, fuel pressure in the injector is low, and the spray tip is not opened. Fuel simply flows through the injector for cooling and lubrication purposes and returns to the fuel tank via the return passage in the fuel manifold (**Figures 22-20 and 22-21**). When the ECM energizes the EUI solenoid, it closes the spill valve and traps fuel in the injector (**Figure 22-22**). A 100 V signal is sent by the ECM to energize the solenoid. When the cam-operated rocker arm pushes the injector plunger downward, pressure on the trapped fuel rises high enough to open the needle valve in the injector tip (**Figure 22-23**). This pressure may reach from 17,250 to 31,000 psi (2500 to 4500 kPa). Injection continues until the ECM deenergizes the injector solenoid, allowing the

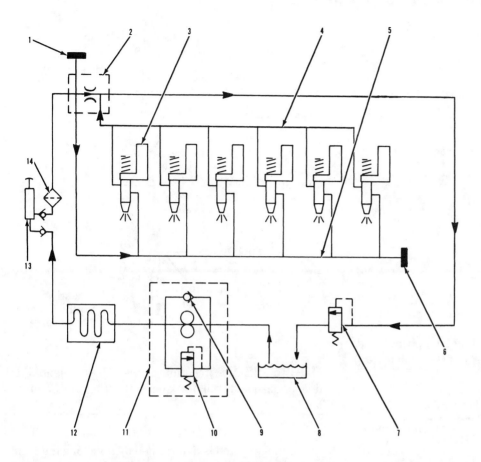

FIGURE 22-19 (Courtesy of Caterpillar, Inc.)

Fuel System Schematic
(1) Vent plug. (2) Adapter (siphon break). (3) Electronically controlled unit injectors. (4) Fuel manifold (return path). (5) Fuel manifold (supply path). (6) Drain plug. (7) Pressure regulating valve. (8) Fuel tank. (9) Check valve. (10) Pressure relief valve. (11) Fuel transfer pump. (12) Electronic control module (ECM). (13) Fuel priming pump. (14) Fuel filter (secondary).

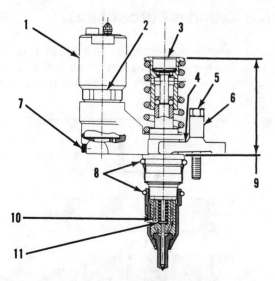

FIGURE 22-20 3176 electronic unit injector (EUI). 1. Solenoid valve assembly. 2. Valve assembly nut. 3. Fuel injection group. 4. Injector clamp. 5. Clamp bolt. 6. Spacer. 7. High-pressure plug. 8. O-rings. 9. Fuel injector group distance specification. 10. Nozzle group. 11. Spacer. (Courtesy of Caterpillar, Inc.)

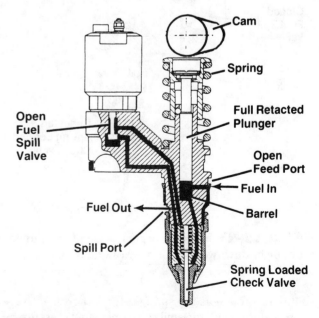

FIGURE 22-21 EUI fuel flow with spill valve open and feed port open. (Courtesy of Caterpillar, Inc.)

527

528 Chapter 22 CATERPILLAR ELECTRONIC CONTROL SYSTEMS

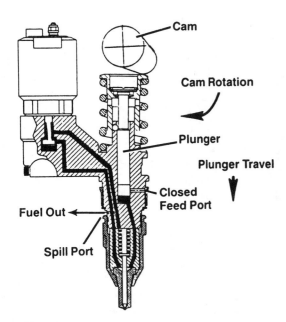

FIGURE 22-22 With spill port and feed port closed, fuel is trapped in injector. (Courtesy of Caterpillar, Inc.)

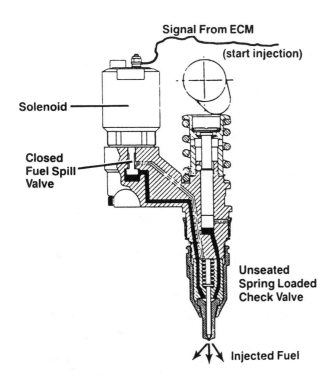

FIGURE 22-23 Injection begins as cam forces injector plunger downward. (Courtesy of Caterpillar, Inc.)

spill port to open as the spring returns the valve to its open position. With the spill valve open, fuel pressure in the injector drops, allowing spring pressure to close the needle valve in the injector tip **(Figure 22-24)**. The start of injection is determined by the ECM. The quan-

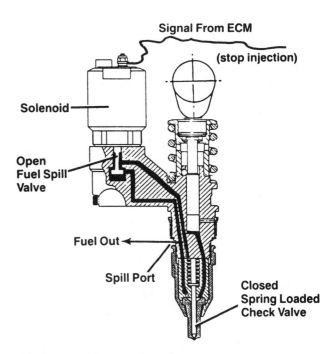

FIGURE 22-24 Injection stops when solenoid is deenergized and spill port opens. (Courtesy of Caterpillar, Inc.)

tity of fuel injected is determined by how long the ECM keeps the EUI solenoid energized. The amount of time the injector remains energized is known as the *pulse width modulated* (PWM) *time signal* **(Figure 22-25)**.

Three-Cylinder Cutout Mode

When the engine is operated at no load, high rpm, the pulse width time is of very short duration. Under these conditions the ECM does not inject fuel into cylinders

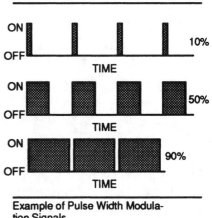

Example of Pulse Width Modulation Signals

FIGURE 22-25 (Courtesy of Caterpillar, Inc.)

4, 5, and 6. This should not be viewed as an engine problem, and it does not require servicing. As soon as more power is needed, all the cylinders again receive fuel and produce full power.

DIAGNOSTIC PROCEDURES

The diagnostic process includes the following steps (Figure 22–26):

1. Obtain as much information as possible about the problem, preferably from the driver, and the conditions under which it occurs.

2. Verify that the problem does in fact exist. If possible, repeat the conditions under which the problem occurs.

3. Determine the cause of the problem. Active codes must be repaired first. Then check any logged diagnostic codes. Use all the information from the codes, your knowledge of the system, and your experience with similar problems in the past.

4. Rather than going through all the possible test procedures, inspect and/or test the most likely cause first. Perform a careful visual inspection, paying special attention to electrical connections. Every pin and socket must be checked to ensure that a good electrical connection is present.

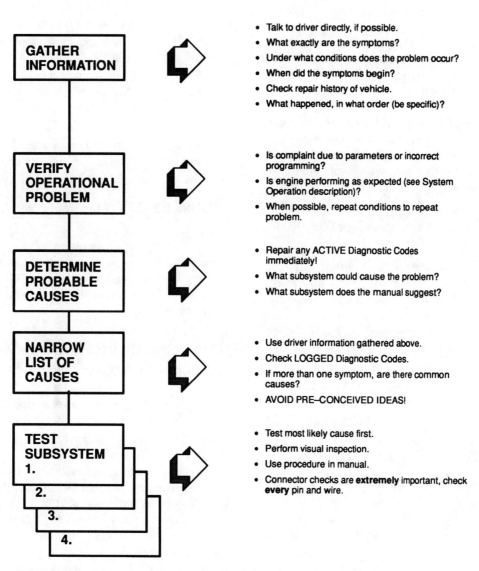

FIGURE 22–26 (Courtesy of Caterpillar, Inc.)

The following troubleshooting and diagnostic procedures are typical for a 3176B engine system. Refer to the appropriate Caterpillar service manual for specific procedures and specifications for the engine model and system being serviced.

TROUBLESHOOTING WITH A DIAGNOSTIC CODE

(Courtesy of Caterpillar, Inc.)

Active Diagnostic Codes

Diagnostic codes are used by the 3176 system to warn the vehicle operator of a problem and indicate to the service technician the nature of the problem. Some codes are used only to record an event and do not indicate problems.

An *active* diagnostic code represents a problem that should be investigated and corrected *as soon as possible*. Repairing the cause of an active code will cause the code to be cleared from the active diagnostic code screen.

When an active code is generated, the 3176 diagnostic lamp will turn on and remain on, blinking every 5 seconds. If the condition generating the fault occurs for only a brief moment, the lamp will go off after 5 seconds and the fault will be *logged*.

NOTE: After a problem has been investigated/corrected, the related diagnostic code should be cleared from memory.

Active codes may be viewed using either of the electronic service tools (ECAP or DDT) and may also be viewed using the diagnostic lamp and cruise control switches. The diagnostic lamp should not be confused with the warning lamp that is used with 3176 Caterpillar engine protection.

NOTE: Some troubleshooting procedures, such as disconnection of a sensor, may cause diagnostic codes. Make sure all diagnostic codes are noted prior to the start of the troubleshooting procedure to avoid confusion.

Using the ECAP or DDT to Display Active Codes (Figures 22-27 and 22-28)

1. With the key off, install an ECAP or DDT into the 3176 system.

2. Turn the key on (engine does not need to be started to view codes).

3176 SYSTEM FAULT CODES

Fault Code	Description
01	Override of idle shutdown throttle
25	Boost pressure sensor fault
27	Coolant temperature sensor fault
28	Check throttle sensor adjustment
31	Loss of vehicle speed signal
32	Throttle position sensor fault
34	Loss of engine rpm signal
35	Engine overspeed warning
36	Vehicle speed signal out of range
37	Fuel pressure sensor fault
41	Vehicle overspeed warning
42	Check sensor calibrations
47	Idle shutdown timed out
51	Intermittent battery power to ECM
52	ECM or personality module fault
53	ECM fault
55	No detected faults
56	Check customer or system parameters
57	Parking brake switch fault
63	Low fuel pressure warning
72	Cylinder 1 or cylinder 2 fault
73	Cylinder 3 or cylinder 4 fault
74	Cylinder 5 or cylinder 6 fault

FIGURE 22-27 (Courtesy of Caterpillar, Inc.)

3. Refer to the operating manual and special instructions for the service tool (listed under 3176 Service Tools) to read the code(s). The ECAP display menu will direct you to the proper screen to display diagnostic messages.

Using the Cruise Control Switches to Display Active Codes

1. Turn the key on (engine does not need to be started to view codes).

2. The 3176 Diagnostic Lamp will turn on for 5 seconds, blink off, turn on again for 5 seconds, then off for 5 seconds. At the end of that time, the lamp will begin to flash the first number of the two-digit code (count the flashes). After 2 seconds off, it will flash the second digit. If two or more codes are present, they will follow the first after a few seconds and be displayed in the same manner.

3. Active diagnostic codes may be displayed at any time by using cruise control switches. With the key on

> Monitor Throttle Position Sensor Signal
>
> Duty Cycle
> 37 %
>
> Valid keys are PREV SCREEN, MAIN MENU, and HELP.

- For the remote mounted sensor the DUTY CYCLE should be between 15 and 20% at the low idle stop and increase to between 80 and 85% at the high idle stop.
- For the pedal mounted sensor the DUTY CYCLE should be between 10 and 22% with the pedal released and increase to between 75 and 90% with the pedal fully depressed.

FIGURE 22-28 (Courtesy of Caterpillar, Inc.)

or the engine running, turn the cruise control ON/OFF switch to OFF, and move the SET/RESUME switch to the RESUME position. Once the codes begin to flash, the switch may be released. The 3176 Diagnostic Lamp will flash out all the codes that are currently active or intermittent codes that have occurred since the key was turned on.

Logged Diagnostic Codes

When an ECM generates a diagnostic code, it usually logs the code in permanent memory within the ECM. The time the code occurred (in hours on the internal diagnostic clock) is logged along with the code. The logged codes can then be later retrieved or erased using an ECAP or DDT service tool. They can be a valuable indicator when troubleshooting intermittent problems.

When investigating logged diagnostic codes, keep in mind:

NOTE: Some troubleshooting procedures, such as disconnection of a sensor, may cause diagnostic codes. Make sure all diagnostic codes are noted prior to the start of the troubleshooting procedure to avoid confusion.

- Some codes may be sensitive and may log occurrences that did not result in driver complaints. If the time the code was logged does not correlate with a complaint, *there may be nothing to fix.*
- *The most likely cause of an intermittent problem is connections or damaged wiring.* Next likely is the component (sensor or injector, for instance). Least likely is the ECM.
- Some codes represent "events," not failures. These codes are 71-00 Idle shutdown override (01), 71-01 Idle shutdown occurrence (47), 84-00 Vehicle overspeed warning (41), and 190-00 Engine overspeed warning (35). The following engine protection codes are events *and* faults: 100-01 Low oil pressure warning (46), 100-11 Very low oil pressure (46), 110-00 High coolant temperature warning (61), 110-11 Very high coolant temperature (61), 111-01 Low coolant level warning (62), 111-11 Very low coolant level (62). Passwords are required to clear these faults from the log.
- The ECM will automatically delete any logged diagnostic code after 100 ECM hours, except for those requiring passwords to clear (see previous paragraph).

To troubleshoot a logged diagnostic code, refer to the procedure in the manual for troubleshooting the specific code first (Troubleshooting with a Diagnostic Code). If symptoms continue, use the procedure for troubleshooting symptoms (see Troubleshooting without a Diagnostic Code).

USING THE ECAP OR DDT TO DISPLAY LOGGED CODES

With the key on or the engine running, follow the directions of the Operating Manual and Special Instructions for ECAP or DDT (listed under 3176 Service Tools) to view the codes. The ECAP menu will direct you to the proper screen to display logged diagnostic messages.

USING THE ECAP OR DDT TO CLEAR LOGGED CODES

Follow directions in the Operating Manual and Special Instructions for ECAP or DDT to clear logged codes.

The ECAP menu will direct you to the proper screen to clear logged diagnostic codes.

EUI SERVICE

Electronic Unit Injector Removal

1. Drain the fuel supply manifold to avoid fuel spillage.
2. Clean the exterior of the rocker arm cover and remove the cover.
3. Loosen the locknuts on the injector rocker arm and the valve rocker arms, and back out the adjusting screws several turns.
4. Remove the rocker arm assembly.
5. Disconnect the EUI control wiring.
6. Remove the injector hold-down.
7. Carefully pry the injector upward using a rollhead pry bar.

EUI Testing

Testing of the electronic unit injectors requires the use of the Caterpillar Pop Tester Group 1U6661 with the 1U6663 Holding Block and a 1U6665 Power Supply. Follow the test equipment instructions and the Caterpillar service manual for proper procedures and specifications.

EUI Installation

1. If the sealing area in the injector sleeve in the cylinder head is rough or grooved, the sleeve must be reworked or replaced. Follow service manual instructions for this procedure. Remove any loose carbon.
2. Use new O-ring seals on the injector body and lubricate them with clean engine oil.
3. Push the injector down into the injector sleeve by hand until it bottoms.
4. Install the bolt in the hold-down clamp, start it by hand, and tighten it to specified torque.
5. Connect the EUI wiring to the injector.
6. Install the rocker arm assembly and position the rocker arm sockets into the pushrod seats.
7. Start the rocker arm hold-down bolts by hand and then tighten them evenly to specified torque.
8. Adjust the valve rocker arms and injector rocker arm to specifications (Figure 22–29).

Injector Mechanism
(1) Rocker arm. (2) Adjusting screw. (3) Locknut.

FIGURE 22–29 (Courtesy of Caterpillar, Inc.)

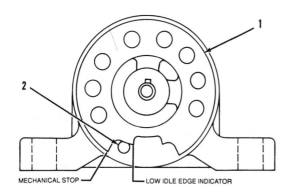

Figure 1 - Correct Low Idle Position Adjustment
(1) Rotary disc. (2) Roll pin.

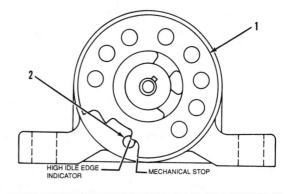

Figure 2 - Correct High Idle Position Adjustment
(1) Rotary disc. (2) Roll pin.

FIGURE 22–30 Proper TPS adjustment positions. (Courtesy of Caterpillar, Inc.)

SPEED/TIMING SENSOR REMOVAL, INSTALLATION, AND CALIBRATION

THROTTLE POSITION SENSOR TEST AND ADJUSTMENT

This requires the use of the ECAP or DDT tool with the appropriate service program module and adapters. When correctly adjusted, the remote-mounted TPS will produce a duty cycle of 15 to 20% at the low-idle position and 80 to 85% at the maximum throttle position. To correct the low-idle duty cycle adjust the pedal stop or throttle linkage until the correct duty cycle reading is achieved on the ECAP or DDT. To correct the maximum throttle duty cycle adjust the pedal stop or throttle linkage until the 80 to 85% duty cycle reading is displayed on the ECAP or DDT.

CAUTION:

Never allow the sensor roll pin stop to act as a pedal stop. This can cause the sensor shaft to break! **(Figure 22–30).**

The pedal-mounted TPS is not adjustable and is automatically calibrated by the ECM.

SPEED/TIMING SENSOR REMOVAL, INSTALLATION, AND CALIBRATION

1. Disconnect the speed/timing wiring connector at the sensor and inspect it for corrosion or damage.

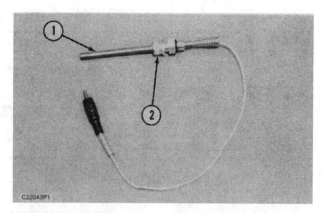

Magnetic Pick-up Sensor
(1) Magnetic pick-up sensor. (2) Adapter sleeve for magnetic pick-up.

Timing Adapter Module Installed On ECAP
(3) ECAP. (4) Timing adapter module. (5) Cable for magnetic pick-up.

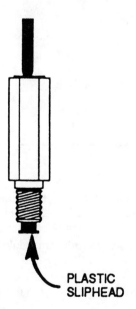

FIGURE 22–31 Engine speed/timing sensor showing plastic slip head. (Courtesy of Caterpillar, Inc.)

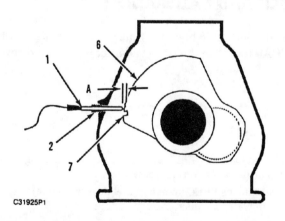

Installed Magnetic Pick-up Sensor
(1) Magnetic pick-up sensor. (2) Adapter sleeve for magnetic pick-up. (6) Counterweight surface. (7) TDC slot. (A) 1.00 mm (.040 in).

FIGURE 22–32 (Courtesy of Caterpillar, Inc.)

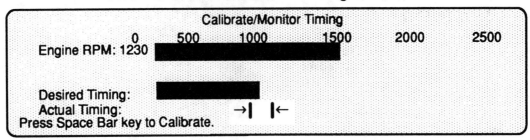

FIGURE 22-33 (Courtesy of Caterpillar, Inc.)

2. Remove the sensor (unscrew) the sensor from the timing cover.

3. Inspect the plastic end of the sensor for wear or contamination **(Figure 22-31)**.

4. Carefully pry the plastic sensor end (slip head) to the fully extended position with a screwdriver. Gently push the plastic end in to the retracted position. Resistance to movement should be firm. If there is no resistance, replace the sensor.

5. Install a new O-ring on the sensor. Pull the sensor tip out to the extended position.

6. Position the no. 1 engine cylinder at TDC.

7. Screw the sensor into the gear cover by hand and tighten to specifications.

8. Reconnect and lock the sensor wiring connector.

9. Calibrate the electronic injection timing as described.

ELECTRONIC INJECTION TIMING CHECK AND CALIBRATION

To check electronic injection timing requires the use of the ECAP tool with the timing adapter module and several other tools and adapters. The procedure includes connecting the ECAP, timing adapter module, and magnetic pickup sensor **(Figure 22-32)**. It is critical that the no. 1 piston be positioned between 10° and 90° BTDC for installation of the magnetic pickup sensor, otherwise the sensor will be damaged when the engine is started. Follow the instructions provided with the test equipment and in the Caterpillar service manual to ensure safety and accuracy. All active faults must be corrected before calibration is attempted. If the DESIRED TIMING bar is not located between the two vertical timing tolerance lines on the ECAP display, the calibration is incorrect. Causes for an incorrect display include incorrect installation of test equipment, a bent magnetic pickup sensor, a faulty timing reference ring, or incorrect calibration. To calibrate injection timing run the engine at 1500 rpm and press the SPACE key on the ECAP until timing is calibrated (about 15 seconds) **(Figure 22-33)**.

REVIEW QUESTIONS

1. Caterpillar has _____ _____ of electronic fuel injection control systems.

2. Caterpillar's PEEC system of fuel delivery and injection timing are controlled _____ .

3. A Caterpillar scroll pump has no _____ governor.

4. The rack position sensor provides information on the position of the _____ _____ _____ to the ECM to help control fuel delivery.

5. Should certain PEEC system components fail, a _____ _____ allows the vehicle to be driven to a repair facility.

6. A sealed pressure transducer module is mounted below the _____ _____ _____ _____ .

7. When a problem occurs in the PEEC system, the _____ _____ _____ stays on until the problem is corrected.

8. The 3406E/3176B electronic control system uses _____ operated, electronically controlled unit injectors.
9. The ECM is mounted on the engine and is cooled by fuel from the _____ _____ as it circulates through passages in the cooling plate.
10. The ECM adjusts fuel injection and _____ _____ in relation to changes in boost pressure.
11. The engine protection system monitors engine _____ _____, _____ _____, and _____ level.
12. The amount of time the injector remains energized is known as the _____ _____ _____ (PWM) time signal.
13. The pedal-mounted TPS is not adjustable and is _____ _____ by the ECM.

TEST QUESTIONS

1. Caterpillar has three types of engine control systems:
   a. PEEC, HEUI, and the 3176
   b. ECM, the 3176, and the 3180
   c. PEEC, the 3176, and the 3180
   d. ECM, PEEC, and the 3196
2. The Caterpillar PEEC systems of fuel delivery and injection timing are controlled
   a. mechanically
   b. electronically
   c. manually
   d. by a crankshaft system
3. The timing advance solenoid on the Caterpillar PEEC system has a
   a. standard solenoid
   b. timing control
   c. special relay
   d. brushless torque motor
4. The Caterpillar PEEC system has a vehicle speed buffer that amplifies vehicle speed output signals and provides the signal for the
   a. engine timing   c. vehicle speedometer
   b. cruise control   d. driver's convenience
5. Should the Caterpillar system fail, it will provide
   a. a driver warning signal
   b. the means to automatically stop the engine
   c. a backup system to allow the driver to get to the repair shop
   d. a code system to identify the problem
6. The Caterpillar inlet air pressure sensor is used to protect the engine during operation
   a. at high altitude
   b. at high altitude or with a restricted air cleaner
   c. with a restricted air cleaner
   d. when the ambient temperature is below −20°C
7. The ECAP tool can be used to access fault codes even with the engine
   a. shut down   c. idling
   b. running   d. running at high rpm
8. The ECM is mounted on the engine and cooled by
   a. fuel from the transfer pump
   b. coolant from the coolant system
   c. fuel from the fuel tank supply pump
   d. coolant directly from the radiator
9. The engine protection system monitors engine
   a. oil pressure   c. coolant level
   b. coolant temperature   d. all of the above
10. The amount of time the injector remains energized is known as the
    a. PWM signal   c. modulated signal
    b. PMW signal   d. open time signal

CHAPTER 23

HYDRAULIC ELECTRONIC UNIT INJECTOR (HEUI) FUEL SYSTEM

INTRODUCTION

The HEUI system is used on several makes of midsize truck engines. The system uses hydraulically actuated unit injectors that are electronically controlled. Engine lubricating oil is used as the hydraulic fluid. The engine oil pump feeds oil to an oil reservoir on top of the engine front cover. A high-pressure pump is mounted on the engine. The pump feeds high-pressure oil to oil rails machined into the cylinder heads. The injection pressure regulator valve, injection control pressure sensor (ICP), and the electronic control module (ECM) regulate pump pressure at between 500 to 3000 psi (3447 to 20,684 kPa). Excess oil is dumped through a check valve back to the oil pan (**Figure 23–1**).

PERFORMANCE OBJECTIVES

After thorough study of this chapter and the appropriate training models and service manuals, you should be able to do the following:

1. List the four major systems of the HEUI system.

2. Describe the HEUI fuel supply system operation.

3. Describe the operation of the hydraulic injection control pressure system.

4. Describe the operation of the hydraulic electronic unit injector during the fill cycle, fuel injection, and end of injection.

5. Describe the function of the HEUI electronic control system.

6. Describe the function and operation of the glow plug system.

7. Diagnose HEUI system faults using the on-board self-diagnostics and the tester connected to the diagnostic link.

TERMS YOU SHOULD KNOW

Look for these terms as you study this chapter, and learn what they mean.

HEUI	fill cycle
tandem fuel transfer pump	injection cycle
diaphragm stage	end of injection
plunger stage	DTC
hydraulic control pressure system	DLC
IPR	self-tests
ICP	KOER tests
ICP sensor	retrieve/clear continuous DTCs
HEUI solenoid	KOEO tests
poppet valve	
intensifier piston and plunger	

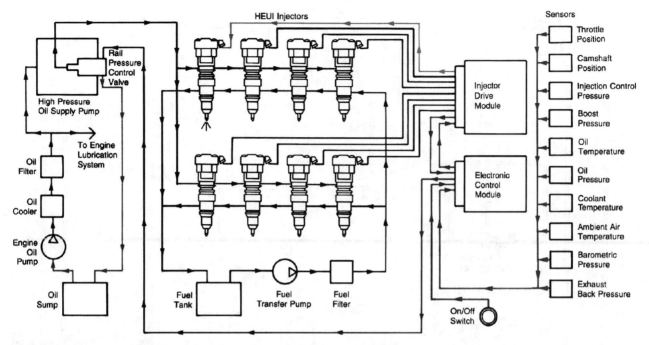

FIGURE 23-1 HEUI fuel injection system. (Courtesy of Navistar International Transportation Corporation.)

HEUI SYSTEM OVERVIEW

The HEUI fuel injection system comprises four interacting systems:

1. *Fuel supply system:* supplies fuel to the unit injectors.
2. *Hydraulic control pressure system:* provides the hydraulic pressure to activate the unit injectors.
3. *Unit injectors:* hydraulic electronic unit injectors inject atomized fuel into the combustion chambers.
4. *Electronic control system:* controls fuel metering and fuel injection timing.

HEUI FUEL SUPPLY SYSTEM COMPONENTS AND OPERATION

The HEUI fuel system fuel tank, fuel strainer, and supply line provide fuel to the tandem fuel transfer pump. (See the fuel flow diagram in **Figure 23-2**.) Fuel flows from the fuel tank through the fuel strainer and fuel line to the engine camshaft-operated tandem fuel transfer pump. The tandem transfer pump consists of a diaphragm stage and a piston stage. The diaphragm stage pumps the fuel to the primary fuel filter at 4 to 6 psi (28 to 41 kPa) **(Figure 23-3)**. Air trapped in the primary filter is vented through a screened orifice back to the fuel tank via the fuel return line. In some applications an electric heating element is located in the base of the filter housing to keep the fuel at a temperature above the wax formation level. A water trap and sensor may be used to illuminate a lamp on the instrument panel to indicate that water should be drained from the trap.

From the filter, fuel flows to the piston stage of the transfer pump The piston stage of the pump raises fuel pressure to 40 psi (276 kPa). From there fuel flows through fuel lines to drilled passages in the cylinder heads to provide fuel to the injectors. Excess fuel from the fuel galleries is routed to the pressure regulator on the side of the filter housing. The pressure regulator contains a spring-loaded spool valve to control fuel supply pressure at 40 psi (276 kPa). Return fuel is warmed by cylinder head heat and is returned to the fuel tank **(Figure 23-4)**.

HYDRAULIC INJECTION CONTROL PRESSURE SYSTEM

The HEUI system uses hydraulic pressure to actuate the unit injectors. In this system engine oil is pumped into a reservoir by the engine oil pump. The reservoir is mounted on top of the front cover of the en-

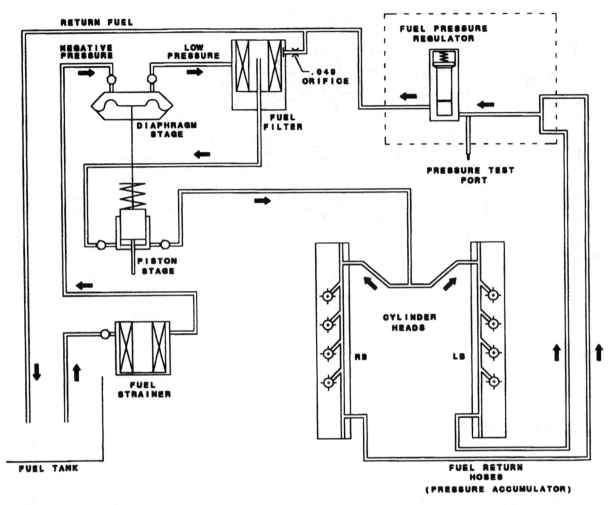

FIGURE 23-2 HEUI fuel system schematic showing fuel flow. (Courtesy of Navistar International Transportation Corporation.)

gine **(Figure 23-5)**. A high-pressure hydraulic pump takes oil from the reservoir and delivers it to the oil rails in the cylinder heads. The pump is a nine-plunger swash plate design that produces very high hydraulic pressure. Pressure is controlled by the rail pressure control valve (RPCV) **(Figures 23-6 and 23-7)**, programmed information in the electronic control module (ECM), and signals from the injection control pressure sensor (ICPS). Injection control pressure ranges from 500 to 3000 psi (3447 to 20,685 kPa). The regulator valve dumps excess oil through a check valve back to the engine oil sump. The ICP sensor is a ceramic disc pressure sensor that converts pressure to an analog voltage signal ranging from 0 to 5 V. The ECM uses this signal and programmed information to determine rail pressure. A pressure relief valve dumps excess pressure into the front cover in the event that pressure exceeds 4000 psi (27,580 kPa).

RAIL PRESSURE CONTROL VALVE OPERATION

The components of the electronically controlled rail pressure control valve (RPCV) are shown in **Figure 23-6**. The RPCV regulates pump output pressure between 450 and 3000 psi (3103 and 20,685 kPa).

With the engine off, the return spring holds the valve spool to the right, and the drain ports are closed. Around 1500 psi (10,342 kPa) is required to start a warm engine. A cold engine requires 3000 psi (20,685 kPa) of oil pressure to start. An electronic signal from the ECM causes the solenoid to generate a magnetic field, which pushes the armature to the right against the pin and poppet, holding it closed while pressure builds in the spool chamber. Spool chamber pressure and spool spring force hold the spool to the right, closing the drain ports. All oil is

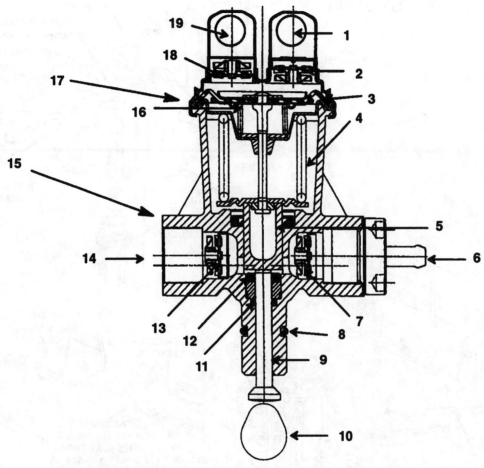

1. Fuel Vacuum Inlet
2. Inlet Check Valve
3. Diaphragm
4. Spring
5. Piston
6. Low Pressure Inlet
7. Inlet Check Valve
8. O–Ring
9. Tappet
10. Camshaft Lobe
11. Oil Seal
12. Fuel Seal
13. Outlet Check Valve
14. High Pressure Outlet
15. Piston Stage
16. Spring
17. Diaphragm Stage
18. Outlet Check Valve
19. Low Pressure Outlet

FIGURE 23–3 HEUI tandem fuel supply pump. (Courtesy of Navistar International Transportation Corporation.)

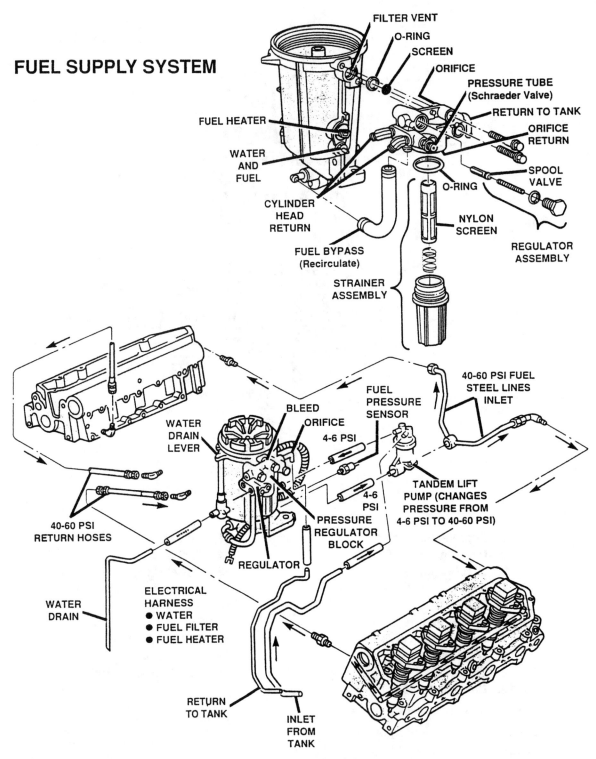

FIGURE 23–4 Fuel supply system components. (Courtesy of Ford Motor Company.)

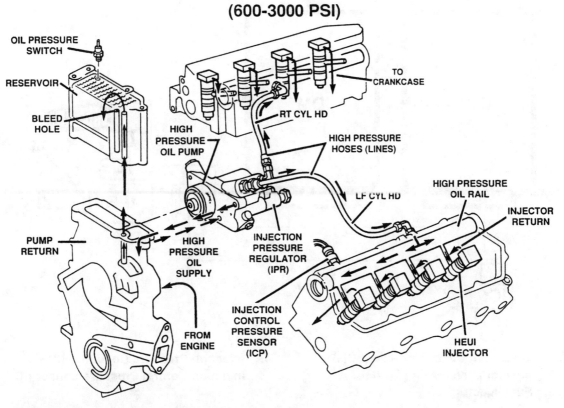

FIGURE 23-5 (Courtesy of Ford Motor Company.)

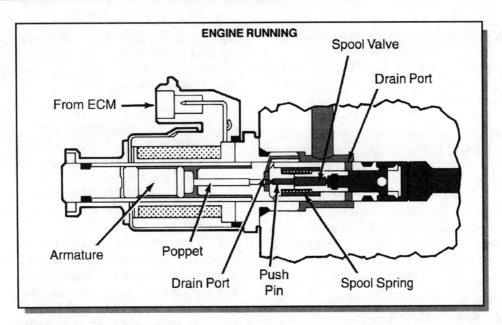

FIGURE 23-6 Cross section of rail pressure control valve (RPCV). (Courtesy of Navistar International Transportation Corporation.)

541

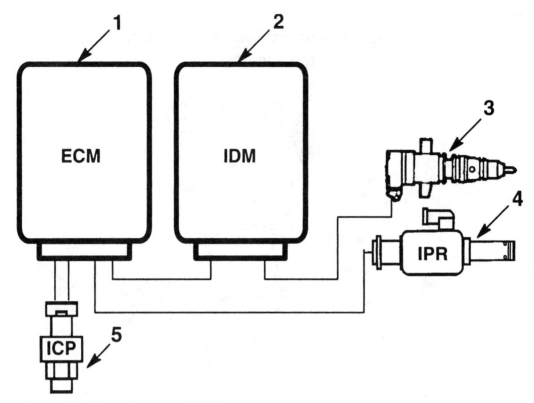

1. Electronic Control Module (ECM)
2. Injector Driver Module (IDM)
3. Fuel Injector
4. Injection Pressure Regulator (IPR)
5. Injection Control Pressure Sensor (ICP)

FIGURE 23-7 (Courtesy of Navistar International Transportation Corporation.)

sent to the pressure rail until the required pressure is attained.

As soon as the engine starts, the ECM signals the RPCV to provide the required rail pressure. The ECM compares the actual rail pressure to the required rail pressure and adjusts the ECM signal to the RPCV to obtain the required rail pressure. The spool position determines how far the drain ports are open and how much oil is bled off from the pump outlet, thereby controlling rail pressure. Poppet position is controlled by the strength of the solenoid magnetic field on signal from the ECM. Spool position is determined by a balance between forces on the left and right sides of the spool. Pump outlet pressure is infinitely variable between 450 and 3000 psi (3103 and 20,685 kPa).

HYDRAULIC ELECTRONIC UNIT INJECTOR COMPONENTS

Fuel is supplied to the injectors by passages drilled into the cylinder head. These passages intersect with fill ports in the injectors. Fuel transfer pump pressure fills the area under the injector plunger. The hydraulic electronic unit injector consists of the following major components (**Figure 23-8**):

1. *Solenoid:* a fast-acting electromagnet that, when energized, pulls the poppet valve off its seat against spring pressure.

2. *Poppet valve:* The poppet valve is lifted off its seat when the injector solenoid is energized by the ECM. This closes off the drain and opens the inlet for high hydraulic pressure to enter the area above the intensifier piston.

3. *Intensifier piston and plunger:* When the injector solenoid is energized, high hydraulic pressure acts on the top of the intensifier piston, pushing it and the plunger down. Because the surface area of the plunger top is seven times greater than that of the plunger, a sevenfold force multiplication is produced and transmitted to the fuel under the plunger. This highly pressurized fuel lifts the nozzle valve off its seat and allows fuel to be forced out of the nozzle orifices.

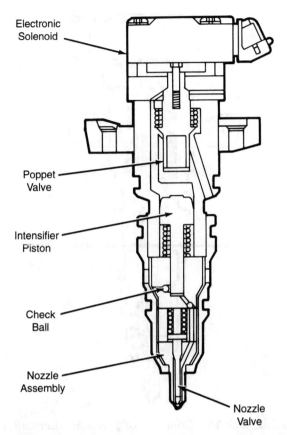

FIGURE 23-8 (Courtesy of Navistar International Transportation Corporation.)

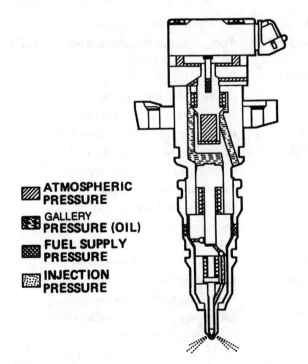

FIGURE 23-9 (Courtesy of Navistar International Transportation Corporation.)

FUEL INJECTOR OPERATION

4. *Nozzle assembly:* consists of a conventional inward-opening needle valve and seat. Spring pressure keeps the valve seated until fuel pressure lifts it off its seat during injection. The valve controls the opening and closing of the nozzle orifices. A check ball that closes during injection (due to fuel pressure) prevents the escape of trapped fuel. After injection the check valve opens to allow the plunger cavity to fill with fuel.

FUEL INJECTOR OPERATION

The three stages of injector operation are **(Figure 23-9)** (1) the fill cycle, (2) the fuel injection cycle, and (3) end of injection.

FILL CYCLE

During the fill cycle the poppet valve is closed, blocking off hydraulic pressure to the injector. The plunger and intensifier are kept at the top of their travel limit by spring pressure, and the plunger cavity is full of fuel at supply system pressure of 40 psi (276 kPa).

FUEL INJECTION CYCLE

When the ECM determines that injection should take place, the following events occur:

1. The ECM signals the IDM to energize the injector solenoid (115 V).
2. Energizing the solenoid pulls the poppet valve off its seat.
3. A land on the upper part of the poppet valve closes off the path to the drain.
4. A land on the lower part of the poppet valve opens the poppet chamber to high hydraulic pressure.
5. High hydraulic pressure flows into the area on top of the intensifier piston.
6. Pressure on top of the intensifier piston forces it down along with the plunger.
7. The resulting high-fuel pressure lifts the nozzle valve off its seat and allows fuel to be forced out of the nozzle openings **(Figure 23-10)**.

END OF INJECTION

The end of injection occurs when the ECM and IDM end the electrical signal to the injector solenoid, triggering the following events:

1. Spring tension seats the poppet valve. This opens the path to the drain and closes off the poppet chamber to hydraulic pressure.

544 Chapter 23 HYDRAULIC ELECTRONIC UNIT INJECTOR (HEUI) FUEL SYSTEM

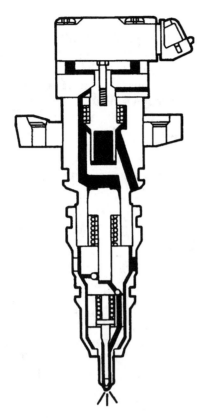

FIGURE 23–10 (Courtesy of Navistar International Transportation Corporation.)

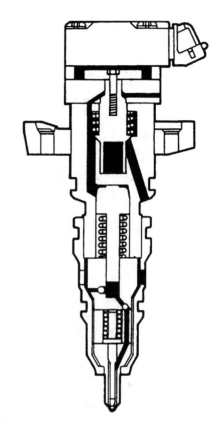

FIGURE 23–11 (Courtesy of Navistar International Transportation Corporation.)

2. Oil in the intensifier and poppet chambers flows out through vents in the poppet valve sleeve and out the adapter drain hole.

3. When oil pressure in the area above the intensifier drops below the pressure under the plunger, the plunger returns to its upper position, the nozzle valve seats, closing the orifices, and injection stops **(Figure 23–11)**.

ELECTRONIC CONTROL SYSTEM

The ECM continuously monitors powertrain operation. When a fault is detected, a diagnostic trouble code (DTC) is stored in the ECM memory. The code remains stored even when the fault no longer exists. These faults must be retrieved with the use of a special tester connected to the data link connector (DLC). Retrieving these faults helps in the diagnosis of intermittent faults. The ignition key–activated self-tests can retrieve current fault codes only.

The electronic control system typically includes the ECM, wiring harness, DLC, and the following input sensors, switches, and outputs.

ACC	Air conditioner clutch switch
APS (TPS)	Accelerator pedal position sensor (throttle position sensor)
BOO	Brake ON/OFF switch
BPA	Brake pedal applied switch
BPS	Barometric pressure sensor
CCC	Converter clutch control switch
CEL	Check engine light (yellow)
CID	Cylinder identification sensor
CMP	Camshaft position sensor
CPP	Clutch pedal position switch
DLC	Data link connector, SAE J1587, ATA compatible
EBP	Exhaust back-pressure sensor
ECM	Electronic control module
EOT	Engine oil temperature sensor
EPC	Electronic pressure control sensor
GPR	Glow plug relay switch
HEUI	Hydraulic electronic unit injector
IAT	Intake air temperature sensor
ICP	Injection control pressure sensor

IDM	Injection driver module
IVS	Idle validation switch
MAP	Manifold absolute pressure sensor
SCCS	Speed control command switches
SEL	Stop engine light (red)
SS1	Shift solenoid 1
SS2	Shift solenoid 2
STO	Self-test output
TAC	Tachometer
TCIL	Transmission control indicator lamp
TCS	Transmission control switch
TFT	Transmission fluid temperature
TR	Transmission range

Data Storage

The ECM is able to store the following data:

NOTE: Features used vary with engine make and model.

1. Total fuel consumed
2. Total idle fuel consumed
3. Total idle time
4. Total PTO time
5. Total miles (kilometers) driven
6. Total engine hours
7. Maintenance data
8. Diagnostic data

These data can be retrieved with the use of service tools that are compatible with the SAE J1587 data link.

Parameters that are customer programmable within the range specified by the engine manufacturer include the following:

1. Cruise control
2. Vehicle speed limiting
3. Customer password protection
4. Engine and maintenance monitoring
5. Idle-speed control
6. Idle shutdown timer (with override)
7. Engine brake control
8. PTO control

GLOW PLUG SYSTEM

The glow plug system warms up the combustion chambers to improve cold-engine starting and reduce exhaust emissions during engine warm-up. The ECM is programmed to turn on the WAIT to start light and energize the glow plugs through the relay each time the ignition switch is turned on. The ECM uses information from the EOT (engine oil temperature) sensor and the BARO (barometric) sensor to determine the length of the WAIT period and the glow plug ON time separately. When the engine oil temperature is low or the barometric pressure is low (as at high altitude), the glow plug ON time is longer. If the battery voltage is high, the ECM modulates the glow plugs by turning the relay ON/OFF at programmed intervals to protect the glow plugs. When the ECM turns the WAIT light off, the engine is ready to start. However, the glow plugs may remain on for up to 120 seconds after the engine starts to reduce exhaust emissions during engine warm-up (**Figure 23–12**).

HEUI SYSTEM DIAGNOSIS AND SERVICE

CAUTION:

Very high voltage is used to energize the injectors. This voltage is high enough to cause serious injury or possibly death if contacted. Always follow service manual procedures to avoid injury or damage.

CAUTION:

Very high hydraulic pressures (500 to 3000 psi/3447 to 20,685 kPa) are used to actuate the injectors. Always follow service manual procedures when servicing this system to avoid injury or damage.

Checking Electrical Connectors

All HEUI system electrical connectors are protected from environmental damage and corrosion. Never puncture or pierce the insulation on any wiring, because this will allow moisture to enter and cause corrosion. Disconnect any suspected connector in the manner prescribed in the service manual. Look for damaged locking tabs or screws, and dirty, bent, or broken pins or sockets. Use only the recommended solvent and tools to clean and straighten faulty items. Replace any badly damaged wiring or connectors. Reconnect the wiring and make sure the connectors are securely locked and environmentally protected as intended. Recheck system operation to ensure the repair is successful.

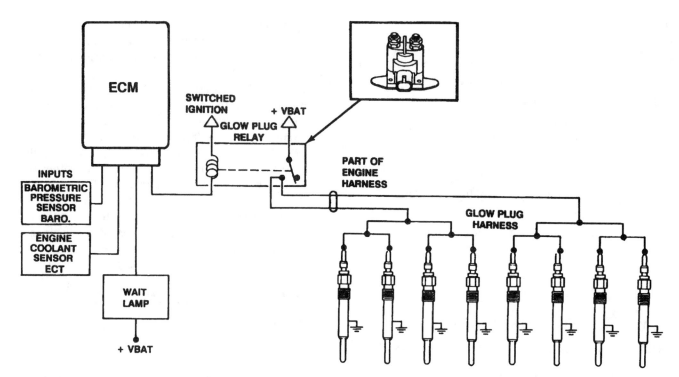

FIGURE 23–12 Glow plug system. (Courtesy of Navistar International Transportation Corporation.)

1. Check Engine/Diagnostic Lamp

(Items 1 through 4 appear here courtesy of Caterpillar, Inc. and apply to the 3100 series engine.)

The check engine/diagnostic lamp (amber color) that is located on the truck dashboard is used as a diagnostic lamp to communicate status or operation situations of the electronically controlled engine and control system.

When the ignition key is first turned ON, the check engine/diagnostic lamp illuminates for 5 seconds (as a lamp check). The lamp also illuminates and blinks every 5 seconds whenever the ECM detects an active fault or condition. If the lamp comes on and continues to blink every 5 seconds after initial start-up, the ECM has detected a system problem.

The check engine/diagnostic lamp is also used for the idle shutdown timer. Ninety seconds before the programmed idle time expires, the lamp starts to flash at a rapid rate. The electronically controlled engine will shut down after the 90-second interval. To disable the idle shutdown timer, the clutch pedal or service brake pedal must be depressed during this final 90 seconds (check engine/diagnostic lamp flashing). Diagnostic code 01, override of idle shutdown timer, will be logged if the electronic control module (ECM) has been programmed to allow override of the timer.

Some codes record events or indicate that a mechanical system needs attention rather than logging an electronic system problem. Troubleshooting is not required for codes 35, 41, 47 and 55. Code 01 will not flash out. Some codes will limit operation or performance of the engine while the code is active (check engine/diagnostic lamp is flashing). Acceleration rates with these codes may be significantly slower **(Figure 23–13)**.

2. Operation with Intermittent Diagnostic Codes

If the check engine/diagnostic lamp illuminates during engine operation and then shuts off, an intermittent fault may have occurred. If this occurs, the fault will be logged into the permanent memory of the ECM.

In most cases it is not necessary to stop the engine and vehicle because of an intermittent code, however, the operator should retrieve the code(s) and refer to the chart **(Figure 23–13)** to identify the nature of the event.

NOTE: Figure 23–13 indicates the potential effect on engine performance with active diagnostic codes. The circumstances (low power, vehicle/engine speed limits, excessive smoke, etc.) involved during the time the lamp was on should be noted or documented by the operator for future reference and to help troubleshoot the situation.

Diagnostic Flash Code	Effect on Engine Performance				Suggested Driver Action		
	Engine Misfire	Low Power	Engine Speed Reduced	Engine Shutdown	Shut Down Vehicle	Service ASAP	Schedule Service
01—Idle shutdown override							
15—Injection actuation pressure sensor fault	X	✓	✓			✓	
17—Excessive injection actuation pressure fault[b]		✓	✓			✓	
18—Injection actuation pressure control valve driver fault	X	X	X	X		✓	
19—Injection actuation pressure system fault	X	✓	✓			✓	
21—Sensor supply voltage fault[a,b]		✓				X	✓
25—Boost pressure sensor fault[a]		X					✓
27—Coolant temperature sensor fault[a,b]	X					X	✓
28—Check throttle sensor adjustment			✓			✓	
31—Loss of vehicle speed signal			✓				✓
32—Throttle position sensor fault			✓			✓	
34—Engine rpm signal fault				X		X	X
35—Engine overspeed warning							
36—Vehicle speed signal fault			✓				✓
38—Inlet air temperature sensor fault[a,b]	✓	✓					✓
41—Vehicle overspeed warning							
42—Check sensor calibrations		X					✓
47—Idle shutdown occurrence				✓			
49—Inlet air heater driver fault[a]							✓
51—Intermittent battery power to ECM	✓			X		✓	
55—No detected faults							
56—Check customer/system parameters		X	X			X	✓
59—Incorrect engine software				✓		✓	
61—High/very high coolant temperature warning		X	X			✓	
71—Cylinder 1 fault	✓	✓				✓	
72—Cylinder 2 fault	✓	✓				✓	
73—Cylinder 3 fault	✓	✓				✓	
74—Cylinder 4 fault	✓	✓				✓	
75—Cylinder 5 fault	✓	✓				✓	
76—Cylinder 6 fault	✓	✓				✓	

Note:—An "X" indicates that the effect on engine performance *will* occur if the code is active. A ✓ (check mark) indicates that the effect on engine performance *may* occur if the code is active, depending on the exact failure.

[a]These diagnostic flash codes may affect the system only under specific environmental conditions, such as engine start-up at cold temperature, cold-weather operation at high altitudes, etc.

[b]These diagnostic flash codes reduce the effectiveness of the engine monitoring feature when active.

Shut Down Vehicle: Drive the vehicle cautiously off the road and get immediate service. Severe engine damage may result.
Service ASAP (As Soon As Possible): Go to the nearest qualified service location.
Schedule Service: Investigate the problem when convenient.

FIGURE 23–13 Diagnostic codes and their relationship to engine performance for the 3100 series Caterpillar engine. (Courtesy of Caterpillar, Inc.)

3. Engine Operation with Active Diagnostic Codes

If the check engine/diagnostic lamp stays on during normal engine operation, first check to ensure that the engine has proper oil pressure, coolant temperature, and coolant level. If the engine has the proper oil pressure and the check engine/diagnostic lamp is on, this indicates that the system has identified a situation that is not within specification.

To retrieve the code(s), refer to the following procedure and to **Figure 23–13,** and follow the Suggested Driver Action (i.e., Shut Down Engine/Vehicle, Service ASAP or Schedule Service) provided.

The active code should be investigated and corrected as soon as possible. Repairing the cause of the code will shut the lamp off (if there is only one active code).

4. Diagnostic Flash Code Retrieval

Use the following procedure to retrieve diagnostic codes with the check engine/diagnostic lamp that have occurred since the engine has started.

1. Turn the cruise control ON/OFF switch to the OFF position.

2. Move the SET/RESUME switch to either position and hold that position until the lamp begins to flash.

The check engine/diagnostic lamp will flash to indicate a two-digit flash code, and the cruise control switch may be released. The sequence of flashes represents the system diagnostic message. Count the first sequence of flashes to determine the first digit of the diagnostic code. After a 2-second pause, the second sequence of flashes will identify the second digit of the diagnostic code.

Any additional diagnostic codes will follow (after a pause) and will be displayed in the same manner. Flash code 55 signals that no detected faults have occurred since the ignition key switch was turned on.

NOTE: This procedure can be used to determine **active** codes (a constantly flashing lamp indicates an active code) or any codes occurring since the engine was started (but not currently active). It cannot be used to get **logged** codes occurring before the ignition switch was turned on (e.g., from yesterday or a week ago). An electronic service tool is necessary to retrieve these codes. For further information or assistance with repairs, refer to Truck Engine Test Procedures and/or the service manual for troubleshooting, or contact an authorized Caterpillar dealer.

REVIEW QUESTIONS

1. The HEUI system is used on several _____ _____ _____ engines.
2. The system uses _____ actuated injectors that are _____ controlled.
3. Engine lubricating oil is used as the _____ _____.
4. Name and describe the four interacting systems of the HEUI system.
5. The HEUI fuel system, fuel tank, fuel strainer, and supply line provide fuel to the _____ _____ _____ pump.
6. The HEUI system uses a _____ _____ to actuate the injectors.
7. Name the three stages of fuel injector operation.
8. The ignition key–activated self-test can retrieve only _____ _____ _____.
9. Explain the purpose of the glow plugs.
10. Explain the operation of a glow plug.

TEST QUESTIONS

1. The HEUI system used on several diesel engines uses hydraulically actuated injectors that are controlled
   a. mechanically
   b. by air pressure
   c. electronically
   d. by a high-pressure main pump

2. Excess oil pressure is dumped through the check valve back to the
   a. oil pan   c. fuel supply system
   b. fuel tank   d. overflow valve

3. The electronic control module regulates pump pressure between
   a. 600 and 1000 psi
   b. 600 and 3000 psi
   c. 100 and 500 psi
   d. 6000 and 30,000 psi
4. The hydraulic control pressure system provides the hydraulic pressure to
   a. supply the main pressure pump
   b. help the electrical solenoid
   c. keep oil at a specific pressure
   d. activate the unit injectors
5. The HEUI system injects atomized fuel into the combustion chamber by using
   a. hydraulic electronic unit injectors
   b. a fuel rail
   c. mechanical electronic unit injectors
   d. mechanical unit injectors
6. Fuel from the fuel filter flows
   a. to the transfer pump
   b. to the piston stage of the transfer pump
   c. to the fuel rail
   d. directly to the unit injector
7. The piston stage of the pump raises pressure to
   a. 10 psi
   b. 40 to 60 psi
   c. 40 psi
   d. 120 psi
8. Fuel is supplied to the injectors by
   a. steel fuel lines
   b. high-pressure flexible lines
   c. passages drilled in the cylinder block
   d. passages drilled in the cylinder heads
9. The stage(s) of injector operation is (are)
   a. fill cycle
   b. fuel injection cycle
   c. end of injection
   d. all of the above
10. The glow plugs must remain on for up to 120 seconds after the engine starts to
    a. reduce exhaust emissions during warm-up
    b. avoid glow plug damage
    c. assist in fuel consumption
    d. allow better cold-start performance

♦ CHAPTER 24 ♦

AMBAC INTERNATIONAL DISTRIBUTOR PUMPS

INTRODUCTION

AMBAC International single-plunger distributor pumps of the PS type include the PSJ, PSU, and Model 100 pumps. These pumps are compact units designed particularly for applications with relatively broad speed requirements such as military vehicles, medium- and heavy-duty trucks, farm equipment, and tractors. In addition to its small size, other design features are ease of servicing, low cost, and simplified calibration. All pumps are designed for flange mounting to the engine **(Figures 24–1 to 24–4)**.

PERFORMANCE OBJECTIVES

After thorough study of this chapter and the appropriate training models and service manuals, and with the necessary tools and equipment, you should be able to do the following with respect to diesel engines equipped with AMBAC International distributor-type injection pumps:

1. Describe the basic construction and operation of the AMBAC International distributor injection pumps, including the following:
    a. method of fuel pumping and metering
    b. delivery valve action
    c. lubrication
    d. injection timing control
    e. governor types

2. Diagnose basic fuel injection system problems and determine the correction needed.

3. Remove, clean, inspect, test, measure, repair, or replace fuel injection system components as required to meet manufacturer's specifications.

TERMS YOU SHOULD KNOW

Look for these terms as you study this chapter, and learn what they mean.

- single-plunger distributor pump
- reciprocating and rotating plunger
- four-lobe cam
- three-lobe cam
- head cavity
- inlet ports
- outlet ports
- distributing slot
- metering sleeve
- metering hole
- shut-off position
- delivery valve
- externally mounted timing mechanism
- Intravance
- governor friction clutch
- droop screw
- torque cam
- aneroid
- excess-fuel starting device
- electronically controlled distributor pump
- pump flushing
- pump calibration
- port closure
- cam nose angle

(This chapter is based on and contains information provided courtesy of AMBAC International Corporation.)

FIGURE 24–1 Model 100 pump. (Courtesy of AMBAC International Corp.)

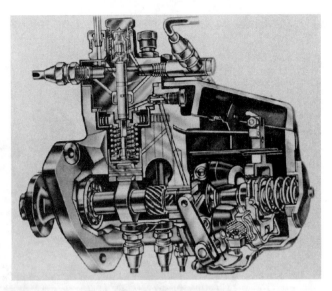

FIGURE 24–2 PSB pump cross section. (Courtesy of AMBAC International Corp.)

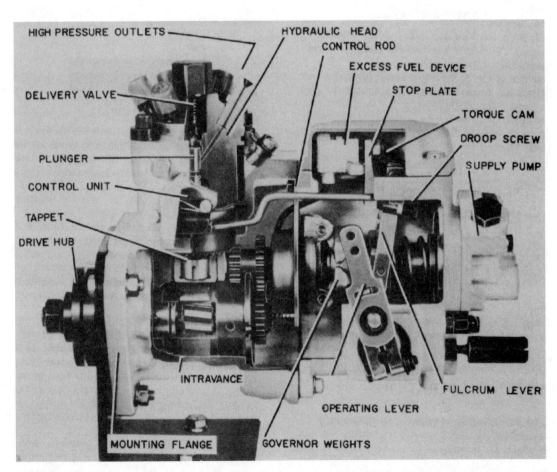

FIGURE 24–3 Partial cutaway view of Model 100 distributor pump. (Courtesy of AMBAC International Corp.)

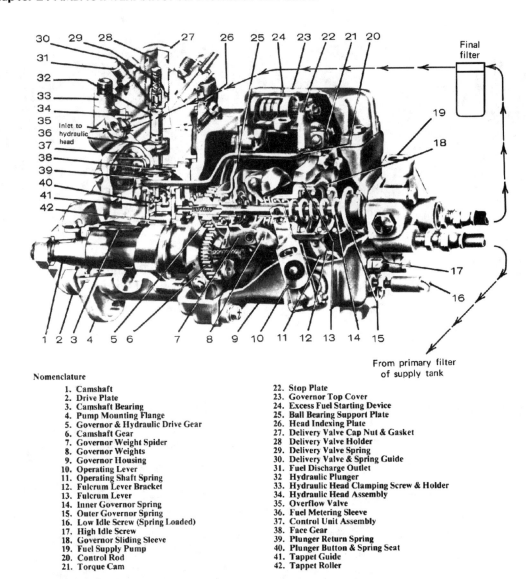

FIGURE 24-4 Model 100 pump components. (Courtesy of AMBAC International Corp.)

Nomenclature

1. Camshaft
2. Drive Plate
3. Camshaft Bearing
4. Pump Mounting Flange
5. Governor & Hydraulic Drive Gear
6. Camshaft Gear
7. Governor Weight Spider
8. Governor Weights
9. Governor Housing
10. Operating Lever
11. Operating Shaft Spring
12. Fulcrum Lever Bracket
13. Fulcrum Lever
14. Inner Governor Spring
15. Outer Governor Spring
16. Low Idle Screw (Spring Loaded)
17. High Idle Screw
18. Governor Sliding Sleeve
19. Fuel Supply Pump
20. Control Rod
21. Torque Cam
22. Stop Plate
23. Governor Top Cover
24. Excess Fuel Starting Device
25. Ball Bearing Support Plate
26. Head Indexing Plate
27. Delivery Valve Cap Nut & Gasket
28. Delivery Valve Holder
29. Delivery Valve Spring
30. Delivery Valve & Spring Guide
31. Fuel Discharge Outlet
32. Hydraulic Plunger
33. Hydraulic Head Clamping Screw & Holder
34. Hydraulic Head Assembly
35. Overflow Valve
36. Fuel Metering Sleeve
37. Control Unit Assembly
38. Face Gear
39. Plunger Return Spring
40. Plunger Button & Spring Seat
41. Tappet Guide
42. Tappet Roller

full-load delivery
high idle
peak torque
governor cutoff
low idle
internal leakage check
static timing
flow timing

PS MODEL PUMP OPERATION

The prominent feature of the PS series is the single precision plunger, which is actuated by the pump camshaft. It reciprocates for pumping action and rotates continuously for distribution of fuel via discharge outlets to individual cylinders of the engine. The precise rotating and reciprocating action of the single plunger distributes a uniform volume of fuel per cylinder for smooth power and fuel economy. There is no need to make fuel delivery adjustments for any one cylinder.

The plunger is lapped and fitted into the head and metering sleeve, thereby making these three components an inseparable assembly. The lower extension of the plunger is specially machined for positioning with the plunger face gear by a drive that fixes the angular position of the plunger in relation to the face gear (**Figure 24-5**).

The plunger makes one complete revolution for every two revolutions of the pump camshaft, which is operating at engine crankshaft speed (**Figures 24-6 and 24-7**). The plunger rotates continuously while moving vertically through the pumping cycle; therefore, on an eight-cylinder engine, the four-lobe cam ac-

PS MODEL PUMP OPERATION

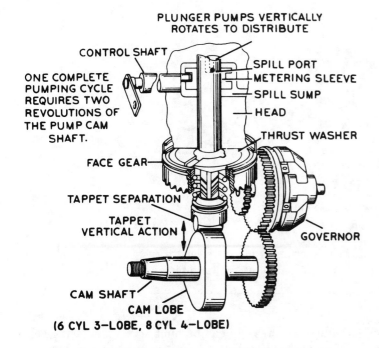

FIGURE 24-5 (Courtesy of AMBAC International Corp.)

tuates the plunger eight times for every two revolutions of the pump camshaft. On a six-cylinder engine, the three-lobe cam actuates the plunger six times for every two revolutions of the pump camshaft.

Pump Lubrication

All PS series pumps are engine-oil lubricated. Internal ducting allows an adequate amount of lubricating oil to be supplied under pressure to the tappet assembly, camshaft bushings, thrust washer faces, and governor shaft. All other parts are splash-lubricated. The internal timing device (Intravance) is actuated by lubricating oil pressure as is the excess-fuel starting device. Lubricating oil is returned to the crankcase through the pump housing and the external timing device, when used.

Fuel Flow

Fuel enters the pump from the supply system through the pump housing inlet connection. This action fills the sump area and the head cavity between the top of the plunger and the bottom of the delivery valve when the plunger is at the bottom of its stroke (**Figures 24–6** and **24–7**).

As the continuously rotating plunger moves upward under cam action, it closes two horizontal galleries, which contain the inlet ports. This traps the fuel and builds up pressure until the spring-loaded delivery

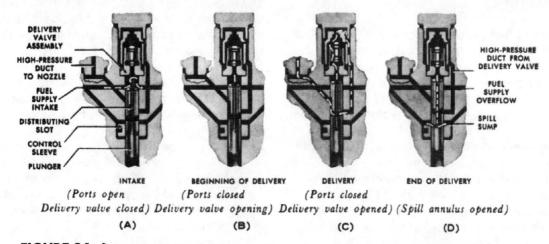

FIGURE 24-6 Details of distributor pumping action and high-pressure passage to cylinder. (Courtesy of AMBAC International Corp.)

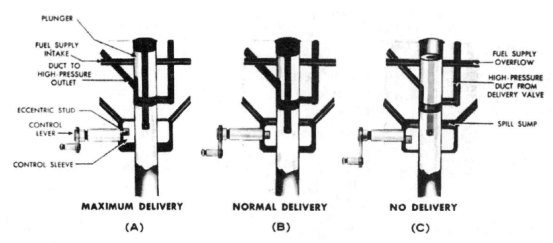

FIGURE 24-7 Position of sleeve (up or down) determines effective length of plunger stroke (metering) and therefore the amount of fuel delivered to the injector. (Courtesy of AMBAC International Corp.)

valve is caused to open. Further movement of the plunger forces the fuel through the delivery valve and is carried through the intersecting duct to the annulus and distributing slot in the plunger. The vertical distributing slot on the plunger connects to the outlet duct, which is then registered as the plunger rotates.

The rotary and vertical movement of the plunger are so placed in relation to the outlet ports that the vertical distributing slot overlaps only one outlet duct during the effective portion of each stroke.

After sufficient vertical movement of the plunger, its metering port passes the edge of the metering sleeve, and the fuel under pressure is forced down the center of the plunger and into the sump surrounding the metering sleeve. This completes a pumping cycle for one cylinder.

Fuel Metering

The quantity of fuel delivered per stroke is controlled by the position of the metering sleeve in relation to the fixed port closing position. That is the point at which the top of the plunger covers the top of the horizontal galleries or *fill ports*, as they are called **(Figure 24-7)**.

As the horizontal metering hole, during the plunger stroke, passes over the top edge of the metering sleeve, pumping pressure is relieved down through the center hole of the plunger and out into the sump surrounding the metering sleeve. Fuel delivery terminates despite the continued upward movement of the plunger.

When the metering sleeve is in its lowest position, the metering hole in the plunger is uncovered by the top edge of the sleeve before the upper end of the plunger can cover the horizontal galleries. In this condition, no pressure can be built up even after the galleries are covered. As a result, no fuel can be delivered. This is the shut-off position.

If the metering sleeve is moved to midposition, the hole in the plunger is uncovered later in the stroke by the sleeve; hence, the effective stroke of the plunger is longer and fuel is delivered. If the metering sleeve position is raised further, the horizontal hole in the plunger remains covered by the sleeve until relatively late in the plunger stroke, thereby increasing the effective stroke of the plunger and fuel delivery.

To sum up, the upward movement of the metering sleeve increases, and its down movement decreases the quantity of fuel pumped per stroke.

Delivery Valve Operation

The delivery valve assembly is located directly above the plunger. Its function is to assist the injection of fuel by preventing irregular loss of fuel from the delivery to the supply side of the system between pumping strokes **(Figure 24-8,** item 14).

The delivery valve assembly consists of a valve with a conical seat and a mating valve body. Opening pressure is controlled by the delivery valve spring that sits on top of the valve.

Pressure is created after the plunger on the upward stroke closes the horizontal galleries (scallops), which form the inlet ports. When the pressure overcomes the force of the valve spring, the valve opens and fuel under pressure flows through the distributing passage into the injection tubing.

When the horizontal hole in the plunger passes the top edge of the metering sleeve, there is a sudden drop

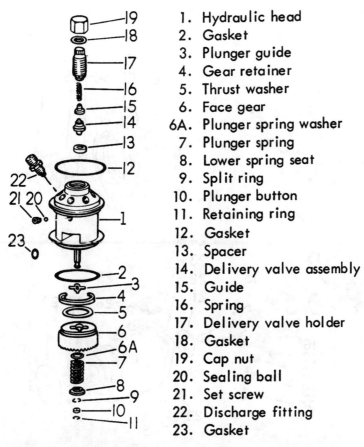

FIGURE 24–8 Hydraulic head components. (Courtesy of AMBAC International Corp.)

1. Hydraulic head
2. Gasket
3. Plunger guide
4. Gear retainer
5. Thrust washer
6. Face gear
6A. Plunger spring washer
7. Plunger spring
8. Lower spring seat
9. Split ring
10. Plunger button
11. Retaining ring
12. Gasket
13. Spacer
14. Delivery valve assembly
15. Guide
16. Spring
17. Delivery valve holder
18. Gasket
19. Cap nut
20. Sealing ball
21. Set screw
22. Discharge fitting
23. Gasket

in fuel pressure below the delivery valve; the force of the valve spring, combined with the differential in pressure, starts to return the valve to its seat.

As the valve starts down into its body, the lower edge of the retraction piston enters the valve bore and, at that moment, the flow of fuel to the bore in the head stops. Further downward movement of the piston increases the volume on the high-pressure side by the amount of piston movement (its displacement volume) and consequently reduces the residual pressure in the affected injection tubing, nozzle, and nozzle holder. This lowered pressure assists the rapid closing of the injection nozzle valve and diminishes the effect of hydraulic pressure waves that exist in the tubing between injections. As a result, the possibility of the nozzle's opening is reduced before the next regular delivery cycle.

PS SERIES DISTRIBUTOR PUMP TIMING

An external or internal timing mechanism may be incorporated in PS series distributor-type fuel injection pumps. These timing mechanisms provide a means for automatically varying injection timing in relation to TDC as engine speed increases to obtain optimum engine combustion efficiency. These devices are adaptable to various fuel injection pumps, provide a timing advance of up to 20°, and become operative at speeds from 300 to 3200 rpm, depending on application and the type used **(Figure 24–9)**.

There are three types of the externally mounted timing mechanisms, which, except for some parts differences, are similar and function in the same manner.

In operation, centrifugal force, due to the rotation of the shaft and spindle, tends to open the weights and is resisted by the pressure of the springs. At a predetermined speed, the pressure of the springs and the centrifugal force are balanced. When that speed is exceeded, the movement of the weights causes the sliding gear to move longitudinally. This, in turn, causes the pump camshaft to rotate slightly. In so doing, the timing of the pump is advanced the desired amount.

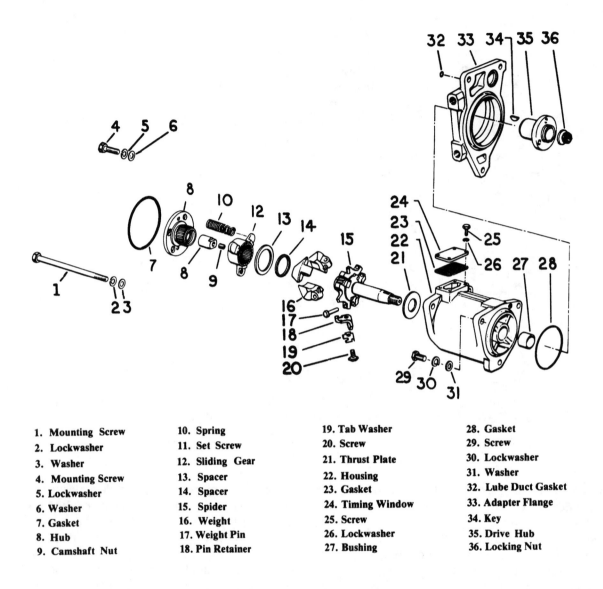

FIGURE 24-9 Exploded view of external timing device. (Courtesy of AMBAC International Corp.)

1. Mounting Screw
2. Lockwasher
3. Washer
4. Mounting Screw
5. Lockwasher
6. Washer
7. Gasket
8. Hub
9. Camshaft Nut
10. Spring
11. Set Screw
12. Sliding Gear
13. Spacer
14. Spacer
15. Spider
16. Weight
17. Weight Pin
18. Pin Retainer
19. Tab Washer
20. Screw
21. Thrust Plate
22. Housing
23. Gasket
24. Timing Window
25. Screw
26. Lockwasher
27. Bushing
28. Gasket
29. Screw
30. Lockwasher
31. Washer
32. Lube Duct Gasket
33. Adapter Flange
34. Key
35. Drive Hub
36. Locking Nut

Internal Advance Mechanism (Intravance)

An advance mechanism is built inside the cam to alter the relationship between the input shaft and the cam surface. The camshaft is supported between two bearings, and the drive shaft is supported inside the camshaft. The camshaft gear is machined on the camshaft. The amount of cam advance is determined by the longitudinal position of the sliding double-splined sleeve. This sleeve has helical splines on both the inner and outer surfaces to engage with corresponding helical splines on the drive shaft and the camshaft. The sleeve position is de-

PS SERIES DISTRIBUTOR PUMP TIMING

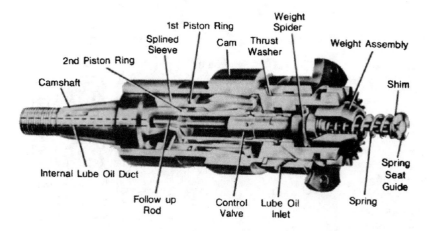

FIGURE 24-10 American Bosch distributor pump automatic injection timing device known as *Intravance*. This device provides up to 20° of injection timing advance. The device is effective from 300 to 3200 rpm, depending on engine application. (Courtesy of AMBAC International Corp.)

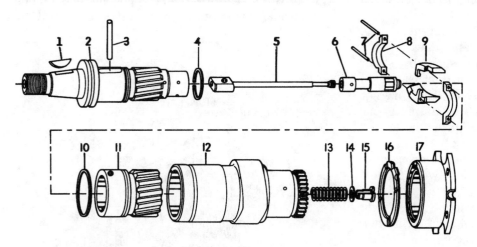

FIGURE 24-11 Exploded view of internal advance mechanism. (Courtesy of AMBAC International Corp.)

1. Woodruff key
2. Drive shaft
3. Cross pin
4. Seal ring
5. Follow-up rod
6. Control valve
7. Weight pins
8. Spiders
9. Weights
10. Seal ring
11. Splined sleeve
12. Cam
13. Spring
14. Shim
15. Nut
16. Thrust washer
17. Bearing plate assembly

termined by a hydraulic servo valve (**Figures 24-10** and **24-11**).

The control action is as follows, assuming that the entire mechanism is initially at equilibrium at some intermediate speed. If the speed increases, the weights move outward and the hydraulic servo valve moves to the right. Lube oil, which is supplied through the right bearing, flows through the valve to the working chamber and forces the sleeve to the left. The sleeve is sealed on both inner and outer surfaces by piston rings. As the sleeve moves to the left, the inner follow-up rod also moves to the left, increasing the spring force.

Movement of the sleeve and follow-up rod continues until the spring force overcomes the higher centrifugal force. The movement of the sleeve causes the cam to advance with respect to the drive shaft by the helix angle of the splines and the spring rate. The spring rate and preload are tailored to the engine. The advance curve is dependent on speed only and is not affected by fuel delivery, lube oil temperature, viscosity, etc.

When the speed decreases, the opposite occurs. The weights move inward, causing the hydraulic servo valve to move to the left. Lube oil is released from the working chamber through the left-hand valve annulus and out through the drain hole in the shaft. As the oil leaves, the sleeve and the follow-up rod move to the right to reduce the spring force. This action continues until the force is reduced and the weights again resume their equilibrium position.

GOVERNOR OPERATION

The variable governor portion of single-plunger distributor fuel injection pumps is a mechanical centrifugal device that is assembled as a unit in the governor housing. The primary purpose of the governor is to serve as a means of presetting and maintaining within close limits the diesel engine over a nominal speed range, irrespective of engine load. In addition, the governor controls engine idling speed to prevent stalling, and maximum speed to prevent overspeeding.

The governor is driven at engine speed by a gear mounted on the governor drive shaft that engages a gear mounted on the pump camshaft. The governor drive shaft assembly consists of the weights, pivot pins, spider, shaft, and friction clutch **(Figures 24–12 and 24–13)**. (See also **Figures 24–3 and 24–4**).

Friction clutch slippage adjustment is accomplished by adding or removing spacers from between the leaf springs and the spider.

The design of the friction clutch permits the weight and spider assembly to momentarily slip on its hub whenever sudden speed or load changes occur. This action dampens torsional vibrations in the pump camshaft and drive, and protects the pump and governor components. It also helps minimize wear.

As the weights of the governor revolve, the centrifugal force tends to move them outward. The movement is opposed by the governor springs acting through the sliding sleeve assembly. If the speed is decreased, the spring force exceeds the centrifugal force of the weights and moves the sliding sleeve toward the injection pump. Thus, any change in speed changes position of the sliding sleeve assembly.

The fulcrum lever bracket is attached to the fulcrum lever by means of a pin. The fulcrum lever bracket fits over the operating lever shaft but is not directly connected to it. The torsion spring hub is firmly secured to the operating shaft and connected through the torsion spring to the fulcrum lever bracket.

The ends of the spring straddle the tongue of the spring hub and fulcrum lever bracket with the result that the spring tension tends to keep these two parts in line with each other. It is normal for them to become separated during operation of the governor at full-load position.

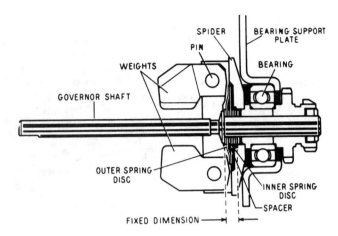

FIGURE 24–12 Governor components and friction clutch adjustment dimension. (Courtesy of AMBAC International Corp.)

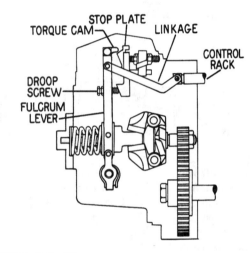

FIGURE 24–13 Governor at medium speed under load. (Courtesy of AMBAC International Corp.)

When the engine returns to normal speed as determined by foot-pedal position, the torsion spring brings the tongue of the spring hub and the fulcrum lever bracket in line.

A movement of the operating lever rotates the spring hub and the fulcrum lever bracket, thereby causing the fulcrum lever to turn about the pivot pin. This movement changes the position of the pump control unit because the upper end of the fulcrum lever is connected by linkage to the control unit lever. The fuel quantity is increased or decreased, depending on the direction of operating lever movement.

When the engine is operating under load and the load is removed, the speed increases, the weights move outward, forcing the sleeve away from the pump, and the fulcrum lever pulls the control rod to a position of less fuel delivery. This condition is known as "high idle."

An adjustable maximum-fuel-delivery stop plate is provided. An extension of torque cam at the upper end of the fulcrum lever contacts the stop plate during periods when full-load fuel quantities are required by the engine. The control lever travel is therefore limited by the position of the stop plate.

A speed variation from full load to low load is known as *speed droop*, and a droop screw adjustment is provided.

The angle of the torque cam in relation to the fulcrum lever affects the quantity of fuel increase at the lower speeds. It is very important that, once this angle has been set by the factory, no changes are made in the angle, because damage to the engine might result due to excess fuel delivery.

With the governor at medium speed under full load the reduced speed causes the weight to collapse and the sleeve to move toward the pump by the spring forces. The fulcrum lever connected to the sleeve also moves toward the pump. Movement at the upper end is limited by the stop plate and therefore the fulcrum lever pivots about the droop screw with the movement being absorbed by the rotation of the fulcrum lever bracket against the torsion spring.

Smoke Limiter (Aneroid)

The fuel aneroid is mounted on top of the governor housing. It is also called a *smoke limiter*. It reduces smoke anytime the engine goes from a light-load to a full-load condition. During light-load conditions, fuel delivery is cut by the governor to meet engine requirements. If the engine load is suddenly increased, the governor reacts and increases fuel delivery. In the case of a naturally aspirated engine, fuel delivery is limited by the governor to ensure that a specific air/fuel ratio will not be exceeded. Because a naturally aspirated engine is an air pump, at a given speed it will pump only so much air. If the air/fuel ratio gets too rich, smoke will occur as the result of incomplete combustion. If a turbocharger is added to the engine, increased fuel delivery causes the turbocharger to force the engine to breathe more air, so the air/fuel ratio is not exceeded; however, during sudden load changes, the turbocharger does not come up to speed quickly enough, and the engine smokes, acting momentarily like an overfueled naturally aspirated engine. The fuel aneroid limits fuel delivery to approximate naturally aspirated engine levels until boost pressure from the turbocharger working against a spring-loaded diaphragm moves the stop plate in the direction of increased fuel delivery. This momentary lag in fuel delivery does not affect engine performance; all it does is eliminate the puff of smoke often associated with turbocharged engines.

Excess-Fuel Starting Device

In the normal start position, the spring holds the piston back in the housing, thereby allowing more fuel to be metered to the engine. The lube oil that is supplied to the device is normally taken from the rear camshaft bushing area and piped to the underside of the excess-fuel starting device housing. With the engine stopped and no lube oil pressure within the device, the spring force moves the piston and stop plate back to the excess-fuel starting position.

With the cranking motor engaged and the accelerator pedal depressed, an additional pressure is exerted on the stop plate by the fulcrum lever and cam. When the engine has started, lube oil pressure overcomes the spring force and the cam pressure, and the stop plate is literally pushed into "run" position. The stop plate remains in the "run" position until the engine is shut down.

ELECTRONICALLY CONTROLLED DISTRIBUTOR PUMP

Components and Operation

The supply pump transfers the fuel from the fuel tank through a combination water separator and filter and produces a hydraulic pressure that varies as a function of pump rpm. This is controlled by the pressure regulator (**Figures 24–14** and **24–15**).

Fuel at supply pressure flows to the charging sump and is conducted through holes in the porting sleeves and spill to move the plungers out to the base circle of the internal cam ring.

As the distributor shaft rotates, the angled slots admitting fuel to fill the pumping chamber close off the ports in the spill sleeve. At a selected angular position of the distributor shaft, the plungers are forced inward by the cam. At this angle, or a slightly later angle, the angled slots close off the ports in the porting sleeve. This is called *port closing*. The plungers now create

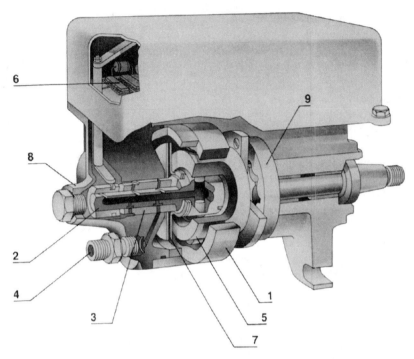

FIGURE 24–14 Cutaway view of the United Technologies Model 75 electronically controlled distributor injection pump. (Courtesy of AMBAC International Corp.)

pressure in the pumping chamber. This pressure is directed by the distributor shaft to one of the outlet ports, through the snubber or delivery valve, thus opening the nozzle.

At a further angle of the distributor shaft, the metering sleeve angled slot opens the axial passage of the distributor shaft to spill the remainder of the pumping stroke, thus ending the injection.

The electronic control module controls the fuel quantity by adjusting the axial position of the metering sleeve. It also programs the start of injection over the speed range by driving a servo valve, which in turn specifies the correct hydraulic pressure for the advance mechanism function.

This servo valve controls the angular location of the injection cam and also, through ramped adjusted plates, controls the axial location of the distributor shaft. The angled port closing slot in the distributor shaft maintains the proper alignment of port closing with the cam.

The axial location of the distributor shaft can also be adjusted by the electronic control, independently of the cam, moving port closing to a different slope on the cam profile and thereby changing the injection rate.

A pulse dampener is built into the pressure gallery to minimize spill pulsations and aid in uniform filling.

Microprocessor-Based Control System

- Control capabilities for governing, timing and rate of injection during engine operation are much more precise with electronic controls.
- Sleeve metering with constant beginning of injection
- Integral supply pump
- Fuel lubrication
- Pressure capability for direct-injection and indirect-injection engines up to 12,500 psi (850,000 kPa)
- Pump displacement rate capability up to 3.75 mm^3 per crank angle degree
- System self-diagnostic feature with visual display on vehicle dashboard
- On-engine service diagnostic access port

Starting and Warm-Up

With microprocessor controls, engine starting and warm-up is achieved by modulating fuel flow and injection timing based on sensed engine parameters at no extra cost with no extra components.

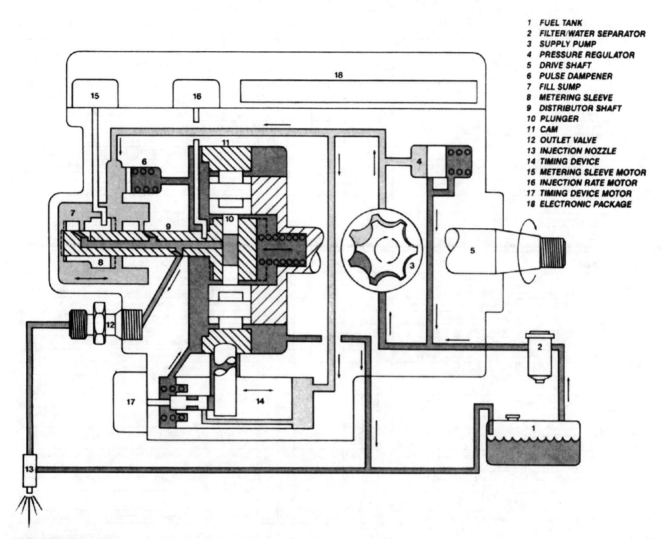

FIGURE 24–15 Model 75 electronically controlled injection pump schematic. (Courtesy of AMBAC International Corp.)

Hydraulic/Electronic Timing Advance

The hydraulic timing advance mechanism operates on supply pump pressure acting on a hydraulic piston to rotate the cam ring. Electronic modulation of cam ring rotation is controlled by sensing speed, load, or other inputs.

FUEL SYSTEM TROUBLESHOOTING

The fuel system troubleshooting guide in **Figure 24–16** lists common operating symptoms that may be encountered, probable causes for each symptom, and recommended remedies. It is good practice to check out those items in the list of causes that do not require injection pump removal first. If these items are not the cause of the problem, then the injection pump may be at fault. In this case the pump must be removed, cleaned, and serviced. The following procedures are typical. Always refer to the appropriate service manual for exact procedures and specifications.

AMBAC DISTRIBUTOR INJECTION PUMP SERVICE

Pump Cleaning and Flushing

CLEANING PROCEDURE

1. Thoroughly wash the external surfaces of the injection pump with cleaning fluid. Make certain that all foreign matter is removed.

TROUBLE - SHOOTING GUIDE

Cause	Excessive Engine Smoke - Black	Rough Idle, Engine Vibrating	Poor Performance, Surge, Erratic Action	Low Power	Excessive Engine Smoke - White - Blue	Lube Oil Dilution	Engine Overspeeds	Hard Starting	Recommended Remedy
1. FUEL SUPPLY PRESSURE									
a. Primary Filter Clogged or Restricted.	x	x							Remove and Clean or Replace.
b. Final Stage Filter Clogged or Restricted.	x	x							Remove and Replace.
c. Air Leaks in Fuel Suction System.	x	x	x					x	Inspect and Correct. Replace Parts where Required.
d. Reducers or Restrictions in Fuel Suction Lines.	x	x	x					x	Clean and Repair or Replace Parts as Required.
e. Fuel Tank Improperly Vented.	x	x	x					x	Clean or Replace Parts as Required.
f. Supply Pump Worn.	x	x	x					x	Remove Supply Pump. Repair by Authorized Personnel.
g. Supply Pump Regulating Valve Worn or Stuck.	x	x	x					x	Remove Supply Pump. Repair by Authorized Personnel.
h. Check Valve Leaking or Blocked.	x	x	x					x	Clean or Replace
i. Fuel Return Line Restricted.		x	x						Clean and Flush. Replace if Necessary.
2. TIMING									
a. Injection Pump to Engine Timing Incorrect.	x	x	x	x				x	Re-Time to Engine Manual Specifications.
b. Pump Drive Worn - Drive Gear, Keyway.	x	x	x						Remove, Inspect and Replace Worn Parts.
c. Plunger Drive Broken.		x	x	x				x	Remove Injection Pump. Repair by Authorized Personnel.
3. GOVERNOR									
a. Throttle Linkage Mis-Adjusted.	x	x	x						Adjust to Engine Manual Specifications.
b. Throttle Linkage Sticking or Binding.	x	x	x						Check for Binding, Worn or Loose Parts, Foreign Particles.
c. Shut-Off Lever Restricting Governor Operation.	x	x	x						Adjust to Engine Manual Specifications.
d. High Idle Screw Adjustment Incorrect.	x						x		Re-Adjust to Manual Specifications.
e. Low Idle Screw Adjustment Incorrect.		x							Re-Adjust to Manual Specifications.
f. Incorrect Governor Spring or Setting.	x	x	x				x		Re-Adjust to American Bosch Specifications.
g. Incorrect Friction Clutch Slippage Torque.	x	x					x		Re-Adjust by Authorized Personnel.
h. Control Rod Binding or Sticking.	x	x	x						Repair by Authorized Personnel.
4. CALIBRATION									
a. Fuel Delivery Incorrect.	x	x		x				x	Re-Calibrate by Authorized Personnel.
b. High Pressure Tubings Clogged or Restricted.	x	x	x						Remove, Check Inside Diameters, Drill or Ream, Clean and Flush.
c. Inadequate Lube Oil Pressure to Excess Fuel Device.		x	x	x					Clean Lines, Repair Injection Pump or Lube Pump as Required.
d. Excess Fuel Device Sticking in "Start" Position.		x	x						Remove, Repair or Replace, and Re-Set by Authorized Personnel.
e. Excess Fuel Device Sticking in "Run" Position.								x	Remove, Repair or Replace, and Re-Set by Authorized Personnel.
5. HYDRAULIC HEAD									
a. Injection Pump Plunger Worn or Scored.								x	Remove Pump. Repair by Authorized Personnel.
b. Injection Pump Plunger Sleeve Sticking or Binding.	x	x	x					x	Remove Pump. Repair by Authorized Personnel.
c. Injection Pump Control Unit Sticking or Binding.	x	x	x					x	Remove Pump. Repair by Authorized Personnel.
d. Delivery Valve Sticking or Leaking.	x	x	x					x	Remove, Clean, Repair or Replace as Required.
e. Delivery Valve Spring Broken.	x	x	x	x				x	Replace and Re-Set Opening Pressure.
6. NOZZLES									
a. Nozzles Defective - Leaking - Worn.	x	x	x	x				x	Remove, Replace or Repair, Reassemble, Test, Set Opening Pressure.
b. Incorrect Nozzle Opening Pressure.	x	x	x	x				x	Re-Set to American Bosch Specifications.
c. Nozzle Cap Nut Incorrectly Torqued.		x	x	x				x	Remove, Retighten Cap Nut, Replace Copper Gasket, Clean Engine Recess and Re-Install in Engine.
d. Nozzle Incorrectly Torqued in Engine.	x	x	x	x				x	Remove, Replace Copper Gasket, Clean Recess, Reassemble to Engine (Tighten Evenly to Required Torque).
e. Nozzle Valve Sticking.	x	x	x	x					Remove, Clean, Repair or Replace as Required.
f. Nozzle Spray Holes Plugged or Partially Plugged.	x	x	x						Remove Nozzle. Clean Holes or Replace Nozzle as Required.
7. OIL LEAKAGE									
a. Defective Supply Pump Seal.					x				Replace by Authorized Personnel.
b. Defective Hydraulic Head "O" Ring.					x				Replace by Authorized Personnel.
c. Defective Control Unit "O" Ring.					x				Replace by Authorized Personnel.
d. Loose Control Unit Fit.					x				Remove Pump. Repair by Authorized Personnel.
8. GENERAL									
a. Air Cleaner Restriction.	x	x	x	x					Remove and Clean.
b. Excessive Lube Oil in Air Cleaner.					x		x		Drain, Clean and Refill per Engine Manual Specifications.
c. Defective Tachometer.	x	x	x				x		Recheck with Master Tachometer. Replace if Necessary.
d. Cranking R.P.M. too Low (Cold or Hot).								x	Check Battery and Starter. Replace if Necessary.
e. Low Engine Compression.	x	x						x	Check Compression and Correct as Required.
f. Incorrect Engine Valve Lash.	x	x	x						Check and Correct as Required.
g. Water in Fuel.	x	x	x					x	Drain Water from Filters and Fuel Tank. If Necessary, Drain Fuel Tank and Refill.

FIGURE 24–16 (Courtesy of AMBAC International Corp.)

2. Remove the inspection window cover or timing window cover. If applicable, remove or open drain plugs or fittings.

3. Remove the governor top cover.

4. Flush lube oil compartments in the pump and governor with approved cleaning fluid, such as Varsol, to remove all traces of dirt, grease, carbon and other foreign matter.

NOTE: After removing the inspection covers, check the governor components for damage or binding of parts. Make certain the fulcrum lever operates freely. Also check pump components for faulty or damaged parts, including broken spring, broken or worn control unit or sleeves, binding control rod or rack, broken plunger(s) or camshaft, cracked or damaged housing, etc.

5. Re-install the covers.

FLUSHING PROCEDURE

1. Mount the pump on an approved test stand.

NOTE: It may be necessary to replace the original pump drive hub and/or coupling with a coupling and/or hub that will fit the particular test stand being used.

2. Lubricate the pump as follows: PSJ and PSM types—use a pressurized lube oil system (system and ducted adapters and fixtures available from Bacharach Instrument Co., Pittsburgh, Pa.) filled with SAE 30 oil, or equivalent, to properly lubricate the pump. The lube oil pressure must be 20 psi (138 kPa) minimum, or if the pump includes lube oil pressure adjusting devices, a pressure of 35–40 psi (241–276 kPa) is required.

NOTE: Do not splash-lubricate PSJ or PSM type pumps, as this provides insufficient lubrication to certain components.

IMPORTANT: Do not use calibrating oil or fuel oil as a lubricant.

CAUTION:

Do not allow lube oil to mix with or contaminate calibrating or fuel oil.

---

3. Connect a set of high-pressure tubings to the pump outlets and direct the opposite ends of the tubings into a pail, thereby bypassing the nozzle and holder assemblies and test stand tank.

4. To flush the pump sump, proceed as follows: PSJ and PSM types—operate at a speed of about 1000 rpm with the operating lever in the full-load position for approximately 3 minutes.

This procedure removes any fuel oil and foreign material from the pump sump and eliminates the possibility of contaminating the test oil or damaging the nozzles during the calibration check.

5. Stop the test stand and remove the high-pressure tubing.

Calibration Check

TEST SETUP

1. Install a good, matched set of calibration nozzle and holder assemblies or the engine nozzle and holder assemblies where specified on the test stand. Refer to calibration data for the applicable pump to determine which type nozzle and holder assemblies are to be used.

2. Install required high-pressure tubings. Refer to applicable calibration data for required tubing dimensions. Use a drill of the proper size to check the ends of each tubing. Drill and flush tubings that have crimped ends.

3. Install a transparent plastic suction hose from the test oil tank to the inlet side of the fuel supply pump.

CALIBRATING

1. Observe the fuel (test) oil entering the supply pump. It must be free of air bubbles. Be sure to eliminate suction air leaks.

2. Check to be sure that the supply pressure is adequate.

3. Check the pump calibration in accordance with the data for the applicable pump. Record the fuel delivery as received and also the final fuel deliveries. Use a calibration form to provide a record of test performance for customer's information and agency's files. Identify the pump by serial number.

PUMP ADJUSTMENTS

The following settings must be checked and adjusted if necessary:

1. Port closure
2. Cam nose angle **(Figure 24–17)**
3. Full-load delivery

4. High idle (**Figure 24–18**)
5. Peak torque
6. Droop screw (**Figure 24–19**)
7. Governor cutoff
8. Low idle (**Figure 24–20**)
9. Internal leakage (hydraulic head)

Obtain all the adjustment and flow data for the particular pump being checked from the service manual.

Figure 24–21 shows how to check and correct pump shaft end play. **Figure 24–22** shows how to install the hydraulic head, and **Figure 24–23** shows how to install the sleeve control unit.

Perform all the checks, adjustments, and repairs in the sequence specified and to the specifications given in the service manual. Install the tamper-proof seal, and cap all pump openings.

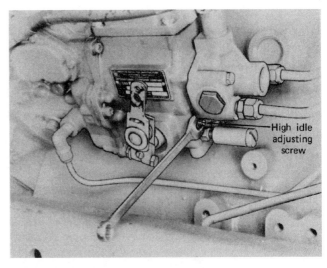

FIGURE 24–18 Adjusting the high-idle screw. (Courtesy of AMBAC International Corp.)

FIGURE 24–19 Droop screw adjustment. (Courtesy of AMBAC International Corp.)

FUEL INJECTION PUMP INSTALLATION AND STATIC TIMING PROCEDURE

1. Rotate the engine until the exhaust valve is just closing and the intake valve is just opening on the No. 6 cylinder and the "TDC MARK" on the pulley is in line with the pointer. The No. 1 and No. 2 push rods should be free.

2. Turn the crank pulley counterclockwise past TDC to 45° BTDC on the pulley, then clockwise to degree setting per engine specification; do this firmly without "bumping" to avoid gear bounce, which would create a backlash error.

FIGURE 24–17 Cam nose angle adjustment. (Courtesy of AMBAC International Corp.)

FUEL INJECTION PUMP INSTALLATION AND STATIC TIMING PROCEDURE 565

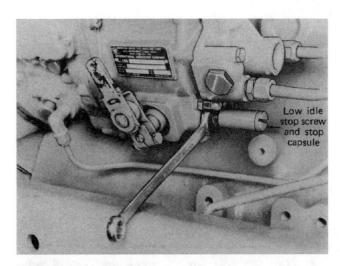

FIGURE 24-20 Adjusting the low-idle stop screw. (Courtesy of AMBAC International Corp.)

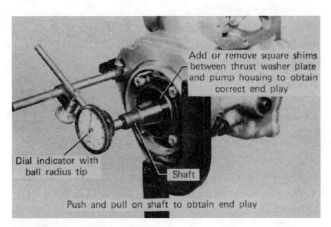

FIGURE 24-21 Checking pump shaft end play. (Courtesy of AMBAC International Corp.)

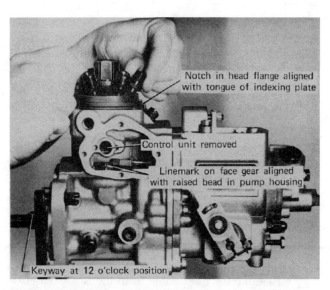

FIGURE 24-22 Hydraulic head installation. (Courtesy of AMBAC International Corp.)

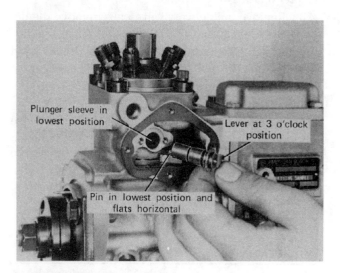

FIGURE 24-23 Installing the sleeve control unit. (Courtesy of AMBAC International Corp.)

3. Install the fuel injection pump (FIP) with capscrews, flat washers, lock washers and nuts, and tighten to specs. Verify that the timing pointer and the mark on the FIP gear drive hub are aligned.

NOTE: Service replacement FIPs may be friction locked in the No. 1 cylinder position. Do not turn the pump shaft except to correct minor errors (less than 0.12 in.), then rotate the hub so that the mark on the hub and pointer are exactly aligned.

4. Carefully install the pump drive gear onto the drive hub in the approximate midposition of the adjustment slots. Remember, the gear will rotate slightly while engaging teeth. Assure that the timing marks are still in alignment.

5. Install capscrews and flat washers through the gear into the pump drive hub, and tighten one at a time finger tight, checking alignment after each. Tighten all capscrews to specs. Assure correct alignment of pointer and FIP drive hub visible through the observation hole.

6. Remove the friction locking screw with aluminum pin from the FIP housing that is located directly behind the FIP mounting flange on the outboard side of the FIP (throttle lever side) and replace with a ¼ in. pipe plug. Assure that the gear backlash is within specifications.

7. As a final timing confirmation, turn the crank pulley counterclockwise past the specified timing to 45°

BTDC, then return (clockwise rotation) to the specified timing and verify that timing marks are aligned. If not aligned, repeat steps 4 and 5 until alignment is reached.

8. Install the gear cover and observation hole cover with capscrews and lock washers and tighten to specs.

9. Adjust throttle lines and idle speed to specifications.

If the replacement pump is not locked in the No. 1 position, No. 1 position will have to be determined by one of the following methods:

Flow Timing Method of Positioning Pump on the No. 1 Cylinder

1. Remove the nozzle line from the No. 1 outlet on the pump.

2. Fabricate a port-closing adapter. Install the adapter on the No. 1 outlet.

3. Remove the delivery valve caps, remove the delivery valve and spring, and replace the cap.

4. Attach a clean fuel supply to the hydraulic head inlet and plug the return outlet.

5. Place the throttle in half- to wide-open position.

6. Turn the pump drive hub (or engine) clockwise (viewing drive end) until a bubbleless stream of fuel is flowing from the adapter. Continue turning until the flow stops. This is port closing. The pump timing pointer and mark on the drive hub should now be in alignment. The pump is now in the No. 1 position.

7. Replace the delivery valve and torque per the torque chart.

NOTE: If there is no flow prior to coming up on mark the pump is probably coming up on No. 6.

NOTE: The pump drive hub turns at engine speed.

Face Gear Timing Mark Method of Positioning Pump on the No. 1 Cylinder

NOTE: The flow timing method should be used if the technician has no method of resealing the pump.

1. Cut the seal wire between the control cover screw and the top governor cover screw.

2. Remove the control cover.

3. Turn the pump until the line on the face gear comes into the opening.

4. Align the pointer and mark and pump drive hub while maintaining eye contact with the line on the face gear.

NOTE: The line has to be visible only in the window (usually near the front). It does not have to align with any mark in the housing.

The pump is now in the No. 1 position.

5. Install the pump drive gear while maintaining hub mark to pointer alignment.

6. Replace the control cover and seal wire in appropriate attaching screws.

REVIEW QUESTIONS

1. AMBAC International single-plunger distributor pumps of the PS type include the _____ _____ and _____ _____ _____ _____.

2. All pumps are designed for _____ _____ to the engine.

3. All PS series pumps are _____ - _____ lubricated.

4. Fuel enters the pump from the supply system through the pump _____ _____ _____.

5. The quantity of fuel delivered per stroke is controlled by the position of the _____ _____ in relation to the fixed port-closing position.

6. The delivery valve assembly is located directly above the _____.

7. An _____ or _____ timing mechanism may be incorporated in PS series distributor-type fuel injection pumps.

8. The variable governor portion of single-plunger distributor fuel injection pumps is a _____ _____ device that is assembled as a unit in the governor housing.

9. The fuel _____ is mounted on top of the governor housing.

10. The electronic control module controls the fuel quantity by adjusting the _____ _____ of the metering sleeve.

TEST QUESTIONS

1. AMBAC International single-plunger distributor-type pumps of the PS type include the
   a. PSJ and PSU
   b. model 100 and PSU
   c. PSJ, PSU, and the model 100 pumps
   d. PSJ, PSU, and PSI

2. All pumps are designed to be mounted to the engine by
   a. a flange mounting
   b. a universal coupling
   c. a bolt or flex coupling
   d. any of the above

3. The prominent feature of the PS series is the single precision plunger that is activated by the
   a. engine camshaft   c. throttle lever
   b. governor   d. pump camshaft

4. Fuel enters the pump from the supply system through the
   a. pump housing inlet connection
   b. first- and second-stage filters
   c. water trap and filters
   d. all of the above

5. The delivery valve assembly consists of a valve with a
   a. 45° angle seat and mating valve body
   b. conical seat and mating valve body
   c. spring-controlled seat and fuel pressure
   d. conical seat controlled by fuel pressure

6. An advance mechanism is built inside the cam to alter the relationship between the
   a. fuel rack and the cam surface
   b. internal shaft and the cam surface
   c. output shaft and the cam surface
   d. input shaft and the cam surface

7. The governor is driven at engine speed by a gear mounted on the
   a. governor drive shaft and engine camshaft
   b. governor drive shaft to the pump camshaft
   c. engine crankshaft
   d. fuel pump and shaft

8. The fuel aneroid reduces smoke anytime the engine goes from
   a. cold to warm temperatures
   b. warm to cold temperatures
   c. light load to full load
   d. full load to light load

9. An injection pump calibration check includes the following:
   a. port closure and cam nose angle
   b. full-load fuel delivery
   c. peak torque setting
   d. all of the above

◆ CHAPTER 25 ◆

ROBERT BOSCH TYPE VE DISTRIBUTOR PUMP

INTRODUCTION

Due to their great flexibility, type VE distributor injection pumps meet the needs of a wide range of applications. A diesel engine's rated speed, power, and configuration set the parameters for a particular pump design. Distributor pumps of the VE type are used in passenger cars, trucks, tractors, and stationary engines.

PERFORMANCE OBJECTIVES

After thorough study of this chapter and the appropriate training models and service manuals, you should be able to do the following with respect to diesel engines equipped with Robert Bosch Type VE distributor injection pumps:

1. Describe the basic construction and operation of the Robert Bosch Type VE distributor injection pump, including the following:
    a. method of pumping and metering
    b. delivery valve action
    c. lubrication
    d. injection timing control
    e. governor types

2. Diagnose basic fuel injection system problems and determine the required corrections.

3. Remove, clean, inspect, test, measure, repair, replace, and adjust fuel injection system components as needed to meet manufacturer's specifications.

TERMS YOU SHOULD KNOW

Look for these terms as you study this chapter, and learn what they mean.

high-pressure pump
distributor
mechanical governor
hydraulic timing mechanism
drive shaft
roller ring
cam plate
distributor plunger
distributor head
vent screw
pressure valve
governor drive gear set
fuel metering control collar
overflow restriction
cutoff bore
high-pressure chamber
speed droop
variable-speed governor
idle- and maximum-speed governor
torque control
manifold pressure compensator
altitude pressure compensator
cold-start module
electronic diesel control
injection pump timing

(This chapter is based on and contains information provided courtesy of Robert Bosch Canada Ltd., Cummins Engine Company, Inc., and Mack Trucks Inc.)

MAJOR COMPONENTS OF THE VE PUMP

The Bosch VE type distributor injection pump has only one pumping unit (cylinder and plunger) regardless of the number of engine cylinders. Fuel is delivered by way of a distributor groove to a number of ports corresponding to the number of engine cylinders. The pump may be driven by gears, toothed belt, or chain. The single-plunger distributor injection pump consists of the following major components (**Figures 25-1 to 25-3**):

1. *Drive shaft:* turns at one-half engine crankshaft speed. It operates the supply pump and drives the cam plate plunger.

2. *Supply pump:* supplies fuel under controlled pressure to the injection pump.

3. *Roller ring:* contains four rollers that push against the cam plate as the roller ring rotates.

4. *Cam plate:* disc with lobes on one side that are acted on by the roller ring, causing the pump plunger to reciprocate to create pumping action.

5. *Pump plunger:* plunger is moved back and forth by the cam plate to create high-pressure pumping action. The plunger has fuel passages that align with other fuel ports in the hydraulic head as the plunger rotates in its plunger barrel.

6. *Metering sleeve:* fits around the pump plunger and is connected through a linkage to the accelerator pedal. Pedal movement slides the metering sleeve along the pump plunger to control the amount of fuel allowed to enter the pumping chamber.

7. *Hydraulic head:* contains the pumping plunger, passages for filling the pumping chamber, and passages to the injection line fittings on the hydraulic head.

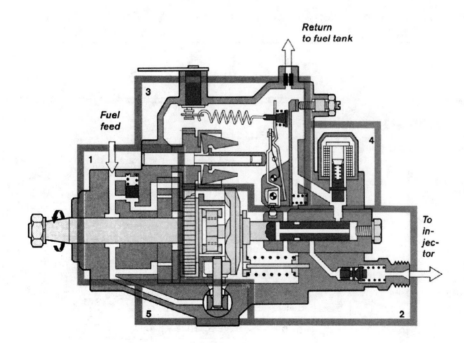

1 Vane-type supply pump
Supplies fuel from tank to injection-pump cavity.

2 High-pressure pump with distributor
Produces injection pressure, moves and distributes fuel to cylinders.

3 Mechanical governor
Controls engine speed, varies fuel delivery over control range.

4 Electromagnetic shutoff valve
Interrupts fuel delivery to stop engine.

5 Injection-timing unit
Adjusts beginning of injection according to engine speed.

FIGURE 25-1 Functional groups of the VE distributor pump. (Courtesy of Robert Bosch Canada Ltd.)

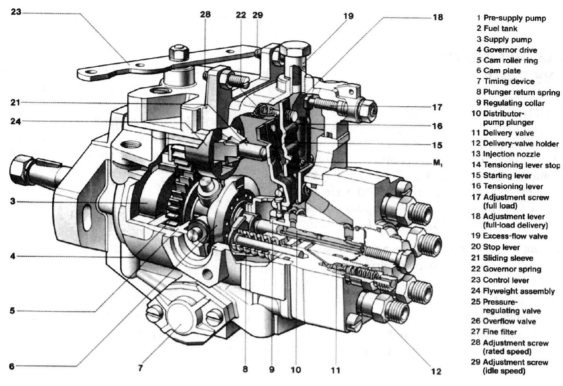

FIGURE 25–2 VE pump cutaway view showing pump components. (Courtesy of Robert Bosch Canada Ltd.)

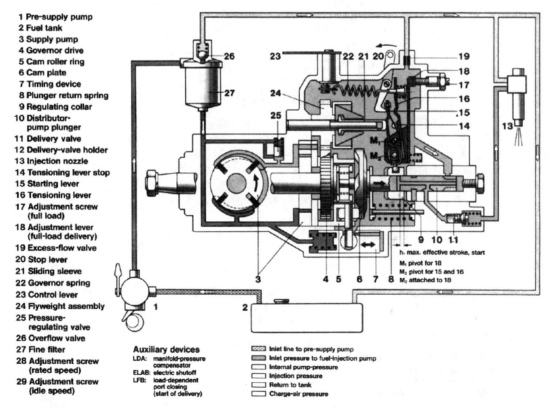

FIGURE 25–3 VE pump components and fuel flow. (Courtesy of Robert Bosch Canada Ltd.)

VE PUMP OPERATION

Fuel Supply Pump and Pressure Regulation

A vane-type fuel supply pump draws fuel from the fuel tank and delivers it to the fuel injection pump cavity. The supply pump surrounds the input shaft with its impeller concentric with the shaft. The impeller runs inside a fixed eccentric ring. Centrifugal action forces the vanes to run in contact against the ring. As the vanes follow the eccentric surface of the ring they slide in and out in their slots in the impeller. Fuel enters the inlet chamber as the vanes slide outward. Rotary motion carries the fuel between the vanes until it is forced into the injection pump cavity while some fuel flows through a bore to the pressure-regulating valve **(Figure 25–3)**.

When fuel supply pressure exceeds a preset value, the valve opens to allow some fuel to flow back to the suction side of the pump. The opening pressure of the valve is determined by the valve spring tension **(Figure 25–4)**. An overflow restriction passage allows a small, variable amount of fuel to return to the fuel tank. This provides cooling and self-ventilation of the pump.

High-Pressure Pumping Action

The rotation of the input shaft is transferred to the pumping plunger in the distributor by dogs on the input shaft and cam plate meshing with a yoke **(Figure 25–5)**. The cam plate surface rides on the rollers during rotation. This converts input shaft rotation to rotary reciprocating motion of the plunger. The plunger is set into the cam plate and is positioned by a locating lug. The cam surface forces the cam plate and plunger to move back and forth to create the pumping action. Precision-matched return springs keep the cam plate in contact with the rollers at all times.

As the plunger rotates, the plunger port aligns with the fill port in the hydraulic head. This allows the supply pump to fill the pumping chamber in front of the plunger with fuel. As the plunger turns, the ports are no longer aligned and the fuel is trapped in front of the plunger. At the same time the cam plate lobe pushes the plunger to force the fuel out of the injection port to the injector. This action is repeated for each injector during one revolution of the drive shaft and plunger **(Figure 25–6)**.

The design of the cam plate surface influences fuel injection pressure and duration. Cam and plunger stroke are individually tailored to combustion chamber design and engine type.

Fuel Metering

A metering sleeve that slides on the pump plunger controls the amount of fuel allowed to enter the pumping chamber. The sleeve meters fuel by closing the spill port to increase fuel delivery and opening it to decrease delivery. The throttle position and the governor control the position of the metering sleeve.

The distributor-plunger motion phases shown in **Figure 25–6** illustrate the metering of fuel to an engine cylinder. With a four-cylinder engine the plunger turns a quarter of a turn as it moves from BDC to TDC with a six-cylinder engine a sixth of a turn.

Fuel pressure opens the pressure valve, and fuel is forced through the pressure line to the injection nozzle.

The effective stroke is completed as soon as the transverse cutoff bore of the plunger reaches the edge of the control collar. After this point, no more fuel is delivered to the injector, and the pressure (delivery) valve closes the line. Fuel is returned through the cutoff bore to the pump cavity as the plunger completes its travel to TDC.

As the plunger returns, its transverse bore is closed at the same time the next control slit opens the fuel inlet bore. The high-pressure chamber is once again filled with fuel, and the cycle repeats for the next cylinder.

Pressure (Delivery) Valve (Figure 25–7)

The pressure valve closes off the injection line from the pump. It removes a specified volume of fuel at the end

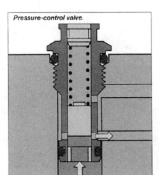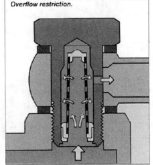

FIGURE 25–4 (Courtesy of Robert Bosch Canada Ltd.)

FIGURE 25–5 1. Yoke. 2. Roller ring. 3. Cam plate. (Courtesy of Robert Bosch Canada Ltd.)

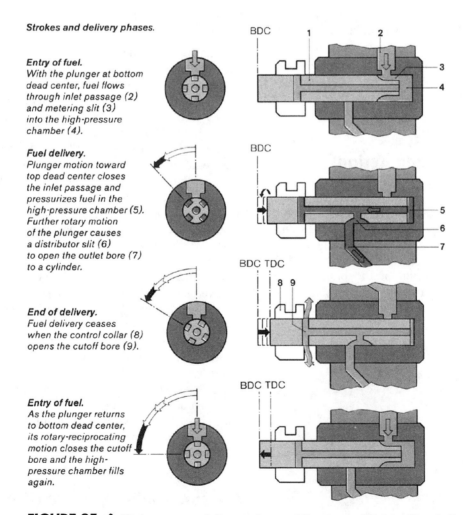

Strokes and delivery phases.

Entry of fuel.
With the plunger at bottom dead center, fuel flows through inlet passage (2) and metering slit (3) into the high-pressure chamber (4).

Fuel delivery.
Plunger motion toward top dead center closes the inlet passage and pressurizes fuel in the high-pressure chamber (5). Further rotary motion of the plunger causes a distributor slit (6) to open the outlet bore (7) to a cylinder.

End of delivery.
Fuel delivery ceases when the control collar (8) opens the cutoff bore (9).

Entry of fuel.
As the plunger returns to bottom dead center, its rotary-reciprocating motion closes the cutoff bore and the high-pressure chamber fills again.

FIGURE 25–6 High-pressure delivery phases. (Courtesy of Robert Bosch Canada Ltd.)

GOVERNOR OPERATION

Pressure valve.

a closed,
b open.
1 Valve holder,
2 Valve seat,
3 Valve spring,
4 Valve body,
5 Shaft,
6 Relief piston,
7 Ring groove,
8 Longitudinal groove.

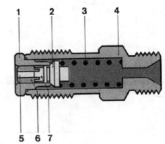

a) closed.

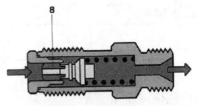

b) open.

FIGURE 25-7 (Courtesy of Robert Bosch Canada Ltd.)

of fuel delivery and determines the precise point at which the injector nozzle closes. The valve is opened under fuel pressure and is closed by spring pressure. Postinjection of fuel (due to pressure waves) is prevented by the addition of a restriction bore, valve plate, and pressure spring. As fuel tries to return, it imposes a flow resistance.

GOVERNOR OPERATION

Mechanical Governor

The governor is gear driven, and the entire assembly rotates in bearings inside the pump housing. The governor influences the position of the control collar on the plunger. The governor forms the top of the injection pump and contains the full-load adjusting screw, overflow (excess flow) restriction, and the engine speed adjusting screw **(Figures 25-1 to 25-3 and 25-8)**. Governor functions include idle-speed regulation, maximum no-load (high-idle) speed control, intermediate-speed control (in variable-speed governors), controlling extra fuel for starting, and changing full-load fuel delivery according to engine speed (torque control).

Variable-Speed Governor

The variable-speed governor controls all engine speeds including low idle, maximum high idle and full load, and everywhere in between. **Figure 25-8** shows the variable-speed governor components. **Figure 25-9** shows the governor in the starting and idle positions. Note the difference in the position of the springs, levers, and control collar. The position of the control collar determines the effective stroke of the plunger and therefore the amount of fuel delivered to the injectors. At engine speeds above idle the starting and idle springs are collapsed and do not affect governor action. Governor action is now controlled by the governor spring as its tension changes with changes in throttle position and engine speed (centrifugal weight position).

Idle- and Maximum-Speed Governor (Figure 25-10)

The idle- and maximum-speed governor controls only the idle and maximum engine speeds. Between these two, engine speed is controlled by accelerator pedal position. With the governor weights at rest (engine stopped) the starting spring holds the starting lever against the sliding sleeve. This puts the control collar in the starting position. Once the engine starts and the accelerator pedal is released, the speed control lever is pulled back to the idle position by the return spring. As the accelerator pedal is depressed and engine speed increases, the control sleeve position is determined by the balance of forces between the centrifugal force acting on the weights and spring pressure. When the operator depresses the accelerator pedal, the starting and idle springs are collapsed, and the intermediate spring becomes effective. Further pedal movement in the full-load direction collapses the intermediate spring. Pedal position is now transferred directly to the sliding sleeve. This allows the operator to control fuel delivery directly. To accelerate or climb a hill the operator presses down on the accelerator pedal. When governor spring tension is overcome by sleeve force, maximum engine speed sets in. If all engine load is relieved, engine speed is limited to its high-idle speed.

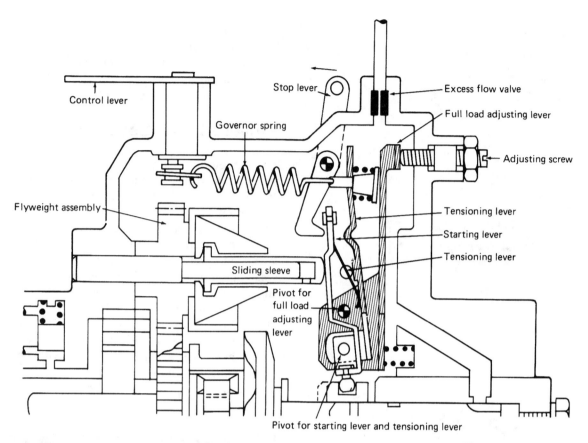

FIGURE 25-8 Robert Bosch distributor pump governor. (Courtesy of Robert Bosch Canada Ltd.)

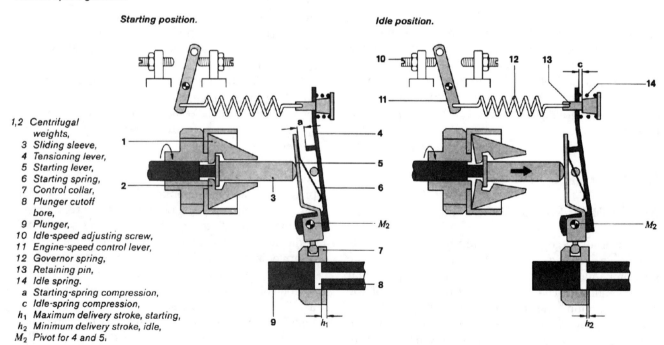

FIGURE 25-9 (Courtesy of Robert Bosch Canada Ltd.)

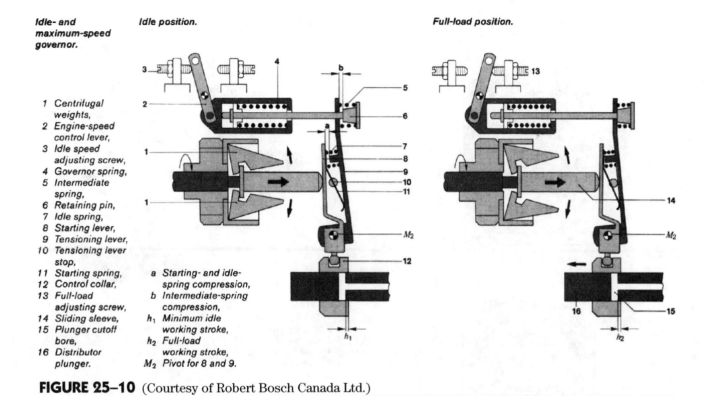

FIGURE 25-10 (Courtesy of Robert Bosch Canada Ltd.)

Injection Timing Control (Figure 25–11)

A hydraulic timing advance device is built into the underside of the injection pump. Injection timing is advanced with increasing engine speed to compensate for injection and combustion lag time. The unit consists of a housing, timing piston, actuation pin, sliding block, and spring.

The timing piston is held in its at-rest position by the preload spring. Fuel pressure proportional to engine speed acts on the side of the piston opposite the spring. The position of the piston during engine operation is determined by the spring force on one side of the piston and by fuel pressure on the other. As engine speed increases, fuel pressure increases and pushes the piston and sliding block against spring pressure. This action is transmitted to the injection pump roller ring through the actuation pin. This causes the roller ring and rollers to rotate slightly in a direction opposite cam plate rotation to advance injection timing. Timing advance may be as much as 24 crankshaft degrees.

Shut-Off Devices

The VE pump may be equipped with a mechanical shut-off device or with an electrical solenoid shut-off. The mechanical shut-off is operated by the driver inside the cab and consists of a control linkage that connects to the outer stop lever (**Figure 25–12,** top). The internal linkage moves the control collar in the pump to the STOP position. With the control collar in this position the cut-off bore in the distributor plunger remains open, and the plunger is unable to pump fuel.

The electric shut-off (**Figure 25–12,** bottom) uses the key-operated start switch to activate a solenoid to hold the fuel valve open when the key is in the START and RUN positions. When the key switch is turned off, the solenoid is deenergized, and the return spring closes the valve to shut off the fuel supply. With electronic diesel control (EDC) the engine is shut off by the "injected quantity actuator" when the signal is "injected fuel quantity zero." In this application the solenoid serves only as a safety shut-off device in case of an actuator defect.

Torque Control

Torque control in the VE pump is accomplished by a pressure valve in the distributor pump or by additions to the governor lever assembly. A second collar with one or two flats ground on it is added to the pressure valve. This provides a restriction known as *positive*

Operation of timing-advance device.

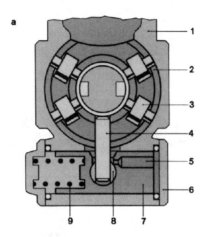

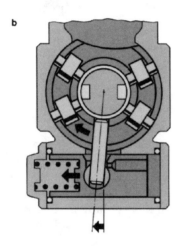

a Position with engine at rest,
b Operational position.
1 Pump housing,
2 Roller ring,
3 Rollers,
4 Actuation pin,
5 Bore in timing piston,
6 Cover,
7 Timing piston,
8 Sliding block,
9 Spring.

FIGURE 25-11 (Courtesy of Robert Bosch Canada Ltd.)

Mechanical shutoff.

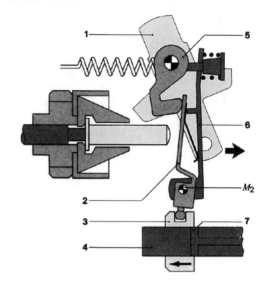

1 Outer stop lever,
2 Starting lever,
3 Control collar,
4 Distributor plunger,
5 Inner stop lever,
6 Tensioning lever,
7 Cutoff bore,
M_2 Pivot point for 2 and 6.

Electrical shutoff (with pull solenoid).

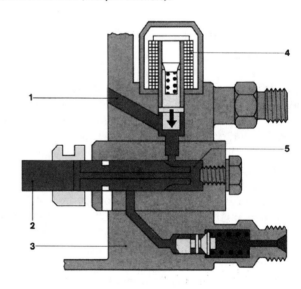

1 Inlet bore,
2 Distributor plunger,
3 Distributor head,
4 Pull (or push) solenoid,
5 High-pressure chamber.

FIGURE 25-12 (Courtesy of Robert Bosch Canada Ltd.)

torque control that reduces fuel delivery as engine speed increases **(Figure 25-13)**. With the governor lever assembly method of torque control, preloading of the torque control spring **(Figure 25-14)** determines the engine speed at which torque control begins. At the specified speed, sleeve force (P_M) and spring preload are in balance. The torque control lever (6) rests against the lug (5) on the tensioning lever (4). The free end of the lever rests against the torque control pin (7). As engine speed increases, sleeve force acting on the starting lever (1) increases. As the pivot point (M_4) moves, the torque control lever forces the torque control pin toward the stop and moves the control collar (8) to provide less fuel delivery. When the pin collar (10) contacts the starting lever (1), torque control ends.

Torque control by pressure valve.
1 Relief collar,
2 Torque-control collar,
3 Flat,
4 Restriction cross-section.

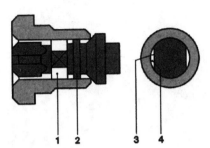

FIGURE 25–13 (Courtesy of Robert Bosch Canada Ltd.)

Manifold-Pressure Compensator (Figure 25–15)

The manifold-pressure compensator adjusts fuel delivery on turbocharged engines in proportion to inlet manifold pressure. As boost pressure increases, fuel delivery increases. When manifold boost pressure decreases, fuel delivery decreases.

The manifold-pressure compensator is mounted on top of the VE pump. The unit consists of a housing divided into two chambers by a diaphragm. A sliding pin with a conical control surface is connected to the diaphragm. A guide pin at right angles to the sliding pin and in contact with the conical surface transfers sliding pin movement to the stop lever, thereby altering the full-load stop position. An adjusting screw at the top of the unit sets the initial position of the sliding pin.

At the top of the compensator are a relief bore and a connection to inlet manifold pressure that acts on the top of the diaphragm. A spring acts on the other side of the diaphragm and is seated on an adjustable nut.

At low engine speeds, boost pressure is not high enough to overcome spring pressure, and the diaphragm, sliding pin, and guide pin remain in the initial position. As engine speed is increased boost pressure increases until spring pressure is overcome and the sliding pin begins to move downward with its conical surface acting on the sliding pin. The sliding pin acts on the tensioning lever and stop lever. This moves the control collar in the direction of greater fuel delivery. Fuel delivery is increased with increasing boost pressure and decreased with decreasing boost pressure. Should the turbocharger fail, the pressure compensator assumes its initial position and operates in that position. A full-load stop screw provides adjustment of full-load fuel delivery.

Load-Dependent Injection Timing (Figure 25–16)

Load-dependent injection timing retards the beginning of injection when the load decreases from full load to part load without any change in the engine's speed control lever position. When the load increases, injection timing is advanced. Modifications to the sliding sleeve (2), governor shaft (7), and pump housing as well as an additional transverse bore in the sliding sleeve (2) and a further bore in the pump housing between the pump cavity and the pump's inlet side achieve the desired results. These modifications result in a pressure drop-off in the pump cavity. This causes the injection timing piston to retard injection timing. When engine speed

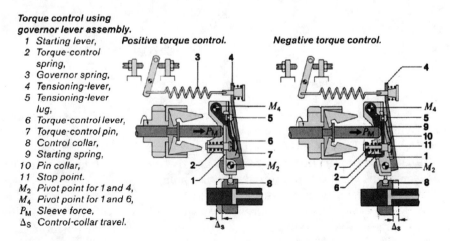

Torque control using governor lever assembly.
1 Starting lever,
2 Torque-control spring,
3 Governor spring,
4 Tensioning-lever,
5 Tensioning-lever lug,
6 Torque-control lever,
7 Torque-control pin,
8 Control collar,
9 Starting spring,
10 Pin collar,
11 Stop point.
M_2 Pivot point for 1 and 4,
M_4 Pivot point for 1 and 6,
P_M Sleeve force.
Δ_S Control-collar travel.

FIGURE 25–14 (Courtesy of Robert Bosch Canada Ltd.)

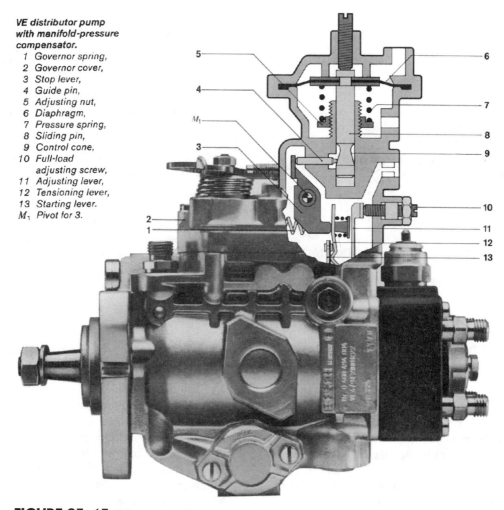

VE distributor pump with manifold-pressure compensator.
1. Governor spring,
2. Governor cover,
3. Stop lever,
4. Guide pin,
5. Adjusting nut,
6. Diaphragm,
7. Pressure spring,
8. Sliding pin,
9. Control cone,
10. Full-load adjusting screw,
11. Adjusting lever,
12. Tensioning lever,
13. Starting lever.
M_1 Pivot for 3.

FIGURE 25–15 (Courtesy of Robert Bosch Canada Ltd.)

drops, as when load is increased, fuel from the pump cavity can no longer flow to the pump inlet, and pressure in the cavity again rises and injection timing is advanced.

Altitude Compensator

At higher altitudes, where air is less dense, fuel delivery is decreased to match the reduced air mass entering the engine. The altitude compensator is similar in construction to the manifold pressure compensator described earlier and shown in **Figure 25–15** and is also mounted on the governor cover. Instead of a diaphragm the altitude compensator has an aneroid capsule that senses changes in atmospheric pressure, causing it to expand at lower atmospheric pressure and contract when pressure rises. The aneroid is connected to the spring-loaded sliding pin and acts on it much like the diaphragm in the manifold pressure compensator. At higher altitudes and lower atmospheric pressures fuel delivery is decreased proportionally.

Cold-Start Modules

Cold-starting modules are designed to improve cold-engine starting by advancing injection timing (**Figures 25–17** and **25–18**). The VE pump may be equipped with a manual or an automatic device. The manual control is located inside the vehicle, and the automatic control is mounted on the injection pump.

The manual control actuates the stop lever, shaft, and inner lever with its ball pin to rotate the roller ring to the advanced position. The ball pin's spherical head engages an oblong slot in the roller ring

ELECTRONIC DIESEL CONTROL (VE PUMP) 579

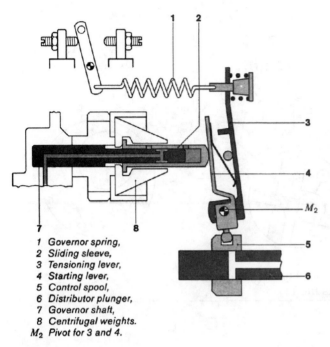

Governor-lever assembly with load-dependent injection timing (LFB).

1. Governor spring,
2. Sliding sleeve,
3. Tensioning lever,
4. Starting lever,
5. Control spool,
6. Distributor plunger,
7. Governor shaft,
8. Centrifugal weights.
M_2 Pivot for 3 and 4.

FIGURE 25-16 (Courtesy of Robert Bosch Canada Ltd.)

that allows the timing piston to take over injection timing when engine speed reaches a certain point without interference from the cold-start device (**Figure 25-17**).

The automatic device operates in the same way but is automatically controlled by a temperature-responsive expansion element. The temperature-sensitive device converts coolant temperature changes into longitudinal motion, which acts on the roller ring to advance injection timing when engine coolant temperature is below the specified value (**Figure 25-18**). This device is used with the V-MAC electronic control system. A temperature-dependent fast-idle device is used with the cold-start device shown in **Figure 25-17**. This device is also actuated by the cold-start control mechanism. With a cold engine the ball pin presses against the engine speed control lever to increase idle speed. When the engine is warm, the idle speed returns to normal.

ELECTRONIC DIESEL CONTROL (VE PUMP) (FIGURE 25-19)

The VE pump electronic control provides more precise control of the following:

1. The quantity of fuel injected (fuel metering)
2. The start of injection (injection timing)
3. The quantity of exhaust gas recirculation (EGR)
4. Charge-air pressure (turbocharger boost pressure)

The EDC consists of four system blocks:

1. Sensors that monitor operating conditions and send this information to the electronic control unit (ECU) as electrical signals.

2. An ECU with its microprocessors that processes incoming data according to stored data to produce outgoing electrical signals to operate system actuators.

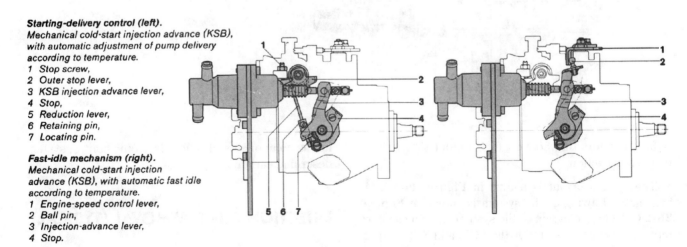

Starting-delivery control (left).
Mechanical cold-start injection advance (KSB), with automatic adjustment of pump delivery according to temperature.
1 Stop screw,
2 Outer stop lever,
3 KSB injection advance lever,
4 Stop,
5 Reduction lever,
6 Retaining pin,
7 Locating pin.

Fast-idle mechanism (right).
Mechanical cold-start injection advance (KSB), with automatic fast idle according to temperature.
1 Engine-speed control lever,
2 Ball pin,
3 Injection-advance lever,
4 Stop.

FIGURE 25-17 (Courtesy of Robert Bosch Canada Ltd.)

580 Chapter 25 ROBERT BOSCH TYPE VE DISTRIBUTOR PUMP

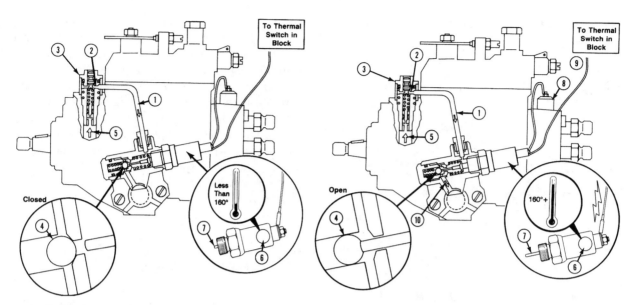

FIGURE 25–18 Cold-start module (KSB) operation with engine below 160° F (left) and above 160° F (right). (Courtesy of Mack Trucks Inc.)

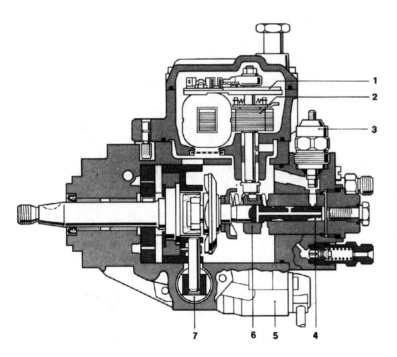

FIGURE 25–19 Distributor-type fuel injection pump with electronic governor. 1. Control-collar position sensor, 2. Injected fuel quantity actuator, 3. Electromagnetic shut-off valve (ELAB), 4. Supply plunger, 5. Solenoid valve for start of injection, 6. Control collar, 7. Timing device. (Courtesy of Robert Bosch Canada Ltd.)

3. Actuators that convert ECU output signals into mechanical action.

The system layout is shown in **Figure 25–20.** A schematic drawing of the system is shown in **Figure 25–21.** A block diagram of the system showing sensor inputs, information stored in the ECU, and ECU output signals with system actuators is shown in **Figure 25–22.** System components operate in the same manner as components of other electronic control systems described earlier.

INJECTION PUMP REMOVAL (TYPICAL)

Always refer to the service manual for exact procedures and specifications.

INJECTION PUMP REMOVAL (TYPICAL)   581

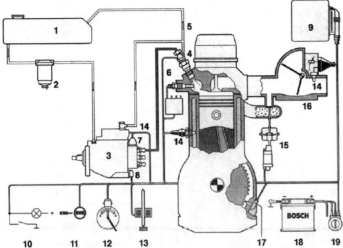

FIGURE 25-20 (Courtesy of Robert Bosch Canada Ltd.)

Fuel-injection system with electronically-controlled distributor-type fuel-injection pump.
1 Fuel tank, 2 Fuel filter, 3 VE pump, 4 Injection nozzle with needle-motion sensor, 5 Fuel return line, 6 Sheathed-element glow plug, glow control unit, 7 Shut-off device, 8 Solenoid valve, 9 ECU, 10 Diagnosis indicator, 11 Speed selector lever, 12 Accelerator-pedal sensor, 13 Road-speed sensor, 14 Temperature sensor (water, air, fuel), 15 EGR valve, 16 Air-flow sensor, 17 Engine-speed and TDC sensor, 18 Battery, 19 Glow-plug and starter switch.

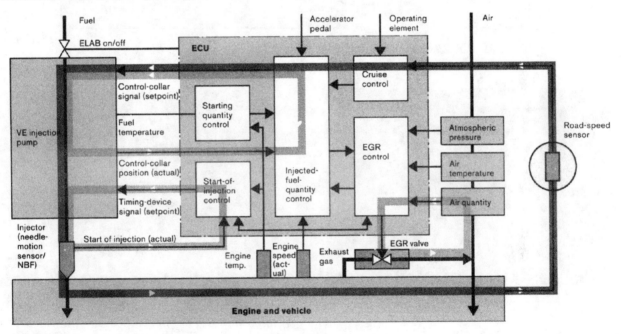

FIGURE 25-21 (Courtesy of Robert Bosch Canada Ltd.)

1. Thoroughly clean the area around the pump, especially around the fittings and inspection cover.

2. Disconnect and remove all lines—high-pressure, fuel inlet, and return—from the pump. Cap all open fittings and lines to prevent contamination.

3. Remove the inspection cover in front of the pump drive gear.

4. Check drive gear backlash with a dial indicator **(Figure 25-23)**.

5. Position the No. 1 piston at TDC on the compression stroke **(Figure 25-24)**.

6. Lock the injection pump driveshaft in position **(Figure 25-25)**.

7. Remove the pump drive gear **(Figure 25-26)**.

8. Remove the injection pump support bracket, pump mounting capscrews, and pump.

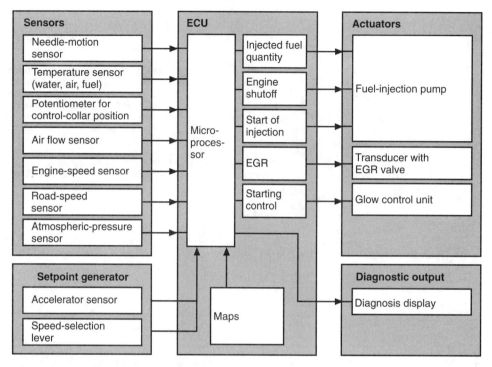

FIGURE 25-22 (Courtesy of Robert Bosch Canada Ltd.)

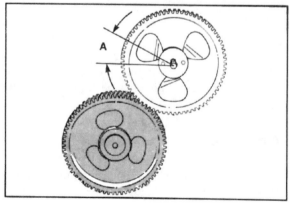

FIGURE 25-23 Checking pump drive gear backlash. (Courtesy of Cummins Engine Company, Inc.)

FIGURE 25-24 Barring the engine over while pressing in on the lock pin to position the No. 1 piston at TDC on compression. (Courtesy of Cummins Engine Company, Inc.)

INJECTION PUMP REPAIRS

Replacing the Shaft Seal (Figure 25-27)

1. Use a seal puller to remove the seal.
2. Inspect the seal seating area for nicks and burrs. Remove any minor nicks and burrs, then clean the area thoroughly.

INJECTION PUMP REPAIRS 583

FIGURE 25-25 Loosening the pump lockscrew to remove the spacer washer. Tightening the lockscrew then locks the pump. (Courtesy of Cummins Engine Company, Inc.)

3. Install the new seal using a protective sleeve on the pump shaft. Drive the seal into place until it bottoms in the seal bore.

Replacing the Delivery Valve (Figure 25-28)

1. Remove the delivery valve holder (if used), delivery valve, and sealing washer.
2. Inspect the sealing surfaces on the high-pressure head, delivery valve, and valve holder.
3. Install the new delivery valve with a new sealing washer and tighten to specifications.

Replacing the Shutdown Solenoid (Figure 25-29)

1. Remove the solenoid and O-ring.
2. Remove the plunger and spring.
3. Place a new O-ring on the replacement solenoid.
4. Install the new solenoid, plunger, and spring into the distributor head and tighten to specifications.

Replacing the Shutdown Lever/Spring (Figure 25-30)

CAUTION:
Mark the shutdown lever position so it can be installed in the same position.

FIGURE 25-26 Removing the pump drive gear after the gear cover is removed. (Courtesy of Cummins Engine Company, Inc.)

1. Disconnect the return spring.
2. Remove the lever and spring.
3. Using the removed lever as a pattern, mark the replacement lever so it can be installed in the same position as the one removed.
4. Align the marks and install the spring and lever. Install the lock washer and nut and tighten to specifications.

Replacing the Overflow Adapter/Seal Ring (Figure 25-31)

1. Remove the overflow adapter.
2. Inspect the sealing surfaces. Check to make sure the orifice is open.
3. Install a new sealing washer and the adapter and tighten to specifications.

The fuel inlet adapter/seal can be replaced using the same procedure.

584 Chapter 25 ROBERT BOSCH TYPE VE DISTRIBUTOR PUMP

FIGURE 25-27 Pump seal removal (top) and installation (bottom). (Courtesy of Cummins Engine Company, Inc.)

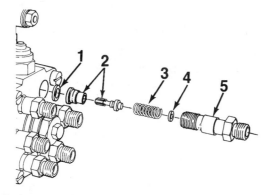

FIGURE 25-28 Delivery valve components. 1. Shim, 2. Delivery valve, 3. Spring, 4. Seal, 5. Holder. (Courtesy of Cummins Engine Company, Inc.)

Cold-Start Module Electrical Check (Figure 25-32)

1. Loosen and remove the electrical element from the module housing.

2. Ground the hex portion of the element and apply 12 V to the electrical terminal. Note whether the

FIGURE 25-29 Shutdown solenoid replacement. (Courtesy of Cummins Engine Company, Inc.)

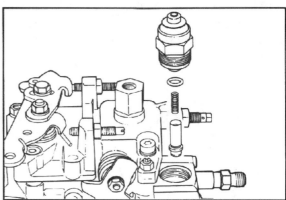

plunger extends. If it does not move after about 1 minute, recheck your connections. If the connections are good and the plunger does not move, replace the element.

3. Install the element and tighten to specifications.

VE INJECTION PUMP TIMING

1. Secure the pump in a vise with soft jaws.

CAUTION:

Do not overtighten the vise or position the pump in the vise in a manner that could damage the pump.

2. Remove the access plug from the rotary head central screw assembly.

3. Thread the timing tool (part no. 3377259, Cummins) extension into the access plug hole finger tight.

4. Thread the dial indicator tip extension into the dial indicator finger tight **(Figure 25-33)**.

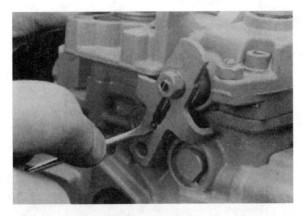

FIGURE 25–30 Shutdown lever/spring replacement. (Courtesy of Cummins Engine Company, Inc.)

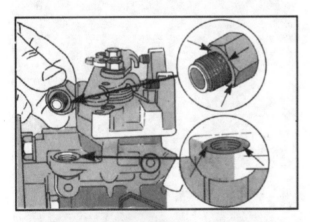

FIGURE 25–31 Fuel inlet adapter seal replacement. (Courtesy of Cummins Engine Company, Inc.)

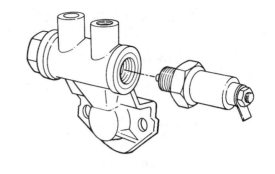
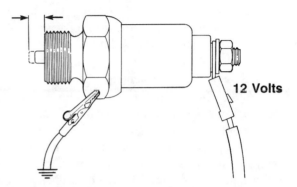

FIGURE 25–32 Cold-start module element check and replacement. (Courtesy of Cummins Engine Company, Inc.)

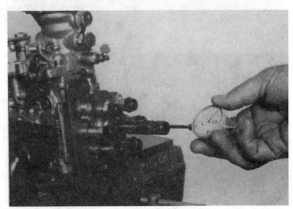

FIGURE 25–33 Installing the timing tool (top) and dial indicator (bottom). (Courtesy of Cummins Engine Company, Inc.)

5. Set the indicator to allow at least 0.125 in. (3.0 mm) travel. Tighten the lock sleeve finger tight.

6. At the point of injection the keyway of the shaft will align with the delivery valve receiving the injection and with the hash mark on the seal housing.

NOTE: The illustrated mark is for reference only and should not be used for setting the pump timing.

7. The No. 1 delivery valve is marked as illustrated. Four cylinder = A, with a firing order of 1-3-4-2. Six cylinder = D, with a firing order of 1-5-3-6-2-4.

8. Install the drive gear retaining nut on the pump drive shaft. Make sure the pump is unlocked.

9. Rotate the pump drive shaft clockwise. The gauge will also rotate in a clockwise direction.

10. As injection is completed to the respective ports the pump will snap **(Figure 25–34)**. At this point the gauge will reverse direction to counterclockwise. Zero the gauge at the point where the needle stops and turns clockwise again.

11. Continue to rotate the pump clockwise until the keyway prepares to align with the No. 1 delivery valve. Verify that the gauge is properly zeroed.

12. Continue rotating the pump clockwise while watching the gauge. Count the revolutions. Each revolution equals 0.50 mm. Three revolutions equals 1.50 mm.

13. Lock the pump as the required plugs lift.

14. Remove the nut from the pump drive shaft by striking the wrench with a sharp blow in a counterclockwise direction. Remove the dial indicator. Install the access plug and tighten to specifications.

FIGURE 25–34 (Courtesy of Cummins Engine Company, Inc.)

VE INJECTION PUMP INSTALLATION

1. Install the injection pump gasket.

2. Install the key in the pump shaft keyway. To prevent the key from falling out, use a small center punch to create a small swelling on one side of the key. Use a hammer to tap the key into place **(Figure 25–35)**.

3. Position the No. 1 piston at TDC by barring the engine over while pushing in on the engine timing pin until it engages and stops rotation.

4. Orient the wide end of the tapered bore of the pump drive gear toward the engine (timing marks away).

5. Use the information from the engine data plate (Cummins critical parts list number) and the control parts list manual to determine the engine year and the agency under which it is certified (EPA or CARB). Use this information to determine which letter on the fuel injection pump drive gear is to be aligned with the camshaft gear.

6. Align the timing marks and set the gear into the housing **(Figure 25–36)**.

FIGURE 25–35 Installing the pump gasket (top) and drive shaft key (bottom). (Courtesy of Cummins Engine Company, Inc.)

VE INJECTION PUMP INSTALLATION 587

FIGURE 25-36 Aligning gear timing marks. (Courtesy of Cummins Engine Company, Inc.)

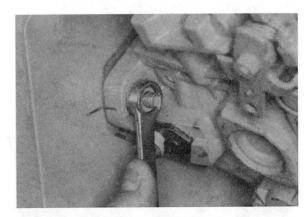

FIGURE 25-38 Scribe mark alignment. (Courtesy of Cummins Engine Company, Inc.)

7. A new or reconditioned pump is locked in a position corresponding to the drive gear keyway when the No. 1 piston is at TDC on compression **(Figure 25-37)**.

8. Position the pump on the engine and finger tighten the mounting screws. The pump must be free to move.

9. Install the drive gear mounting nut and spring washer. Tighten the nut to the specified *preliminary* torque. This nut is not tightened to the final torque until after the pump is unlocked.

10. Rotate the pump to align the scribe marks, and tighten the mounting nuts to specified torque **(Figure 25-38)**.

NOTE: If the new or rebuilt pump is without scribe marks take up the gear lash by rotating the pump against the direction of drive rotation, making sure the engine remains at TDC. Tighten the mounting nuts to specifications.

11. Unlock the pump by loosening the lockscrew. Place the special washer behind the lockscrew head and tighten the screw to specifications.

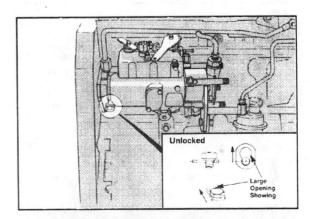

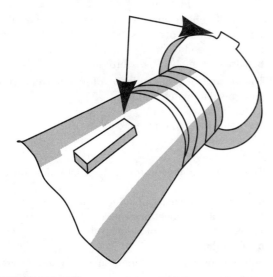

FIGURE 25-37 Pump drive shaft key and pump drive gear keyway alignment. (Courtesy of Cummins Engine Company, Inc.)

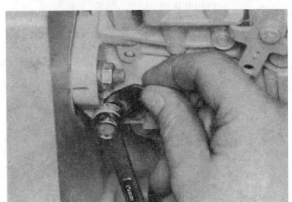

FIGURE 25-39 Unlocking the injection pump. (Courtesy of Cummins Engine Company, Inc.)

12. Disengage the engine timing pin (**Figure 25–39**).

13. Tighten the pump drive gear mounting nut to final torque specifications.

14. Use a chisel and hammer and carefully mark the injection pump flange to match the mark on the engine gear housing.

15. Install the gear train cover.

REVIEW QUESTIONS

1. Name the VE distributor pump's functional groups.
2. The VE distributor fuel pump is driven at _____ speed.
3. Bosch VE fuel injection systems have a _____ _____ fuel supply pump.
4. If supply pump fuel pressure exceeds a preset value, the valve piston opens and allows the fuel back to the _____ _____.
5. The pressure necessary for injection into the cylinder is generated by the _____.
6. The pressure valve determines the point at which the injector nozzle closes at the end of _____.
7. The governor ensures that the engine does not fall below the controlled _____ _____.
8. The desirable governor speed _____ depends on engine application.
9. In the timing advance operation, the timing piston is held in its initial position by a _____ _____.
10. A manifold-pressure compensator is used on _____ diesel engines.
11. Explain the purpose of altitude compensation.
12. Name the electronic diesel control (VE pump) components.
13. Explain how to replace the injection pump shaft seal.

TEST QUESTIONS

1. The distributor fuel injection pump is engine driven at
   a. crankshaft speed
   b. camshaft speed
   c. twice camshaft speed
   d. steady rpm
2. The fuel supply pump draws fuel from the tank and delivers it to the
   a. injection pump cavity
   b. primary filter
   c. secondary filter
   d. high-pressure steel lines
3. Besides driving the distributor pump plunger the cam plate also influences the
   a. fuel injection pressure
   b. duration of fuel injection
   c. fuel injection pressure and duration
   d. governor operation
4. The pressure necessary for injection into the engine cylinders is generated by the
   a. injectors
   b. supply pump
   c. governor
   d. plunger
5. The pressure valve is a fluid-controlled
   a. plunger type
   b. vane type
   c. poppet type
   d. over mechanical
6. Speed droop is the percentage of engine speed increase when
   a. traveling down grade
   b. load is removed
   c. operator decelerates
   d. all of the above
7. The variable-speed governor controls all engine speed
   a. regardless of the operator's demands
   b. regardless of engine workload
   c. between starting and maximum value
   d. for highway operation
8. The hydraulic timing device is built into the
   a. governor assembly
   b. distributor pump drive mechanism
   c. distributor pump upper right side
   d. distributor pump underside
9. On the distributor injection pump the mechanical shutoff is in the form of a
   a. lever assembly
   b. cable assembly
   c. solenoid and lever
   d. throttle
10. Torque control is the variation of fuel delivery with engine speed to match it to the
    a. engine rpm
    b. engine fuel requirement curve
    c. altitude
    d. air humidity
11. The altitude compensator, at high altitudes, lower air density
    a. increases the mass of induced air
    b. keeps air volume the same
    c. reduces the mass of induced air
    d. all of the above
12. Starting noise, fuel consumption, and emissions are significantly influenced by
    a. the glow plug
    b. the start of injection
    c. increased injection pressure
    d. correct operator starting procedure

◆ CHAPTER 26 ◆

STANADYNE (ROOSA MASTER) AND LUCAS CAV DISTRIBUTOR PUMPS

INTRODUCTION

The first distributor pump produced by Stanadyne/Hartford (formerly Roosa Master) was the model A. The basic design of this pump was carried over into the later D, B, and DB models. The JDB pump produced for John Deere is a modified version of the DB model. The DM pump, a heavy-duty model capable of 10,000 psi (68,950 kPa) line pressure is also available in a four-plunger model. The four-plunger DB4 pump is similar in design to the DM model. Stanadyne Diesel System pumps are in widespread use in automotive, industrial, and agricultural equipment.

Lucas CAV first produced the DPA distributor pump under a 1956 licensing agreement with Roosa Master. The DPA was modeled after the Roosa Master A model pump. The more recent Roto-Diesel pump is similar to the DPA model with some additional features.

PERFORMANCE OBJECTIVES

After thorough study of this chapter and the appropriate training models and service manuals, and with the necessary tools and equipment, you should be able to do the following with respect to diesel engines equipped with Stanadyne and Lucas CAV distributor-type injection pumps:

1. Describe the basic construction and operation of Stanadyne and Lucas CAV distributor injection pumps, including the following:
   a. method of fuel pumping and metering
   b. delivery valve action
   c. lubrication
   d. injection timing control
   e. governor types
2. Diagnose basic fuel injection system problems and determine the correction needed.
3. Remove, clean, inspect, test, measure, repair, or replace fuel injection system components as required to meet manufacturer's specifications.

TERMS YOU SHOULD KNOW

Look for these terms as you study this chapter, and learn what they mean.

drive shaft	cam ring
distributor rotor	cam rollers and shoes
hydraulic head	vane-type transfer pump
pumping plungers	end cap

(The information in this chapter is based on and contains information provided courtesy of Stanadyne Diesel Systems, Lucas CAV Ltd., Cummins Engine Company, Inc., and Deere and Company.)

pressure regulator
charge ports
axial bore
discharge port
metering valve bore
metering valve
charge cycle
discharge cycle
delivery valve
fuel-return circuit
timing advance unit
all-speed governor
governor weights
thrust sleeve
governor spring
torque control screw
aneroid fuel control
electrical shut-off
electronic fuel control system
pump testing and calibration procedures

STANADYNE (ROOSA MASTER) DISTRIBUTOR INJECTION PUMPS

The Stanadyne distributor injection pumps use a helix-type metering valve. The radial position of the valve is controlled by throttle position and the governor. Cam-ring-operated pumping plungers pump high-pressure fuel for injection to the distributor head, where fuel is distributed to individual discharge ports for each engine cylinder.

The microprocessor-controlled injection pump uses a number of engine sensors to provide information to the microprocessor. This information is processed to provide the best possible fuel delivery for every engine operating condition and mode.

D Series Pump Operation

The main rotating components of the D series Stanadyne distributor pump are the drive shaft, transfer pump blades, distributor rotor, and governor.

With reference to **Figure 26–1,** the drive shaft (1) engages the distributor rotor (2) in the hydraulic head (6). The drive end of the rotor incorporates two or four pumping plungers (4), depending on pump model.

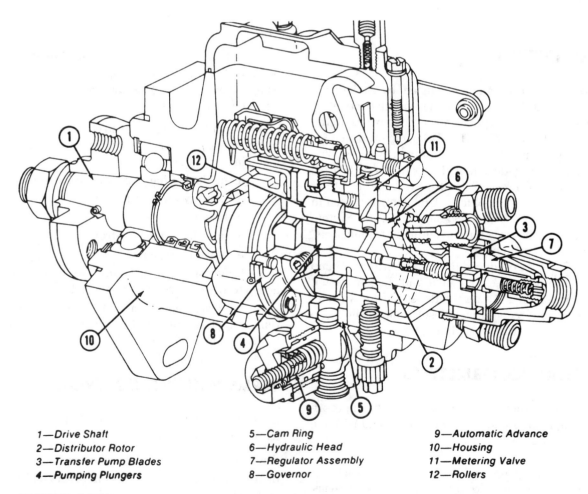

1—Drive Shaft
2—Distributor Rotor
3—Transfer Pump Blades
4—Pumping Plungers
5—Cam Ring
6—Hydraulic Head
7—Regulator Assembly
8—Governor
9—Automatic Advance
10—Housing
11—Metering Valve
12—Rollers

FIGURE 26–1 Cutaway view of Roosa Master distributor-type injection pump. Note pumping plungers and metering valve. (Courtesy of Stanadyne Diesel Systems.)

The plungers are actuated toward each other simultaneously by an internal cam ring (5) through rollers and shoes, which are carried in slots at the drive end of the rotor. The number of cam lobes normally equals the number of engine cylinders.

The transfer pump (3) at the rear of the rotor is of the positive-displacement vane type and is enclosed in the end cap (7). The end cap also houses the fuel inlet strainer and transfer pump pressure regulator. The face of the regulator assembly is compressed against the liner and distributor rotor and forms an end seal for the transfer pump. The injection pump is designed so that the end thrust is against the face of the transfer pump pressure regulator. The distributor rotor incorporates two charging ports and a single axial bore with one discharge port to service all head outlets to the injection lines.

The hydraulic head contains the bore in which the rotor revolves, the metering valve bore, the charging ports, and the head discharge outlets. The high-pressure injection lines to the nozzles are fastened to these discharge outlets.

A high-pressure relief slot is incorporated in some regulators as part of the pressure-regulating slot to prevent excessively high transfer pump pressure, if the engine or pump is accidentally overspeeded.

The transfer pump works equally well with different grades of diesel fuel and varying temperatures, both of which affect fuel viscosity. A unique and simple feature of the regulating system offsets pressure changes caused by viscosity differences. Located in the spring adjusting plug is a thin plate incorporating a sharp-edged orifice **(Figure 26–2)**. The orifice allows fuel leakage past the piston to return to the inlet side of the pump. Flow through a short orifice is virtually unaffected by viscosity changes. The pressure exerted against the back side of the piston is determined by the leakage through the clearance between the piston and the regulator bore and the pressure drop through the sharp-edged orifice. With cold or viscous fuels, very little leakage occurs past the piston. The additional force on the back side of the piston from the viscous fuel pressure is slight. With hot or light fuels, leakage past the piston increases. Fuel pressure in the spring cavity increases also, since the flow past the piston must equal the flow through the orifice. Pressure rises due to increased piston leakage, and pressure rises to force more fuel through the orifice. This variation in piston position compensates for the leakage that would occur with thin fuels, and design pressures are maintained over a broad range of viscosity changes.

Charging Cycle

As the rotor revolves, the inlet passages in the rotor register with the ports of the circular charging passage. Fuel under pressure from the transfer pump, controlled by the opening of the metering valve, flows into the pumping chamber, forcing the plungers apart **(Figures 26–3 to 26–7)**.

The plungers move outward at a distance proportionate to the amount of fuel required for injection on the following stroke. If only a small quantity of fuel is admitted into the pumping chamber, as at idling, the plungers move out a short distance. Maximum plunger travel and, consequently, maximum fuel delivery are limited by the leaf spring, which contacts

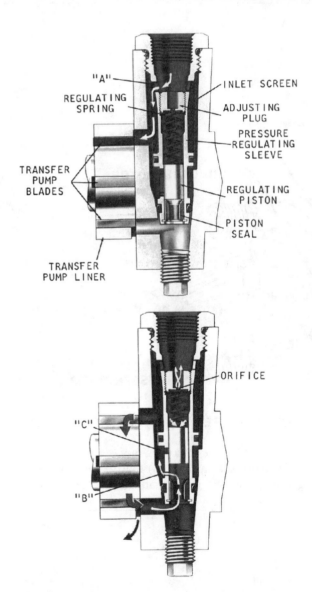

FIGURE 26–2 Pressure-regulating valve operation. Building pressure (top) and regulating (bottom). (Courtesy of Stanadyne Diesel Systems.)

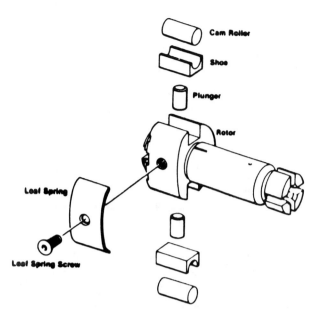

FIGURE 26-3 Exploded view of rotor assembly. (Courtesy of Stanadyne Diesel Systems.)

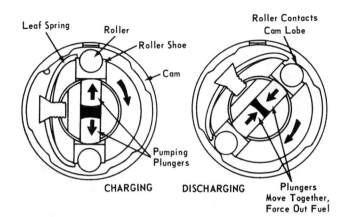

FIGURE 26-4 Cam roller and pumping plunger action during charging and discharging. (Courtesy of Deere and Company.)

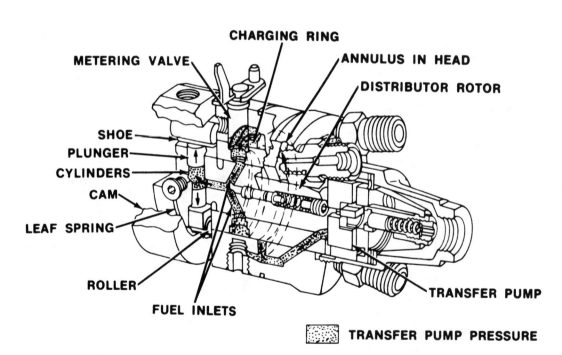

FIGURE 26-5 Roosa Master pump metering is achieved by three controlling factors: (1) transfer pump fuel pressure, (2) metering valve position, and (3) leaf spring tension keeping the pumping plungers together. (Courtesy of Stanadyne Diesel Systems.)

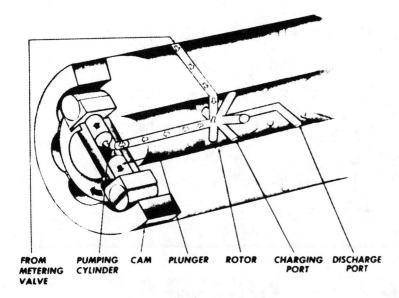

FIGURE 26–6 Fuel pressure force between plungers forces plungers apart against leaf-spring pressure during metering. (Leaf spring not shown here.) (Courtesy of Stanadyne Diesel Systems.)

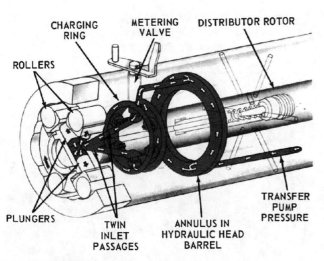

FIGURE 26–7 Charging-cycle fuel flow. (Courtesy of Stanadyne Diesel Systems.)

the edge of the roller shoes. Only when the engine is operating at full load will the plungers move to the most outward position. Note that while the angled inlet passages in the rotor are in registry with the ports in the circular charging passage, the rotor discharge port is not in registry with a head outlet. Note also that the rollers are off the cam lobes. Compare their relative positions.

Discharge Cycle

As the rotor continues to revolve, the inlet passages move out of registry with the charging ports. The rotor discharge port opens to one of the head outlets. The rollers then contact the cam lobes, forcing the shoes in against the plungers, and high-pressure pumping begins **(Figures 26–8 to 26–10).**

The beginning of injection varies according to load (volume of charging fuel), even though rollers may always strike the cam at the same position. Further rotation of the rotor moves the rollers up the cam lobe ramps, pushing the plungers inward. During the discharge stroke the fuel trapped between the plungers flows through the axial passage of the rotor and discharge port to the injection line. Delivery to the injection line continues until the rollers pass the innermost point on the cam lobe and begin to move outward. The pressure in the axial passage is then reduced, allowing the nozzle to close. This is the end of delivery.

The delivery valve operates in a bore in the center of the distributor rotor. Note that the valve requires no seat—only a stop to limit travel. Sealing is accomplished by the close clearance between the valve and bore into which it fits. Because the same delivery valve performs the function of retraction for each injection line, the result is a smooth-running engine at all loads and speeds.

Fuel-Return Circuit

A fuel-return circuit is provided in the pump to prevent air from entering the pumping chambers. A cavity built into the pump collects any air present and returns it via the fuel-return circuit to the fuel tank **(Figure 26–11).** A wire in the air vent passage helps break up any large air bubbles and restricts the flow of return fuel. Differ-

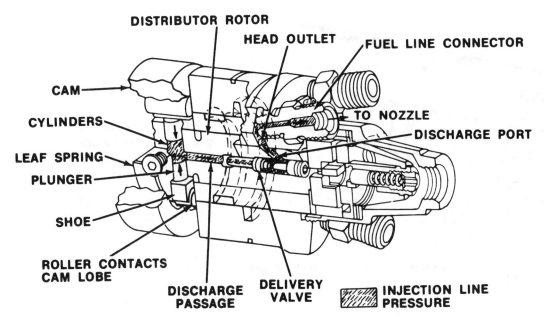

FIGURE 26-8 Continued rotation of the rotor causes the plungers to be forced closer together, forcing fuel out of the high-pressure discharge passage to the injector. The inlet passage is closed when the rotor is in this position. (Courtesy of Stanadyne Diesel Systems.)

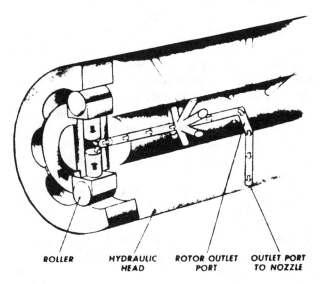

FIGURE 26-9 High-pressure fuel passages. Rollers force the plungers together every time they roll over the high parts of the cam ring. The cam ring has the same number of cams as the engine has cylinders. (Courtesy of Stanadyne Diesel Systems.)

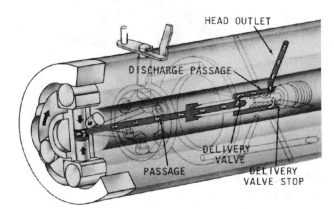

FIGURE 26-10 Discharge-cycle fuel flow. (Courtesy of Stanadyne Diesel Systems.)

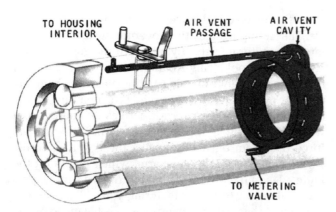

FIGURE 26–11 Fuel-return circuit. (Courtesy of Stanadyne Diesel Systems.)

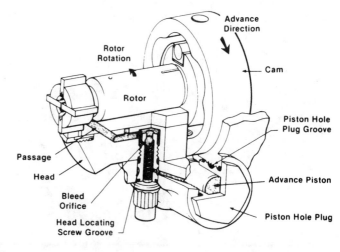

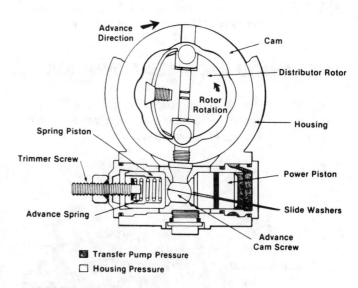

FIGURE 26–12 Roosa Master distributor pump automatic injection timing device. Fuel oil transfer pump pressure is the medium used to operate this device. (Courtesy of Stanadyne Diesel Systems.)

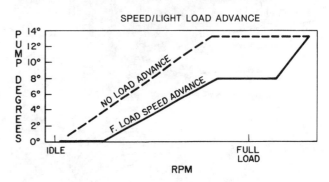

FIGURE 26–13 No-load and full-load advance graph. (Courtesy of Stanadyne Diesel Systems.)

ent size vent wires are available to achieve the specified quantity of return fuel.

STANADYNE (ROOSA MASTER) INJECTION TIMING

Roosa Master basic injection timing is achieved by proper alignment of timing marks at the pump drive mounting. An automatic hydraulic advance mechanism controls timing in relation to engine speed (**Figures 26–12 to 26–14**).

The Roosa Master design permits the use of a simple hydraulic servomechanism, powered by oil from the transfer pump, to rotate the normally stationary cam ring to advance injection timing. Transfer pump pressure, increasing with speed, operates the servoadvance piston against spring pressure as required along a predetermined timing curve.

The automatic advance may also be used to retard injection, as an aid in cold starting, by starting the cam advance from a retard position.

Controlled movement of the cam in the pump housing is induced and limited by the action of the power piston and spring of the automatic advance against the cam advance screw.

During cranking, the cam is in the retard position, since the force of the advance spring is greater than transfer pump pressure. As the engine rpm and transfer pump pressure increase, oil entering the advance housing behind the power piston moves the cam. Any amount of advance may be provided, but the limit is 14 pump degrees. A ball check valve is provided to offset the normal tendency of the cam to return to the retard position during injection.

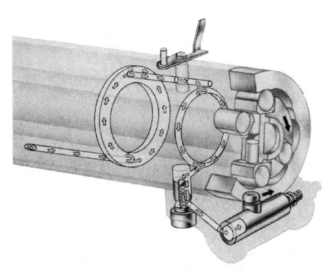

FIGURE 26–14 Fuel flow in the model DC pump advance unit. (Courtesy of Stanadyne Diesel Systems.)

MECHANICAL ALL-SPEED GOVERNOR

The governor serves the purpose of maintaining the desired engine speed within the operating range under various load settings **(Figure 26–15)**.

In the mechanical governor the movement of the weights acting against the governor thrust sleeve rotates the metering valve by means of the governor arm and linkage hook. This rotation varies the registry of the metering valve opening to the passage from the transfer pump, thereby controlling the quantity of fuel to the plungers. The governor derives its energy from weights pivoting in the weight retainer. Centrifugal force tips them outward, moving the governor thrust sleeve against the governor arm, which pivots on the knife edge of the pivot shaft and through a simple, positive linkage, rotates the metering valve. The force of the weights against the governor arm is balanced by the governor spring force, which is controlled by the manually positioned throttle lever and vehicle linkage for the desired engine speed.

In the event of a speed increase due to a load reduction, the resultant increase in centrifugal force of the weights rotates the metering valve clockwise to reduce fuel. This limits the speed increase (within the operating range) to a value determined by the governor spring rate and setting of the throttle.

When the load on the engine is increased, the speed tends to reduce. The lower speed reduces the force generated by the weights, permitting the spring force to

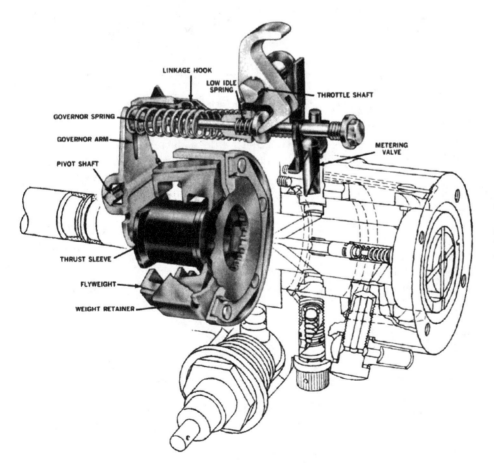

FIGURE 26–15 Mechanical governor components. (Courtesy of Stanadyne Diesel Systems.)

rotate the metering valve in the counterclockwise direction to increase fuel. The speed of the engine at any point within the operating range is dependent on the combination of load on the engine and the governor spring rate and setting as established by the throttle position. A light idle spring is provided for more sensitive regulation when weight energy is low in the low end of speed range. The limits of throttle travel are set by adjusting screws for proper low-idle and high-idle positions.

A light tension spring on the linkage assembly takes up any slack in the linkage joints and also allows the shut-off mechanism to close the metering valve without having to overcome the governor spring force. Only a very light force is required to rotate the metering valve to the closed position.

Torque Control Screw

Torque is commonly defined as the turning moment or "lugging ability" of an engine. Maximum torque varies at each speed in the operating range for two reasons: (1) as engine speed increases, friction losses progressively increase and (2) combustion chamber efficiency drops due to loss of volumetric efficiency (breathing ability of an engine), and due to reduction of time necessary to completely and cleanly burn the fuel in the cylinder. Since the torque increases with increased load conditions, a predetermined point at which maximum torque is desired may be selected for any engine. Thus, as engine rpm decreases, the torque generally increases toward this preselected point. This desirable feature is called "torque backup." Three basic factors affect torque backup:

1. metering valve opening area
2. time allowed for charging
3. transfer pump pressure curve

Of these, the only control among engines for purposes of establishing a desired torque curve is the transfer pump pressure curve and metering valve opening, since the other factors involved are common to all engines. Torque control in Roosa Master fuel injection pumps is accomplished in the following manner:

The manufacturer determines at what speed for a specific application the engine is to develop its maximum torque. The maximum fuel setting is then adjusted for required delivery during dynamometer testing. This delivery must provide acceptable fuel economy. The engine is then brought to full-load governed speed. The fuel delivery is then reduced from that determined by the maximum fuel setting by turning in an adjustment or "torque screw" that moves the metering valve toward the closed position. The engine is then running at full-load governed speed. When the engine is operating at high-idle speed, no load, the quantity of fuel delivered is controlled only by governor action through the metering valve. At this point, the torque screw and maximum fuel adjustment have no effect. As load is applied, the quantity of fuel delivered is controlled only by governor action and metering valve position until full-load governed speed is reached. At this point, further opening of the metering valve is prevented by its contact with the previously adjusted torque screw. Thus, the amount of fuel delivered at full-load governed speed is controlled by the torque screw and not by the roller-to-roller dimension. As additional load is applied and engine rpm decreases, a greater quantity of fuel is allowed to pass into the pumping chamber due to the increased time of registration of the charging ports. During this phase of operation the metering valve position remains unchanged, still being held from further rotation by the torque screw. As engine rpm continues to decrease under increasing load, the rotor charging ports remain in registry for a longer time period, allowing a larger quantity of fuel into the pumping chamber. Fuel delivery increases until the predetermined point of maximum torque is reached **(Figure 26–16)**.

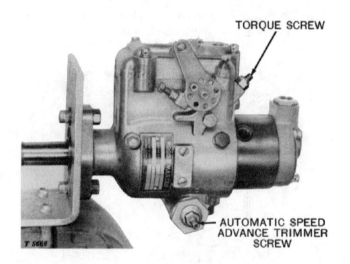

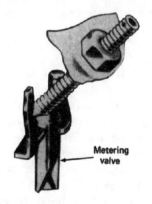

FIGURE 26–16 Torque screw and metering valve. (Courtesy of Stanadyne Diesel Systems.)

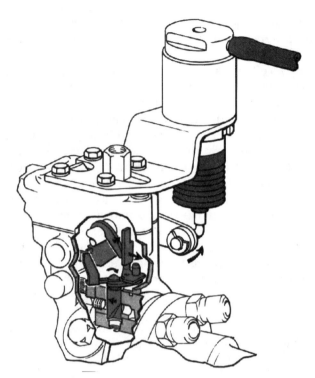

FIGURE 26-17 Aneroid control used on DM pump. (Courtesy of Stanadyne Diesel Systems.)

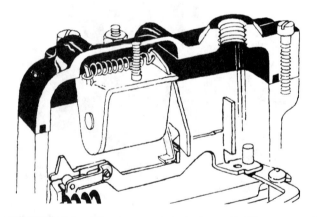

FIGURE 26-18 Electrical fuel shut-off. (Courtesy of Stanadyne Diesel Systems.)

Aneroid Fuel Control

The aneroid is designed to reduce exhaust smoke during acceleration by temporarily limiting fuel delivery on turbocharged engines. The aneroid consists of a spring-loaded diaphragm that pushes upward on the shut-off lever during engine idle, thereby restricting fuel delivery. As the engine is accelerated turbocharger boost pressure directed to the diaphragm gradually overcomes spring pressure as boost pressure increases and pushes the shut-off to the RUN position. When the turbocharger reaches full speed, maximum fuel delivery is provided **(Figure 26-17)**.

Electrical Shut-Off

Electrical shut-off devices are available as an option in both energized to run (ETR) and energized to shut-off (ETSO) models. These solenoids are included in various applications to control the run and stop functions of the engine. They accomplish this by positively stopping fuel flow to the plungers, thereby interrupting injection when the switch is turned off **(Figure 26-18)**.

STANADYNE ELECTRONIC FUEL INJECTION PUMP

The Stanadyne electronic fuel control pump uses a plunger control mechanism in place of a metering valve. Electronic governing eliminates the mechanical governing system from the pump, making it a smaller and lighter package **(Figure 26-19)**.

Plunger control is achieved by physically limiting the radial displacement of the plungers. A yoke, located and guided by a slot at the driven end of the rotor, has a set of fingers that straddle the plungers. Ramps ground into the plungers make contact with these fingers during the charging sequence, thereby limiting the amount of fuel available for the pumping sequence. The axial position of the yoke controls the outward radial displacement of the plungers. Disengagement of the plunger ramps from the yoke fingers begins when pumping commences.

Controlling plunger motion in this manner is advantageous because of lower dynamic loads on the control mechanisms.

Yoke positioning is accomplished by a cam follower that rotates against a cam profile ground into the pilot tube. The follower displacement is rod to the yoke. Low force levels on the cam follower permit the use of actuators to rotate this part. A potentiometer measuring the position of the cam follower provides feedback to the ECU for fuel delivery accuracy.

The advance system is a servo type actuated by a stepper motor. A sensor in the nozzle provides a start-of-injection signal for timing feedback.

An eight-bit, single-chip microcomputer, engine-mounted sensors, and the Model PCF diesel fuel injection pump are the major components of the electronic system. Engine-mounted sensors analyze engine speed, top dead center, start of injection, water

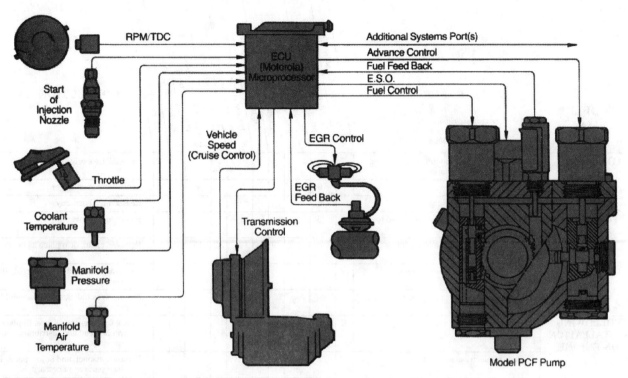

FIGURE 26–19 Stanadyne electronic control system. (Courtesy of Stanadyne Diesel Systems.)

temperature, manifold pressure, and exhaust gas recirculation valve position. Start of injection is determined by a special nozzle that incorporates a Hall-effect needle lift sensor. The ECU, based on data received from the sensors, determines the desired advance and fuel delivery levels best suited to engine needs.

Benefits of Electronic Control of Fuel Injection in the PCF Injection Pump

1. Improved emission control, fuel economy, and engine performance
2. Fueling control functions
   - Maximum-load fuel shaping
   - Altitude fuel compensation
   - Turbocharge boost compensation
   - Excess fuel at cranking speeds
   - Transient fuel trimming
   - Throttle progression tailoring
   - Minimum/maximum or all-speed governing
3. Timing control functions
   - Speed/load advance
   - Cold-start advance
   - Altitude timing compensation
   - Pump installation error
   - Drive shaft wear compensation
   - Timing map flexibility
4. Optional control functions available
   - Cruise control
   - Optimum transmission control
   - EGR (exhaust gas recirculation) control

STANADYNE DISTRIBUTOR PUMP TROUBLESHOOTING

The distributor pump troubleshooting guide in **Figure 26–20** lists common operating problems that may be encountered, the probable cause for each problem, and the required correction. Always refer to the appropriate service manual for exact procedures and specifications for each pump and any particular application.

STANADYNE DISTRIBUTOR PUMP TESTING AND CALIBRATION

Any test is only as good as the testing equipment employed. Incorporation of quality test equipment

PROBLEM

PROBLEM MAY OCCUR	CAUSE *Numbers in "Problem" check chart indicate order in which to check possible "Causes" of problem.*	A. Fuel not reaching pump	B. Fuel delivered from transfer pump but not to nozzles.	C. Fuel reaching nozzles but engine won't start.	D. Engine starts hard.	E. Engine starts and stops.	F. Erratic engine operation—surge, misfiring, poor governor regulation.	G. Engine idles imperfectly.	H. Engine does not develop full power or speed.	I. Engine smokes black.	J. Engine smokes blue or white.	CORRECTION
ON TEST STAND FOLLOWING OVERHAUL	Transfer pump liner locating pin in wrong hole for correct rotation	7										Reinstall properly.
	Plunger missing		9									Assemble new plunger.
	Cam backward in housing		8									Reassemble correctly.
	Metering valve incorrectly assembled to metering valve arm		6									Reassemble correctly.
	Delivery valve sticking, missing, or assembled backward				19		21	17	15			Remove, clean, or replace as needed.
	Hydraulic head vent wires missing		13						24			Install as indicated in reassembly instructions.
	Idling spring missing or incorrect						15	8				Assemble as indicated in reassembly instructions.
FOLLOWING INSTALLATION ON ENGINE	Seizure of distributor rotor	2										Check for cause of seizure. Replace hydraulic head and distributor rotor assembly.
	Failure of electrical shut-off		2		8							Remove, inspect, and adjust parts. Replace parts as necessary.
	Fuel supply lines clogged, restricted, wrong size, or poorly located	9	7	7	5	1	2	13	4			Blow out all fuel lines with filtered air. Replace if damaged. Remove and inspect all flexible lines.
	Air leaks on suction side of system	11			6	7	8	3	5			Troubleshoot the system for air leaks. See Supplementary Inspections in manual.
	Transfer pump blades worn or broken	8			12		18	5	13			Replace.
	Delivery valve retainer screw loose and leaking or incorrectly installed			25			20		21			Inspect delivery valve stop seat for erosion, tighten retainer screw, or replace head and rotor assembly as needed.
DURING OPERATION	Transfer pump regulating piston sticking	10			13		19	12				Remove piston and regulator assembly and inspect for burrs, corrosion, or varnishes. Replace if necessary.
	Shut-off device at STOP position		1									Move to RUN position.
	Plungers sticking			10		21	10		18	17		Disassemble and inspect burrs, corrosion, or varnishes.
	Metering valve sticking or closed		3		14	9	12	7	10			Check for governor linkage binding, foreign matter, burrs, missing metering valve shim, etc.
	Passage from transfer pump to metering valve clogged with foreign matter		11									Disassemble and flush out Hydraulic Head.
	Tank valve closed	1										Open valve.
	Fuel too heavy at low temperature	6			8							Add kerosene as recommended for 0° F (−18° C), −15° F (−26° C), and −30° F (−34° C) temperatures.
	Cranking speed too low			1	2						7	Charge or replace batteries
	Lube oil too heavy at low temperature			18	9							See engine manual.
	Engine engaged with load			2	1							Disengage load.
	Nozzles faulty or sticking			10	17		9	10		5		Replace or correct nozzles.
	Intake air temperature low			5	3							Provide starting aids. See engine manual.
	Engine compression poor			17	10				25	9	8	Correct compression. See engine manual.
	Pump timed incorrectly to engine			3	4		4	4	7	4	3	Correct timing.
	Excessive fuel leakage past plungers (worn or badly scored)			15	22			16	18			Replace rotor and hydraulic head assembly.
	Filters or inlet strainer clogged	5			7	6	3		6			Remove and replace clogged elements. Clean strainer.

FIGURE 26–20A (Courtesy of Stanadyne Diesel Systems.)

PROBLEM

PROBLEM MAY OCCUR	CAUSE *Numbers in "Problem" check chart indicate order in which to check possible "Causes" of problem.*	A. Fuel not reaching pump	B. Fuel delivered from transfer pump but not to nozzles.	C. Fuel reaching nozzles but engine won't start.	D. Engine starts hard.	E. Engine starts and stops.	F. Erratic engine operation—surge, misfiring, poor governor regulation.	G. Engine idles imperfectly.	H. Engine does not develop full power.	I. Engine smokes black.	J. Engine smokes blue or white.	CORRECTION
DURING OPERATION CONT.	Cam, shoes, or rollers worn			14	20				16			Remove and replace.
	Automatic advance faulty or not operating			11	24		11	11	12	7	4	Remove, inspect, correct, and reassemble.
	Governor linkage out of adjustment or broken				16		14	9	11			Adjust governor linkage hook.
	Governor not operating: parts or linkage worn, sticking or binding, or incorrectly assembled		4		26		13	6	9			Disassemble, inspect parts, replace if necessary, and reassemble.
	Maximum fuel setting too low			12	18				14			Reset to pump specifications.
	Engine valves faulty or out of adjustment				28		10	15		6	9	Correct valves or valve adjustment as in engine manual.
	Water in fuel					2	5	1	23			Drain fuel system and pump housing, provide new fuel, prime system.
	Return oil line or fittings restricted				29	4	23	19	8			Remove line, blow clean with filtered air
	Engine rotation wrong		4									Check engine rotation. See engine manual.
	Air intake restricted				3				26	2		Check. See engine manual.
	Wrong governor spring						17		27			Remove and replace with proper spring as in pump specifications.
	Pump housing not full of fuel					7	2					Operate engine for approximately 5 minutes until pump fills with fuel.
	Low cetane fuel			13	11		6	14	20	8		Provide fuel per engine specifications.
	Fuel lines incorrect, leaking, or connected to wrong cylinders			6			1		28			Relocate fuel lines for correct engine firing sequence.
	Tang drive excessively worn						22		19		5	Remove and install new head and rotor assembly and drive shaft as necessary.
	Governor sleeve binding on drive shaft						16					Remove, inspect for burrs, dirt, etc. Correct and reassemble.
	Shut-off device interfering with governor linkage			8	15			2				Check and adjust governor linkage dimension.
	Governor high-idle adjustment incorrect							3				Adjust to pump specifications.
	Torque screw incorrectly adjusted			5	9	23			22	11		Adjust to specification.
	Throttle arm travel not sufficient				4			1				Check installation and adjust throttle linkage.
	Rotor excessively worn			12	16	27						Replace hydraulic head and rotor assembly.
	Maximum fuel setting too high									10		Reset to pump specifications.
	Engine overheating					5			3			Correct as in engine manual.
	Exceeding rated load								1			Reduce load on engine
	Engine cold				30						1	Check thermostats or shutter controls, warm to operating temperature. See engine manual.
	Lube oil pumping past valve guides or piston rings in engine										6	Correct as in engine manual.
	Excess lube oil in engine air cleaner										2	Correct as in engine manual.

FIGURE 26–20B

and adherence to specifications and the following test procedures will reduce testing inaccuracies to a minimum.

CALIBRATION NOZZLES

Several different types of calibrating nozzles are required for testing the various DB2 pump models. Be sure to use only the type of nozzle called for in the individual specification. Some of the permissible types are listed:

TYPE	OPENING PRESSURE—PSI
DN12SD12	2500 (170 ATS)
AMBAC PCU25D050.5	
Orifice plate (SAE Std.)	3000 (204 ATS)
AMBAC TSE77110-5/8	
0.5 Orifice plate	1700 (116 ATS)

NOTE: ATS-atmospheres. 1 atmosphere = 14.7 psi.

Use of the SAE/150 orifice plate nozzle is described in SAE recommended practices J968c and J969b.

INJECTION LINES

Several injection line sizes (length and inner diameter) have been released for service use. Refer to the individual specification for proper size, and see the applicable Roosa Master service bulletin for preparation and maintenance instructions.

CALIBRATING OIL

Guidelines for calibrating oil are listed in SAE recommended practice J969d. Refer to Service Bulletin 201 for brands of calibrating oil approved for use. Calibrating oil should be changed every 3 months or 200 pumps (whichever comes first).

CALIBRATING OIL TEMPERATURE

The temperature of the oil in the test bench must be maintained within 110° to 115° F (43°–46° C) while testing Roosa Master fuel injection equipment.

NOTE: This reading should be taken as close to the inlet as possible. The test bench should be equipped with a heater and thermostatic control to maintain this temperature. Refer to SAE recommended practice J969b.

TEST BENCH

Mount and drive the DB2 pump models according to the test bench manufacturer's instructions. In addition, the test stand coupling should be of the self-aligning, zero backlash type, similar to the Thomas coupling types or Robert Bosch.

GENERAL TEST PROCEDURE

1. Install the applicable transfer pump inlet connector, using two wrenches so that the pump outlet fitting does not get moved at the same time. Install transfer pressure gauge connector 21900. Install a shut-off valve to isolate the gauge when not in use. Connect a pressure gauge to the 21900 adapter. Some automotive DB2 pumps require 0.25° setting accuracy. If so specified, remove the timing line cover and replace it with the 21734 advance gauge. If the 21734 gauge is not required, replace the timing line cover with the 19918 advance window.

2. Determine the proper direction of pump rotation from the specification. Rotation is determined as viewed from the drive end of the pump.

3. If the pump is equipped with an energized-to-run electric shut-off device, energize it at the lowest speed with the specified voltage (see Service Bulletin 108). Move the pump throttle lever to the full-load position. When the transfer pump is primed, allow fuel to bleed for several seconds from loosened injection line nuts at the nozzles. Tighten the line nuts securely.

NOTE: Pump specifications list fuel delivery in cubic millimeters per stroke. Some test benches measure fuel flows in cubic centimeters (milliliters). To convert from cubic millimeters per stroke to cubic centimeters, use the following formula:

$$\text{Delivery in cc} = \frac{\text{mm}^3/\text{stroke} \times \text{no. of strokes}}{1000}$$

Example. If the specification calls for 72 mm³/stroke, and the test stand counter has been set for 500 strokes, simply substitute these numbers into the formula and calculate as follows:

$$\text{Delivery in cc} = \frac{72 \times 500}{1000} = 36 \text{ cc}$$

Bear in mind when testing stanadyne pumps that the specifications refer to engine rpm (erpm) and that most test bench tachometers register pump rpm, which is half of engine speed for four-stroke-cycle engines.

4. Operate the pump at 1000 erpm wide-open throttle (WOT) for 10 minutes. Dry the pump off completely with compressed air. Observe it for leaks and correct as

necessary. Back out the high-idle, low-idle, and torque screws (if equipped).

NOTE: Refer to pump specification for correct sequence of test stand adjustments. Pressurize the transfer pump inlet to the amount indicated on the specification or to a maximum of 5 psi, if not otherwise indicated.

5. Vacuum check: Close the valve in the fuel supply line. At 400 erpm, the transfer pump must be capable of creating a vacuum of at least 18 in. of mercury. If it does not, check for air leaks between the pump inlet and shut-off valve or deficiency in transfer pump components.

6. Fill graduates to bleed air from the test stand and to wet graduates.

7. Check the return oil quantity by directing the return oil flow into an appropriately calibrated graduate for a given time. See individual specifications for allowable quantity and erpm at which to make the check.

8. Operate the pump at the specified speeds with wide-open throttle and observe transfer pump pressure. Adjust the pressure-regulating spring plug to raise or lower transfer pump pressure.

CAUTION:
Under no circumstances should 130 psi be exceeded.

To adjust pressure, remove the line to the transfer pump inlet connector and use a $5/32$ in. hex key wrench 13316 to adjust the plug. Clockwise adjustment increases pressure. Do not overadjust.

NOTE: The transfer pump pressure gauge must be isolated by the shut-off valve at the injection pump when checking fuel delivery and advance movement.

9. Check for minimum delivery at cranking speed.

10. Operate at high speed and adjust the high-idle screw to obtain the specified delivery. Recheck transfer pump pressure on completion of this adjustment.

11. Adjust the low-idle screw to the correct low-idle delivery.

12. Automatic advance: Check the cam position at specified points in the speed range. Adjust the trimmer screw, as required, to obtain proper advance operation. Each line on the advance gauge 19918 equals 2 pump degrees. After setting the advance, check to see that the cam returns to its initial position at zero rpm. Recheck the transfer pump pressure after setting the advance, and correct if necessary.

13. Record the fuel delivery at the checkpoints shown on the pump specification. *Roller settings should not be readjusted on the test bench.* Experience has proved that micrometer and dial indicator settings provide more consistent accurate results in performance. Variations in test bench drives, instrumentation, nozzle lines, and fuels in different areas sometimes result in nonconforming fuel flow readings.

14. While operating at full-load governed speed, set the torque screw (if employed) to specified delivery. Recheck the transfer pump pressure and advance movement on completion of this adjustment.

15. Recheck delivery at the lowest speed checkpoint.

16. Check the governor cutoff at specified speed.

17. Check the electric shut-off (if equipped) at speeds indicated on the specification.

18. Remove the pump from the test stand and assemble all sealing wires. The pump is then ready for installation on the engine.

LUCAS CAV DPA DISTRIBUTOR PUMP DESIGN AND OPERATION

The Lucas CAV DPA distributor pump is very similar in design and operation to the Stanadyne distributor pump. The pump is flange-mounted to the engine. It is oil-tight, and during operation all moving parts are lubricated by fuel oil under pressure, so that no additional lubrication system is required. Pressure maintained within the pump housing prevents the entry of dust, water and other foreign matter (**Figures 26–21** and **26–22**).

Fuel injection is effected by a single pumping element having twin opposed plungers. These plungers

FIGURE 26–21 Lucas CAV DPA injection pump. (Courtesy of Lucas CAV Ltd.)

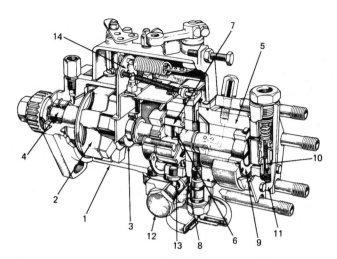

FIGURE 26–22 Lucas CAV DPA injection pump components. 1. Pump housing. 2. Governor weight retainer. 3. Governor thrust sleeve. 4. Drive shaft. 5. Hydraulic head and rotor. 6. Pumping plungers. 7. Rollers and shoes. 8. Adjusting plates. 9. Transfer pump. 10. Pressure-regulating valve. 11. End plate. 12. Advance mechanism. 13. Cam advance stud. 14. Governor. (Courtesy of Lucas CAV Ltd.)

are located within a transverse bore in a central rotating member that acts as a distributor and revolves in a stationary member known as the *hydraulic head*. The pump plungers are actuated by lobes on an internal cam ring. Fuel is accurately metered to the pumping element, and the high-pressure charges are distributed to the engine cylinders at the required timing intervals through ports in the rotor and the hydraulic head. The single pumping element ensures equal delivery of fuel to each engine cylinder.

Most pumps are fitted with a device capable of automatically changing the timing of the fuel injection pulse.

The internally lobed cam ring, mounted in the pump housing, normally has as many lobes as there are engine cylinders and operates the opposed pump plungers through cam rollers carried in shoes sliding in the rotor body. The plungers are forced inward simultaneously as the rollers contact the diametrically opposed cam lobes. This is the injection stroke. The plungers are returned by pressure of the inflowing fuel, and this forms the charging stroke **(Figure 26–23)**.

The pump rotor is driven by the engine through a splined shaft or other drive to suit the engine manufacturer's requirements. Pumps may be mounted horizontally, vertically, or at any convenient angle.

The accurate spacing of cam lobes and delivery ports ensures the exact equality of the timing interval between injections, and the components that affect

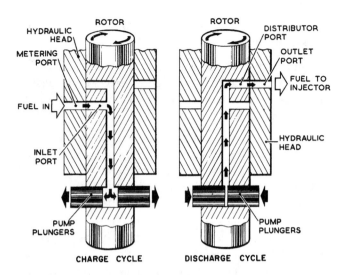

FIGURE 26–23 Charge and discharge cycle fuel flow. (Courtesy of Lucas CAV Ltd.)

timing are designed with one assembly position only to ensure precision.

Fuel entering the pump through the main inlet connection is pressurized by a sliding vane transfer pump carried on the rotor inside the hydraulic head. The pressure rise is controlled by a regulating valve assembly located in the pump end plate. The fuel then flows

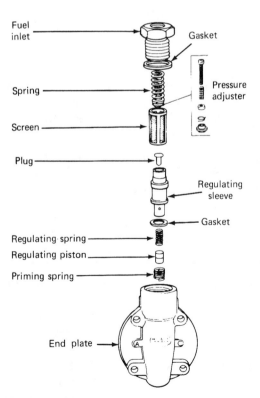

FIGURE 26–24 End plate and pressure regulator components. (Courtesy of Lucas CAV Ltd.)

LUCAS CAV DPA DISTRIBUTOR PUMP DESIGN AND OPERATION

through the passages to the pumping elements **(Figures 26–24 and 26–25)**.

Excess fuel and air are returned to the fuel tank through the fuel return circuit via a fitting at the top of the housing **(Figure 26–26)**.

The outward travel of the opposed pumping plungers is determined by the quantity of fuel metered, which varies in accordance with the setting of the metering valve. The outward travel of the roller assemblies that are displaced centrifugally is always limited by the maximum fuel adjusting plates.

The maximum amount of fuel delivered is regulated by limiting the outward travel of the plungers by means of an adjustable stop.

As the rotor turns, the inlet port is cut off, and the single distributor port in the rotor registers with an outlet port in the hydraulic head. At the same time, the plungers are forced inward by the rollers contacting the cam lobes, and fuel under injection pressure passes to the central bore of the rotor through the aligned ports to one of the injectors. The rotor normally has as many inlet ports as the engine has cylinders, and a similar number of outlet ports in the hydraulic head.

The cam lobes are contoured to provide pressure relief in the injector lines at the end of the injector cycle. This gives a sharp cutoff of fuel and prevents "dribble" at the nozzles.

The governor is contained in a small housing mounted on the pump body, the metering valve being operated by fuel at transfer pressure.

The DPA pump is available with a hydraulic or a mechanical governor. The DPS pump has a two-speed mechanical governor.

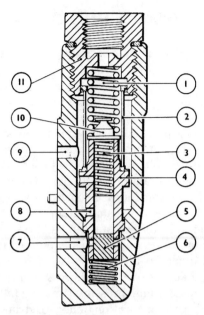

1. Retaining Spring.
2. Nylon Filter.
3. Regulating Spring.
4. Valve Sleeve.
5. Piston.
6. Priming Spring.
7. Fuel passage to Transfer Pump Outlet.
8. Regulating Port.
9. Fuel passage to Transfer Pump Inlet.
10. Spring Guide.
11. Fuel Inlet Connection.

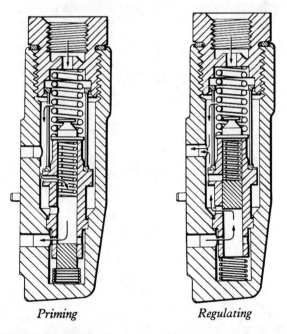

FIGURE 26–25 Cross section of end plate containing pressure-regulating valve. (Courtesy of Lucas CAV Ltd.)

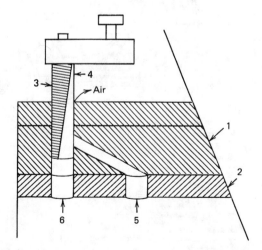

FIGURE 26–26 Fuel-return circuit. 1. Barrel. 2. Sleeve. 3. Metering valve. 4. Air bleed passage. 5. Delivery passage. 6. Transfer pressure. (Courtesy of Lucas CAV Ltd.)

MECHANICAL GOVERNOR OPERATION (FIGURE 26-27)

The governor weights acting on the thrust sleeve cause the governor lever to pivot and the metering valve to rotate. Rotating the metering valve admits more or less fuel to the pumping cylinder. The main governor spring and the idle spring oppose the governor weights. The balance between the governor weight force and spring force determines the position of the metering valve and the amount of fuel delivery. The metering valve can be rotated to the no-fuel position by the fuel shut-off at any engine speed.

At idle speed the main governor spring is extended and provides little force, allowing the weights to fly outward. This turns the metering valve to the idle position. The idle spring provides the necessary speed control. At full-load rated speed the throttle position causes full spring force to be applied to the governor lever in opposition to flyweight action. The metering valve is then rotated to the full-fuel position. At high idle (no load) the throttle still causes full spring force to be applied. However, when the load is removed at full throttle, the engine speeds up, the flyweights move outward to overcome spring pressure and rotate the metering valve to the closed position. The reduced fuel flow reduces engine speed to the point where the flyweight force and the spring force are in balance, thereby maintaining the high-idle engine speed.

TIMING ADVANCE UNIT OPERATION

The timing advance unit uses a hydraulic servomechanism to rotate the normally stationary cam ring against pump rotation to advance injection timing. Fuel from the transfer pump is routed to one side of the advance piston against spring pressure on the other side of the advance pin. The advance pin is screwed into the cam ring with its head positioned between the hydraulic piston and the spring. A ball check valve in the circuit keeps the cam ring stationary as the rollers impact against the cam lobes. Leakage between the piston and cylinder allows the cam to return to its normal position when transfer pressure drops. A damper is used on some pumps to dampen transfer pump pressure pulsations (**Figure 26-28**).

Advance occurs over the entire speed range of the engine. Transfer pump pressure rises and falls with engine speed, causing injection timing to advance more or less. There is no change in advance as long as engine speed remains constant. The advance can be adjusted by replacing the spring with one that has different tension or by using shims to change spring pressure.

A special groove machined into the metering valve allows additional fuel to flow through it and into the advance unit during light loads. At full load the valve is rotated by the governor so that the groove is not aligned with the passage to the advance unit and there is no advance (**Figure 26-29**).

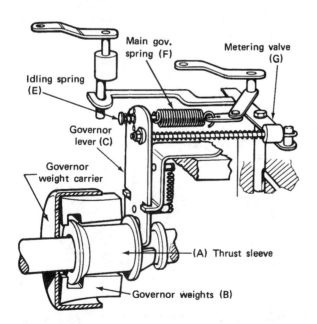

FIGURE 26-27 CAV DPA mechanical governor components and metering valve. (Courtesy of Lucas CAV Ltd.)

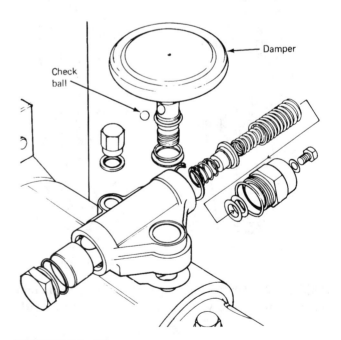

FIGURE 26-28 CAV DPA pump damper. (Courtesy of Lucas CAV Ltd.)

LUCAS DPA PUMP SERVICE

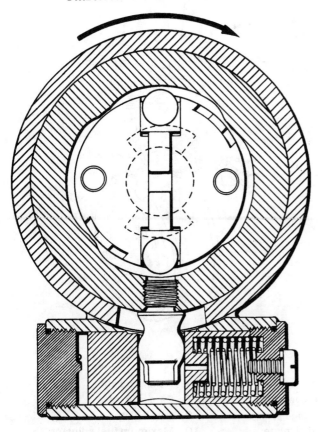

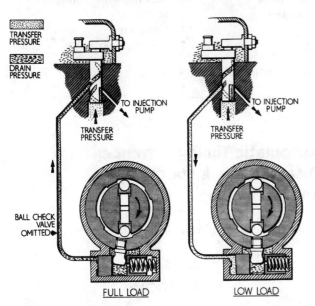

FIGURE 26-29 Timing advance unit operation. (Courtesy of Lucas CAV Ltd.)

Maximum Fuel Delivery Control (Figure 26-30)

The addition of scroll plates on DPS pumps to either side of the cam ring allows maximum fuel delivery to be adjusted with the engine running. It also makes possible the adjustment of excess fuel for starting. With the throttle lever (2) closed the scroll plates (9) are in the excess-fuel position. This allows the rollers (7) to move out farther and allow more fuel to enter the pumping chamber. Once the engine is started and running, the throttle lever (2) is rotated to the normal run position with the link plate (11) held firmly against the maximum fuel adjustment screw (10). In this position the higher profile of the scroll plates (9) limits outward movement of the rollers and pumping plungers and reduces fuel delivery to that set by the adjustment screw.

LUCAS DPA PUMP SERVICE

Back-Leakage Valve Replacement

Remove the back-leakage valve and sealing washer. Inspect the valve to make sure it is not stuck. Inspect the sealing surfaces for possible leak paths and correct if necessary. Assemble the valve with new sealing washers and tighten to specifications (**Figure 26-31**).

Shutdown Solenoid Replacement

Remove the solenoid, O-ring, and plunger. Replace the O-ring. Use the protective sleeve to prevent damage to the O-ring. Inspect the plunger tip. If the tip is damaged or deformed, replace the solenoid assembly. Install the plunger, spring, solenoid, and O-ring and tighten to specifications (**Figure 26-32**).

Sealing Washer Replacement

If there is any evidence of leakage at any of the sealing washers at the bleed screws, vent fitting, or fuel inlet fitting, the sealing washer should be replaced. Inspect the sealing surfaces for leak paths and correct if necessary. Use new sealing washers and tighten the bleed screws and fittings to specifications (**Figure 26-33**).

Control Lever Replacement

Remove the locknut. Inspect the condition of the components. Replace any faulty items. Assemble the spring guide, torsion spring, washer, and stop arm. The stop

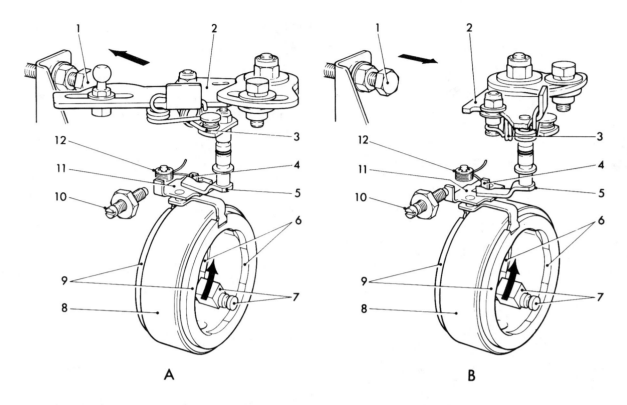

A. Scroll plates in excess fuel position with throttle lever closed
B. Scroll plates in maximum fuel position with throttle lever open

1. Anti-stall stop
2. Throttle control-lever
3. Excess fuel linkage spring
4. Inner tongue-link plate
5. Excess fuel shaft and lever
6. Scroll profiles
7. Roller and shoe
8. Cam ring
9. Scroll plates
10. Maximum fuel adjuster screw
11. Link plate
12. Link plate spring

FIGURE 26-30 Scroll plate mechanism for DPS pump. (Courtesy of Lucas CAV Ltd.)

arm must slide over the flats of the shaft. Install the lever, bushing, and locknut, and tighten the nut to specifications.

Shutdown Lever/Spring Replacement

Remove the locknut and washer. Lift off the lever while allowing the return spring to unwind. Position the return spring with one end of the return spring contacting the boss on the governor cover. Hook the free end of the spring around the shutdown lever and rotate the lever in a clockwise direction until it engages with the flats on the shutdown shaft. Install the nut with a new lockwasher and tighten to specifications **(Figure 26–34)**.

Automatic Timing Advance Disassembly, Inspection, and Assembly

Remove the small plug and washer.

CAUTION:

The spring cap is under tension; remove the cap slowly **(Figure 26–35)**.

Remove the springs and shims. Remove the pressure end plug and O-ring. Remove the cap nut and sealing washer. Remove the head-locating fitting assembly.

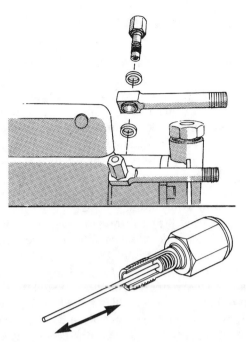

FIGURE 26–31 (Courtesy of Cummins Engine Company, Inc.)

FIGURE 26–32 Shutdown solenoid removal and plunger inspection. (Courtesy of Cummins Engine Company, Inc.)

FIGURE 26–33 (Courtesy of Cummins Engine Company, Inc.)

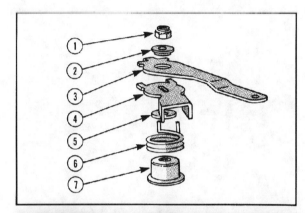

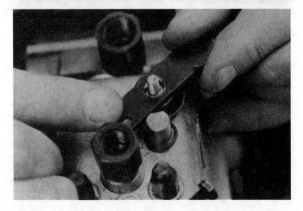

FIGURE 26–34 (Courtesy of Cummins Engine Company, Inc.)

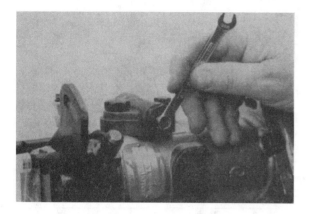

FIGURE 26-35 Small plug removal (top), cap removal (middle), spring removal (bottom). (Courtesy of Cummins Engine Company, Inc.)

CAUTION:

Do not lose the check ball. Remove the housing and slide the advance piston from the bore.

Inspect the advance piston and housing for scoring. Inspect the check ball and seat for erosion. Make sure the ball can move freely on the seat. Make sure the orifice in the side of the seat in the head-locating fitting is open. Check that the cam ring is free to move in the fuel pump **(Figure 26-36)**. Replace any faulty components.

To assemble the unit, position a new gasket on the injection pump housing. Insert the advance piston into the housing with the blank end toward the oil feed hole in the bore. Position the advance housing over the stud in the injection pump with the cam advance screw positioned into the center bore in the piston. Install new O-rings on the head-locating fitting. Position the check ball in the head-locating fitting **(Figure 26-37)**. Position the head-locating fitting through the advance housing and hand tighten. Install the cap nut and a new washer. Tighten the cap nut and head-locating fitting progressively and evenly to specified torque. Check that the piston is able to move freely in the bore. Install the springs and shims into the pocket end of the advance piston. Install a new O-ring on the spring cap and place the shims in the pocket. Install and tighten the spring cap on the advance housing and tighten to specifications. Use a new washer and install the spring cap plug. Tighten to specified torque. Install a new O-ring on the pressure end cap. Use the protective sleeve to avoid O-ring damage. Install the cap and tighten to specifications.

PUMP TESTING AND CALIBRATION

Pressure Testing

All pumps must undergo a pressure test *before* and *after* machine testing using the following method **(Figure 26-38)**.

1. Drain all fuel from the pump and connect an air line to the pump inlet connection. Ensure that the air supply is clean and free from water.

2. Seal off the low-pressure outlet connection on the pump and completely immerse it in a bath of clean test oil.

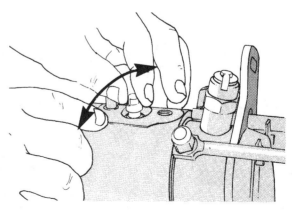

FIGURE 26-36 (Courtesy of Cummins Engine Company, Inc.)

PUMP TESTING AND CALIBRATION   611

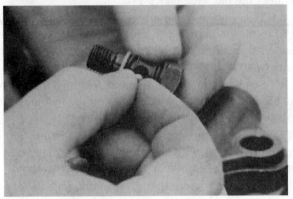

FIGURE 26-37 (Courtesy of Cummins Engine Company, Inc.)

3. Raise the air pressure in the pump to 20 psi (138 kPa). Leave the pump immersed in oil for 10 minutes to allow any trapped air to escape.

4. Observe for leaks after the pump has been immersed for 10 minutes; if the pump is not leaking, reduce the air pressure to 2 psi (14 kPa) for 30 seconds; if there is still no leak, increase the pressure to 20 psi (138 kPa) and if the pump is still leak free after 30 seconds, it can be passed as satisfactory.

5. On pumps without a drive shaft oil seal it is necessary to stop the oil from leaking past the drive shaft during pressure testing. Tool part no. 7144-760 can be used, but it is necessary to blank off the 12×1.5 mm threaded connection.

6. All leaks must be rectified before testing and setting the pump.

Test Procedure

General explanatory notes and an individual test plan, quoting the numbers of the range of pumps to which it may be applied, are published for each different model manufactured. The sequence of operations listed in the test data for the particular type of pump gives the test performance requirements at various pump speeds, the timing procedure, and any special precautions necessary to safeguard the pump.

1. Pressure gauge, tool Part No. AT20/9
2. Pipe, tool Part No. AT20/2
3. Indexing plate
4. Adaptor, tool Part No. 7244-382
5. Pressure adjuster, tool Part No. 7244-410
6. Throttle control bracket
7. Maximum speed screw
8. Throttle control lever
9. Ball screw
10. Maximum fuel adjuster screw
11. Timing cover plate
12. Anti-stall screw
13. Idling adjustment screw
14. Screw, tool Part No. 7244-275/3
15. Backleak connection; banjo, tool Part No. ALP143 and bolt, tool Part No. ALP144

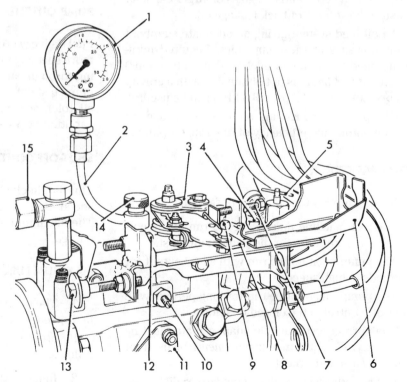

FIGURE 26-38 Pressure test gauge hookup, adjustment points, and special tools list for DPS pump. (Courtesy of Lucas CAV Ltd.)

NOTE: All pump tests, calibrations, and settings must be made using the specified test oils given in the CAV Test Plans. Where the word "fuel" is used in the following test instructions this means the approved test oil and **not** diesel fuel.

TEST MACHINE

A suitable test machine must incorporate the following features:

1. A mounting bracket, for holding the pump.
2. A splined drive coupling for rotating the pump in either direction at all speeds specified in the test data.

NOTE: The pump **must** be rotated in the direction given in the test data and indicated on the pump nameplate. Incorrect rotation will cause serious damage.

3. A set of high-pressure pipes, 863.6 mm long × 2 mm bore × 6 mm outside diameter (34 in. long × 0.080 in. bore × 0.236 outside diameter), for coupling the pump outlet connections to a matched set of injectors, type BDN 12SD12, set at 175 atm (2572 psi) opening pressure.
4. An automatic trip mechanism that directs test oil from the injectors into graduated glasses during the period stated in the test data and then diverts the oil into a drain.
5. A set of graduated glasses for measuring the output from each injector, and one glass of larger capacity to measure the volume of back-leakage oil.
6. An oil feed system giving an adequate supply at constant pressure at the pump inlet. Required minimum fuel feed to the pump inlet is 61 in.^{3} (1000 cm^3) per minute. If this figure is not obtainable with a gravity feed, a pressure feed of 2 psi (13.8 kPa) can be used.
7. One pressure gauge and one vacuum gauge for testing the output and efficiency of the transfer pump.

GENERAL PROCEDURE

The following precautions must be observed:

1. The test machine must be set to run in the correct direction of rotation for the pump under test.
2. The pump must not be run with a low output for long periods at high speed.
3. The pump must not be run for long periods with the shut-off control in the closed position.
4. The correct test machine adapter plate must be used. A plate with a 50 mm hole must *never* be used for a pump with a 46 mm spigot.
5. Unless otherwise stated, standard radial high-pressure connections must be fitted prior to testing. Information is given in the test data and explanatory notes.
6. Prime the pump thoroughly before testing and at all times indicated in the test plan.

PRIMING

Variations for certain pumps are explained in the test data.

1. Slacken both the vent screw on the governor control casing and the head vent screw.
2. Connect the oil feed pipe to the pump inlet and connect the back-leakage pipe.
3. Turn on the oil supply to feed pressure 2 psi (14 kPa) to fill the pump. Run the pump at 100 rpm. When test oil free of air bubbles issues from the vents, retighten the vent valve and the head locking screw.
4. Slacken the connections at the injector end of the high-pressure pipes, or if fitted on the test machine, open the bleeder valves at the injectors.
5. Run the pump at 100 rpm. When test oil free of air bubbles issues from all high-pressure pipes, retighten the connections or close the bleeder valves.
6. Examine the pump after priming, for oil leaks at all jointing faces, connections, and oil seals. Pumps must be free from leaks both when running and when stationary.

PUMP OUTPUT

Fuel delivery is checked at full-throttle setting at one or more speeds of rotation by measuring the volume passing through each injector during 200 pump cycles. The pump test data quotes the maximum fuel delivery, overall tolerance and the maximum permissible delivery variation between injectors.

SHUT-OFF CONTROL

This is checked by running the pump at a specified speed (see test plan) with the shut-off control closed. The maximum fuel delivery permitted on this setting is quoted.

MAXIMUM FUEL SETTING

1. The maximum fuel delivery is checked at a specified speed, with throttle and shut-off controls fully open.

If output is not within the specified limits, adjust as follows:

2. Remove the inspection cover.
3. Slacken the two drive plate screws.

4. Engage the 7144-875 tool with the slot in the periphery of the adjusting plate.

5. Adjust the plate by lightly tapping the knurled end of the tool. The direction in which the drive plate is turned to increase or to decrease fueling depends on the type of adjusting plates fitted.

When looking at the drive end of the pump, if the top adjusting plate has a shallow slot 3 mm deep, the fueling can be increased by turning the adjusting plate in a counterclockwise direction, and decreased by turning it in a clockwise direction.

If the top adjusting plate has a deeper slot of 5.5 mm depth, the fueling can be increased by turning the top adjusting plate in a clockwise direction, and decreased by turning it in a counterclockwise direction.

6. Tighten the drive plate screws *evenly* to the listed torque value, using the 7144-482 Adapter, the 7144-511A Spanner, and a torque wrench.

7. Replace and secure the inspection cover, refill the pump, vent as necessary, and recheck the maximum fuel delivery. Repeat until the fuel delivery is within the specified limits.

GOVERNOR TESTING

Run the pump at more than half the maximum permissible speed of the engine to which it will be fitted, and adjust the maximum speed stop until the specified fuel delivery is obtained. This specified volume is less than at the maximum fuel setting. Reduce the speed of rotation, whereupon the fuel delivery should increase to a specified volume approximately equal to the maximum fuel delivery.

NOTE: Final governor setting must be carried out with the pump fitted to the engine, and in accordance with the engine manufacturer's instructions **(Figure 26–39)**.

TRANSFER PUMP SETTING

Transfer pump vacuum is checked while the pump is running at a low speed, with the two-way cock in the fuel feed line turned to the position that cuts off the fuel supply and connects the pump inlet to the vacuum gauge. A given depression must be obtained in a specified time.

NOTE: The pump may need repriming after this test.

Transfer Pressure Adjustment

To adjust the transfer pressure the operation of the regulating valve is modified as follows.

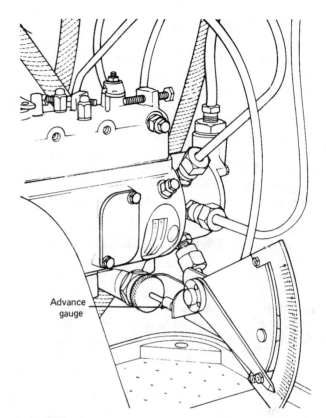

FIGURE 26–39 Advance gauge installation on CAV DPA pump. (Courtesy of Lucas CAV Ltd.)

On certain pumps with aluminum end plates, the transfer pressure can be adjusted within the limits of an individual specification in one or both of two ways: (a) by changing the end plate sleeve plug and (b) by the adjustment of the transfer pressure adjusting screw passing through the plug.

Several plugs with different step thicknesses are available; the stepped portion in contact with the regulating spring determining the spring compression. The screwed transfer pressure adjuster limits the movement of the regulating piston and so controls the maximum uncovered area of the sleeve port. Method (a) modifies transfer pressure characteristics over the lower and middle speed ranges; method (b) controls the pressure rise over the upper speed range.

A third type of transfer pressure adjuster varies the preloading of the regulating spring and maximum lift of the piston. It has a spring peg fitted between the adjusting screw and regulating spring. This has the same effect as method (a) but allows more variation and gives easier adjustment.

Additional information is given in the individual pump test data.

PRESSURIZED CAM BOXES

On some pumps the cam box is pressurized during running by a spring-loaded ball pressurizing valve located in the back-leakage connection in the inspection cover.

The test pressure gauge is fitted to the governor housing vent screw hole.

Pressure limits are given in the appropriate test data. If the pressure is incorrect, check that the pump is not leaking and there is no restriction in the back-leakage passages. If the pressurizing valve is faulty, a new inspection cover complete with valve must be fitted.

Testing of Advance Devices

The appropriate test plan for the individual pump specifies the type of advance device fitted and details the tests for the particular unit.

All advance devices are tested using the special part no. 7244-50 Tool, which consists of a gauge with a scale covering 0 to 18° and a 7244-70 Feeler Pin. To fit this tool, proceed as follows:

1. Remove the small screw from the piston spring cap on the advance device.

2. Pass the threaded bushing of the feeler pin assembly through the hole in the tool bracket.

3. Insert the end of the plunger into the hole in the spring cap and screw the bushing into the spring cap hole. This will clamp the bracket between the spring cap and the shoulder on the threaded bushing.

4. Zero the gauge by moving the scale relative to the pointer **(Figure 26–40)**.

NOTE: The pump must be reprimed after fitting the tool. After priming, operate the throttle and press inward and release the advance gauge pin a few times with the pump running at 100 rpm.

SPEED ADVANCE DEVICE

The tests outlined in the test data must be applied to ensure that the degree of advance obtained is within the stated limits at the speeds specified. To adjust the degree of advance, increase or decrease the thickness of the shims between the piston spring and the spring cap (see Test Data). When the tests are satisfactory, remove the special tool and prime the pump.

COMBINED LOAD AND SPEED ADVANCE

These tests, at different speeds and fuel deliveries, check the movement of the outer piston in response to

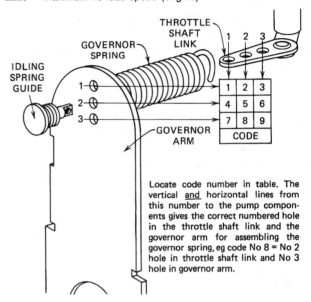

FIGURE 26–40 Nameplate and governor spring position code. (Courtesy of Lucas CAV Ltd.)

changes of speed, and of the inner piston to changes of load. Adjustment is made by altering the thickness of the shims beneath inner and outer piston springs. When the tests are satisfactory, remove the special tool and prime the pump.

Timing

All pumps require timing (pump phasing).

After completion of tests remove the pump from the test machine and drain by slackening the inspection cover screws. Remove the inspection cover. For internal or external timing connect the 7144-262A Stirrup Pipe to the fuel outlet specified on the Test Plan and to the outlet diametrically opposite. Fit the 7144 Relief Valve to the stirrup pipe, and connect the complete tool through a high-pressure pipe to a nozzle testing unit.

For three-cylinder pumps, connect one branch of the stirrup pipe to the specified outlet, and arrange the other branch to face away from the pump and seal it off with the blanking plug.

Normally, a pressure of 30 atm with a relief valve fitted in the system is specified in the test data but sometimes a higher pressure is quoted. To obtain the higher pressures, adjust the relief valve appropriately, or connect the stirrup pipe directly to the nozzle testing unit.

WARNING:

Do not exceed the specified pressure. Excess pressure can cause damage to the shoe assemblies and adjusting plates.

Turn the pump drive shaft in the direction indicated on the nameplate until resistance to further movement is felt. At this point, the master spline on the drive shaft should be in the same plane as the outlet quoted in the test plan. This is the timing position. Access to the timing ring may be obtained through the inspection cover aperture. Move the timing ring until the straight edge of the timing circlip—or the line scribed on the ring in the case of old-type clips—aligns with the mark on the drive plate as specified in the test plan. Circlips with two straight ears are only for spacing, and the circlip ends are positioned remote from the inspection aperture. After carrying out this operation refit the inspection cover and tighten the screws.

The test data gives specific information about the timing mark on the pump flange. A 7244-27 Flange Marking Gauge is available. It consists of a cast aluminum body with a lock screw, around which slides a ring carrying the scribing plate. The ring is held in position by a scale plate on which direct readings in degrees are taken from the edge of the scribing plate.

Interchangeable spigot plates, one with a 46 mm bore and one with a 50 mm bore, are held in position by a cap-head screw, and these accommodate the different pump spigots. Four interchangeable inserts adapt the gauge to any type of pump drive. The inserts are held by two screws and positioned relative to the scale zero by a dowel pin.

To mark the flange, hold the pump in the timed position, and fit the marking gauge, with the appropriate insert and spigot plate, to the pump drive, and set to the indexing figure specified in the test data. Using the gauge as a template, scribe a line on the flange between the plates of the scribing gauge.

REVIEW QUESTIONS

1. The Stanadyne distributor injection pumps use a _____ _____ metering valve.
2. The main rotating parts of the D series Stanadyne distributor pump are the _____ _____, transfer pump blades, _____ _____, and governor.
3. Explain the charging cycle.
4. A fuel-return circuit is provided in the pump to prevent _____ from entering the pumping chamber.
5. Roosa Master basic injection timing is achieved by proper alignment of _____ _____ at the _____ _____ mounting.
6. The governor serves the purpose of maintaining the desired _____ _____ within the operating range under various _____ _____.
7. The aneroid is designed to reduce _____ _____ during acceleration.
8. List the benefits of electronic control of fuel injection.
9. The temperature of the oil on the test bench must be maintained at _____ to _____ while testing Roosa Master fuel injection equipment.
10. CAV-DPA fuel injection is effected by a single pumping element having _____ _____ _____.
11. The DPA pump is available with _____ or _____ governor.
12. Explain control lever replacement in a Lucas DPA pump.

TEST QUESTIONS

1. The JBD pump produced by John Deere is a modified version of the
   a. A model   c. DM model
   b. DB model   d. DB4 model

2. The microprocessor-controlled injection pump uses a number of engine sensors to provide information to the
   a. microprocessor   c. computer
   b. modulator   d. injection pump

3. In the discharge cycle the delivery valve operates
   a. in the distributor rotor
   b. in the center of the pump
   c. outside the distributor rotor
   d. in a bore in the center of the distributor rotor
4. The fuel-return circuit is provided in the pump
   a. to prevent excess fuel from entering
   b. to assist in cooling
   c. to prevent air from entering the pumping chamber
   d. to prevent overconsumption of fuel
5. The Stanadyne automatic hydraulic advance controls timing in relation to
   a. engine speed   c. engine load
   b. road speed   d. engine temperature
6. Torque is commonly defined as the
   a. turning moment of the engine
   b. lugging ability of the engine
   c. twisting motion of the engine
   d. all of the above
7. The aneroid is designed to reduce smoke during acceleration on
   a. naturally aspirated engines
   b. engines over 200 HP
   c. highway vehicles
   d. turbocharged vehicles
8. The Stanadyne electronic fuel injection pump advance system is a servo actuated by a(n)
   a. mechanical lever   c. accelerator rod
   b. vacuum valve   d. stepper motor
9. Lucas CAV DPA pumps may be mounted
   a. horizontally
   b. vertically
   c. at any convenient angle
   d. all of the above
10. All pumps after reconditioning must undergo a
    a. pressure test
    b. pressure test before and after a machine test
    c. heat test
    d. fuel output test

CHAPTER 27

ENGINE DIAGNOSIS, PERFORMANCE TESTING, AND TUNE-UP

INTRODUCTION

Accurate diagnosis of engine performance problems requires a thorough understanding of the function, design, and operation of the engine and its support systems. Incorrect diagnosis is costly, time consuming, and may result in a loss of customer confidence. This chapter discusses the diagnosis, performance testing, and tune-up of diesel engines and engine systems.

PERFORMANCE OBJECTIVES

After thorough study of this chapter and sufficient practical work on diesel engines and with the appropriate service manuals you should be able to:

1. Diagnose engine performance problems using a dynamometer and other test equipment.
2. Perform an engine compression test and interpret the results.
3. Perform a cylinder leakage test and interpret the results.
4. Perform a diesel engine tune-up.

TERMS YOU SHOULD KNOW

Look for these terms as you study this chapter, and learn what they mean.

engine diagnosis
engine tune-up
compression testing
cylinder leakage testing
engine performance testing
water manometer
mercury manometer
tachometer
engine dynamometer
chassis dynamometer
dynamometer worksheet

ENGINE DIAGNOSIS

The course of action to be taken in any repair work is determined by following a systematic approach to problem diagnosis. The following diagnostic chart will assist in the process of identifying the source of the problem and the extent of work required to correct the problem. The diagnostic guide and chart are provided here courtesy of Mack Trucks Inc. The letters under the ITEM column refer to lettered sections in the diagnostic chart where causes and corrections are listed for the symptoms opposite each letter. The chart includes engine mechanical and engine support system problem diagnosis. In some cases reference is made to specific Mack Truck

DIAGNOSTIC CHART

Symptom	Probable Cause	Remedy
A. Engine will not crank	1. Batteries have low output	1. Check batteries, charge or replace if required.
	2. Loose or corroded battery connection	2. Clean and tighten battery connections.
	3. Broken or corroded wires	3. Check voltage at connections—switch to starter and battery to starter, replace defective parts.
	4. Faulty starter solenoid, starter	4. Check operation of starter solenoid/starter as outlined under *Electrical Section* repair. Replace defective parts.
	5. Faulty key switch	5. Replace key switch.
	6. Internal seizure	6. Bar the engine over at least one complete revolution. If the engine cannot be rotated a complete revolution, internal damage is indicated, and the engine must be disassembled.
B. Engine cranks—will not start	1. Slow cranking speed	1. Check items listed for "Engine will not crank."
	2. Chassis equipped with Mack Puff Limiter. Puff limiter air cylinder preventing full starting fuel to be delivered by the injection pump.	2. Check emergency spring brakes. They must be set prior to starting. Also check all service lines for air leaks and operation of the air cylinder and reversing valve. Replace all defective parts.
	3. Emergency shut-off valve closed or partly closed	3. Check emergency shut-off system and make necessary corrections.
	4. Low ambient air temperature	4. Use starting aid [20° F (6° C) or below].
	5. No fuel to engine	5. Check for empty fuel tank, plugged fuel tank connections, obstructed or kinked fuel suction lines, fuel transfer pump failure, or plugged fuel filters.
	6. Governor throttle shaft linkage binding; improper setting of accelerator linkage	6. Check accelerator linkage; free all binding parts/make necessary adjustments.
	7. Defective fuel transfer pump	7. Check transfer pump for minimum output pressure. If low, change fuel filters. Look for air leaks and recheck pressure. If still below minimum, pump is defective. Replace fuel transfer pump.
	8. Poor-quality fuel or water in fuel	8. Drain fuel from tank. Install new fuel filters and fill tank with recommended Mack-specified diesel fuel.
	9. Improper lubricating oil viscosity; oil too thick for free crankshaft rotation	9. Drain oil. Install new oil filters and fill crankcase with Mack-specified lubricating oil.
	10. Low compression	10. Check cylinder compression. If low, see symptom "Low compression."

Symptom	Probable Cause	Remedy
B. Engine cranks—will not start	11. Engine improperly timed	11. Check engine timing. The timing setting is protected by seals; therefore, this item must be checked at a factory-authorized service station.
C. Engine misfiring	1. Poor-quality fuel; water or dirt in fuel	1. Drain fuel from tank. Install new fuel filters and fill tank with recommended Mack-specified diesel fuel.
	2. Air in fuel system	2. Check system for air leaks and correct. Air will generally get into the fuel system on the suction side of the fuel transfer pump.
	3. Broken or leaking high-pressure fuel lines	3. Check for fuel leaks and replace defective parts.
	4. Restrictions in fuel lines or drain lines	4. Check for fuel flow. If no flow, replace lines.
	5. Low fuel supply pressure	5. Check to be sure there is no fuel in fuel tank. Look for leaks or sharp bends or kinks in fuel line between fuel tank and fuel transfer pump; clogged suction pipe in tank; and plugged fuel suction hose. Look for air in the system. Check fuel pressure. If lower than specified, change filters and recheck. If still low, replace or repair transfer pump.
	6. Improper valve lash	6. Check and make necessary adjustments.
	7. Defective fuel injection nozzle or fuel pump	7. Run engine at speed that gives maximum misfiring or rough running. Loosen fuel line nut, one at a time, on injection pump, cutting fuel flow to cylinder. When fuel is cut from a given cylinder and running speed does not change, it is an indication cylinder is not firing. Remove nozzle and check. If defective, repair or replace. If nozzle checks ok, check for low compression. However, if no fuel is evident when nut is loosened, it is an indication pump is defective. Repair or replace defective parts.
	8. Engine improperly timed	8. Check engine timing.
	9. Cylinder head gasket leakage	9. Check for visible signs of leakage, coolant in the lubricating oil, and oil traces in the coolant. Use a compression tester to check each cylinder. Replace cylinder head gasket.
	10. Worn camshaft lobe	10. Check rocker arm movement. If not within specification, replace defective parts.

continued

Symptom	Probable Cause	Remedy
D. Engine stalls at low speeds	1. Idle speed too low	1. Check idle setting and make the necessary adjustments.
	2. Fuel tank vent plugged or partly plugged	2. Check vent arrangement and make the necessary repairs.
	3. Low fuel supply	3. Check to be sure fuel is in fuel tank. Look for leaks, sharp bends, or dents in fuel supply lines. Check for air in fuel system. Check that fuel pressure is within the recommended specification. If it is not, check filters, replace, and recheck pressure. If still low, repair or replace transfer pump.
	4. Injection pump overflow valve leaky, stuck open, or closed	4. Repair or replace valve.
	5. Defective fuel injection nozzle	5. Isolate defective fuel injection nozzle and replace (refer to C item 7).
	6. Defective fuel injection pump	6. Remove injection pump. Test and make repairs as required. Reinstall pump.
	7. High parasitic load	7. Check for excessive loading due to engaged auxiliary attachments.
E. Erratic engine speed	1. Air leaks in fuel suction line	1. Check for air leaks and make necessary corrections.
	2. Throttle linkage loose or out of adjustment	2. Check all throttle linkages and make necessary adjustments.
	3. Injection pump governor failure	3. Remove injection pump and look for damaged or broken springs or other components. Check for free travel of fuel rack. Check for correct governor spring. Install new parts in place of damaged or defective ones. Recalibrate injection pump and install.
F. Low power **NOTE:** When diagnosing low-power complaints, it is possible the trouble can be traced to chassis components other than the engine. Make sure the chassis rolls freely when the brakes are released.	1. Restrictions in the air intake system; clogged air filter(s), etc.	1. Check the air pressure in air intake manifold. Replace air filter and make necessary corrections to air system.
	2. Poor-quality fuel	2. Drain fuel tank(s), clean and bleed fuel system. Replace fuel filters. Fill tank with the recommended Mack-specified diesel fuel.
	3. Damage or restrictions in the accelerator/shut-off cable linkage	3. Check linkage, adjust to obtain sufficient travel. Replace if damaged or bent.

Symptom	Probable Cause	Remedy
F. Low power	4. Low fuel pressure	4. Check fuel supply lines for leaks, kinks, restrictions, air in system, etc. Check fuel pressure. If low, replace filters and recheck; if still low, replace or repair fuel transfer pump. Also check for sticking, binding, or defective fuel overflow valve. Repair or replace valve.
	5. Improper valve lash	5. Set valve lash to specified clearance.
	6. Air cylinder improperly shimmed	6. Remove air cylinder. Check operation and "PLE" dimension. Make corrections and install on unit.
	7. Puff limiter air cylinder stuck in puff-limiting position	7. Check operation of air cylinder and replace if defective.
	8. Defective reversing relay permitting excessive air pressure to be directed to the air cylinder	8. Check air pressure at relay outlet; if over specified limit, replace relay.
	9. Restrictions in the tip turbine air cleaner/air induction system	9. Check air induction system. Replace air cleaner and make any necessary repairs.
	10. Air leak in tip turbine fan, bleed air connection	10. Check all air connections. They must be tight.
	11. Tip turbine fan exhaust port plugged	11. Remove blockage. Exhaust port must be open at all times.
	12. Excessive dirt buildup in tip turbine fan housing or wheel	12. Clean, repair, or replace unit.
	13. Tip turbine wheel rub	13. Repair or replace unit.
	14. Intercooler core blocked	14. Remove core. Clean or replace. Reinstall core.
	15. Intercooler gasket leakage	15. Disassemble intercooler and replace gasket.
	16. Incorrect fuel injection timing; plugged fuel tank vents	16. Adjust timing.
	17. Fuel injection nozzle failure	17. Isolate defective nozzle and replace.
	18. Turbocharger has carbon deposits or other causes of friction	18. Inspect the turbocharger. Clean, repair, or replace unit.
	19. Internal fuel injection pump wear preventing full rack travel	19. Remove injection pump. Make necessary repairs, recalibrate pump, and install on engine.
	20. High-altitude operation	20. Engine loses horsepower with increases in altitude. The percentage of power loss is governed by the altitude at which the engine is operated. Make the necessary adjustments.
	21. Low compression	21. Check items listed for "Low compression."
	22. Exhaust restriction	22. Check exhaust system for restrictions.

continued

Symptom	Probable Cause	Remedy
	NOTE: Items 23 and 24 relate to chassis equipped with CMCAC.	
F. Low power	23. Restrictions in cooler	23. Perform restriction pressure test. Clean out restriction.
	24. Restrictions in cooler inlet and outlet tubes	24. Disconnect and clean obstructions.
G. Engine will not reach no-load governed rpm	1. Air in fuel system	1. Check system for air leaks and correct. Air will generally get into the fuel system on suction side of fuel transfer pump.
	2. Accelerator linkage loose or out of adjustment	2. Check linkage and make necessary adjustment.
	3. Restricted fuel lines; stuck overflow valve	3. Check flow in fuel lines and overflow valve for defective spring, poor valve, valve setting, or sticking. Make all necessary repairs.
	4. High-idle adjustment set too low	4. Check high-idle adjustment. Make necessary adjustment.
	5. Chassis equipped with the Mack Maxi-Miser—air leaks in the air supply line or defective control valve	5. Should the air supply to the transmission or air line from the transmission to the fuel injection pump governor leak, the maximum engine rpm would automatically be restricted in all gears and not just fifth gear. Check all connecting hoses and make necessary repairs. Replace control valve if required.
	6. Fuel injection pump calibration incorrect	6. Remove injection pump and nozzle assemblies from engine. Check calibration. Make necessary adjustments and install on engine.
	7. Internal fuel pump governor wear	7. Remove injection pump from engine. Make all necessary repairs, recalibrate, and install on engine.
H. Excessive engine vibration	1. Loose vibration damper hub nut/bolt.	1. Check condition of mounting. Make necessary corrections and retorque.
	2. Defective/damaged vibration damper	2. Replace damaged/defective parts.
	3. Fan blade not in balance	3. Loosen/remove fan belts and operate engine for a short time at the speed at which the vibration was present. If vibration disappears, replace fan assembly.
	4. Engine supports are loose, worn, or defective	4. Tighten all mounting bolts. Install new components as required.
	5. Misfiring or running rough	5. Check items listed for "Engine misfiring."

ENGINE DIAGNOSIS

Symptom	Probable Cause	Remedy
I. Excessive smoke during acceleration **NOTE:** This item is geared to chassis equipped with the Mack Puff Limiter.	1. Plugged or broken outlet line to air cylinder	1. Check and replace broken/deteriorated air line.
	2. Plugged or broken pressure-sensing line	2. Check and replace broken/deteriorated air line.
	3. Inoperative air cylinder	3. Check operation of air cylinder; replace if defective.
	4. Air cylinder improperly shimmed	4. Remove air cylinder and check operation, "PLE" dimension, and shim pack. Make necessary corrections and install in injection pump.
	5. Defective or stuck reversing relay valve	5. Correct sticking valve or replace it.
J. Excessive black or gray smoke.	1. Insufficient air for combustion	1. Check for air cleaner restrictions. Check inlet manifold pressure and inspect turbocharger for correct operation. Make necessary repairs.
	2. High exhaust back pressure	2. Check for faulty exhaust piping or muffler obstruction. Repair/replace defective parts.
	3. Improper grade of fuel	3. Drain fuel from tank. Install new fuel filters and fill tank with recommended Mack-specified diesel fuel.
	4. Faulty injection nozzle	4. Isolate faulty nozzle and replace. Refer to "Engine misfiring," item 7.
	5. Improper engine timing	5. Check timing and make necessary corrections.
K. Excessive blue or white smoke.	1. Engine lubricating oil level too high	1. Remove excess lubricating oil. If oil is contaminated with either fuel or coolant, completely drain system. Change oil filters, locate source of leak and correct. Fill with Mack-specified lubricating oil. Check oil level with dipstick. *Do not overfill.*
	2. Failure of turbocharger oil seals.	2. Check inlet manifold for oil and repair turbocharger as required.
	3. Worn piston rings	3. Check cylinder walls for scuffing. Clean up or replace as required. Install new piston rings.
	4. Engine misfiring or running rough	4. Check all items listed under "Engine misfiring."
	5. Engine-to-pump timing	5. Check and reset timing.
L. Excessive fuel consumption	1. Restrictions in air induction system	1. Inspect system and remove restrictions. Replace defective parts.
	2. External fuel system leakage	2. Check fuel system external piping for signs of fuel leakage. Make necessary corrections.
	3. Incorrect injection timing	3. Check engine timing and make corrections.
	4. Defective injection nozzle assembly	4. Isolate defective nozzle assembly. Remove and replace defective parts.

continued

Symptom	Probable Cause	Remedy
L. Excessive fuel consumption	5. Fuel injection pump calibration incorrect	5. Remove injection pump and nozzle assemblies from engine. Check calibration and make necessary adjustments.
	6. Internal engine wear	6. Overhaul engine.
M. Excessive oil consumption	1. External oil leaks	1. Check engine for visible signs of oil leakage. Look for loose/stripped oil drain plug, broken gaskets (cylinder head cover, etc.), front and rear oil seal leakage. Replace all defective parts.
	2. Clogged crankcase breather/pipe	2. Remove obstruction.
	3. Excessive exhaust back pressure	3. Check exhaust pressure and make necessary corrections.
	4. Worn valve guides	4. Replace valve guides.
	5. Air compressor passing oil	5. Repair or replace air compressor.
	6. Failure of seal rings in turbocharger	6. Check inlet manifold for oil and make necessary repairs.
	7. Internal engine wear	7. Overhaul engine.
N. Engine overheats	1. Coolant level low	1. Determine cause. Replace leaking gaskets and hoses. Tighten connections and add coolant.
	2. Loose or worn fan belts	2. Adjust belt tension, replace belts.
	3. Air flow through radiator restricted	3. Remove all obstructions from outer surface of radiator.
	4. Radiator pressure cap defective	4. Check pressure release of radiator cap. Replace cap if defective.
	5. Defective coolant thermostat/water gauge	5. Check thermostat for proper opening temperature and correct installation. Check temperature gauge. Repair if necessary.
	6. Fan improperly positioned; viscous fan drive not operating properly	6. Check operation of fan and make necessary adjustment.
	7. Chassis with shutters—shutters not opening properly	7. Check shutter operation and make necessary repairs.
	8. Combustion gases in coolant	8. Determine point at which gases enter the system. Repair or replace components as required.
	9. Plugged oil cooler	9. Remove oil cooler from engine. Disassemble, remove restrictions, replace defective parts.
	10. Defective water pump	10. Remove pump and make necessary repairs.
O. High exhaust temperature	1. Operating chassis in wrong gear ratio for load, grade, and altitude	1. Select the correct gear ratio for load and grade conditions.
	2. Restrictions in air induction system	2. Inspect system and remove restrictions. Replace defective parts.
	3. Air leaks in air induction system	3. Check pressure in the air intake manifold. Look for restrictions at the air cleaner/intercooler. Make necessary corrections.

Symptom	Probable Cause	Remedy
O. High exhaust temperature	4. Leaks in exhaust system (preturbo)	4. Check exhaust system for leaks and make necessary repairs.
	5. Fuel injection timing incorrect	5. Adjust injection timing.
	6. Restriction in the exhaust system	6. Inspect system and make necessary repairs.

NOTE: In addition to the preceding items, the following (items 7 through 12) relate to chassis equipped with intercooler engine.

	7. Plugged tip turbine fan air cleaner	7. Check air cleaner and remove restriction.
	8. Leaking tip turbine fan, bleed air connection	8. Check all air connections. They must be airtight for efficient operation.
	9. Tip turbine fan exhaust port plugged	9. Remove blockage. Exhaust port must be open at all times.
	10. Excessive dirt buildup in fan housing or wheel	10. Clean, repair, or replace unit.
	11. Turbine wheel rub	11. Repair or replace the unit.
	12. Intercooler gasket leaking	12. Disassemble intercooler and replace gasket.
	13. Improper valve lash	13. Set valve lash to specified clearance.
	14. Defective injection nozzle assembly	14. Isolate defective nozzle assembly. Remove and replace defective parts.
	15. Fuel injection pump calibration incorrect	15. Remove injection pump and nozzle assemblies from engine. Check calibration and make necessary adjustment.

NOTE: Items 16 through 23 relate to chassis equipped with CMCAC.

High Pyrometer—Normal Boost

	16. Loose ducting	16. Repair loose connections.
	17. Core fin obstructions	17. Clean core fins.

High Pyrometer—Low Boost

	18. Blockage in ducting between air cleaner and turbo	18. Check for blockage and clean.
	19. Dirty turbocharger	19. Remove turbo and clean.
	20. Leaks in the pressurized side of the induction system	20. Check for and repair leaks.
	21. Core leak or inlet manifold leak	21. Check for damaged core gaskets.
	22. Open petcock	22. Close petcock.
	23. Core leakage	23. Pressure test core. Remove, repair, or replace core if test results are not satisfactory.
P. Low lubricating oil pressure	1. Oil leak—line, gasket, etc. Insufficient lubricating oil level	1. Check oil level and add make-up oil as required. Oil must be to Mack-recommended specifications. Check for oil leaks.
	2. Wrong oil viscosity	2. Drain lubricating oil. Change oil filters and fill with oil meeting Mack specifications.

continued

Symptom	Probable Cause	Remedy
P. Low lubricating oil pressure	3. Defective oil pressure gauge	3. Check operation of oil gauge. If defective, replace gauge.
	4. Plugged oil filter(s)	4. Install new oil filter elements. Clean or install new oil cooler core. Drain oil from engine and install oil meeting Mack specifications.
	5. Lubricating oil diluted with fuel oil	5. Check fuel system for leaks. Make necessary repairs. Drain diluted lubricating oil. Install new filter elements and fill crankcase with recommended oil meeting Mack specifications.
	6. Defective oil pump relief valve	6. Remove valve. Check for seat condition and sticking relief valve spring, proper spring tension, and cap. Check assembly parts. The use of incorrect parts will result in improper oil pressure. Make necessary repairs or install new relief valve.
	7. Incorrect meshing of oil pump gears	7. Check mounting arrangement. If engine has been rebuilt, check for proper gear ratio combination of the oil pump–driven gear and drive gear. Check for correct oil pad gasket. Incorrect gear combinations will result in immediate gear failure and possible engine damage.
	8. Excessive clearance between crankshaft and bearings	8. Overhaul engine and replace worn and defective parts.
Q. Low oil pressure at idle, good oil pressure at high rpm	1. Improperly functioning piston cooling oil pressure control valve	1. Check control valve for proper seating; check spring for proper tension and length; check for correct cap.
R. Oil in cooling system	1. Defective oil cooler O-ring(s)	1. Disassemble and replace O-ring(s).
	2. Defective oil cooler core	2. Remove oil cooler. Disassemble and repair/replace oil cooler core.
	3. Blown head gasket	3. Replace head gasket.
S. Coolant in lubricating oil	1. Defective oil cooler core	1. Disassemble and repair/replace oil cooler core.
	2. Blown head gasket	2. Replace head gasket.
	3. Defective water pump oil seal(s)	3. Remove water pump, disassemble and replace defective parts.
	4. Mack V8 engines only—cylinder sleeve seal (O-rings) failures.	4. Replace cylinder sleeve seals (O-rings).
	5. Mack 6-cylinder engines only—leaking or loose plugs behind valve lifter cover.	5. Replace or tighten loose plugs.
T. Low compression	1. Improper valve lash	1. Set valve lash to specified clearance.
	2. Blown head gasket	2. Replace head gasket.
	3. Broken or weak valve spring	3. Check for and replace defective parts.
	4. Burned valves or seat and parts	4. Remove head from engine and recondition head.
	5. Piston rings stuck, worn, broken/improperly seated	5. Overhaul engine.
	6. Camshaft or valve lifters worn	6. Replace camshaft and/or valve lifters. Overhaul engine if required.

engine equipment. The chart is similar to those found in service manuals for other engines. Always refer to the appropriate service manual for any specific engine.

For engines with electronically controlled fuel systems the on-board diagnostics and the appropriate electronic diagnostic tools should be used to retrieve diagnostic faults stored in computer memory. These are described in other chapters in this book.

Diagnostic Guide (to pp. 618-626)

ITEM	SYMPTOM
A	Engine will not crank
B	Engine cranks—will not start
C	Engine misfiring
D	Engine stalls at low speed
E	Erratic engine speed
F	Low power
G	Engine will not reach no-load governed rpm
H	Excessive engine vibration
I	Excessive smoke during acceleration
J	Excessive black or gray smoke
K	Excessive blue or white smoke
L	Excessive fuel consumption
M	Excessive oil consumption
N	Engine overheats
O	High exhaust temperature
P	Low lubricating oil pressure
Q	Low oil pressure at idle, good oil pressure at high rpm
R	Oil in cooling system
S	Coolant in lubricating oil
T	Low compression

DIESEL ENGINE TUNE-UP

A diesel engine tune-up is performed when the engine has operated the number of hours specified in the service manual. It may also be performed when a performance problem has developed. The number and type of procedures performed varies with vehicle owners' requirements, different service shops, and engine make and model. Services performed may include the following:

1. Clean and inspect the engine.
2. Change the engine oil and filters.
3. Service the cooling system.
4. Check for loose fasteners, lines, fittings, supports, and mounting brackets.
5. Check the air intake system.
6. Check the exhaust system.
7. Service the fuel system.
8. Perform a compression test and cylinder leakage test if needed.
9. Adjust valve clearances.
10. Check throttle linkage.
11. Check engine monitoring gauges.
12. Check engine performance with a dynamometer.

Tune-Up Procedure

1. *Clean the engine exterior.* The importance of cleanliness when working on diesel engines and fuel injection systems must be strongly emphasized. It takes only a small particle of dirt to cause serious damage. It is therefore good practice to clean the exterior of the engine before any tune-up work is begun. Be sure to use a face shield and protective clothing when using a steam cleaner or high-pressure water cleaner.

CAUTION:

Both the steam cleaner and the high-pressure water cleaner work well; however, some precautions must be observed. Be sure that water or steam does not enter the engine, induction system, or fuel system. Cover any vents or openings to prevent moisture contamination.

CAUTION:

Do not steam clean any fuel injection components while the engine is running. If fuel injection components are steam cleaned with the engine stopped, do not start the engine until you are sure that the temperature of the injection system components has completely stabilized. Steam cleaning fuel injection components causes considerable temperature differences in injection system components, which can cause scoring of parts or can result in actual seizure if done while the engine is running or if the engine is started soon after steam cleaning. Due to the extremely close tolerances between certain injection system components, uneven temperatures between these parts can cause serious damage if the engine is operated under these conditions.

Make sure all areas of the engine are properly cleaned, particularly places where disassembly is required. These areas are especially prone to the entry of dirt when covers or plates are removed to gain access to other parts of the engine.

Run the engine at operating temperature and check for oil, fuel, and coolant leaks or other problems.

2. *Change the engine oil and oil filters.* It is important that the engine oil and oil filters be changed at the recommended time or mileage intervals. See Chapter 6 for details on service and viscosity ratings of engine lubricating oils and filter changing.

3. *Service the cooling system.* The cooling system must be in good operating condition for it to be able to keep the engine at its most efficient operating temperature. A good engine tune-up includes a careful inspection of the cooling system and the correction of any faults that may be discovered. See Chapter 5 for details of cooling system service.

Check the coolant level. Check the coolant condition before adding antifreeze or water to ensure that coolant quality is maintained. Check for coolant leakage and correct if necessary. Check the condition of all hoses and clamps. Check the condition and tension of all belts. Check the pulleys, fan, fan shroud, fan drive, and radiator shutter operation. Change the coolant filter and conditioner as required. Replace any parts and make any adjustments that inspection procedures indicate are needed.

4. *Check for loose fasteners, lines, fittings, supports, and mounting brackets.* If there are any loose bolts or nuts, make sure that both the male and female threads are in good condition before tightening them to specified torque. Faulty fasteners must be replaced. Make sure that any required lock washers, cotter pins, or flat locks are in place. Missing or damaged locking devices must be replaced. Make sure that all lines and tubing are routed properly and are in good condition. Lines and tubing should be routed properly to prevent chafing and have sufficient distance from hot exhaust system components. All support brackets should be in place and in good condition.

5. *Check the air intake system.* Check all induction system hoses, tubes, and connections. The entire induction system should be airtight to prevent the entry of dust and dirt. Check the turbocharger connections as well. Make sure that piping is not damaged or dented, which could cause air intake to be restricted. Service the air cleaner as required. See Chapter 4 for details on induction system service.

6. *Check the exhaust system.* Make sure there are no exhaust leaks or damaged pipes that could restrict the exhaust. Exhaust leaks are usually evidenced by streaked lines of black extending from the leak. Repair as needed. See Chapter 4 for details on exhaust system service.

7. *Service the fuel system.* An adequate supply of clean fuel of the recommended type is absolutely essential to good engine performance. Incorrect fuel, contaminated fuel, or restricted fuel flow can all cause serious engine performance problems. Check the fuel tank and filler cap (be sure the vent is open), the water trap, and the fuel lines. Drain any water from the water trap. Make sure fuel lines are not dented or kinked to restrict flow in both supply and return lines. Clean any fuel screens that may be in the fuel supply system. Check the fuel supply pump pressure if necessary.

Remove, test, and calibrate the injectors if required. Injectors must perform in a uniform manner for all engine cylinders to ensure proper power balance between cylinders. On engines with engine cam-operated injectors, the engine valve clearances must be adjusted before injectors are adjusted. Unit injectors also require timing and fuel rack control adjustments. If an engine compression test is to be performed, it should be done before the injectors are installed.

8. *Perform an engine cylinder compression test if needed.* Cylinder compression must be adequate and must be even between cylinders as specified for all cylinders to produce equal power. Compression testing and cylinder leakage tests are discussed in this chapter.

9. *Adjust engine valve clearance.* The valve bridges do not normally need adjustment during a tune-up; however, they must be checked. If adjustment of any amount is required, it is an indication of valve and seat problems. Adjust valve clearance after checking valve bridge adjustment.

10. *Check the throttle linkage.* Make sure the throttle linkage is able to move through the entire range of travel and does not bind throughout that range. If the throttle linkage is not able to move to the full-fuel position, the engine will not be able to produce its full power. A bent linkage or obstructions can be the cause of interference with linkage movement. Make sure connecting pins and clevises are not worn excessively. Correct any condition that interferes with free and full throttle movement. Adjust engine low-idle and high-idle speed to specifications.

11. *Check engine monitoring gauges mounted in the instrument panel.* Check any abnormal condition to determine whether the monitored function is in error or whether the gauge or sensor is at fault. Correct as needed.

12. *Check engine performance with a dynamometer.* Performance testing with a dynamometer is discussed in this chapter.

COMPRESSION TESTING

To perform a compression test on a diesel engine, proceed as follows with a tester capable of 500 psi (3450 kPa) **(Figures 27–1 and 27–2).**

COMPRESSION TESTING 629

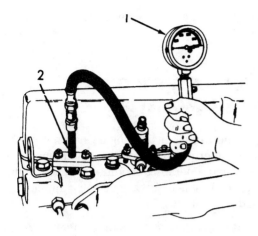

Checking Compression Pressure

1. Compression Tester Gauge Assembly
2. Compression Tester Adapter

FIGURE 27-1 Typical compression testing equipment. Different adapters are required for various engines. (Courtesy of Allis-Chalmers.)

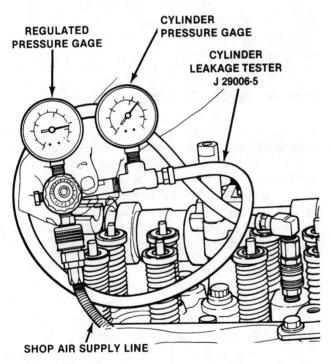

FIGURE 27-2 This arrangement allows for compression testing and cylinder leakage testing. (Courtesy of Detroit Diesel Corporation.)

1. The engine should be at operating temperature.
2. The batteries and cranking system should be in good condition.
3. Remove the air cleaner element.

4. Disable the fuel system and glow plug system as specified by the vehicle manufacturer. This may require disconnecting a fuel solenoid lead on some models, as well as glow plug connections. Refer to the service manual for the procedure to follow.

5. Remove either glow plugs or injectors to allow installation of a compression tester. Some engines are compression tested through the glow plug holes, while others are tested through the injector holes. Compression testers may be equipped with screw-in or clamp-in adapters, depending on application. Be sure to follow engine and equipment manufacturer's instructions for use of the compression tester.

6. Install the compression tester into the No. 1 cylinder.

7. Crank the engine through at least six compression strokes, and note the highest compression reading and the number of strokes required to obtain the reading. Repeat the test on each of the remaining cylinders, cranking the engine the same number of strokes as were required to obtain the highest reading for the No. 1 cylinder.

8. Record the results from all cylinders.

9. Analyze the test results as follows:

Normal—Compression builds up quickly and evenly to specified compression pressure on all cylinders.

Piston ring leakage—Compression low on first stroke but tends to build up on following strokes. Does not reach normal cylinder pressure.

CAUTION:

Due to the high compression ratio of a diesel engine and the very small combustion chamber volume, do not add oil to any cylinder for compression testing. Extensive engine damage can result from this procedure.

Comparison Between Cylinders

In order for cylinders to produce relatively even power output, compression pressures should be similar within certain limits. The comparison of compression pressures between cylinders is just as important as actual compression pressures. Since engine temperature, oil, viscosity, and engine cranking speed all have a bearing on test results, some variation from specified pressures may be expected. However, compression pressures should be comparable between cylinders. Some manufacturers allow a maximum difference in pressures between cylinders of 15%. A greater than 15% difference requires repair to the affected cylinders.

CYLINDER LEAKAGE TESTING

Cylinder leakage tests are performed to determine whether compression or combustion pressures are able to leak past the rings into the crankcase, past the exhaust valves into the exhaust system, past the intake valves into the induction system, or past the head gasket into the engine coolant.

A cylinder leakage test may be performed using a shop air line adapter made from a discarded injector of the appropriate type—a male shop air coupler is welded to the injector main body after the plunger etc. have been removed. This allows the adapter to be installed into the injector hole, and shop air to be coupled to the adapter **(Figure 27–2)**.

Shop air pressure of 150 psi (1000 kPa) is required for this test.

To perform a cylinder leakage test proceed as follows:

1. The engine must be at operating temperature.
2. Disconnect the battery ground cable.
3. Remove the air cleaner element.
4. Disable the fuel system and glow plug system and remove injectors as specified by the vehicle manufacturer. This may require disconnecting a fuel solenoid lead on some models, as well as glow plug connections. Refer to the service manual for the procedure to follow.
5. Install an air line adapter into the No. 1 cylinder.
6. Remove the crankcase oil dipstick, oil filler cap, and radiator cap.
7. Turn the crankshaft to position the No. 1 piston at TDC on the compression stroke. Make sure that the piston is exactly at TDC on the upstroke of the compression stroke. This is important for three reasons. First, the piston will be forced down by shop air if not exactly at the TDC position. Second, the piston rings should be at the bottom of their grooves for this test. Moving the piston up will do this. Third, valves are closed with the piston in this position.
8. Now connect shop air to the adapter in the No. 1 cylinder.
9. Listen for air leakage into the exhaust system at the exhaust pipe. If present, this indicates exhaust valve leakage. Listen for air leakage at the air intake. If present, this indicates intake valve leakage. Listen for air leakage at the oil filler cap or dipstick tube. If present, this indicates leakage past the rings. Listen for leakage at injector holes of cylinders adjacent to the one being tested. If present, this indicates cylinder head gasket leakage between cylinders. Watch for air bubbles in coolant in radiator. If present, this indicates cylinder head gasket leakage to cooling system.
10. Disconnect the shop air line from the cylinder adapter and repeat the test procedure for all cylinders.

PERFORMANCE TESTING WITH A DYNAMOMETER

Diesel engine performance and engine systems' performance can be determined by using a dynamometer designed to simulate operating conditions. Several types are used, each designed for a particular application. These include the following:

1. Chassis dynamometer used to test highway vehicles.
2. Engine dynamometer used to test engines removed from the vehicle or equipment.
3. Power take-off dynamometer used to test agricultural and industrial tractors equipped with a power take-off shaft.

All three types use a power absorption unit to which engine power is applied for testing. The degree of load applied can be varied by a hand-held control device. The power absorption unit may be of the hydraulic or electrical type. See Chapter 2 for a description of dynamometer operation.

Dynamometer Test Instruments

A complete set of engine monitoring instruments is used to provide the operator with the necessary information during the test procedure. An electronic test instrument capable of all functional tests is used on engines with electronic control systems. The tester plugs into the engine data link, which is connected to the electronic control module. Electronic control systems and system tests are described in other chapters of this book.

1. A *pressure gauge* to determine engine lubricating oil pressure is connected to the engine main oil gallery.
2. An *oil temperature gauge* is inserted into the engine oil through the dipstick tube or hole.
3. A *temperature gauge* to determine engine coolant temperature is usually connected to the engine side of the thermostat housing.
4. A *pressure gauge* to measure cooling system pressure is usually connected to the water manifold.
5. A *water manometer* or *low-reading pressure gauge* to measure crankcase pressure is connected to the crankcase above the oil level **(Figures 27–3 and 27–4)**.

6. A *water manometer* or *low-reading vacuum gauge* to determine negative inlet air pressure (restriction) is connected to the air inlet to the engine **(Figure 27–5)**.

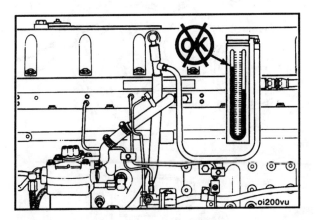

FIGURE 27–3 Checking crankcase pressure (blowby) with a water manometer. (Courtesy of Cummins Engine Company, Inc.)

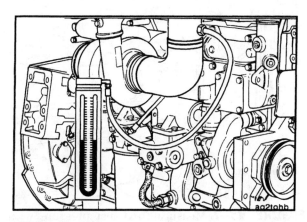

FIGURE 27–5 Measuring inlet air restriction with a water manometer. (Courtesy of Cummins Engine Company, Inc.)

7. A *mercury manometer* or *low-reading pressure gauge* to measure boost pressure on turbocharged engines is connected to the intake air manifold **(Figure 27–6)**.

8. A *water manometer* or *low-reading pressure gauge* to measure exhaust back pressure is connected to the exhaust manifold flange **(Figure 27–7)**.

9. A *contact pyrometer* is used to determine individual cylinder exhaust temperatures. Temperature is checked at the manifold port near the cylinder head at each cylinder to determine whether exhaust temperatures are within limits and are relatively even among cylinders **(Figure 27–8)**.

10. A *tachometer* is used to determine engine crankshaft speed in revolutions per minute **(Figure 27–9)**.

11. A *power gauge* (horsepower and/or kilowatts) indicates power produced by the engine.

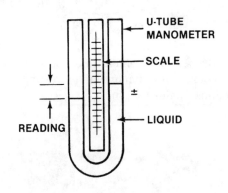

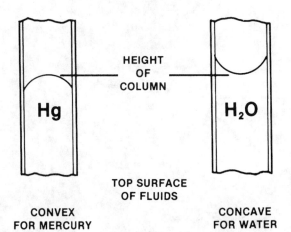

FIGURE 27–4 U-tube manometer and illustration of where to take the readings. (Courtesy of Detroit Diesel Corporation.)

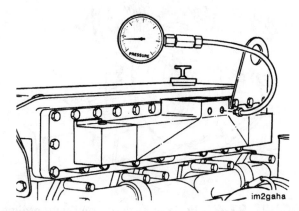

FIGURE 27–6 Measuring turbo boost pressure at the intake manifold. (Courtesy of Cummins Engine Company, Inc.)

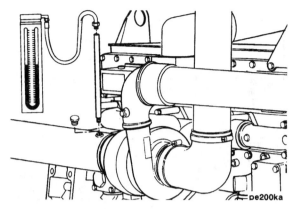

FIGURE 27-7 Measuring exhaust pressure with a mercury manometer. (Courtesy of Cummins Engine Company, Inc.)

FIGURE 27-8 Pyrometer used to test exhaust temperature. (Courtesy of Caterpillar, Inc.)

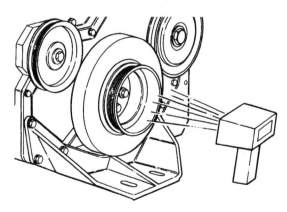

FIGURE 27-9 Checking engine speed with a digital optical tachometer. (Courtesy of Cummins Engine Company, Inc.)

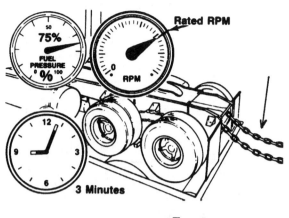

FIGURE 27-10 Vehicle must be securely anchored and blocked for dynamometer testing. (Courtesy of Cummins Engine Co., Inc.)

FIGURE 27-11 Power take-off (PTO) dynamometer. (Courtesy of Deere and Company.)

Dynamometer Worksheet

Date _____ Repair Order No. _____ Operator _____

ESN _____ CPL _____ Fuel Pump Code _____

Complaint _____ SC Code _____

PARAMETER	CODE SPECIFICATIONS	ACTUAL READING
Fuel Pressure (psi @ RPM)		
Fuel Rate (pph)		
Check Point 1 (psi @ RPM)	(Reference Only)	
Intake Mfd. Pressure (in.Hg)		
Intake Mfd. Temperature	76°C [170°F]	
Governor Break RPM		
No-Air Setting (psi @ RPM)		
*Intake Air Restriction	25 in. H_2O, Maximum	
*Exhaust Air Restriction	3 in. Hg, Maximum	
*Fuel Inlet Restriction	8 in. Hg (Dirty Filter), Maximum	
CELECT™ Fuel Inlet Restriction	9 in Hg (Dirty Filter), Maximum	
*Fuel Drain Line Restriction	2.5 in. Hg with Check Valves, Max. 6.5 in. Hg without Check Valves, Max.	
Engine Blowby	12 in. H_2O New Engines, Max. 18 in. H_2O Used Engines, Max.	

*Recorded at Maximum Horsepower Speed and Full Load

Road Speed Limit _____ Engine High Speed Limit _____

Check Oil Level ____ Low ____ High ____ OK   Fuel Quality ____ OK ____ Not OK

ENGINE SPEED (RPM)	*FUEL RATED (pph)	FUEL PRESSURE (psi)	INTAKE MANIFOLD PRESSURE (in. Hg)	INTAKE MANIFOLD TEMP	† ENGINE BLOWBY (in.H_2O)	LUBRICATING OIL PRESSURE (psi)	HORSE-POWER OR TORQUE

* Be Sure That the Fuel Rate is Corrected for Temperature.

Fuel Temperature	Correction for Flow Rate
Less than 7°C [45°F]	Flow meter **not** accurate
7 to 13°C [45 to 55°F]	Subtract 2% from flow rate reading
13.0 to 20.0°C [55 to 68°F]	Subtract 1% from flow rate reading
20.0 to 29°C [68 to 85°F]	No Correction
29 to 42°C [85 to 108°F]	Add 1% to flow rate reading
42 to 56°C [108 to 132°F]	Add 2% to flow rate reading
56°C above [132°F]	Flow meter **not** accurate.

Pressure Conversions

1 in. H_2O = 0.074 in. Hg = 0.036 psi
1 in. Hg = 13.514 in. H_2O = 0.491 psi
1 psi = 2.036 in. Hg = 27.7 in. H_2O

FIGURE 27-12 (Courtesy of Cummins Engine Co., Inc.)

12. A *fuel consumption meter* is connected to the fuel supply to the engine to measure fuel consumption in pounds per brake horsepower hour (lbs/BHP/hr).

DANGER:

Do not perform a dynamometer test with radial ply, recapped, or snow-tread tires. Tread damage and tread separation may occur.

CAUTION:

Do not attempt to operate a chassis or engine dynamometer unless you are trained and authorized to do so. Incorrect procedures endanger the operator and the equipment.

Figure 27–10 shows a chassis dynamometer. A PTO dynamometer is shown in **Figure 27–11**. A typical dynamometer worksheet is shown in **Figure 27–12**.

REVIEW QUESTIONS

1. Incorrect engine diagnosis is _____, time _____, and may result in a loss of customer _____.
2. How do you retrieve fault codes stored in the memory of the ECM?
3. It is a good practice to clean the engine exterior before any _____ _____ is begun.
4. When cleaning with steam or high-pressure water, be sure it does not enter the _____ _____ or _____ _____.
5. To check for oil, fuel, and coolant leaks run the engine at operating temperature. (T) (F)
6. Check all induction system _____, _____, and _____.
7. Cylinder compression must be _____ and _____ to produce equal power.
8. Explain the purpose of cylinder leakage testing.
9. The chassis dynamometer is used to test _____.
10. An electronic test instrument capable of all functional tests is used on engines with electronic control systems. (T) (F)

TEST QUESTIONS

1. If an engine cranks normally but does not start,
   a. perform a dynamometer test
   b. observe the injector spray while cranking the engine
   c. both of the above
   d. neither of the above
2. A cylinder balance test is performed to determine whether all cylinders
   a. are the same size
   b. are the same weight
   c. produce equal power
   d. produce equal emissions
3. A cylinder leakage test does *not* check for
   a. leakage past the rings
   b. leakage past the valves
   c. leakage past gaskets
   d. leakage past the oil pump
4. Compression pressures that are too low are the result of
   a. carbon buildup
   b. excessive cranking speeds
   c. high compression ratios
   d. cylinder leakage
5. To obtain maximum performance, minimum fuel consumption, and lowest emissions, there must be
   a. adequate cylinder compression
   b. proper fuel atomization
   c. adequate fuel injection at the correct time
   d. all of the above
6. Avoid all contact with fuel under injection pressure because it
   a. will penetrate the skin, causing injury
   b. is hot and will cause burns
   c. will cause skin blisters
   d. could cause a fire
7. Injection lines should be
   a. made of copper tubing
   b. all the same length and size
   c. routed close to the exhaust manifold
   d. all of the above

8. Technician A says diesel injection fuel lines should never be bent. Technician B says you can bend a line in order to remove it. Who is correct?
   a. Technician A   c. both are correct
   b. Technician B   d. both are incorrect
9. The air/fuel ratio of a diesel engine is approximately
   a. 60 to 1   c. 18 to 1
   b. 30 to 1   d. 9 to 1
10. The diesel fuel injection system that does not require high-pressure fuel injection lines is the
    a. unit injector system
    b. distributor pump system
    c. electric pump system
    d. hydraulic pump system
11. Diesel engine starting aids do not include
    a. a glow plug   c. block heaters
    b. fuel heaters   d. a choke plate
12. Testing injector pressure will identify opening and closing pressures. (T) (F)
13. Injection pump timing is preset and cannot be changed. (T) (F)
14. Two adjustments to be made after the injection pump is installed are low-idle speed and maximum speed. (T) (F)
15. Injector parts are all prefitted for a specific injector and should never be mixed with other injector parts. (T) (F)
16. Injector fuel lines must never be bent and can be replaced only with original manufacturer's parts. (T) (F)

CHAPTER 28

CYLINDER HEAD AND VALVE SERVICE

INTRODUCTION

Cylinder head and valve train service may be performed without engine removal. This is the case when valve or cylinder head gasket problems occur on an engine that is otherwise in good condition. When the engine is removed for a complete overhaul, the cylinder head and valve train are normally reconditioned. Procedures for cylinder head and valve service vary somewhat depending on engine design. A two-stroke-cycle engine, for example, does not have any intake valves. Differences in valve train components exist between engines with two or four valves per cylinder and with in-block or overhead camshafts. In spite of these differences the reconditioning procedures are very similar, though the specifications may differ. This chapter discusses these procedures and provides examples used for different engines. Always follow the procedures and specifications given in the appropriate service manual for the engine being serviced.

PERFORMANCE OBJECTIVES

After thorough study of this chapter, sufficient practice on the appropriate training components, and with the appropriate shop manuals, tools and equipment, you should be able to do the following:

1. Follow the general precautions outlined in this chapter and in the service manual.

2. Clean, inspect, and accurately measure all cylinder head and valve components to determine their serviceability.

3. Recondition and replace all cylinder head and valve components as needed to meet manufacturer's specifications.

4. Correctly assemble and adjust all cylinder head and valve components to meet manufacturer's specifications.

TERMS YOU SHOULD KNOW

Look for these terms as you study this chapter, and learn what they mean.

valve spring compressor
mushroomed valve stem tips
cylinder head pressure testing
crack detection methods
cylinder head warpage or flatness
valve stem-to-guide clearance
loose valve seat insert
oversize seat insert
counterboring
interference fit
valve head–to–cylinder head clearance
valve seat grinding
roughing stone
finishing stone
tapered pilot
expanding pilot
stone holder
stone driver
stone dresser
stone chatter
overcutting
topping

undercutting
throating
valve seat concentricity
valve grinder
valve refacing
valve grinding angle
crossfeed
depth feed
valve spring tester
valve spring pressure
valve spring height
valve opening pressure
guide studs
valve adjustment

CYLINDER HEAD REMOVAL PRECAUTIONS

The entire cylinder head area should be thoroughly cleaned using a pressure washer or steam cleaner. Use a face mask and gloves during cleaning.

CAUTION:

Never steam clean the fuel injection system components while the engine is running. The resulting difference in component temperatures can cause injection plungers to score or seize.

DANGER:

Be aware that engine parts and fluids may be hot enough to cause severe burns if contacted. Avoid contact with hot parts or fluids.

DANGER:

Be aware that some engine systems are under high pressure and that disconnecting them improperly could result in injury. This includes hydraulic systems, fuel injection lines, air lines, and cooling systems.

CAUTION:

Disconnect the battery ground cable at the battery before beginning disassembly procedures. This avoids possible damage to the electrical system components.

Cab-Over and Cab-Forward Trucks

Servicing cab-over and cab-forward truck engines requires the following special safety precautions:

BEFORE TILTING CAB

- Check clearance above and in front of truck cab.
- Keep tool chests and workbenches away from front of cab.
- Inspect sleeper and cab interior for loose luggage, tools, and liquid containers that could fall forward.
- Check "buddy" or "jump" seats on right side of cab to be sure that they are secured in place.

WHILE TILTING CAB

- Engine must not be running.
- Check position of steering shaft U-joint to prevent binding.
- Never work under a partially tilted cab.
- When cab is tilted past the overcenter position, use cables or chains to prevent it from falling. Do not rely on the cab hydraulic lift mechanism to retain or break the fall of the cab in any position beyond the overcenter point.
- After the cab has been tilted make sure the safety catch locks the cab in place.

WHILE CAB IS TILTED

- If it is necessary to open a door, take care to avoid damage to door hinges and/or window glass.

WHILE LOWERING CAB

- Engine must not be running.
- Be sure that the cab lowers properly on mounting and locating pins. Twisted or misaligned cabs may miss locating pins, and cab latch will not secure cab in locked position.
- Lock cab-latching mechanism when the cab is all the way down. Failure to lock the cab will allow it to swing forward when the truck is stopped suddenly.

Industrial Attachments

- Be sure that all hydraulic, mechanical, or air-operated equipment such as loaders, buckets, and the like are lowered to their at-rest position on the floor or are safely supported by other means before you attempt to do any engine work.

CYLINDER HEAD REMOVAL

Problems that require cylinder head removal for correction include the following:

- head gasket failure
- cracked cylinder head
- worn valve guides
- worn valve seats
- worn valve face and stem

Procedures for cylinder head removal vary among the different engine makes and models as well as the type of vehicle or equipment in which they are used. The following procedures although typical are therefore general in nature and should be used in conjunction with the appropriate service manual.

1. Remove or tilt the hood, tilt the cab, or remove the engine cover panels depending on the type of vehicle or equipment being serviced.

2. Thoroughly clean the entire cylinder head area with a pressure washer or steam cleaner.

3. Drain the coolant from the radiator and engine block drains into suitable drain pans. (See Chapter 5 for cooling system service.)

4. Disconnect any coolant hoses, oil lines, fuel lines, air lines, and electrical connections that are connected to the cylinder head.

5. Remove any item such as the air compressor, alternator, air conditioning compressor, or other item that may be mounted on the cylinder head.

NOTE: These may be removed without disconnecting them from the system but should be tied up, away from possible interference during cylinder head removal.

6. Remove the connections to the intake and exhaust manifolds including the turbocharger if so equipped.

7. Remove any other items that could interfere with cylinder head removal. This may include the intake and exhaust manifolds, water manifold, and turbocharger. (See Chapter 4 for intake, exhaust, and turbocharger service.)

8. Remove the rocker arm covers. They may be one-, two-, or three-piece design **(Figure 28–1)**.

9. Remove the Jacobs brake (or C brake) control wire connections, the mounting bolts, and the entire unit if so equipped **(Figure 28–2)**.

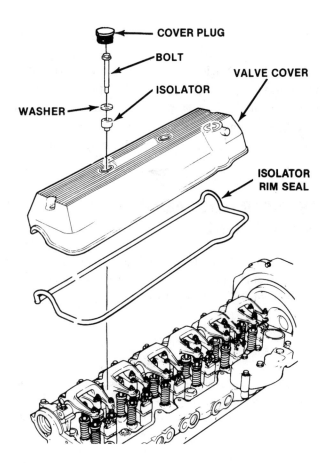

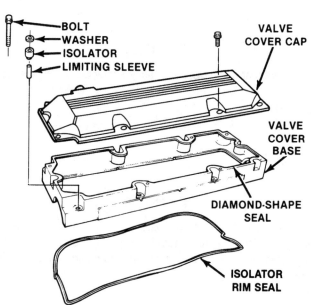

FIGURE 28–1 One-piece (top) and two-piece (bottom) valve cover designs. (Courtesy of Detroit Diesel Corporation.)

CYLINDER HEAD REMOVAL

FIGURE 28–2 Jacobs brake removal. (Courtesy of Caterpillar, Inc.)

10. Loosen the locknuts on the injector adjusting screws (if so equipped) and loosen the screws several turns to aid in later assembly **(Figure 28–3)**.

11. Loosen the locknuts on the valve bridge and valve adjusting screws. Loosen the screws several turns to aid in later assembly.

12. Loosen the rocker arm (or rocker box) mounting screws evenly to avoid distortion of the rocker shaft (if so equipped). Remove the bolts, rocker arms, and push rods, keeping them in the proper order **(Figures 28–4 and 28–5)**.

13. On engines with injection nozzles and injection pump disconnect the injection lines from the pump and nozzle holders. Cap all lines and fittings to prevent the entry of dirt or moisture.

14. On engines with unit injectors disconnect any fuel lines that connect the injector with the fuel gallery in the head. Cap all lines and fittings.

Nozzle-Type Injector Removal

a. On engines with an injection pump and nozzles loosen the injection line fittings just enough to relieve fuel pressure, then remove the lines from the injectors and pump.

CAUTION:

Be careful not to bend the injection lines in any way, since this changes the inside diameter of the line and affects performance. (See Chapter 12 for more detailed procedures.)

b. Remove the hold-down nuts or cap screws. Turn the nozzle back and forth while pulling on it to remove it from the head. If the nozzle is stud-mounted, use a roll-head prybar or an injector slide hammer puller to remove the injector.

CAUTION:

Be very careful not to damage or bend the nozzles during removal. Pencil nozzles are most easily damaged.

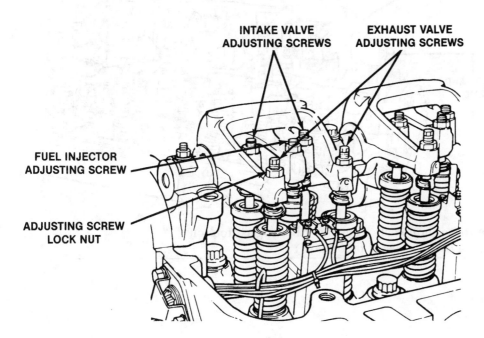

FIGURE 28–3 Fuel injector and valve adjusting screws and locknuts on 60 series engine. (Courtesy of Detroit Diesel Corporation.)

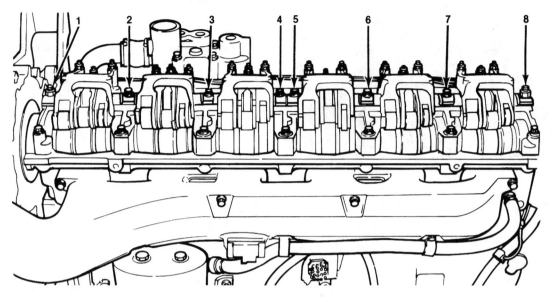

FIGURE 28-4 Rocker shaft bolts and nuts (typical). (Courtesy of Detroit Diesel Corporation.)

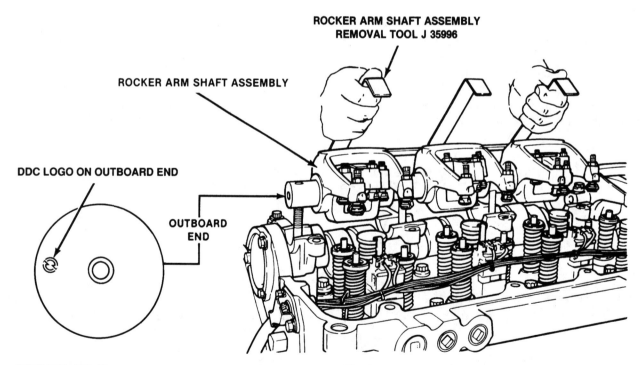

FIGURE 28-5 Removing the rocker arm shaft assembly using special tool on DDC series 60 engine. (Courtesy of Detroit Diesel Corporation.)

Unit Injector Removal (See Chapters 18, 20, 22, and 23 for more detailed procedures.)

a. Disconnect the wires from the electronic injectors.
b. Remove the hold-down nut or cap screw.
c. Use a roll-head prybar or slide hammer injector puller to remove the injector.

15. On overhead camshaft engines the camshaft may be gear- or belt-driven. To remove the cylinder head, the camshaft drive must first be disconnected. The camshaft may also have to be removed. To gain access requires that the camshaft drive cover or cover plate be removed, after which the camshaft can be removed **(Figures 28–6 and 28–7)**. To remove the camshaft remove the camshaft bearing caps and then lift the camshaft from the head.

16. Loosen the cylinder head bolts in the reverse order of the tightening sequence (in a circular pattern from both ends toward the middle), one turn each. Then remove all the bolts and the cylinder head. Note the location of any special bolts or bolts of different lengths for proper assembly later. Use lift equipment on larger heads **(Figure 28–8)**.

CYLINDER HEAD DISASSEMBLY

Preliminary cleaning can be done without the valves being removed. This protects the valve seats from damage when removing carbon deposits from the combustion chamber. Scrapers and a rotary wire brush are used for this purpose. Any deposits in the ports should also be removed.

Cylinder head disassembly includes removal of rocker arms and pivots, valve bridges or crossheads, valve springs, valves, and locks.

All parts should be kept in order for correct reassembly. A C-clamp type or lever type of valve-spring compressor is used to compress the valve springs far enough to allow the locks to be removed **(Figures 28–9 and 28–10)**. Sometimes the locks and retainer are stuck to the valve stem, and the spring cannot be compressed with the spring compressor. In this case the spring retainer should be tapped with a soft-faced hammer while the spring compressor is applying some pressure to the spring. Tapping the retainer with the hammer will allow the locks and retainer to pop loose.

CAUTION:
Always keep compressed springs under full control.

If the valve stem tips are mushroomed from rocker arm action, they should be dressed before valve removal to avoid damage to the valve guides. A hand file can be used for this purpose.

Care must be exercised when using the valve-spring compressor to avoid slipping off the compressed spring. The compressed spring releases a powerful punch if it snaps out of place and can cause serious injury.

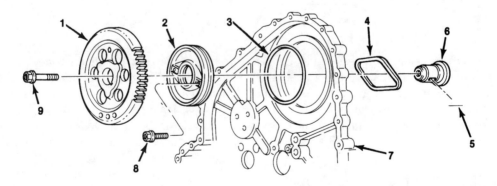

1. Gear, Camshaft Drive
2. Thrust Plate, Camshaft
3. O-Ring, Camshaft Thrust Plate
4. Seal, Camshaft Thrust Plate
5. Key, Camshaft Drive Gear Hub
6. Hub, Camshaft Drive Gear
7. Gear Case
8. Bolt, Camshaft Thrust Plate Retaining (2)
9. Bolt, Camshaft Retaining

FIGURE 28–6 Drive gear and related parts for overhead camshaft on DDC series 60 engine. (Courtesy of Detroit Diesel Corporation.)

642 Chapter 28 CYLINDER HEAD AND VALVE SERVICE

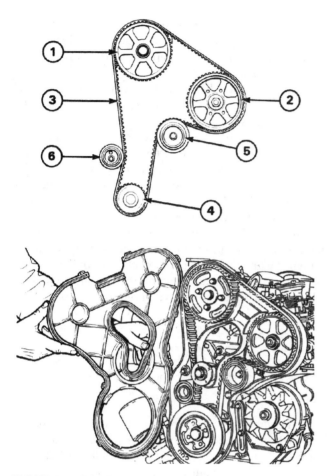

FIGURE 28–7 Fiberglass-reinforced timing belt drives the overhead camshaft and the fuel injection pump. (A) 1. Camshaft drive. 2. Injection pump drive. 3. Toothed belt. 4. Crankshaft drive sprocket. 5. Fixed idler pulley. 6. Adjustable tensioner pulley. (C) Timing belt cover. (Courtesy of Detroit Diesel Corporation.)

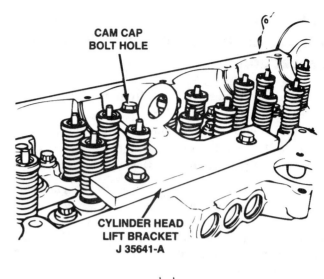

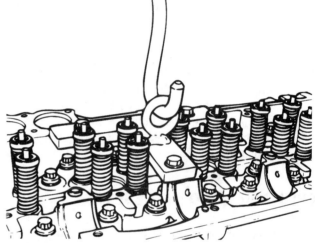

FIGURE 28–8 Cylinder head removal using lift attachment. (Courtesy of Detroit Diesel Corporation.)

After disassembly, the cleaning process can be completed. Any remaining deposits that were not accessible while the valves were in place must be removed **(Figures 28–11** and **28–12)**. The cylinder head should then be pressure tested or tested for cracks by the magnetic detection method.

CYLINDER HEAD CRACK DETECTION

If careful visual inspection has not revealed any cracks, the cylinder head must be tested further. The following methods are typical.

1. *Pressure testing:* All coolant passages in the cylinder head are sealed off with plugs and plates designed for the purpose. An adapter to allow shop air line to be connected is installed in one of the coolant openings. With air pressure applied to the head, it is immersed in a tank of water. Air bubbles escaping indicate points of leakage. Pinpoint the source of the bubbles and mark the location. Do not mistake leakage from the plugs or plates for cracks in the head. Mark any leaking injector sleeves for later replacement. Welding of cracks in the head is allowed in some cases. Refer to the service manual for instructions. The pressure-testing method reveals leakage in areas that are not visible. The magnetic crack detection and dye penetrant methods are limited to accessible and visible areas.

2. *Magnetic crack detection:* A u-shaped electromagnet is placed on the cylinder head surface. Metal filings are sprinkled around the area being checked. The filings are attracted to any crack, making it visible. The

CYLINDER HEAD WARPAGE

FIGURE 28-9 Using a C-clamp spring compressor. (Courtesy of Deere and Company.)

FIGURE 28-11 Cleaning the cylinder head. (Courtesy of Mack Trucks Inc.)

detector is repositioned several times to ensure accurate results (**Figures 28-13** and **28-14**).

3. *Dye penetrant:* A dye penetrant is sprayed over the suspected area, and excess penetrant is wiped off. A developer is sprayed on the area. The developer draws the penetrant out of any crack, making it visible.

CYLINDER HEAD WARPAGE

Cylinder head warpage (flatness) should be checked with a straightedge and feeler gauge longitudinally and across the surface. Typical maximum allowable warpage is 0.003 in. (0.075 mm) across and 0.003 in. (0.075 mm) over any two cylinders of cylinder head length with 0.006 in. (0.150 mm) overall on a four-cylinder head. Refer to the service manual for actual specifications for any particular engine (**Figures 28-15** and **28-16**). Measure cylinder head height with an outside micrometer to determine whether the head has previously been machined or surface ground. If head warpage is excessive and if head height allows, correct the warpage by surface grinding or machining. Record the amount of stock removed by stamping the amount on the cylinder head (**Figures 28-17** and **28-18**).

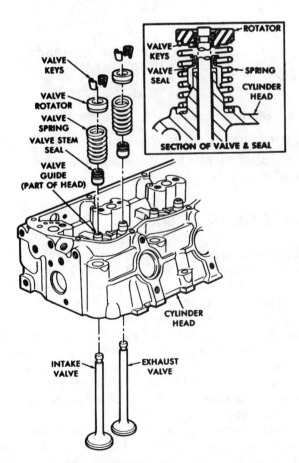

FIGURE 28-10 Valves and related parts. (Courtesy of Detroit Diesel Corporation.)

644 Chapter 28 CYLINDER HEAD AND VALVE SERVICE

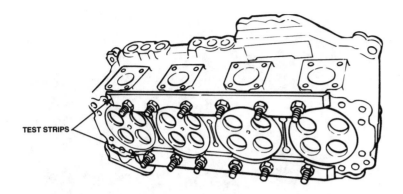

FIGURE 28–12 Cylinder head ready for pressure test. (Courtesy of Detroit Diesel Corporation.)

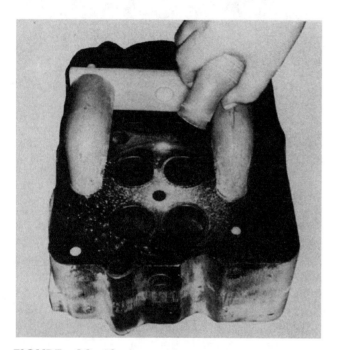

FIGURE 28–13 Electromagnetic crack detection. (Courtesy of Cummins Engine Company, Inc.)

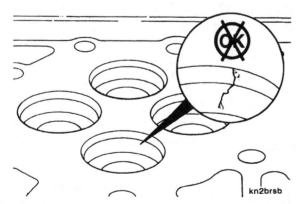

FIGURE 28–14 Inspecting valve seat and port area for cracks. (Courtesy of Cummins Engine Company, Inc.)

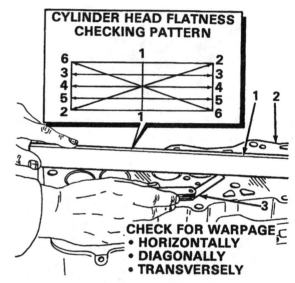

FIGURE 28–15 Measuring cylinder head flatness. (Courtesy of Navistar International Transportation Corporation.)

VALVE GUIDE SERVICE

Clean the valve guides with a scraper type of cleaner to remove all the carbon. Follow this with a rifle type of bristle brush to remove any loose material, and polish the guide **(Figure 28–19)**. Measure valve stem-to-guide clearance to determine guide wear. Use a small-bore gauge and outside micrometer to measure guide diameter **(Figures 28–20 and 28–21)**. Take measurements at the top, middle, and bottom of the guide. If wear exceeds allowable limits in the service manual, the condition must be corrected to ensure good valve-to-seat sealing.

Another method is to use an inside micrometer to measure the inside diameter of the valve guide. The valve stem diameter measurement is subtracted from the valve guide diameter measurement to determine valve stem-to-guide clearance.

VALVE BRIDGE OR CROSSHEAD AND GUIDE   645

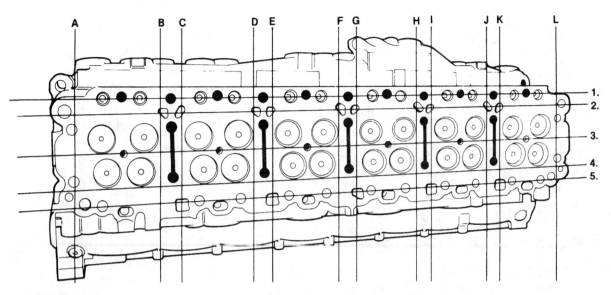

FIGURE 28–16 Cylinder head warpage measurement locations. (Courtesy of Detroit Diesel Corporation.)

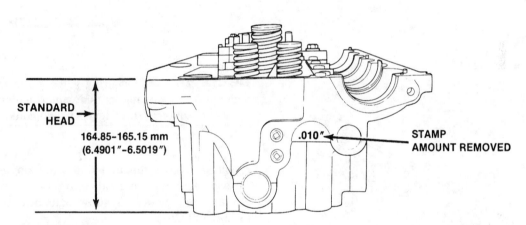

FIGURE 28–17 Typical cylinder head height dimension and recorded amount of stock removed. (Courtesy of Detroit Diesel Corporation.)

When valve guide reconditioning or replacement is required, this must always be done before attempting to recondition the valve seats. This is necessary to maintain the correct relationship between the valve guides and seats.

Several methods of restoring correct valve stem-to-guide clearance are employed. On engines with replaceable guides, guides are removed with a special press or driver, and new guides installed to the correct depth **(Figures 28–22 to 28–24)**. Valves with standard-diameter stems can then be used.

On cylinder heads with integral guides, the guides can be reamed to an oversize for which oversized stem valves are available. Another method is to machine the guides to allow new valve guide inserts of standard diameter to be installed.

VALVE BRIDGE OR CROSSHEAD AND GUIDE

Inspect crossheads or valve bridges for wear, cracks, or damage. The magnetic crack detection method should be used. The guide bore in the crosshead or bridge should be measured for wear with a small-bore gauge or inside micrometer. If wear exceeds limits or if cracks are evident, the bridge or crosshead should be

646 Chapter 28 CYLINDER HEAD AND VALVE SERVICE

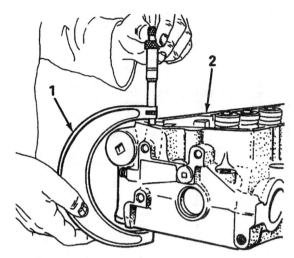

FIGURE 28–18 Measuring cylinder head height. (Courtesy of Navistar International Transportation Corporation.)

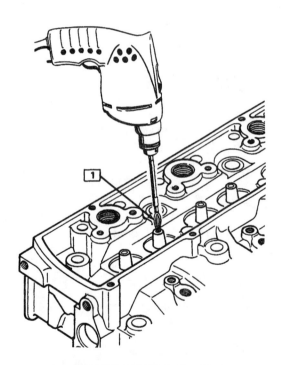

FIGURE 28–19 Cleaning valve guides. (Courtesy of Deere and Company.)

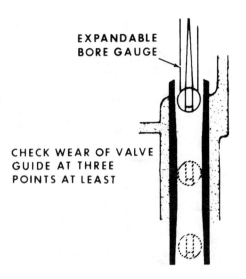

FIGURE 28–20 Measuring valve guide wear with a small-bore gauge and outside micrometer. Measurements are taken at three levels as indicated.

replaced. Inspect the adjusting screw threads and nut. Threads should be in good condition. The rocker lever contact area should not show signs of excessive wear **(Figure 28–25)**.

The bridge or crosshead guide should be measured for wear with an outside micrometer **(Figure 28–26)**. If wear exceeds maximum allowable limits specified in the service manual, replace the guide. The guide should also be checked for straightness; if it is not at right angle to the head surface, it should be replaced **(Figure 28–27)**.

VALVE SEAT INSERT REPLACEMENT

Valve seats may be replaceable inserts or an integral part of the head. Damaged, loose, or worn seat inserts are replaced with new inserts. Integral seats may be repaired by machining and installing seat repair inserts. Seat inserts may be hardened cast-iron alloy, chrome steel, or stellite material. Stellite seats are the hardest and provide high corrosion resistance and a low wear rate.

FIGURE 28–21 Using a dial-type small-bore gauge to measure guide wear. (Courtesy of Cummins Engine Company, Inc.)

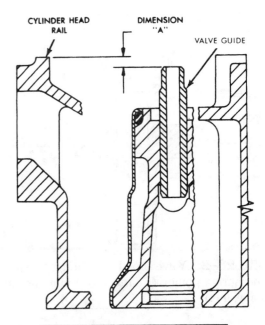

Tool No.	Cyl. Head	DIMENSION "A"
J 7560	2 Valve	.010"–.040"
J24519	4 Valve	.150"–.018"

FIGURE 28–23 Example of valve guide installation dimensions. (Courtesy of Detroit Diesel Corporation.)

FIGURE 28–24 Measuring valve guide installed height with a depth micrometer. (Courtesy of Mack Trucks Inc.)

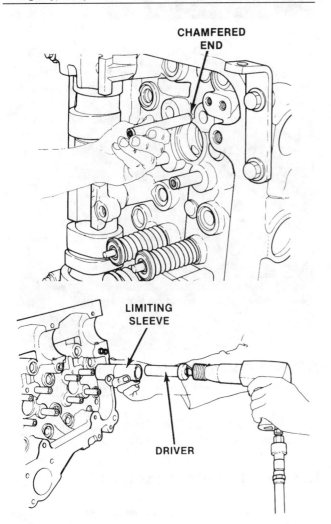

FIGURE 28–22 Installing a valve guide. (Courtesy of Detroit Diesel Corporation.)

648 Chapter 28 CYLINDER HEAD AND VALVE SERVICE

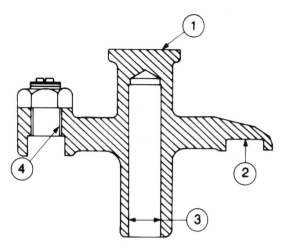

FIGURE 28-25 Valve crosshead (bridge) inspection and measuring points. (Courtesy of Cummins Engine Company, Inc.)

FIGURE 28-26 Measuring the bridge guide for wear. (Courtesy of Cummins Engine Company, Inc.)

FIGURE 28-27 Bridge guide pin removal tool (top) and installer (bottom). (Courtesy of Mack Trucks Inc.)

Oversized valve seat inserts can be used in heads where original seat inserts have loosened. Use a hammer and blunt nose punch to check for seat insert looseness in the head **(Figure 28-28).** The head is machined to provide an interference (negative) fit, and the oversized seat installed. Seat inserts are generally available in the following inch oversizes: 0.002, 0.005, 0.010, 0.015, 0.020, 0.030, 0.040, and 0.060 in. Similar oversize inserts in metric dimensions are also available. If a seat insert is to be installed to repair a damaged integral seat, the old seat area must be counterbored by a machining process. If a damaged seat insert is to be replaced, the old insert must be removed, the counter-bore cleaned up and machined to oversize, and the new insert installed.

Removing a Valve Seat Insert

One way to remove a seat insert is to use a roll-head pry bar. The tool tip is inserted to engage the lower edge of

VALVE SEAT INSERT REPLACEMENT

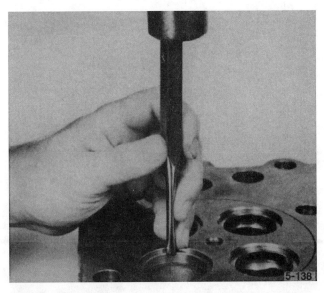

FIGURE 28-28 To check valve seat insert for looseness use a blunt-nose punch on top of the insert and tap the punch with a hammer. A ringing sound indicates the insert is tight. A dull sound indicates a loose insert. (Courtesy of Mack Trucks Inc.)

Straight Pull

FIGURE 28-29 Valve seat insert removal. (Courtesy of Winona Van Norman Machine Company.)

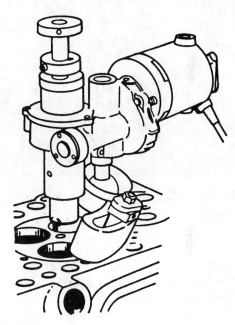

FIGURE 28-30 Valve seat insert counterboring tool. (Courtesy of Mack Trucks Inc.)

the seat insert while force is applied to the other end to pry the seat out. The seat may crack during this procedure, so eye protection should be used. Another method is to use a puller specially designed for seat insert removal, as shown in **Figure 28-29**.

Counterboring for a Valve Seat Insert

Counterboring equipment ranges from hand-operated devices to state-of-the-art head reconditioning equipment with counterboring capability (**Figure 28-30**). The hand-operated equipment is relatively inexpensive and can be used by small shops; however, the procedure is more time consuming than when more sophisticated equipment is used.

The diameter of the counterbore is determined from the insert diameter minus the interference fit required (**Figure 28-31**). Typical interference fit requirements for different insert sizes are given here.

INSERT OD (IN.)	INTERFERENCE FIT (IN.)	(MM)
1–2	−0.002–0.004	−0.05– −0.10
2–3	−0.003–0.005	−0.075– −0.125
3–4	−0.004–0.006	−0.10– −0.150

Although these are typical, the recommendations of the insert manufacturer should be followed, taking into consideration the insert material and design. The counterbore depth must also be considered, depending on the seat insert height dimension (**Figure 28-32**).

Installing a Valve Seat Insert

1. Insert the proper-size pilot into the valve guide.
2. Shrink the seat insert in dry ice or in a freezer. Shrinking the insert makes it easier to install.

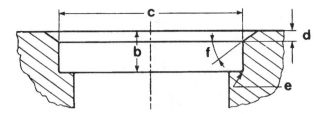

FIGURE 28-31 Valve seat insert and counterbore dimensions. (Courtesy of Deere and Company.)

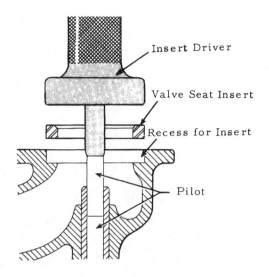

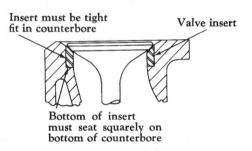

FIGURE 28-33 Installing a valve seat insert. (Courtesy of Cummins Engine Company, Inc.)

FIGURE 28-32 Measuring counterbore depth. (Courtesy of Mack Trucks Inc.)

3. If the insert is to be installed in an aluminum head, the head should be preheated to about 250°F (120°C) to expand the counterbore.

4. Place the insert on the counterbore with the chamfered outer edge down. The chamfer aids installation and helps prevent shaving metal from touching the counterbore sides.

5. Place the insert driver onto the pilot.

6. With the insert centered over the counterbore, drive or press it into place. Make sure that it bottoms in the counterbore. Remove the insert driver from the pilot **(Figure 28-33)**.

7. If required, the seat insert may be staked, peened, or rolled in place. This prevents the insert from dropping out in case it loosens. A seat insert that drops into the cylinder causes severe piston, head, and valve damage. Place the rolling tool into the drive head of the head reconditioning equipment. Place enough downward pressure on the tool to roll the cylinder head metal over the edge of the seat insert. An alternative method is to peen metal over the edge of the insert.

8. After the insert is installed, the seat must be ground to the proper angle and seat width and the valve head–to–cylinder head surface clearance measured **(Figures 28-34 and 28-35)**. If this dimension is less than specified, grind the seat to lower the valve.

VALVE SEAT GRINDING

Valve seat service is required when the valve seats no longer provide a good seal. They may be pitted, worn, burned, cracked, or recessed. Seats that have only minor surface damage can be ground or milled to restore sealing. Badly burned, pitted, cracked, or recessed seats must be replaced.

Special equipment is required to grind valve seats. A high-speed driver (8000 to 12,000 rpm) is used

VALVE SEAT GRINDING

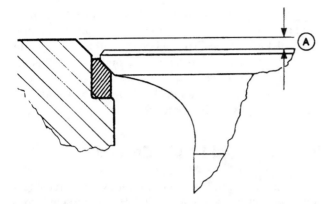

FIGURE 28-34 Valve head–to–cylinder head surface clearance dimension. (Courtesy of Deere and Company.)

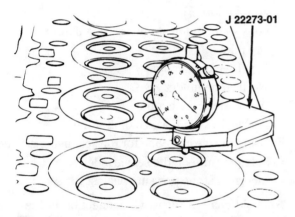

FIGURE 28-35 Measuring valve head–to–cylinder head surface dimension. (Courtesy of Detroit Diesel Corporation.)

FIGURE 28-36 Valve seat grinding equipment includes driver, stone holders, stones, pilots, and stone dressing fixture. (Courtesy of Sioux Tools Inc.)

to drive a grinding stone mounted on a pilot inserted tightly in the valve guide **(Figure 28-36)**. Roughing stones, finishing stones, and special stones for induction-hardened or stellite seats are available. When seats require only a little grinding, only the finishing stone is used. Seats that require more grinding may require the use of a roughing stone first and then a finishing stone to complete the job. Seat grinding stones are available in a number of diameters and cutting angles, to suit the various valve seat diameters and angles. Special diamond wheels are available that maintain their angle and do not need dressing. Three-angle seat milling equipment is used in another method of seat reconditioning. Valve seats (both integral and replaceable) that are worn down too far or are too badly damaged can be restored by installing new valve seat inserts as described earlier.

Valve Seat Grinding Pilots

Valve seat grinding pilots are required to support the seat grinder when grinding the valve seats. The pilot is inserted firmly into the guide, and the seat grinder is then placed on the pilot. There are two basic types of pilots: tapered and expanding. The tapered pilot wedges tightly into the guide when properly installed. The expanding pilot is tapered at the top end and expands at the bottom. It is inserted tightly into the guide at the top and then expanded at the bottom by tightening the adjusting screw at the top. This locks the pilot into place. Valve seat grinding pilots are available in a wide range of fractional inch, decimal inch, and metric sizes.

Seat Grinding Stones

Seat grinding stones or wheels are available in a variety of sizes and types. Different valve seat diameters require seat grinding stones of different diameters. Valve seat angles determine the stone angle required. The grit size required depends on the valve seat material to be ground.

Seat grinding stones have a threaded metal insert in the center for assembly to the stone holder. The hole diameter and thread type of the stone must match that of the stone holder. Common thread sizes are ½ in. × 20, 9/16 in. × 16 and 11/16 in. × 16. Stone diameters range from 1 to 2 in. in 1/16 in. steps. Stones over 2 in. in diameter are usually available in 1/8 in. steps. The correct stone size to use is determined by the valve seat diameter and the clearance around the seat. The proper size stone should be 1/16 to 1/8 in. larger than the valve head diameter. The stone is usually either 30 or 45°, depending on the valve seat angle. For topping (overcutting) the valve seat a 15° stone should be used. For undercutting (throating) a 65° stone is required **(Figure 28–37)**.

Seat grinding stones come in rough, medium, and fine or finish grit. Roughing stones are used to rough grind hard valve seats after which they are finish ground with a finishing stone. For stellite valve seats a stone with a special grit compound is required, after which they are finish ground. Finishing stones are made of very fine grit and produce a good surface finish on the valve seat. Medium-grit stones are sometimes used as a compromise instead of using both a roughing and finishing stone. The success of this method depends on the type of seat material being ground. A finishing stone produces a smoother finish.

Both sides of a stone may be dressed to provide two different angles on the same stone.

How to Avoid Stone Chatter

Stone chatter may be encountered when dressing a seat grinding stone or when grinding a valve seat. Stone chatter must be avoided to ensure accuracy in seat grinding. If stone chatter is allowed, it can destroy a valve seat in seconds. Stone chatter may be caused by any of the following.

- Loose pilot
- Loosely mounted stone dressing fixture
- Dry pilot
- Dry drive connection
- Worn stone holder
- Worn drive connection
- Poorly dressed seat grinding stone
- Using the wrong stone for the seat being ground
- Side pressure on the seat grinder
- Incorrect stone pressure against the seat

To avoid chatter, the drive connection in the stone holder and the driver must be in good condition. The ID of the stone holder must not be worn excessively. The pilot must be tight in the guide. The stone dresser must be mounted securely. A drop of oil on the drive connection and on the pilot will reduce friction and sticking. Avoid getting any oil on the seat or stone. The seat grinding stone must be dressed carefully and accurately. Select a hard seat grinding stone for hardened seats. Do not exert any side pressure on the seat grinder while grinding. Vary the downward pressure on the seat to establish the pressure required to avoid chatter.

Seat Grinding Procedure

To grind the seats, select the proper-diameter pilot and install it tightly in the valve guide. Select a seat grinding stone of the correct angle (same angle as the seat), usually 45 or 30°. The stone should be 1/16 or 1/8 in. larger in diameter than the valve head. Frequently dress the stone in the stone dressing fixture **(Figure 28–38)** during the seat grinding procedure, to ensure a good seat finish. Use eye protection, a shield, and a dust collector.

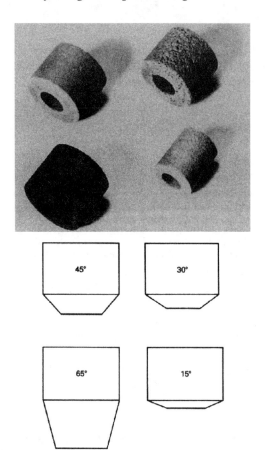

FIGURE 28–37 Seat grinding stones (top) from roughing to finishing types. Common seat grinding stone angles. (Courtesy of Sioux Tools Inc.)

VALVE SEAT GRINDING

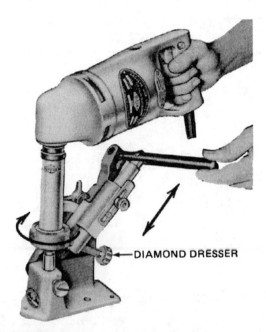

FIGURE 28-38 Dressing a seat grinding stone. Set the dresser to the angle of the stone, then move the dresser smoothly and evenly across the face of the stone using very light cuts. (Courtesy of Sioux Tools Inc.)

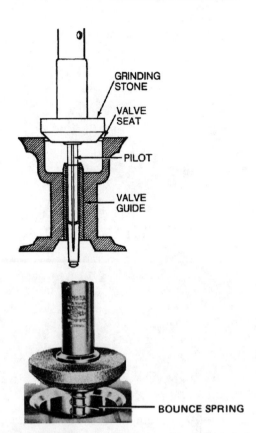

FIGURE 28-39 Valve seat grinding. Bounce spring lifts stone off seat for quick, easy inspection of work. (Courtesy of Sioux Tools Inc.)

It is good practice to wipe the seats clean by using a piece of fine emery cloth between the stone and the seat and giving it a good hard rub. This avoids contaminating the seat grinding stone with any oil or carbon residue that may be on the valve seat. Seat grinding stones should be handled in a manner that will keep them clean. Stones will soak up oil like a blotter. This causes them to become glazed and ineffective for seat grinding. Remove only as much material from the seat as required to provide a good finish of sufficient width all the way around the seat. Avoid any side pressure during the grinding process.

Grind the seat with short bursts only, checking frequently to inspect progress. Pressure of the stone against the seat must be precisely controlled to avoid chatter and to provide a good seat finish (**Figures 28-39 to 28-42**).

NOTE: If the valve seat is too wide after grinding, it must be narrowed to specifications, usually 1/16 to 3/32 in. (1.6 to 2.3 mm), with the exhaust seat being the wider for better heat transfer. This will require topping (overcutting), throating (undercutting), or both.

Narrowing a Valve Seat

The objective is twofold: (1) The seat should be the correct width, and (2) it should contact the center of the valve face. To determine whether topping or throating is required to narrow the valve seat, a new or reconditioned valve must be used.

Overcutting is done in the same manner as seat grinding, except that a 15° stone is used for the purpose. This narrows the seat from the combustion chamber side and lowers the point of contact on the valve face (farther from the margin). Undercutting is done similarly with a 65° stone, which narrows the seat from the port side.

Valve Seat-to-Face Contact and Concentricity

To determine where the seat contacts the valve face, use a new or reconditioned valve. Mark the valve face with a series of pencil marks across the face of the valve all around the valve. Insert the valve in the guide, press down on the valve, and turn it one-quarter turn and back. Remove the valve and check the pencil marks. The pencil marks will be wiped out at the point where the seat contacts the valve. (Because of the interference angle, only the edge of the seat on the combustion cham-

654 Chapter 28 CYLINDER HEAD AND VALVE SERVICE

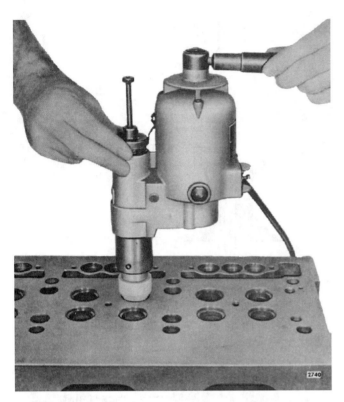

FIGURE 28–40 Grinding a valve seat. (Courtesy of Detroit Diesel Corporation.)

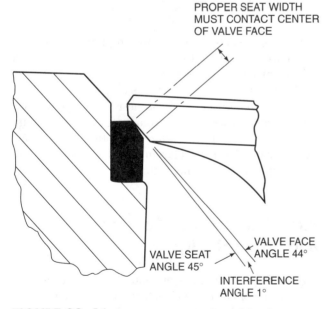

FIGURE 28–41 Typical seat grinding objectives.

ber side will wipe out the pencil marks.) This should be about one-third of the way down on the face of the valve away from the margin to center the seat on the valve face **(Figure 28–43)**. Turning the valve only one-quarter turn while in contact with the seat provides the

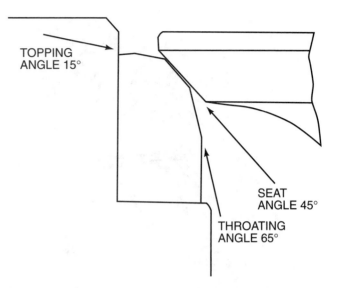

FIGURE 28–42 Topping and throating angles for a 45° seat.

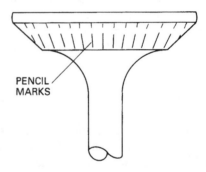

FIGURE 28–43 Pencil marks across valve face wiped out at area of valve face-to-seat contact.

means for checking whether or not the seat is concentric. If the pencil marks are wiped out at one point all the way around the valve, seat concentricity is within limits. A dial-type seat concentricity gauge may be used to check seat runout **(Figure 28–44)**. Prussian blue may also be used to check seat contact and concentricity. If the guide, seat, and valve have all been properly reconditioned, they will be concentric (centered in relation to each other). Concentricity is required to provide a good seal between the valve and seat.

VALVE SERVICE

Cleaning the Valves

Valves that do not appear to be worn or damaged upon visual inspection should be cleaned on a wire wheel buffer to remove all the carbon. Badly damaged or worn valves must be replaced.

VALVE SERVICE 655

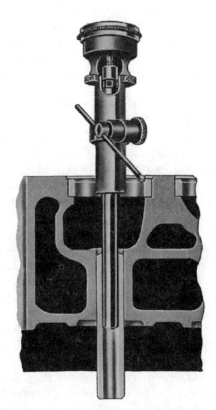

FIGURE 28–44 Cross section of valve seat runout tool in place. (Courtesy of Mack Trucks Inc.)

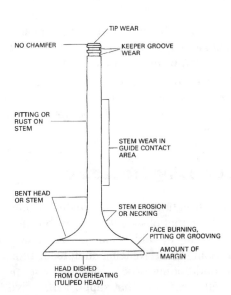

FIGURE 28–45 Valve inspection and measuring points.

CAUTION:
Wear a face shield and protective gloves when cleaning valves on a wire wheel buffer.

CAUTION:
To avoid damaging the valve face and stem do not press the valve too hard against the wire wheel while buffing.

Other methods used to clean valves include placing the valves in a suitable basket and immersing them in a chemical cleaning tank for about 30 minutes, then rinsing and drying them. Glass bead blasting is also a very effective cleaning method.

Inspecting and Measuring Valves

Valves should be inspected and measured for wear or damage as follows. See **Figure 28–45** for valve inspection areas.

1. *Margin.* Minimum margin width after grinding should be $1/32$ in. (0.8 mm). A valve with too little margin has reduced heat capacity and will warp and burn.

2. *Stem wear.* Stem wear is measured with a micrometer in the area of contact with the valve guide. Maximum wear usually occurs at each end of this area. Maximum valve stem wear should be no more than half the maximum guide clearance specification given in the service manual **(Figure 28–46)**.

3. *Bent valve.* A bent valve has a head that is tilted off square from the valve stem. Bend usually occurs in the area between the head and the guide contact area

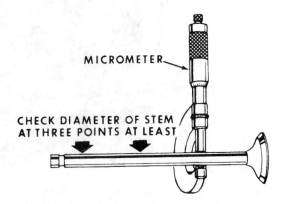

FIGURE 28–46 Measuring valve stem wear. (Courtesy of J. I. Case, a Tenneco company.)

of the stem. Valve bend is most easily checked with the valve chucked in the valve refacer. Rotating the valve will reveal head wobble or runout if the valve is bent.

4. *Keeper groove wear.* Keeper grooves that are worn have the potential for valve breakage. Valves with worn keeper grooves should be replaced and new keepers used.

Valve Grinding Procedure

After the valves have been thoroughly cleaned and inspected, those that passed the inspection should be reconditioned in a valve refacing machine **(Figure 28–47)**. Some valve refacing equipment requires that valve reconditioning procedures follow a specific sequence. Valve refacers that support the valve tip in a coned shaft require the valve tip to be dressed and chamfered before it is refaced. If this is not done, the valve face will be ground off center. Equipment manufacturers' instructions for procedures and sequence should be followed. To recondition a valve, the tip should be dressed and chamfered and the valve refaced. Remove only enough material from the valve face or tip to produce the desired results. If too much material is removed from the valve tip, there may be interference between the rocker arm and spring retainer or valve rotator. Do not remove more than the maximum allowed in the appropriate service manual. A bent or distorted valve will easily be noticed as valve head wobble as the valve rotates in the machine. Replace any valves that are bent.

To grind a valve, first dress the stones on the refacing machine. Set the dressing screw depth with the machine off **(Figure 28–48)**. Try a few passes across the stone, and set the dressers so they almost touch the stones. Dress the side of the stem grinding stone and the face of the valve refacing stone. With the machine running, slowly feed the dresser toward the stone to take a very light cut only. Use a face mask for protection. Move the dresser slowly across the stone. Do not remove excessive material from the stone.

After dressing the stones, clamp the valve in the stem grinder holding fixture so that it nearly touches the stone. Carefully feed the valve stem closer to the stone until contact is made while moving the stem across the side of the stone **(Figure 28–49)**. Remove only enough material to provide a smooth finish. Never remove more than the amount specified in the service manual. Next, chamfer the edges of the valve stem in the fixture provided on the machine **(Figure 28–50)**.

To reface a valve, insert it in the grinder chuck to the proper depth and make sure it is tight in the chuck **(Figure 28–51)**. With the machine off, adjust the

FIGURE 28–47 Valve grinding equipment. (Courtesy of Sioux Tools Inc.)

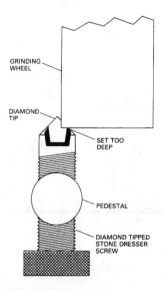

FIGURE 28–48 Setting the stone dresser too deep can cause the diamond tip to be destroyed.

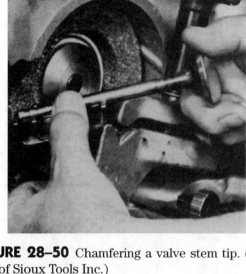

FIGURE 28–50 Chamfering a valve stem tip. (Courtesy of Sioux Tools Inc.)

FIGURE 28–49 Dressing a worn valve stem tip. (Courtesy of Sioux Tools Inc.)

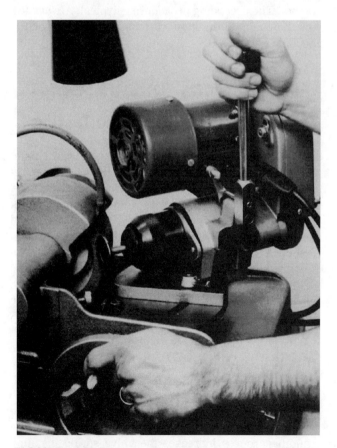

FIGURE 28–51 Valve face grinding. (Courtesy of Tobin-Arp Manufacturing Company.)

657

chuck head to the correct angle for the valve. With a 45° valve set the chuck head at 44° to provide a 1° interference angle. With a 30° valve set it at 29°. Use the control lever to move the valve toward the grinding wheel. Set the depth feed so that the valve face nearly touches the stone face. Turn on the machine and adjust the coolant flow to direct it at the valve face. Adjust the coolant flow rate. With one hand on the cross feed and the other on the depth feed, move the valve face across the full width of the grinding stone. At the same time adjust the depth feed to take a very light cut only. Do not allow the valve face to go off the edge of the stone in either direction. This rounds off the valve face and the stone. Remove only enough material to provide a smooth finish over the entire valve face. Never allow the valve stem to touch the grinding stone. A valve nicked in this way must be replaced due to possible breakage at that point. To inspect the valve face back the valve away from the stone. Do not slide it off the edge of the stone. If there is insufficient margin on the valve after grinding, the valve must be replaced, since it would then run too hot. The preceding procedure is typical. Always follow the equipment manufacturer's instructions for the machine being used.

VALVE TRAIN COMPONENTS SERVICE

Valve Spring Inspection and Testing

Inspect valve springs for acid etching. Discard damaged or bent springs. Check all the springs for squareness. Rotate the spring and check the height variance. Test the spring pressures with a spring tester. Check the service manual for spring free height, valve-closed spring height, and valve-open spring height pressures. Test each spring at both heights and compare to specifications. Replace springs that do not meet standards **(Figures 28–52** and **28–53).** Inspect spring retainers, seats, valve rotators, and keepers for wear or damage and replace faulty parts.

Checking Valve Spring Installed Height

The spring installed height must be checked, because removing metal from the valves and seats during the grinding process increases the valve spring installed height and results in less spring pressure. To check spring installed height, insert the appropriate valve in the guide. Install the spring retainer and keepers (not the spring). While pulling squarely on the spring retainer to keep the valve and keepers seated, measure

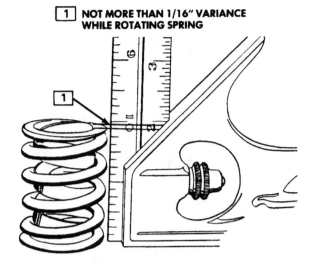

FIGURE 28–52 Checking valve spring squareness. (Courtesy of General Motors Corporation.)

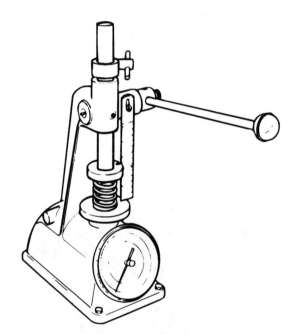

FIGURE 28–53 Checking valve spring pressure. (Courtesy of Detroit Diesel Corporation.)

the distance from the top of the spring seat to the spring contact surface of the spring retainer. To correct the spring installed height replace the valve, the seat, or both.

Checking Push Rods

All push rods should be checked for wear or ridging on the ends. Worn push rods must be replaced. Check whether the push rod is bent by rolling it on a flat machined surface and checking for clearance anywhere between the push rod and the flat surface. Bent push rods should be replaced. Inspect and measure cam followers and components for cracks and wear. Replace faulty parts. See **Figures 28–54** to **28–62** for typical procedures.

Checking Rocker Arms and Shafts

Inspect the rocker arms for wear at the push rod end and at the valve stem end **(Figure 28–63)**. If any obvious wear is present, the rocker arm must be replaced. In some cases the valve stem end of the rocker arm can be reground to restore the wear surface. Check the service manual to determine whether this can be done.

Inspect the pivot area of the rocker arm for wear and scoring. Check for galling of the metal on the bottom of the rocker shaft hole. Check for wear in the push rod seat area of rocker arms. Inspect the rocker arm shafts in all wear areas. Look for any damage in other areas as well. Look for scoring, galling, stepped wear, or ridging. If wear is smooth and shiny and there is no evidence of stepped wear, measure the shaft diameter in the wear areas with a micrometer to determine the amount of wear. Replace the shaft if wear is excessive.

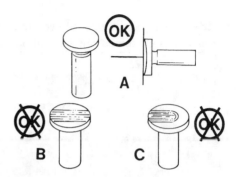

FIGURE 28–56 Checking cam follower wear. (Courtesy of Cummins Engine Company, Inc.)

FIGURE 28–54 Push rod wear inspection. Push rod ball ends and seat wear must be smooth and even over the entire contact area. (Courtesy of Cummins Engine Company, Inc.)

FIGURE 28–57 Measuring cam follower shaft wear. (Courtesy of Cummins Engine Company, Inc.)

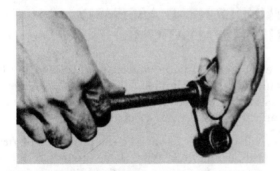

FIGURE 28–55 Checking push tube and push tube seat contact pattern using Prussian blue. There must be a minimum of 80% surface contact. (Courtesy of Cummins Engine Company, Inc.)

FIGURE 28–58 Measuring cam follower bushing diameter. (Courtesy of Cummins Engine Company, Inc.)

660 Chapter 28 CYLINDER HEAD AND VALVE SERVICE

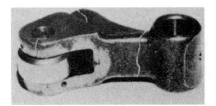

FIGURE 28–59 Cracked cam follower must be replaced. (Courtesy of Cummins Engine Company, Inc.)

FIGURE 28–60 Measuring cam follower roller inside diameter for wear. (Courtesy of Cummins Engine Company, Inc.)

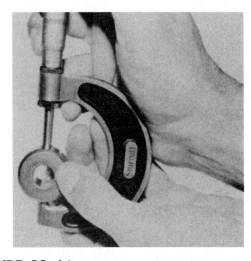

FIGURE 28–61 Measuring cam follower roller outside diameter for wear. (Courtesy of Cummins Engine Company, Inc.)

INJECTOR SLEEVE TESTING AND REPLACEMENT

Any indication of coolant leakage while the cylinder head was pressure tested as outlined earlier

FIGURE 28–62 Installing the cam follower roller pin using a feeler gauge to prevent collapse of the cam follower yoke. (Courtesy of Cummins Engine Company, Inc.)

requires sleeve replacement. The following procedure is typical.

1. Using the special tool required, remove the injector sleeve.
2. Thoroughly clean the injector bore in the cylinder head. Emery cloth may be used. Pay special attention to the O-ring seal groove if so equipped.
3. Install new O-ring seals (if used). Lubricate the seal as recommended.
4. Tap the sleeve into place until fully seated.
5. On some designs the installation must be completed by flaring, reaming, and spot facing procedures. Follow service manual instructions for these procedures (**Figures 28–64** to **28–69**).

CYLINDER HEAD ASSEMBLY AND INSTALLATION

All cylinder head parts, valves, springs, and the like, should be absolutely clean for assembly. Valves and seats should be lubricated with engine oil during assembly.

- Valve stem seals may be the positive type or the rubber umbrella type. Some engines rely on tapered valve guides to deflect oil away from the valve stems and do not have any valve stem seals. Positive-type stem seals are sometimes installed on these engines or on those originally equipped with umbrella seals. This usually requires ma-

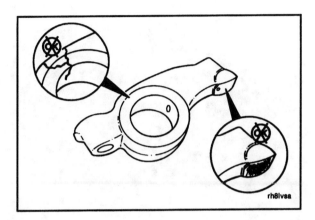

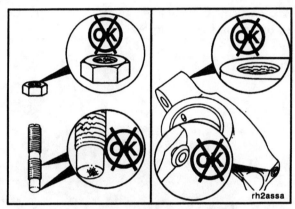

FIGURE 28–64 Removing an injector nozzle sleeve. (Courtesy of Mack Trucks Inc.)

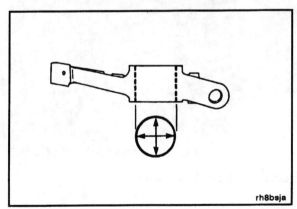

FIGURE 28–63 Rocker arm inspection and measuring points. (Courtesy of Cummins Engine Company, Inc.)

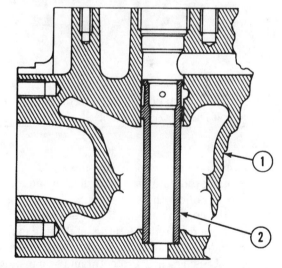

FIGURE 28–65 Cross section of injector nozzle sleeve in place in head. (Courtesy of Mack Trucks Inc.)

662 Chapter 28 CYLINDER HEAD AND VALVE SERVICE

FIGURE 28–66 Removing a unit injector tube. (Courtesy of Detroit Diesel Corporation.)

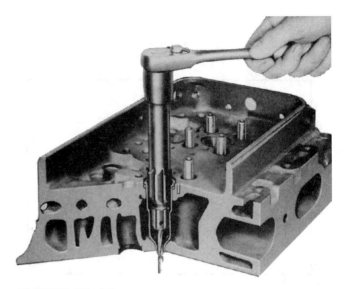

FIGURE 28–68 Reaming a unit injector tube. (Courtesy of Detroit Diesel Corporation.)

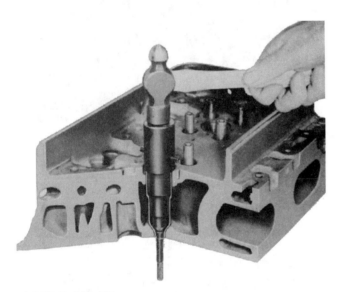

FIGURE 28–67 Installing a unit injector tube. (Courtesy of Detroit Diesel Corporation.)

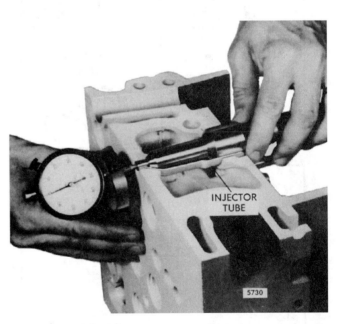

FIGURE 28–69 Measuring injector tube bevel seat–to–cylinder head surface dimension. (Courtesy of Detroit Diesel Corporation.)

chining the guides to accept the seals. A guide machining tool and electric drill are used for this purpose. Follow the tool manufacturer's instructions to machine the guides. Remove all metal chips from the guides and the intake and exhaust ports before installing the seals. Install the valves and seals using a seal protector and seal installing tool **(Figure 28–70)**.

- Install the valve rotators either above or below the spring as specified by the manufacturer. Valve spring seats (if used) are installed between the spring and cylinder head. Variable-rate springs are usually installed with the closest spaced coils toward the cylinder head **(Figure 28–71)**.

- After assembling the valves in the cylinder head, check the valve opening spring pressure if required by the service manual **(Figure 28–72)**. Check the valves for leakage with a vacuum-type tester. Correct any valves and seats that fail the vacuum test **(Figure 28–73)**.

- On engines with an overhead camshaft the assembly sequence will be different and must be fol-

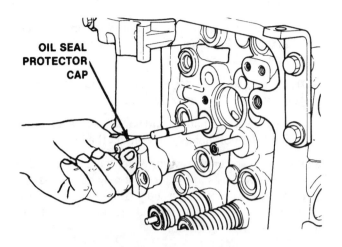

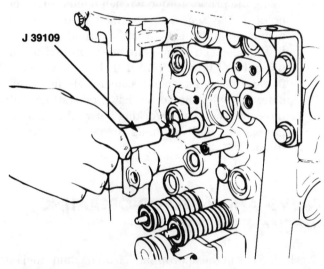

FIGURE 28–70 Installing a valve stem seal using a seal protector (top) and installing tool (bottom). (Courtesy of Detroit Diesel Corporation.)

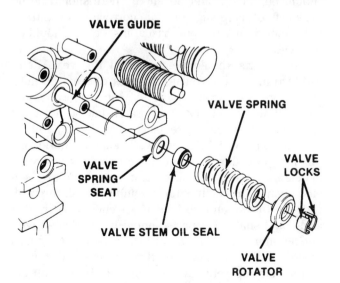

FIGURE 28–71 Valve spring and related parts assembly. (Courtesy of Detroit Diesel Corporation.)

FIGURE 28–72 Checking pressure required to open valve. (Courtesy of Detroit Diesel Corporation.)

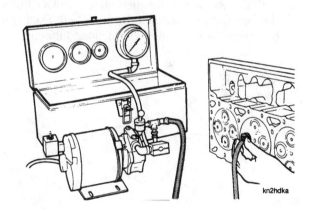

FIGURE 28–73 Checking valve sealing with a vacuum pump tester. (Courtesy of Cummins Engine Company, Inc.)

lowed as outlined in the appropriate service manual.

- Gasket surfaces, bolts, and threaded bolt holes must be absolutely clean. Make sure there is no fluid or grease in any blind bolt holes in the block. Grease or fluid in a blind bolt hole can cause the block to crack when the cylinder head bolts are tightened due to the very high hydraulic pressures created.
- Install guide studs in the engine block to aid in keeping the gaskets in place and in guiding the head into position. Guide studs can be made by sawing off the heads of spare head bolts and tapering the end on a grinder **(Figure 28–74)**.
- Whereas some head gaskets are installed dry, others must be coated on both sides with the recommended sealer. Some head gaskets are of the single-piece design; others consist of several gaskets, O-rings, and fire rings. Be sure to place all of them in the proper position with the correct side up. Install them dry or coated with the correct sealer as specified in the appropriate service manual.
- Carefully lower the head or heads into place while avoiding moving the gaskets and O-rings out of place and avoiding damaging them. Thoroughly clean all head bolts or cap screws and inspect them for damage by corrosion or pitting in both the shank and threaded areas. Replace bolts that are pitted, corroded, or stretched. Coat the bolt threads with the recommended lubricant or sealer. Replace the head bolt washers (if used). Place the bolts (with washers if used) into the bolt holes.

NOTE: Make sure that special bolts and bolts of different lengths are placed in the correct location.

- Using a speed handle and socket turn each bolt in until the bolt heads just contact the head.

NOTE: On in-line engines with two or three heads that do not have dowel pins in the block, it may be necessary to align the heads by placing a straightedge across the intake or exhaust manifold surfaces and keeping them aligned as the head bolts are tightened.

- Using the proper torque wrench tighten the head bolts in the sequence specified in the service manual in increments of half a turn at a time until the specified torque has been reached for all the bolts **(Figure 28–75)**. If the engine manufacturer's recommended sequence is not available, the bolts should be tightened starting with the center bolts and working outward in a circular pattern until all bolts are at the specified torque.

VALVE- AND INJECTOR-OPERATING MECHANISM SERVICE

Wash all parts in the appropriate cleaning fluid (fuel oil or solvent) as recommended in the service manual. Blow dry with compressed air. Inspect all parts visually for abnormal wear patterns. Wear patterns should be smooth and polished. There should be no evidence of chipping, cracks, corrosion, or pitting. Parts that do not pass the visual inspection should be replaced.

Other parts should be measured for wear. Refer to illustrations for typical examples of required measurements for various types of cam followers and cam follower mounting parts. Parts that do not meet specifications must be replaced. Install cam followers thoroughly lubricated with engine lubricating oil. The procedure and sequence to follow will vary, depending on engine make and model. Follow service manual procedures for assembly. Install the push rods or tubes (and springs if so equipped), rocker arms, and shafts. All parts should be lubricated with engine lubricating oil at all wear points **(Figures 28–76 to 28–78)**. Make sure that all valve adjustments and injector-operating adjustments are backed off before attempting to turn the crankshaft. This is necessary to avoid damage to valves and injectors. Assemble

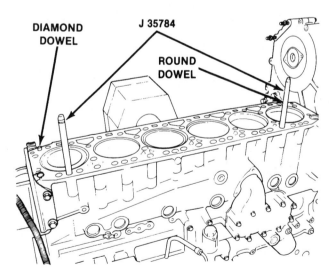

FIGURE 28–74 Cylinder head guide studs (J35784) aid in cylinder head installation. (Courtesy of Detroit Diesel Corporation.)

CYLINDER HEAD

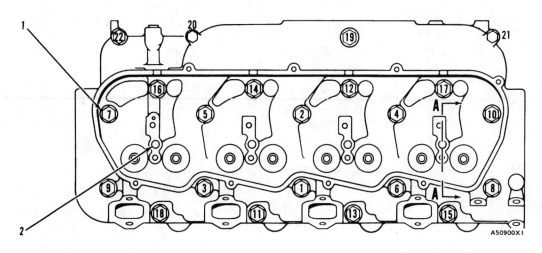

(1) Put 4S9416 Anti-Seize Compound on bolt threads and tighten bolts according to the following HEAD BOLT CHART:

HEAD BOLT CHART		
Tightening Procedure	***EARLIER BOLTS** (with six dash marks)	***LATER BOLTS** (with seven dash marks)
Step 1. Tighten bolts 1 thru 18 in number sequence to:	60 ± 10 lb. ft. (80 ± 14 N·m)	60 ± 10 lb. ft. (80 ± 14 N·m)
Step 2. Tighten bolts 1 thru 18 in number sequence to:	95 ± 5 lb. ft. (130 ± 7 N·m)	110 ± 5 lb. ft. (150 ± 7 N·m)
Step 3. Again tighten bolts 1 thru 18 in number sequence to:	95 ± 5 lb. ft. (130 ± 7 N·m)	110 ± 5 lb. ft. (150 ± 7 N·m)
*See BOLT HEAD IDENTIFICATION pictures for EARLIER and LATER identification.		
Torque for head bolts 19 thru 22 (tighten in number sequence to)	32 ± 5 lb. ft. (43 ± 7 N·m)	

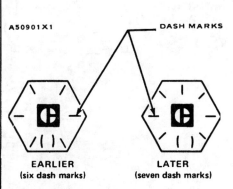

BOLT HEAD IDENTIFICATION

FIGURE 28–75 Typical cylinder head bolt tightening sequence and specifications. Follow the sequence and specifications given in the appropriate service manual. (Courtesy of Caterpillar, Inc.)

666 Chapter 28 CYLINDER HEAD AND VALVE SERVICE

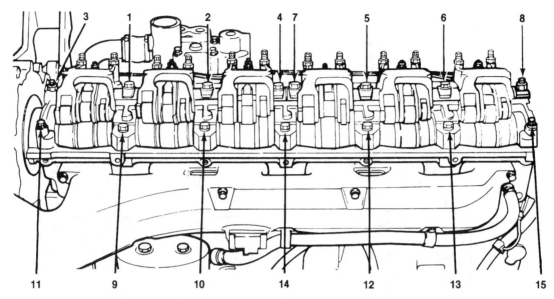

FIGURE 28–76 Bolt and nut tightening sequence for camshaft bearing caps and rocker arm shaft bolts and nuts on DDC series 60 engine. (Courtesy of Detroit Diesel Corporation.)

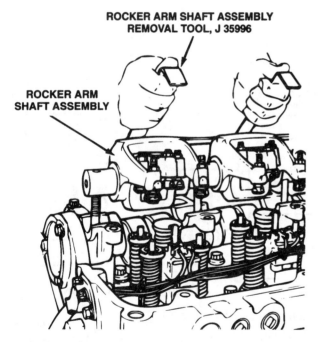

FIGURE 28–77 Installing rocker shaft assembly on overhead cam engine. (Courtesy of Detroit Diesel Corporation.)

and install the compression release mechanism if so equipped. Install the injector control rack and linkage, if so equipped, making sure that all timing marks are properly aligned according to service manual specifications.

ADJUSTING THE VALVE TRAIN (FIGURES 28–79 TO 28–82)

The objective in adjusting the valve train is to provide sufficient valve lash (clearance) to allow for any expansion of parts due to heat and still ensure that valves will be fully seated when closed. At the same time there must not be excessive lash, which would retard valve timing and cause rapid wear of valve train parts.

The static valve adjustment is made with the engine cold. A cold engine is an engine that has reached a stabilized temperature to within 10°F (6°C) of normal room temperature, 70°F (21°C). Further valve- and injector-operating mechanism adjustment may be required on some engines after the engine has reached operating temperature. Due to extremely close piston-to-valve tolerance, some engine valve trains must not be adjusted with the engine running. Inserting the feeler gauge could cause the pistons to strike the valves, causing major damage.

The static valve adjustment is required after an engine overhaul to ensure proper engine starting and prevent damage to the valves from pistons hitting the valves. The valve bridge or crosshead adjustment must be made first, then the valve lash adjustment. On some engines the valve lash adjustment is made on one cylinder while the injector adjustment is made on another. Follow the procedures outlined in the appropriate service manual.

ADJUSTING THE VALVE TRAIN

FIGURE 28–78 Lubricate all valve train contact points. (Courtesy of Mack Trucks Inc.)

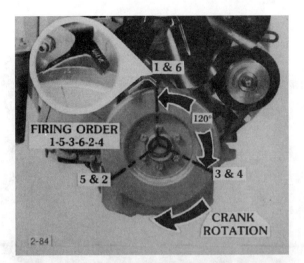

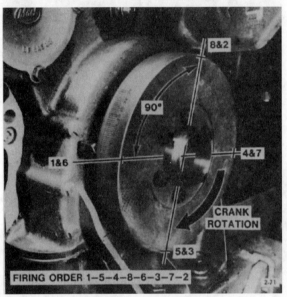

FIGURE 28–79 Crankshaft position and sequence of valve train adjustment for in-line six (top) and V8 engines (bottom). (Courtesy of Mack Trucks Inc.)

Valve adjustment is sometimes done as a routine service procedure when there is no engine overhaul involved. This is often done with the engine running at a slow idle speed. When this is done, provision must be made to prevent oil from squirting and spraying over other engine parts.

Adjustment should be made only within the limits and sequence prescribed in the manufacturer's shop manual.

The valve train usually has an adjusting screw and locknut provided at the push rod or the rocker arm. Clearance or lash is measured between the valve stem and rocker arm or crosshead or bridge. With the locknut loosened off several turns, turning the adjusting screw will increase or decrease the amount of lash, depending on which direction the screw is being turned. Valves should be adjusted to specifications, then the locknut should be tightened. Make sure that the adjusting screw does not turn while tightening the locknut. Recheck the clear-

FIGURE 28–80 Adjusting clearance between valve bridge and valve stem tips. (Courtesy of Mack Trucks Inc.)

FIGURE 28–81 Adjusting clearance between rocker arm and valve bridge. (Courtesy of Mack Trucks Inc.)

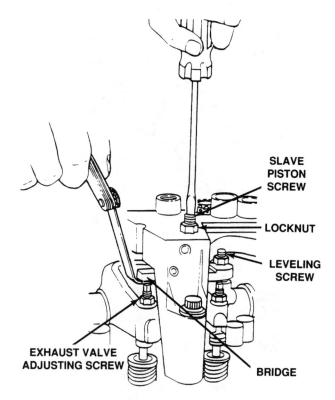

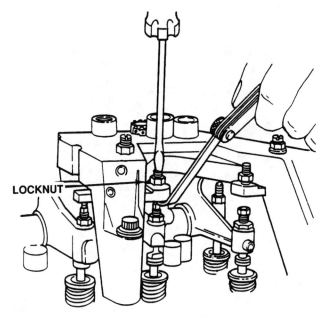

FIGURE 28–82 Adjusting the slave piston screw (top) and leveling screw (bottom) on a Jacobs brake. (Courtesy of Detroit Diesel Corporation.)

ance with a feeler gauge and correct the adjustment if needed.

The proper sequence must be followed when adjusting valves in order to ensure that adjustment is being made while the cam follower is on the base circle or heel of the cam lobe.

On engines equipped with camshaft-operated injectors, adjust the injectors according to the recommended setting and sequence prescribed in the appropriate chapters in this book and in the service manual.

REVIEW QUESTIONS

1. Cylinder head and valve train service may be performed without _____ _____.
2. A two-stroke-cycle engine does not have any _____ valves.
3. Engines may have _____ or _____ valves per cylinder.
4. Once the cylinder head has been removed, carefully remove the _____ at the top of the cylinders.
5. When disassembling a cylinder head, keep all parts in _____.
6. Valves with mushroomed stem tips can be forced out. (T) (F)
7. What cylinder head checks must be performed after disassembly?
8. Measure valve _____ -to- _____ clearance to determine guide wear.
9. How can valve guide clearance be corrected if worn?
10. Valve seats may be _____ or an _____ part of the head.
11. When installing a valve seat, shrink the seat in _____ _____.
12. Valve seat service is required when the valve seats no longer have a _____ _____.
13. Describe valve seat grinding procedures.
14. The required valve seat grinding stone grit size depends on the valve _____ _____.
15. Name four items to inspect on a valve.
16. Valve springs must be checked for _____ _____.
17. Inspect the rocker arms for wear at the _____ end and at the _____ stem end.
18. A valve train that has excessive lash would _____ valve timing.

TEST QUESTIONS

1. Excessive cylinder head warpage may be corrected by
   a. retightening
   b. surface grinding
   c. sandblasting
   d. valve grinding
2. Valve seats are usually ground by the manufacturer to angles of
   a. 60° or 70°
   b. 10° or 15°
   c. 30° or 45°
   d. 60° or 90°
3. Cylinder heads that are equipped with pressed-in type valve guides must have all the guides installed
   a. in the same direction
   b. in the opposite direction
   c. to the same approximate position
   d. to the proper depth
4. Cracked valve seat inserts may be repaired by
   a. welding
   b. installing plugs in cracks
   c. installing crack sealer
   d. replacement
5. Remove the rocker arm and shaft assembly by
   a. loosening each bracket all the way before going to the next one
   b. loosening each bracket, in turn, a little until all are loose
   c. loosening the front bracket first
   d. leaving one bracket tight until all others are loosened
6. When a valve seat needs grinding, it must be
   a. done before the guide is reconditioned
   b. ground down the width of the valve face
   c. done after the guide is reconditioned
   d. ground when the head is warm
7. A valve seat that is ground and left too wide will
   a. coat with carbon, leak, and burn
   b. break the valve stem
   c. be difficult to open
   d. stick closed

8. Excessive clearance between the valve stem and guide can cause
   a. poor engine performance
   b. high oil consumption
   c. valve breakage
   d. all of the above
9. It is considered good practice to keep all the
   a. push rods in order
   b. head bolts at the right temperature
   c. the small parts in one large container
   d. valve springs under pressure
10. Which of the following is *not* true when installing cylinder heads? The head bolts
    a. may be tightened in any sequence
    b. should be cleaned and lubricated
    c. should be tightened in the proper sequence
    d. must be tightened to the correct torque
11. On most engines, the intake manifold bolts should be
    a. tightened from the center out
    b. tightened in any order
    c. tightened evenly from the ends
    d. none of the above
12. Technician A says cylinder head bolts should be tightened in sequence from front to back. Technician B says each cylinder head bolt should be tightened to full torque, one after the other, then retorqued. Who is correct?
    a. Technician A   c. both are correct
    b. Technician B   d. both are incorrect
13. Grinding the valve face at a different angle than the seat produces a(n)
    a. resistance to spring pressure
    b. interference angle
    c. conference angle
    d. clearance volume
14. Excessive valve spring installed height can cause
    a. slow valve timing   c. valve float
    b. fast valve timing   d. leaking stem seals
15. When grinding the face of the valve, always keep
    a. the valve in the center of the stone face
    b. the valve in one place after it touches the stone
    c. moving the valve back and forth, staying on the stone
    d. the valve head on the left side of the stone
16. Valve springs must be square on both ends to prevent
    a. up-and-down motion of the valve
    b. end-to-end wear of the valve stem
    c. side-to-side wear of the valve stem and guide
    d. none of the above
17. Oil seals are used on valves to keep the oil from getting into the
    a. intake manifold and cooling system
    b. rocker cover and intake manifold
    c. cylinder and exhaust manifold
    d. cooling system and exhaust manifold
18. With the valve installed in the head, the stem-to-guide clearance may be checked by use of a(n)
    a. inside micrometer   c. dial indicator
    b. small-bore gauge   d. outside micrometer
19. Valve springs should be tested for
    a. roundness and length
    b. roundness and strength
    c. height and weight
    d. squareness and length
20. Valve springs that do not test within 10 lb of the recommended pressure should be
    a. re-heat-treated   c. discarded
    b. stretched   d. shimmed
21. In order to ensure correct valve operation, the valve springs should be checked for
    a. squareness   c. installed height
    b. tension   d. all of the above
22. Technician A says only the valve seat must be concentric for good valve sealing. Technician B says the valve, seat, and guide must be concentric. Who is correct?
    a. Technician A   c. both are correct
    b. Technician B   d. both are incorrect
23. When the face of a valve is ground, the amount of material to be removed must be
    a. a small amount at a time
    b. at the same machine setting
    c. as much as the machine can cut
    d. with the machine oscillating
24. Valve grinding stones must be kept dressed because they
    a. wear longer
    b. look better
    c. will do more accurate work
    d. will cut faster
25. Technician A says a good valve seat must be within specified width limits. Technician B says a good valve seat must contact the center of the valve face. Who is correct?
    a. Technician A   c. both are correct
    b. Technician B   d. both are incorrect

APPENDIX

WEIGHT, MEASUREMENT, AND TEMPERATURE CONVERSION

Description	Multiply	By	For Metric Equivalent
ACCELERATION	Feet/sec^2	0.3048	meters/sec^2 (m/s^2)
	Inches/sec^2	0.0254	meters/sec^2 (m/s^2)
TORQUE	Pound-inches	0.11298	newton-meters (N · m)
	Pound-feet	1.3558	newton-meters (N · m)
POWER	horsepower	0.746	kilowatts (kw)
PRESSURE or STRESS	inches of water	0.2488	kilopascals (kPa)
	pounds/sq. in.	6.895	kilopascals (kPa)
ENERGY or WORK	BTU	1055	joules (J)
	foot-pounds	1.3558	joules (J)
	kilowatt-hours	3,600,000 or 3.6 × 10^6	joules (J = one Watt)
LIGHT	foot candles	10.76	lumens/meter^2 (lm/m^2)
FUEL PERFORMANCE	miles/gal	0.4251	kilometers/liter (km/l)
	gal/mile	2.3527	liters/kilometer (l/km)
VELOCITY	miles/hour	1.6093	kilometers/hour (km/h)
LENGTH	inches	25.4	millimeters (mm)
	feet	0.3048	meters (m)
	yards	0.9144	meters (m)
	miles	1.609	kilometers (km)
AREA	inches^2	645.2	millimeters^2 (mm^2)
		6.45	centimeters^2 (cm^2)
	feet^2	0.0929	meters^2 (m^2)
	yards^2	0.8361	meters^2 (m^2)
VOLUME	inches^3	16,387	(mm^3)
	inches^3	16.387	(cm^3)
	inches^3	0.0164	liters (l)
	quarts	0.9464	liters (l)
	gallons	3.7854	liters (l)
	yards^3	0.7646	meters^3 (m^3)
MASS	pounds	0.4536	kilograms (kg)
	tons	907.18	kilograms (kg)
	tons	0.90718	tonne
FORCE	kilograms	9.807	newtons (N)
	ounces	0.2780	newtons (N)
	pounds	4.448	newtons (N)
TEMPERATURE	degrees Fahrenheit	0.556 (°F −32)	degrees Celsius (°C)

INCH CAP SCREW TORQUE VALUES

SAE Grade	Head Markings	SAE Grade	Nut Markings
SAE GRADE 1, SAE GRADE 2	No Mark	2	No Mark

SAE Grade	Head Markings	SAE Grade
SAE GRADE 5, SAE GRADE 5.1, SAE GRADE 5.2		5 Nut Markings

SAE Grade	Nut Markings	SAE Grade
SAE GRADE 8, SAE GRADE 8.2		8 Nut Markings

DIA.	WRENCH SIZE	SAE GRADE 1 OIL N•m(lb-in)	SAE GRADE 1 DRY N•m(lb-in)	*SAE GRADE 2 OIL N•m(lb-in)	*SAE GRADE 2 DRY N•m(lb-in)	SAE GRADE 5 OIL N•m(lb-in)	SAE GRADE 5 DRY N•m(lb-in)	SAE GRADE 8 OIL N•m(lb-in)	SAE GRADE 8 DRY N•m(lb-in)
#6		0.5(4.5)	0.7(6)			1.4(12)	1.7(15)		
#8		0.9(8)	1.2(11)			2.4(21)	3.2(28)		
#10		1.4(12)	1.8(16)			3.4(30)	4.6(41)		
#12		2(19)	2.8(25)			5.4(48)	7.3(65)		
		N•m(lb-ft)	N•m(lb-ft)	N•m(lb-ft)	N•m(lb-ft)	N•m(lb-ft)	N•m(lb-ft)	N•m(lb-ft)	N•m(lb-ft)
1/4	7/16	3.5(2.5)	4(3)	5(4)	7(5)	8(6)	11(8)	12(8.5)	16(12)
5/16	1/2	7(5)	9(6.5)	10(7.5)	14(10)	16(12)	23(17)	24(18)	33(24)
3/8	9/16	12(8.5)	16(12)	19(14)	24(18)	30(22)	41(30)	41(30)	54(40)
7/16	5/8	19(14)	26(19)	30(22)	41(30)	47(35)	68(50)	68(50)	95(70)
1/2	3/4	24(21)	41(30)	47(35)	61(45)	75(55)	102(75)	102(75)	142(105)
9/16	13/16	41(30)	54(40)	68(50)	88(65)	108(80)	142(105)	149(110)	203(150)
5/8	15/16	54(40)	75(55)	88(65)	122(90)	149(110)	197(145)	203(150)	278(205)
3/4	1-1/8	102(75)	136(100)	163(120)	217(160)	258(190)	353(260)	366(270)	495(365)
7/8	1-5/16	163(120)	244(165)	163(120)	224(165)	414(305)	563(415)	590(435)	800(590)
1	1-1/2	244(180)	332(245)	244(180)	332(245)	624(460)	848(625)	881(650)	1193(880)
1-1/8	1-11/16	346(255)	468(345)	346(255)	468(345)	780(575)	1058(780)	1248(920)	1695(1250)
1-1/4	1-7/8	488(360)	664(490)	488(360)	665(490)	1098(810)	1492(1100)	1763(1300)	2393(1765)
1-3/8	2-1/16	637(470)	868(640)	637(470)	868(640)	1438(1061)	1953(1440)	2312(1705)	3140(2315)
1-1/2	2-1/4	848(625)	1153(850)	848(625)	1153(850)	1912(1410)	2590(1910)	3065(2260)	4163(3070)

*For SAE Grade 2 fasteners 152 mm (6 in.) or less in length, use torque values for SAE Grade 2. For fasteners longer than 152 mm (6 in.), use SAE Grade 1 torque values.

Do not use these values if a different torque value or tightening procedure is listed for a specific application. Torque values listed are for general use only. Check tightness of cap screws periodically.

Shear bolts are designed to fail under predetermined loads. Always replace shear bolts with identical grade.

Fasteners should be replaced with the same or higher grade. If higher grade fasteners are used, these should be tightened only to the strength of the original.

Make sure fastener threads are clean and you properly start thread engagement. This will prevent them from failing when tightening.

Tighten plastic insert or crimped steel–type locknuts to approximately 50% of amount shown in the table. Tighten toothed or serrated-type locknuts to full torque value.

Source: Cummins Engine Company, Inc.

METRIC CAP SCREW TORQUE VALUES

Property Class and Head Markings	4.6	4.8	8.8	9.8	10.9	12.9
Property Class and Nut Markings	5	5	10	10	10	12

DIA.	WRENCH SIZE	4.6 OIL N·m(lb-ft)	4.6 DRY N·m(lb-ft)	4.8 OIL N·m(lb-ft)	4.8 DRY N·m(lb-ft)	8.8 or 9.8 OIL N·m(lb-ft)	8.8 or 9.8 DRY N·m(lb-ft)	10.9 OIL N·m(lb-ft)	10.9 DRY N·m(lb-ft)	12.9 OIL N·m(lb-ft)	12.9 DRY N·m(lb-ft)
M3	5.5mm	0.4(0.2)	0.5(0.3)	0.5(0.4)	0.7(0.5)	1(0.8)	1.3(1)	1.5(1)	2(1.5)	1.5(1)	2(1.5)
M4	7mm	0.9(0.6)	1.1(0.8)	1(0.9)	1.5(1)	2.5(1.5)	3(2)	3.5(2.5)	4.5(3)	4(3)	5(4)
M5	8mm	1.5(1)	2.5(1.5)	2.5(1.5)	3(2)	4.5(3.5)	6(4.5)	6.5(4.5)	9(6.5)	7.5(5.5)	10(7.5)
M6	10mm	3(2)	4(3)	4(3)	5.5(4)	7.5(5.5)	10(7.5)	11(8)	15(11)	13(9.5)	18(13)
M8	13mm	7(5)	9.5(7)	10(7.5)	13(10)	18(13)	25(18)	25(18)	35(26)	30(22)	45(33)
M10	16mm	14(10)	19(14)	20(15)	25(18)	35(26)	50(37)	55(41)	75(55)	65(48)	85(63)
M12	18mm	25(18)	35(26)	35(26)	45(33)	65(48)	85(63)	95(70)	130(97)	110(81)	150(111)
M14	21mm	40(30)	50(37)	55(41)	75(55)	100(74)	140(103)	150(111)	205(151)	175(129)	240(177)
M16	24mm	60(44)	80(59)	85(63)	115(85)	160(118)	215(159)	235(173)	315(232)	275(203)	370(273)
M18	27mm	80(59)	110(81)	115(85)	160(118)	225(166)	305(225)	320(236)	435(321)	375(277)	510(376)
M20	30mm	115(85)	160(118)	165(122)	225(166)	320(236)	435(321)	455(356)	620(457)	535(395)	725(535)
M22	33mm	160(118)	215(159)	225(167)	305(225)	435(321)	590(435)	620(457)	840(620)	725(535)	985(726)
M24	36mm	200(148)	275(203)	285(210)	390(288)	555(409)	750(553)	790(583)	1070(789)	925(682)	1255(926)
M27	41mm	295(218)	400(295)	415(306)	565(417)	810(597)	1100(811)	1155(852)	1565(1154)	1350(996)	1835(1353)
M30	46mm	400(295)	545(402)	565(417)	770(568)	1100(811)	1495(1103)	1570(1158)	2130(1571)	1835(1353)	2490(1837)
M33	51mm	545(402)	740(546)	770(568)	1050(774)	1500(1106)	2035(1500)	2135(1575)	2900(2139)	2500(1844)	3390(2500)
M36	55mm	700(516)	950(700)	990(730)	1345(992)	1925(1420)	2610(1925)	2740(2021)	3720(2744)	3205(2364)	4355(3212)

CAUTION:

Use only metric tools on metric hardware. Other tools may not fit properly. They may slip and cause injury.

Do not use these values if a different torque value or tightening procedure is listed for a specific application. Torque values listed are for general use only. Check tightness of cap screws periodically.
Shear bolts are designed to fail under predetermined loads. Always replace shear bolts with identical grade.
Fasteners should be replaced with the same or higher grade. If higher grade fasteners are used, these should be tightened only to the strength of the original.
Make sure fastener threads are clean and you properly start thread engagement. This will prevent them from failing when tightening.
Tighten plastic insert or crimped steel-type locknuts to approximately 50% of amount shown in the table. Tighten toothed or serrated-type locknuts to full torque value.
Source: Cummins Engine Company, Inc.

PIPE PLUG TORQUE VALUES

Size		Torque In Aluminum Components		Torque In Cast-Iron or Steel Components	
Thread in.	Actual Thread O.D. in.	N·m	ft-lb	N·m	ft-lb
1/16	0.32	5	45 in.-lb	15	10
1/8	0.41	15	10	20	15
1/4	0.54	20	15	25	20
3/8	0.68	25	20	35	25
1/2	0.85	35	25	55	40
3/4	1.05	45	35	75	55
1	1.32	60	45	95	70
1-1/4	1.66	75	55	115	85
1-1/2	1.90	85	65	135	100

TORQUE VALUES FOR STRAIGHT THREAD FITTINGS AND PLUGS

Thread Size in.	Steel Fitting (In Cast Iron or Steel[a])		Brass Fitting	
5/16–24	5 to 6 N·m	(45 to 50 in.-lb)	3.1 to 3.4 N·m	(27 to 30 in.-lb)
3/8–24	10 to 11 N·m	(90 to 100 in.-lb)	5.1 to 5.7 N·m	(45 to 50 in.-lb)
7/16–20	15 to 17 N·m	(135 to 150 in.-lb)	6.5 to 7.3 N·m	(58 to 65 in.-lb)
1/2–20	15 to 18 N·m	(11 to 13 ft-lb)	8.2 to 9.0 N·m	(72 to 80 in.-lb)
9/16–18	30 to 34 N·m	(22 to 25 ft-lb)	12.7 to 14.1 N·m	(112 to 125 in.-lb)
3/4–16	50 to 56 N·m	(37 to 41 ft-lb)	24 to 28 N·m	(18 to 21 ft-lb)
7/8–14	71 to 79 N·m	(52 to 58 ft-lb)	38 to 39 N·m	(26 to 29 ft-lb)
1 1/16–12	102 to 113 N·m	(75 to 83 ft-lb)	50 to 56 N·m	(37 to 41 ft-lb)
1 3/16–12	122 to 136 N·m	(90 to 100 ft-lb)	64 to 71 N·m	(47 to 52 ft-lb)
1 5/16–12	142 to 157 N·m	(105 to 116 ft-lb)	71 to 79 N·m	(52 to 58 ft-lb)
1 5/8–12	213 to 237 N·m	(157 to 175 ft-lb)	91 to 102 N·m	(67 to 75 ft-lb)
1 7/8–12	305 to 339 N·m	(225 to 250 ft-lb)	102 to 115 N·m	(75 to 85 ft-lb)
2 1/2–12	457 to 509 N·m	(337 to 375 ft-lb)	126 to 141 N·m	(93 to 104 ft-lb)

[a]Steel fittings in brass or aluminum *must* use comparable brass fitting torque for the thread size.
Source: Cummins Engine Company, Inc.

TORQUE WRENCH USAGE*

Turn Tighter Chart

Turn (Bolt or Nut)	Turn Degrees
1	360°
½	180°
⅓	120°
¼	90°
⅙	60°
1/12	30°

NOTE: The side of a nut or bolt head can be used for reference if a mark cannot be put on.

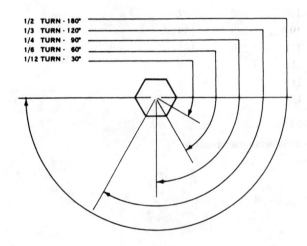

Torque Wrench Extension

When a torque wrench extension is used with a torque wrench, the torque indication on the torque wrench will be less than the real torque.

Source: Caterpillar Inc.

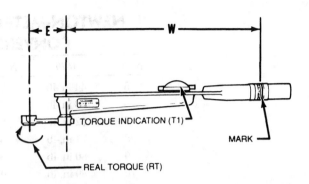

(E) Torque wrench drive axis–to–torque wrench extension drive axis. **(W)** Mark on handle-to–torque wrench drive axis.

1. Put a mark on the handle. Measure the handle from the mark to the axis of the torque wrench drive (W).

2. Measure the torque wrench extension from the torque wrench drive to the axis of the torque wrench extension drive (E).

3. To get correct torque indication (TI) when the real torque (RT) is known:

$$TI = \frac{RT \times W}{W + E}$$

Example: W = 304.8 mm (12 in.); E = 65.0 mm (2.56 in.); RT (from specifications) = 17 N · m (125 lb-ft).

$$TI = \frac{(170 \times 304.8)}{(304.8 + 65.0)}$$

$$= 140 \text{ N} \cdot \text{m}$$

$$TI = \frac{125 \times 12}{12 + 2.56}$$

$$= 103 \text{ lb-ft}$$

4. Hold the torque wrench handle with the longest finger of the hand over the mark on the handle to get the real torque (RT) with low torque indication (TI) on the torque wrench.

NEWTON-METER TO FOOT-POUND CONVERSION CHART

$N \cdot m$	ft-lb	$N \cdot m$	ft-lb	$N \cdot m$	ft-lb
5	44 in.-lb	70	52	170	125
6	53 in.-lb	75	55	175	129
7	62 in.-lb	80	59	180	133
8	71 in.-lb	85	63	185	136
9	80 in.-lb	90	66	190	140
10	89 in.-lb	95	70	195	144
12	9	100	74	200	148
14	10	105	77	205	151
15	11	110	81	210	155
16	12	115	85	215	159
18	13	120	89	220	162
20	15	125	92	225	165
25	18	130	96	230	170
30	22	135	100	235	173
35	26	140	103	240	177
40	30	145	107	245	180
45	33	150	111	250	184
50	37	155	114		
55	41	160	118		
60	44	165	122		
65	48				

NOTE: To convert from newton-meters to kilogram-meters, divide newton-meters by 9.803.

Source: Cummins Engine Company, Inc.

TAP-DRILL CHART—U.S. CUSTOMARY AND METRIC

NOTE ON SELECTING TAP-DRILL SIZES: The tap-drill sizes shown in this table give the theoretical tap-drill size for approximately 60% and 75% of full thread depth. Generally, it is recommended that drill sizes be selected in the 60% range, as these sizes will provide about 90% of the potential holding power. Drill sizes in the 75% range are recommended for shallow hole tapping (less than 1-½ times the hole diameter) in soft metals and mild steel.

Tap Size 60%	Tap Size 75%	Drill Size	Tap Size 60%	Tap Size 75%	Drill Size	Tap Size 60%	Tap Size 75%	Drill Size	Tap Size 60%	Tap Size 75%	Drill Size
		48			4.40mm			7.50mm			13.25mm
		1.95mm	12–24		16			¹⁹⁄₆₄		⅝–11	¹⁷⁄₃₂
		⁵⁄₆₄			4.50mm			7.60mm		M15×1.5	13.50mm
	3–48	47		M5.5×.9	15			N	M15×1.5		13.75mm
		2.00mm		12–28	4.60mm			7.70mm			³⁵⁄₆₄
	M2.5×.45	2.05mm	12–24		14		M9×1.25	7.75mm	⅝–11	M16×2	14.00mm
		46			13			7.80mm			14.25mm
3–48	3056	45		M5.5×.9	4.70mm			7.90mm			⁹⁄₁₆
		2.10mm			4.75mm		⅜–16	⁵⁄₁₆	M16×2	⅝–18	14.50mm
M2.5×.45	M2.6×.45	2.15mm	12–28		³⁄₁₆	M9×1.25	M9×1	8.00mm	⅝–18	M16×1.5	³⁷⁄₆₄
3–56	4–36	44			12			O			14.75mm
		2.20mm			4.80mm			8.10mm	M16×1.5		15.00mm
M2.6×.45		2.25mm			11	M9×1		8.20mm			19.32
4–36	4–40	43			4.90mm			P			15.25mm
		2.30mm			10			8.25mm			³⁹⁄₆₄
		2.35mm			9			8.30mm		M17×1.5	15.50mm
4–40	4–48	42		M6×1	5.00mm	⅜–16	⅛–27NPT	²¹⁄₆₄	M17×1.5	M18×2.5	15.75mm
		³⁄₃₂			8			8.40mm			⅝
	M3×.6	2.40mm			5.10mm		⅜–24	Q	M18×2.5	M18×2	16.00mm
4–48		41	¼–20		7		M10×1.5	8.50mm	M18×2		16.25mm
		2.45mm			¹³⁄₆₄			8.60mm		¾–10	⁴¹⁄₆₄
		40			6			R		M18×1.5	16.50mm
M3×.6	M3×.5	2.50mm	M6×1		5.20mm	⅜–24		8.70mm	¾–10	M19×2.5	²¹⁄₃₂
		39			5	⅛–27NPT		¹¹⁄₃₂	M18×1.5		16.75mm
	5–40	38		M6×.75	5.25mm		M10×1.25	8.75mm	M19×2.5		17.00mm
M3×.5		2.60mm			5.30mm	M10×1.5		8.80mm			⁴³⁄₆₄
5–40	5–44	37	¼–20		4			S			17.25mm
		2.70mm			5.40mm			8.90mm	¾–16	¾–16	¹¹⁄₁₆
5–44	6–32	36	M6×.75	¼–28	3	M10×1.25	M10×1	9.00mm		M20×2.5	17.50mm
		2.75mm			5.50mm			T			17.75mm
		⁷⁄₆₄			⁷⁄₃₂			9.10mm	M20×2.5	M20×2	18.00mm
		35			5.60mm			²³⁄₆₄	M20×2		18.25mm
		2.80mm	¼–28		2	M10×1		9.20mm			²³⁄₃₂
		34			5.70mm			9.30mm		M20×1.5	18.50mm
6–32	6–40	33			5.75mm		⁷⁄₁₆–14	U			⁴⁷⁄₆₄
	M3.5×.6	2.90mm			1			9.40mm	M20×1.5		18.75mm
		32			5.80mm		M11×1.5	9.50mm			19.00mm
M3.5×.6		3.00mm			5.90mm			⅜			¾
6–40		31			A			V			19.25mm
		3.10mm			¹⁵⁄₆₄			9.60mm		⁷⁄₈–9	⁴⁹⁄₆₄
		⅛	M7×1		6.00mm			9.70mm		M22×2.5	19.50mm
		3.20mm			B			9.75mm	⁷⁄₈–9		²⁵⁄₃₂
	M4×.75	3.25mm			6.10mm	M11×1.5		9.80mm			19.75mm
		30			C	⁷⁄₁₆–14		W	M22×2.5	M22×2	20.00mm
	M4×.7	3.30mm	M7×1		6.20mm			9.90mm		⁷⁄₈–14	⁵¹⁄₆₄
M4×.75		3.40mm			D		⁷⁄₁₆–20	²⁵⁄₆₄	M22×2		20.25mm
M4×.7	8–32	29		M7×.75	6.25mm			10.00mm		M22×1.5	20.50mm
		3.50mm			6.30mm	⁷⁄₁₆–20		X	⁷⁄₈–14		¹³⁄₁₆
	8–36	28			E		M12×1.75	10.20mm			20.75mm
8–32		⁹⁄₆₄			¼			Y	M22×1.5	M24×3	21.00mm
		3.60mm	M7×.75		6.40mm			¹³⁄₃₂			⁵³⁄₆₄
8–36		27			6.50mm			Z			²¹⁄₂₅mm
		3.70mm		⁵⁄₁₆–18	F	M12×1.75	M12×1.5	10.50mm			²⁷⁄₃₂
		26			6.60mm		½–13	²⁷⁄₆₄	M24×3		21.50mm
	M4.5×.75	3.75mm			G	M12×1.5	M12×1.25	10.75mm			21.75mm
	10–24	25			6.70mm	M12×1.25		11.00mm			⁵⁵⁄₆₄
		3.80mm			¹⁷⁄₆₄	½–13		⁷⁄₁₆		M24×2	22.00mm
		24		M8×1.25	6.75mm	¼–18NPT		11.25mm	M24×2	1″–8	⅞
		3.90mm	⁵⁄₁₆–18		H			11.50mm			22.25mm
		23			6.80mm			²⁹⁄₆₄		M24×1.5	22.50mm
M4.5×.75		⁵⁄₃₂			6.90mm			11.75mm	1″–8		⁵⁷⁄₆₄
10–24		22		5.16–24	I			11.50mm	M24×1.5	M25×2	22.75mm
	M5×1	4.00mm	M8×1.25	M8×1	7.00mm		½–20	²⁹⁄₆₄		1″–12	23.00mm
	10–32	21			J		⁹⁄₁₆–12	¹⁵⁄₃₂			²⁹⁄₃₂
		20			7.10mm		M14×2	12.00m	M25×2	1″–12	23.25mm
	M5×.9	4.10mm	⁵⁄₁₆–24		K			12.25mm	1″–12	1″–14	⁵⁹⁄₆₄
M5×1	M5×.8	4.20mm		M8×1	⁹⁄₃₂	⁹⁄₁₆–12		³¹⁄₆₄		M25×1.5	23.50mm
10–32		19			7.20mm	M14×2	M14×1.5	12.50mm	M20×1.5		23.75mm
M5×.9		4.25mm			7.25mm		⁹⁄₁₆–18	½	1″–14		¹⁵⁄₁₆
M5×.8		4.30mm			7.30mm	M14×1.5	M14×1.25	12.75mm			
		18			L	M14×1.25		13.00mm			
		¹¹⁄₆₄			7.40mm	⁹⁄₁₆–18		³³⁄₆₄			
		17			M						

(Courtesy of Cummins Engine Company, Inc.)

FRACTION, DECIMAL, MILLIMETER CONVERSIONS

8 THS	16 THS	32 NDS	64 THS	inches	mm	8 THS	16 THS	32 NDS	64 THS	inches	mm
			1	0.0156	0.397				33	0.5156	13.097
		1		0.0313	0.794			17		0.5313	13.494
			3	0.0469	1.191				35	0.5469	13.891
	1			0.0625	1.588		9			0.5625	14.288
			5	0.0781	1.984				37	0.5781	14.684
		3		0.0938	2.381			19		0.5938	15.081
			7	0.1094	2.778				39	0.6094	15.478
1				0.1250	3.175	5				0.6250	15.875
			9	0.1406	3.572				41	0.6406	16.272
		5		0.1563	3.969			21		0.6563	16.669
			11	0.1719	4.366				43	0.6719	17.066
	3			0.1875	4.763		11			0.6875	17.463
			13	0.2031	5.159				45	0.7031	17.859
		7		0.2188	5.556			23		0.7188	18.256
			15	0.2344	5.953				47	0.7344	18.653
¼				0.2500	6.350	¾				0.7500	19.050
			17	0.2656	6.747				49	0.7656	19.447
		9		0.2813	7.144			25		0.7813	19.844
			19	0.2969	7.541				51	0.7969	20.241
	5			0.3125	7.938		13			0.8125	20.638
			21	0.3281	8.334				53	0.8281	21.034
		11		0.3438	8.731			27		0.8438	21.431
			23	0.3594	9.128				55	0.8594	21.828
3				0.3750	9.525	7				0.8750	22.225
			25	0.3906	9.922				57	0.8906	22.622
		13		0.4063	10.319			29		0.9063	23.019
			27	0.4219	10.716				59	0.9219	23.416
	7			0.4375	11.113		15			0.9375	23.813
			29	0.4531	11.509				61	0.9531	24.209
		15		0.4688	11.906			31		0.9688	24.606
			31	0.4844	12.303				63	0.9844	25.003
½				0.5000	12.700	1 in.				1.0000	25.400

Conversion factor: 1 inch = 25.4 mm

Source: Cummins Engine Company, Inc.

DRIVE BELT TENSION

SAE Belt Size	Belt Tension Gauge Part No.		Belt Tension New		Belt Tension Range Used[a]	
	Click-type	Burroughs	N	lbf	N	lbf
0.380 in.	3822524		620	140	270 to 490	60–110
0.440 in.	3822524		620	140	270 to 490	60–110
1/2 in.	3822524	ST-1138	620	140	270 to 490	60–110
11/16 in.	3822524	ST-1138	620	140	270 to 490	60–110
3/4 in.	3822524	ST-1138	620	140	270 to 490	60–110
7/8 in.	3822524	ST-1138	620	140	270 to 490	60–110
4 rib	3822524	ST-1138	620	140	270 to 490	60–110
5 rib	3822524	ST-1138	670	150	270 to 530	60–120
6 rib	3822525	ST-1293	710	160	290 to 580	65–130
8 rib	3822525	ST-1293	890	200	360 to 710	80–160
10 rib	3822525	3823138	1110	250	440 to 890	100–200
12 rib	3822525	3823138	1330	300	530 to 1070	120–240

[a]A belt is considered used if it has been in service for 10 minutes or longer.
If used belt tension is less than the minimum value, tighten the belt to the maximum used belt value.
Source: Cummins Engine Company, Inc.

INDEX

Aftercoolers, 37
Air box, 9
Air change pressure, 220
Air charge temperatures, 220
Air cleaner (dry type), 31, 32
Air cleaner functions, 31
Air cleaner restriction indicator, 32
Air cleaner types, 31
Air-cooling system, 61
Air density, 24
Air/fuel ratio, 5
Air induction system, 299
Air molecules, 7
Air-operated, 69
Air-operated starting systems, 174
Air shutdown, 33, 37
Air shutdown valve, 35
Air starter Diagnostic chart, 197
Air-starting system, 195, 196
Air-to-air aftercooler, 38
Air-to-air intercooler, 38
Air velocity, 34
Alternative, fuels, 207
Alternator, 148
Alternator assembly, 169
Alternator bearing service, 168
Alternator bench test, 171
Alternator brushless, 158
Alternator circuit, 151
Alternator components, 148
Alternator current control, 153
Alternator disassembly, 166
Alternator drive belt, 148, 171
Alternator installation, 172

Alternator operation, 150
Alternator output test, 166
Alternator problem diagnosis, 159
Alternator removal, 166
Alternator types, 148
Altitude-pressure compensator (ADA), 258, 260
AMBAC, 311
   AMBAC distributor injection pump service, 56
   AMBAC International Corp., 279, 304
   calibration check, 563
   delivery valve operation, 554
   distributor pump timing, 555
   electronically controlled distributor pump, 559
   fuel flow, 553
   fuel injection pump installation, 564
   fuel metering, 554
   fuel starting device, 559
   fuel system troubleshooting, 561
   hydraulic/electronic timing advance, 561
   internal advance mechanism, 556
   microprocessor-based control system, 560
   pump adjustment, 563
   pump lubrication, 553
   PS model pump operation, 552
   smoke limiter, 559
Ammeter, 125, 155, 157
Analyzing exhaust smoke, 58
Aneroid capsule, 258
Antifreeze additives, 71
Antifreeze coolant, 71
Antifreeze inhibitors, 71

Antimony (phosphorous), 120
Application of injection nozzles, 222
Armature, 176, 180, 193
Ash content, 204, 207
Aspirated air, 30
Atmospheric pressure, 5
Atom, 104
Atomization, 220
Automatic disengagement, 187
AVR tester, 163

Balancer shafts, 2
Ballast resistor, 119
Barometer, 24
Base, 121
Base current, 12
Battery, 148, 175
Battery cable connections (removing), 140
Battery cables, 134
Battery capacity test, 144
Battery cell voltage test, 142
Battery charging, 132, 138, 143
Battery cleaning, 140
Battery construction, 130, 131
   cell connections, 131
Battery discharging, 132
Battery drain test, 144
Battery electrolytes, 132
Battery function, 130
Battery inspection, 138
Battery installation, 145
Battery maintenance free, 132
Battery polarity, 133
Battery quick charge test, 142

INDEX

Battery quick charge test, *continued*
    cold cranking amperes, 135
    reserve capacity, 135
Battery rating, 135
Battery selection, 145
Battery service precautions, 136
Battery state of charge, 141
Battery state of charge testing, 142
Battery-systems multiple, 134
Battery terminal, 133
Battery terminal cleaning, 140
Battery testing, 138
Battery voltage, 142
Bendix drive, 183
Bendix drive check, 195
Bendix drive components, 185
Bendix drive sleeve, 183
Bendix drive spring, 183
Bendix-type drive, 196
Belt driven alternator, 148
Belt tension adjustment, 77
Belt tension chart, 78
Binary digital voltage signals, 268
Black exhaust smoke, 47
Bleeding air from the fuel system, 215
Bleeding the injection system, 309
Blower, 9
Blower assembly, 54
Blower cleaning, 53
Blower components, 38
Blower design, 36
Blower diode, 121
Blower disassembly, 53
Blower drive shaft, 54
Blower function, 33
Blower inspection (on engine), 53
Blower installation, 54
Blower operation, 36
Blower removal, 53
Blower rotor, 36
Blower rotor gears, 54
Blower service, 52
Blue exhaust smoke, 47
Bore, 12
Boron (indium), 120
Bosch injection nozzle, 222
Bosch injection pump, 280
Bosch PES pump timing, 308
Bosch preventive maintenance, 297
Bosch pump, 306
Brake power, 22
Brush assembly, 176
Brush holder assembly, 149
Brushless alternator, 158
Brushless charging operating principles, 159

Cable connection test, 138
Cable and wiring, 175
Cam follower, 2
Camshaft, 2
Canister-type element, 92
Cap tester, 79
Capacitor (condenser), 119, 151
Capacitor C1, 159

Carbon monoxide, 40
Caterpillar Electronic Control System, 512
    component description, 522
    component replacement, 521
    diagnosis procedures, 529
    3406E electronic control system, 524
    3406E/3176B electronic control system, 521
    electronic injection timing, 534
    electronic unit injector, 527
    engine protection system, 526
    EUI service, 532
    fault codes, 518
    fuel flow system, 526
    PEEC limp-home system, 514
    PEEC system components, 513
    PEEC system tests, 517
    pressure transducer module, 516
    programmable electronic engine control, 513
    rack actuator, 517
    speed/timing sensor, 533
    three-cylinder cutout mode, 528
    throttle position sensor, 533
    timing advance unit operation, 517
    troubleshooting diagnostic chart, 530
Caterpillar fuel injection system, 451
    alternative method, 471
    automatic timing advance, 460
    automatic timing advance operation, 477
    automatic timing advance unit, 458, 466
    caterpillar mechanical unit injector fuel system, 493
    checking the set point, 470, 487
    control rack, 494
    crossover lever adjustment, 486
    dashpot governor, 478
    dashpot operation, 456, 498
    engine checking, 500
    engine timing by the timing pin method, 467
    engine timing check, 465
    fuel injection nozzle, 479
    fuel injection pump, 463
    fuel injection pump removal, 462
    fuel injection pump service, 489
    fuel injection pump timing, 481
    fuel pressure, 501
    fuel pump calibration, 489, 492
    fuel ratio control, 459, 468, 477, 485, 489
    fuel ratio control adjustment, 472
    fuel ratio control operation, 457
    fuel setting, 483, 504
    fuel timing, 506
    fuel transfer pump, 493
    governor, 494, 500
    governor check, 468
    governor components, 456, 457
    governor dashpot operation, 476
    governor operation, 452, 475
    governor servo, 498
    governor servo operation, 455
    injection pump operation, 452
    injector synchronization, 502, 503

Caterpillar Fuel Injection System, *continued*
    lifter, 463
    load stop adjustment, 484
    low-idle adjustment, 470, 487
    mechanical unit injector system diagnosis, 500
    mechanical unit injector system service, 500
    plunger, 463
    scroll fuel system, 453
    scroll fuel system diagnosis, 462,
    scroll fuel system service, 462
    scroll injection pump, 454
    scroll system transfer pump, 454
    sealed unit fuel ratio control, 458
    sleeve-metering fuel injection, 473
    sleeve-metering fuel system fuel flow, 473
    sleeve-metering injection pump operation, 474
    sleeve-metering system diagnosis, 480
    sleeve-metering system service, 480
    TDC magnetic transducer, 466
    top center compression position, 464
    torque control, 455
    unit injector, 493
CELECT ECI (electronically controlled injection), 429
    ambient air pressure sensor, 430, 434
    boost pressure sensor, 430
    brake switch, 433
    clutch switch, 433
    compression brake control switch, 436
    coolant level sensor, 429
    coolant temperature sensor, 429
    cruise control switches, 436
    electronic control module, 429
    electronically controlled injector, 435
    electronically controlled unit injectors, 435
    engine position sensor, 429
    fuel pump installation, 421
    fuel pump testing, 421
    gear pump regulator, 433
    idle-speed adjust switch, 436
    intake air temperature sensor, 430, 433
    oil pressure sensor, 430
    oil temperature sensor, 430
    pressure regulator, 433
    throttle position sensor, 430
    turbo boost pressure sensor, 434
    vehicle speed sensor, 430, 434
    warning lights, 435
    wiring harness, 429
CELECT ECM outputs, 433
CELECT exhaust side, 431
CELECT fuel pump side, 431
CELECT system components, 432
Centistokes, 205
Centrifugal compressor, 34
Centrifugal force, 204
Centrifugal type of oil cleaner, 94
Cetane number, 204, 205
Charge indicator, 148, 155
Charging principles, 147

682 INDEX

Charging service, 147
Charging system, 147
Charging system analyzer, 162
Charging system diagnostic chart, 160
Charging system function, 147
Charging system preliminary inspection, 161
Charging system service precautions, 159
Charging system tests, 161
Charging system tester, 163
Charging system warning light, 155
Chassis dynamometer, 27
Checking oil level, 95
Checking oil level conditions, 95
Chemical analysis, 97
Chemical energy, 2
Chemically treated filter element, 70
Circuit breaker, 117, 124
Circuit protection, 117
    fuses, 117
    fusible link, 117
Circuit resistance test, 163
Cloud point, 204, 205
Clutch drive, 68
Codes (active), 273
Codes (inactive), 273
Combustion chamber design, 218
Combustion stages, 220
Common ground principle, 110
Commutator, 176
Commutator end cap (plate), 178
Commutator plate (end plate), 178
Compressed air, 196
Compressed natural gas (CNG), 207
Compressed nitrogen gas, 198
Compression ratio, 7, 13
Compression stroke, 5
Compressor, 35
Compressor wheel, 34, 35
Compressor wheel damage, 49
Computer chip (integrated circuit), 268
Computer controlled actuators (output device), 272
Computer memory, 269
Condenser (capacity), 119
Conduction, 61
Conductor, 104
Connecting rod, 2
Control rack (rod), 327
Control rod (rack), 237
    automatic full-load control-rod stop, 273
    control-lever stops, 256
    excess fuel stops, 253
    reduced delivery stop, 256
    spring-loaded-control-rod stop, 255
    spring-loaded-idle-speed stop, 256
Convection, 60
"Convectional" theory, 105
Coolant circulation, 62
Coolant conditioner, 70, 80
Coolant filter, 70, 80
Coolant inspection, 78
Coolant reserve tank, 66
Coolant temperature problems, 71
Coolant testing, 78

Cooled hole-type nozzle, 224
Cooling system, 81
Cooling system cleaning, 79
Cooling system diagnosis, 72, 74
Cooling system filtering, 81
Cooling system flushing, 79
Cooling system overheating, 72
Cooling system principles, 60
Cooling system service, 60, 72
Cooling system visual inspection, 72
Crankcase pressure, 9, 49
Crankcase ventilation, 94
Crankcase ventilation service, 99
Cranking motor relay, 183
Crank pin travel, 14
Crankshaft, 2, 3
Crankshaft counterweight, 2
Cross-flow radiator, 67
Cummins C brake, 15
Cummins Comulink diagnostic tool, 273
Cummins Electronic Control Systems, 424
    ECM inputs, 433
    engine position sensor, 426
    fault codes, 428
    fuel control valve, 427
    fuel flow, 427
    Pace electronic control system, 425
    Pacer system fault codes, 429
    PT pacer electronic system, 425
    PT system operation, 427
    vehicle speed sensor, 426
Cummins fault code, 273
Cummins pressure time, 205
Cummins PT Fuel Injection System, 384
    air/fuel control, 393
    constant-speed governor, 392
    Cummins injection timing control, 398
    engine speed adjustment, 420
    fixed-time and STC engines, 414
    fuel drain line restriction check, 417
    fuel injector operation, 388, 393
    fuel pump air leak, 418
    fuel pump drive, 385
    fuel pump function, 388
    fuel pump major components, 385
    fuel pump removal, 421
    fuel rail pressure, 418
    fuel supply line restriction check, 417
    fuel supply pump operation, 389
    fuel system adjustment, 417
    fuel system check, 417
    fuel system diagnosis, 403
    gear pump, 385
    governor operations, 390
    governor types, 390
    HVT oil flow (advanced timing), 399
    HVT oil flow (retarded timing), 399
    hydraulic governor, 393
    hydraulic variable timing, 398, 399
    injector adjustment, 410
    injector adjustment sequence, 412
    injector installation, 408
    injector removal, 407
    injector service, 403, 407

Cummins PT Fuel Injection System, *continued*
    injector timing (typical for L-10 series) 414
    injector timing check, 414
    mechanical variable-speed (MVS) governor, 391
    mechanical variable timing, 398
    (MVS) governor, 392
    MVT actuator, 398
    pressure regulation, 388
    PT-D injector, 395
    PT-D injector metering, 394
    PT-D top-stop injector, 397
    PT (type D) injector, 393
    PT (Type D) injector pumping, 394
    PT (Type D) injector top-stop, 394
    PTG-AFC fuel rate adjustment, 419
    PTG-AFC low-idle-speed adjustment, 420
    PTG-AFC-No-air valve adjustment, 421
    PTG-AFC-VS high-idle-speed adjustment, 421
    PTG-AFC pump fuel ratio, 419
    pulsation damper, 385
    shutdown valve, 387
    special variable-speed (SVS) governor, 392
    STC injector adjustment, 413
    step-timing control system, 402
    suction side air leak, 418
    testing injectors, 407
    throttle, 387
    throttle leakage check, 420
    top-stop injector, 402
    troubleshooting charts, 404, 405, 406, 407
Current-carrying conductor, 104
Current output test, 162
Cycle efficiency, 25
Cylinder balance, 26
Cylinder block, 2, 4
Cylinder bore, 12
Cylinder head, 2
Cylinder head and valve service, 636
    adjusting the valve train, 666
    cab-forward trucks, 637
    checking push rods, 658
    counterboring valve seat, 649
    cylinder head crack detection, 642
    cylinder head disassembly, 641
    cylinder head installation, 660
    cylinder head removal, 637, 638
    cylinder head warpage, 643
    injector service, 664
    inspecting valves, 655
    inspector sleeve testing, 660
    installing a valve seat, 649
    measuring valves, 655
    narrowing a valve seat, 653
    nozzle-type injector removal, 639
    removing a valve seat, 648
    rocker arms, 659
    seat grinding, 652
    seat grinding stones, 651
    shafts, 659
    unit injector removal, 641

INDEX

Cylinder head and valve service, *continued*
    unit injector tube, 662
    valve adjustment, 667
    valve bridge, 645
    valve crosshead, 641
    valve grinding procedure, 656
    valve seat grinding, 650
    valve seat grinding pilots, 651
    valve seat inserter, 646
    valve seat-to-face contact, 653
    valve service, 654, 664
    valve spring height, 658
    valve spring testing, 658
    valve stem seal, 663
    valve train service, 658
Cylinder liner, 2, 3
Cylinder misfire test, 441
    CELECT ECI, 130
    CELECT exhaust side, 431
    CELECT fuel pump side, 431
    CELECT injector puller installer, 444
    CELECT injector replacement, 443
    CELECT system components, 432
    compulink, 442
    fuel pump installation, 421
    fuel pump testing, 421
    gear pump rest, 448
    injector and valve adjustment, 446
    injector installation, 443
    injector O-rings, 444
    injector removal, 443, 444
    shut-off valve tests, 448, 449
    supply pump, 448
    valve adjustment, 448

Data Link information, 275
Dead battery, 159
Delcotron 400 (alternator), 158
Delta-connector stator, 152
Delta stator, 153
Delta-wound stator, 156
Design of injection nozzle, 222
Detroit Diesel, 3
Detroit Diesel diagnosis
    checking fuel flow, 335
    tailored torque governor, 336
Detroit Diesel electronic control (DDEC), 267
Detroit Diesel electronic control system, 270, 368
    checking electrical connectors, 378
    clearing trouble codes, 378
    Data Communication Links, 374
    DDEC development, 369
    DDEC diagnosis, 375
    DDEC diagnostic codes, 377
    DDEC engine protection system, 373
    DDEC fuel flow and EUI operation, 374
    DDEC function, 369
    DDEC 111 component description, 370
    DDEC operating features, 370
    DDEC testing, 375
Detroit Diesel injector chart, 344–349
Detroit Diesel injector removal, 338
    cleaning injector parts, 343

Detroit Diesel injector removal, *continued*
    control rack and plunger, 339
    fuel output test, 341
    injector disassembly, 342
    injector high pressure test, 341
    injector troubleshooting, 342
    installing fuel injector in tester, 339
    pressure holding test, 341
    spray pattern test, 340
    spray tip test, 341
    tailored torque governor, 336
    valve opening test, 340
Detroit Diesel inspecting injector parts, 351
    assembly follower, 353
    assembly injector filters, 352
    assembly plunger, 353
    assembling rack and gears, 353
    assembling spray tip, 353
    assembling spring cage, 353
    Belleville spring adjustment, 360
    buffer screw adjustment, 364
    checking spray tip concentricity, 354
    checking valve assembly, 353
    engine tune up, 355
    fuel modulator adjustment, 366
    governor gap adjustment, 357
    idle speed adjustment, 364
    injector assembly, 352
    injector installation, 354
    no-load engine speed adjustment, 363
    positioning injector rack control levers, 360
    setting injector timing (two-stroke-cycle engine), 355
    setting injector timing height (8.2L four-stroke-cycle engine), 356
    starting aid screw, 362
    testing reconditioned tester, 354
    throttle delay adjustment, 365
    timing fuel injector, 356
    tune-up for mechanical governor, 355
    tune-up sequence for hydraulic governor, 355
    Type A governor spring, 363
    Type B governor spring, 364
Detroit Diesel insufficient fuel, 338
Detroit Diesel mechanical governor, 332
    fuel injector rack, 334
    fuel modulator, 333
    hydraulic governor assembly, 334
    tailored torque (TT) governor operation, 334
    throttle delay mechanism, 334
    variable-speed governor, 333
Detroit Diesel mechanical unit injector, 324
    barrel construction, 326
    crown valve injectors, 329
    crown valve operation, 327
    fast idle cylinder, 331
    governor types, 331
    helix construction, 326
    injector control tube, 330
    injector timing, 328
    needle valve injector, 329
    needle valve operation, 327

Detroit Diesel mechanical unit injector, *continued*
    plunger construction, 326
    unit injector function, 325
    unit injector identification, 326
    unit injector operation, 327
Diagnostic procedures using the CEL, 375
Diagnostic procedures using the DDL Reader (DDR), 375
Diaphragm pump, 210
Diaphragm pump operation, 211
Diesel cooling system, 60
Diesel engine, 1, 2
Diesel engine systems, 17
Diesel fuel, 203
Diesel fuel classification, 20
Diesel fuel injection, 217
Diesel fuel injection pump, 208
Diesel fuel injection tubing, 225
Diesel fuel quality, 204
Diesel fuel specifications, 208
Diesel fuel supply system, 203
Digital system, 268
Digital Volt-ohmmeter (DVOM), 126
Diode D5, 120, 159
Diode testing, 168
Diode trio, 150, 156, 169
Direct injection combustion chamber, 218
Dirt ejector, 31
Dished piston head, 219
Displacement, 12
Double-acting fuel supply pump operation, 295
Drive end frame, 14
Drive end housing, 17
Drive pinion, 181
Drive pinion clearance check, 195
Dry element cleaning, 46
Dry-type air cleaning service, 44
Dynamometer testing, 630
Dynamometer worksheet, 633
Dynamometers, 26

Earth's magnetic poles, 105
ECM (electronic control module) input and output devices, 271
ECM output devices, 271
    lighted display, 271
    power module, 271
    relay, 271
    servo motor, 271
    solenoid, 271
    transistor, 271
EDC system schematic diagram, 295
Electric speed-control device, 262
Electric starting motor, 174, 175
Electric starting system, 175
Electric starting system service, 187
Electrical circuits, 109
    parallel circuits, 109
    series circuits, 109
    series parallel circuits, 110
Electrical circuit diagnosis, 123
Electrical circuit service, 123
Electrical common ground principle, 110

Electrical current, 104
Electrical power, 109
Electrical principles, 103
Electrical problems, 122
   corroded connections, 122
   electrical feedback, 122
   grounds, 123
   incorrect resistance, 122
   loose connections, 122
   opens, 123
   shorts, 123
   voltage drop, 123
Electrical service basics, 103
Electrical symbols, 113
   automatic switches, 114
   electrical switches, 114
   manually operated switches, 114
   relay switch, 114
Electrical wiring, 111
   straight wire, 111
   twisted wires, 111
   twisted/shielded wires, 111
   wiring diagram, 111
Electricity, 104
Electricity flows through a conductor, 107
   atoms, 106
   conductor, 107
   electromotive force (EMF), 107
   electron drift, 108
   electrons, 107
   free electrons, 107
   ground, 107
Electricity, measured in, 108
   amperes (current), 108
   current (amperes), 108
   Ohms (resistance), 108
   Ohm's Law, 108
   resistance, 108
   voltage (volts), 108
Electromagnetic induction, 106
Electromagnets, 105, 176
Electromechanical regulator, 154
Electron flow, 104
Electronic control advantages, 275
Electronic control features, 274, 436
   CELECT electronic injector operation, 437
   CELECT system basic fuel flow, 437
   cooling fan with clutch count, 436
   cruise control, 436
   Cummins Echeck tool, 440
   data link, 437
   electronically controlled injector, 438
   engine protection system, 437
   fault code diagnosis, 439
   fuel metering, 438
   gear down protection, 436
   governor options, 437
   idle shutdown, 437
   pinpoint testing, 441
   power takeoff, 436
   progressive shifting, 436
   road relay and sensor plus, 437
   using the compulink tool, 441
Electronic control module, 69

Electronic control module (ECM) components, 268
   Analog converter, 268
   circuit board, 269
   clock, 268
   digital converter, 268
   housing, 269
   memory, 269,
   microprocessor, 268
   multiple connector, 269
   power transistor, 269
   voltage regulator, 268
Electronic control module (ECM) operation, 267
Electronic control of diesel fuel injection, 266
Electronic control of diesel operating principles, 266
Electronic control system, 277
Electronic unit injector replacement, 378
Electronic voltage regulator, 157
Electrons, 104
Emitter, 121
End frame, 150
Energy cell, 219
Energy check light (CEL), 272
Engine balance, 25
Engine braking system, 15
Engine classification, 15
Engine components, 2
Engine diagnosis, performance testing and tune up, 617
   comparison between cylinders, 629
   compression test, 628
   cylinder leakage testing, 630
   diagnostic chart, 618–626
   diagnostic guide, 627
   diesel engine tune up, 627
   engine diagnosis, 617
Engine displacement, 12
Engine driven charging pump, 198
Engine dynamometer, 26
Engine efficiency, 19, 22
Engine emission standards, 40
Engine lubricating oil, 86
   antiwear, 86
   ashless dispersants, 86
   bearing corrosion inhibitors, 86
   foam depressants, 86
   metallic depressants, 86
   oxidation inhibitors, 86
   pour point depressants, 86
   rust inhibitors, 86
   viscosity index improvers, 87
Engine oil analysis, 97
Engine oil classification, 87
   API service classification, 87
   multigrade oils, 87
   SAE viscosity classification, 87
Engine oil cooler, 92
Engine oil data book, 88
Engine oil functions, 85
   bearing protection, 85
   control of combustion chamber deposits, 85
   control of valve deposits, 86

Engine oil functions, *continued*
   cooling, 85
   lubrication, 85
   rust control, 85
   scuff protection, 85
   sealing, 85
   sludge control, 85
   varnish control, 85
   wear control, 85
Engine oils, 83
Engine performance, 19
Engine power, 19, 20, 21
Engine protection system, 272
Engine speed, 14
Engine torque, 22
Engine won't crank properly, 187
Ethylene glycol-based antifreeze, 71
EUI installation, 379
EUI solenoid replacement, 380
Exhaust emissions, 39, 41
Exhaust smoke opacity meter, 58
Exhaust stroke, 5
Exhaust system, 29, 37
Exhaust system service, 43, 55
Exhaust valve timing, 10

Fan, 68
Fan capacity, 68
Fan drive, 68
Fan drive belt tension, 77
Farads (F), 120
Fault codes, 273
Field brush testing, 167
Field coil(s), 151, 176, 193
Field frame, 175
Field windings, 178, 180
Filter change intervals, 97
Filter elements, 31
Filtered air, 31
Filtered drive hub, 36
Five-vaned air motor, 195
Flash point, 204, 207
Floating bearing, 35
Fluid weep hole, 4
Flyweights, 240
Four-stroke cycle, 1, 5
Four-stroke-cycle engine, 2
Friction, 22, 107
   static negative charge, 107
   static positive charge, 107
Friction power, 21
Fuel atomization, 41
Fuel efficiency, 25
Fuel filter, 213
Fuel filter/water separator, 214
   depth type, 214
   final (primary) filters, 214
   final (secondary) filters, 214
   primary (final) filters, 214
   secondary (final) filters, 214
   surface filters, 214
Fuel fluidity, 205
Fuel heaters, 205
Fuel injection control (FIC) module, 315
Fuel injection nozzle, 2, 221

Fuel injection pressure, 218
Fuel injection principles, 217
Fuel pump design, 210
Fuel supply pump, 292
    AMBAC APE pump, 293
    majormec pump, 294
    single-acting fuel, 292
Fuel supply pump operation, 210
Fuel tank, 297
Fuel tank design, 209
Full-load speed regulation, 249

Gear-driven alternator, 148
Gear pump, 211
Gear pump operation, 212
Gear type, 88
Gear type fuel supply pump, 212
Gear type oil pump, 89
Governor fuel delivery, 235
Governor functions, 238
    maximum speed, 238
    minimum-maximum-speed, 238
    variable speed, 238
Governor maximum-speed regulations, 247
Governor operations, 236
Governor springs, 24
Governor terminology, 238
    high idle speed, 238
    hunting, 238
    low-idle speed, 238
    overrun, 238
    promptness, 238
    rated fuel load, 238
    sensitivity, 238
    speed droop, 238
    stability, 238
    underrun, 238
Governor types, 238
    hydraulic, 238
    isochronous, 238
    mechanical, 238
    pneumatic, 238
Grounded armature, 190
Grounded field coil, 190
Grounds, 193

Hand priming pump, 299
Heat, 107
Heat coil, 62
Heat energy, 2, 5, 61
Heat exchanger (radiator), 61, 63, 71, 81
Heat radiator (exchanger), 61, 63, 71, 81
Heat valve, 204, 205
Helix, 328
Hermetically sealed, 120
High pressure lines, 299
Hole-type nozzle, 224
Horsepower, 20
Hydraulic cranking system diagnostic chart, 200
Hydraulic Electronic Unit Injector (HEUI) fuel system, 536
    diagnostic codes, 546, 548
    diagnostic flash codes, 548

Hydraulic Electronic Unit Injector (HEUI) fuel system, *continued*
    electronic connectors, 545
    electronic control system, 544
    engine/diagnostic lamp, 546
    fuel injector operation, 543
    fuel supply pump, 539
    fuel supply system, 540
    glow plug system, 545
    HEUI system diagnosis, 545
    HEUI system service, 545
    hydraulic electronic unit injector components, 542
    hydraulic injection control pressure system, 537
    rail pressure control valve, 538, 541
Hydrocarbons, 40
Hydrometer readings, 138
Hydrostarter, 199
Hydrostarter accumulation, 199
Hydrostarter hand pump, 199
Hydrostarter operation, 198
Hydrostarter reservoir, 198
Hydrostarter system, 198
Hydrostarter torque, 198

Ignition delay period, 205, 218
Indicated power, 21
Indicators, 45
Indium (boron), 120
Induction system, 29
Induction system service, 43
Inertia, 22
Inertia air-starting system, 196
Injection nozzle, 305
Injection nozzle function, 221
Injection nozzle operation, 221
Injection nozzle service, 226
Injection pump calibration, 305
Injection pump design and operation (Bosch), 280
    Bosch smoke limiter, 286
    delivery valve, 282
    constant-pressure type, 285
    control rack operation, 284
    control-rod stop, 286
    excess fuel device, 286
    helix, 281
    pressure compensators, 286
    pump element, 280, 282
    pump plunger, 283
    pump plunger stroke, 284
    pump size, 282
    single-port element, 281
    starting groove, 282
Injection pump disassembly, 301
    cleaning of parts, 301
    inspection of parts, 301
    reassembly, 302
Injection pump flushing, 300
Injection pump mounting, 299
Injection pump removal, 300
Injection pump timing, 299, 307
Injection tubing, 225
Injector assembly removal, 378

Injector problems, 226
    incorrect spray pattern, 226
    leaky nozzle, 226
    nozzle dribble, 226
    opening pressure too high, 226
    opening pressure too low, 226
Input sensors (functions), 271
Installing injection nozzles, 231
Insulators, 104
Intake manifold, 33, 34
Intake port, 10
Intake stroke, 5
Integral charging system, 158
Integral electronic regulator, 150
Integral regulator, 150
Integrated circuit (computer chip), 268
Intercoolers, 45
Intermediate bearing, 178
Intermediate-speed regulations, 239
Internal actuator, 197

Jacobs brake, 15
Journal bearing (turbocharger), 35

Keel cooling system, 61
Key switch, 175
Kilometers, 21
Kinetic inertia, 10, 22

Lanova combustion chamber, 219
Light-emitting diodes (LEDS), 121
Limp-in mode, 270
Liner packing rings, 2
Liquid cooling system, 61
Liquid cooling system function, 60
Lubrication system, 84
Lubrication system components, 83
Lubrication system correction chart, 95
Lubrication system diagnosis, 95
Lubrication system principles, 83
Lubrication system services, 83, 95
Lucas CAV DPA Distributor Cap, 603
    automatic timing advance, 608
    governor testing, 613
    Lucas ignition pump, 604
    Lucas pump service, 607
    maximum fuel delivery control, 607
    mechanical governor, 606
    pump calibration, 610
    pump testing, 610
    testing of advance devices, 614
    timing, 614
    timing of advance unit, 606
    transfer pressure adjustment, 613
    transfer pump setting, 613
Lucas CAV Ltd., 305

Mack Dynatard, 15
Mack Trucks diagnostic blink codes, 316
Mack Trucks diagnostic computer, 322
Mack Trucks diagnostic procedures, 314
Mack Trucks engine protection V-MAC, 320
Mack Trucks fuel control, 319
    FIC, 319
    V.MAC, 319

INDEX

Mack Trucks idle shutdown, 321
Mack Trucks injection timing V-MAC, 320
Mack Trucks intermittent problems, 319
Mack Trucks no active codes, 319
Mack Trucks road speed timing V-MAC, 321
Mack Trucks shutdown V-MAC, 321
Mack Trucks speed control V-MAC, 320
Mack Trucks troubleshooting procedure, 314
Magnetic field, 105, 178, 180
Magnetic flux line, 105
Magnetic force, 105
Magnetic switch, 183
Magnetic switch test, 191
Magnetism, 105
Magnetite, 105
Magnets, 105
Manifold-pressure compensation, 256, 260
Marine test exchange, 70
Maximum-speed governors, 239
Maximum-speed regulations, 239
Mechanical driven blower, 33
Mechanical efficiency, 24
Mechanical energy, 2
Mechanical gauges, 45
Mechanical governor, 240, 241
Metering fuel delivery, 237
Metering sleeve, 218
Metering units, 241
Microfarad, 120
Minimum-maximum-speed governor RQ, 24
  construction, 241
  operating characteristics, 242
  torque-control mechanism, 243
Minimum-maximum-speed governors, 239
M-type chamber, 219
Muffler types, 40
Multicylinder, 127
Multimeter, 127
Multiplunger injection pump, 204

Negative rectifier assembly, 152
Nitrogen, 5
Noisy turbo charger, 49
Nozzle shapes, 224
Nozzle testing, 230
Nucleus, 104

Ohmmeter, 126
Ohm's Law, 108
Ohms (resistance), 108
Oil bath air cleaner, 32, 33
Oil bath air cleaner service, 47
Oil change intervals, 97
Oil consumption, 47, 95, 97
Oil consumption without exhaust smoke, 49
Oil contamination, 97
Oil cooler, 61, 93
Oil cooler coil, 101
Oil cooler service, 99
Oil filter, 91
  bypass valve, 91
  centrifugal type, 92

Oil filter, *continued*
  depth type, 91
  filter bypass, 92
  filter elements, 91
  full flow filtering, 91
  relief valve, 92
  shunt type, 91
  surface type, 91
Oil filtering system, 93
Oil leaks, 95
Oil pan, 2
Oil pressure indicators, 93
  automatic alarms, 93
  automatic shutdown devices, 93
  electric gauges, 93
  oil warning lights, 93
Oil pressure regulation, 88
Oil pressure testing
  oil pickup, 98
  oil pressure, 98
  oil pump, 98
Oil pump, 88
Oil pump operation, 90
Oil pump service, 98
Oil seal sleeves, 54
Oil temperature, 73
One-way clutch, 178
Open armature coils, 190
Open field circuits, 190
Opens, 193
Operation, 5
Overcharged battery, 159
Overcooling, 73
Overhead camshaft, 4
Oxides of nitrogen, 40
Oxygen, 5

Particulate matter, 40
Password, 270
Phosphorous (antimony), 120
Pinion gear retainer ring, 194
Pinion teeth, 193
Pinpoint testing, 274
Pintle nozzle, 224, 225
Piston, 2
Piston pin, 2
Piston pin travel, 14
Piston positions, 11
Piston rings, 2
Piston speed, 14
Piston stroke, 12
Plunger pump, 210
Pneumatic governor, 262, 263
  variable speed governor EP/M, 262
Pneumatic governor diaphragm, 263
Pneumatic governor torque control, 264
Polarity, 105, 106
Ports, 9
Positive drive, 184
Positive rectifier assembly, 151
Power absorption unit, 27
Power stroke, 5
Precleaners, 31
Precombustion chamber, 219, 225

Preengaged air-starting system, 197
Pressure, 107
  piezoelectric effect, 107
Pressure cap, 66
Pressure cap testing, 79
Pressure sensor, 34
Pressure time system, 203
Primary filter(s), 299
Priming pumps, 212, 309
Priming the lubrication system, 99
Producing electricity, 107
Prony brake, 20
Protons, 104
P.T.O. Dynamometers, 27
Pulley, 150
Pulley alignment, 171
Pump lubricating oil, 297
Pump operation (hand), 290
Pump phasing, 302, 305
Pump size, 288
  design and construction, 291
  governor type, 291
  injection time control device, 290
  operation, 291
  PF pump design, 290
  PF pump operation, 290
  size MW pump, 288
Pumping element, 280
Push rod, 2

Radiation, 60
Radiator (heat exchange), 61, 63
Radiator backflushing, 80
Radiator cap, 66, 67
Radiator cap operation, 68
Radiator capacity, 67
Radiator clamp, 63
Radiator core construction, 67
Radiator fan, 67
Radiator fins, 61
Radiator hoses, 63, 65
Radiator inspection, 78
Radiator shroud inspection, 73
Radiator shutter service, 78
Radiator shutters, 69
Radiator testing, 78
Radiator tubes, 6
Raw sea water, 71
Raw (sea) water pump, 63, 71
Rectifier (or bridge), 48
Rectifier (slip ring), 150
Rectifier bridge tests, 168
Rectifying the current, 151
Reference voltage, 270
Regulator bypass test, 163
Regulator service, 166
Regulator valve, 88
Removing injection nozzles, 226
Resistance (ohms), 188
Resistor, 117
Resistor R5, 159
Resistor R7, 159
Rheostat, 119
Road horsepower, 27
Robert Bosch, 279, 311

INDEX 687

Robert Bosch electronic diesel control (EDC) system, 292
Robert Bosch Type VE Distributor Pump, 568
  altitude compensation, 578
  cold-start module electrical check, 584
  cold-start modules, 578
  electronic diesel control, 579
  fuel metering, 571
  governor operation, 573
  idle maximum speed governor, 573
  injection pump removal, 580
  injection pump repairs, 582
  injection timing control, 575
  load-dependent injection timing, 577
  major components, 569
  manifold-pressure compensator, 577
  mechanical governor, 573
  pressure valve, 571
  pumping action, 571
  shut-off devices, 575
  supply pump, 577
  tamper control, 575
  variable-speed governor, 573
  VE injection pump installation, 586
  VE injection pump timing, 584
  VE pump operation, 577
Rocker arm, 2
Rocker arm shaft, 2
Rotor, 149
Rotor shafts, 54
Rotor testing, 167
Rotor timing gear, 36
Rotor type, 88
Rotor type of oil pump, 90

SAE power, 21
Scavenge efficiency, 25
Scavenge line, 31
Scavenge pump, 88, 91
Scavenging, 9, 10
Seal, 80
Secondary filter, 299
Self-powered test light, 124
Semiblocking thermostats, 65
Semi-parallel circuits, 110
Sensor (active), 271
Sensor (passive), 271
Service precautions, 44
Shorted armature, 190
Shorted circuit, 193
Shorted diode, 121
Shutter control, 69
Shutters, 69
Silicon crystal, 120
Silicon wafer, 120
Single phase stator, 150
"Single-port elements", 281
Slip ring (rectifier), 150, 151
Society of Automotive Engineers, 21
Solenoid, 175, 181
Solenoid-operated air control valve, 69
Solenoid switch, 18
Solenoid test, 191
Speed pattern, 225

Spill timing method of pump phasing, 303
Spill timing method of pump timing, 303
Sprag clutch, 185, 186
Sprag clutch check, 195
Sprag clutch drive, 185
Sprag-drive starter, 196
Spray pattern, 225
Spray pattern analysis, 228
Stanadyne (Roosa Master)
  aneroid fuel control, 598
  calibration nozzles, 602
  calibration oil, 602
  charging cycle, 591
  D series pump operation, 590
  discharge cycle, 593
  electrical shut-off, 598
  fuel return circuit, 593
  general test procedure, 602
  injection lines, 602
  mechanical all-speed governor, 596
  pressure-regulating valve, 591
  Roosa Master pump metering, 592
  Stanadyne distributor pump testing, 599
  Stanadyne distributor pump troubleshooting, 599
  Stanadyne electronic fuel injection pump, 598
  Stanadyne (Roosa Master) distributor injection pump, 590
  Stanadyne (Roosa Master) injection timing, 595
  torque control screw, 597
Starter disassembly, 191
Starter driver, 178
Starter free-speed test, 189
Starter magnetic switch, 186
Starter motor bench test, 189
Starter operation, 199
Starter preliminary inspection, 187
Starter protection, 175
Starter protection device, 186
Starter removal, 188
Starter solenoid, 182
Starting motor, 175
Starting motor circuits, 180
Starting motor drives, 182
Starting motor solenoid circuits, 18
Starting system diagnosis chart, 188
Starting system principles, 174
Starting system service, 174
Static inertia, 10, 22
Stator assembly, 149
Stator interpreting test results, 190
Stator parts inspection, 191
Stator testing, 167
Stator torque test, 189
Stop linkage, 299
Stroke, 12
Suction hoses and fitting, 297
Sulfur content, 204, 207
Sulfur dioxide, 40
Supercharger, 23, 33
Swirl chamber (turbulence chamber), 219
Swivel-mounted cam plate, 258

Synthetic oils, 88
  formulated synthetic oils, 88

Tappet adjustment, 306
Tension of brush spring, 195
Test lamp, 124
Testing a thermostat, 79
Testing control, 240, 249
Testing injection nozzle, 228
Thermal efficiency, 25
Thermistor, 119
Thermo-modulated fan, 69
Thermostat, 61, 63, 80
Thermostat design, 65
Thermostat service, 79
Thermostatic sensor circuit, 186
Three-phase, 152
Three-phase stator, 150
Throttle linkage, 299
Throttling pintle nozzle, 225
Thrust bearing, 35
Top dead center, 5
Torque control, 240, 249
Transducers, 120
Transistorized regulator, 154
Transistors, 121
Tubes, 66
Turbine, 35
Turbine wheel, 35
Turbine wheel damage, 49
Turboboost, 34
Turbocharger, 23, 30, 33
Turbocharger cleaning, 51
Turbocharger components, 36
Turbocharger disassembly, 50
Turbocharger function, 33
Turbocharger installation, 52
Turbocharger operation, 34
Turbocharger overhaul, 50
Turbocharger reassembly, 51
Turbocharger removal, 50
Turbocharger service, 47
Turbocharger system service, 43
Turbocharger systems, 29
Turbulence chamber engines, 225
Turbulence chamber (swirl chamber), 219
Two-segment commutator, 180
Two-stroke-cycle, 1
Two-stroke-cycle diesel operation, 9
Two-stroke-cycle engine operation, 8
Typical nozzle, 221
Typical oil pump, 100

Ultrasonic cleaning, 229
Unit injector, 218
Unit injector function, 325
Unit injector identification, 326
Unit injector system, 203

Vacuum, 5
Valve, 2
Valve overlap, 10
Valve spring, 2
Valve timing, 9
Vane pump, 21

Variable-pitch fans, 68
Variable-speed governor EP/RSUV, 253
Variable-speed governor RQV, 244
    construction, 249
    operating characteristics, 250
Variable speed governors, 239, 247
V-belt service, 73
V-belt tension, 172
Vented filler cap, 209
Vernatherm shutter control system, 70
Viscosity, 204
Viscous drive, 68
Viscous friction, 22
V-MAC blink code, 318
V-MAC diagnostic tool, 322
V-MAC electronic tool, 311
V-MAC module, 313
V-MAC performance features, 311
V-MAC system, 311, 312
V-MAC system components and functions, 313
    coolant level sensor, 314
    coolant temperature sensor, 313
    econovance timing advance unit, 313
    engine oil pressure sensor, 314
    fuel injection control module, 313

V-MAC system components and functions, *continued*
    fuel injection pump, 313
    fuel rack actuator, 313
    fuel shut-off solenoid, 313
    intake manifold air temperature sensor, 314
    MPH (road speed) sensor, 313
    RPM/TDC engine position sensor, 313
    switches, 314
    throttle position sensor, 314
    timing event marker, 313
V-MAC system diagnostics, 319
Volatility, 204, 205, 206
Voltage adjustment cap, 155
Voltage drop, 188
Voltage drop test, 164
Voltage drop testing, 165
Voltage efficiency, 22
Voltage indicator, 155, 157
Voltage limiter, 117
Voltage output factors, 155
Voltage production, 152
Voltage regulation, 154
Voltage regulator, 148, 151

Voltage-sensing ADLD circuits, 187
Voltmeter, 125
Voltmeter tests, 161

Warning devices, 45
Warning light, 155
Water manifold, 63
Water manometers, 45
Water pump, 61, 62
Water pump service, 73
Watt's formula, 20
Wire conductors, 111
Wire harness, 111, 148
Wire terminals, 111
Wiring and cables, 175
Wiring harness, 148
Wiring repairs, 127
Wiring test chart, 127
Work, 19

Y-connected stator, 152
Y stator, 153

Zinc diode, 121
Zinc electrodes, 71
Zinc plug, 81